MODERN NONLINEAR OPTICS
Part 2
Second Edition

ADVANCES IN CHEMICAL PHYSICS

VOLUME 119

EDITORIAL BOARD

MODERN NONLINEAR OPTICS

Part 2

Second Edition

ADVANCES IN CHEMICAL PHYSICS
VOLUME 119

Edited by

Myron W. Evans

Series Editors

I. PRIGOGINE

Center for Studies in Statistical Mechanics and Complex Systems
The University of Texas
Austin, Texas
and
International Solvay Institutes
Université Libre de Bruxelles
Brussels, Belgium

and

STUART A. RICE

Department of Chemistry
and
The James Franck Institute
The University of Chicago
Chicago, Illinois

AN INTERSCIENCE® PUBLICATION
JOHN WILEY & SONS, INC.

This book is printed on acid-free paper. ♾

Published simultaneously in Canada.

For ordering and customer service, call 1-800-CALL-WILEY

Library of Congress Catalog Number: 58-9935

ISBN 0-471-38931-5

Printed in the United States of America.

10 9 8 7 6 5 4 3 2 1

CONTRIBUTORS TO VOLUME 119
Part 2

CARL E. BAUM, Air Force Research Laboratory, Kirtland Air Force Base, NM

THOMAS E. BEARDEN, Fellow Emeritus, Alpha Foundation Institute for Advanced Study and Director, Association of Distinguished American Scientists, CEO, CTEC Inc., and Magnetic Energy Limited, Huntsville, AL

BOGUSŁAW BRODA, Department of Theoretical Physics, University of Łódź, Lódź, Poland

PATRICK CORNILLE, Advanced Electromagnetic Systems, S.A., St. Rémy-Lès-Chevreus, France; CEA/DAM/DIE, Bryeres le Chatel, France

J. R. CROCA, Departamento de Fisica, Faculdade de Ciências, Universidade de Lisboa, Lisboa, Portugal

M. W. EVANS, 50 Rhyddwen Road, Craigcefnparc, Swansea, Wales, United Kingdom

K. GRYGIEL, Nonlinear Optics Division, Adam Mickiewicz University, Institute of Physics, Poznań, Poland

V. I. LAHNO, Department of Theoretical Physics II, Complutense University, Madrid, Spain and State Pedagogical University, Poltava, Ukraine

B. LEHNERT, Alfven Laboratory, Royal Institute of Technology, Stockholm, Sweden

P. SZLACHETKA, Nonlinear Optics Division, Adam Mickiewicz University, Institute of Physics, Poznań, Poland

R. Z. ZHDANOV, Department of Theoretical Physics II, Complutense University, Madrid, Spain

INTRODUCTION

Few of us can any longer keep up with the flood of scientific literature, even in specialized subfields. Any attempt to do more and be broadly educated with respect to a large domain of science has the appearance of tilting at windmills. Yet the synthesis of ideas drawn from different subjects into new, powerful, general concepts is as valuable as ever, and the desire to remain educated persists in all scientists. This series, *Advances in Chemical Physics*, is devoted to helping the reader obtain general information about a wide variety of topics in chemical physics, a field that we interpret very broadly. Our intent is to have experts present comprehensive analyses of subjects of interest and to encourage the expression of individual points of view. We hope that this approach to the presentation of an overview of a subject will both stimulate new research and serve as a personalized learning text for beginners in a field.

I. PRIGOGINE
STUART A. RICE

PREFACE

This volume, produced in three parts, is the Second Edition of Volume 85 of the series, *Modern Nonlinear Optics*, edited by M. W. Evans and S. Kielich. Volume 119 is largely a dialogue between two schools of thought, one school concerned with quantum optics and Abelian electrodynamics, the other with the emerging subject of non-Abelian electrodynamics and unified field theory. In one of the review articles in the third part of this volume, the Royal Swedish Academy endorses the complete works of Jean-Pierre Vigier, works that represent a view of quantum mechanics opposite that proposed by the Copenhagen School. The formal structure of quantum mechanics is derived as a linear approximation for a generally covariant field theory of inertia by Sachs, as reviewed in his article. This also opposes the Copenhagen interpretation. Another review provides reproducible and repeatable empirical evidence to show that the Heisenberg uncertainty principle can be violated. Several of the reviews in Part 1 contain developments in conventional, or Abelian, quantum optics, with applications.

In Part 2, the articles are concerned largely with electrodynamical theories distinct from the Maxwell–Heaviside theory, the predominant paradigm at this stage in the development of science. Other review articles develop electrodynamics from a topological basis, and other articles develop conventional or U(1) electrodynamics in the fields of antenna theory and holography. There are also articles on the possibility of extracting electromagnetic energy from Riemannian spacetime, on superluminal effects in electrodynamics, and on unified field theory based on an SU(2) sector for electrodynamics rather than a U(1) sector, which is based on the Maxwell–Heaviside theory. Several effects that cannot be explained by the Maxwell–Heaviside theory are developed using various proposals for a higher-symmetry electrodynamical theory. The volume is therefore typical of the second stage of a paradigm shift, where the prevailing paradigm has been challenged and various new theories are being proposed. In this case the prevailing paradigm is the great Maxwell–Heaviside theory and its quantization. Both schools of thought are represented approximately to the same extent in the three parts of Volume 119.

As usual in the *Advances in Chemical Physics* series, a wide spectrum of opinion is represented so that a consensus will eventually emerge. The prevailing paradigm (Maxwell–Heaviside theory) is ably developed by several groups in the field of quantum optics, antenna theory, holography, and so on, but the paradigm is also challenged in several ways: for example, using general relativity, using O(3) electrodynamics, using superluminal effects, using an extended electrodynamics based on a vacuum current, using the fact that longitudinal waves may appear in vacuo on the U(1) level, using a reproducible and repeatable device, known as the *motionless electromagnetic generator*, which extracts electromagnetic energy from Riemannian spacetime, and in several other ways. There is also a review on new energy sources. Unlike Volume 85, Volume 119 is almost exclusively dedicated to electrodynamics, and many thousands of papers are reviewed by both schools of thought. Much of the evidence for challenging the prevailing paradigm is based on empirical data, data that are reproducible and repeatable and cannot be explained by the Maxwell–Heaviside theory. Perhaps the simplest, and therefore the most powerful, challenge to the prevailing paradigm is that it cannot explain interferometric and simple optical effects. A non-Abelian theory with a Yang–Mills structure is proposed in Part 2 to explain these effects. This theory is known as O(3) *electrodynamics* and stems from proposals made in the first edition, Volume 85.

As Editor I am particularly indebted to Alain Beaulieu for meticulous logistical support and to the Fellows and Emeriti of the Alpha Foundation's Institute for Advanced Studies for extensive discussion. Dr. David Hamilton at the U.S. Department of Energy is thanked for a Website reserved for some of this material in preprint form.

Finally, I would like to dedicate the volume to my wife, Dr. Laura J. Evans.

Myron W. Evans

Ithaca, New York

CONTENTS

MODERN NONLINEAR OPTICS
Part 2
Second Edition

ADVANCES IN CHEMICAL PHYSICS

VOLUME 119

OPTICAL EFFECTS OF AN EXTENDED ELECTROMAGNETIC THEORY

B. LEHNERT

Alfvén Laboratory, Royal Institute of Technology, Stockholm, Sweden

CONTENTS

Modern Nonlinear Optics, Part 2, Second Edition, Advances in Chemical Physics, Volume 119,
Edited by Myron W. Evans. Series Editors I. Prigogine and Stuart A. Rice.
ISBN 0-471-38931-5

I. INTRODUCTION

Conventional electromagnetic field theory based on Maxwell's equations and quantum mechanics has been very successful in its application to numerous problems in physics, and has sometimes manifested itself in an extremely good agreement with experimental results. Nevertheless, in certain areas these joint theories do not seem to provide fully adequate descriptions of physical reality. Thus there are unsolved problems leading to difficulties with Maxwell's equations that are not removed by and not directly associated with quantum mechanics [1,2].

Because of these circumstances, a number of modified and new approaches have been elaborated since the late twentieth century. Among the reviews and conference proceedings describing this development, those by Lakhtakia [3], Barrett and Grimes [4], Evans and Vigier [5], Evans et al. [6,7], Hunter et al. [8], and Dvoeglazov [9] can be mentioned here. The purpose of these approaches can be considered as twofold:

- To contribute to the understanding of so far unsolved problems
- To predict new features of the electromagnetic field

The present chapter is devoted mainly to one of these new theories, in particular to its possible applications to photon physics and optics. This theory is based on the hypothesis of a nonzero divergence of the electric field in vacuo, in combination with the condition of Lorentz invariance. The nonzero electric field divergence, with an associated "space-charge current density," introduces an extra degree of freedom that leads to new possible states of the electromagnetic field. This concept originated from some ideas by the author in the late 1960s, the first of which was published in a series of separate papers [10,12], and later in more complete forms and in reviews [13–20].

As a first step, the treatment in this chapter is limited to electromagnetic field theory in orthogonal coordinate systems. Subsequent steps would include more advanced tensor representations and a complete quantization of the extended field equations.

II. UNSOLVED PROBLEMS IN CONVENTIONAL ELECTROMAGNETIC THEORY

The failure of standard electromagnetic theory based on Maxwell's equations is illustrated in numerous cases. Here the following examples can be given.

1. Light appears to be made of waves and simultaneously of particles. In conventional theory the individual photon is on one hand conceived to be a massless particle, still having an angular momentum, and is on the other hand regarded as a wave having the frequency ν and the energy $h\nu$, whereas the angular momentum is independent of the frequency. This dualism of the wave and particle concepts is so far not fully understandable in terms of conventional theory [5].
2. The photon can sometimes be considered as a plane wave, but some experiments also indicate that it can behave like a bullet. In investigations on interference patterns created by individual photons on a screen [21], the impinging photons produce dot-like marks on the latter, such as those made by needle-shaped objects.
3. In attempts to develop conventional electrodynamic models of the individual photon, it is difficult to finding axisymmetric solutions that both converge at the photon center and vanish at infinity. This was already realized by Thomson [22] and later by other investigators [23].
4. During the process of total reflection at a vacuum boundary, the reflected beam has been observed to be subject to a parallel displacement with respect to the incident beam. For this so-called Goos–Hänchen effect, the displacement was further found to have a maximum for parallel polarization of the incident electric field, and a minimum for perpendicular polarization [24,25]. At an arbitrary polarization angle, however, the displacement does not acquire an intermediate value, but splits into the two values for parallel and perpendicular polarization. This behaviour cannot be explained by conventional electromagnetic theory.
5. The Fresnel laws of reflection and refraction of light in nondissipative media have been known for over 180 years. However, these laws do not apply to the total reflection of an incident wave at the boundary between a dissipative medium and a vacuum region [26].
6. In a rotating interferometer, fringe shifts have been observed between light beams that propagate parallel and antiparallel with the direction of rotation [4]. This Sagnac effect requires an unconventional explanation.
7. Electromagnetic wave phenomena and the related photon concept remain somewhat of an enigma in more than one respect. Thus, the latter concept should in principle apply to wavelengths ranging from about 10^{-15} m of gamma radiation to about 10^5 m of long radiowaves. This leads to an as yet not fully conceivable transition from a beam of individual photons to a nearly plane electromagnetic wave.
8. As the only explicit time-dependent solution of Cauchy's problem, the Lienard–Wiechert potentials are claimed be inadequate for describing

the entire electromagnetic field [2]. With these potentials only, the implicitly time-independent part of the field is then missing, namely, the part that is responsible for the interparticle long-range Coulomb interaction. This question may need further analysis.

9. There are a number of observations which seem to indicate that superluminal phenomena are likely to exist [27]. Examples are given by the concept of negative square-mass neutrinos, fast galactic miniquasar expansion, photons tunneling through a barrier at speeds greater than c, and the propagation of so called X-shaped waves. These phenomena cannot be explained in terms of the purely transverse waves resulting from Maxwell's equations, and they require a longitudinal wave component to be present in the vacuum [28].
10. A photon gas cannot have changes of state that are adiabatic and isothermal at the same time, according to certain studies on the distribution laws for this gas. To eliminate such a discrepancy, longitudinal modes, which do not exist in conventional theory, must be present [29,30].
11. It is not possible for conventional electromagnetic models of the electron to explain the observed property of a "point charge" with an excessively small radial dimension [20]. Nor does the divergence in self-energy of a point charge vanish in quantum field theory where the process of renormalization has been applied to solve the problem.

III. BASIS OF PRESENT APPROACH

The present modified form of Maxwell's equations in vacuo is based on two mutually independent hypotheses:

- The divergence of the electric field may differ from zero, and a corresponding "space-charge current" may exist in vacuo. This concept should not become less conceivable than the earlier one regarding introduction of the displacement current, which implies that a nonvanishing curl of the magnetic field and a corresponding current density can exist in vacuo. Both these concepts can be regarded as intrinsic properties of the electromagnetic field. The nonzero electric field divergence can thereby be interpreted as a polarization of the vacuum ground state [13] which has a nonzero energy as predicted by quantum physics [5], as confirmed by the existence of the Casimir effect. That electric polarization can occur out of a neutral state is also illustrated by electron–positron pair formation from a photon [18].
- This extended form of the field equations should remain Lorentz-invariant. Physical experience supports such a statement, as long as there are no results that conflict with it.

A. Formulation in Terms of Electromagnetic Field Theory

1. Basic Equations

On the basis of these two hypotheses the extended field equations in vacuo become

$$\operatorname{curl}\frac{\mathbf{B}}{\mu_0} = \mathbf{j} + \frac{\varepsilon_0 \partial \mathbf{E}}{\partial t} \tag{1}$$

$$\operatorname{curl}\mathbf{E} = \frac{-\partial \mathbf{B}}{\partial t} \tag{2}$$

$$\mathbf{j} = \bar{\rho}\,\mathbf{C} \tag{3}$$

in SI units. Here $\mathbf{B}$ and $\mathbf{E}$ are the magnetic and electric fields, $\mathbf{j}$ is the current density, and $\bar{\rho}$ the charge density arising from a nonzero electric field divergence in vacuo. As a consequence of the divergence of equations (1) and (2),

$$\operatorname{div}\mathbf{E} = \frac{\bar{\rho}}{\varepsilon_0} \tag{4}$$

$$\operatorname{div}\mathbf{B} = 0 \qquad \mathbf{B} = \operatorname{curl}\mathbf{A} \tag{5}$$

and

$$\mathbf{E} = -\nabla\phi - \frac{\partial \mathbf{A}}{\partial t} \tag{6}$$

The space-charge current density in vacuo expressed by Eqs. (3) and (4) constitutes the essential part of the present extended theory. To specify the thus far undetermined velocity $\mathbf{C}$, we follow the classical method of recasting Maxwell's equations into a four-dimensional representation. The divergence of Eq. (1) can, in combination with Eq. (4), be expressed in terms of a four-dimensional operator, where $(\mathbf{j}, ic\bar{\rho})$ thus becomes a 4-vector. The potentials $\mathbf{A}$ and ϕ are derived from the sources $\mathbf{j}$ and $\bar{\rho}$, which yield

$$-\left(\nabla^2 - \frac{1}{c^2}\frac{\partial^2}{\partial t^2}\right)\left(\mathbf{A}, \frac{i\phi}{c}\right) \equiv \square\left(\mathbf{A}, \frac{i\phi}{c}\right) = \mu_0(\mathbf{j}, ic\bar{\rho}) = \mu_0\bar{\rho}(\mathbf{C}, ic) \equiv \mu_0\mathbf{J} \tag{7}$$

when being combined with the condition of the Lorentz gauge. The Lorentz condition is further discussed in Appendix A.

It should be observed that Eq. (7) is of a "Proca type," here being due to generation of a space-charge density $\bar{\rho}$ in vacuo (free space). Such an equation can describe a particle with the spin value unity [31].

Returning to the form (3) of the space-charge current density, and observing that $(\mathbf{j}, ic\bar{\rho})$ is a 4-vector, the Lorentz invariance thus leads to

$$j^2 - c^2\bar{\rho}^2 = \bar{\rho}^2(C^2 - c^2) = \text{const} = 0 \qquad C^2 = c^2 \tag{8}$$

where $j^2 = \mathbf{j}^2$ and $C^2 = \mathbf{C}^2$. The constant in this relation has to vanish because it should be universal to any inertial frame, and because the charge density varies from frame to frame. This result is further reconcilable with the relevant condition that the current density $\mathbf{j}$ of Eq. (3) should vanish in absence of the space-charge density $\bar{\rho}$. In this way Eqs. (1)–(6) and (8) provide an extended Lorentz invariant form of Maxwell's equations that includes all earlier treated electromagnetic phenomena but also contains new classes of time-dependent and steady solutions, as illustrated later.

Concerning the velocity field $\mathbf{C}$, the following general features can now be specified:

- The vector $\mathbf{C}$ is time-independent.
- The direction of the unit vector of $\mathbf{C}$ depends on the geometry of the particular configuration to be analyzed, as is also the case for the unit vector of the current density $\mathbf{j}$ in any configuration treated in terms of conventional electromagnetic theory. As will be shown later, the direction of $\mathbf{C}$ thus depends on the necessary boundary conditions.
- Both $\operatorname{curl}\mathbf{C}$ and $\operatorname{div}\mathbf{C}$ can differ form zero, but here we restrict ourselves to

$$\operatorname{div}\mathbf{C} = 0 \tag{9}$$

We finally observe that a combination of Eqs. (1) and (4) leads to the classical relation

$$\operatorname{div}\mathbf{j} = -\frac{\partial\bar{\rho}}{\partial t} \tag{10}$$

of the 4-vector $(\mathbf{j}, ic\bar{\rho})$.

The introduction of the current density (3) in 3-space is, in fact, less intuitive than what could appear at first glance. As soon as the charge density (4) is permitted to exist as the result of a nonzero electric field divergence, the Lorentz invariance of a 4-current (7) with the time part $ic\bar{\rho}$ namely requires the associated space part to adopt the form (3), that is, by necessity.

The degree of freedom introduced by a nonzero electric field divergence leads both to new features of the electromagnetic field and to the possibility of satisfying boundary conditions in cases where this would not become possible in conventional theory.

In connection with the basic ideas of the present approach, the question may be raised as to why only div **E**, and not also div **B**, is permitted to be nonzero. This issue can be considered to be both physical and somewhat philosophical. Here we should remember that the electric field is associated with an equivalent "charge density" $\bar{\rho}$ considered as a source, whereas the magnetic field has its source in the current density **j**. The electric field lines can thereby be "cut off" by ending at a corresponding "charge," whereas the magnetic field lines generated by a line element of the current density are circulating around the same element. From the conceptual point of view it thus appears more difficult to imagine how these circulating magnetic field lines could be cut off to form magnetic poles by assuming div **B** to be nonzero, than to have electric field lines ending on charges with a nonzero div **E**.

Some investigators have included magnetic monopoles in extended theories [32,33], also from the quantum-theoretic point of view [20]. According to Dirac [34], the magnetic monopole concept is an open question. In this connection it should finally be mentioned that attempts have been made to construct theories based on general relativity where gravitation and electromagnetism are derived from geometry, as well as theories including both a massive photon and a Dirac monopole [20].

2. *The Momentum and Energy Balance*

We now turn to the momentum and energy balance of the electromagnetic field. In analogy with conventional deductions, Eq. (1) is multiplied vectorially by **B** and Eq. (2), by $\varepsilon_0\mathbf{E}$. The sum of the resulting equations is then rearranged into the local momentum balance equation

$$\operatorname{div}{}^2\mathbf{S} = \bar{\rho}(\mathbf{E} + \mathbf{C} \times \mathbf{B}) + \varepsilon_0 \frac{\partial}{\partial t}(\mathbf{E} \times \mathbf{B}) \tag{11}$$

where ${}^2\mathbf{S}$ is the electromagnetic stress tensor [35] and Eq. (3) has been employed. The integral form of Eq. (11) becomes

$$\int {}^2\mathbf{S} \cdot \mathbf{n}\, dS = \mathbf{F}_e + \mathbf{F}_m + \frac{\partial}{\partial t} \int \mathbf{g}\, dV \tag{12}$$

where dS and dV are surface and volume elements, respectively,

$$\mathbf{F}_e = \int \bar{\rho}\mathbf{E}\, dV \quad \mathbf{F}_m = \int \bar{\rho}\mathbf{C} \times \mathbf{B}\, dV \tag{13}$$

are the electric and magnetic volume forces, and

$$\mathbf{g} = \varepsilon_0 \mathbf{E} \times \mathbf{B} = \frac{1}{c^2} \mathbf{S} \tag{14}$$

can be interpreted as an electromagnetic momentum with $\mathbf{S}$ denoting the Poynting vector. Here the component S_{jk} of the tensor ${}^2\mathbf{S}$ is the momentum that in unit time crosses in the j- direction for a unit element of surface whose normal is oriented along the k axis [35]. The difference in the present results (11) and (12) as compared to conventional theory is in the appearance of the terms, which include the nonzero charge density $\bar{\rho}$ in vacuo.

In a similar way scalar multiplications of Eq. (1) by $\mathbf{E}$ and Eq. (2) by $\mathbf{B}/\mu_0$ yields, after subtraction of the resulting equations, the local energy balance equation

$$-\operatorname{div} \mathbf{S} = -\left(\frac{1}{\mu_0}\right) \operatorname{div}(\mathbf{E} \times \mathbf{B}) = \bar{\rho} \mathbf{E} \cdot \mathbf{C} + \frac{1}{2} \varepsilon_0 \frac{\partial}{\partial t} (\mathbf{E}^2 + c^2 \mathbf{B}^2) \tag{15}$$

This equation differs from that of the conventional Poynting theorem, due to the existence of the term $\bar{\rho}\mathbf{E} \cdot \mathbf{C}$ in vacuo. That there should arise a difference has also been emphasized by Evans et al. [6] as well as by Chubykalo and Smirnov-Rueda [2]. These investigators note that the Poynting vector in vacuo is only defined in terms of transverse plane waves, that the case of a longitudinal magnetic field $\mathbf{B}^{(3)}$ leads to a new form of the Poynting theorem, and that the Poynting vector can be associated only with the free magnetic field. We shall return to this question later, when considering axisymmetric wavepackets and the photon interpreted as a particle with an associated pilot wave. It will also be seen later in this context that $\mathbf{F}_e$, $\mathbf{F}_m$, and the integral of $\bar{\rho}\mathbf{E} \cdot \mathbf{C}$ can disappear in the special case of axisymmetric wavepackets, and that $\bar{\rho}\mathbf{E} \cdot \mathbf{C}$ disappears for plane waves.

3. *The Energy Density*

The last term in Eq. (15) includes the local "field energy density"

$$w_f = \frac{1}{2} \left(\varepsilon_0 \mathbf{E}^2 + \frac{\mathbf{B}^2}{\mu_0} \right) \tag{16}$$

interpreted in terms of the electromagnetic field strengths $\mathbf{E}$ and $\mathbf{B}$. An alternative form [35], which at least holds for steady states and for waves where the field quantities vary as $\exp(-i\omega t)$ and have the same phases, is given by the local "source energy density"

$$w_s = \frac{1}{2} (\bar{\rho}\phi + \mathbf{j} \cdot \mathbf{A}) = \frac{1}{2} \bar{\rho}(\phi + \mathbf{C} \cdot \mathbf{A}) \tag{17}$$

interpreted in terms of the sources $\bar{\rho}$ and $\mathbf{j}$, which generate the electromagnetic field, and where the form (17) is a direct measure of the local work performed on the electric charges and currents. The total field energy becomes

$$W = \int w_f \, dV = \int w_s \, dV \tag{18}$$

provided it leads to surface integrals that vanish at infinity, and at the origin. Thus, Eq. (18) does not hold when the field quantities become divergent at the origin or at infinity.

In the present approach a physically relevant expression for the local energy density is sometimes needed. In such a case we shall prefer the form (17) to that of Eq. (16). Thus there are situations where the moment has to be taken of the local energy density, with some space-dependent function f. Since w_f and w_s represent entirely different spatial distributions of energy, it is then observed that

$$\int f \cdot w_f \, dV \neq \int f \cdot w_s \, dV \tag{19}$$

A further feature of physical interest is that the *local* energy density (17) can become positive as well as *negative* in some regions of space, even if the total energy W becomes positive as long as relation (18) holds. It is, however, not clear at this stage whether the form (17) could open up a possibility of finding negative energy states.

When considering the energy density of the form (17), it is sometimes convenient to divide the electromagnetic field into two parts when dealing with charge and current distributions that are limited to a region in space near the origin. This implies that the potentials are written as

$$\mathbf{A} = \mathbf{A}_s + \mathbf{A}_v \quad \phi = \phi_s = \phi_v \tag{20}$$

Here $\text{curl}^2\, \mathbf{A}_s \neq 0$ and $\nabla^2 \phi_s \neq 0$ refer to the "source part" of the field that is nonzero within such a limited region, whereas $\text{curl}^2\, \mathbf{A}_v = 0$ and $\nabla^2 \phi_v = 0$ refer to the "vacuum part" outside the same region [13,20], and the notation $\text{curl}^2 \equiv \text{curl curl}$ is used henceforth. For a model of a charged particle such as the electron, the potentials $\mathbf{A}_v$ and ϕ_v would thus be connected with its long-distance magnetic dipole field and electrostatic Coulomb field, respectively [20]. The total energy becomes

$$\begin{aligned} W = \frac{1}{2} \varepsilon_0 \int \left(c^2 \mathbf{A}_s \cdot \text{curl}^2\, \mathbf{A}_s - \phi_s \nabla^2 \phi_s \right) dV + \frac{1}{2} \varepsilon_0 \int \mathbf{n} \cdot \left[c^2 (\mathbf{A}_s \times \text{curl}\, \mathbf{A}_v \right. \\ \left. - \mathbf{A}_v \times \text{curl}\, \mathbf{A}_s) + \phi_s \nabla \phi_v - \phi_v \nabla \phi_s \right] dS \end{aligned} \tag{21}$$

where S now stands for the bounding surfaces to be taken into account. There are, in principle, two possibilities:

- When there is a single bounding surface S that can be extended to infinity where the electromagnetic field vanishes, only the space-charge parts $\mathbf{A}_s$ and ϕ_s will contribute to the energy (21). This possibility is of special interest in this context, which concentrates mainly on photon physics.
- When there is also an inner surface S_i enclosing the origin and at which the field diverges, special conditions have to be imposed for $\mathbf{A}_s$ and ϕ_s to represent a total energy, and for convergent integrated expressions still to result from the analysis [13,20]. These conditions will apply to a model of charged particle equilibrium states, such as those representing charged leptons discussed in Section V.A and Appendix B.

B. Formulation in Terms of Quantum Mechanics

An adaptation of quantum mechanics implies that a number of constraints are imposed on the system as follows.

- The energy is given in terms of the quantum $h\nu$, where ν is the frequency.
- The angular momentum (spin) of a particle-like state becomes $h/2\pi$ for a boson and $h/4\pi$ for a fermion.
- The magnetic moment of a charged particle, such as the electron, is quantized according to the Dirac theory of the electron [36], including a small modification according to Feynman [37], which results in an excellent agreement with experiments. As based on a tentative model of "self-confined" (bound) circulating radiation [11,13,20], the quantization of energy and its alternative form mc^2 can also be shown to result in an angular momentum equal to about $h/4\pi$, and a magnetic moment of the magnitude obtained in the theory by Dirac. One way to obtain exact agreement with the results by Dirac and Feynman is provided by different spatial distributions of electric charge and energy density. This is possible within the frame of the present theory [13,20]. However, it has also to be observed that these results apply to an electron in an electromagnetic field, and they could therefore differ from the result obtained for a free electron.
- With e as a given elementary electric charge, there is also a condition on the quantization of magnetic flux. This could be reinterpreted as a subsidiary condition in an effort to quantize the electron charge and deduce its absolute value by means of the present theory [13,18,20], but the details of such an analysis are not yet available. Magnetic flux quantization is discussed in further detail in Appendix B.

In a first step, these conditions can be imposed on the general solutions of the present electromagnetic field equations. At a later stage the same equations should be quantized by the same procedure as that applied earlier in quantum electrodynamics to Maxwell's equations [39].

C. Derivation from Gauge Theory

It should finally be mentioned that the basic equations (1)–(8) have been derived from gauge theory in the vacuum, using the concept of covariant derivative and Feynman's universal influence [38]. These equations and the Proca field equations are shown to be interrelated to the well-known de Broglie theorem, in which the photon rest mass m_0 can be interpreted as nonzero and be related to a frequency $\nu' = m_0 c^2/h$. A gauge-invariant Proca equation is suggested by this analysis and relations (1)–(8). It is also consistent with the earlier conclusion that gauge invariance does not require the photon rest mass to be zero [20,38].

IV. MAIN CHARACTERISTICS OF MODIFIED FIELD THEORIES

Before turning to the details of the present analysis, we describe and compare the main features of some of the modified and extended theories that have been proposed and elaborated on with the purpose of replacing Maxwell's equations. This description includes a Proca-type equation as a starting point. Introducing the 4-potential $A_\mu = (\mathbf{A}, i\phi/c)$ and the 4-current J_μ, the latter equation can be written as

$$\Box A_\mu = \mu_0 J_\mu \tag{22}$$

A. Electron Theory by Dirac

According to the Dirac [36] electron theory, the relativistic wavefunction Ψ has four components in spin-space. With the Hermitian adjoint wave function $\bar{\Psi}$, the quantum mechanical forms of the charge and current densities become [31,40]

$$\bar{\rho} = e\bar{\Psi}\Psi \tag{23}$$

and

$$\mathbf{j} = ce(\bar{\Psi}\alpha_i\Psi) \qquad i = 1, 2, 3 \tag{24}$$

where α_i are the Dirac matrices of the three spatial directions (x, y, z). There is more than one set of choices of these matrices [41].

Expressions (23) and (24) could be interpreted as the result of the electronic charge being "smeared out" over the volume of an electron with a very small but nonzero radius. The 4-current of the right-hand side of equation (22) thus becomes

$$J_{\mu} = ce(\bar{\Psi}\alpha_i\Psi, i\bar{\Psi}\Psi) \tag{25}$$

in this case.

B. Photon Theory by de Broglie, Vigier, and Evans

At an early stage Einstein [42] as well as Bass and Schrödinger [43] considered the possibility for the photon to have a very small but nonzero rest mass m_0. Later de Broglie and Vigier [44] and Evans and Vigier [5] derived a corresponding form of the 4-current in the Proca-type equation (22) as given by

$$J_{\mu} = \left(\frac{1}{\mu_0}\right)\left(\frac{2\pi m_0 c}{h}\right)^2\left(\mathbf{A}, \frac{i\phi}{c}\right) \tag{26}$$

As a consequence, the solutions of the field equations were also found to include longitudinal fields. Thereby Evans [45] was the first to give attention to a longitudinal magnetic field part, $\mathbf{B}^{(3)}$, of the photon in the direction of propagation.

C. Present Nonzero Electric Field Divergence Theory

The present approach of Eqs. (1)–(8) includes the four-current

$$J_{\mu} = \bar{\rho}(\mathbf{C}, ic) = \varepsilon_0(\operatorname{div}\mathbf{E})(\mathbf{C}, ic) \tag{27}$$

The solutions of the corresponding field equations have a wide area of application. They can be integrated to yield such quantities as the electric charge of a steady particle-like state, as well as a nonzero rest mass in a dynamic state representing an individual photon that also includes longitudinal field components in the direction of propagation. Thereby application of de Broglie's theorem for the photon rest mass links the concepts of expressions (26) and (27) together, as well as those of the longitudinal magnetic fields. This point is illuminated further in the following sections.

The present theory should be interpreted as microscopic in nature, in the sense that it is based only on the electromagnetic field itself. This applies to both free states of propagating wavefronts and the possible existence of bound steady axisymmetric states in the form of self-confined circulating radiation. Consequently, the extended theory does not need to include the concept of an initial particle rest mass. The latter concept does not enter into the differential equations of the electromagnetic field, simply because a rest mass should first originate from a spatial integration of the electromagnetic energy density, such as in a bound state [11–13].

When further relating the present approach to Eqs. (23) and (24) of the Dirac theory, we therefore have to consider wavefunctions that only represent states without a rest mass. One functions of this special class is given by [40]

$$\Psi = u(x, y, z)\begin{bmatrix} U \\ 0 \\ \pm U \\ 0 \end{bmatrix} \tag{28}$$

where u is an arbitrary function and U a constant. This form yields a charge density

$$\bar{\rho} = 2e\bar{U}U\bar{u}u \tag{29}$$

and the corresponding current density components

$$j_z = c\bar{\rho};\ j_x = 0 \qquad \text{and} \qquad j_y = 0 \tag{30}$$

where a bar over U and u indicates the complex conjugate value. Other forms analogous to the wavefunction (28) can be chosen to correspond to the cases

$$j_y = \pm c\bar{\rho}: \qquad j_z = j_x = 0 \tag{31}$$

$$j_x = \pm c\bar{\rho}; \qquad j_y = j_z = 0 \tag{32}$$

This result, as well as the form of expressions (23) and (24), shows that the charge and current density relations (3), (4), and (8) of the present extended theory become consistent with and related to the Dirac theory. It also implies that this extended theory can be developed in harmony with the basis of quantum electrodynamics.

The introduced current density $\mathbf{j} = \varepsilon_0(\operatorname{div}\mathbf{E})\mathbf{C}$ is thus consistent with the corresponding formulation in the Dirac theory of the electron, but this introduction also applies to electromagnetic field phenomena in a wider sense.

D. Nonzero Conductivity Theory by Bartlett, Harmuth, Vigier, and Roy

Bartlett and Corle [46] proposed modification of Maxwell's equations in the vacuum by assigning a small nonzero electric condictivity to the formalism. As pointed out by Harmuth [47], there was never a satisfactory concept of propagation velocity of signals within the framework of Maxwell's theory. Thus, the equations of the latter fail for waves with nonnegligible relative frequency bandwidth when propagating in a dissipative medium. To resolve this problem, a nonzero electric conductivity σ and a corresponding current density

$$\mathbf{j}_\sigma = \sigma\mathbf{E} \tag{33}$$

were thus introduced into a modified form of Maxwell's equations in vacuo. In the same system of equations, a magnetic current density given by a nonzero magnetic field divergence was introduced as well [47].

This electric conductivity concept was later reconsidered by Vigier [48], who showed that the introduction of the current density (33) is equivalent to adding a related nonzero photon rest mass to the system, such as in the Proca-type equation represented by expressions (22) and (26). The dissipative "tired light" mechanism underlying this conductivity can be related to a nonzero energy of the vacuum ground state, as predicted by quantum physics [5,49]. That the current (33) is related to the form (26) of a 4-current can be understood from the conventional field equations for homogeneous conducting media [35].

The effects of the nonzero electric conductivity were further investigated by Roy et al. [20,50–52]. They have shown that the introduction of a nonzero conductivity yields a dispersion relation that results in phase and group velocities depending on a corresponding nonzero photon rest mass, due to a tired-light effect.

In principle, this nonzero conductivity effect could also be included in the present theory of a nonzero electric field divergence.

E. Single-Charge Theory by Hertz, Chubykalo, and Smirnov-Rueda

A set of first-order field equations was proposed by Hertz [53–55], who substituted the partial time derivatives in Maxwell's equations by total time derivatives

$$\frac{d}{dt} = \frac{\partial}{\partial t} + \mathbf{v}_d \cdot \nabla \tag{34}$$

Here $\mathbf{v}_d$ denotes a constant velocity parameter that was interpreted as the velocity of the ether. Hertz' theory was discarded and forgotten at that time, because it spoiled the spacetime symmetry of Maxwell's equations.

Chubykalo and Smirnov-Rueda [2,56] have presented a renovated version of Hertz' theory, that is in accordance with Einstein's relativity principle. For a single point-shaped charged particle moving at the velocity $\mathbf{v}$, the displacement current in Maxwell's equation is modified into a "convection displacement current"

$$\mathbf{j}_{\text{disp}} = \varepsilon_0 \frac{\partial \mathbf{E}}{\partial t} + \varepsilon_0(\mathbf{v} \cdot \nabla)\mathbf{E} \tag{35}$$

The approach by Chubykalo and Smirnov-Rueda further includes longitudinal modes and Coulomb long-range electromagnetic fields that cannot be described by the Lienard–Wiechert potentials [2,57].

V. NEW FEATURES OF PRESENT APPROACH

The extra degree of freedom introduced into the present theory by the nonzero electric field divergence gives rise to new classes of phenomena such as "bound" steady electromagnetic equilibria and "free" dynamic states, including wave phenomena. These possibilities are demonstrated by Fig. 1.

A. Steady Equilibria

The form of the current density term in Eq. (1), as given by expressions (3) and (8), predicts *steady* electromagnetic *equilibria* to exist in vacuo. For such equilibria, Eq. (1)–(6) and (8) combine to

$$c^2 \mathrm{curl}^2 \mathbf{A} = -\mathbf{C}(\nabla^2 \phi) = \frac{\mathbf{C}\bar{\rho}}{\varepsilon_0} \tag{36}$$

Detailed analyses of these equilibria and their applications are given elsewhere [13,15,18,20]. Here we only summarize those parts of the theory that are of interest in connection with wave phenomena, photon physics, and long-range interaction. We later return to Eqs. (36) when discussing the concepts of

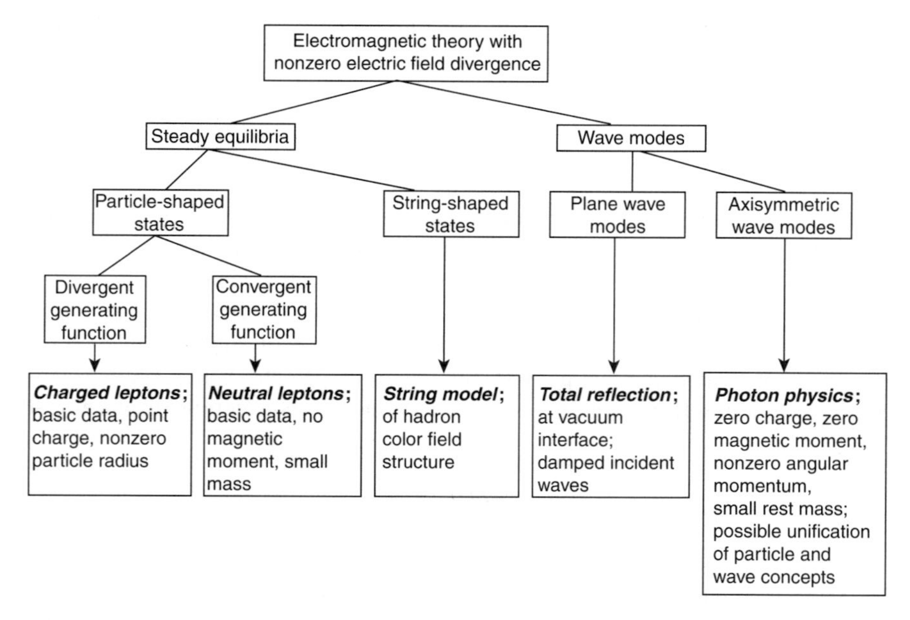

Figure 1. New features introduced by the concept of nonzero electric field divergence in vacuum space. The arrows point to possible areas of application.

instantaneous interaction and long-range forces. A more detailed description of the theory on the equilibrium state is given in Appendix B.

Among the steady states, axisymmetric equilibria are of special interest. These states can be subdivided into two classes: (1) those of "particle-shaped" geometry, where the geometric configuration varies in the axial direction and becomes bounded in both this and the radial directions; and (2) those of "string-shaped" geometry, where the geometric configuration is uniform in the axial direction.

For both these classes the general solution of the electromagnetic field is given in terms of differential operators acting on a *generating function* $CA - \phi$, where the particle-shaped equilibria are treated in a frame (r, θ, φ) of spherical coordinates, with a current density $\mathbf{j} = (0, 0, C\bar{\rho})$, a magnetic vector potential $\mathbf{A} = (0, 0, A)$, and $C = \pm c$. Analogously, the string-shaped equilibria are treated in a frame (r, φ, z) of cylindrical coordinates, with $\mathbf{j} = (0, C\bar{\rho}, 0)$, $\mathbf{A} = (0, A, 0)$, and no dependence on z. The analysis has been limited to separable generating functions

$$F = CA - \phi = G_0 G \qquad G = R(\rho) \cdot T(\theta) \tag{37}$$

where G_0 is a characteristic amplitude, $\rho = r/r_0$ with r_0 as a characteristic radius, R and T as parts of the dimensionless normalized generating function G, and $T(\theta) = 1$ in the case of string-shaped geometry.

1. *Particle-Shaped States*

From the general solutions for particle-shaped states, integrated field quantities

$$q_0 = 2\pi\varepsilon_0 r_0 G_0 J_q \tag{38}$$

$$M_0 = \pi\varepsilon_0 C r_0^2 G_0 J_M \tag{39}$$

$$m_0 = \pi\left(\frac{\varepsilon_0}{c^2}\right) r_0 G_0^2 J_m \tag{40}$$

$$s_0 = \pi\left(\frac{\varepsilon_0 C}{c^2}\right) r_0^2 G_0^2 J_s \tag{41}$$

are obtained where q_0 is the net electric charge, M_0 the magnetic moment, m_0 the mass, s_0 the angular momentum (spin), and (J_q, J_M, J_m, J_s) the corresponding integrals with respect to ρ and θ. These integrals include the charge and mass densities; the latter are given by Einstein's relation for the energy divided by c^2. Here the source energy density w_s of expression (17), and not the field energy density w_f of expression (16), is used when forming the integrals of the mass m_0 and the angular momentum s_0.

Imposition of the quantum condition

$$s_0 = \frac{h}{4\pi} \tag{42}$$

on a model for leptons can in a simple physical picture be regarded as an application of a corresponding periodicity condition for "self-confined" (bound) electromagnetic radiation that circulates around the axis of symmetry.

Depending on the form of the radial part $R(\rho)$ of the generating function, there are two subclasses of particle-shaped axisymmetric equilibria as follows.

a. Convergent Case. A part R that converges at the origin $\rho = 0$ leads to zero net charge q_0 and magnetic moment M_0. Such a result can provide a model for the neutrinos. The solution that is obtained after imposing the spin condition (42) leads to a very small but nonzero value of the quantity $m_0 r_0$, thus allowing for a small mass. Concerning such a model, it has to be pointed out that neutrinos in the laboratory frame move nearly at the speed of light, and that their interaction with the surroundings is weak. The neutrino is neutral and has no color charge.

b. Divergent Case. A part R that diverges at the origin $\rho = 0$ leads to nonzero values of all integrated quantities (38)–(41). These can still become finite when permitting the radius r_0 to shrink to the value of a "point charge," thereby outbalancing the divergence in the integrals (J_q, J_M, J_m, J_s). This applies also to a very small but nonzero radius r_0. One further has to impose the spin condition (42) and a condition on the magnetic moment. In presence of an electromagnetic field the latter becomes

$$H_0 \equiv \frac{M_0 m_0}{q_0 s_0} = \frac{J_M J_m}{2 J_q J_s} = 1 + \delta_F \tag{43}$$

as being related to the Bohr magneton and Feynman's [37] small correction $\delta_F = e^2/4\pi\varepsilon_0 hc = 0.00115965246$. The experimental values of δ_F are 0.00115965221 for the electron and about 0.00116 for the muon. An alternative is to relate the magnetic moment to a free electron, thereby corresponding to half the value given by Dirac.

The present configuration could become a model for charged leptons. With these conditions imposed, the integrated charge q_0 has been given by [20]

$$\frac{|q_0|}{e} = \left(\frac{2\varepsilon_0 ch J_q^2}{e^2 J_s} \right)^{1/2} \tag{44}$$

and is determined by a rather restricted range in parameter space. Thus, detailed analysis shows that there are choices of the generating function by which the value $|q_0| = e$ is covered within such a limited range. To investigate whether it is possible to obtain the exact result $|q_0| = e$, an additional condition has to be imposed. The flux quantization mentioned in Section III.B may provide a candidate for this, combined with variational analysis [13,18,20]. A corresponding electron model is described in Appendix B.

If the result $|q_0| = e$ would come out of a pure theoretical deduction, then the electronic charge would no longer be an independent constant of nature, but would become a quantized charge determined by Planck's constant and the velocity constant c of light, as indicated by Eq. (44). According to relation (43), this would then also apply to the product $M_0 m_0$, whereas all quatitities M_0 and m_0 have thus far not been deduced theoretically for the electron, but have been determined by measurements.

On purely physical grounds it appears to be unacceptable to have a charged particle whose characteristic radius r_0 is strictly equal to zero, and where the particle has no internal structure. Even if experiments as well as the present theory are reconcilable with an extremely small radius, this does not exclude r_0 from being nonzero. In the present model of a steady equilibrium one can conceive electromagnetic radiation to be forced to propagate in circular orbits around the axis of symmetry. This leads to the question of whether such a model has to be modified to include a correction due to general relativity. When passing by a gravitational mass, light is known to be deflected. This effect is proposed here to be "inverted," in the sense that the circular orbit is assumed to give rise to an additional kind of centrifugal force that modifies the steady balance of the bound state represented by Eq. (36). Using the expression for the deflection of a light ray given by Weber [58], this extra force has been introduced into the same equations as a small correction [15,20]. As a result, an equilibrium can be established for a very small but nonzero radius r_0, with a small shift of the equilibrium parameters.

2. *String-Shaped States*

The string-shaped equilibria that result from Eqs. (36) can serve as an analogous model that reproduces several desirable features of the earlier proposed string configuration of the hadron color field structure. These equilibria have a constant longitudinal stress that tends to pull the ends of the configuration toward each other. The magnetic field is thereby located to a narrow channel, and the system has no net electric charge. Since the divergence of the magnetic field is zero, no model based on magnetic poles is needed.

B. Wave Phenomena

The basic equations (1)–(8) also predict the existence of free time-dependent states, in the form of *nontransverse* wave phenomena in vacuo. Combination of the same equations yields

$$\left(\frac{\partial^2}{\partial t^2} - c^2\nabla^2\right)\mathbf{E} + \left(c^2\nabla + \mathbf{C}\,\frac{\partial}{\partial t}\right)(\operatorname{div}\mathbf{E}) = 0 \tag{45}$$

for the electric field. The magnetic field can be determined from the electric field by means of Eq. (2). A divergence operation on Eq. (1) further gives

$$\left(\frac{\partial}{\partial t} + \mathbf{C}\cdot\nabla\right)(\operatorname{div}\mathbf{E}) = 0 \tag{46}$$

In some cases this equation will become useful for the analysis, but it does not introduce more information than that already contained in Eq. (45). As will be shown later, Eq. (46) leads to the same dispersion relation for $\operatorname{div}\mathbf{E} \neq 0$ as Eq. (45) for the wave as a whole.

Three limiting cases can be identified on the basis of Eq. (45):

- When $\operatorname{div}\mathbf{E} = 0$ and $\operatorname{curl}\mathbf{E} \neq 0$, the result is a conventional *transverse* electromagnetic wave, henceforth denoted as an "EM wave."
- When $\operatorname{div}\mathbf{E} \neq 0$ and $\operatorname{curl}\mathbf{E} = 0$, a purely *longitudinal* electric space-charge wave arises, denoted here as an "S wave."
- When both $\operatorname{div}\mathbf{E} \neq 0$ and $\operatorname{curl}\mathbf{E} \neq 0$, a hybrid *nontransverse* electromagnetic space-charge wave appears, denoted here as an "EMS wave." The S wave can be considered as a special degenerate form of the EMS wave.

A general form of the electromagnetic field can be obtained from a superposition of various EM, S, and EMS modes. Thereby it should be observed that the EMS modes can have different velocity field vectors **C**. These wave concepts provide new possibilities in the study of problems in optics and photon physics, both when considering plane waves and axisymetric modes with associated wavepackets.

It should finally be noted that many authors use the term "longitudinal waves" for all modes having at least one field component in the direction of propagation. This would then apply as a common term to both the S and EMS waves.

VI. PLANE WAVES

Because of their relative simplicity, plane waves provide a convenient first demonstration of the wave types defined in the previous section.

A. General Features

The nontransverse plane waves that arise from the present approach are treated in the case of a constant velocity vector **C** and where any field component Q is assumed to have the form

$$Q(x, y, z, t) \equiv Q_0 \exp(i\Theta) \qquad \Theta = -\omega t + \mathbf{k} \cdot \mathbf{r} \tag{47}$$

and ω and $\mathbf{k} = (k_x, k_y, k_z)$ are the frequency and wavenumbers in a rectangular frame with $\mathbf{r} = (x, y, z)$. Equations (1)–(8) then yield

$$c^2 \mathbf{k} \times \mathbf{B} = (\mathbf{k} \cdot \mathbf{E})\mathbf{C} - \omega \mathbf{E} \tag{48}$$

$$\omega \mathbf{B} = \mathbf{k} \times \mathbf{E} \tag{49}$$

There are three types of modes as demonstrated by Fig. 2, with **k** chosen in the z direction and the velocity vector **C** located in the plane perpendicular to **B**.

1. *The Conventional Electromagnetic Mode*

When $\mathbf{k} \cdot \mathbf{E} = 0$ and $\mathbf{k} \times \mathbf{E} \neq 0$, there is a conventional EM wave with a magnetic field according to Eq. (49), and a dispersion relation

$$\omega = \pm kc \tag{50}$$

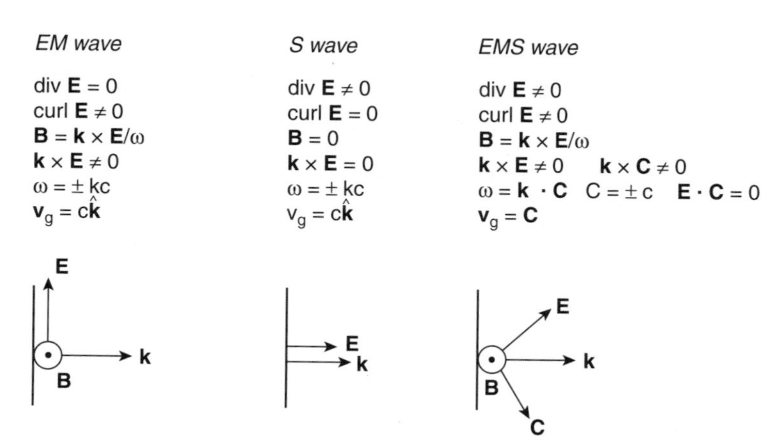

Figure 2. The three fundamental wave types of an extended electromagnetic theory with nonzero electric field divergence in the vacuum, as demonstrated by the simple case of plane waves.

The phase and group velocities are

$$v_p = \pm c \qquad \mathbf{v}_g = \pm c\hat{\mathbf{k}} \qquad \hat{\mathbf{k}} = \frac{\mathbf{k}}{k} \tag{51}$$

where k stands for the modulus of the wavenumber and $\hat{\mathbf{k}}$ for its unit vector. All components of the electric and magnetic fields are perpendicular to the direction of propagation that is along the wave normal.

2. *The Pure Electric Space-Charge Mode*

When $\mathbf{k} \cdot \mathbf{E} \neq 0$ and $\mathbf{k} \times \mathbf{E} = 0$, there is a purely longitudinal S wave without a magnetic field. Thus $\mathbf{C} \times \mathbf{E} = 0$ and $\mathbf{k} \times \mathbf{C} = 0$ due to Eq. (48). The dispersion relation and the phase and group velocities are the same as (51) for the EM wave. The field vectors **E** and **C** are parallel with the wave normal. Possibly this mode may form a basis for telecommunication without induced magnetic fields.

3. *The Electromagnetic Space-Charge Mode*

When both $\mathbf{k} \cdot \mathbf{E} \neq 0$, and $\mathbf{k} \times \mathbf{E} \neq 0$, there is a nontransverse EMS wave with a magnetic field due to Eq. (49). This is the mode of most interest to this context. Here $\mathbf{k} \times \mathbf{C}$ differs from zero, and Eqs. (48) and (49) combine to

$$\left(\omega^2 - k^2c^2\right)\mathbf{E} + (\mathbf{k} \cdot \mathbf{E})\mathbf{F} = 0 \qquad (\mathbf{B} \neq 0) \tag{52}$$

and

$$\mathbf{F} = c^2\mathbf{k} - \omega\mathbf{C} \tag{53}$$

which corresponds to Eq. (45). Scalar multiplication of Eq. (52) by **k**, combined with the condition $\omega \neq 0$, leads to the dispersion relation

$$\omega = \mathbf{k} \cdot \mathbf{C} \qquad (\mathbf{k} \times \mathbf{C} \neq 0) \tag{54}$$

This relation could as well have been obtained directly from Eq. (46). Since **k** and **C** are not parallel in a general case, the phase velocity becomes

$$v_p = \frac{\omega}{k} = \hat{\mathbf{k}} \cdot \mathbf{C} \tag{55}$$

and the group velocity becomes

$$\mathbf{v}_g = \frac{\partial \omega}{\partial \mathbf{k}} = \mathbf{C} \tag{56}$$

Thus the phase and group velocities of the EMS wave differ from each other and also from those of the EM and S waves. The field vectors **E** and **C** have components that are both perpendicular and parallel to the wave normal.

From Eq. (49) we have $\mathbf{k} \cdot \mathbf{B} = 0$ and $\mathbf{E} \cdot \mathbf{B} = 0$. Scalar multiplication of Eq. (53) by **C** in combination with relation (54) further yields $\mathbf{C} \cdot \mathbf{F} = 0$. Combining this result with the scalar product of Eq. (52) with **C**, we obtain $\mathbf{E} \cdot \mathbf{C} = 0$. Finally scalar multiplication of Eq. (48) by **E** results in $\mathbf{E}^2 = c^2\mathbf{B}^2$ when combined with Eq. (49).

4. *Relations between the Plane-Wave Modes*

For the EMS mode it is thus seen that **k** and **E** are localized to a plane perpendicular to **B**, and that **E** and **C** form a right angle. We can introduce the general relation

$$\mathbf{k} \cdot \mathbf{E} = kE(\cos\chi) \tag{57}$$

Conventional theory is then represented by the angle $\chi = \pi/2$ and leads to a single EM mode. Here the same angle stands for the extra degree of freedom introduced by the nonzero electric field divergence, as a result of which a set of possible plane wave solutions is being generated. The set thus ranges for decreasing χ, from the EM mode given by $\chi = \pi/2$, via the EMS modes for $\pi/2 > \chi > 0$, to the S mode where $\chi = 0$. Thus the choice of χ, wave type, and the velocity vector **C** will depend on the boundary conditions and the geometry of the special problem to be considered. An example of this is given later in the discussion of total reflection in Section VI.B.

We finally turn to the momentum and energy balance equation (11)–(15) of Section III.A.2. Since $\bar{\rho}$ is nonzero for the S and EMS modes, these equations will differ from those of the conventional EM mode in vacuo:

- For the S mode both balance equations contain a contribution from $\bar{\rho}\mathbf{E}$ but have no magnetic terms.
- For the EMS mode the momentum balance equation includes the additional forces $\mathbf{F}_e$ and $\mathbf{F}_m$. Because of the result $\mathbf{E} \cdot \mathbf{C} = 0$ the energy balance equation (15) of a plane EMS wave will on the other hand be the same as for the EM wave.

Poynting's theorem for the energy flow of plane waves in vacuo thus applies to the EM and EMS modes, but not to the S mode. Vector multiplication of Eqs. (52) and (53) by **k**, and combination with Eq. (49) and the result $\mathbf{E} \cdot \mathbf{C} = 0$, is easily shown [16,20] to result in a Poynting vector that is parallel with the group velocity **C** of Eq. (56). Later in Section VII.C.3 we shall return to Poynting's theorem in the case of axisymmetric photon wavepackets.

B. Total Reflection at a Vacuum Interface

The process of total reflection of an incident wave in an optically dense medium against the interface of an optically less dense medium turns out to be of particular and renewed interest with respect to the concepts of *nontransverse* and *longitudinal* waves. In certain cases this leads to questions not being fully understood in terms of classical electromagnetic field theory [26]. Two crucial problems that arise at a vacuum interface can be specified as follows:

1. Because of the classical theory of total reflection, the excited electromagnetic field within the less dense medium consists of a nontransverse wave confined to the immediate neighborhood of the bounding surface [35]. When the less dense medium becomes a vacuum region, this may be expected to cause complications. At first glance, matching at a vacuum interface then appears to become impossible by a transmitted electromagnetic (EM) wave with a vanishing electric field divergence. Analysis has shown, however, that such a matching is possible, but only in a dissipation-free case [16,19,20].
2. Additional complications arise when the EM wave in a *dissipative* medium approaches a vacuum interface at an oblique angle [26]. The incident and reflected wave fields then become inhomogeneous (damped) in the direction of propagation. As a consequence the matching at the interface to a conventional undamped electromagnetic wave in vacuo becomes impossible.

Case 2 of a dissipative medium is now considered where $x = 0$ defines the vacuum interface in a frame (x, y, z). The orientation of the xy plane is chosen such as to coincide with the plane of wave propagation, and all field quantities are then independent on z as shown in Fig. 3. In the denser medium (region I) with the refractive index $n_{\mathrm{I}} = n > 1$ and defined by $x < 0$, an incident (i) EM wave is assumed to give rise to a reflected (r) EM wave. Here φ is the angle between the normal direction of the vacuum boundary and the wave normals of the incident and reflected waves. Vacuum region (II) is defined by $x > 0$ and has a refractive index of $n_{\mathrm{II}} = 1$. The wavenumber [35] and the phase (47) of the weakly damped EM waves then yield

$$\Theta_{i,r} = \left(\frac{\omega}{c}\right)[-ct \pm n(\cos\varphi)x + n(\sin\varphi)y + i\bar{\delta}\left(\frac{\omega}{c}\right)n[\pm(\cos\varphi)x + (\sin\varphi)y] \tag{58}$$

with the upper and lower signs corresponding to (i) and (r), and where the damping factor $\bar{\delta} = 1/2\omega\eta\bar{\varepsilon} \ll 1$ with $\bar{\varepsilon}$ denoting the electric permittivity and η

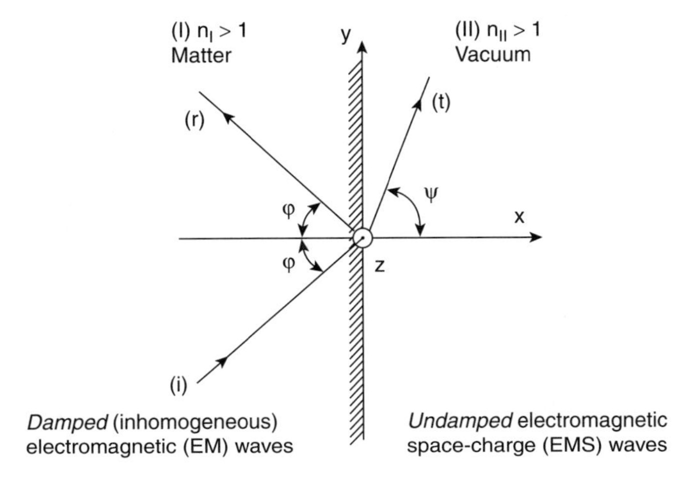

Figure 3. Total reflection of a plane incident damped (inhomogeneous) conventional EM wave at the boundary $x = 0$ between a dissipative medium (I) and a vacuum region (II). The incident and reflected EM waves can be matched at $x = 0$ to undamped transmitted EMS waves in the limit $\pi/2$ of the angle Ψ, but not by an undamped transmitted EM wave in vacuo.

the electric resistivity of medium I. For the phase of a transmitted wave we further adopt the notation

$$\Theta_t = \frac{\omega}{c}(-ct + p_t x + r_t y) + i\left(\frac{\omega}{c}\right)(q_t x + s_t y) \tag{59}$$

where (p_t, r_t, q_t, s_t) are real.

The possibility of matching a transmitted EM wave to the incident and reflected waves is first investigated. This requires the phases (58) to be matched at every point of the interface $x = 0$ to the phase (59). This condition becomes

$$r_t = n_\varphi > 0 \qquad s_t = \bar{\delta} n_\varphi > 0 \qquad n_\varphi = n(\sin\varphi) \tag{60}$$

where total reflection corresponds to $n_\varphi > 1$. For the transmitted EM wave in vacuo, combination of Eqs. (45) and (59) results in

$$1 = p_t^2 + r_t^2 - (q_t^2 + s_t^2) \tag{61}$$

$$\frac{q_t}{s_t} = -\frac{r_t}{p_t} \tag{62}$$

The transmitted wave should further travel in the positive x direction, into region II, and this also applies in the limit where the angle of its wave normal with the vacuum interface approaches the zero value of total reflection. Thus $p_t > 0$. Equations (62) and (60) then yield the condition

$$q_t = -\frac{n_\varphi^2 \bar{\delta}}{p_t} < 0 \tag{63}$$

For total reflection, however, there should be no flow of energy into medium II, and the transmitted wave then must represent an energy flow directed parallel to the interface, thereby limited in amplitude to a narrow layer at the vacuum side of the interface [35]. This excludes the negative value of q_t given by Eq. (63) and the form (59). It does therefore become impossible to match the inhomogeneous (damped) EM waves in region I by a homogeneous (undamped) EM wave in region II. This agrees with an earlier statement by Hütt [26].

Turning instead to the possibility of matching the incident and reflected waves to EMS waves in the vacuum region, we consider the two cases of parallel and perpendicular polarization of the electric field of the incident wave. For an EMS wave the velocity $\mathbf{C}$ is now expressed by

$$\mathbf{C} = c(\cos\beta\cos\alpha,\ \cos\beta\sin\alpha,\ \sin\beta) \tag{64}$$

In combination with the definitions (47) and (59), the dispersion relation (54) of this wave type yields

$$\frac{1}{\cos\beta} = p_t\cos\alpha + r_t\sin\alpha \tag{65}$$

$$q_t\cos\alpha = -s_t\sin\alpha \tag{66}$$

For the factor q_t in Eq. (66) matching of the phases by a transmitted EMS wave then becomes possible when

$$q_t = -s_t tg\alpha = -n_\varphi\bar{\delta} tg\alpha \tag{67}$$

When there is total reflection, the velocity vector $\mathbf{C}$ of Eq. (64) and the corresponding current density (3) can then be directed almost parallel with the interface $x = 0$, that is, when $|\cos\alpha| \ll 1$, $|\sin\alpha| \cong 1$, $tg\alpha < 0$. The EMS wave can then be matched to the slightly decreasing amplitudes of the EMS waves in the positive y direction of the interface. This also implies that q_t in Eq. (67) can be made positive and large for weakly damped EM waves in medium I. Even a moderately large $q_t > 0$ provides the possibility of having a transmitted energy flow along the interface, within a narrow boundary layer, and for an EMS wave amplitude to drop steeply with increasing distance x from the interface. This possibility becomes consistent with the observed physical behavior during total reflection.

As a next step the electric and magnetic fields have to be matched at the interface. This raises three questions that must be faced, in common with those of conventional theory [35]:

1. The first issue is due to the expectation that the transmitted and reflected waves are no longer in phase at the surface $x = 0$ with the incident wave.
2. The second question concerns the amplitude ratios between the reflected and incident waves. For both homogeneous and inhomogeneous incident waves these ratios must have a modulus equal to unity, because no energy loss through the instantaneous reflection process at $x = 0$ is expected.
3. Question 2 leads to a third issue that concerns the energy flow of the transmitted wave in medium II. This flow should be directed along the surface $x = 0$, and be localized to a narrow region near the same surface.

To meet these requirements we first observe that the wavenumber and the phase are coupled to the angles of the velocity **C** given by expression (64). In this way the angle of any transmitted EMS wave in medium II can be expressed in terms of the angles α and β. In analogy with the classical analysis on total reflection, which includes phase differences [35], we introduce a complex form of the angle α of an EMS wave. The definitions

$$\cos\alpha = g_0 \exp(i\bar{\gamma}) = g_0 \cos\bar{\gamma} + ig_0 \sin\bar{\gamma} = (1 - \sin^2\alpha)^{1/2} \tag{68}$$

$$\sin^2\alpha = 1 - g_0^2 \cos 2\bar{\gamma} - ig_0^2 \sin 2\bar{\gamma} \tag{69}$$

are therefore adopted where g_0 and $\bar{\gamma}$ are real and $g_0 > 0$.

The details of the deductions are given elsewhere [16,19,20]; the results can be summarized and discussed as follows:

- For inhomogeneous (damped) incident EM waves the necessary matching of the phases at the vacuum interface can be provided by the nontransverse EMS waves, but not by conventional EM waves in the vacuum region.
- The reflected EM wave arising from an incident inhomogeneous EM wave of plane polarization at an arbitrary angle has a nearly plane polarization when being associated with transmitted EMS waves.
- In the cases of both homogeneous (undamped) and inhomogeneous (damped) incident waves, the transmitted nontransverse EMS waves become confined to a narrow layer at the vacuum side of the interface, and no energy is extracted from the reflection process. The inclusion of EMS waves in a dissipation-free case is, of course, unnecessary and questionable.
- A far-from-simple question concerns the value of the damping factor $\bar{\delta}$, which in physical reality forms the limit between the analysis of

homogeneous and inhomogeneous incident waves. In most experimental situations there is a very large ratio $1/\bar{\delta}$ between the damping length and the wavelength of the incident wave, and this makes it difficult to decide which results on the homogeneous and inhomogeneous cases would be physically relevant. The results on inhomogeneous waves should first become applicable at large enough values of the damping factor $\bar{\delta}$, but this would require large initial amplitudes of the incident wave to give rise to a detectable reflected wave.

VII. AXISYMMETRIC WAVE MODES

As discussed for several decades by a number of authors, the nature of light and photon physics is related not only to the propagation of plane wavefronts but also to axisymmetric wavepackets, the concepts of a rest mass, a magnetic field in the direction of propagation, and an associated angular momentum (spin).

The analysis of plane waves is straightforward in several respects. As soon as we begin to consider waves varying in more than one space dimension, however, we will encounter new phenomena that further complicate the analysis. This also applies to the superposition of elementary modes to form wavepackets. In this section an attempt is made to investigate dissipation-free axially symmetric modes in presence of a nonzero electric field divergence [16,20]. Such a wavepacket configuration could provide a model for the individual photon [19].

In analogy with the treatment of axisymmetric equilibria, we will also seek a model where the entire vacuum space is treated as one entity, without internal boundaries and boundary conditions, thereby also avoiding divergent solutions.

A. Elementary Normal Modes

A cylindrical frame of reference (r, φ, z) is introduced where φ is an ignorable coordinate. In this frame the velocity vector is now assumed to have the form

$$\mathbf{C} = c(0, \cos\alpha, \sin\alpha) \tag{70}$$

with a constant α. We further define the operators

$$D_1 = \frac{\partial^2}{\partial r^2} + \frac{1}{r}\frac{\partial}{\partial r} + \frac{\partial^2}{\partial z^2} - \frac{1}{c^2}\frac{\partial^2}{\partial t^2} \tag{71}$$

$$D_2 = \frac{\partial}{\partial t} + c(\sin\alpha)\frac{\partial}{\partial z} \tag{72}$$

$$D_3 = \frac{\partial^2}{\partial z^2} - \frac{1}{c^2}\frac{\partial^2}{\partial t^2} \tag{73}$$

The basic equations then reduce to

$$\left(D_1 - \frac{1}{r^2}\right)E_r = \frac{\partial}{\partial r}(\operatorname{div}\mathbf{E}) \tag{74}$$

$$\left(D_1 - \frac{1}{r^2}\right)E_\varphi = \frac{1}{c}(\cos\alpha)\frac{\partial}{\partial t}(\operatorname{div}\mathbf{E}) \tag{75}$$

$$D_1 E_z = \left[\frac{\partial}{\partial z} + \frac{1}{c}(\sin\alpha)\frac{\partial}{\partial t}\right](\operatorname{div}\mathbf{E}) \tag{76}$$

and

$$D_2(\operatorname{div}\mathbf{E}) = 0 \tag{77}$$

for the vector field $\mathbf{E}$.

Using the operator (73) we have from Eq. (74)

$$D_3 E_r = \frac{\partial^2 E_z}{\partial r \partial z} \tag{78}$$

Since D_2 commutes with $\partial/\partial r$, combination of Eqs. (76) and (78) yields

$$D_2\left(D_1 - \frac{1}{r^2}\right)E_r = 0 \tag{79}$$

and

$$D_2 D_1 E_z = 0 \tag{80}$$

when steady states defined by $\partial/\partial t = 0$ are excluded. Equation (75) is further combined with Eq. (78) to yield

$$D_3\left(D_1 - \frac{1}{r^2}\right)E_\varphi = \frac{1}{c}(\cos\alpha)\frac{\partial^2}{\partial z\,\partial t}D_1 E_z \tag{81}$$

The set (77) and (79)–(81) of equations corresponds to two branches of solutions:

1. When $D_2 E_r$ and $D_2 E_z$ differ from zero, Eqs. (77)–(80) can be satisfied only when $\operatorname{div}\mathbf{E} = 0$. This, in turn, implies that the right-hand members of Eqs. (74)–(76) all disappear. Consequently, this branch represents a classical electromagnetic (EM) mode with vanishing electric field divergence.

2. When $D_2E_r = D_2E_z = 0$, Eqs. (77) and (79)–(81) can all be satisfied when $\operatorname{div} \mathbf{E} \neq 0$. This branch represents an electromagnetic space-charge (EMS) mode with nonzero electric field divergence in vacuo.

These branches are now discussed for propagating modes depending on z and t as $\exp[i(-\omega t + kz)]$. In fact, there are a number of choices with respect to the form (70) as represented by $\pm\cos\alpha$ and $\pm\sin\alpha$, and that satisfy the condition $C^2 = c^2$ of Eq. (8), thereby corresponding to the two directions along z and φ. From now on we also introduce the normalized radial coordinate $\rho = r/r_0$, where r_0 stands for a characteristic radial dimension.

1. *Conventional Case of a Vanishing Electric Field Divergence*

For branch 1 of a vanishing electric field divergence, the corresponding axisymmetric EM mode is obtained from Eqs. (45) and (74)–(76). Since no dispersion relation for such a mode is available at this point of the deductions, we first introduce the notation

$$\bar{\theta}^2 = \left[k^2 - \left(\frac{\omega}{c}\right)^2\right] r_0^2 \tag{82}$$

with $\bar{\theta}^2 \geq 0$ for phase velocities ω/k that at least do not exceed the limit c. The general solution of the electric field would then become

$$(E_r, E_\varphi) = [(c_{r1}, c_{\varphi 1}) I_1(\bar{\theta}\rho) + (c_{r2}, c_{\varphi 2}) K_1(\bar{\theta}\rho)] \times \exp[i(-\omega t + kz)] \tag{83}$$

$$E_z = [c_{z1} I_0(\bar{\theta}\rho) + c_{z2} K_0(\bar{\theta}\rho)] \times \exp[i(-\omega t + kz)] \tag{84}$$

where c_{r1}, c_{r2}, $c_{\varphi 1}$, $c_{\varphi 2}$, c_{z1}, c_{z2} are arbitrary constants and I_1, K_1, I_0, K_0 are Bessel functions with imaginary argument. At the origin $\rho = 0$ it is known that I_1 vanishes, I_0 becomes finite, and both K_1 and K_0 become infinite. For large ρ, both I_1 and I_0 tend to infinity, whereas K_1 and K_0 tend to zero. Consequently, a nonzero form of the solutions (83) and (84) becomes infinite either at $\rho = 0$ or at large values of ρ.

As a next step we assume the value $\bar{\theta}^2 = 0$ corresponding to a phase velocity where $|\omega/k| = c$. Then Eqs. (74)–(76) reduce to

$$\left[D_\rho - \left(\frac{1}{\rho^2}\right)\right](E_r, E_\varphi) = 0 \qquad D_\rho E_z = 0 \tag{85}$$

where

$$D_\rho = \frac{\partial^2}{\partial \rho^2} + \frac{1}{\rho}\frac{\partial}{\partial \rho} \tag{86}$$

The solutions then have the form

$$(E_r, E_\varphi) \propto k_1\rho + \frac{k_2}{\rho} \qquad E_z \propto k_3 \ln \rho + k_4 \tag{87}$$

where k_1, k_2, k_3, k_4 are constants. This result agrees with that of Eqs. (83) and (84) in the limit $\bar{\theta}^2 = 0$. We then recover the result that the nonzero form of the electric field becomes infinite either at the origin or at infinity, and this also applies to the magnetic field through Eq. (2). This divergent behavior of the EM field in vacuo was realized by Thomson [22] and further by Heitler [59] as well as by Hunter and Wadlinger [23].

Conventional theory thus results in axisymmetric modes in vacuo having the following properties:

- The electric and magnetic fields have components in all three spatial directions, and thus also in the longitudinal direction of propagation.
- The nonzero solutions of these field components either diverge at the origin or become divergent at large distances from the axis of symmetry. Such solutions are therefore not physically relevant to configurations that are extended over the entire vacuum space. The introduction of artificial internal boundaries within the vacuum region would also become irrelevant from the physical point of view, nor would it remove the difficulties with the boundary conditions.

2. *Present Case of a Nonzero Electrical Field Divergence*

From now on we therefore consider branch 2 of the axisymmetric EMS mode.

a. Field Components in the Laboratory Frame. The dispersion relation is obtained from Eqs. (77) and (78), which yield

$$\omega = kc(\sin\alpha) \qquad v = c(\sin\alpha) \tag{88}$$

where the phase and group velocities ω/k and $\partial\omega/\partial k$ are both equal to v. Equation (81) then takes the form

$$\left[\frac{\partial^2}{\partial r^2} + \frac{1}{r}\frac{\partial}{\partial r} - \frac{1}{r^2} - k^2(\cos\alpha)^2\right]E_\varphi = -(tg\alpha)\left[\frac{\partial^2}{\partial r^2} + \frac{1}{r}\frac{\partial}{\partial r} - k^2(\cos\alpha)^2\right]E_z \tag{89}$$

We introduce the function

$$G_0 \cdot G = E_z + (\cot\alpha)E_\varphi \qquad G = R(\rho)\exp[i(-\omega t + kz)] \tag{90}$$

where G_0 is an amplitude factor and $R(\rho)$ is a dimensionless function of ρ. The operator

$$D = D_\rho - \theta^2(\cos\alpha)^2 \qquad D_\rho = \frac{\partial^2}{\partial\rho^2} + \frac{1}{\rho}\frac{\partial}{\partial\rho} \qquad \theta = kr_0 \tag{91}$$

is further defined, and the parameter θ of this last equation should not be confused with the polar coordinate of the spherical frame of reference used in Section V.A and Appendix B. Using Eqs. (89), (90) and (78), the electric and magnetic field components become

$$E_r = -iG_0[\theta(\cos\alpha)^2]^{-1}\frac{\partial}{\partial\rho}[(1-\rho^2 D)G] = -\frac{1}{r_0}\frac{\partial\phi}{\partial\rho} + i\omega A_r \tag{92}$$

$$E_\varphi = G_0(tg\alpha)\rho^2 DG = i\omega A_\varphi \tag{93}$$

$$E_z = G_0\big(1-\rho^2 D\big)G = -ik\phi + i\omega A_z \tag{94}$$

and

$$B_r = -G_0[c(\cos\alpha)]^{-1}\rho^2 DG = -ikA_\varphi \tag{95}$$

$$B_\varphi = -iG_0(\sin\alpha)[\theta c(\cos\alpha)^2]^{-1}\frac{\partial}{\partial\rho}[(1-\rho^2 D)G]$$
$$= ikA_r - \frac{1}{r_0}\frac{\partial A_z}{\partial\rho} \tag{96}$$

$$B_z = -iG_0[\theta c(\cos\alpha)]^{-1}\left(\frac{\partial}{\partial\rho} + \frac{1}{\rho}\right)(\rho^2 DG) = \frac{1}{r_0}\frac{1}{\rho}\frac{\partial}{\partial\rho}(\rho A_\varphi) \tag{97}$$

Consequently, the function G can be considered as a *generating function* from which the entire electromagnetic field of an elementary axisymmetric EMS mode can be determined, in analogy with the generating function (37) of a steady equilibrium state. It should also be observed that G and its derivatives can be chosen to become finite at $\rho = 0$ and zero at $\rho = \infty$. Such a choice then makes it possible for the EMS modes to remain finite and physically acceptable within the entire range of ρ. Insertion into the basic equations confirms the result (92)–(97).

Expressions for the charge density $\bar{\rho}$ and the potentials $\mathbf{A}$ and ϕ are readily obtained from relations (92)–(97) as shown in detail elsewhere [19]. These relations are thus given in the laboratory frame, and they can be considered to correspond to the Lorentz gauge, which is discussed further in Appendix A.

As seen from Eqs. (90)–(97), the elementary axisymmetric EMS mode consists of a three-dimensional propagating configuration that periodically repeats itself along the z axis with the wavelength $2\pi c(\sin\alpha)/\omega$, and where each of the fields **E** and **B** has three nonzero components. This mode can be considered as the axisymmetric correspondence to the plane-wave mode in rectangular geometry.

The result (92)–(97) is reconcilable with that of Evans and Vigier [5–7], in the sense that all field components are nonzero, and thereby also the axial magnetic field B_z. A vanishing component B_z of an axisymmetric field would also be in contradiction with the basic equations. The three axisymmetric magnetic and electric field components of Eqs. (92)–(97) form a helical structure similar to that by Evans and Vigier [5–7] with its cyclic field relations. The present result is, however, not identical with that of Evans and Vigier, because it originates from equations leading to a Proca-type relation (7), which differs from the forms (22) and (26) used by de Broglie, Vigier, and Evans.

b. Field Components in the Rest Frame. The intrinsic properties of a photon, such as a possibly existing rest mass, should be related to a rest frame K', which follows the phase and group velocity v of Eq. (88). In the present case where $v = c(\sin\alpha) < c$, such a "rest frame" becomes physically relevant, but not in the case where $v = c$. For this purpose we make a transformation from the laboratory frame K to the rest frame K'. Introducing

$$\varepsilon \equiv \left[1 - \left(\frac{v}{c}\right)^2\right]^{1/2} = \cos\alpha \qquad v = [0, 0, c(\sin\alpha)] \tag{98}$$

the Lorentz transformation yields

$$r' = r \qquad z' = \frac{z - c(\sin\alpha)t}{\varepsilon} \equiv \frac{\bar{z}}{\varepsilon} \tag{99}$$

where a prime refers to the rest frame henceforth. Thus

$$\frac{\partial}{\partial z'} = \varepsilon\left(\frac{\partial}{\partial z}\right) \qquad k' = \varepsilon k \qquad \theta' = k' r_0 = \varepsilon\theta \tag{100}$$

The normalized generating function then becomes

$$G = R(\rho)\exp[i(-\omega t + kz)] = R(\rho)\exp(ik'z') \equiv G' \tag{101}$$

which is time-independent in the rest frame. Further

$$D = D_\rho - \theta^2(\cos\alpha)^2 = D_\rho - (\theta')^2 \equiv D' \qquad DG = D'G' \tag{102}$$

holds for the operator D' in K'.

The Lorentz transformation is further applied to the electric and magnetic fields, which become

$$\mathbf{E}' = \frac{1}{\varepsilon}\mathbf{E} - \left(\frac{1}{\varepsilon} - 1\right)(\hat{\mathbf{z}} \cdot \mathbf{E})\hat{\mathbf{z}} + \frac{1}{\varepsilon}\mathbf{v} \times \mathbf{B} \tag{103}$$

$$\mathbf{B}' = \frac{1}{\varepsilon}\mathbf{B} - \left(\frac{1}{\varepsilon} - 1\right)(\hat{\mathbf{z}} \cdot \mathbf{B})\hat{\mathbf{z}} - \frac{1}{\varepsilon}\mathbf{v} \times \frac{\mathbf{E}}{c^2} \tag{104}$$

where $\hat{\mathbf{z}} = (0, 0, 1)$. This results in the components

$$E'_r = -iG_0(\theta')^{-1}\frac{\partial}{\partial\rho}[(1 - \rho^2 D')G'] \tag{105}$$

$$E'_\varphi = 0 \tag{106}$$

$$E'_z = G_0(1 - \rho^2 D')G' \tag{107}$$

and

$$B'_r = -G_0 c^{-1}\rho^2 D'G' \tag{108}$$

$$B'_\varphi = 0 \tag{109}$$

$$B'_z = -iG_0(c\theta')^{-1}\left(\frac{\partial}{\partial\rho} + \frac{1}{\rho}\right)(\rho^2 D'G') \tag{110}$$

Here we observe that the axial components $E_z = E'_z$ and $B_z = B'_z$ are invariant during the Lorentz transformation, as given by Eqs. (103) and (104).

The result obtained can also be interpreted in the way that the field components in K' are obtained from those in K by replacing the angle α by $\alpha' = 0$, specifically, by replacing the velocity vector $\mathbf{C}$ in expression (70), which refers to the frame K by the vector

$$\mathbf{C}' = c(0, 1, 0) \qquad (\alpha' = 0) \tag{111}$$

which refers to the rest frame K'. This further supports the adopted form for the velocity vector $\mathbf{C}$. In the frame K' the current density (3) has a component in the φ direction only, and it circulates around the axis of symmetry, thereby generating purely poloidal fields $\mathbf{E}'$ and $\mathbf{B}'$, that is, where $E'_\varphi = 0$ and $B'_\varphi = 0$. This situation is similar to that of the "bound" steady equilibrium state described in Section V.A.

From relations (105)–(110) expressions are also obtained for the charge density $\bar{\rho}'$ and the potentials $\mathbf{A}'$ and ϕ', as given elsewhere [19]. These expressions in the rest frame K' can be considered to correspond to the Coulomb gauge.

B. Wavepackets

To form a photon-like particle, the elementary normal EMS modes now have to be superimposed to create a wavepacket of finite axial extensions and of finite linewidth in wavelength space. Here we are free to choose an amplitude factor G_0 of the generating function (90) having the form

$$G_0 = g_0(\cos\alpha)^2 \tag{112}$$

where g_0 is constant.

The normal modes are further superimposed to form a wavepacket having the amplitude

$$A_k = \left(\frac{k}{k_0}\right)\exp[-z_0^2(k-k_0)^2] \tag{113}$$

within the wavenumber interval dk and centered around the wavenumber k_0. Integration of the modes given by Eqs. (92)–(97) is then represented by the integrals

$$P_\mu = \int_{-\infty}^{+\infty} k^\mu A_k \exp[ik(z-vt)]\,dk \qquad v = c(\sin\alpha) \tag{114}$$

Introducing the variable

$$p = z_0(k-k_0) + \frac{i\bar{z}}{2z_0} \qquad \bar{z} = z - vt \tag{115}$$

the integral (114) can be written as

$$P_\mu = \left(\frac{k_0^\mu\sqrt{\pi}}{z_0}\right)\exp\left[-\left(\frac{\bar{z}}{2z_0}\right)^2 + ik_0\bar{z}\right]\cdot\left[1 + f\left(\frac{\bar{z}}{z_0}, \frac{1}{k_0 z_0}\right)\right] \tag{116}$$

where f is a polynomial in terms of the quantities $\bar{z}/z_0$ and $1/k_0z_0$. The case $k_0z_0 \gg 1$ is of most physical interest, because it represents a small linewidth and will be adopted in the following deductions. Then the contribution from f in Eq (116) can be dropped with good approximation. The case of small linewidth is demonstrated by the form (113), where the amplitude in k space drops to $1/e$ of its maximum value for $\Delta k = k - k_0 = 1/z_0$. Then the linewidth $\Delta k/k_0 = 1/k_0z_0$ becomes small for $k_0z_0 \gg 1$.

Applying Eqs. (114) and (116) to the field quantities (92)–(97) and introducing the notation

$$E_0 = E_0(\bar{z}) = \left(\frac{g_0}{r_0}\right)\left(\frac{\sqrt{\pi}}{k_0z_0}\right)\exp\left[-\left(\frac{\bar{z}}{2z_0}\right)^2 + ik_0\bar{z}\right] \tag{117}$$

the wavepacket field components now become

$$E_r = -iE_0[R_5 + (\theta_0')^2 R_2] \tag{118}$$

$$E_\varphi = E_0\theta_0(\sin\alpha)(\cos\alpha)[R_3 - (\theta_0')^2 R_1] \tag{119}$$

$$E_z = E_0\theta_0(\cos\alpha)^2[R_4 + (\theta_0')^2 R_1] \tag{120}$$

$$B_r = -\left(\frac{1}{c}\right)(\sin\alpha)^{-1}E_\varphi \tag{121}$$

$$B_\varphi = \left(\frac{1}{c}\right)(\sin\alpha)E_r \tag{122}$$

$$B_z = -i\left(\frac{1}{c}\right)E_0(\cos\alpha)[R_8 - (\theta_0')^2 R_7] \tag{123}$$

and

$$\bar{\rho} = -i\left(\frac{\varepsilon_0}{r_0}\right)E_0[R_6 + (\theta_0')^2 R_9 - (\theta_0')^4 R_1] \tag{124}$$

$$\psi \equiv \phi + \mathbf{C}\cdot\mathbf{A} = ir_0E_0[2R_4 - R(\cos\alpha)^2 + 2(\theta_0')^2 R_1] \tag{125}$$

where

$$\theta_0 = k_0 r_0 \qquad \theta_0' = \theta_0(\cos\alpha) \tag{126}$$

and

$$R_1 = \rho^2 R \qquad R_2 = \frac{d}{d\rho}(\rho^2 R) \tag{127}$$

$$R_3 = \rho^2 D_\rho R \qquad R_4 = (1 - \rho^2 D_\rho)R \tag{128}$$

$$R_5 = \frac{d}{d\rho}[(1 - \rho^2 D_\rho)R] \quad R_6 = D_\rho[(1 - \rho^2 D_\rho)R] \tag{129}$$

$$R_7 = \left(\frac{d}{d\rho} + \frac{1}{\rho}\right)(\rho^2 R) \qquad R_8 = \left(\frac{d}{d\rho} + \frac{1}{\rho}\right)(\rho^2 D_\rho R) \tag{130}$$

$$R_9 = D_\rho(\rho^2 R) - (1 - \rho^2 D_\rho)R \tag{131}$$

In the rest frame K' analogous deductions can be made for small linewidths because

$$k_0' z_0' = k_0 z_0 \gg 1 \tag{132}$$

according to the Lorentz transformation of Eqs. (99) and (100). Likewise, we introduce

$$E'_0 = E'_0(z') = \left(\frac{g_0}{r_0}\right)\left(\frac{\sqrt{\pi}}{k'_0 z'_0}\right)\exp\left[-\left(\frac{z'}{2z_0}\right)^2 + ik'_0 z'\right] \tag{133}$$

and obtain the field quantities $E'_\varphi = B'_\varphi = 0$ and

$$E'_r = -iE'_0(\cos\alpha)\,[R_5 + (\theta'_0)^2 R_2] \tag{134}$$

$$E'_z = E'_0\theta'_0(\cos\alpha)[R_4 + (\theta'_0)^2 R_1] \tag{135}$$

$$B'_r = -\left(\frac{1}{c}\right)E'_0\theta'_0(\cos\alpha)[R_3 - (\theta'_0)^2 R_1] \tag{136}$$

$$B'_z = -i\left(\frac{1}{c}\right)E'_0(\cos\alpha)[R_8 - (\theta'_0)^2 R_7] \tag{137}$$

$$\bar{\rho}' = -i\left(\frac{\varepsilon_0}{r_0}\right)E'_0(\cos\alpha)[R_6 + (\theta'_0)^2 R_9 - (\theta'_0)^4 R_1] \tag{138}$$

$$\psi' \equiv \phi' + \mathbf{C}' \cdot \mathbf{A}' = ir_0 E'_0(\cos\alpha)[2R_4 - R + 2(\theta'_0)^2 R_1] \tag{139}$$

We recall that these results are approximate, valid for small linewidths, where $(z'/2z_0)^2$ varies much more slowly with z' than $k'_0 z'$ in expression (133). In this case conditions such as $\text{div}'\mathbf{B}' = \text{div}\,\mathbf{B} = 0$ will be satisfied only approximately by expressions (136) and (137).

In the limit of zero linewidth of a needle-shaped axisymmetric wavepacket, having infinite length z_0 in the direction of propagation, the present deduction would reduce "backward" to the elementary normal mode investigated in Section VII.A.2.

Here we also notice that, for a photon with nonzero rest mass, the *intrinsic* properties of the wavepacket are expected to be clearly visible in the rest frame K'. Among other things, this applies to the components (134)–(137), which then represent *steady* electric and magnetic fields being entirely localized to the $r'z'$ plane, thereby also having strong components in the axial direction of propagation. This property clearly supports the photon model with a *static* magnetic field part $\mathbf{B}^{(3)}$ as deduced by Evans and Vigier [5–7,45], and where the axial electric and magnetic field components of Eqs. (107), (110), (135), and (137) are invariant to the Lorentz transformation.

C. Integrated Field Quantities

Considered as a particle, the photon constitutes a unique concept, because it may never be seen at rest in the laboratory frame. Nevertheless, integrated field quantities such as the total charge q, magnetic moment M, rest mass m_0, and angular momentum (spin) s are properties that could be attributed to the photon wavepacket model and the "rest frame" K' as shown by the present and following deductions.

To proceed with the analysis, we first observe that the wavepacket components (E_φ, E_z, B_r) in the laboratory frame are in phase with the generating function G, whereas the components (E_r, B_φ, B_z) are 90° out of phase with G according to Eqs. (118)–(123). Similarly, (E'_z, B'_r) in the rest frame are in phase with G', whereas (E'_r, B'_z) are 90° out of phase with G' as shown by Eqs. (134)–(137). In the analysis that follows we choose the normalized generating functions (90) and (101) to be symmetric with respect to the axial centra $\bar{z} = 0$ and $z' = 0$ of the wavepackets. With $\bar{z} = z - c(\sin\alpha)t$, we thus have

$$G = R(\rho)\cos k\bar{z} \qquad G' = R(\rho)\cos(k'z') \tag{140}$$

where the real parts of expressions (90) and (101) have been adopted.

1. *Charge and Magnetic Moment*

In the laboratory frame the integrated electric charge is given by

$$q = \varepsilon_0 \int \operatorname{div}\mathbf{E}\, dV = \varepsilon_0 \int \mathbf{n}\cdot\mathbf{E}\, dS = 0 \tag{141}$$

where dV and dS are volume and surface elements, respectively, and the integration is extended over entire space. Here the surface integral vanishes because of the choice of the generating function, which should vanish at infinity as well as all its derivatives. The analogous result $q' = 0$ is obtained in the rest frame.

The integrated magnetic moment in the laboratory frame becomes

$$\begin{aligned} M &= \frac{1}{2}\varepsilon_0 c \int r(\operatorname{div}\mathbf{E})\, dV \\ &= \pi\varepsilon_0 c \int_{-\infty}^{+\infty}\int_0^{\infty}\left[r\frac{\partial}{\partial r}(rE_r) + r^2\frac{\partial}{\partial z}E_z\right]dr\, d\bar{z} \\ &= \pi\varepsilon_0 c \int_0^{\infty}\left\{r\frac{\partial}{\partial r}\left[r\int_{-\infty}^{+\infty}E_r\, d\bar{z}\right] + r^2[E_z]_{-\infty}^{+\infty}\right\}dr = 0 \end{aligned} \tag{142}$$

because E_z vanishes at $\bar{z} = \pm\infty$, and E_r of Eq. (118) has an antisymmetric form due to the factor $\sin k\bar{z}$ when the choice (140) is made for the generating function G. The analogous result $M' = 0$ is obtained in the rest frame. Observe that the *local* magnetic fields $\mathbf{B}$ and $\mathbf{B}'$ of Eqs. (121)–(123) and (136)–(137) are *nonzero* even when the total magnetic moments M and M' vanish. For

$M = M' = 0$, the fields $\mathbf{B}$ and $\mathbf{B}'$ will only decrease more rapidly toward zero at an increasing distance from the origin than in cases where M and M' would become nonzero.

As in the case of the neutrino model of Section V.A.I.a, the photon could have a structure with both positive and negative local contributions to the charge and magnetic moment that add up to zero when integrated over the total volume.

2. *Mass*

To deduce expressions for the integrated mass and angular momentum, the radial part $R(\rho)$ of the generating function (140) has to be specified. Here we make the choice

$$R = \rho^{-\gamma} \exp\left(-\frac{1}{\rho}\right) \qquad \rho = \frac{r}{r_0} \qquad \gamma \gg 1 \tag{143}$$

which is convergent both at the axis $\rho = 0$ and at infinity. We further consider integrals of the form

$$J_\mu = \int_0^\infty F_\mu \, d\rho \qquad F_\mu = \rho^{-\mu} \exp\left(-\frac{2}{\rho}\right) \tag{144}$$

The integrand F_μ increases from zero at $\rho = 0$ to a maximum at $\rho = \hat{\rho}_\mu$ where

$$\hat{\rho}_\mu = \frac{\hat{r}_\mu}{r_0} = \frac{2}{\mu} \tag{145}$$

and it then drops very steeply toward zero for increasing values of ρ beyond $\rho = \hat{\rho}_\mu$ when $\mu \gg 1$. Here

$$\frac{\hat{r}_\mu}{\hat{r}_{\mu-1}} = \frac{\mu - 1}{\mu} \rightarrow 1 \qquad (\mu \gg 1) \tag{146}$$

The integral (144) is evaluated by making the substitution $p = 2/\rho$, and then becomes

$$J_\mu = 2^{-(\mu-1)}(\mu - 2)! \tag{147}$$

which yields

$$\frac{J_\mu}{J_{\mu-1}} = \frac{\mu - 2}{2} \rightarrow \frac{\mu}{2} \qquad (\mu \gg 1) \tag{148}$$

The integrated total mass in the laboratory frame K is now given by

$$m = \frac{W}{c^2} \tag{149}$$

where expressions (17), (124), and (125) are used to perform the integration of Eq. (18). For a volume element $dV = 2\pi r\, dr\, d\bar{z}$ we first carry out the integration with respect to $\bar{z}$. With $E_0(\bar{z})$ given by Eq. (117) this leads to the integrals

$$J_z = \int_{-\infty}^{+\infty} \exp\left[-2\left(\frac{\bar{z}}{2z_0}\right)^2\right] \left\{ \begin{matrix} (\sin k_0\bar{z})^2 \\ (\cos k_0\bar{z})^2 \end{matrix} \right\} d\bar{z} = z_0 \left(\frac{\pi}{2}\right)^{1/2} \tag{150}$$

for a small line width. Introducing the notation

$$a_0 = \frac{\varepsilon_0 \pi^{5/2} g_0^2 z_0}{c^2 (k_0 z_0)^2 \sqrt{2}} \tag{151}$$

the mass becomes

$$m = a_0 \bar{J}_m \qquad \bar{J}_m = \int_0^{\infty} W_{ms}\, d\rho \tag{152}$$

where

$$W_{ms} = \rho[R_6 + (\theta_0')^2 R_9 - (\theta_0')^4 R_1] \cdot [2R_4 - R(\cos\alpha)^2 + 2(\theta_0')^2 R] \tag{153}$$

In the rest frame K' the corresponding results are

$$J_z' = z_0' \left(\frac{\pi}{2}\right)^{1/2} \tag{154}$$

and

$$a_0' = \frac{\varepsilon_0 \pi^{5/2} g_0^2 z_0'}{c^2 (k_0' z_0')^2 \sqrt{2}} \qquad k_0' z_0' = k_0 z_0 \tag{155}$$

The integrated mass in this frame becomes

$$m' = a_0' \bar{J}_m' \qquad \bar{J}_m' = \int_0^{\infty} W_{ms}'\, d\rho \tag{156}$$

where

$$W_{ms}' = \rho[R_6 + (\theta_0')^2 R_9 - (\theta_0')^4 R_1] \cdot [2R_4 - R + 2(\theta_0')^2 R] \tag{157}$$

From expressions (128) and (143) combined with relations (144) and (148), we can easily see that the integrals (152) and (156) of W_{ms} and W'_{ms} become equal in the limit of large γ. The same result comes out if we use the alternative form w_f of Eq. (16) for the energy density in combination with the field components (118)–(123) and (134)–(137). According to Eqs. (151), (152), (155), and (156) and the Lorentz transformation (99), we now have

$$m' = \varepsilon m \qquad \varepsilon = \cos\alpha \tag{158}$$

On the other hand, when introducing the phase and group velocities v of expressions (88), the energy relations due to Planck, Einstein, and de Broglie result in

$$h\nu = \frac{h\omega}{2\pi} = mc^2 = m_0c^2\left[1 - \left(\frac{v}{c}\right)^2\right]^{-1/2} = \frac{m_0c^2}{\cos\alpha} \tag{159}$$

where m_0 is the rest mass. Consequently, the mass m' in the rest frame K' of the present theory becomes identical with the conventional form for the rest mass m_0. This can be taken as an additional confirmation of the performed deductions.

3. *Momentum and Energy Balance in an Axisymmetric Case*

The momentum balance of the electromagnetic field is governed by Eqs. (11)–(15). Here the local forces

$$\mathbf{f}_e = \bar{\rho}\,\mathbf{E} \qquad \mathbf{f}_m = \mathbf{j} \times \mathbf{B} = \bar{\rho}\,\mathbf{C} \times \mathbf{B} \tag{160}$$

and the Poynting vector **S** are considered in the laboratory frame K. Because of the axial symmetry of the present configuration, all integrated components in the r and φ directions of the cylindrical frame (r, φ, z) will vanish. To investigate the contributions to the volume integrals of Eqs. (12) and (13), we can thus restrict ourselves in studying the symmetry properties with respect to the axial direction, that is, to the center $\bar{z} = 0$ of the propagating wavepacket. With the chosen symmetric form (140) for G, it is then seen from Eqs. (118)–(124) that $(\bar{\rho}, E_r, B_\varphi, B_z)$ become antisymmetric with respect to $\bar{z}$, whereas (E_φ, E_z, B_r) become symmetric.

In addition, Eqs. (118)–(124) show that $(\bar{\rho}, E_r, B_\varphi)$ are of zero order in the smallness parameter ε of the definition (98), (E_φ, B_r, B_z) are of first order in ε, and E_z of second order.

Because of the symmetry conditions integration of the local forces (160) results in vanishing electric and magnetic volume forces $\mathbf{F}_e$ and $\mathbf{F}_m$ as given by Eqs. (13). Of the Poynting vector **S** and the electromagnetic momentum vector **g** of the propagating wavepacket, only the component due to $E_r B_r - E_r B_z$ is of interest in forming the integral of expression (12), that is, in the z direction of propagation.

Concerning the local energy balance of Eq. (15), it is finally seen that the term

$$\bar{\rho}\,\mathbf{E}\cdot\mathbf{C} = \bar{\rho}(E_\varphi C_\varphi + E_z C_z) \tag{161}$$

is antisymmetric and has a vanishing volume integral. Therefore Poynting's conventional energy theorem holds for the integrated work of the electromagnetic forces in the present axisymmetric configuration.

4. *Angular Momentum*

As a consequence of the results just obtained, it is now possible to use the conventional form

$$\mathbf{s} = \mathbf{r} \times \frac{\mathbf{S}}{c^2} \tag{162}$$

for the density of the angular momentum as given by Schiff [39] and Morse and Feshbach [31], where $\mathbf{r}$ represents the radius vector from the origin.

In the laboratory frame $E_r B_z$ is of first order in ε; and $E_z B_r$ of third order. Further, $cB_r \cong -E_\varphi$ for small ε according to Eq. (121). Therefore the density of angular momentum can be written as

$$\mathbf{s} = \pm\left(\frac{r}{c}\right) w_f (\cos\alpha)\hat{\mathbf{z}} \tag{163}$$

where $\hat{\mathbf{z}} = (0, 0, 1)$,

$$w_f \cong \varepsilon_0 E_r^2 \tag{164}$$

and the two spin directions in equation (163) correspond to the two alternative choices of $\mathbf{C}$ as given by Eq. (8). This is analogous to the two spin directions in a steady particle-shaped state as given by expression (41). The modulus of the integrated angular momentum now becomes

$$s = \left(\frac{1}{c}\right)(\cos\alpha)\int r w_f \, dV \tag{165}$$

For the integrated mass the alternative expression

$$m = \left(\frac{1}{c^2}\right)\int w_f \, dV \tag{166}$$

can then be used as given by Eq. (18), and will be equivalent to expression (152).

5. *Quantum Conditions*

According to Eqs. (165), (166), (118), and (122) and the quantum conditions for the photon to behave as a boson, the mass and angular momentum now become

$$m = a_0 \bar{J}_m = \frac{h\nu}{c^2} \qquad \bar{J}_m = \int_0^\infty W_{mf}\, d\rho \tag{167}$$

$$s = a_0 r_0 c \bar{J}_s = \frac{h}{2\pi} \qquad \bar{J}_s = \int_0^\infty \rho(\cos\alpha) W_{mf}\, d\rho \tag{168}$$

where

$$W_{mf} = \rho[R_5 + (\theta_0')^2 R_2]^2 \tag{169}$$

Combination of Eqs. (167) and (168) results in

$$\omega = 2\pi\nu = \frac{c\bar{J}_m}{r_0 \bar{J}_s} \tag{170}$$

Insertion of the radial function (143) into expressions (129) and (127) for R_5 and R_2 yields a corresponding form for W_{mf} of Eq. (169), which then consists of 10 terms. Integration of each of these terms by means of Eqs. (144) and (147) yields expressions for $\bar{J}_m$ and $\bar{J}_s$, which in the limit of large γ give rise to the relation

$$\frac{\bar{J}_m}{\bar{J}_s}(\cos\alpha) = \gamma \quad (\gamma \gg 1) \tag{171}$$

The radii $\hat{r}_\mu$ of Eq. (145), which define the maximum of each term in W_{mf} thus converge for increasing γ values toward a common value $\hat{r}$, as given by Eq. (146). Consequently

$$\frac{\bar{J}_m}{r_0 \bar{J}_s}(\cos\alpha) = \frac{\gamma}{r_0} = \frac{1}{\hat{r}} \tag{172}$$

in the limit of large μ, that is, where $\mu \cong 2\gamma \gg 1$. Combination of relations (172) and (170) finally yields

$$2\pi\hat{r}(\cos\alpha) = \frac{c}{\nu} = \frac{\lambda_0}{\sin\alpha} \simeq \lambda_0 \tag{173}$$

with very good approximation for $(\cos\alpha)^2 \ll 1$, and where $\lambda_0 = 2\pi/k_0$ is the mean wavelength of the wavepacket that has a small linewidth, namely,

$k_0 z_0 \gg 1$ and $\nu \cong \nu_0 = ck_0/2\pi$. This result thus applies to large γ values of the generating function, where the integrands W_{mf} and ρW_{mf} of the mass (167) and angular momentum (168) have sharply peaked maxima located at the radius $\hat{r}$.

The result (173) applies to a photon model with the angular momentum $h/2\pi$ of a boson, whereas the photon radius $\hat{r}$ would become half as large for the angular momentum $h/4\pi$ of a fermion. Moreover, the present analysis on superposition of EMS normal modes is applicable not only to narrow linewidth wavepackets but also to a structure of short pulses and soliton-like waves. In these latter cases the radius in Eq. (173) is expected to be replaced by an average value resulting from a spectrum of broader linewidth.

VIII. FEATURES OF PRESENT INDIVIDUAL PHOTON MODEL

The present axisymmetric model of the individual photon becomes associated with a number of important questions discussed in the current literature.

A. The Nonzero Rest Mass

The question of the possible existence of a nonzero photon rest mass was raised by Einstein [42], Bass and Schrödinger [43], de Broglie and Vigier [44], and further by Evans and Vigier [5], among others. It includes such crucial points as the relation to the Michelson–Morley experiment, and the so far undetermined value of such a mass and its experimental determination.

1. *Comparison with the Michelson–Morley Experiment*

The velocity of the earth in its orbit around the sun is about $10^{-4}c$. If this would also turn out to be the velocity with respect to a stationary ether, and massive photons would move at the velocity $v = c(\sin\alpha)$ in the same ether, then the velocity u of photons recorded at the earth's surface would become

$$u = \frac{v + w}{1 + (vw/c^2)} \tag{174}$$

with $w = \delta \cdot c$ and $\delta = \pm 10^{-4}$. Introducing

$$\sin\alpha = 1 - \frac{1}{2}(\cos\alpha)^2 - \frac{1}{8}(\cos\alpha)^4 - \cdots \equiv 1 - f \tag{175}$$

the departure from c of the recorded photon velocity would be given by

$$\begin{aligned} 1 - \frac{u}{c} &= f\frac{[1 - \delta(2 - f) + \delta^2(1 - f)]}{1 - \delta^2(1 - f)^2} \\ &\cong \frac{1}{2}(\cos\alpha)^2(1 \pm 2\delta) \end{aligned} \tag{176}$$

From Eq. (176) the following conclusions can be drawn regarding recorded velocities:

- For $\cos\alpha \leq 10^{-4}$, corresponding to a photon rest mass $m_0 < 0.74 \times 10^{-39}\,\text{kg} \cong 10^{-9} m_e$, a change in the eighth decimal of the recorded velocity of light can hardly be detected.
- With the same assumption of $\cos\alpha \leq 10^{-4}$, and when turning from a direction where $\delta = +10^{-4}$ to the opposite direction where $\delta = -10^{-4}$, the change in $1 - (u/c)$ would even become much less; 10^{-12}. Also such a value hardly becomes detectable.

Consequently, there should be no noticeable departure in recorded velocity from the Michelson–Morley experiments when the photon rest mass is changed from zero to about $10^{-39}\,\text{kg} \cong 10^{-9} m_e$ or less. For a photon rest mass in the range $10^{-68} < m_0 < 10^{-45}\,\text{kg}$ as considered by Evans and Vigier [5], this departure would become extremely small and very hard to detect. Still the physics with such a nonzero mass becomes fundamentally different from that being based on a photon mass, which is exactly equal to zero. This is also clearly demonstrated by the present analysis and its results in the laboratory and "rest" frames, as given by the Lorentz transformation.

In all approaches with a nonzero photon rest mass the velocity c should be considered as an asymptotic limit at infinite energy that can never be fully approached in physical reality by a single photon *in vacuo*.

2. *The Undetermined Value of the Rest Mass*

In the case of the EMS mode of Eq. (45), the limit of zero rest mass corresponds to $\cos\alpha = 0$. In this limit where $\cos\alpha$ and m_0 are *exactly* equal to zero, the result is like that of a conventional axisymmetric EM mode that either diverges at the axis or at infinity, and must be discarded as pointed out in Section VII.A.1 on branch 1 of solutions. Therefore the present results hold only for a nonzero rest mass, but this mass can be allowed to become very small. This implies that the quantum conditions $mc^2 = h\nu$ for the total energy and $s = h/2\pi$ for the angular momentum are satisfied for a whole class of small values of $\cos\alpha$ and the corresponding rest mass.

As pointed out by de Broglie and Vigier [25], this indeterminableness of the photon rest mass appears to be a serious objection to the underlying theory. The problem is that the derivations depend simply on the *existence* of the nonzero rest mass, but not on its *magnitude*. To this mass in their analysis, de Broglie and Vigier use examples of other "macroscopic quantum effects" that have been considered in theoretical physics, such as ferromagnetism and the possibility of indefinite precision in measuring Planck's constant. Additional examples are given in this context, such as those of the electron and neutrino masses and of the electric charge discussed in Section V.A.1, and where independent theoretical deductions of their absolute values have not been possible to make so far.

Thus, the uncertainty in the absolute value of the nonzero photon rest mass does not necessarily imply that the corresponding theory is questionable, but rather could be due simply to some so far "hidden" extra condition or refinement that may have to be added at a later stage to obtain such a value.

3. *Possible Methods for Determination of the Photon Rest Mass*

For the determination of the photon rest mass, the following considerations can be of importance:

- From the hypothesis of a nonzero electrical conductivity in the vacuum and the corresponding dispersion relation [20,48, 50–52], the concepts of "tired light" and the observed cosmical redshift could be interpreted and associated with a nonzero photon mass of about 10^{-68} kg. The related frequency dependence can also become a measure of the mass.
- Anisotropic effects of the recorded frequency of cosmic microwave background radiation have been proposed for photon rest mass determination [20].
- The Vigier mass of the photon being associated with the de Broglie wavelength λ_B is

$$m_0 = \frac{h}{\lambda_B c} \tag{177}$$

  where $\lambda_B = 10^{26}$ m is put equal to the radius of the universe [48,60].
- Possibly a deeper understanding of the details in certain interference phenomena, such as the Goos–Hänchen effect and the Sagnac effect mentioned in Section II, list items 4 and 6 could provide estimates of the photon rest mass. As shown by de Broglie and Vigier [44] and Vigier [61], these effects can have explanations in terms of such a mass.

B. The Photon as a Particle with an Associated Wave

An essential feature of de Broglie's picture of the wave-particle duality consists of regarding the particle and the associated wave as simultaneously existing physically real entities. Relations (70), (111), (118)–(123), (134)–(137), (158), (159), and (170) of the present wavepacket model in the frames K and K' are consistent with such a picture, which could be considered as a "hybrid" system of unified wave and particle nature.

In the laboratory frame K this electromagnetic field configuration has the total energy $h\nu = mc^2$. Thereby the fraction $(m - m_0)c^2$ can be regarded as the energy of a "free" pilot wave of radiation, and the fraction m_0c^2 as the energy of a "bound" particle state of "self-confined" radiation. The rest mass m_0 thus represents an integrating part of the total field energy. In the φ direction the "bound" radiation "moves" around the z axis of symmetry at the velocity $C'_\varphi = c$ according to Eq. (111). With the sharply defined radius $\hat{r}$ of Eq. (146), this part of the radiation field becomes associated with a frequency

$$\nu_0 = \frac{c}{2\pi\hat{r}} \tag{178}$$

of revolution around the z axis in the rest frame K'. In combination with relations (173) and (159), and for $\sin\alpha \cong 1$, this yields

$$\nu_0 = \nu(\cos\alpha) = \nu\left(\frac{m_0}{m}\right) \tag{179}$$

as also being supported by the idea that all parts of the EM field energy should be included in the same way in the total energy $h\nu$. The relation

$$h\nu_0 = m_0 c^2 \tag{180}$$

by de Broglie is then recovered through Eq. (179). In K the pilot wave thus becomes associated with the component $C_z = c(\sin\alpha)$, and the rest mass with the component $C_\varphi = c(\cos\alpha)$.

Provided that m_0/m is independent of the frequency ν, the results (178)–(180) and (88) thus permit the angle α also to be independent of ν. This would have two important consequences:

- For a constant angle α the phase and group velocities and the velocity vector $\mathbf{C}$ of Eqs. (88) and (70) become independent of frequency. Then there are no dispersion effects that would cause signals of different wavelengths to have different propagation times over large cosmical distances. Even if there would arise a frequency dependence of α, the extremely small rest mass mentioned in Section VIII.A.1 would lead to small dispersion effects, even at large cosmical distances.
- The two-frequency paradox [6] by de Broglie can be resolved, in the sense that the frequency $\nu_0 = \nu'$ is coupled to the frequency ν by relation (179).

If Eqs. (178)–(180) would not become fully satisfied, thereby leading to a frequency dependence for α, then Eq. (159) would result in a dispersion relation that depends on the rest mass m_0 as discussed in Section VIII.A.3.

C. The Electric Charge, Angular Momentum, and Longitudinal Field

The local electric charge density of the present theory can have either sign. In the photon model, however, the boundary conditions on the electric field cause the total integrated charge to vanish as given by Eq. (141). A way out of the discussed problem earlier of zero net charge of the photon is to assert that the photon is its own antiphoton [5]. The result of Eq. (141) provides an alternative to this.

At this point a question arises as to the possibility of having an expression for the angular momentum also in the rest frame K'. However, such a question is not simple, because the definition of a Poynting vector in the rest frame of the wavepacket is not straightforward.

As in the case of the electric charge, there are local contributions of various signs to the total integrated magnetic moment. The latter vanishes in the present analysis where these contributions outbalance each other. In spite of this, there is a nonzero local magnetic field that has a nonzero component in the axial direction that is invariant to Lorentz transformations in the same direction, and is associated with a nonzero photon rest mass. This intrinsic axial magnetic field is reconcilable but not identical with that of Evans and Vigier [5,6,45], who have based their approach on a different form of 4-current (26) in the Proca-type equation (22). As expected, the intrinsic current system and magnetic field are purely poloidal in the rest frame, whereas the magnetic field in the laboratory system is helical. The helicity of the photon field has also been considered in an analysis by Evans [28] and Dvoeglazov [62].

In this connection it should finally be mentioned that Comay [63] and Hunter [64] have discussed the $\mathbf{B}^{(3)}$ field concept by Evans and Vigier on the basis of conventional electromagnetic theory where the 4-current of Eq. (22) vanishes in the vacuum. Their analysis leads to the obvious conclusion that the $\mathbf{B}^{(3)}$ field vanishes in such a case. This does, however, not rule out the existence of $\mathbf{B}^{(3)}$ when there is a nonzero 4-current of the type (26) introduced by de Broglie, Vigier and Evans. Thus, without a Proca-type equation (22), no steady-state magnetic "spin field" can exist in a rest frame K'.

D. The Photon Radius

The present photon model is partly supported by experiments performed. Thus far in earlier investigations on two-slit interference phenomena of individual photons in the ultraviolet range [21], dot-shaped marks were observed at a screen, as in a similar case with electrons. Such a concentrated local energy behavior is hard to reconcile with a plane wave, even when the distribution of photon impacts exhibits a probability character given by quantum theory. In the case of impinging individual photons, these marks seem to be consistent with a limited radial extension of the wavepacket as given by Eq. (173), at least when $\cos\alpha \geq 10^{-5}$. Apart from the factor $\cos\alpha$, Eq. (173) agrees with a form by Hunter and Wadlinger [23] which appears to be consistent with microwave transmission experiments. Even if both these forms sometimes lead to a rather limited photon diameter, however, there remains the question how a photon can release a single electron at atomic dimensions, such as in the phtoelectric effect.

E. The Thermodynamics of a Photon Gas

With a nonzero rest mass one would at a first glance expect a photon gas to have three degrees of freedom: two transverse and one longitudinal. This would alter Planck's radiation law by a factor of $\frac{3}{2}$, in contradiction with experience [20]. A detailed analysis based on the Proca equation shows, however, that the $\mathbf{B}^{(3)}$ spin field cannot be involved in a process of light absorbtion [5]. This is also made plausible by the present model of Sections VII and VIII, where the spin field is "carried away" by the pilot field. As a result, Planck's law is recovered in all practical cases [20]. In this connection it has also to be observed that transverse photons cannot penetrate the walls of a cavity, whereas this is the case for longitudinal photons which would then not contribute to the thermal equilibrium [43].

The equations of state of a photon gas have been considered by Mézáros [29] and Molnar et al. [30]. It has thereby been found that Planck's distribution and the Wien and Rayleigh–Jeans laws cannot be invariant to an adiabatic change of state occurring in an ensemble of photons. The dilemma that arises is due to the fact that the changes of state cannot be adiabatic and isothermal at the same time. It is probable that the cause of this contradiction can be the lack of a longitudinal magnetic flux density, such as the field $\mathbf{B}^{(3)}$ by Evans and Vigier, in the original and standard treatments. Thus, in an adiabatically deformed photon gas the intensity will change in time, and so will the field $\mathbf{B}^{(3)}$. These important questions may require further investigation.

F. Tests of the Present Model

The results achieved so far that support the present individual photon model can be summarized as follows:

- The total reflection of an incident wave in a dissipative medium that is bounded by a vacuum region
- The convergence of the axisymmetric solutions, which has no counterpart in conventional theory
- The needle-shaped wavepacket solutions of the individual photon model, which agree at least sometimes with the dot-shaped marks on a screen in interference experiments
- The nonzero angular momentum which does not exist for circularly and plane polarized conventional electromagnetic waves
- The contradictions that arise in the thermodynamics of a photon gas, and that are likely to be clarified by having a longitudinal magnetic field part

At this stage there is an open question as to whether other radial parts of the generating function, exist being similar to that of Eq. (143) and giving results same as or similar to those just mentioned. Thus the photon model and its internal structure could also be tested in observations of various scattering processes.

IX. NONLOCALITY AND SUPERLUMINOSITY

Another class of electromagnetic phenomena beyond the concepts of conventional theory, has generated an increasing interest, namely, when considering instantaneous long-range interaction as well as signals propagating at velocities greater than c. Investigations within this new field are still at a preliminary stage, also including questionable concepts and interpretations. Suggestions are made here as to how the present approach could be included in the analysis.

A. General Questions

It was pointed out by Dirac [34] that, as long as we are dealing only with transverse waves, we cannot bring in the Coulomb interactions. There must then also arise longitudinal interactions between pairs. In fact, as already argued by Faraday and Newton and further stressed by Chubykalo and Smirnov-Rueda [2], among others, instantaneous long-range interaction takes place not instead of but along with the short-range interaction in classical field theory. This point of view has also been expressed by Pope [65] in stating that instantaneous "action at a distance" and the finite speed of light are generally considered as antithetical, but it is well known that in relativistic physics light has both finite and infinite speed. Thus c is not a velocity but is a spacetime constant having the dimensions of velocity. In this connection Argyris and Ciubotariu [66] point out that the unquantized longitudinal–scalar part of the field yields the Coulomb potential, and that transverse photons transport energy whereas longitudinal (virtual) photons do not carry energy away. Thus, there is direct interaction between a transverse photon and the gravitation field of a black hole, but not with a longitudinal photon. The Coulomb field is therefore able to cross the event horizon of a black hole. The $\mathbf{B}^{(3)}$ field concept should also be understood to be related to this nonlocalized action at a distance [28].

In addition to the long-range interaction, there is also the phenomenon of superluminosity, defined by signals that can propagate and possibly carry energy at a finite speed but that exceeds the value c. There are a number of observations reviewed by Recami [27] indicating that such motions could exist and can be associated with tachyon particles.

B. Instantaneous Long-Range Interaction

The problem of nonlocality due to instantaneous long-range interaction is now considered both in electromagnetic and in gravitational theory.

1. The Electromagnetic Case

In their considerations on the field generated by a single moving charged particle, Chubykalo and Smirnov-Rueda [2,56,57] have claimed the Lienard–Wiechert potentials to be incomplete. These potentials are then not able to describe long-range instantaneous Coulomb interaction. However, in a modified theory by Chubykalo and Smirnov-Rueda such interaction is included. The applicability of these potentials is, however, still under discussion [9].

A further analysis of the concept of long-range instantaneous interaction has been presented [2,56,57] in which the Proca field equation is divided into two pairs. The first of these manifests the instantaneous and longitudinal aspect of electromagnetic nature, as represented by functions $f[\mathbf{R}(t)]$ of an implicit time dependence. For a single charge system this would then lead to the form $\mathbf{R}(t) = \mathbf{r} - \mathbf{r}_q(t)$, where $\mathbf{r}$ is a fixed vector from the point of observation to the origin, and $\mathbf{r}_q(t)$ is the position of the moving charge. The implicit time dependence then implies that all explicit time derivatives disappear from the basic equations of the first part of the divided pairs. The second part is responsible for transverse wave phenomena, as represented by functions $\mathbf{g}(\mathbf{r}, t)$ of an explicit time dependence.

The basic equations (1)–(8) of the present extended formulation also include Coulomb interaction. Likewise, these equations can be split into one part representing long-range instantaneous interaction having an implicit time dependence, and another part that represents signals having a nonzero propagation time and an explicit time dependence. The first part is then given by Eqs. (1)–(8) without partial time derivatives, thereby reducing to a form being analogous to the general equations (36) of a steady state. The second part defined by Eqs. (1)–(8) with an explicit time dependence includes propagating transverse EM waves, as well as S and EMS waves.

The subdivision of the solutions into parts of implicit and explicit time dependence could in fact be interpreted in two ways:

1. For sufficiently slow time variations, the explicit time derivatives can be neglected and the solutions of a quasisteady equilibrium are then obtained with good approximation. Such a situation arises when the sources of the EM field, as given by the charges and currents, vary slowly as compared to the time required for a wave to pass from its source to the field point in question. Then a corresponding interpretation of an implicit time dependence would not be reconcilable with the absolute concept of instantaneous long-range interaction. It would merely be due to the fact that the time variations of the sources have been chosen slow enough for the propagation time of the wave to appear infinitely short, somewhat in analogy with the situation of Achilles and the turtle.

2. There are, however, a number of arguments that could support the existence of long-range instantaneous interaction. Thus, superluminal phenomena

cannot be explained from Maxwell's equations without longitudinal wave concepts [28]. The energy of "longitudinal modes" cannot be stored locally in space but can be spread by an arbitrary velocity [62]. Moreover, nonlocality behaviour is supported by observations as a fundamental property of the universe [67]. There are also several quantum-mechanical arguments in favor of long-range interaction, such as that of the Aharonov–Bohm effect [68] and those raised in connection with the Einstein–Podolsky–Rosen thought experiment. With these arguments, and with the similarity between EM and gravitational long-range interaction described in the next subsection (IX.B.2), it should be justified to use the part of implicit time dependence in the basic equations as a theoretical model for instantaneous long-range interaction.

Instantaneous action at a distance, represented here by longitudinal components, can thus be interpreted as a classical equivalent of nonlocal quantum interactions [2].

2. *The Gravitational Case*

There are some similarities between electromagnetism and gravitation that could be important to the investigation of long-range interaction. Thus, there is a resemblance between the Coulomb and Newton potentials [66]. A holistic view of electromagnetism and gravitation would imply that action at a distance occurs in a similar way in gravitation [28], and vice versa. This is supported by the general principle that there exists no screen against gravitational forces acting between distant massive sources [48]. In other words, the position, velocity, and acceleration of a source of gravity would then be felt by the target body in much less than the light–time between them.

C. Superluminosity

In addition to the long-range interaction at infinite speed, there are proposals for the existence of superluminal phenomena that should manifest themselves in propagation at a finite speed, but being larger than c. There now seems to be observational evidence for such phenomena, as well as indications thereof in the theoretical analysis, but further investigation is needed.

1. *Observational Evidence*

The longstanding idea of superluminal motion has become subject to renewed interest, due to a number of recent discoveries and observations, as described in a survey by Recami [27]. Thus the squared mass of muon–neutrinos is found to be negative. There are further observations that can be interpreted as superluminal expansions inside quasars, in some galaxies and in galactic objects. Also, so called "X-shaped waves" have been observed [69] to propagate at a velocity larger than c. Finally Nimtz et al. [70] have performed an experiment where a wavepacket was used to transmit Mozart's Symphony No. 40 at a speed of $4.7c$ through a tunnel formed by a barrier of 114 mm length. There is a difficulty in the interpretation of tunneling experiments, because no group velocity can be defined in the barrier traversal that is associated with evanescent waves.

2. *Theoretical Analysis*

Reviews on the theoretical analysis of superluminal phenomena have been presented by Barut et al. [71] on tachyons and by Olkovsky and Recami [72] on tunneling processes. A special study has been conducted by Walker [73] on the propagation speed of a longitudinally oscillating electric field, generated along the axis of vibration of an electric charge and based on Lienard–Wiechert potentials. It was shown that both the phase and group velocities of the field were infinite next to the charge, and then decayed rapidly to the speed of light in one wavelength. In the zero-frequency limit this analysis gave results of a nonlocal character, and their relation to or contradiction with the theory by Chubykalo and Smirnov-Rueda [2,56,57] is not clear at this stage.

The central assumption underlying the standard approach to tachyon theory is that the usual Lorentz transformation also applies to the superluminal case. One therefore simply takes the Lorentz factor $[1-(v/c)^2)]^{1/2}$ and substitutes $v > c$ into it [27,74]. This leads directly to an imaginary rest mass and propagation time for tachyons, with many difficulties of interpretation [74].

These central concepts of tachyon theory also come out of the present approach. An alternative way to satisfy the condition (8) of Lorentz invariance is thus to replace the form (70) of the velocity vector $\mathbf{C}$ by

$$\mathbf{C} = c(0, i \sinh \alpha, \cosh \alpha) = c(0, C_\varphi, C_z) \tag{181}$$

in an axisymmetric case. For propagating normal modes of the form $\exp[i(-\omega t + kz)]$, Eq. (46) for an EMS-like tachyon mode then yields the dispersion relation

$$\omega = kC_z = kv \qquad v = c(\cosh \alpha) \tag{182}$$

which replaces relation (88). Thus the phase and group velocities are both equal to $v > c$ for a given value of α.

If we further assume that relations (159) for a photon can be adopted and modified to apply to the tachyon case, the result in combination with Eq. (182) would become

$$h\nu = mc^2 = m_0 c^2 \left[1 - \left(\frac{v}{c}\right)^2\right]^{-1/2} = \frac{m_0 c^2}{i(\sinh \alpha)} \tag{183}$$

or

$$m_0 = \frac{i(\sinh\alpha)h\nu}{c^2} = i(\sinh\alpha)m \qquad m_0^2 < 0 \tag{184}$$

Consequently, this leads to what could be considered as an imaginary rest mass, as is also obtained in current tachyon theory.

The introduction of the form (181) for $\mathbf{C}$ into the basic equation would lead to formal solutions analogous to those of Section VII for the axisymmetric EMS mode and wavepacket. Physical interpretations of such solutions are assumed to be difficult and have not been attempted at this stage. Here we shall only end the discussion with some preliminary and speculative points related to the imaginary rest mass. It is first noticed that an ansatz

$$m_0 c^2 = h\nu' \tag{185}$$

analogous to that by de Broglie for the photon would lead to an imaginary frequency

$$\nu' = \frac{m_0 c^2}{h} = i\nu(\sinh\alpha) = \frac{ic(\sinh\alpha)(\cosh\alpha)}{\lambda_0} \tag{186}$$

with λ_0 as the average wavelength of a propagating tachyon wavepacket. This frequency and rest mass are related to the component C_φ of the velocity vector $\mathbf{C}$, and can be associated with a balance condition in the φ direction. Second, when returning to expression (173) for the photon wave packet, and by comparing the forms (70) and (181), an analogous expression

$$2\pi\hat{r} = \frac{\lambda_0}{\cosh\alpha} \tag{187}$$

could be attempted for the tachyon radius. With the imaginary velocity C_φ of "revolution" in the φ direction, the corresponding imaginary frequency would then become

$$\nu' = \frac{C_\varphi}{2\pi\hat{r}} = \frac{ic(\sinh\alpha)(\cosh\alpha)}{\lambda_0} \tag{188}$$

which is the same result as that of Eq. (186). The interpretation of this imaginary frequency is difficult, and it is not clear that it could become associated with a damping factor and a limited tachyon life time.

X. THE WAVE AND PARTICLE CONCEPTS OF A LIGHT BEAM

At this point it is appropriate to return to the general wave and particle concepts of light. To be more specific, we can ask whether an individual photon can be treated both as a plane wave and as a particle, and whether a broad light beam can be considered both as a stream of individual photon particles and as a plane wave. For a closer examination of these questions it becomes necessary to investigate every particular case with respect to the following points:

- The choice of theoretical representation has to be reconcilable with the geometric configuration to be studied.
- The boundary conditions impose certain constraints.
- The initial conditions by which light is being generated will influence the choice of representation.

The analysis that follows is based partly on a preliminary and *tentative* approach [75].

A. The Individual Photon

In many cases the photon can be represented by the two alternative models of a plane wave and a particle-like wavepacket. This should also apply to interference phenomena with individual photons [21]. For a given point at the screen of an experiment with two apertures, the resulting interference pattern obtained from individual photon impacts could thus be interpreted in two alternative ways:

- The photon is as a plane wave divided into two parts that pass through either of the apertures, and then interfere at the screen.
- The photon is as an axisymmetric wavepacket divided into two parts that pass the apertures and interfere with each other when ending up at the screen to form a common dot-shaped mark.

Irrespective of whether the photon is considered as a plane wave or a wavepacket of narrow radial extension, it must thus be divided into two parts that pass each aperture. In both cases interference occurs at a particular point on the screen. When leading to total cancellation by interference at such a point, for both models one would be faced with the apparently paradoxical result that the photon then destroys itself and its energy $h\nu$. A way out of this contradiction is to interpret the dark parts of the interference pattern as regions of forbidden transitions, as determined by the conservation of energy and related to zero probability of the quantum-mechanical wavefunction.

The photon has the energy $h\nu$ both at its source and at the screen. The division into two parts of the photon during its flight would at first sight be in conflict with the quantization of energy. A possible solution of this problem can, however, be found in terms of the Heisenberg uncertainty principle. A randomness in phase can be assumed to arise when the plane wave parts are transferred into dot-shaped geometry when hitting the screen, or when the axisymmetric wavepacket parts are formed and reunited during their flight. Since both photon models include long wavetrains in the direction of propagation, the average uncertainty in phase of the interference process is expected to be of the order of half a wavelength λ. At the surface of the screen this can in its turn be interpreted as an uncertainty $\Delta t \cong \lambda/2c = 1/(2\nu)$ in time. This finally leads to an uncertainty $\Delta E \cong h/(2\pi\Delta t) = h\nu/\pi$ in energy. The latter would then become of the order of half the energy $h\nu$, thus being carried by each photon part that passes through the apertures.

These questions appear to be understandable in terms of both photon models. The wavepacket axisymmetric model has, however, an advantage of being more reconcilable with the dot-shaped marks finally formed by an individual photon impact on the screen of an interference experiment. If the photon would have been a plane wave just before the impact, it would then have to convert itself during the flight into a wavepacket of small radial dimensions, and this becomes a less understandable behavior from a simple physical point of view. Then it is also difficult to conceive how a single photon with angular momentum (spin) could be a plane wave, without spin and with the energy $h\nu$ spread over an infinite volume. Moreover, with the plane-wave concept, each individual photon would be expected to create a continuous but weak interference pattern that is spread all over the screen, and not a pattern of dot-shaped impacts.

An individual axisymmetric photon wavepacket that propagates in vacuo and meets a mirror surface, should be reflected in the same way as a plane wave, on account of the matching of the electromagnetic field components at the surface. Inside a material with a refraction index greater than that in vacuo, the transmission of the wavepacket is affected by interaction with atoms and molecules, in a way that is outside the scope of the discussion here.

Whenever applicable, the present wavepacket model of Section VII can be regarded as a unification of the wave and particle concepts of an individual photon. It propagates as a wave, and at the same time has a static part that is associated with the intrinsic properties of spin, a nonzero rest mass, and a static magnetic and electric field part. Although it is never at rest, the photon thus has many features in common with other particles such as leptons, which, in turn, also can behave as waves according to de Broglie. The geometric structure of the individual "localized" photon wavepacket of the present simplified theoretical model is thereby not in conflict with the quantum-theoretical wavefunction that represents the probability distribution of a photon before its position has been localized through a measurement.

B. Density Parameters of a Broad Beam of Wavepackets

A light beam is now being considered that consists of a stream of individual axisymmetric photon wavepackets of narrow linewidth and where the macroscopic breadth of the beam is much larger than the individual photon radius $\hat{r}$ of Eq. (173). The volume density of the wavepackets is assumed to be uniform in space. Then the mean distance between the centra of the wavepackets becomes

$$d = \left(\frac{1}{n_p}\right)^{1/3} \tag{189}$$

With the energy $h\nu = hc/\lambda_0$ of each photon, the energy flux per unit area is given by

$$\psi_p = \frac{n_p\, \nu hc}{\lambda_0} \cong \frac{n_p hc^2}{\lambda_0} \qquad \lambda_0 = \frac{2\pi}{k_0} \tag{190}$$

according to Eq. (88) when $(\cos\alpha)^2 \ll 1$. Combination of relations (189), (190), and (173) then yields a ratio

$$\theta_\perp = \frac{d}{2\hat{r}} = \pi\left(\frac{hc^2}{\psi_p \lambda_0^4}\right)^{1/3}(\cos\alpha) \tag{191}$$

of the mean transverse separation distance between the photons and the diameter $2\hat{r}$ of a single photon. Multiphoton states [23] are not considered here, but could somewhat modify the analysis.

Since $\hat{r}$ represents a sharply defined radius according to the theory described earlier, there is a critical value $\theta_{\perp c} \cong 1$ of the ratio (191). It corresponds to

$$\left(\psi_p \lambda_0^4\right)_{\perp c}(\cos\alpha)^{-3} \cong \pi^3 hc^2 \cong 1.85 \times 10^{-15}\,[\mathrm{W} \cdot \mathrm{m}^2] \tag{192}$$

Equation (191) is related to an earlier analysis of multiphoton phenomena [23] for which its left-hand member would have to be replaced by the value $\theta_\perp = (\pi/2)^{2/3}$.

In analogy with Eq. (191) there is also a ratio of the mean longitudinal separation distance between two photon wavepackets situated on a common axis and the individual wavepacket length $2z_0$, as given by

$$\theta_\parallel = \frac{d}{2z_0} \ll \theta_\perp \qquad (z_0 \gg \hat{r}) \tag{193}$$

for small linewidths. Here $\theta_{\|c} \cong 1$ is the corresponding critical value of $\theta_{\|}$.

The regimes of the parameters (191) and (193) are now subdivided with respect to overlapping of the individual photon fields:

(I) $\theta_{\perp} > \theta_{\perp c}$: There is no transverse overlapping, and there exist two subregimes.

- (Ia) $\theta_{\perp} \gg \theta_{\|} > \theta_{\|c}$: There is no longitudinal overlapping.
- (Ib) $\theta_{\perp} \gg \theta_{\|} < \theta_{\|c}$: There is longitudinal overlapping.

(II) $\theta_{\perp} < \theta_{\perp c}$: There is both transverse and longitudinal overlapping.

1. Longitudinal Field Overlapping

When $\theta_{\|} < \theta_{\|c}$ the photon wavepackets that are lined up after each other on the same axis can match their phases and combine into more elongated packets. As a result the linewidth of a single photon is expected to become larger than that of a dense photon beam. The local energy flux density ψ_p should at the same time be given by the number of photons passing a cross-sectional area, regardless of the value of the ratio $\theta_{\|}$.

2. Transverse Field Overlapping

The case of transverse field overlapping is far more complicated than that of longitudinal overlapping. According to the previous analysis, the field energy is limited mainly within a well-defined radius $\hat{r}$. As long as the ratio $\theta_{\perp}$ exceeds the critical value $\theta_{\perp c}$, the photon wavepackets of the beam will not overlap and will have hardly any mutual interaction. The beam then behaves as a stream of individual photon particles, that is, when representing each photon as an axisymmetric wavepacket.

When $\theta_{\perp}$ decreases beyond $\theta_{\perp c}$, however, a rapidly increasing overlapping of the individual photon fields would take place. An extensive analysis of moderately large overlapping in the range near $\theta_{\perp c}$ is expected to become complicated, and will not be undertaken here, where we limit the discussion to the range $\theta_{\perp} \ll \theta_{\perp c}$ of strong overlapping. For this latter range the following points apply:

- If the individual wavepacket solutions of the present theory could be superimposed, this would imply that the field vectors become *multivalued* at every point inside the photon beam. The individual photon fields would then have to cancel each other. This implies that the axisymmetric small-scale wavepacket solution of Section VII does not apply and cannot satisfy the basic Eqs. (1)–(8) in the case of a nearly plane (one-dimensional) and broad photon-dense beam configuration.
- From this mutual cancellation of the individual photon fields an apparently paradoxical conclusion would follow, namely, that the beam energy gradually vanishes as $\theta_{\perp}$ decreases beyond $\theta_{\perp c}$. To preserve the energy

flux, such a conclusion then must be combined with an additional hypothesis, as will be shown in the following section.

When considering the model of a beam that consists of axisymmetric photon wavepackets, it can finally be seen from relation (191) that transverse overlapping would not occur for visible light at very small energy fluxes, whereas such overlapping should arise at the energy fluxes of strong laser light in the visible regime as well as for electromagnetic waves in the radiofrequency regime.

C. Energy Flux Preservation

Within any parameter regime, the quantization and preservation of the beam energy flux has to be imposed as a necessity. Thus, to preserve this flux we propose a tentative approach by assuming that the increasing deficit of beam energy due to overlapping and cancellation of the axisymmetric EMS wavepacket fields is compensated by the energy contribution due to a simultaneously appearing and increasing plane wave of the EM or EMS type as being defined earlier. This assumption is also supported by the requirement of having a wave system with phase and group velocities in the direction of the beam, and with wavefronts perpendicular to the direction of propagation. Consequently we make the ansatz

$$\psi_p = \psi_{\text{EMS}} + \psi_{\text{PL}} \tag{194}$$

where ψ_{EMS} and ψ_{PL} are the contributions to the total energy flux from the individual EMS fields and from the plane wave, respectively. This proposed scenario can thus be described as follows:

- In the regime $\theta_\perp \gg \theta_{\perp c}$ of negligible field overlapping the beam consists of a stream of individual EMS wavepackets, each with an energy $h\nu$ and rest mass m_0, propagating at the velocity $v = c(\sin\alpha)$ and giving rise to the energy flux $\psi_p = \psi_{\text{EMS}}$.
- When $\theta_\perp$ decreases to values $\theta_\perp \cong \theta_{\perp c}$, field overlapping begins to influence the beam and a rather complex hybrid state is established. In the case of a plane EM wave, EMS and EM fields are then superimposed and coupled, with a reduced rest mass, and propagating at a velocity v in the range between $c(\sin\alpha)$ and c, thereby giving rise to a combined energy flux (194). The coupling of these two wave fields has been discussed to some detail elsewhere [75].
- When finally reaching the regime $\theta_\perp \ll \theta_{\perp c}$, the EMS wave fields of axisymmetric wavepackets would be canceled by overlapping, and only the plane wave therefore remains as a possible solution of the basic equations in the present macroscopic beam geometry. A plane EM wave propagates at the velocity $v = c$, thereby giving rise to an energy flux

$\psi_p = \psi_{PL}$ for a beam consisting of photons, each having the energy $h\nu$. In this regime we are thus back to the plane-wave representation, here applying to the collective behavior of photons in a beam. Plane EMS waves could also be involved under these conditions, such as in the special process of total reflection described in Section VI.B.

A question that can be raised in this tentative approach concerns the angular momentum. In the case of a low-density beam of negligible EMS field overlapping, all the individual axisymmetric photon wavepackets will carry their own spin. For a high-density beam that has been converted into a plane EM wave, however, there are two reasons why it is difficult to see how there could exist a total nonzero spin related to the beam energy flux ψ_p: (1) the plane geometry makes it hard to picture how an angular momentum could arise for a wave system of "flat" appearance and infinite spatial extension; and (2) the extended approaches that are based on a nonzero 4-current in the Proca-type equation (22) lead to photon models with intrinsic spin, whereas this does not result from the transverse wave solutions of conventional electromagnetic theory [5,20,25]. Such a conclusion is also consistent with that by Heitler [59], who points out that a plane wave of infinite extension in the transverse direction cannot have any angular momentum about an axis in the direction of propagation. Further, for an axisymmetric circularly polarized beam of finite circular cross section, there arises an angular momentum that can have contributions from individual photons at its periphery, somewhat like the contributions from the gyro motion of charged particles at the boundary of a magnetized plasma body. In any case, these questions require further analysis in which quantum-mechanical aspects may also have to be included.

D. Beam Conditions for Wave or Particle Representation

The preceding discussion thus indicates that the basic equations, with their initial and boundary conditions, will determine whether a ray of comparatively large transverse dimensions is to be represented by a plane wave or by a beam of particle-like photon wavepackets. Thus two general cases should be observed:

- The condition $\theta_\perp \ll \theta_{\perp c}$ of strong transverse overlapping is both necessary and sufficient for the ray to be represented by a plane wave, because the basic equations do not permit axisymmetric individual photon wavepacket solutions to exist in such a case.
- The condition $\theta_\perp \gg \theta_{\perp c}$ of negligible transverse overlapping is necessary but not sufficient for the ray to be represented by a beam of axisymmetric wavepackets, because the basic equations can then in principle be satisfied both in plane wave and axisymmetric wavepacket geometry.

Energy flux preservation in the sense of the ansatz of Section X.C and Eq. (194) should hold in both these cases, that is, irrespective of the value of $\theta_\perp$.

1. *Initial Conditions*

The initial conditions of the source can contribute to the character of the emitted electromagnetic radiation. Thus emission from excited atoms occurs in the form of individual photon wavepackets and can give rise to particle-like photon beams under appropriate conditions. Electromagnetic radiation that is excited by the current system of a macroscopic antenna is, on the other hand, expected to produce nearly plane waves at large distances from the source.

2. *Boundary Conditions*

The boundary conditions can also have a decisive influence on the type of representation. A first example is given by the transmitted wave at a vacuum boundary, as discussed in Section VI.B. Here the incident and reflected plane waves can be matched at the interface by a plane transmitted wave, but hardly by a transmitted beam of axisymmetric photon wavepackets.

A second example is given by the microwave transmission experiments [23] mentioned at the end of Section VIII.D. These were performed in presence of an adjustable aperture. A cutoff of the transmitted power was observed at a certain aperture size, in agreement with theory [23]. The question then arises how the earlier discussed overlapping effects become reconcilable with such a result. The answer is that microwave transmission in a device of limited size differs in many respects from the behavior of a broad beam. An important feature of the former case is due to the boundary conditions imposed by the aperture, in which an essential part of the overlapping fields is being "scraped off." As a result of this, the single photon picture could still be relevant to the experimentally observed behavior. But it should then also be remembered that light transmission through a circular hole becomes strongly reduced according to conventional thory when the wavelength exceeds the hole diameter [76].

Large wavelengths and low energies $h\nu$ favour transverse overlapping according to Eq. (191). To avoid such overlapping, and thereby satisfy the boundary conditions of an axisymmetric individual photon, the energy flux ψ_p has to be chosen at an extremely low level. Thus, to observe a "giant" axisymmetric photon wavepacket in the radiofrequency regime, the energy flux of equation (192) would have to become extremely small.

XI. CONCLUDING REMARKS

The present theory has been developed in terms of an extended Lorentz invariant form of the electromagnetic field equations, in combination with an addendum of necessary basic quantum conditions. From the results of such a simplified approach, theoretical models have been obtained for a number of physical systems. These models could thus provide some hints and first approximations to the properties of the individual photon, light beams, and charged and neutral leptons. However, at least certain parts of the same models must, in a future and more rigorous approach, be substituted by a quantization of the extended field equations from the beginning. Possibly the present models may still represent the "most probable states" of a corresponding rigorous quantum-theoretical approach.

Last but not least, it must be further investigated whether light manifests itself differently under different conditions. One of these manifestations is represented by an axisymmetric solution of the present theory, which has the nonzero angular momentum of a boson particle. Another is represented by a plane-polarized wave having zero angular momentum.

As pointed out by Sarfatti [77], the basic form (7) of the four-current is related to the result of anholonomic electrodynamics.

APPENDIX A: THE LORENTZ CONDITION

With the introduction of the two potentials $\mathbf{A}$ and ϕ, the basic equations (1)–(6) can be recast into the system

$$\Box \mathbf{A} + \nabla L = -\mu_0 \mathbf{j} \tag{A.1}$$

$$\Box \phi - \frac{\partial}{\partial t} L = -\frac{\bar{\rho}}{\varepsilon_0} \tag{A.2}$$

where

$$L = \operatorname{div} \mathbf{A} + \frac{1}{c^2}\frac{\partial \phi}{\partial t} = \left[\nabla, -\left(\frac{i}{c}\right)\frac{\partial}{\partial t}\right] \cdot \left[\mathbf{A}, \frac{i\phi}{c}\right] \tag{A.3}$$

In terms of O(3) covariant derivatives, Evans et al. [78] have shown that the Proca-type equation represented by the form (22) can be derived without imposing the conventional Lorentz condition $L = 0$. This result is supported by the following two considerations.

When the two square brackets in the right-hand member of Eq. (A.3) both transform as 4-vectors, their "scalar" product becomes invariant in spacetime. The quantity L is then equal to an *arbitrary* constant. Consequently the terms containing L in Eqs. (A.1) and (A.2) vanish regardless whether the Lorentz condition $L = 0$ is being satisfied.

A second confirmation is obtained from a gauge transformation

$$\mathbf{A}' = \mathbf{A} + \nabla \chi \qquad \phi' = \phi - \frac{\partial \chi}{\partial t} \tag{A.4}$$

with an arbitrary scalar function $\chi = \chi(x, y, z, t)$. From Eq. (A.3) we then have

$$L = \operatorname{div} \mathbf{A}' + \frac{1}{c^2}\frac{\partial \phi'}{\partial t} - \left(\nabla^2 - \frac{1}{c^2}\frac{\partial^2}{\partial t^2}\right)\chi = L'(x, y, z, t) - \Box\chi \tag{A.5}$$

A choice of χ is now made where

$$\chi = \chi_1 + \chi_2 \tag{A.6}$$

with

$$\Box\chi_1 = L' \tag{A.7}$$

$$2\chi_2 = c_0 c^2 t^2 - (c_1 x^2 + c_2 y^2 + c_3 z^2) \tag{A.8}$$

From equation (A.5) it is further seen that L' is canceled by the choice of χ_1, as in conventional theory. We then have

$$L = -\Box\chi_2 = c_0 + c_1 + c_2 + c_3 = \text{const} \tag{A.9}$$

This also confirms that the Proca-type equation does not require L to vanish.

Moreover, Anastasovski et al. [79] have interpreted the basic EM field equations in a new and different way. Thereby the Lorentz condition $L = 0$ has been discarded, and Eqs. (A.1) and (A.2) have been modified to the form

$$\Box\mathbf{A} = -\nabla L = \mu_0 \mathbf{j}_A \tag{A.10}$$

$$\Box\phi = \frac{\partial}{\partial t}L = \frac{\bar{\rho}_A}{\varepsilon_0} \tag{A.11}$$

where the new 4-current thus becomes

$$(\mathbf{j}_A, ic\bar{\rho}_A) = -\left(\frac{1}{\mu_0}\right)\left(\nabla, -\left(\frac{i}{c}\right)\frac{\partial}{\partial t}\right)L \tag{A.12}$$

which is Lorentz-invariant, as expected. This can be interpreted as a vacuum current and charge. It further raises the question of a possible extraction of electromagnetic energy from the vacuum [79].

APPENDIX B: ELECTRON MODEL OF PRESENT THEORY

The present model of charged particle states is connected with two fundamental questions in EM field theory. The first of these concerns the electric charge quantization, partly through its relation with the magnetic flux in non-Abelian electrodynamics [5]. Such a quantization has been discussed in terms of the magnetic monopole theories by Dirac [34] and t'Hooft and Polyakov [80]. A second question of fundamental importance concerns quantum field theory in general. It has been stated by Ryder [80] that, despite the numerous success of the latter theory, the question of how to describe the basic matter fields of nature has remained unanswered, except through the introduction of quantum numbers and symmetry groups. Thus, as far as field theory goes, the matter fields are treated as point objects. Even in classical field theory these present us with unpleasant problems, in the shape of the infinite self-energy of a point charge. In the comparative success of renormalization theory the feeling remains that there ought to be a more satisfactory way of doing things [80].

B.1. General Equations of the Equilibrium State

Turning now to a more detailed description of the theory on steady equilibria described in Section V.A, the basic equations (36) in spherical coordinates with the adopted axisymmetrix geometry can be written as

$$\frac{(r_0\rho)^2\bar{\rho}}{\varepsilon_0} = D\phi = \left[D + (\sin\theta)^{-2}\right](CA) \tag{B.1}$$

where the operator is given by

$$D = D_\rho + D_\theta; \qquad D_\rho = -\frac{\partial}{\partial\rho}\left(\rho^2\frac{\partial}{\partial\rho}\right); \qquad D_\theta = -\frac{\partial^2}{\partial\theta^2} - \frac{\cos\theta}{\sin\theta}\frac{\partial}{\partial\theta} \tag{B.2}$$

Combination of Eqs. (B.1) and (37) yields

$$CA = -(\sin\theta)^2 DF \tag{B.3}$$

$$\phi = -[1 + (\sin\theta)^2 D\,]F \tag{B.4}$$

$$\bar{\rho} = -(\varepsilon_0/r_0^2\rho^2)D[1 + (\sin\theta)^2 D]F \tag{B.5}$$

According to Eq. (17), the source energy density now becomes

$$w_s = \frac{1}{2}(\bar{\rho}\phi + \mathbf{j}\cdot\mathbf{A}) = \frac{1}{2}\bar{\rho}(\phi + CA) \tag{B.6}$$

Introducing the functions

$$f(\rho,\theta) = -(\sin\theta)D[1 + (\sin\theta)^2 D]G \tag{B.7}$$

$$g(\rho,\theta) = -[1 + 2(\sin\theta)^2 D]G \tag{B.8}$$

and the normalized integrals

$$J_\nu = \int_{\varepsilon_\nu}^{\infty} \int_0^{\pi} I_\nu \, d\theta \, d\rho \qquad (\nu = q, M, m, s) \tag{B.9}$$

the integrated charge q_0, magnetic moment M_0, mass m_0, and angular momentum s_0 of expressions (38)–(41) become

$$q_0 = 2\pi\varepsilon_0 r_0 G_0 J_q \qquad I_q = f \tag{B.10}$$

$$M_0 = \pi\varepsilon_0 C r_0^2 G_0 J_M \qquad I_M = \rho(\sin\theta) f \tag{B.11}$$

$$m_0 = \pi\left(\frac{\varepsilon_0}{c^2}\right) r_0 G_0^2 J_m \qquad I_m = fg \tag{B.12}$$

$$s_0 = \pi\left(\frac{\varepsilon_0 C}{c^2}\right) r_0^2 G_0^2 J_s \qquad I_s = \rho(\sin\theta) fg \tag{B.13}$$

Here ε_ν are small dimensionless radii of circles centered at the origin $\rho = 0$ in the case when the radial part R of the generating function (37) is divergent at $\rho = 0$, and $\varepsilon_\nu = 0$ when R is convergent at $\rho = 0$. The former case will be shown to correspond to an electrically charged particle state, whereas the latter leads to a neutral particle state. The former state will be mainly considered here. For the same state we notice that q_0 and M_0 depend on the sign of G_0 but not m_0 and s_0. The sign of $C = \pm c$ affects M_0 and s_0 but not q_0 and m_0. The two spin directions are related to the two signs of C. The integrals J_m and J_s include the energy density and should be positive in this analysis, which excludes negative energy states.

With the separable normalized generating function G of Eq. (37), the integrands I_ν of the normalized expressions (B.9) become

$$I_q = \tau_0 R + \tau_1 (D_\rho R) + \tau_2 D_\rho (D_\rho R) \tag{B.14}$$

$$I_M = \rho(\sin\theta) I_q \tag{B.15}$$

$$I_m = \tau_0\tau_3 R^2 + (\tau_0\tau_4 + \tau_1\tau_3) R (D_\rho R) + \tau_1\tau_4 (D_\rho R)^2 + \tau_2\tau_3 R D_\rho (D_\rho R) + \tau_2\tau_4 (D_\rho R) \cdot [D_\rho (D_\rho R)] \tag{B.16}$$

$$I_s = \rho(\sin\theta) I_m \tag{B.17}$$

where

$$\tau_0 = -(\sin\theta)(D_\theta T) - (\sin\theta) D_\theta [(\sin\theta)^2 (D_\theta T)] \tag{B.18}$$

$$\tau_1 = -(\sin\theta) T - (\sin\theta) D_\theta [(\sin\theta)^2 T] - (\sin\theta)^3 (D_\theta T) \tag{B.19}$$

$$\tau_2 = -(\sin\theta)^3 T \tag{B.20}$$

$$\tau_3 = -T - 2(\sin\theta)^2 (D_\theta T) \tag{B.21}$$

$$\tau_4 = -2(\sin\theta)^2 T \tag{B.22}$$

To demonstrate the principal difference between the charged and neutral particle states, we now consider the case of a convergent radial part R at $\rho = 0$, and where we can put $\varepsilon_v = 0$. For the integral J_q of the electric charge (B.10) integration by parts with respect to ρ then yields

$$J_q = \int_0^\infty \int_0^\pi \tau_0 R \, d\rho + \int_0^\pi \left\{ -\tau_1 \left[\rho^2 \frac{dR}{d\rho} \right]_0^\infty + \tau_2 \left[\rho^2 \frac{d^2}{d\rho^2} \left(\rho^2 \frac{dR}{d\rho} \right) \right]_0^\infty \right\} d\theta \tag{B.23}$$

When R and its derivatives vanish at infinity and are finite at $\rho = 0$, and when the integrals of τ_1 and τ_2 given by Eqs. (B.19) and (B.20) are finite, Eq. (B.23) reduces to

$$J_q = \int_0^\infty R \, d\rho \cdot \int_0^\pi \tau_0 \, d\theta \equiv J_{q\rho} \cdot J_{q\theta} \tag{B.24}$$

For convergent integrals $J_{q\rho}$ we thus have to analyze the integral $J_{q\theta}$. Partial integration with respect to θ yields

$$J_{q\theta} = \left\{ (\sin\theta) \frac{d}{d\theta} [(\sin\theta)^2 (D_\theta T)] + (\sin\theta) \frac{dT}{d\theta} \right\}_0^\pi \tag{B.25}$$

For all finite functions T with finite derivatives at $\theta = (0, \pi)$, it is then seen that J_q and q_0 vanish in general.

Turning next to the magnetic moment (A.11) and its integral, partial integration with respect to ρ yields

$$J_{M\rho} \equiv \int_0^\infty \rho R \, d\rho = -\left(\frac{1}{2}\right) \int_0^\infty \rho (D_\rho R) \, d\rho = \frac{1}{4} \int_0^\infty \rho D_\rho (D_\rho R) \, d\rho \tag{B.26}$$

Combining expressions (B.26) with Eqs. (B.15) and (B.18)–(B.20), partial integration with respect to θ finally results in

$$J_M = J_{M\rho} \left\{ (\sin\theta)^3 \frac{d}{d\theta} [(\sin\theta)(D_\theta T - 2T)] \right\}_0^\pi \tag{B.27}$$

where it is seen that finite functions T with finite derivatives at $\theta = (0, \pi)$ lead to $J_M = 0$, and to a vanishing magnetic moment M_0.

Concerning the mass and angular momentum given by expressions (A.12), (A.13), (A.16), and (A.17), the convergent normalized integrals J_m and J_s

cannot vanish in general. A simple test with $R = e^{-\rho}$ and $T = 1$ can be used to illustrate this.

It is thus seen that radial functions R that are convergent at the origin lead to a class of neutral particle states where q_0 and M_0 both vanish, whereas m_0 and s_0 are nonzero.

For the present theory to result in electrically charged particle states it therefore becomes necessary to look into radial functions R that are *divergent* at the origin. This leads to the subsequent question whether the corresponding integrals (B.9) would then be able to form the basis of an equilibrium having *finite* and nonzero values of all the quantities q_0, M_0, m_0, and s_0. In the next section we will shown how this question can be answered.

B.2. The Charged-Particle State

The questions just being raised can be settled by studying a generating function $F = G_0 R(\rho) \cdot T(\theta)$ of the form [13,50]

$$R = \rho^{-\gamma} \exp(-\rho) \tag{B.28}$$

$$\begin{aligned} T &= 1 + \sum_{\nu=1}^{n} \{a_{2\nu-1} \sin[(2\nu - 1)\theta] + a_{2\nu} \cos[2\nu\theta]\} \\ &= 1 + a_1 \sin\theta + a_2 \cos 2\theta + a_3 \sin 3\theta + \cdots \end{aligned} \tag{B.29}$$

where γ is a positive constant. A neutral particle state [20], as mentioned in Section V.A.I.a, can instead be represented by $\gamma < 0$ in the form (B.28). In the same expression a change to a variable $\rho^* = \text{const}\,\rho$ only results in a change of the amplitude G_0 in expression (37). The form (B.28) diverges at $\rho = 0$ and tends strongly to zero as ρ tends to infinity. The part T of (B.29) is chosen to have top–bottom symmetry with respect to the "equatorial" plane $\theta = \pi/2$.

In the radial integration with the form (B.28), expressions (B.9) have dominating contributions from the largest negative powers of ρ. Therefore a single term is chosen in (B.28), and not a negative power series. Consequently we can write [13,20]

$$J_\nu = J_{\nu\rho} \cdot J_{\nu\theta} \qquad J_{\nu\theta} = \int_0^\pi I_{\nu\theta}\, d\theta \qquad (\nu = q, M, m, s) \tag{B.30}$$

where

$$J_{q\rho} = \frac{1}{\gamma - 1} \varepsilon_q^{-(\gamma-1)} \qquad J_{M\rho} = \frac{1}{\gamma - 2} \varepsilon_M^{-(\gamma-2)} \tag{B.31}$$

$$J_{m\rho} = \frac{1}{2\gamma - 1} \varepsilon_m^{-(2\gamma-1)} \qquad J_{s\rho} = \frac{1}{2(\gamma - 1)} \varepsilon_s^{-2(\gamma-1)} \tag{B.32}$$

and

$$I_{q\theta} = \tau_0 - \gamma(\gamma - 1)\tau_1 + \gamma^2(\gamma - 1)^2\tau_2 \tag{B.33}$$

$$I_{M\theta} = (\sin\theta)I_{q\theta} \tag{B.34}$$

$$I_{m\theta} = \tau_0\tau_3 - \gamma(\gamma - 1)(\tau_0\tau_4 + \tau_1\tau_3) \quad + \gamma^2(\gamma - 1)^2(\tau_1\tau_4 + \tau_2\tau_3) - \gamma^3(\gamma - 1)^3\tau_2\tau_4 \tag{B.35}$$

$$I_{s\theta} = (\sin\theta)I_{m\theta} \tag{B.36}$$

For all the integrands (B.31)–(B.32) to diverge at small values of all ε_ν, the condition $\gamma > 2$ has to be satisfied.

Finite and nonzero values of the integrated quantities (B.31)–(B.32) can now be obtained by shrinking the characteristic radius r_0 to very small values, as represented by

$$r_0 = c_0 \cdot \varepsilon^{\gamma - 1} \qquad (0 \leq \varepsilon \ll 1 \quad \text{and} \quad \gamma > 2) \tag{B.37}$$

where c_0 is a positive constant having the dimension of length. Then it is further necessary to make the choice

$$\varepsilon_q \equiv \varepsilon; \qquad \varepsilon_M = \varepsilon^{[2(\gamma-1)/(\gamma-2)]}; \qquad \varepsilon_m = \varepsilon^{[(\gamma-1)/(2\gamma-1)]}; \qquad \varepsilon_s = \varepsilon \tag{B.38}$$

which converts the radial integrals (B.31)–(B.32) into

$$J_{q\rho} = \frac{1}{\gamma - 1}\,\varepsilon^{-(\gamma-1)} \qquad J_{M\rho} = \frac{1}{\gamma - 2}\,\varepsilon^{-2(\gamma-1)} \tag{B.39}$$

$$J_{m\rho} = \frac{1}{2\gamma - 1}\,\varepsilon^{-(\gamma-1)} \qquad J_{s\rho} = \frac{1}{2(\gamma - 1)}\,\varepsilon^{-2(\gamma-1)} \tag{B.40}$$

In connection with expressions (B.38) we also notice that a choice $\varepsilon_M = \varepsilon_m = \varepsilon$ would lead to a form for $J_{M\rho}J_{m\rho}$ due to equations (B.31) and (B.32), which makes the product $M_0 m_0$ finite and nonzero [18].

The integrated quantities (B.10)–(B.13) now become

$$q_0 = \frac{2\pi\varepsilon_0 c_0 G_0 J_{q\theta}}{\gamma - 1} \tag{B.41}$$

$$M_0 = \frac{\pi\varepsilon_0 C c_0^2 G_0 J_{M\theta}}{\gamma - 2} \tag{B.42}$$

$$m_0 = \frac{\pi(\varepsilon_0/c^2) c_0 G_0^2 J_{m\theta}}{2\gamma - 1} \tag{B.43}$$

$$s_0 = \frac{\pi(\varepsilon_0 C/c^2) c_0^2 G_0^2 J_{s\theta}}{2(\gamma - 1)} \tag{B.44}$$

For the integrated charge q_0 combination of equations (B.10), (B.13), and (42) finally yields

$$\frac{|q_0|}{e} = \left[\frac{2\varepsilon_0 c h J_{q\theta}^2}{(\gamma - 1) e^2 J_{s\theta}} \right]^{1/2} \simeq 11.71 \frac{J_{q\theta}}{\left[(\gamma - 1) J_{s\theta} \right]^{1/2}} \tag{B.45}$$

for $C = +c$ and where the charge is normalized in respect to the measured value e. In equation (B.45) the coefficient $2\varepsilon_0 ch/e^2$ is equal to the inverted value of the fine structure constant. For $\gamma = 2$ the result of Eq. (44) is recovered.

B.3. The Point Charge Concept and the Related Divergence

It has thus been shown that the present theory of charged particle equilibria necessarily leads to point-like configurations with an excessively small characteristic radius r_0, permitted, in principle, even to approach the limit $r_0 = 0$. In this way the integrated field quantities can be rendered finite and nonzero. As pointed out in Section V.A.1.b, a strictly vanishing radius would not become physically acceptable, whereas a nonzero but very small radius is reconcilable both with experiments and with the present analysis. It would leave space for some form of internal particle structure. A small but lower limit of the radius would also be supported by considerations based on general relativity [15,20].

In this connection there is an important question concerning the infinite self-energy of a point charge in classical as well as in quantum field theory. The latter uses a renormalization process to solve the problem, namely, by subtracting two "infinities" to end up with a finite result. Despite the success of such a procedure, a more physically satisfactory way is needed [80]. Possibly the present theory may provide such an alternative, by tackling the divergence problem in a more surveyable manner. The finite result of a difference between two "infinities" due to renormalization theory would then be replaced by a finite result obtained from the product of an "infinity" and a "zero," as demonstrated by the present analysis.

B.4. Quantized Charged Equilibrium

To specify the equilibrium state more in detail, the quantum conditions of Section III.B now have to be imposed.

B.4.1. Conditions on Spin and Magnetic Moment

The imposed spin condition is represented by Eq. (42), which has to be combined with expression (B.13) for the angular momentum.

Concerning the magnetic moment, relation (43) is combined with Eqs. (B.41)–(B.44) to result in the form

$$H_0 = \overline{\Gamma} \frac{J_{M\theta} J_{m\theta}}{J_{q\theta} J_{s\theta}}; \qquad \overline{\Gamma} = \frac{(\gamma - 1)^2}{(\gamma - 2)(2\gamma - 1)} \tag{B.46}$$

In presence of an imposed external magnetic field, the quantity $H_0 = 1 + \delta_F$ stands for the right-hand member of Eq. (43), where e has been replaced by q_0 in the Feynman correction δ_F. On the other hand, if the influence of the magnetic field has a decisive effect on the internal balance of the equilibrium state, the value for a free electron may instead have to be written as $H_0 = \frac{1}{2} + \delta^*$, where δ^* would become a corresponding as yet undeduced small correction. This point is, however, an open question.

As a simple example, and to illustrate the ranges and orders of magnitude of involved parameters, a polar part of the generating function

$$T = 1 - N(\sin\theta)^{\bar{\alpha}} \tag{B.47}$$

was earlier chosen in the case where γ approaches the value 2 from above and N and $\bar{\alpha}$ are the independent variables [13,20]. Since $(\sin\theta)^{\bar{\alpha}}$ can be expanded into a series of $\sin\mu\theta$ and $\cos\mu\theta$ terms with an integer μ, the form (B.47) becomes a special case of the general symmetric form (B.29). The result obtained from (B.47) shows that the value $|q_0| = e$ then obtained from Eq. (B.45) is included within a relatively limited parameter range of the variables N and $\bar{\alpha}$.

B.4.2. Condition on Magnetic Flux

According to Eq. (B.3), the magnetic flux function becomes

$$\Gamma = 2\pi r A\,(\sin\theta) = -2\pi r_0 \left(\frac{G_0}{C}\right) \rho(\sin\theta)^3 DG \tag{B.48}$$

where

$$DG = T(D_\rho R) + R(D_\theta T) \tag{B.49}$$

and

$$D_\rho R = -[\gamma(\gamma-1) + 2(\gamma-1)\rho + \rho^2]R \tag{B.50}$$

after insertion of the radial part (B.28) and with the operators (B.2). The magnitude of the flux Γ increases strongly as ρ approaches small values from above. For such values relations (B.48)–(B.50) and (B.37) combine to

$$\Gamma \cong -2\pi \left(\frac{c_0 G_0}{C}\right) (\sin\theta)^3 [D_\theta T - \gamma(\gamma-1)T]\varepsilon^{\gamma-1} \cdot \rho^{-(\gamma-1)} \qquad (\rho \ll 1) \tag{B.51}$$

We now choose the limiting radius ε_Γ for this flux in analogy with the radii ε_ν of Eq. (B.9), that is, where $\varepsilon_\nu = \varepsilon$ for q_0, s_0 and $M_0 m_0$ according to Section B.2. With $\rho = \varepsilon_\Gamma = \varepsilon$, it is then seen from expression (B.51) that there is a resulting finite and nonzero magnetic flux Γ.

From the general expression (B.48) and the result (B.51), there is thus a magnetic flux that vanishes at large radial distances ρ, and that for a decreasing ρ grows in magnitude, to reach a steeply pronounced maximum at $\rho = \varepsilon_\Gamma$. Such a configuration resembles that of a thin current loop with almost all current at $\rho = \varepsilon$. For such a loop the magnetic field lines cut the equatorial plane $\theta = \pi/2$ at right angles, and there is a positive "upward" flux within an inner region $0 < \rho < \varepsilon$ that is equal to a negative "downward" flux within the outer region $\varepsilon < \rho$. The upward flux in the equatorial plane thus becomes

$$\Gamma_0 \equiv -\Gamma\left(\varepsilon, \theta = \frac{\pi}{2}\right) = 2\pi\left(\frac{c_0 G_0}{C}\right) A_\Gamma \tag{B.52}$$

where

$$A_\Gamma = -[\gamma(\gamma - 1)T - D_\theta T]_{[\theta=\pi/2]} \tag{B.53}$$

It has, however, to be stressed that expressions (B.52) and (B.53) are the result of a first and preliminary deduction of the relevant magnetic flux. Thus, magnetic island formation could somewhat increase its total value.

The quantization of the magnetic flux is based on the following line of thought. The quantized value s_0 of the angular momentum depends on the type of configuration being considered. Thus it becomes $s_0 = h/4\pi$ for fermions and $s_0 = h/2\pi$ for bosons. The electron, which is a fermion, can be considered as a system that also has the quantized charge q_0. If the total magnetic flux Γ_{tot} associated with such a system should be quantized as well, it is likely to be given by the two quantized concepts s_0 and q_0, in a relation having the dimension of a magnetic flux:

$$\Gamma_{\text{tot}} = \left|\frac{s_0}{q_0}\right| = \frac{h}{4\pi|q_0|} \tag{B.54}$$

Using the spin condition (42) in combination with expressions (B.41), (B.44), and (B.54), the condition for magnetic flux quantization is reduced to

$$8\pi f_m A_\Gamma J_{q\theta} = J_{s\theta} \tag{B.55}$$

where $f_m > 1$ when there is magnetic island formation, and $f_m = 1$ in absence of such formation. With the definition (B.53), the quantities A_Γ and $J_{q\theta}$ will have the same signs.

B.4.3. *Available Parameters of the Equilibrium State*

With the adopted form (B.28)–(B.29) of the generating function, the independent variables consist of the exponential factor γ in the radial part (B.28) and $2n$ amplitudes a_μ in the polar part (B.29) with $\mu = 2\nu - 1$ or 2ν. There are two resulting quantum conditions: (1) the combined spin and magnetic moment condition (B.46) with a given value of H_0, the choice of which could have two possible options; and (2) the magnetic flux condition (B.55). As the present theory stands, there is thus a considerable degree of freedom for the resulting charged particle equilibria, and this freedom could be even greater with the possible existence of forms of the generating function other than (B.28)–(B.29) that satisfy the basic conditions of a physically relevant charged equilibrium state.

B.5. The Possible Extremum of the Electric Charge

In Section V.A.1.b the question was raised as to whether the magnitude of the elementary electric charge e could be due to a unique quantized value that ought to emerge from a variational analysis [13,20]. Here we shall outline a correspo- nding analysis to some detail. The notation

$$A_\nu \equiv J_{\nu\theta} \qquad (\nu = q, M, m, s) \tag{B.56}$$

is introduced for polar integrals that are obtained from Eqs. (B.30) and (B.33)–(B.36). Relation (B.45) can then be written as

$$\left(\frac{q_0}{e}\right)^2 = 2\left(\frac{\varepsilon_0 ch}{e^2}\right)S \qquad S = \frac{A_q^2}{(\gamma - 1)A_s} \tag{B.57}$$

The combined condition (B.46) for spin and magnetic moment has the further form

$$Q = H_0 - \frac{1}{2}g_s - \delta_F = 0 \tag{B.58}$$

with the notation

$$H_0 = \frac{(\gamma - 1)^2 A_M A_m}{(\gamma - 2)(2\gamma - 1)A_q A_s} \tag{B.59}$$

Here the Landé factor $g_s = 2$ corresponds to the electron with an imposed magnetic field according to Dirac and Feynman, and $g_s = 1$ may represent a possible option for a free electron. Further

$$\delta_F = \frac{q_0^2}{4\pi\varepsilon_0 hc} = \frac{S}{2\pi} \tag{B.60}$$

is the correction by Feynman in terms of expressions (B.41), (B.44), and (B.57), whereas δ_F may have a modified value in the case of a free electron. In Eq. (B.59) for values of γ that approach 2 from above, it has been shown [13,20] that $A_M/(\gamma - 2)$ still has a finite value, earlier [20] denoted by the symbol A_M.

The condition (B.55) on magnetic flux quantization finally becomes

$$V = \frac{8\pi f_m A_\Gamma A_q}{A_s} - 1 = 0 \tag{B.61}$$

The proposed variational analysis now implies that an extremum of the normalized square (B.57) of the electric charge is being searched for, under the subsidiary constraints (B.58) and (B.61). With $2n$ terms in the expansion (B.29), the quantities S, Q, and V will all depend on $2n+1$ variables $(\gamma, a_1, a_2, \ldots, a_{2n})$. Defining the function

$$P = S + L_Q \cdot Q + L_V \cdot V \tag{B.62}$$

where L_Q and L_V are Lagrange multipliers, the variational analysis then has the form of a system with $2n+1$ equations

$$\frac{\partial P}{\partial \gamma} = \frac{\partial S}{\partial \gamma} + L_Q \frac{\partial Q}{\partial \gamma} + L_V \frac{\partial V}{\partial \gamma} = 0 \tag{B.63}$$

$$\frac{\partial P}{\partial a_\mu} = \frac{\partial S}{\partial a_\mu} + L_Q \frac{\partial Q}{\partial a_\mu} + L_V \frac{\partial V}{\partial a_\mu} = 0 \qquad (\mu = 1, 2, \cdots, 2n) \tag{B.64}$$

$$\frac{\partial P}{\partial L_Q} = Q = 0 \qquad \frac{\partial P}{\partial L_V} = V = 0 \tag{B.65}$$

One way of eliminating L_Q and L_V is to use the first two equations (B.64), which yield

$$L_Q = \frac{\frac{\partial S}{\partial a_1} \cdot \frac{\partial V}{\partial a_2} - \frac{\partial S}{\partial a_2} \cdot \frac{\partial V}{\partial a_1}}{\frac{\partial Q}{\partial a_2} \cdot \frac{\partial V}{\partial a_1} - \frac{\partial Q}{\partial a_1} \cdot \frac{\partial V}{\partial a_2}} \tag{B.66}$$

$$L_V = \frac{\frac{\partial S}{\partial a_2} \cdot \frac{\partial Q}{\partial a_1} - \frac{\partial S}{\partial a_1} \cdot \frac{\partial Q}{\partial a_2}}{\frac{\partial Q}{\partial a_2} \cdot \frac{\partial V}{\partial a_1} - \frac{\partial Q}{\partial a_1} \cdot \frac{\partial V}{\partial a_2}} \tag{B.67}$$

This results in a system of $2n+1$ equations for the $2n+1$ unknown independent variables $(\gamma, a_1, a_2, \ldots, a_{2n})$ as given by

$$\left(\frac{\partial Q}{\partial a_2} \cdot \frac{\partial V}{\partial a_1} - \frac{\partial Q}{\partial a_1} \cdot \frac{\partial V}{\partial a_2}\right) \frac{\partial S}{\partial \gamma} + \left(\frac{\partial S}{\partial a_1} \cdot \frac{\partial V}{\partial a_2} - \frac{\partial S}{\partial a_2} \cdot \frac{\partial V}{\partial a_1}\right) \frac{\partial Q}{\partial \gamma} + \left(\frac{\partial S}{\partial a_2} \cdot \frac{\partial Q}{\partial a_1} - \frac{\partial S}{\partial a_1} \cdot \frac{\partial Q}{\partial a_2}\right) \frac{\partial V}{\partial \gamma} = 0 \tag{B.68}$$

$$\left(\frac{\partial Q}{\partial a_2} \cdot \frac{\partial V}{\partial a_1} - \frac{\partial Q}{\partial a_1} \cdot \frac{\partial V}{\partial a_2}\right) \frac{\partial S}{\partial a_\mu} + \left(\frac{\partial S}{\partial a_1} \cdot \frac{\partial V}{\partial a_2} - \frac{\partial S}{\partial a_2} \cdot \frac{\partial V}{\partial a_1}\right) \frac{\partial Q}{\partial a_\mu} + \left(\frac{\partial S}{\partial a_2} \cdot \frac{\partial Q}{\partial a_1} - \frac{\partial S}{\partial a_1} \cdot \frac{\partial Q}{\partial a_2}\right) \frac{\partial V}{\partial a_\mu} = 0 \qquad (\mu = 3, 4, \ldots, 2n) \tag{B.69}$$

$$Q = 0 \qquad V = 0 \tag{B.70}$$

As seen from expressions (B.30), (B.33)–(B.36), (B.18)–(B.22), and (B.56)–(B.61), the system (B.68)–(B.70) leads to highly nonlinear relations between the independent variables.

Since a high degree of accuracy requires the inclusion of a large number $2n$ of amplitudes a_μ in the expansion (B.29), a numerical analysis is not expected to become trivial. The objective is thus to determine whether S and q_0 of Eq. (B.57) will converge at large n toward an asymptotic value, and how this value will be related to the experimentally determined elementary charge e. Results of such an analysis are not available at this stage, and parts of it include open questions such as those concerning the detailed spatial distribution of the magnetic flux near the origin $\rho = 0$ and its influence on the values of A_Γ and f_m in Eq. (B.55).

Without using the detailed variational equations, a first hint can be obtained by applying the special polar part (B.47) of the generating function. A simple numerical analysis in the case $\gamma = 2$ and $f_m = 1$ then yields the result $|q_0/e| \cong 1.6$ according to conditions (B.70), the form (B.55), and expression (B.45). The corresponding extremum value should at least become somewhat lower, because the special function (B.47) is not likely to be that function that in a variational analysis should result in the lowest possible value of S and q_0.

References

1. R. P. Feynman, *Lectures on Physics: Mainly Electromagnetism and Matter*, Addison-Wesley, Reading, MA, 1964.
2. A. E. Chubykalo and R. Smirnov-Rueda, in M. W. Evans, J.-P. Vigier, S. Roy, and G. Hunter (Eds.), *The Enigmatic Photon*, Kluwer, Dordrecht, 1998, Vol. 4, p. 261.
3. A. Lakhtakia (Ed.), *Essays on the Formal Aspects of Electromagnetic Theory*, World Scientific, Singapore, 1993.
4. T. Barrett and D. M. Grimes (Eds.), *Advanced Electromagnetics*, World Scientific, Singapore, 1995.
5. M. Evans and J.-P. Vigier, *The Enigmatic Photon*, Kluwer, Dordrecht, Vols. 1 (1994), 2 (1995).
6. M. W. Evans, J.-P. Vigier, S. Roy, and S. Jeffers, *The Enigmatic Photon*, Kluwer, Dordrecht, 1996, Vol. 3.
7. M. W. Evans, J.-P. Vigier, S. Roy, and G. Hunter, *The Enigmatic Photon*, Kluwer, Dordrecht, Vol. 4.
8. G. Hunter, S. Jeffers, and J.-P. Vigier, *Causality and Locality in Modern Physics*, Kluwer, Dordrecht, 1998.
9. V. V. Dvoeglazov (Ed.), *Contemporary Fundamental Physics*, Nova Science Publishers, Huntington, NY, 2000.
10. B. Lehnert, *Electron and Plasma Physics*, Royal Institute of Technology, Stockholm, Report TRITA-EPP-79-13 (1979).
11. B. Lehnert, *Spec. Sci. Technol.* **9**, 177 (1986); **11**, 49 (1988).
12. B. Lehnert, Dept. Plasma Physics, Royal Institute of Technology, Stockholm, Report TRITA-EPP-86-16 (1986) and Report TRITA-EPP-89-05 (1989).

13. B. Lehnert, *Spec. Sci. Technol.* **17**, 259; 267 (1994).
14. B. Lehnert, *Optik* **99**, 113 (1995).
15. B. Lehnert, *Physica Scripta* **19**, 204 (1996).
16. B. Lehnert, in M. W. Evans, J.-P. Vigier, S. Roy, and G. Hunter (Eds.), *The Enigmatic Photon*, Kluwer, Dordrecht, 1998, Vol. 4, p. 219.
17. B. Lehnert, in G. Hunter, S. Jeffers, and J.-P. Vigier (Eds.), *Causality and Locality in Modern Physics*, Kluwer, Academic, Dordrecht, 1998, p. 105.
18. B. Lehnert, *Physica Scripta* **T82**, 89 (1999).
19. B. Lehnert, in V. V. Dvoeglazov (Ed.), *Contemporary Fundamental Physics* Nova Science Publishers, Huntington, NY, 2000, Vol. 2, pp. 3–20.
20. B. Lehnert and S. Roy, *Extended Electromagnetic Theory*, World Scientific, Singapore, 1998.
21. Y. Tsuchiya, E. Inuzuka, T. Kurono and M. Hosoda, *Adv. Electron. Electron Phys.* **64A**, 21 (1985).
22. J. J. Thomson, *Nature*, 232 (Feb. 8, 1936).
23. G. Hunter and R. L. P. Wadlinger, *Phys. Essays* **2**(2), 154 (1989) [with an experiment by F. Engler (unpublished)].
24. A. Mazet, C. Imbert, and S. Huard, *C. R. Acad. Sci., Ser. B* **273**, 592 (1971).
25. L. de Broglie and J.-P. Vigier, *Phys. Rev. Lett* **28**, 1001 (1972).
26. R. Hütt, *Optik* **78**, 12 (1987).
27. E. Recami, in G. Hunter, S. Jeffers, and J.-P. Vigier (Eds.), *Causality and Locality in Modern Physics*, Kluwer, Dordrecht, 1998, p. 113.
28. M. W. Evans, in M. W. Evans, J.-P. Vigier, S. Roy, and G. Hunter (Eds.), *The Enigmatic Photon*, Kluwer, Dordrecht, 1998, Vol. 4. p. 51.
29. M. Mézáros, in M. W. Evans, J.-P. Vigier, S. Roy, and G. Hunter (Eds.), *The Enigmatic Photon*, Kluwer, Dordrecht, 1998, Vol. 4, p. 147.
30. P. R. Molnár, T. Borbély, and B. Fajszi, in M. W. Evans, J.-P. Vigier, S. Roy, and G. Hunter (Eds.), *The Enigmatic Photon*, Kluwer, Dordrecht, 1998, Vol. 4, p. 205.
31. P. M. Morse and H. Feshbach, *Methods of Theoretical Physics*, McGraw-Hill, New York, 1953, Part I, pp. 221,260.
32. K. Imaeda, *Prog. Theor. Phys.* **5**, 133 (1950).
33. T. Ohmura (Kikuta), *Prog. Theor. Phys.* **16**, 684,685 (1956).
34. P. A. M. Dirac, *Directions in Physics*, Wiley-Interscience, New York, 1978.
35. J. A. Stratton, *Electromagnetic Theory*, McGraw-Hill, New York, 1941, Sec. 1.10, Chap. II and Sec. 1.23, 2.19, 5.15.2, 9.4–9.8.
36. P. A. M. Dirac, *Proc. Roy. Soc.* **117**, 610 (1928); **118**, 351(1928).
37. R. Feynman, *QED: The Strange Theory of Light and Matter*, Penguin, London, 1990.
38. P. K. Anastasovski, T. E. Bearden, C. Ciubotariu, W. T. Coffey, L. B. Crowell, G. J. Evans, M. W. Evans, R. Flower, S. Jeffers, A. Labounsky, D. Leporini, B. Lehnert, M. Mézáros, J. K. Mosciki, P. R. Molnár, H. Munera, E. Recami, D. Roscoe, and S. Roy, *Found. Phys. Lett.* **13**(2), 179 (2000); see also G. Feldman and P. T. Mathews, *Phys. Rev.* **130**, 1633 (1963).
39. L. Schiff, *Quantum Mechanics*, McGraw-Hill, New York, 1949, Chaps. XIV and X, Sec. 36.
40. R. B. Leighton, *Principles of Modern Physics*, McGraw-Hill, New York, 1959, Chap. 20.
41. C. W. Sherwin, *Introduction to Quantum Mechanics*, Holt, Rinehart Winston, New York, 1960, Chap. II.

42. A. Einstein, *Ann. Phys.* (Leipzig) **7**, 132 (1905); **18**, 121 (1917).
43. L. Bass and E. Schrödinger, *Proc. Roy. Soc. A* **232**, 1 (1955).
44. L. de Broglie and J.-P. Vigier, *Phys. Rev. Lett.* **28**, 1001 (1972).
45. M. W. Evans, *Physica B* **182**, 227,237 (1992).
46. D. F. Bartlett and T. R. Corle, *Phys. Rev. Lett.* **55**, 99 (1985).
47. H. F. Harmuth, *IEEE Trans.* **EMC-28**(4), 250,259,267 (1986).
48. J.-P. Vigier, *IEEE Trans. Plasma Sci.* **18**, 64 (1990).
49. B. Haisch and A. Rueda, in G. Hunter, S. Jeffers, and J.-P. Vigier (Eds.), *Causality and Locality in Modern Physics*, Kluwer, Dordrecht 1998, p. 71.
50. S. Roy and M. Roy, in J.-P. Vigier, S. Roy, S. Jeffers, and G. Hunter (Eds.), *The Present State of Quantum Theory of Light*, Kluwer, Dordrecht, 1996.
51. G. Kar, M. Sinha, and S. Roy, *Int. J. Theor. Phys.* **32**, 593 (1993).
52. S. Roy, G. Kar, and M. Roy, *Int. J. Theor. Phys.* **35**, 579 (1996).
53. H. Hertz, *Wied. Ann.* **41**, 369 (1890).
54. H. Hertz, *Ges. Werke* **2**, 256 (1894).
55. H. Hertz, *Electric Waves*, transl. by D. E. Jones, Dover, New York, 1962.
56. A. E. Chubykalo and R. Smirnov-Rueda, *Modern Phys. Lett. A* **12**, 1 (1997).
57. A. E. Chubykalo and R. Smirnov-Rueda, *Phys. Rev. E* **53**, 5373 (1996).
58. J. Weber, *General Relativity and Gravitational Waves*, Interscience, New York, 1961, Chap. 5.4.
59. W. Heitler, *The Quantum Theory of Radiation*, 3rd ed., Clarendon Press, Oxford, 1954, p. 401.
60. P. Ecimovic, K. Kafatos, and R. Amoroso, in G. Hunter, S. Jeffers, and J.-P. Vigier (Eds.), *Causality and Locality in Modern Physics*, Kluwer, Dordrecht, 1998, p. 165.
61. J.-P. Vigier, *Phys. Lett. A* **234**, 75 (1997).
62. V. V. Dvoeglazov, in M. W. Evans, J.-P. Vigier, S. Roy, and G. Hunter (Eds.), *The Enigmatic Photon*, Kluwer, Dordrecht, 1998, Vol. 4, p. 305.
63. E. Comay, *Chem. Phys. Lett.* **261**, 601 (1996).
64. G. Hunter, *Chem. Phys.* **242**, 331 (1999).
65. N. V. Pope, in G. Hunter, S. Jeffers, and J.-P. Vigier (Eds.), *Causality and Locality in Modern Physics*, Kluwer, Dordrecht, 1998, p. 187.
66. J. Argyris and C. Ciubotariu, in G. Hunter, S. Jeffers, and J.-P. Vigier (Eds.), *Causality and Locality in Modern Physics* Kluwer, Dordrecht 1998, p. 143.
67. M. Kafatos, in G. Hunter, S. Jeffers, and J.-P. Vigier (Eds.), *Causality and Locality in Modern Physics*, Kluwer, Dordrecht, 1998, p. 29.
68. Y. Aharonov and D. Bohm, *Phys. Rev.* **115**, 485 (1959).
69. P. Saari and K. Reivelt, *Phys. Rev. Lett.* **79**, 4135 (1997).
70. G. Nimtz, A. Enders, and H. Spieker, in A. van der Merwe and A. Garuccio (Eds.), *Wave and Particle in Light and Matter, Proc. Trani Workshop* (Italy, Sept. 1992), Plenum, New York (H. Aichmann and G. Nimtz: "Tunnelling of a FM-Signal: Mozart 40").
71. A. O. Barut, G. D. Maccarone, and E. Recami, *Il Nuovo Cimento*, **71A**(4), 509 (1982).
72. V. S. Olkhovsky and E. Recami, *Phys. Reports* **214**(6), 339 (1992).
73. W. D. Walker, in G. Hunter, S. Jeffers, and J.-P. Vigier (Eds.), *Causality and Locality in Modern Physics*, Kluwer, Dordrecht 1998, p. 127.
74. K. A. Peacock, in G. Hunter, S. Jeffers, and J.-P. Vigier (Eds.), *Causality and Locality in Modern Physics*, Kluwer, Dordrecht 1998, p. 227.

75. B. Lehnert, Royal Institute of Technology, Stockholm, Report TRITA-ALF-1999-02 (1999).
76. J. D. Jackson, *Classical Electrodynamics*, John Wiley & Sons, Inc., New York, London, Sydney, 1962, Ch. 9.8.
77. J. Sarfatti, Private communication at Vigier III Symposium, Berkeley, California, August 21–25, 2000.
78. M. W. Evans, unpublished comment (2000); P. K. Anastasovski, T. E. Bearden, C. Ciubotariu, W. T. Coffey, L. B. Crowell, G. J. Evans, M. W. Evans, R. Flower, A. Labuonsky, B. Lehnert, M. Mézáros, P. R. Molnár, S. Roy, and J.-P. Vigier, *Found. Phys.* **30**, 1123 (2000).
79. P. K. Anastasovski, T. E. Bearden, C. Ciubotariu, W. T. Coffey, L. B. Crowell, G. J. Evans, M. W. Evans, R. Flower, S. Jeffers, A. Labounsky, B. Lehnert, M. Mézaŕos, P. R. Molnaŕ, J.-P. Vigier, and S. Roy, *Physica Scripta* **61**, 513 (2000).
80. L. H. Ryder, *Quantum Field Theory*, Cambridge Univ. Press, 1996, Chap. 10.

O(3) ELECTRODYNAMICS

M. W. EVANS

CONTENTS

I. TOPOLOGICAL BASIS FOR HIGHER-SYMMETRY ELECTRODYNAMICS

Topology is the study of geometrical configurations invariant under transformation by continuous mappings. It provides what is probably the most fundamental known framework for the description of physical models using the mathematical techniques of group theory [1] and gauge theory [2]. A study of the topology of a

Modern Nonlinear Optics, Part 2, Second Edition, Advances in Chemical Physics, Volume 119, Edited by Myron W. Evans. Series Editors I. Prigogine and Stuart A. Rice.
ISBN 0-471-38931-5

given experiment can be used to decide whether that experiment is possible or not, and the decision is made in the language of group theory. Topological considerations can be applied to the vacuum itself, so that the vacuum becomes structured, or has a given configuration. On the basis of the fact that topology is a fundamental description, then it is also a fundamental description of the vacuum itself, and decides the structure of physical objects such as electrodynamic field equations [3,4] in the vacuum. The group-theoretic description of the received equations of classical electrodynamics, the Maxwell–Heaviside equations [5], is U(1), homomorphic with O(2) ($U(1) \approx O(2)$). The latter is the group of rotations in two dimensions, and the former is the group of all numbers of the form $e^{i\phi} = \cos\phi + i\sin\phi$, whose group space is a circle. The two groups are homomorphic or similar in form. Each element of O(2) is given uniquely [6] by an angle $\boldsymbol{\alpha}$, the angle of rotation in a plane. The group space of both O(2) and U(1) is therefore a circle. The received view [5,6] asserts that the classical electromagnetic field is a gauge field invariant under local U(1) gauge transformations. In other words, Maxwell–Heaviside theory is a U(1) symmetry Yang–Mills gauge field theory. Unified field theory proceeds on this assertion, specifically, that the electromagnetic sector has U(1) symmetry. The topological basis for this conclusion in the received view is given by such phenomena as the Aharonov–Bohm effect [6], where the classical vacuum is deduced to have a nontrivial topology [6]. This is combined with the view that electrodynamics is a U(1) gauge theory to give the received explanation of the Aharonov–Bohm effect [3,4,6]. In gauge theory in general, however, the vacuum has a rich topological structure, and this structure is not confined to U(1). Other groups may be used, and each has physical, or measurable, gauge-invariant, consequences. Therefore, the most fundamental basis for the development of field equations, such as those of classical electrodynamics, is the topology of the vacuum itself. In order to understand this further, some topological concepts must be introduced and defined.

Basic to the understanding of topology are simply and non-simply connected spaces. The relevant topological space is the vacuum itself. A simply connected space is one in which all closed curves may be shrunk to a point; and in a non-simply connected space, this is not true in general. In a non-simply connected space, a function may be many-valued, for example $\cos(\phi \pm 2\pi n)$. In this view therefore, the Aharonov–Bohm effect can exist physically if and only if the vacuum *itself* is not simply connected. The group theoretic description of the Aharonov–Bohm effect follows from these considerations. The $U(1) \approx O(2)$ group is not simply connected because its group space (denoted S^1) is a circle. The group space S^1 itself is not simply connected [6]. In the received view, this argument is used to show that the Aharonov–Bohm effect is supported by a vacuum topology described by the group U(1).

In the 1990s, however, there have been several attempts to extend the received view of classical electrodynamics, for example, the work of Barrett

[3,4], Lehnert et al. [7–10], Evans et al. [11–20] and Harmuth et al. [21,22]. These attempts stem from anomalies and self-inconsistencies in classical electrodynamics viewed as a U(1) gauge field theory. Some of these are reviewed in Section III of this chapter. The basis for these developments resides, as it must, in vacuum topology and its subsidiary languages of group and gauge theory. In other words, it may be possible to describe classical electrodynamics with groups other than U(1) in a non-simply connected vacuum, the relevant topological space. Once a particular group is chosen, general gauge field theory [3,4,6] may be used to write down the physical field equations of electrodynamics and the field tensor [3,4,11–20]. The results of the hypothesis are compared with empirical data as usual, and cross compared with the U(1) description. This method is developed and reviewed in this chapter. The basis of our development, therefore, is the topology of the vacuum, which ultimately decides which set of field equations is the more accurate in its description of data. The basis for gauge theory is fiber bundle theory, which is briefly reviewed in Section II.

We will be concerned in this article with the non-simply connected vacuum described by the group O(3), the rotation group. The latter is defined [6] as follows. Consider a spatial rotation in three dimensions of the form

$$\begin{bmatrix} X' \\ Y' \\ Z' \end{bmatrix} = (R) \begin{bmatrix} X \\ Y \\ Z \end{bmatrix} \qquad \text{or} \qquad r' = Rr \tag{1}$$

where R is a rotation matrix. Rotations have the property

$$X'^2 + Y'^2 + Z'^2 = X^2 + Y^2 + Z^2 \tag{2}$$

which can be written

$$r'^T r' = r^T r \tag{3}$$

where T denotes "transpose." Therefore

$$\begin{aligned} r^T R^T R r &= r^T r \\ R^T R &= 1 \end{aligned} \tag{4}$$

where R is an orthogonal 3×3 matrix. These matrices form a group. If R_1 and R_2 are orthogonal, then so is $R_1 R_2$:

$$(R_1 R_2)^T R_1 R_2 = R_2^T R_1^T R_1 R_2 = 1 \tag{5}$$

This group is denoted O(3) in three dimensions, and O(n) in n dimensions. The rotation group O(3) is a Lie group (i.e., is a continuous group), and is non-Abelian (i.e., its rotation matrices do not commute) [6]. A simple example of an O(3) group is the one formed by the unit vectors of a Cartesian frame in three-dimensional space:

$$\begin{aligned} \boldsymbol{i} \times \boldsymbol{j} &= \boldsymbol{k} \\ \boldsymbol{j} \times \boldsymbol{k} &= \boldsymbol{i} \\ \boldsymbol{k} \times \boldsymbol{i} &= \boldsymbol{j} \end{aligned} \tag{6}$$

Therefore, we can adopt as our fundamental hypothesis that the topological space under consideration (i.e., the vacuum) is described by O(3) rather than U(1) and work out the consequences [11–20]. Some of the latter are reviewed in this chapter. An O(3) group can also be formed by the complex unit vectors defined by

$$\begin{aligned} \boldsymbol{e}^{(1)} &= \frac{(\boldsymbol{i} - i\boldsymbol{j})}{\sqrt{2}} \\ \boldsymbol{e}^{(2)} &= \frac{(\boldsymbol{i} + i\boldsymbol{j})}{\sqrt{2}} \\ \boldsymbol{e}^{(3)} &= \boldsymbol{k} \end{aligned} \tag{7}$$

so that

$$\begin{aligned} \boldsymbol{e}^{(1)} \times \boldsymbol{e}^{(2)} &= i\boldsymbol{e}^{(3)*} \\ \boldsymbol{e}^{(2)} \times \boldsymbol{e}^{(3)} &= i\boldsymbol{e}^{(1)*} \\ \boldsymbol{e}^{(3)} \times \boldsymbol{e}^{(1)} &= i\boldsymbol{e}^{(2)*} \end{aligned} \tag{8}$$

forms an O(3) group suitable for the description of circularly polarized radiation, and therefore of radiation in general [11–20]. Here, an asterisk (*) denotes complex conjugate. There are several other ways of defining the O(3) group, one of which is that it is the little group of the Poincaré group of special relativity [6]. A little group with structure O(3) is the group of a particle with mass. So if O(3) is adopted as the group describing classical electrodynamics, the photon, on quantization, may have a tiny mass (empirically estimated [23] as less than 10^{-68} kg). The little group for the massless photon in the received view is unphysical, it is the Euclidean E(2) [6,11–20]. This means that a particle without mass is an unphysical object. The photon without mass is obtained by quantizing a classical U(1) theory, suggesting that the received view is also unphysical. We do not have to search far to find some unphysical properties of the U(1) Yang–Mills gauge field theory of classical electromagnetism. For example, the electromagnetic phase is random, the 4-potential A^{μ} is unphysical as the result of

Heaviside's development of Maxwell's original concept of a physical vector potential, which was based, in turn, on Faraday's electrotonic state. Barrett [2] has reviewed extensive empirical evidence for a physical classical A^{μ}, in contradiction to U(1) theory, hereinafter described as "U(1) electrodynamics." The vacuum for the Aharonov–Bohm effect is non-simply connected, and therefore supports a *physical* A^{μ} [3,6]. The potential A^{μ} has no physically discernible effect if and only if the space is simply connected. Since U(1) is non-simply connected, there is a self-contradiction in the received view, [3,6] and since A^{μ}, by definition, is unphysical in U(1) electrodynamics, we must search for a new type of classical electrodynamics. In this chapter, we base this search on the group O(3), and hereinafter describe it as "O(3) electrodynamics." The basic topological space is that of the vacuum, and is described by the O(3) group and gauge theory based on this group. One consequence is that the potential is physical as required, another is that the unphysical random phase of U(1) electrodynamics is replaced by a gauge-invariant physical phase *factor* of O(3) electrodynamics. These changes are shown to have foundational consequences in interferometry and aspects of physical optics, for example. Furthermore, several of the well-developed techniques of non-Abelian gauge field theory [3,4,6] may be brought to bear on classical electrodynamics, because the group O(3) is a non-Abelian group, as argued already. This enriches and develops the subjects of classical and quantum electrodynamics and unified field theory.

The group space of O(3) is doubly connected (i.e., non-simply connected) and can therefore support an Aharonov–Bohm effect (Section V), which is described by a physical inhomogeneous term produced by a rotation in the internal gauge space of O(3) [24]. The existence of the Aharonov–Bohm effect is therefore clear evidence for an extended electrodynamics such as O(3) electrodynamics, as argued already. A great deal more evidence is reviewed in this article in favor of O(3) over U(1). For example, it is shown that the Sagnac effect [25] can be described accurately with O(3), while U(1) fails completely to describe it.

The O(3) group is homomorphic with the SU(2) group, that of 2×2 unitary matrices with unit determinant [6]. It is well known that there is a two to one mapping of the elements of SU(2) onto those of O(3). However, the group space of SU(2) is simply connected in the vacuum, and so it cannot support an Aharonov–Bohm effect or physical potentials. It has to be modified [26] to $SU(2)/Z2 \approx SO(3)$.

Therefore, this is a statement of our fundamental hypothesis, specifically, that the topology of the vacuum defines the field equations through group and gauge field theory. Prior to the inference and empirical verification of the Aharonov–Bohm effect, there was no such concept in classical electrodynamics, the ether having been denied by Lorentz, Poincaré, Einstein, and others. Our development of O(3) electrodynamics in this chapter, therefore, has a well-defined basis in fundamental topology and empirical data. In the course of the development of this chapter, several misconceptions and inconsistencies of U(1) electrodynamics are brought to light, and these are remedied straightforwardly by changing the gauge group from U(1) to O(3). The implications are briefly reviewed for quantum electrodynamics and unified field theory, starting with electroweak theory. One major result of the latter is the existence of a novel massive boson, the existence of which is consistent with novel empirical data as discussed in Section XII. The gradual and consistent accumulation of evidence leads in this chapter to the conclusion that an O(3) gauge group is to be preferred over a U(1) gauge group in classical electrodynamics.

Some by-products of the development emerge, such as the fact that the acceptance of a structured vacuum described by an O(3) gauge group leads directly to the existence of novel charges and currents in the vacuum. These are conserved, or Noether, currents and charges and are clearly topological in origin. They spring from the fact that the vacuum is a topological space. Four such entities emerge:

1. A topological vacuum electric charge, also proposed empirically by Lehnert et al. [7–10]
2. A topological vacuum electric current, also proposed empirically by Lehnert et al. [7–10]
3. A topological vacuum magnetic charge, proposed also by Barrett [3,4] and Harmuth [21,22]
4. A vacuum topological magnetic current, proposed also by Barrett [3,4] and Harmuth [21,22].

Each of these four objects can provide energy, which can be loosely termed "vacuum energy:" energy coming from the topology of the vacuum.

In well-defined limits, the field equations of O(3) electrodynamics can collapse to a set of two complex conjugate equations that resemble those of U(1) electrodynamics (Maxwell–Heaviside equations), and a third equation for a novel fundamental spin component of O(3) electrodynamics, the $\boldsymbol{B}^{(3)}$ component [11–20] in the basis ((1),(2),(3)). This component also springs from the topology of the vacuum, described by an O(3) gauge group and is therefore a magnetic flux density that exists in the vacuum because of this choice of gauge group. Clearly, the $\boldsymbol{B}^{(3)}$ component is fundamental to O(3) electrodynamics and is not a static magnetic field of U(1) electrodynamics. The $\boldsymbol{B}^{(3)}$ component is an observable of the third Stokes parameter, topological phases, interferometry, and magneto-optics and is a radiated field that propagates with the radiation. In the laboratory, it propagates for all practical purposes at the speed of light, c, as does the third Stokes parameter to which it is proportional [11–20]. It is a fundamental property of the O(3) electromagnetic field that emanates from the topology of the vacuum. It forms an O(3) group with the plane wave $\boldsymbol{B}^{(1)} = \boldsymbol{B}^{(2)*}$

of magnetic flux density in the vacuum in O(3) electrodynamics, giving the B cyclic theorem [11–20]

$$\begin{aligned} \boldsymbol{B}^{(1)} \times \boldsymbol{B}^{(2)} &= iB^{(0)}\boldsymbol{B}^{(3)*} \\ \boldsymbol{B}^{(2)} \times \boldsymbol{B}^{(3)} &= iB^{(0)}\boldsymbol{B}^{(1)*} \\ \boldsymbol{B}^{(3)} \times \boldsymbol{B}^{(1)} &= iB^{(0)}\boldsymbol{B}^{(2)*} \\ B^{(0)} &= \left|\boldsymbol{B}^{(3)}\right| \end{aligned} \tag{9}$$

which is Lorentz-invariant, as it is, within a common factor on both sides, simply a relation between rotation generators of the O(3) group.

An important by-product of the development in this chapter (Section X) is the possible existence of scalar interferometry, which is interferometry between structured scalar potentials, first introduced by Whittaker [27,28] and that can be defined in terms of $\boldsymbol{B}^{(3)}$. This is a type of interferometry that depends on physically meaningful potentials that can exist self-consistently, as we have argued, only in a non-singly connected O(3) vacuum, because potentials in the nonsingly connected U(1) vacuum are assumed to be unphysical.

In summary of this introduction therefore, we develop a novel theory of electrodynamics based on vacuum topology that gives self-consistent descriptions of empirical data where an electrodynamics based on a U(1) vacuum fails. It turns out that O(3) electrodynamics does not incorporate a monopole, as a material point particle, because it is a theory based on the topology of the vacuum. The next section provides foundational justification for gauge field theory using fiber bundle theory.

II. BASIS IN FIBER BUNDLE THEORY

The gauge concept [3] was introduced by Weyl in 1918. In consequence of gauge theory, the absolute magnitude or norm of a physical vector depends on its location in spacetime. This notion is the basis of all contemporary gauge theory, which is expressed in the language [6] of group theory and has been highly developed mathematically [29–32]. For our purposes, it is sufficient to give a brief account of the elements of gauge theory as used in optics and electrodynamics, including O(3) electrodynamics. A gauge theory is a theory of special relativity in O(3) and U(1) electrodynamics, and in electroweak theory, and borrows concepts [6] from general relativity. For example, the homogeneous field equation of both U(1) and O(3) electrodynamics are Jacobi identities akin to the Bianchi identity in general relativity. Several reviews of contemporary gauge theory are given in Ref. 4, and the theory is firmly rooted in rigorous mathematical concepts such as fiber bundle theory. The latter leads to the field equations of O(3) electrodynamics through concepts [29–32] such as principal bundle, associated vector bundle, connections on principal bundles, covariant derivatives of sections of a vector bundle, exterior covariant derivative, and the curvature of a connection. In optics and electrodynamics however, these mathematical concepts reduce to those of gauge potentials. It is sufficient to know, therefore, that the theory of O(3) electrodynamics is rigorously founded in fiber bundle theory and in the theory of extended Lie algebra [4,15]. The interested reader is referred elsewhere for mathematical details [29–32] because, in natural philosophy, a theory stands or falls by comparison with empirical data, not by mathematical rigor alone. The latter is necessary but not sufficient for a theory in optics and electrodynamics.

A simple example in classical electrodynamics of what is now known as "gauge invariance" was introduced by Heaviside [3,4], who reduced the original electrodynamical equations of Maxwell to their present form. Therefore, these equations are more properly known as the Maxwell–Heaviside equations and, in the terminology of contemporary gauge field theory, are identifiable as U(1) Yang–Mills equations [15]. The subject of this chapter is O(3) Yang–Mills gauge theory applied to electrodynamics and electroweak theory.

The Maxwell–Heaviside field equations are, in SI units

$$\begin{aligned} \nabla\cdot\boldsymbol{E} &= 0; \qquad \nabla\cdot\boldsymbol{B} = 0 \\ \nabla\times\boldsymbol{E}+\frac{\partial\boldsymbol{B}}{\partial t} &= \boldsymbol{0} \\ \nabla\times\boldsymbol{B}-\frac{1}{c^2}\frac{\partial\boldsymbol{E}}{\partial t} &= \boldsymbol{0} \end{aligned} \tag{10}$$

where $\boldsymbol{D}$ is the electric displacement, ρ is the electric charge density, $\boldsymbol{B}$ is magnetic flux density, $\boldsymbol{E}$ is the electric field strength, $\boldsymbol{H}$ is the magnetic field strength, and $\boldsymbol{J}$ is the current density. The received view is to assert that in the vacuum:

$$\boldsymbol{D} = \varepsilon_0\boldsymbol{E}; \qquad \boldsymbol{B} = \mu_0\boldsymbol{H} \tag{11}$$

where ε_0 and $\boldsymbol{\mu}_0$ are permittivity and permeability in vacuo. Equations (12) then reduce to

$$\begin{aligned} \nabla\cdot\boldsymbol{D} &= \rho; \qquad \nabla\cdot\boldsymbol{B} = 0 \\ \nabla\times\boldsymbol{E}+\frac{\partial\boldsymbol{B}}{\partial t} &= 0 \\ \nabla\times\boldsymbol{H} &= \boldsymbol{J}+\frac{\partial\boldsymbol{D}}{\partial t} \end{aligned} \tag{12}$$

The notion of gauge invariance is illustrated on this level by denoting

$$\boldsymbol{B} = \nabla \times \boldsymbol{A} \tag{13}$$

$$\boldsymbol{E} = -\nabla \times \boldsymbol{S} \tag{14}$$

where $\boldsymbol{A}$ is the vector potential of Maxwell and $\boldsymbol{S}$ the vector potential of Stratton. Using the identity

$$\nabla \times \nabla\chi \equiv \mathbf{0} \tag{15}$$

$$\boldsymbol{B} = \nabla \times \boldsymbol{A} = \nabla \times (\boldsymbol{A} + \nabla\chi) \tag{16}$$

it is seen that any gradient $\nabla\chi$ can be added to $\boldsymbol{A}$ or $\boldsymbol{S}$, leaving $\boldsymbol{B}$ and $\boldsymbol{E}$ unchanged. Therefore, in the received view, $\boldsymbol{B}$ and $\boldsymbol{E}$ are gauge-invariant, measurable, and physical, whereas $\boldsymbol{A}$ and $\boldsymbol{S}$ are defined only up to an arbitrary gradient function and are therefore mathematical in nature, are not measurable, and have no physical effect. However, this can be true as argued in Section I only if the vacuum is simply connected, whereas the group spaces of U(1) and O(3) are not simply connected. We find empirically [3,4] several experimental verifications of the fact that $\boldsymbol{A}$ and $\boldsymbol{S}$ are in fact physical quantities, and that $\boldsymbol{A}$ and $\boldsymbol{S}$ cannot be changed arbitrarily by adding a gradient of a scalar. However elaborate the mathematical justification for U(1) electrodynamics becomes, this paradox remains.

During the course of this review chapter, we shall unearth several flaws in U(1) electrodynamics, some of which are discussed in Section III. One consequence of the gauge and metric invariance of the free space Maxwell–Heaviside equations is that they are also invariant under the general Lorentz transformation, consisting of boosts, rotations, and spacetime translations [6]. They are invariant also under the fundamental symmetry operations of motion reversal (T) and parity inversion (P). These properties mean that they are unable to describe interferometry and simple optical properties such as normal reflection without self-contradiction. The Maxwell–Heaviside theory and its gauge invariance is rigidly adhered to in the received view, but nevertheless, these basic flaws are there and are discussed systematically in this chapter. In the course of development of O(3) electrodynamics, a more general form of gauge theory is needed, and this more general form is based on vacuum topology and group theory. Therefore, in our view, O(3) electrodynamical equations apply in the vacuum as well as in field matter interaction [11–20]. In general, they must be solved without approximation using numerical techniques, but with well-defined assumptions, analytical solutions emerge. These include the B cyclic theorem [11–20].

The systematic development of gauge theory relies on a rotation of a n dimensional function ψ of the spacetime coordinate x^μ in special relativity. The rotation is expressed as

$$\psi' = \exp\left(iM^a\Lambda^a(x^\mu)\right)\psi \equiv S(x^\mu)\psi \tag{17}$$

where M^a are group generators, and where Λ^a is an angle that is a function of x^μ through special relativity [6]. In general, M^a are $n \times n$ matrices or tensors. In O(3) electrodynamics, the indices a can be (1), (2), and (3) of the complex basis (7), or Cartesian indices as in the basis (6). From Eq. (17), it is found that

$$\partial_\mu\psi' = S(\partial_\mu\psi) + (\partial_\mu S)\psi \tag{18}$$

that is, that $\partial_\mu\psi$ does not transform covariantly. It is well known that this problem is addressed through the introduction of the covariant derivative:

$$D_\mu\psi \equiv (\partial_\mu - igM^aA_\mu^a)\psi \tag{19}$$

where g is in general a proportionality constant giving the right units, and where A_μ^a is the vector potential, sometimes referred to as the "connection." In U(1) electrodynamics, A_μ^a reduces to the familiar 4-potential A_μ^a of the Maxwell–Heaviside theory, a 4-vector. This means that in U(1) electrodynamics, the internal gauge space is a scalar space in which $M = -1$ and in which the covariant derivative reduces to

$$D_\mu(U(1)) = \partial_\mu + igA_\mu \tag{20}$$

which, in momentum space, is the familiar minimal prescription. In O(3) electrodynamics however, A_μ^a is a 12-vector, and can be expressed as

$$\boldsymbol{A}_\mu = A_\mu^{(1)}\boldsymbol{e}^{(1)} + A_\mu^{(2)}\boldsymbol{e}^{(2)} + A_\mu^{(3)}\boldsymbol{e}^{(3)} \tag{21}$$

in the basis ((1),(2),(3)). Similarly, the familiar field tensor $F_{\mu\nu}$ of U(1) electrodynamics becomes

$$\boldsymbol{G}_{\mu\nu} = G_{\mu\nu}^{(1)}\boldsymbol{e}^{(1)} + G_{\mu\nu}^{(2)}\boldsymbol{e}^{(2)} + G_{\mu\nu}^{(3)}\boldsymbol{e}^{(3)} \tag{22}$$

in O(3) electrodynamics. Since ((1),(2),(3)) is a physical space, each of the tensors $G_{\mu\nu}^{(i)}; i = 1, 2, 3$ is well defined in Minkowski spacetime [11–20].

General gauge field theory emerges when the covariant derivative is applied to ψ [6]:

$$D'_\mu\psi = SD_\mu\psi \tag{23}$$

It is useful to go through this derivation in detail because it produces the inhomogeneous term responsible for the Aharonov–Bohm effect in O(3) electrodynamics. The effect of the rotation may be written as

$$(\partial_\mu - igA'_\mu)\psi' = S(\partial_\mu - igA_\mu)\psi \tag{24}$$

which means that

$$\begin{aligned}
\partial_\mu \psi' &= S(\partial_\mu \psi) + (\partial_\mu S)\psi \\
(\partial_\mu S)\psi - igA'_\mu S\psi &= -igSA_\mu \psi \\
igA'_\mu S &= igSA_\mu + \partial_\mu S \\
A'_\mu SS^{-1} &= SA_\mu S^{-1} - \frac{i}{g}(\partial_\mu S)S^{-1} \\
A'_\mu &= SA_\mu S^{-1} - \frac{i}{g}(\partial_\mu S)S^{-1}
\end{aligned} \tag{25}$$

The end result is that the inhomogeneous term $-(i/g)(\partial_\mu S)S^{-1}$ appears in the vacuum. This term originates in the topology of the vacuum, and it is different for U(1) electrodynamics and O(3) electrodynamics. In U(1) electrodynamics, the gauge transformation (25) reduces to

$$A'_\mu \rightarrow A_\mu + \frac{1}{g}\partial_\mu \Lambda \tag{26}$$

which is the covariant form of Eq. (15). In O(3) electrodynamics however, the inhomogeneous term and the vector potential are both physical quantities, as originally envisaged by Maxwell and Faraday. The 12-vector A_μ is the equivalent of Faraday's electrotonic state and of Maxwell's physical vector potential [3,4]. It follows that the effect (25) on the vector potential in O(3) electrodynamics is produced by a physical rotation, and later in this review, it is shown that this physical rotation is the rotation of the platform in the Sagnac effect [20]. More generally, a rotation in the internal gauge space of O(3) electrodynamics produces a phase difference that is also physical and measurable [3,4]. O(3) electrodynamics is therefore able to describe the Sagnac effect precisely, whereas U(1) electrodynamics has no explanation for the Sagnac effect because of its gauge invariance. Quantities such as the 12-vector potential of O(3) electrodynamics are gauge-covariant, not gauge-invariant, because the inhomogeneous term in O(3) electrodynamics is a physical term, not a random mathematical construct as in U(1) electrodynamics.

In general gauge field theory [6], the field tensor is proportional to the commutator of covariant derivatives. This is the result of a round trip or closed loop with covariant derivatives in Minkowski spacetime, and in condensed notation, the result can be written as

$$G_{\mu\nu} = \frac{i}{g}[D_\mu, D_\nu] \tag{27}$$

$$G_{\mu\nu} = \partial_\mu A_\nu - \partial_\nu A_\mu - ig[A_\mu, A_\nu] \tag{27a}$$

In U(1) electrodynamics, we recover the familiar 4-curl of Maxwell–Heaviside theory because the commutator $[A_\mu, A_\nu]$ is zero. In O(3) electrodynamics, Eq. (22) applies and each component $G_{\mu\nu}^{(1)}$, $G_{\mu\nu}^{(2)}$, and $G_{\mu\nu}^{(3)}$ is defined as

$$G_{\mu\nu}^{(i)} = \partial_\mu A_\nu^{(i)} - \partial_\nu A_\mu^{(i)} - ig\varepsilon_{(i)(j)(k)} A_\mu^{(j)} A_\nu^{(k)}, \qquad (i),(j),(k) = (1),(2),(3) \tag{28}$$

in the complex circular basis ((1),(2),(3)), [11–20]. Whereas $F_{\mu\nu}$ of U(1) electrodynamics is gauge- and Lorentz-invariant, $G_{\mu\nu}$ of O(3) electrodynamics transforms covariantly under rotation in the internal space ((1),(2),(3)), a representation of the physical space of three dimensions:

$$G'_{\mu\nu} = S G_{\mu\nu} S^{-1} \tag{29}$$

The homogeneous field equation of O(3) electrodynamics is inferred from the Jacobi identity of covariant derivatives

$$\sum_{\text{cyclic}} [D_\rho, [D_\mu, D_\nu]] \equiv 0 \tag{30}$$

and can be written as the identity [11–20]

$$D_\mu \tilde{\boldsymbol{G}}^{\mu\nu} \equiv \mathbf{0} \tag{31}$$

The inhomogeneous field equation is not an identity, but an equation of the Yang–Mills type [6]

$$D_\mu \boldsymbol{H}^{\mu\nu} = \boldsymbol{J}^\nu \tag{32}$$

where $\boldsymbol{H}^{\mu\nu}$ is a generalization of $\boldsymbol{G}^{\mu\nu}$ to include polarization and magnetization, and where $\boldsymbol{J}^\nu$ is the charge current 12-vector, defined as

$$J^{\nu(i)} \equiv \left(\rho^{(i)}, \frac{\boldsymbol{J}^{(i)}}{c}\right) \qquad i = 1, 2, 3 \tag{33}$$

where c is the speed of light in vacuo for all practical purposes in the laboratory. Equations (31) and (32) are developed fully in Section (IV) and are compared with the Lehnert, Barrett, and Harmuth equations cited in Section I. These equations extend the symmetry of the electromagnetic sector of unified field theory with many consequences, some of which are discussed in Section (XI) for electroweak theory, and in Part 3 of this three-volume series for grand unified theory.

The development just given illustrates the fact that the topology of the vacuum determines the nature of the gauge transformation, field tensor, and field equations, as inferred in Section (I). The covariant derivative plays a central role in each case; for example, the homogeneous field equation of O(3) electrodynamics is a Jacobi identity made up of covariant derivatives in an internal O(3) symmetry gauge group. The equivalent of the Jacobi identity in general relativity is the Bianchi identity.

Finally, in this section, we develop the concept of electromagnetic phase from U(1) to O(3). This is a nontrivial development [4] that has foundational consequences for interferometry and physical optics for example. In U(1) electrodynamics, the electromagnetic phase is defined up to an arbitrary factor [4] because of gauge invariance. The U(1) phase is therefore

$$\gamma \equiv \omega t - \boldsymbol{\kappa} \cdot \boldsymbol{r} + \alpha \tag{34}$$

where ω is the angular frequency at instant t; $\boldsymbol{\kappa}$ is the wave-vector at coordinate $\boldsymbol{r}$, and $\boldsymbol{\alpha}$ is random. In other words, the U(1) electromagnetic phase factor $\exp\,(i(\omega t - \boldsymbol{\kappa} \cdot \boldsymbol{r}))$ can be multiplied by the factor $e^{i\alpha}$ because gauge transformation in U(1) is a random rotation in the (scalar) internal gauge space. The random rotation is represented by the operator $e^{i\alpha}$ where $\boldsymbol{\alpha}$ is random. This operation leads to Eq. (26), where the gradient function is random as usual in U(1) electrodynamics. Therefore the U(1) electromagnetic phase is unphysical. This is true despite the fact that the theory of U(1) electrodynamics is the received view, adhered to rigidly. Therefore [4], the field tensor in U(1) electrodynamics, is underdetermined because the phase is arbitrary; and the potential 4-vector of U(1) electrodynamics is overdetermined because it is also arbitrary—an infinite number of A^μ corresponds, in the received view, to one physical condition. Dirac attempted to remedy these flaws by introducing a phase *factor*

$$\Phi(C) = \exp\left(i\frac{e}{\hbar}\oint_C A_\mu \, dx^\mu\right) \tag{35}$$

where e is electric charge, and $\hbar$ is the Dirac constant. The Dirac phase factor completely defines [4] the system on the U(1) level. The phase factor in O(3) electrodynamics is obtained by generalizing this concept, as first accomplished by Wu and Yang [33]. The phase factor in O(3) electrodynamics can be written as

$$\Phi^*(C) = P\exp\left(ig\oint_C A_\mu \, dx^\mu\right) \tag{36}$$

$$\Phi^*(C) = P\exp\left(i\oint_C \kappa_\mu \, dx^\mu\right) = P\exp\left(ig\int\!\!\int B^{(3)}\, dAr\right) \tag{36a}$$

where a magnetic flux of topological origin appears on the right hand side, an area integral over the $\boldsymbol{B}^{(3)}$ (Evans–Vigier) field [11–20]. Here, $\Phi^*(C)$ specifies parallel transport over any loop C in rotation, g is the same factor that appears in the definition of the covariant derivative, [Eq. (20)], and P specifies path dependence in the integral [4]. On the left-hand side appears the line integral corresponding to the dynamical phase factor, which is equal through a non-Abelian Stokes theorem to the topological phase defined by the surface integral over $\boldsymbol{B}^{(3)}$. This result is a clear illustration of the topological origin of $\boldsymbol{B}^{(3)}$, and the phase factor is not a random quantity as in U(1) electrodynamics, but a gauge-covariant quantity. It is the holonomy of the connection A_μ in O(3) electrodynamics and plays a central role in interferometry, including the Aharonov–Bohm effect. Consideration of interferometry leads to the conclusion that O(3) electrodynamics provides a self-consistent description of several situations where U(1) electrodynamics either fails (e.g., the Sagnac effect) or is self-inconsistent (e.g., Michelson interferometry).

III. REFUTATION OF U(1) ELECTRODYNAMICS

From the foregoing, U(1) electrodynamics was never a complete theory, although it is rigidly adhered to in the received view. It has been argued already that the Maxwell–Heaviside theory is a U(1) Yang–Mills gauge theory that discards the basic commutator $\boldsymbol{A}^{(1)} \times \boldsymbol{A}^{(2)}$. However, this commutator appears in the fundamental definition of circular polarity in the Maxwell–Heaviside theory through the third Stokes parameter

$$S_3 = |-i\omega^2 A^{(1)} \times A^{(2)}| = \omega^2 A^{(0)2} \tag{37}$$

so there is an internal inconsistency. In O(3) electrodynamics, on the other hand, the fundamental definition of the $\boldsymbol{B}^{(3)}$ field ensures that circular polarity is consistently defined

$$\boldsymbol{B}^{(3)*} = -ig\boldsymbol{A}^{(1)} \times \boldsymbol{A}^{(2)} \tag{38}$$

so that circular polarity in O(3) electrodynamics is due to the $\boldsymbol{B}^{(3)}$ field, which is therefore a foundational physical observable. This argument is a simple and straightforward refutation of U(1) electrodynamics, specifically, of Maxwell–Heaviside theory considered as a U(1) symmetry gauge field theory. The third Stokes parameter is a fundamental signature of circular polarization and was first recognized as such by Stokes in 1852 before the development of Maxwell's original equations in the 1860s [3]. Circular polarization was discovered empirically by Arago in 1811.

There is in effect no circular polarization in U(1) electrodynamics if we choose to define circular polarization in terms of the third Stokes parameter.

This result is inconsistent with the fact that the differential equation developed by Heaviside from Maxwell's original equations describe circular polarization. The root of the inconsistency is that U(1) gauge field theory is made to correspond with Maxwell–Heaviside theory by discarding the commutator $\boldsymbol{A}^{(1)} \times \boldsymbol{A}^{(2)}$. The neglect of the latter results in a reduction to absurdity, because if S_3 vanishes, so does the zero order Stokes parameter:

$$S_0 = \pm S_3 \tag{39}$$

and S_0 describes the intensity of radiation. This result is another self inconsistency of U(1) electrodynamics.

In O(3) electrodynamics, on the other hand, Eq. (38), defining the $\boldsymbol{B}^{(3)}$ field, is consistent with the O(3) field Eq. (31) and (32) because Eq. (38) is part of the definition of the field tensor in O(3) electrodynamics [11–20].

A second simple refutation of U(1) electrodynamics is perfect normal reflection. The explanation of this foundational effect in Maxwell–Heaviside electrodynamics relies on the phase in U(1) electrodynamics, which, as argued already, is a random quantity. If we choose α in Eq. (34) to be zero for simplicity and without loss of generality, then the received view of perfect normal reflection (Fig. 1) is as follows:

$$\exp\left(i(\boldsymbol{\kappa}\cdot\boldsymbol{r} - \omega t)\right) \xrightarrow{R} \exp\left(i(-\boldsymbol{\kappa}\cdot\boldsymbol{r} - \omega t)\right) \tag{40}$$

However, normal reflection, in, for example, the Z axis, is equivalent to the parity inversion operation P. The effect of this operation on the U(1) phase factor is as follows:

$$\exp\left(i(\boldsymbol{\kappa}\cdot\boldsymbol{r} - \omega t)\right) \xrightarrow{P} \exp\left(i(\boldsymbol{\kappa}\cdot\boldsymbol{r} - \omega t)\right) \tag{41}$$

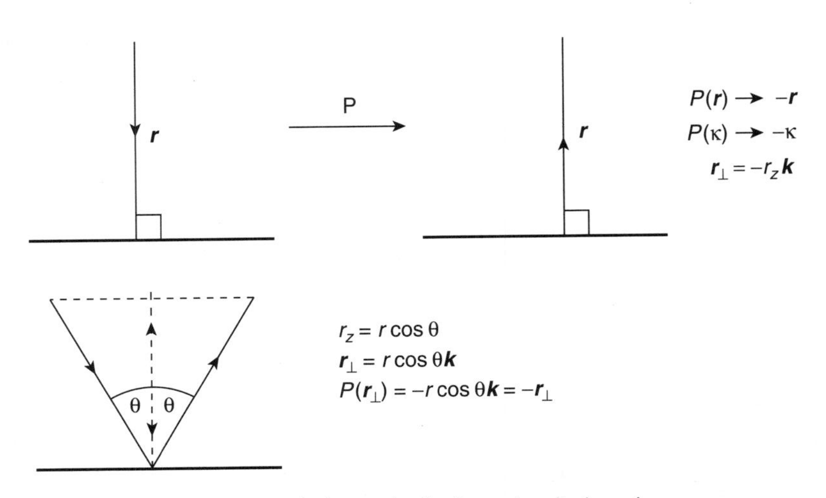

Figure 1. Equivalence of reflection and parity inversion.

Thus the received view of normal reflection (1) in U(1) electrodynamics violates parity. This violation is not allowed in classical physics. For off-normal reflection (Fig. 1), projections on to the normal result in the same paradox using the empirical fact that the angle of reflection is equal to the angle of incidence. In the received view, Eq. (40) is held to rigidly, but is nevertheless in violation of parity. This is true if and only if Snell's law is true. In conclusion, $P(\omega t - \boldsymbol{\kappa}\cdot\boldsymbol{r} = (\omega t - \boldsymbol{\kappa}\cdot\boldsymbol{r})$, which is Snell's law in Maxwell–Heaviside theory.

It is highly significant that this paradox disappears in O(3) electrodynamics through the use of the physical phase factor:

$$\Phi = \exp\left(i\oint \boldsymbol{\kappa}\cdot d\boldsymbol{Z}\right) = \exp\left(ig\int \boldsymbol{B}^{(3)}\cdot d\boldsymbol{S}\right) \tag{42}$$

On the left-hand side appears a line integral, and on the right-hand side, there is an area integral over $\boldsymbol{B}^{(3)}$. If a beam of light originates at an origin O and is normally reflected from a perfectly reflecting mirror at point Z, the line integral is as follows:

$$\oint \boldsymbol{\kappa}\cdot d\boldsymbol{Z} = \int_0^Z \kappa\, dZ - \int_Z^0 \kappa\, dZ = 2\kappa Z \tag{43}$$

Note that this gives, fortuitously, the same change, $2\boldsymbol{\kappa}Z$, as in the U(1) description of normal reflection, which therefore fortuitously describes the empirical result.

The area integral on the right-hand side of Eq. (42) is a topological phase [4], because the origin of $\boldsymbol{B}^{(3)}$ is topological as argued already, that is, $\boldsymbol{B}^{(3)}$ springs from the vacuum configuration. Using the relation [11–20]

$$g = \frac{\kappa^2}{B^{(0)}} \tag{44}$$

the right-hand-side exponent becomes $\boldsymbol{\kappa}^2 S$, where S is an area

$$S = 2\frac{\kappa}{Z} \tag{45}$$

If the distance OZ is n wavelengths, λ, then the area becomes

$$S = \frac{n\lambda^2}{\pi} \tag{46}$$

The outcome of these two very simple examples is that all electrodynamics (classical and quantum) must be upgraded to a gauge theory of higher symmetry, such as O(3). Equation (42) is self-consistent, because under P, both sides are negative. The left-hand side is negative because the line integral changes sign under P, and the right-hand side is negative because the integral is negative under P (product of an axial vector $\boldsymbol{B}$ and a polar vector $\boldsymbol{S}$).

Michelson interferometry is dependent on normal reflection from two mirrors at right angles, and so the same foundational argument as just given can be used to show that U(1) electrodynamics does not describe Michelson interferometry self-consistently. Without loss of generality, we can write Eq. (38) as

$$\pi R \kappa A^{(0)} \boldsymbol{k} \cdot R\boldsymbol{k} = \boldsymbol{B}^{(3)} \cdot A r\boldsymbol{k} \tag{47}$$

which can be integrated straightforwardly to give the non-Abelian Stokes theorem [11–20]

$$2\pi\kappa A^{(0)} \oint \boldsymbol{R} \cdot d\boldsymbol{R} = \int\int \boldsymbol{B}^{(3)} \cdot d\boldsymbol{Ar} \tag{48}$$

where R is given by

$$R = \frac{1}{\kappa} = \frac{\lambda}{2\pi} \tag{49}$$

and where λ is the wavelength. Multiplying both sides by $g = \boldsymbol{\kappa}/A^{(0)}$ defines the required non-Abelian phase factor in terms of a non-Abelian Stokes theorem

$$\Phi = \exp(2\pi i \oint \boldsymbol{\kappa} \cdot d\boldsymbol{R}) = \exp\left(i \frac{\kappa}{A^{(0)}} \int\int \boldsymbol{B}^{(3)} \cdot d\boldsymbol{Ar} \right) \tag{50}$$

which is closely related to Eq. (42). The line integrals must be evaluated over a closed curve [11–20] and have the foundational property

$$\oint_{AO} \boldsymbol{\kappa} \cdot d\boldsymbol{R} = -\oint_{0A} \boldsymbol{\kappa} \cdot d\boldsymbol{R} \tag{51}$$

which is the root cause [34] of Michelson interferometry, and interferometry in general. In U(1) electrodynamics, the change in phase of a light beam originating at the beamsplitter [35] and arriving back at the beamsplitter after normal reflection from either mirror is zero because of the property (41). This is contrary to the empirical observation [35] of the Michelson interferogram, the basis of Fourier transform infrared spectroscopy. In the usual U(1) theory, therefore, the path-dependent part of the electromagnetic phase is the familiar $\boldsymbol{\kappa} \cdot \boldsymbol{r}$, and the complete electromagnetic phase is $\omega t - \boldsymbol{\kappa} \cdot \boldsymbol{r} + \alpha$, where α is random and can be set to zero for simplicity of argument. The phase $\omega t - \boldsymbol{\kappa} \cdot \boldsymbol{r}$ is invariant under both

P and T because it is a dimensionless number, and we shall show that the complete failure of U(1) electrodynamics to describe the Sagnac effect is due to the T invariance of U(1) phase. In Michelson interferometry, as described by O(3) electrodynamics, there is a change in the measurable phase factor after reflection because of the property of line integrals. The phase factors arriving back at the beamsplitter from either mirror are different, and an interferogram appears as observed [35] empirically by changing the length of one arm of the interferometer. The Fourier transformation of this interferogram gives a spectrum.

The inverse Faraday effect depends on the third Stokes parameter empirically in the received view [36], and is the archetypical magneto-optical effect in conventional Maxwell–Heaviside theory. This type of phenomenology directly contradicts U(1) gauge theory in the same way as argued already for the third Stokes parameter. In O(3) electrodynamics, the paradox is circumvented by using the field equations (31) and (32). A self-consistent description [11–20] of the inverse Faraday effect is achieved by expanding Eq. (32):

$$\partial_\mu \boldsymbol{H}^{\mu\nu(1)*} = \boldsymbol{J}^{\nu(1)*} + ig\boldsymbol{A}_\mu^{(2)} \times \boldsymbol{H}^{\mu\nu(3)} \tag{52}$$

$$\partial_\mu \boldsymbol{H}^{\mu\nu(2)*} = \boldsymbol{J}^{\nu(2)*} + ig\boldsymbol{A}_\mu^{(3)} \times \boldsymbol{H}^{\mu\nu(1)} \tag{53}$$

$$\partial_\mu \boldsymbol{H}^{\mu\nu(3)*} = \boldsymbol{J}^{\nu(3)*} + ig\boldsymbol{A}_\mu^{(1)} \times \boldsymbol{H}^{\mu\nu(2)} \tag{54}$$

Using the constitutive relation

$$\boldsymbol{H}^{(3)*} = \frac{1}{\mu}\boldsymbol{B}^{(3)*} \tag{55}$$

gives the magnetic field strength induced in the inverse Faraday effect from first principles of O(3) gauge field theory as

$$\boldsymbol{H}^{(3)*} = -i\frac{g'}{\mu_0}\boldsymbol{A}^{(1)} \times \boldsymbol{A}^{(2)} \tag{56}$$

where

$$g' = \frac{\mu_0}{\mu} g \tag{57}$$

Here, μ is the magnetic permeability of the material in which the inverse Faraday effect is observed. We can write Eq. (52) as

$$\partial_\mu \boldsymbol{H}^{\mu\nu(1)*} = \boldsymbol{J}^{\nu(1)*} + \Delta\boldsymbol{J}^{\nu(1)*} \tag{58}$$

so that the transverse current detected [37] in the inverse Faraday effect is given by

$$\Delta \boldsymbol{J}^{\nu(1)*} = ig\varepsilon \boldsymbol{A}_{\mu}^{(2)} \times \boldsymbol{G}^{\mu\nu(3)} \tag{59}$$

and causes a signal in an induction coil due to the vacuum $\boldsymbol{B}^{(3)}$ field appearing in the O(3) field tensor $\boldsymbol{G}^{\mu\nu(3)}$.

The explanation of the inverse Faraday effect in U(1) electrodynamics relies on the clearly self-inconsistent introduction of $\boldsymbol{A}^{(1)} \times \boldsymbol{A}^{(2)}$ phenomenologically: "self-inconsistent" because U(1) gauge field theory sets $\boldsymbol{A}^{(1)} \times \boldsymbol{A}^{(2)}$ to zero identically. As argued already, the conjugate product $\boldsymbol{A}^{(1)} \times \boldsymbol{A}^{(2)}$ is proportional to the third Stokes parameter in the vacuum and so is a fundamental property of circularly polarized light. As such, it must be considered as a fundamental object in gauge field theory applied to electrodynamics. In U(1) gauge field theory, this is not possible, but it is possible self-consistently in O(3) gauge field theory.

In Maxwell–Heaviside electrodynamics, the field energy, Poynting vector, and Maxwell stress tensor are incorporated in the stress energy momentum tensor [38]. In order to obtain a non-null energy and field momentum (Poynting vector), the method of averaging is used. The conventionally defined Poynting vector, for example, becomes proportional to $\boldsymbol{E} \times \boldsymbol{B}^* = \boldsymbol{E}^{(1)} \times \boldsymbol{B}^{(2)}$. This method is inconsistent with electrodynamics considered as a U(1) gauge field theory, but consistent with O(3) electrodynamics.

Recall that in general gauge field theory, for any gauge group, the field tensor is defined through the commutator of covariant derivatives. In condensed notation [6]

$$G_{\mu\nu} = \partial_{\mu} A_{\nu} - \partial_{\nu} A_{\mu} - ig[A_{\mu}, A_{\nu}] \tag{60}$$

where the commutator is nonzero in general. The connection or generalized potential A_{μ} is defined in general through the gauge group symmetry. The field tensor $G_{\mu\nu}$ is covariant for all gauge groups, and is always compatible with special relativity for all gauge group symmetries. In this general theory therefore, the homogeneous and inhomogeneous Maxwell equations in the vacuum are the U(1) gauge field equations

$$D^{\nu} \tilde{G}_{\mu\nu} \equiv 0 \tag{61}$$

$$D^{\nu} G_{\mu\nu} = 0 \tag{62}$$

where D^{ν} denotes the covariant derivative pertinent to U(1) and where $\tilde{G}^{\mu\nu}$ is the dual of $G_{\mu\nu}$ as usual. In the U(1) gauge theory, the commutator in Eq. (60) vanishes because the U(1) group has only one structure constant and the internal symmetry of the gauge theory is a scalar symmetry. The covariant derivative in U(1) is

$$D^{\nu} = \partial^{\nu} + igA^{\nu} \tag{63}$$

Therefore Eqs. (61) and (62) reduce to

$$(\partial^{\nu} + igA^{\nu})\tilde{F}_{\mu\nu} \equiv 0 \tag{64}$$

$$(\partial^{\nu} + igA^{\nu})F_{\mu\nu} \equiv 0 \tag{65}$$

which become the free-space homogeneous and inhomogeneous Maxwell–Heaviside equations if and only if

$$A^{\nu}\tilde{F}_{\mu\nu} = 0 \tag{66}$$

$$A^{\nu}F_{\mu\nu} \equiv 0 \tag{67}$$

or in vector notation

$$\begin{matrix} \boldsymbol{A}\cdot\boldsymbol{B} = 0 & \boldsymbol{A}\times\boldsymbol{E} = \boldsymbol{0} \\ \boldsymbol{A}\cdot\boldsymbol{E} = 0 & \boldsymbol{A}\times\boldsymbol{B} = \boldsymbol{0} \end{matrix} \tag{68}$$

For plane waves, and using the usual U(1) relation

$$\boldsymbol{B} = \nabla \times \boldsymbol{A} \tag{69}$$

the vector potential is proportional to $\boldsymbol{B}$ and so

$$\boldsymbol{B} \times \boldsymbol{E} = \boldsymbol{0} \tag{70}$$

If we attempt to define the free-space field energy and momentum in terms of the products $\boldsymbol{B}\cdot\boldsymbol{B}$ and $\boldsymbol{B} \times \boldsymbol{E}$, the results are zero in U(1) gauge field theory. In order to obtain the conventional field energy and Poynting vector of the free electromagnetic field, products such as $\boldsymbol{B}^{(1)} \times \boldsymbol{B}^{(2)}$ and $\boldsymbol{B}^{(1)} \times \boldsymbol{E}^{(2)}$ have to be used. This procedure, although common place, and referred to in the literature as "time averaging" [38], introduces phenomenology extraneous to U(1), because it introduces the complex internal gauge space ((1),(2),(3)). These inconsistencies in U(1) gauge field theory applied to electrodynamics are therefore summarized as follows: (1) if the U(1) covariant derivative is used, the field energy, momentum, and third Stokes parameter vanish; and (2) if the phenomenological "time averaging" procedure is used, the resultant Poynting vector is proportional to $\boldsymbol{B}^{(1)} \times \boldsymbol{E}^{(2)}$, and is perpendicular to the plane of definition of

U(1), whose group space is a circle. This result is another internal inconsistency, because the group space of a gauge theory is a circle, there can be no physical quantity in free space perpendicular to that plane. It is necessary but not sufficient, in this view, that the Lagrangian in U(1) field theory be invariant [6] under U(1) gauge transformation.

In O(3) electrodynamics, the stress energy momentum tensor is defined [11–20] as

$$T_{\mu}^{\nu} = \varepsilon_0 \left(\boldsymbol{G}^{\mu\rho} \cdot \boldsymbol{G}_{\rho\mu} - \frac{1}{4} \boldsymbol{G}^{\rho\nu} \cdot \boldsymbol{G}_{\rho\nu} \right) \tag{71}$$

giving the field energy self-consistently as

$$U = \varepsilon_0 (E^{1(1)} E_1^{(2)} + E^{2(1)} E_2^{(2)} + E^{1(2)} E_1^{(1)} + E^{2(2)} E_2^{(1)} + E^{3(3)} E_3^{(3)*}) \tag{72}$$

The Poynting vector is self-consistently defined as

$$T_1^0 = \varepsilon_0 (\boldsymbol{G}^{02} \cdot \boldsymbol{G}_{21} + \boldsymbol{G}^{03} \cdot \boldsymbol{G}_{31}) \tag{73}$$

$$T_2^0 = \varepsilon_0 (\boldsymbol{G}^{01} \cdot \boldsymbol{G}_{12} + \boldsymbol{G}^{03} \cdot \boldsymbol{G}_{32}) \tag{74}$$

$$T_3^0 = \varepsilon_0 (\boldsymbol{G}^{01} \cdot \boldsymbol{G}_{13} + \boldsymbol{G}^{02} \cdot \boldsymbol{G}_{23}) \tag{75}$$

and is finite. The $\boldsymbol{B}^{(3)}$ component is defined through Eq. (38), giving, self-consistently, the result (39).

The root cause of these further problems with electrodynamics considered as a U(1) gauge field theory is that parallel transport [6] must be used when an internal gauge space is present. The internal gauge space of U(1) is a scalar, and parallel transport results in a covariant derivative whose momentum representation is the minimal prescription [6]. This covariant derivative, however, leads self-inconsistently to a null energy density and Poynting vector as just argued. Therefore, in U(1), the Maxwell–Heaviside equations are obtained if and only if the field energy and Poynting vector are identically zero. A null Poynting vector means null energy and a null third Stokes parameter. The root cause of this is the neglect of $\boldsymbol{A}^{(1)} \times \boldsymbol{A}^{(2)}$, and we have come full circle. The only way out is to adopt a gauge field theory of higher symmetry than U(1).

A related problem is that the linear momentum of radiation in U(1) is defined by

$$\langle p \rangle = \varepsilon_0 c \int \boldsymbol{E} \times \boldsymbol{B} \, dV \tag{76}$$

which is again zero. The linear momentum of a photon, however, is nonzero in quantum theory and is $\hbar\kappa$, leading to the Compton effect and Compton scattering. It is well known that there is no classical equivalent of the Compton effect [39], so the correspondence principle is lost in the received view based on U(1) gauge field theory. In O(3) electrodynamics, however [11–20], there exists, in general, the longitudinally directed potential $A^{(3)}$ as part of the definition of the field tensor. The classical quantum equivalence in the Compton effect is then given simply as

$$\mathbf{p}^{(3)} = e\mathbf{A}^{(3)} = \hbar\boldsymbol{\kappa} \tag{77}$$

where e is regarded as the coupling constant in the definition of the constant $g = e/\hbar$, which appears in free space in *both* U(1) and O(3) electrodynamics. This is another characteristic of gauge field theory applied to electrodynamics, that charge e can act as a coupling constant in the covariant derivative. This is true for all internal gauge symmetries, so e need not be defined solely by the charge on the electron. These concepts are discussed further in Ref. 6. Therefore O(3) electrodynamics saves the quantum classical correspondence principle in Planck–Einstein quantization. Equation (77) has the following manifestly covariant form:

$$\begin{aligned} p^{\mu(3)} &= eA^{\mu(3)} = \hbar\kappa^{\mu} \\ A^{\mu(3)} &= \frac{1}{c}(A^{(0)}, cA^{(3)}) \end{aligned} \tag{78}$$

These concepts of O(3) electrodynamics also completely resolve the problem that, in Maxwell–Heaviside electrodynamics, the energy momentum of radiation is defined through an integral over the conventional tensor $T^{\mu\nu}$, and for this reason is not manifestly covariant. To make it so requires the use of special hypersurfaces as attempted, for example, by Fermi and Rohrlich [40]. The O(3) energy momentum (78), in contrast, is generally covariant in O(3) electrodynamics [11–20].

The Maxwell–Heaviside theory seen as a U(1) symmetry gauge field theory has no explanation for the photoelectric effect, which is the emission of electrons from metals on ultraviolet irradiation [39]. Above a threshold frequency, the emission is instantaneous and independent of radiation intensity. Below the threshold, there is no emission, however intense the radiation. In U(1), electrodynamics energy is proportional to intensity and there is, consequently, no possible explanation for the photoelectric effect, which is conventionally regarded as an archetypical quantum effect. In classical O(3) electrodynamics, the effect is simply

$$En = ecA^{(3)} = \text{constant} \times \text{frequency} \tag{79}$$

and in Planck–Einstein quantization, the constant of proportionality is $\hbar$, which turns out to be a universal constant of physics. The concomitant momentum relation, Eq. (77), is shown empirically by the Compton effect as argued already. Equation (77) means that above a given threshold frequency, there is enough energy in the photon to cause electron emission in the photoelectric effect. All the energy and momentum of the photon are transferred to the electron in a collision above a certain threshold frequency because at this point, the potential energy responsible for keeping the electron in place is exceeded. If we attempt to apply this logic to $\langle \boldsymbol{p} \rangle$ in Eq. (76), there is no threshold frequency possible on the classical level because $\langle \boldsymbol{p} \rangle$ cannot be proportional to frequency, only to beam intensity. The momentum $\boldsymbol{p}^{(3)} = e\boldsymbol{A}^{(3)}$ of classical O(3) electrodynamics is not proportional to intensity; it is proportional to frequency through the gauge equation (77), which also leads to the B cyclic theorem [11–20], the fundamental Lorentz invariant angular momentum relation of O(3) electrodynamics.

In the O(3) Compton effect, the observable change of wavelength is

$$\Delta\lambda = 2\left(\frac{eA^{(3)}}{mc}\right)\lambda_0 \sin^2\frac{\theta}{2} \tag{80}$$

where λ is the wavelength of the incident beam, m is the electron mass, and θ is the scattering angle. If Eq. (77) is applied to this result, we recover the usual quantum description of the Compton effect.

The concept of $\boldsymbol{A}^{(3)}$ can also be used to suggest a way out of the Dirac paradox [41] of U(1) electrodynamics, in which Dirac maintains that so long as we are dealing with transverse waves, we cannot bring in the Coulomb interaction between charged particles. In O(3) electrodynamics, there is a force given by

$$F^{\mu(3)} = e\frac{\partial A^{\mu(3)}}{\partial \tau} \tag{81}$$

whenever the beam interacts with an electron. This interaction results in a longitudinal force with a change of wavelength as just described for the Compton effect. This is not a Coulomb force since $\boldsymbol{E}^{(3)}$ is zero in vacuo [11–20].

Similarly, $\boldsymbol{A}^{(3)}$ can be used to suggest a way out of the de Broglie paradox [42], which points out that momentum and energy transform differently under Lorentz transformation from frequency. This paradox led de Broglie to postulate the existence of empty waves, which, however, have never been observed empirically. It can therefore be suggested that the Lorentz frequency transform must always be applied to

$$e\boldsymbol{A}^{(3)} = \frac{\hbar\omega}{c}\boldsymbol{e}^{(3)} \tag{82}$$

because this momentum is proportional to frequency empirically. If this momentum is interpreted as that of a particle traveling at the speed of light, the momentum becomes indeterminate (massless particle) or infinite (massive particle) unless it is always interpreted as being a constant ($\hbar$) multiplied by ω/c, which always exists empirically as the speed of light. The energy must evidently be interpreted in the same way, namely, as a constant multiplied by frequency. The Lorentz transform applied to frequency produces the aberration of light as usual [39] in special relativity. In this interpretation, there is no de Broglie paradox and no need to postulate the existence of empty waves [42].

The Sagnac effect cannot be described by U(1) electrodynamics [4,43] because of the invariance of the U(1) phase factor under motion reversal symmetry *(T)*:

$$\Phi = \exp\left(i(\omega t - \boldsymbol{\kappa}\cdot\boldsymbol{r})\right) \xrightarrow{T} \exp\left(i(\omega t - \boldsymbol{\kappa}\cdot\boldsymbol{r})\right) \tag{83}$$

The T operator generates the counterclockwise (A) loop from the clockwise (C) loop in the Sagnac effect, with the result that there is no difference in phase factor for journeys around the A and C loops, and no interferogram. This is contrary to observation when the Sagnac platform is at rest [43]. When the platform of the Sagnac interferometer [3] is rotated, there is the well-known Sagnac phase shift, which was first detected in 1913. This defies description by U(1) electrodynamics because the Maxwell–Heaviside field equations in the vacuum are invariant to rotation, which is part of the most general type of Lorentz transform [6]. The Maxwell–Heaviside equations in vacuo are also gauge- and metric-invariant, and are not capable of describing the Sagnac effect at all. The O(3) electrodynamics, in contrast, are completely successful in describing the interferogram with platform at rest and with a rotating platform. The details of this important advantage of O(3) electrodynamics are discussed in Section (VI), where a kinematic explanation of the Sagnac effect is also given using O(3) gauge theory. More details of magneto-optical effects are given in Section (VII).

The Aharonov–Bohm effect is self-inconsistent in U(1) electrodynamics because [44] the effect depends on the interaction of a vector potential $\boldsymbol{A}$ with an electron, but the magnetic field defined by $\boldsymbol{B} = \nabla \times \boldsymbol{A}$ is zero at the point of interaction [44]. This argument can always be used in U(1) electrodynamics to counter the view that the classical potential $\boldsymbol{A}$ is physical, and adherents of the received view can always assert in U(1) electrodynamics that the potential must be unphysical by gauge freedom. If, however, the Aharonov–Bohm effect is seen as an effect of O(3) electrodynamics, or of SU(2) electrodynamics [44], it is easily demonstrated that the effect is due to the physical inhomogeneous term appearing in Eq. (25). This argument is developed further in Section VI.

Barrett has argued convincingly that there are several effects in classical electrodynamics [3,4] where the potential must be physical, and Ref. 3 lists empirically observed effects where this is the case. The arguments in this section point to the fact that U(1) electrodynamics, defined as U(1) gauge field theory applied to electrodynamics, is self-inconsistent in the vacuum, as well as in field–matter interaction. In the next section, the field equations of electrodynamics seen as an O(3) gauge field theory applied to electrodynamics are given in full, revealing the presence in free space of conserved topological charges and currents that do not appear in U(1) electrodynamics and that in general are not zero.

IV. FIELD EQUATIONS OF O(3) ELECTRODYNAMICS IN THE COMPLEX CIRCULAR BASIS

In their most condensed form, the field equations are Eqs. (31) and (32), respectively, and, in general, must be solved without approximation on a computer with constitutive relations, as usual in classical electrodynamics. The familiar field tensors $\tilde{G}^{\mu\nu}$ and $H^{\mu\nu}$of the homogeneous and inhomogeneous Heaviside–Maxwell equations [U(1) Yang–Mills gauge field theory] become vectors in the O(3) symmetry internal gauge space of Eqs. (31) and (32), which are equations of O(3) symmetry Yang–Mills gauge field theory. Therefore an object such as $\tilde{\boldsymbol{G}}^{\mu\nu}$ is a vector in the internal gauge space and a tensor in Minkowski spacetime, and an object such as $\boldsymbol{J}^{\mu}$ is a 3-vector in the internal O(3) space and a 4-vector in Minkowski spacetime. The ordinary derivatives of the Maxwell–Heaviside equations are replaced in Eqs. (31) and (32) by covariant derivatives in an internal gauge space, with three rotation generators [11–20]. Eqs. (31) and (32) are gauge-covariant, and not gauge-invariant, under all conditions, including the vacuum. As argued already, the homogeneous Eq. (31) is a Jacobi identity of the O(3) group, and the tilde denotes dual tensor as usual. The homogeneous field equation, Eq. (31), originates in the cyclic identity between O(3) covariant derivatives, Eq. (30), and can be developed by writing out the covariant derivative in terms of the coupling constant g, which has the classical units $\boldsymbol{\kappa}/A^{(0)}$ [11–20]. The coupling constant, as usual in gauge theory [6], couples the dynamical field to its source, so in Eqs. (31) and (32), the dynamical field is never free of its source, and there is no source-free region. This is also true in U(1) electrodynamics on a rigorous level because g also appears in the U(1) covariant derivative as argued already. A field propagating without a source is a violation of causality. On Planck quantization, the coupling constant g has the units $e/\hbar$ in both O(3) and U(1) gauge theory, and for one photon in free space

$$eA^{(0)} = \hbar\boldsymbol{\kappa} \tag{84}$$

signaling that the photon is always coupled to its source. The quantity e has the dual role [6] of coupling constant and charge on the electron. The presence of g in the theory does not mean that the gauge bosons are charged after quantization, anymore than it means that the U(1) gauge bosons are charged after quantization. The role of g is to measure the "strength" with which the dynamical electromagnetic field couples to its source. This aspect of g [6] is a consequence of the gauge principle, and g originates in parallel transport—it is a coefficient needed to ensure that units are balanced [6].

The homogeneous field equation (31) can be expanded in terns of the O(3) covariant derivative [6,11–20]:

$$(\partial_\mu + g\boldsymbol{A}_\mu \times)\tilde{\boldsymbol{G}}^{\mu\nu} \equiv \boldsymbol{0} \tag{85}$$

A particular solution is

$$\partial_\mu \tilde{\boldsymbol{G}}^{\mu\nu} = \boldsymbol{0} \tag{86}$$

the first equation of which gives

$$\boldsymbol{A}_\mu \times \tilde{\boldsymbol{G}}^{\mu\nu} = 0 \tag{87}$$

$$\partial_\mu \tilde{\boldsymbol{G}}^{\mu\nu(i)} = 0; \qquad i = 1, 2, 3 \tag{88}$$

that is, Heaviside–Maxwell-type equations and an equation for $\boldsymbol{B}^{(3)}$, which in vector form is

$$\frac{\partial \boldsymbol{B}^{(3)}}{\partial t} = \boldsymbol{0} \tag{89}$$

The latter equation can be interpreted to mean that the third Stokes parameter does not vary with time in a circularly polarized beam of light. The particular solution (87) gives the B cyclic theorem (9) self-consistently [11–20].

In the vacuum (in the absence of matter), the inhomogeneous O(3) field equation (32) can be interpreted as

$$\partial_\mu \boldsymbol{G}^{\mu\nu} = 0 \tag{90}$$

$$\boldsymbol{J}^\nu = g\varepsilon_0 \boldsymbol{A}_\mu \times \boldsymbol{G}^{\mu\nu} \tag{91}$$

where $\boldsymbol{J}^\nu$ is a conserved vacuum current. Equation (90) gives the component equations:

$$\partial_\mu G^{\mu\nu(i)} = 0; \qquad i = 1, 2, 3 \tag{92}$$

The first two are Maxwell–Heaviside-type equations, and the third, in vector form, is

$$\nabla \times \boldsymbol{B}^{(3)} = \boldsymbol{0} \tag{93}$$

which can be interpreted to mean that the third Stokes parameter is irrotational in the vacuum. It can be shown [17] that the current $\boldsymbol{J}^{\mathrm{v}}$ self-consistently gives the vacuum energy

$$En^{(3)} = \frac{1}{\mu_0}\int \boldsymbol{B}^{(3)}\cdot\boldsymbol{B}^{(3)}dV \tag{94}$$

due to the $\boldsymbol{B}^{(3)}$ field.

In the presence of matter (electrons and protons), the inhomogeneous field equation (32) can be expanded as given in Eqs. (52)–(54) and interprets the inverse Faraday effect self-consistently as argued already. Constitutive relations such as Eq. (55) must be used as in U(1) electrodynamics.

The fundamental field equations (31) and (32) can be expanded out fully in the (1),(2),(3) basis defined by Eqs. (8) to give four field equations: the O(3) equivalents of the Coulomb, Gauss, Ampère–Maxwell, and Faraday laws. This expansion shows clearly that the adoption of an O(3) configuration for the vacuum produces conserved vacuum charges and currents from the first principles of gauge field theory. The vacuum electric charge and vacuum electric current were introduced empirically and developed by Lehnert [7–10]; and the magnetic equivalents were introduced and developed empirically by Harmuth [21,22] and later developed from gauge theory by Barrett [3,4], whose field equations in SU(2) gauge group symmetry are isomorphic with the field equations in O(3) gauge group symmetry given here.

The Gauss law in O(3) electrodynamics is

$$\nabla\cdot\boldsymbol{B}^{(1)*} \equiv ig(\boldsymbol{A}^{(2)}\cdot\boldsymbol{B}^{(3)} - \boldsymbol{B}^{(2)}\cdot\boldsymbol{A}^{(3)}) \tag{95}$$

$$\nabla\cdot\boldsymbol{B}^{(2)*} \equiv ig(\boldsymbol{A}^{(3)}\cdot\boldsymbol{B}^{(1)} - \boldsymbol{B}^{(3)}\cdot\boldsymbol{A}^{(1)}) \tag{96}$$

$$\nabla\cdot\boldsymbol{B}^{(3)*} \equiv ig(\boldsymbol{A}^{(1)}\cdot\boldsymbol{B}^{(2)} - \boldsymbol{B}^{(1)}\cdot\boldsymbol{A}^{(2)}) \tag{97}$$

and allows for the possibility of a topological magnetic monopole originating in the vacuum configuration defined by the O(3) gauge group. Empirical evidence for such a monopole has been reviewed by Mikhailov [4] and interpreted by Barrett [45]. However, the right-hand side of Eqs. (95) to (97) can also be zero for particular solutions [11–20], in which case no magnetic monopole exists. In general, Eqs. (95)–(97) must be solved numerically and simultaneously with the other three equations [Eqs. (98)–(100)] given next. This is not a trivial task, but would give a variety of solutions not present in U(1) electrodynamics, solutions can be compared with empirical data.

The Faraday law on the O(3) level is

$$\nabla \times \boldsymbol{E}^{(1)*} + \frac{\partial \boldsymbol{B}^{(1)*}}{\partial t} \equiv -ig(cA_0^{(3)}\boldsymbol{B}^{(2)} - cA_0^{(2)}\boldsymbol{B}^{(3)} + \boldsymbol{A}^{(2)} \times \boldsymbol{E}^{(3)} - \boldsymbol{A}^{(3)} \times \boldsymbol{E}^{(2)}) \tag{98}$$

$$\nabla \times \boldsymbol{E}^{(2)*} + \frac{\partial \boldsymbol{B}^{(2)*}}{\partial t} \equiv -ig(cA_0^{(1)}\boldsymbol{B}^{(3)} - cA_0^{(3)}\boldsymbol{B}^{(1)} + \boldsymbol{A}^{(3)} \times \boldsymbol{E}^{(1)} - \boldsymbol{A}^{(1)} \times \boldsymbol{E}^{(3)}) \tag{99}$$

$$\nabla \times \boldsymbol{E}^{(3)*} + \frac{\partial \boldsymbol{B}^{(3)*}}{\partial t} \equiv -ig(cA_0^{(2)}\boldsymbol{B}^{(1)} - cA_0^{(1)}\boldsymbol{B}^{(2)} + \boldsymbol{A}^{(1)} \times \boldsymbol{E}^{(2)} - \boldsymbol{A}^{(2)} \times \boldsymbol{E}^{(1)}) \tag{100}$$

and contains on the right-hand sides terms proportional to a conserved topological vacuum magnetic current, which was introduced empirically by Harmuth [21,22] and developed by Barrett [3,4] using SU(2) gauge field theory. This vacuum magnetic current provides energy, in the same way as the current $\boldsymbol{J}^{\text{v}}$ leads to the energy in Eq. (94), and this energy emanates from the vacuum configuration. In principle, therefore, it can be used as a source of mechanical energy provided devices are available to convert the vacuum topological magnetic current into mechanical energy. The same is true of the topological magnetic charge in Eqs. (95)–(97). These charges and currents vanish only in very special cases [11–20], and in general are nonzero. They originate from fundamental topological considerations as argued in Section I.

The O(3) Coulomb law in field–matter interaction is

$$\nabla \cdot \boldsymbol{D}^{(1)*} = \rho^{(1)*} + ig(\boldsymbol{A}^{(2)} \cdot \boldsymbol{D}^{(3)} - \boldsymbol{D}^{(2)} \cdot \boldsymbol{A}^{(3)}) \tag{101}$$

$$\nabla \cdot \boldsymbol{D}^{(2)*} = \rho^{(2)*} + ig(\boldsymbol{A}^{(3)} \cdot \boldsymbol{D}^{(1)} - \boldsymbol{D}^{(3)} \cdot \boldsymbol{A}^{(1)}) \tag{102}$$

$$\nabla \cdot \boldsymbol{D}^{(3)*} = \rho^{(3)*} + ig(\boldsymbol{A}^{(1)} \cdot \boldsymbol{D}^{(2)} - \boldsymbol{D}^{(1)} \cdot \boldsymbol{A}^{(2)}) \tag{103}$$

In the vacuum, the quantities $\rho^{(i)}, i = 1, 2, 3$, disappear, but the topological Noether charges proportional to the remaining right-hand-side terms do not disappear, leaving one of the Lehnert equations [7–10]. Lehnert introduced the vacuum charge empirically. Lehnert and Roy [10] have given clear empirical evidence for the existence of vacuum charge and current. The latter appears in the O(3) Ampère–Maxwell law, which in field–matter interaction is

$$\begin{aligned} &\nabla \times \boldsymbol{H}^{(1)*} - \boldsymbol{J}^{(1)*} - \frac{\partial \boldsymbol{D}^{(1)*}}{\partial t} \\ &\quad = -ig(cA_0^{(2)}\boldsymbol{D}^{(3)} - cA_0^{(3)}\boldsymbol{D}^{(2)} + \boldsymbol{A}^{(2)} \times \boldsymbol{H}^{(3)} - \boldsymbol{A}^{(3)} \times \boldsymbol{H}^{(2)}) \end{aligned} \tag{104}$$

$$\begin{aligned} &\nabla \times \boldsymbol{H}^{(2)*} - \boldsymbol{J}^{(2)*} - \frac{\partial \boldsymbol{D}^{(2)*}}{\partial t} \\ &\quad = -ig(cA_0^{(3)}\boldsymbol{D}^{(1)} - cA_0^{(1)}\boldsymbol{D}^{(3)} + \boldsymbol{A}^{(3)} \times \boldsymbol{H}^{(1)} - \boldsymbol{A}^{(1)} \times \boldsymbol{H}^{(3)}) \end{aligned} \tag{105}$$

$$\nabla \times \boldsymbol{H}^{(3)*} - \boldsymbol{J}^{(3)*} - \frac{\partial \boldsymbol{D}^{(3)*}}{\partial t}$$
$$= -ig(cA_0^{(1)}\boldsymbol{D}^{(2)} - cA_0^{(2)}\boldsymbol{D}^{(1)} + \boldsymbol{A}^{(1)} \times \boldsymbol{H}^{(2)} - \boldsymbol{A}^{(2)} \times \boldsymbol{H}^{(1)}) \quad (106)$$

In the vacuum, the terms $\boldsymbol{J}^{(i)}, i = 1, 2, 3$ disappear, but the topological Noether electric vacuum currents on the right-hand sides of these equations do not. These are the equivalents of the vacuum current introduced empirically by Lehnert [7–10]. These vacuum charges and currents originate in the vacuum configuration and provide energy as argued already. This can loosely be called "vacuum energy." In principle, it can be converted to useful form, and this type of energy does not originate in point electric charge; it originates in the topology of the vacuum itself.

The Lehnert field equations in the vacuum also exist in U(1) form, and were originally postulated [7–10] in U(1) gauge field theory. It can be demonstrated as follows, that they originate from the U(1) gauge field equations when matter is not present:

$$(\partial^{\nu} - igA^{\nu*})F_{\mu\nu} = 0 \quad (107)$$

This equation can also be written as

$$\partial^{\nu}F_{\mu\nu} = igA^{\nu*}F_{\mu\nu} \qquad g = \kappa/A^{(0)} \quad (108)$$

giving the first Lehnert equation in the form

$$\nabla \cdot \boldsymbol{D} = -ig\boldsymbol{A}^* \cdot \boldsymbol{D} \equiv \rho \quad (109)$$

Similarly, Eq. (107) shows that the second Lehnert equation is

$$\nabla \times \boldsymbol{H} - \frac{\partial \boldsymbol{D}}{\partial t} = \boldsymbol{J} = ig(cA^{0*}\boldsymbol{D} + \boldsymbol{A}^* \times \boldsymbol{H}) \quad (110)$$

and vacuum charge and current emanate directly from U(1) gauge field theory as well as from O(3) gauge field theory as just argued. The constant $e/\hbar$ must be regarded as a coupling constant in both cases [6], because it arises from the gauge principle. Similarly, the vacuum magnetic monopole and charge can be obtained from the U(1) gauge equation:

$$(\partial^{\nu} - igA^{\nu*})\tilde{F}_{\mu\nu} \equiv 0 \quad (111)$$

and in vector form are

$$\nabla \cdot \boldsymbol{B} = ig\boldsymbol{A}^* \cdot \boldsymbol{B} \quad (112)$$

$$\frac{\partial \boldsymbol{B}}{\partial t} + \nabla \times \boldsymbol{E} = ig\,(cA^{0*}\boldsymbol{B} + \boldsymbol{A}^* \times \boldsymbol{E}) \quad (113)$$

In both U(1) and O(3), the existence of vacuum charges and currents depends on the existence of the coupling constant g, which is due fundamentally to the notion of covariant derivative, and can be traced, therefore, to the original gauge principle of Weyl, as discussed in Section II. The coupling constant g must be introduced in vacuo if we accept special relativity and the gauge principle. The existence of vacuum charges and currents follows. The arguments in Section III lead us to reject the U(1) gauge theory of electrodynamics in favor of another theory such as O(3) electrodynamics. These vacuum charges and currents are conserved in the sense that they are Noether currents, and therefore do not violate the Noether theorem [6], specifically, conservation of charge/current, energy, and momentum.

It is seen that as the gauge group is changed from U(1) to a higher symmetry, more solutions are allowed for the field equations, and therefore for the vacuum charges and currents. Mikhailov has detected a magnetic monopole in six independent experiments [4], interpreted as a topological magnetic monopole by Barrett [4,5], and a magnetic monopole means the presence of magnetic current. This has also been detected empirically [46]. Both the magnetic charge and the current are topological in origin. In the case of U(1) gauge field theory applied to electrodynamics, the vacuum configuration is described by a U(1) group symmetry, and in O(3) electrodynamics by an O(3) gauge group symmetry.

All gauge theory depends on the rotation of an n-component vector whose 4-derivative does not transform covariantly as shown in Eq. (18). The reason is that $\psi(x)$ and $\psi(x+dx)$ are measured in different coordinate systems; the field ψ has different values at different points, but $\psi(x)$ and $\psi(x)+d\psi$ are measured with respect to different coordinate axes. The quantity $d\psi$ carries information about the nature of the field ψ itself, but also about the rotation of the axes in the internal gauge space on moving from $x\ +dx$. This leads to the concept of parallel transport in the internal gauge space and the resulting vector [6] is denoted $\psi(x)+d\psi$. The notion of parallel transport is at the root of all gauge theory and implies the introduction of g, defined by

$$\delta\psi = igM^a A_\mu^A \, dx^\mu \psi \tag{114}$$

where dx^μ is the distance over which the vector is carried, M^a are group rotation generators, and A_μ^a are generalized vector potentials for the given internal gauge symmetry [e.g., U(1) or O(3)]. The covariant derivative is therefore

$$D_\mu \equiv \partial_\mu - igM^a A_\mu^a \tag{115}$$

and is defined in this way under all conditions, in the presence and absence of matter (electrons and protons). It follows that the electromagnetic field tensor under all conditions for all gauge groups is

$$G_{\mu\nu} \equiv \frac{i}{g}[D_\mu, D_\nu] \tag{116}$$

and if g is zero, the field tensor becomes infinite for any gauge group, including U(1). Here, [,] denotes commutator as usual. The constant g interpreted in this way is neither a property of the source (an electron) nor of the field, but a constant that couples source and field. Note that gauge theory is a necessary condition for the existence of vacuum charges and currents, but not sufficient. The actual existence of these entities must be determined empirically, as in the experiments by Mikhailov [4] and in the work of Lehnert and Roy [10]. The gauge equations on both the U(1) and O(3) levels allow for the fact that vacuum charges and currents may be zero [11–20]. The $\boldsymbol{B}^{(3)}$ field of O(3) electrodynamics, however, is always nonzero in the vacuum, as it is the direct result of a vacuum configuration described by O(3) symmetry. If vacuum charges and currents do exist, however, they provide the possibility of extracting energy from the vacuum as developed in Section XI.

V. FIELD EQUATIONS OF O(3) ELECTRODYNAMICS IN THE CARTESIAN BASIS: REDUCTION TO THE LAWS OF ELECTROSTATICS

In this section, it is shown that the field equations of O(3) electrodynamics written in the Cartesian basis have a substantially different meaning from those written in the complex circular basis of Section IV. The latter basis essentially introduces motion and dynamics, while Eqs. (31) and (32), written in the Cartesian basis, produce the laws of electrostatics self-consistently. This is confirmation of the mathematical and physical correctness of Eqs. (31) and (32).

In the Cartesian basis, the O(3) field tensor is

$$\boldsymbol{G}_{\mu\nu} = G^X_{\mu\nu}\boldsymbol{i} + G^Y_{\mu\nu}\boldsymbol{j} + G^Z_{\mu\nu}\boldsymbol{k} \tag{117}$$

and the O(3) potential is

$$\boldsymbol{A}_\mu = A^X_\mu\boldsymbol{i} + A^Y_\mu\boldsymbol{j} + A^Z_\mu\boldsymbol{k} \tag{118}$$

where the upper indices (X, Y, Z) denote an O(3) internal space defined by the Cartesian unit vectors in Eq. (6). The components of the field tensor are

$$G^{\mu\nu}_X = \partial^\mu A^\nu_X - \partial^\nu A^\mu_X - ig[A^\mu_Y, A^\nu_Z] \tag{119}$$

$$G^{\mu\nu}_Y = \partial^\mu A^\nu_Y - \partial^\nu A^\mu_Y - ig[A^\mu_Z, A^\nu_X] \tag{120}$$

$$G^{\mu\nu}_Z = \partial^\mu A^\nu_Z - \partial^\nu A^\mu_Z - ig[A^\mu_X, A^\nu_Y] \tag{121}$$

where the potentials are real quantities. Therefore the commutators vanish:

$$[A_Y^\mu, A_Z^\nu] = [A_Z^\mu, A_X^\nu] = [A_X^\mu, A_Y^\nu] = 0 \tag{122}$$

The covariant derivative of O(3) electrodynamics in the Cartesian basis is

$$D_\mu = \partial_\mu - igM^a A_\mu^a = \partial_\mu - ig(A_\mu^X + A_\mu^Y + A_\mu^Z) \tag{123}$$

and a rotation in the internal gauge space is denoted by

$$\begin{aligned}\psi' &= e^{i(M^a \Lambda^a (x^\mu))}\psi \\ &= e^{i\Lambda_X(x^\mu)} e^{i\Lambda_Y(x^\mu)} e^{i\Lambda_Z(x^\mu)}\psi\end{aligned} \tag{124}$$

For a rotation about the Z axis

$$\psi' = e^{i\Lambda_Z(x^\mu)}\psi \equiv S\psi \tag{125}$$

producing the gauge transformation:

$$A_Z \rightarrow A_Z + \frac{1}{g}\partial_Z \Lambda \tag{126}$$

This is, self-consistently, the same result as for O(3) electrodynamics in the complex circular basis [11–20] because of the relation $\boldsymbol{k} = \boldsymbol{e}^3$.

The use of Cartesian indices for the internal O(3) gauge space produces the laws of electrostatics as follows. For clarity, the derivation is given in detail. First, the components of the magnetic field disappear:

$$B_X = G_X^{32} = \partial^3 A_X^2 - \partial^2 A_X^3 - ig[A_Y^3, A_Z^2] = 0 \tag{127}$$

$$B_Y = G_Y^{13} = \partial^1 A_Y^3 - \partial^3 A_Y^1 - ig[A_Z^1, A_X^3] = 0 \tag{128}$$

$$B_Z = G_Z^{21} = \partial^2 A_Z^1 - \partial^1 A_Z^2 - ig[A_X^2, A_Y^1] = 0 \tag{129}$$

This means that a magnetic field is always a quantity that depends on motion, or a current. If there is no magnetic field, there is no electric current, that is, no motion of charge. The use of Cartesian indices for the internal O(3) gauge space therefore corresponds to an electrostatic situation where there is no movement of charge. The use of complex circular indices corresponds to electrodynamics.

The nonzero static electric field components are given by equations such as:

$$G_X^{01} = \partial^0 A_X^1 - \partial^1 A_X^0 - ig[A_Y^0, A_Z^1] \tag{130}$$

$$G_X^{10} = \partial^1 A_X^0 - \partial^0 A_X^1 - ig[A_Y^1, A_Z^0] \tag{131}$$

which correspond to

$$-E_X = \frac{1}{c}\frac{\partial}{\partial t}A_X^1 + \frac{\partial}{\partial X}A_X^0 \tag{132}$$

$$E_x = -\frac{1}{c}\frac{\partial}{\partial t}A_X^1 - \frac{\partial}{\partial X}A_x^0 \tag{133}$$

The static electric field is therefore given self-consistently by

$$\boldsymbol{E} = -\nabla A^0 - \frac{1}{c}\frac{\partial}{\partial t}\boldsymbol{A} \tag{134}$$

The vector potential $\boldsymbol{A}$ is zero, however, because the magnetic field is zero, and we arrive at the familiar law of electrostatics:

$$\boldsymbol{E} = -\nabla A^0 \tag{135}$$

Using the vector identity (16), it is found that $\boldsymbol{E}$ is irrotational:

$$\nabla \times \boldsymbol{E} = \boldsymbol{0} \tag{136}$$

In the Cartesian basis, the homogeneous field equation of O(3) electrodynamics can be written out as three component equations:

$$\partial_\mu \tilde{G}_X^{\mu\nu} = ig(A_\mu^Y \tilde{G}_Z^{\mu\nu} - A_\mu^Z \tilde{G}_Y^{\mu\nu}) \tag{137}$$

$$\partial_\mu \tilde{G}_Y^{\mu\nu} = ig(A_\mu^Z \tilde{G}_X^{\mu\nu} - A_\mu^X \tilde{G}_Z^{\mu\nu}) \tag{138}$$

$$\partial_\mu \tilde{G}_Z^{\mu\nu} = ig(A_\mu^X \tilde{G}_Y^{\mu\nu} - A_\mu^Y \tilde{G}_X^{\mu\nu}) \tag{139}$$

For $\nu = 0$

$$\partial_\mu \tilde{G}_X^0 = A_\mu^Y \tilde{G}_Z^{\mu 0} - A_\mu^Z \tilde{G}_Y^{\mu 0} \tag{140}$$

and using

$$\tilde{G}^{10} = B_X \tag{141}$$

this gives the result

$$\partial_X B_X = 0 \tag{142}$$

The complete result for $\nu = 0$ is therefore

$$\nabla \cdot \boldsymbol{B} = 0 \tag{143}$$

which is self-consistent with Eqs. (127)–(129), indicating the absence of a magnetic field because of the absence of moving charges.

For $\nu = 1$, we obtain

$$\begin{aligned}\partial_0 \tilde{G}_X^{01} + \partial_2 \tilde{G}_X^{21} + \partial_3 \tilde{G}_X^{31} \\ = ig(A_0^Y \tilde{G}_Z^{01} + A_2^Y \tilde{G}_Z^{21} + A_3^Y \tilde{G}_Z^{31} - A_0^Z \tilde{G}_Y^{01} - A_2^Z \tilde{G}_Y^{21} - A_3^Z \tilde{G}_Y^{31})\end{aligned} \tag{144}$$

that is, $\partial_0 B_X^X = 0$. Repeating this procedure gives

$$\frac{\partial \boldsymbol{B}}{\partial t} = \boldsymbol{0} \tag{145}$$

which is self-consistent with $\boldsymbol{B} = \boldsymbol{0}$.

The inhomogeneous field equation (32) in the Cartesian basis must be written in the static limit where

$$J^{\nu} = (\rho, \boldsymbol{0}) \tag{146}$$

The component equations look like

$$\partial_\mu H_X^{\mu\nu} + ig(A_\mu^Y H_Z^{\mu\nu} - A_\mu^Z H_Y^{\mu\nu}) = J_X^{\nu} \tag{147}$$

For $\nu = 0$, we obtain

$$\begin{aligned}\partial_1 H_X^{10} + \partial_2 H_X^{20} + \partial_3 H_X^{30} + ig(A_1^Y H_Z^{10} - A_1^Z H_Y^{10} + A_2^Y H_Z^{20} \\ - A_2^Z H_Y^{20} + A_3^Y H_Z^{30} - A_3^Z H_Y^{30}) = J_X^0 = \rho\end{aligned} \tag{148}$$

and this results in the equation

$$\nabla \cdot \boldsymbol{D} = \rho \tag{149}$$

which is the Coulomb law of electrostatics. The Coulomb law is well known to be self-consistent with Eqs. (135) and (136). For $\nu = 1$ and other indices, we obtain the self-consistent result

$$\frac{\partial \boldsymbol{D}}{\partial t} = \boldsymbol{0} \tag{150}$$

which is true for an electrostatic displacement $\boldsymbol{D}$.

In summary, the laws of O(3) electrodynamics in the Cartesian basis reduce to the laws of electrostatics:

$$\begin{aligned}\boldsymbol{E} &= -\nabla A^0 \\ \nabla \times \boldsymbol{E} &= \boldsymbol{0} \\ \nabla \cdot \boldsymbol{D} &= \rho\end{aligned} \tag{151}$$

and this is an indication of the correctness and self-consistency of Eqs. (31) and (32). The need for the complex circular basis now becomes clear: this basis introduces dynamics into the O(3) laws. The Cartesian representation of the gauge space describes a static situation where there is charge but no current (movement of charge). A magnetic field always requires the movement of charge. It has therefore been shown that the laws of electrostatics are laws of a gauge field theory of O(3) internal symmetry. This is another refutation of the received view, that the laws of electrostatics are laws of a gauge field theory of U(1) internal symmetry.

The Gauss and Ampère laws of magnetism are obtained mathematically, and somewhat artificially, from the fact that using a Cartesian basis gives Eq. (143) (the Gauss law); and from the fact that there is no current and no $\boldsymbol{B}$, so we have

$$\boldsymbol{J} = \nabla \times \boldsymbol{B} = \boldsymbol{0} \tag{152}$$

and the Ampère law follows. However, there is a more satisfactory way of obtaining the Gauss and Ampère laws by using the complex circular basis. The latter is needed because magnetism is not a static phenomenon, as evidenced by the both the Ampère and Faraday laws. Magnetism is always a dynamic phenomenon, so we always need complex circular indices. Therefore the Gauss and Ampère laws are obtained from the particular solutions (87) and (91) leading to Eqs. (88) and (92). The phenomenon of radiation is then removed by removing the Maxwell displacement current in Eqs. (88) and (92). This removes the radiated $\boldsymbol{B}^{(3)}$ field and leaves the Gauss, Ampère, Coulomb, and Faraday laws of the received view at the expense of generality. This procedure is a method of obtaining the old laws from O(3) electrodynamics, which is, however, more general and self-consistent. In forcing a reduction of O(3) electrodynamics to the received view, we lose the vacuum charges and currents and a great deal of information.

Information is also lost if we replace the ((1),(2),(3)) basis by the (X, Y, Z) basis for the internal gauge space. The reason is that the former basis is essentially dynamical and the latter is essentially static. This is again a self-consistent result, because electrodynamics, by definition, requires the movement of charge. The misnamed subject of "magnetostatics" also requires the movement of charge, and so is not static.

VI. EXPLANATION OF INTERFEROMETRY AND RELATED PHYSICAL OPTICAL EFFECTS USING O(3) ELECTRODYNAMICS

The explanation of interferometric effects in U(1) electrodynamics is in general self-inconsistent, and sometimes, as in the Sagnac effect, nonexistent. In this section, the theory of interferometry and related physical optical effects is developed with O(3) electrodynamics, which is found to give an accurate and self-consistent explanation, for example, of the Sagnac effect in terms of the fundamental component $\boldsymbol{B}^{(3)}$. The latter is therefore a physical observable in all interferometry.

In order to understand interferometry at a fundamental level in gauge field theory, the starting point must be the non-Abelian Stokes theorem [4]. The theorem is generated by a round trip or closed loop in Minkowski spacetime using covariant derivatives, and in its most general form is given [17] by

$$\exp\left(\oint D_\mu \, dx^\mu\right) = \exp\left(-\frac{1}{2}\int [D_\mu, D_\nu] \, d\sigma^{\mu\nu}\right) \tag{153}$$

where the integral over the closed loop on the left-hand side is related to an integral over the hypersurface $\sigma^{\mu\nu}$ of the commutator of covariant derivatives. The electromagnetic phase factor in O(3) electrodynamics is developed as an exponential from Eq. (153) and is given most generally by

$$\exp\left(g\oint D_\mu \, dx^\mu\right) = \exp\left(-\frac{1}{2}g\int [D_\mu, D_\nu] \, d\sigma^{\mu\nu}\right) \tag{154}$$

The observable phase is the real part of this exponential, specifically, the cosine. Recall that in ordinary U(1) electrodynamics, the phase factor is given by the exponent

$$\phi = \exp\left(i(\omega t - \boldsymbol{\kappa}\cdot\boldsymbol{r} + \alpha)\right) \tag{155}$$

where α is random.

To reduce Eq. (153) to the ordinary Stokes theorem, the U(1) covariant derivative is used

$$D_\mu = \partial_\mu + igA_\mu \tag{156}$$

to give the result

$$\oint A_\mu \, dx^\mu = -\frac{1}{2}\int F_{\mu\nu} \, d\sigma^{\mu\nu} \tag{157}$$

The space part of this expression is the ordinary, or Abelian, Stokes theorem

$$\oint \boldsymbol{A}\cdot d\boldsymbol{r} = \int \boldsymbol{B}\cdot d\boldsymbol{Ar} = \int \nabla \times \boldsymbol{A}\cdot d\boldsymbol{Ar} \tag{158}$$

with the following fundamental property:

$$\oint_{OA} \boldsymbol{A} \cdot d\boldsymbol{r} = -\oint_{AO} \boldsymbol{A} \cdot d\boldsymbol{r} \tag{159}$$

In U(1) electrodynamics in free space, there are only transverse components of the vector potential, so the integral (158) vanishes. It follows that the area integral in Eq. (157) also vanishes, and so the U(1) phase factor cannot be used to describe interferometry. For example, it cannot be used to describe the Sagnac effect. The latter result is consistent with the fact that the Maxwell–Heaviside and d'Alembert equations are invariant under T, which generates the clockwise (C) Sagnac loop from the counterclockwise (A) loop [17]. It follows that the phase difference observed with platform at rest in the Sagnac effect [47] cannot be described by U(1) electrodynamics. This result is also consistent with the fact that the traditional phase of U(1) electrodynamics is invariant under T as discussed already in Section (III). The same result applies for the Michelson–Gale experiment [48], which is a Sagnac effect.

From Eqs. (157) and (158) the integral

$$\text{Int} = -\frac{1}{2}\int F_{\mu\nu}\, d\sigma^{\mu\nu} = 0 \tag{160}$$

vanishes in interferometry as described by U(1) electrodynamics. Therefore, in order to explain interferometry and related optical effects by gauge theory, a non-Abelian Stokes theorem and a non-Abelian phase factor are required. This means that O(3) electrodynamics is capable of describing interferometry but U(1) electrodynamics is not. An area integral is needed that does not vanish, as in Eq. (160), and equated through the theorem (157) to a line integral. It is straightforward to show that the only possible solution for the O(3) phase factor is

$$P \exp\left(ig \oint \boldsymbol{A}^{(3)} \cdot d\boldsymbol{r}\right) = P' \exp\left(ig \int \boldsymbol{B}^{(3)} \cdot d\boldsymbol{Ar}\right) \tag{161}$$

and since $g = \boldsymbol{\kappa}/A^{(0)}$ classically the phase factor reduces to

$$P \exp\left(i \oint \boldsymbol{\kappa}^{(3)} \cdot d\boldsymbol{r}\right) = P' \exp\left(ig \int B^{(3)} \cdot d\boldsymbol{Ar}\right) \tag{162}$$

for all interferometry and physical optics. Equation (162) is nonzero if and only if the Evans–Vigier field $\boldsymbol{B}^{(3)}$ is nonzero, and the latter is therefore responsible for all interferometry and related physical optical effects.

The P on the left-hand side of Eq. (162) denotes path ordering and the P' denotes area ordering [4]. Equation (162) is the result of a round trip or closed loop in Minkowski spacetime with O(3) covariant derivatives. Equation (161) is a direct result of our basic assumption that the configuration of the vacuum can be described by gauge theory with an internal O(3) symmetry (Section I). Henceforth, we shall omit the P and P' from the left- and right-hand sides, respectively, and give a few illustrative examples of the use of Eq. (162) in interferometry and physical optics.

The Sagnac effect with a platform at rest [47] is explained as the phase factor:

$$\exp\left(i\oint_{A-C}\boldsymbol{\kappa}^{(3)}\cdot d\boldsymbol{r}\right) = \exp\left(2i\oint\boldsymbol{\kappa}^{(3)}\cdot d\boldsymbol{r}\right) \tag{163}$$

which is nonzero and gives an observable interferogram, a cosine function:

$$\gamma = \cos\left(2\oint\boldsymbol{\kappa}^{(3)}\cdot d\boldsymbol{r} \pm 2\pi n\right) \tag{164}$$

Using the relation:

$$B^{(0)} = \left|\boldsymbol{B}^{(3)}\right| = gA^{(0)2} \tag{165}$$

the right-hand side of Eq. (162) may be written as

$$\Phi = \exp(i\kappa^2 Ar) \tag{166}$$

and so Eq. (164) becomes

$$\gamma = \cos\left(2\frac{\omega^2}{c^2}Ar \pm 2\pi n\right) \tag{167}$$

This is an expression for the observed phase difference with the platform at rest in the Sagnac experiment [47]; it is a rotation in the internal gauge space. In U(1) electrodynamics, there is no phase difference when the platform is at rest, as discussed already.

When the platform is rotated in the Sagnac effect, there is an additional rotation in the internal gauge space described by

$$\psi' = \exp(iJ_Z\alpha(x^{\mu}))\psi \tag{168}$$

where $\alpha(x^{\mu})$ is an angle in the plane of the Sagnac platform [48]. The effect on the gauge potential $A_{\mu}^{(3)}$ is as follows:

$$A_{\mu}^{(3)} \rightarrow A_{\mu}^{(3)} + \frac{1}{g}\partial_{\mu}\alpha \tag{169}$$

The angular frequency of rotation of the platform is

$$\Omega = \frac{\partial \alpha}{\partial t} \tag{170}$$

and so Eq. (169) implies that the additional rotation of the platform has the effect

$$\omega \rightarrow \omega \pm \Omega \tag{171}$$

on frequency, depending on the sense of rotation of the platform, which therefore produces the phase factor difference

$$\Delta\gamma = \exp\left(i\left(\frac{Ar}{c^2}((\omega+\Omega)^2 - (\omega-\Omega)^2)\right)\right) \tag{172}$$

and an interferogram

$$\mathrm{Re}\,(\Delta\gamma) = \cos\left(4\frac{\omega\Omega Ar}{c^2} \pm 2\pi n\right) \tag{173}$$

as observed [49] to very high accuracy. This formula was first given by Sagnac [50] using kinematic methods. There is no explanation for it in U(1) electrodynamics [4].

The calculation can be repeated using matter waves, because the Sagnac effect exists in electrons [51] as well as in photons. The starting point is the same, namely, the assumption that the vacuum configuration is described by an O(3) gauge group symmetry. The same structured vacuum applies to both electrodynamics and dynamics, wherein the energy momentum tensor is also a vector in the internal gauge space:

$$\begin{aligned} \boldsymbol{p}^{\mu} &= p^{\mu(1)}\boldsymbol{e}^{(1)} + p^{\mu(2)}\boldsymbol{e}^{(2)} + p^{\mu(3)}\boldsymbol{e}^{(3)} \\ &= \hbar(\boldsymbol{\kappa}^{\mu(1)}\boldsymbol{e}^{(1)} + \boldsymbol{\kappa}^{\mu(2)}\boldsymbol{e}^{(2)} + \boldsymbol{\kappa}^{\mu(3)}\boldsymbol{e}^{(3)}) \end{aligned} \tag{174}$$

where

$$\omega^2 = c^2\boldsymbol{\kappa}^2 + \frac{m_0^2 c^4}{\hbar^2} \tag{175}$$

Here, ω is the angular frequency of a matter wave, such as that of an electron, $\boldsymbol{\kappa}$ is its wave number magnitude, and m_0 is the rest mass of the particle corresponding to the matter wave. The rest mass could be the photon's rest mass, estimated to be less than 10^{-68} kg.

Both p_μ and $\boldsymbol{\kappa}_\mu$ are governed by a gauge transformation

$$p_\mu \rightarrow Sp_\mu S^{-1} - i(\partial_\mu S)S^{-1} \tag{176}$$

and similarly for $\boldsymbol{\kappa}_\mu$. The rotation of the Sagnac platform is governed by Eq. (168), from which we obtain

$$\boldsymbol{\kappa}^{0(3)} \rightarrow \boldsymbol{\kappa}^{0(3)} \pm \partial^0 \alpha \tag{177}$$

which is the same as Eq. (171). This is a topological result given by the structure of the vacuum and is valid for all matter waves, including the electromagnetic wave as argued already. The holonomy difference with platform at rest for A and C loops [round trips in Minkowski spacetime with O(3) covariant derivatives] for matter waves is

$$\Delta\gamma = \exp(2i\boldsymbol{\kappa}^2 Ar) \tag{178}$$

where, from Eq. (175)

$$\boldsymbol{\kappa}^2 = \frac{\omega^2}{c^2} - \frac{m_0^2 c^4}{\hbar^2} \tag{179}$$

The extra holonomy difference due to the rotating platform is the same as for electromagnetic waves:

$$\Delta\Delta\gamma = \exp\left(\frac{4i\omega\Omega Ar}{c^2}\right) \tag{180}$$

This result is true for all matter waves and also in the Michelson–Gale experiment, where it has been measured to a precision of one part in 10^{23} [49]. Hasselbach et al. [51] have demonstrated it in electron waves. We have therefore shown that the electrodynamic and kinematic explanation of the Sagnac effect gives the same result in a structured vacuum described by O(3) gauge group symmetry.

The preceding is a result of special relativity precise to one part in 10^{23} [49]. Its explanation in standard special relativity is as follows. Let the tangential velocity of the disk be v_1 and the velocity of the particle be v_2 in the laboratory frame [52]. When the particle and disk are moving in the same direction, the velocity of the particle is $v_2 - v_1 = v_3$ relative to an observer on the periphery of the disk. Vice-versa, the relative velocity is $v_2 + v_1 = v_4$. The special theory of relativity states that time for the two particles will be dilated to different extents, so the time dilation difference relative to the observer on the periphery of the disk is

$$\Delta t = \left(1 - \frac{v_3^2}{c^2}\right)^{1/2} - \left(1 - \frac{v_4^2}{c^2}\right)^{-1/2} \tag{181}$$

using the binomial theorem. When the disk is stationary [53]:

$$t = \frac{2\pi r}{v_2} \tag{182}$$

where r is its radius. So the observable time difference of the Sagnac effect is

$$\Delta\Delta t = \frac{4\pi r v_1}{c^2} = \frac{4\Omega A r}{c^2} \tag{183}$$

as deduced already as a rotation in the O(3) gauge space of a structured vacuum.

The Maxwell–Heaviside theory of electrodynamics has no explanation for the Sagnac effect [4] because its phase is invariant under T, as argued already, and because the equations are invariant to rotation in the vacuum. The d'Alembert wave equation of U(1) electrodynamics is also T-invariant. One of the most telling pieces of evidence against the validity of the U(1) electrodynamics was given experimentally by Pegram [54] who discovered a little known [4] cross-relation between magnetic and electric fields in the vacuum that is denied by Lorentz transformation.

It can be shown straightforwardly, as follows, that there is no holonomy difference if the phase factor (154) is applied to the problem of the Sagnac effect with U(1) covariant derivatives. In other words, the Dirac phase factor [4] of U(1) electrodynamics does not describe the Sagnac effect. For C and A loops, consider the boundary

$$X^2 + Y^2 = 1 \tag{184}$$

of the assumed circular paths of the two light beams of the Sagnac effect. The line integral vanishes around the boundary

$$\begin{aligned} \oint dr &= \int_0^{2\pi} dX + \int_0^{2\pi} dY \\ &= -\int_0^{2\pi} \sin\phi\, d\phi + \int_0^{2\pi} \cos\phi\, d\phi \\ &= 0 = -\int dr \end{aligned} \tag{185}$$

and so

$$\oint \boldsymbol{\kappa} \cdot d\boldsymbol{r} = -\oint \boldsymbol{\kappa} \cdot d\boldsymbol{r} = 0 \tag{186}$$

in U(1) electrodynamics and the relevant holonomy in this symmetry of electrodynamics is the same

$$\exp\left(i\oint_C \boldsymbol{\kappa} \cdot d\boldsymbol{r}\right) = \exp\left(-i\oint_C \boldsymbol{\kappa} \cdot d\boldsymbol{r}\right) = 1 \tag{187}$$

for both beams. There is no interferogram with the platform at rest, contrary to observation.

Furthermore, the only electromagnetic vector present in free space in the Maxwell–Heaviside theory is the plane wave [11–20]:

$$\boldsymbol{A}^{(1)} = \boldsymbol{A}^{(2)*} = \frac{A^{(0)}}{\sqrt{2}}(i\boldsymbol{i} + \boldsymbol{j})e^{i(\omega t - \boldsymbol{\kappa} \cdot \boldsymbol{r})} \tag{188}$$

which is always perpendicular to $\boldsymbol{r}$, so we obtain Eq. (187) self-consistently. Owing to the gauge invariance of the Maxwell–Heaviside theory, there is no extra effect of a moving platform, again contrary to observation. The principle of gauge invariance, and U(1) electrodynamics in general, fail to describe the Sagnac effect.

On the O(3) level, it can be shown that if we write out the commutator of covariant derivatives in Eq. (153) the phase factor becomes [6]

$$\gamma = \exp\left(\int [D_\mu, D_\nu]\, d\sigma^{\mu\nu}\right) \tag{189}$$

$$\gamma = \exp\left(-i\frac{g}{2}\int (\partial_\mu A_\nu - \partial_\nu A_\mu)\, d\sigma^{\mu\nu} - g^2 \int [A_\mu, A_\nu]\, d\sigma^{\mu\nu}\right) \tag{190}$$

ut as just argued, integrals such as

$$I(U(1)) = \int (\partial_\mu A_\nu - \partial_\nu A_\mu)\, d\sigma^{\mu\nu} \tag{191}$$

vanish for both A and C loops, leaving the only source of nonzero holonomy, Eq. (162), leading to the observable interferogram in Eq. (167). This derivation can be self-checked using a closed loop with O(3) covariant derivatives in Minkowski spacetime [6] whereupon the holonomy in one direction is

$$\gamma_A = \exp\left(-i\frac{g}{2}\int G_{\mu\nu}\, dS^{\mu\nu}\right) \tag{192}$$

and in the other direction is

$$\gamma_C = \exp\left(i\frac{g}{2} \int G_{\mu\nu}\, dS^{\mu\nu} \right) \tag{193}$$

where $S^{\mu\nu}$ is the area enclosed by the loop. The holonomy represents a rotation in the internal O(3) gauge space and is a general result for all gauge group symmetries. If the internal basis of the space of O(3) is (a, b, c), the holonomy can be expressed as

$$\gamma = \exp\left(\mp i\frac{g}{2} \int (\partial_\mu A_\nu^a - \partial_\nu A_\mu^a - ig\varepsilon_{abc} A_\mu^b A_\nu^c)\, dS^{\mu\nu} \right) \tag{194}$$

If the internal symmetry is U(1), the holonomy in either direction is

$$\begin{aligned} \gamma(U(1)) &= \exp\left(\mp ig \int (\partial_\mu A_\nu - \partial_\nu A_\mu)\, dS^{\mu\nu} \right) \\ &= \exp\left(\mp ig \oint A_\mu\, dx^\mu \right) = 1 \end{aligned} \tag{195}$$

and the ordinary Stokes theorem can be used to show that there is no holonomy difference.

If the internal group symmetry is O(3) in the basis ((1),(2),(3)), we obtain:

$$\begin{aligned} \exp\left(\mp i\frac{g}{2} \int (\partial_\mu A_\nu^{(1)} - \partial_\nu A_\mu^{(1)})\, dS^{\mu\nu} \right) &= 1 \\ \exp\left(\mp i\frac{g}{2} \int (\partial_\mu A_\nu^{(2)} - \partial_\nu A_\mu^{(2)})\, dS^{\mu\nu} \right) &= 1 \\ \exp\left(\mp i\frac{g}{2} \int (\partial_\mu A_\nu^{(3)} - \partial_\nu A_\mu^{(3)})\, dS^{\mu\nu} \right) &= 1 \end{aligned} \tag{196}$$

and the only source of holonomy difference is the commutator term, which is written in general as [17]

$$\gamma = \exp\left(\mp \frac{g^2}{2} \int \varepsilon_{abc} A_\mu^a A_\nu^c\, dS^{\mu\nu} \right) \tag{197}$$

Considering the special case

$$\gamma = \exp\left(\mp \frac{g^2}{2} \int (A_X^{(1)} A_Y^{(2)} - A_X^{(2)} A_Y^{(1)})\, dS^{XY} \right) \tag{198}$$

and using Eqs. (165) and (188), it is found that the holonomy is

$$\gamma = \exp(\mp i\kappa^2 Ar) \tag{199}$$

The difference in holonomy is Eq. (178), and the interferogram can be written as

$$\gamma = \cos(2\kappa^2 Ar \pm 2\pi n) \tag{200}$$

with the platform at rest.

The Sagnac effect caused by the rotating platform is therefore due to a rotation in the internal gauge space ((1),(2),(3)), which results in the frequency shift in Eq. (171). The frequency shift is experimentally the same to an observer on and off the platform and is independent of the shape of the area Ar. The holonomy difference (172) derived theoretically depends only on the magnitudes ω and Ω, and these scalars are frame-invariant, as observed experimentally. There is no shape specified for the area Ar in the theory, and only its scalar magnitude enters into Eq. (172), again in agreement with experiment.

In the one photon limit, O(3) electrodynamics [11–20] produces the result:

$$eA^{(0)} = \hbar\kappa \tag{201}$$

Substituting this into

$$\gamma = \exp\left(\mp i\frac{e}{\hbar}B^{(3)}Ar\right) \tag{202}$$

for a beam made up of one photon, the flux $B^{(3)}Ar$ becomes $\hbar/e$ and so, in the one photon limit

$$\gamma = \exp(\pm i) \tag{203}$$

The observable phase difference is therefore nonzero for one photon in O(3) electrodynamics. The effect with platform in motion is the same as Eq. (172) for one photon.

Equations leading to Eq. (162) apply in general in O(3) electrodynamics and to interferometry and physical optics in general. They imply the existence of the quantity

$$g_m \equiv \frac{1}{V}\int B^{(3)} dAr \tag{204}$$

in which the units of a topological magnetic monopole are directly dependent on the vacuum configuration. We therefore have the relation

$$\Phi = g g_m V \tag{205}$$

and the observation of phase Φ implies the existence of both $\boldsymbol{B}(3)$ and g_m. The latter must not be confused with the Dirac point magnetic monopole or with the quantities on the right-hand sides of Eqs. (95) to (97).

In the Maxwell–Heaviside theory of electrodynamics, the electromagnetic phase is a product of two 4-vectors together with a random quantity α:

$$\phi = \kappa_\mu x^\mu + \alpha = \omega t - \boldsymbol{\kappa} \cdot \boldsymbol{r} + \alpha \tag{206}$$

Let $\alpha = 0$ without loss of generality, because it is a random number. Then the remaining part of the phase in Eq. (206) is invariant under parity inversion, which is the same as perfect normal reflection as argued in Section III. Therefore the phase arriving back at the beam splitter [55] in one arm of the Michelson interferometer is unchanged for all r, the length of the arm. The same is true for the other arm, and so there is no interferogram, because the phases arriving back from either arm are always the same as the phase in the beam that initially entered the beam splitter. This result is clearly contrary to observation, and U(1) electrodynamics is unable to explain Michelson interferometry, the basis of Fourier transform infrared spectral techniques and instruments.

In O(3) electrodynamics, the interferogram is described by the holonomy

$$\exp\left(i\oint_{1-2} \boldsymbol{\kappa}^{(3)} \cdot d\boldsymbol{r}\right) = \exp\left(2i\int B^{(3)}\, dAr\right) \tag{207}$$

where 1–2 represents a path traversal from beam splitter to mirror and back to beamsplitter. Using the property

$$\oint_1 \boldsymbol{\kappa}^{(3)} \cdot d\boldsymbol{r} = -\oint_2 \boldsymbol{\kappa}^{(3)}\, d\boldsymbol{r} \tag{208}$$

this is nonzero, and the interferogram is the cosine function [17]

$$\mathrm{Re}(\gamma) = \cos(2\boldsymbol{\kappa}^{(3)} \cdot \boldsymbol{r} \pm 2\pi n) \tag{209}$$

which is nonzero and depends on r. By varying r, an interferogram is generated as observed empirically [55]. Its Fourier transform is a spectral function, and in general the beam is polychromatic.

The principle of interferometry in O(3) electrodynamics follows from the fact that it is caused by a rotation in the internal gauge space

$$\exp\left(i\oint_{1-2} \kappa_\mu dx^\mu\right) = \exp(iJ_Z\Lambda(x^\mu))\exp\left(i\oint_1 \kappa_\mu dx^\mu\right) \tag{210}$$

or more succinctly

$$\gamma' = e^{iJ_Z\Lambda(x^\mu)}\gamma \tag{211}$$

In Michelson interferometry, for example, the left-hand-side of Eq. (210) becomes

$$\gamma = \exp(2i\kappa_\mu x^\mu) \tag{212}$$

whose real part is Eq. (209), the interferogram. This result follows from the fact that the rotation (211) in the O(3) internal gauge space results in

$$\begin{aligned} A_\mu^{(3)} &\rightarrow A_\mu^{(3)} + \frac{1}{g}\partial_\mu\Lambda \\ \omega &\rightarrow \omega + \frac{\partial\Lambda}{\partial t} \\ \kappa &\rightarrow \kappa + \frac{1}{c}\frac{\partial\Lambda}{\partial t} \end{aligned} \tag{213}$$

and if $\omega = \partial\Lambda/\partial t$, Eq. (212) follows. We have already applied Eq. (210) to the Sagnac effect.

In U(1) electrodynamics, the equivalent of Eq. (210) is the rotation in the U(1) internal gauge space:

$$e^{i(\omega t - \boldsymbol{\kappa}\cdot\boldsymbol{r} + \Lambda)} = e^{i\Lambda}e^{i(\omega t - \boldsymbol{\kappa}\cdot\boldsymbol{r})} \tag{214}$$

in other words

$$\psi' = e^{i\Lambda}\psi \tag{214a}$$

where Λ is random. The electromagnetic phase in U(1) electrodynamics is defined only up to a random number Λ, whereas the phase in O(3) electrodynamics is fully defined and gives rise to physical effects in interferometry. The details of the effect depend on the geometry of the interferometer.

Another example of a physical effect of this type is the Aharonov–Bohm effect, which is supported by a multiply connected vacuum configuration such as that described by the O(3) gauge group [6]. The Aharonov–Bohm effect is a gauge transform of the true vacuum, where there are no potentials. In our notation, therefore the Aharonov–Bohm effect is due to terms such as $(1/g)\partial_\mu\Lambda$, depending on the geometry chosen for the experiment. It is essential for the Aharonov–Bohm effect to exist such that $(1/g)\partial_\mu\Lambda$ be physical, and not random. It follows therefore that the vacuum configuration defined by the

U(1) group does not support the Aharonov–Bohm effect [26]. The vacuum configuration defined by the SU(2) group cannot support the effect because SU(2) is singly connected [6], leaving O(3) as the only possibility. This is another strong indication of the need for O(3) electrodynamics. Barrett [26] has also reasoned that the U(1) vacuum configuration cannot support the Aharonov–Bohm effect. First, there is a fundamental topological flaw in Heaviside's reduction of the potential to a mathematical convenience because this can apply only in singly connected spaces, whereas U(1) itself is not singly connected, and Maxwell–Heaviside theory is asserted to be a U(1) Yang–Mills gauge field theory. This is another self-inconsistency of the received view. In fact, any polarized classical wave such as a circularly polarized wave has two vectorial components that form the O(3) symmetry basis ((1),(2),(3)) [3]. Another inconsistency of the received view of the Aharonov–Bohm effect is that it depends on the interaction of an assumed physical vector potential A with an electron. However [26], the magnetic field $\boldsymbol{B} = \nabla \times \boldsymbol{A}$ is always zero at the point of interaction, and the effect is described self-inconsistently [6] as an integral over the flux due to $\boldsymbol{B}$. At the point of interaction this flux is always zero. The effect actually depends on the inhomogeneous term generated by the gauge transform of the vacuum [6] into regions where both the magnetic field and the potential are zero. So the effect is an interferometric effect determined by gauge transformed terms such as

$$A'_i = -\frac{i}{g}(\partial_i S_i)S_i^{-1} = \frac{1}{g}\partial_i \Lambda(i); \qquad i = 1, 2, 3 \tag{215}$$

in O(3) electrodynamics, where these terms are physical. The Aharonov–Bohm effect is therefore a rotation in the internal gauge space of a vacuum configuration described by the O(3) group, and not the U(1) group, where terms such as (215) are random.

VII. EXPLANATION OF MAGNETO-OPTICS AND OTHER EFFECTS USING O(3) ELECTRODYNAMICS

The subject of O(3) electrodynamics was initiated through the inference of the $\boldsymbol{B}^{(3)}$ field [11] from the inverse Faraday effect (IFE), which is the magnetization of matter using circularly polarized radiation [11–20]. The phenomenon of radiatively induced fermion resonance (RFR) was first inferred [15] as the resonance equivalent of the IFE. In this section, these two interrelated effects are reviewed and developed using O(3) electrodynamics. The IFE has been observed several times empirically [15], and the term responsible for RFR was first observed empirically as a magnetization by van der Ziel et al. [37] as being proportional to the conjugate product $\boldsymbol{A}^{(1)} \times \boldsymbol{A}^{(2)}$ multiplied by the Pauli matrix

σ in europium-ion-doped glasses. Good agreement was obtained [37] between theory and experiment, implying that the resonance equivalent of this term is present in nature. In other words, resonance can be induced between the states of the Pauli matrix by circularly polarized radiation. This resonance phenomenon is potentially of widespread utility as argued in this section because (1) it has a much higher resolution than ESR or NMR, (2) it has its own spectral fingerprint or chemical shift pattern, and (3) RFR can be observed without the use of superconducting magnets. In O(3) electrodynamics, it is essentially due to the product of the Pauli matrix with the $\boldsymbol{B}^{(3)}$ field and also exists [20] in O(3) quantum electrodynamics.

The IFE was inferred phenomenologically by Pershan [56] in terms of the conjugate product of circularly polarized electric fields, $\boldsymbol{E} \times \boldsymbol{E}^* = \boldsymbol{E}^{(1)} \times \boldsymbol{E}^{(2)}$. In O(3) electrodynamics, it is described from the first principles of gauge field theory by the inhomogeneous field equation (32), which can be expanded as

$$\partial_\mu \boldsymbol{H}^{\mu\nu(1)*} = \boldsymbol{J}^{\nu(1)*} + ig\boldsymbol{A}_\mu^{(2)} \times \boldsymbol{H}^{\mu\nu(3)} \tag{216}$$

$$\partial_\mu \boldsymbol{H}^{\mu\nu(2)*} = \boldsymbol{J}^{\nu(2)*} + ig\boldsymbol{A}_\mu^{(3)} \times \boldsymbol{H}^{\mu\nu(1)} \tag{217}$$

$$\partial_\mu \boldsymbol{H}^{\mu\nu(3)*} = \boldsymbol{J}^{\nu(3)*} + ig\boldsymbol{A}_\mu^{(1)} \times \boldsymbol{H}^{\mu\nu(2)} \tag{218}$$

that is, as three cyclically symmetric equations in the O(3) symmetry basis ((1),(2),(3)) without empiricism. In order to make further progress, a constitutive relation must be used, as follows, but there is no need to assume the existence of $\boldsymbol{E} \times \boldsymbol{E}^*$ empirically. This is proportional to $\boldsymbol{A}^{(1)} \times \boldsymbol{A}^{(2)}$ which is part of the fundamental definition of the O(3) field tensor [11–20]. The constitutive relation used is [20]

$$\boldsymbol{H}^{\mu\nu(3)*} = \varepsilon \boldsymbol{G}^{\mu\nu(3)*} \tag{219}$$

so that

$$\boldsymbol{H}^{(3)*} = -i\frac{g}{\mu}\boldsymbol{A}^{(1)} \times \boldsymbol{A}^{(2)} \tag{220}$$

where ϵ and $\boldsymbol{\mu}$ are the electric permittivity and magnetic permeability of the sample being magnetized by a circularly polarized electromagnetic field whose signature, the third Stokes parameter, is proportional to $\boldsymbol{A}^{(1)} \times \boldsymbol{A}^{(2)}$ and therefore to $\boldsymbol{B}^{(3)}$ (Section III). If the vacuum configuration is assumed to be described by an O(3) group, it follows that the inverse Faraday effect is due to $\boldsymbol{B}^{(3)}$, and is empirical evidence for $\boldsymbol{B}^{(3)}$, leading to the development of O(3) electrodynamics.

The magnetization in the IFE is now defined as

$$\partial_\mu \boldsymbol{H}^{\mu\nu(1)*} = \boldsymbol{J}^{\nu(1)*} + \Delta \boldsymbol{J}^{\nu(1)*} \tag{221}$$

where

$$\Delta \boldsymbol{J}^{\nu(1)*} = ig\varepsilon \boldsymbol{A}_{\mu}^{(2)} \times \boldsymbol{G}^{\mu\nu(3)} \tag{222}$$

It can be worked out precisely [15] in an electron gas for a visible frequency laser such as that used by van der Ziel et al. [37]. The magnetic flux density set up in the electron gas is

$$B_{\text{sample}}^{(3)} = \frac{N}{V}\left(\frac{\mu_0 e^3 c^2 B^{(0)}}{2m^2\omega^3}\right) B_{\text{free space}}^{(3)} \tag{223}$$

where there are N electrons in a volume V, and where m is the mass of the electron. It is inversely proportional to the cube of the angular frequency of the circularly polarized laser. The free-space value of $\boldsymbol{B}^{(3)}$ is

$$B_{\text{free space}}^{(3)} = \left(\frac{\mu_0 I}{c}\right)^{1/2} \tag{224}$$

in terms of the intensity I (W/m^2) of the laser and so

$$B_{\text{sample}}^{(3)} = \frac{N}{V}\left(\frac{\mu_0^2 e^3 c}{2m^2}\right)\frac{I}{\omega^3} e^{(3)} \tag{225}$$

For example, for a pulsed Nd-YaG laser [57] where $I = 5.5 \times 10^{12}$ W/m^2, and $\omega = 1.77 \times 10^{16}$ rad/s, we obtain

$$\begin{aligned} |\boldsymbol{B}_{\text{sample}}^{(3)}| &= 1.06 \times 10^{-35}\frac{N}{V} \\ &\approx 10^{-9}\,\text{T} = 10^{-5}\,\text{G} \end{aligned} \tag{226}$$

which for $N/V = 10^{26}\,\text{m}^{-3}$ (Avogadro's number) is the same order of magnitude as that observed experimentally by van der Ziel et al. [37] in the first inverse Faraday effect experiment. More generally, g/μ is a frequency dependent hyper polarizability [58], giving the possibility of the as yet undeveloped IFE spectroscopy with its characteristic [58] spectral fingerprint. IFE spectroscopy is magnetization near optical resonance caused by the $\boldsymbol{B}^{(3)}$ field in O(3) electrodynamics and is potentially as useful as infrared or Raman spectroscopy.

We can write Eq. (216) as

$$\partial_{\mu}\boldsymbol{H}^{\mu\nu(1)*} = \boldsymbol{J}^{\nu(1)*} + \Delta\boldsymbol{J}^{\nu(1)*} \tag{227}$$

where the transverse current can be developed as

$$\Delta\boldsymbol{J}^{\nu(1)*} = \varepsilon g^2 \boldsymbol{A}_{\mu}^{(2)} \times (\boldsymbol{A}^{\mu(1)} \times \boldsymbol{A}^{\nu(2)}) \tag{228}$$

causes a signal in an induction coil due to the vacuum $\boldsymbol{B}^{(3)}$ field, a component of $\boldsymbol{G}^{\mu\nu(3)}$. This transverse current causes the inverse Faraday effect as observed experimentally in an induction coil [37].

The explanation of the IFE in the Maxwell–Heaviside theory relies on phenomenology that is self-inconsistent. The reason is that $\boldsymbol{A}^{(1)} \times \boldsymbol{A}^{(2)}$ is introduced phenomenologically [56] but the same quantity (Section III) is discarded in U(1) gauge field theory, which is asserted in the received view to be the Maxwell–Heaviside theory. In O(3) electrodynamics, the IFE and third Stokes parameter are both manifestations of the $\boldsymbol{B}^{(3)}$ field proportional to the conjugate product that emerges from first principles [11–20] of gauge field theory, provided the internal gauge space is described in the basis ((1),(2),(3)).

Equation (228) can be developed further using the following result:

$$\boldsymbol{F} \times (\boldsymbol{G} \times \boldsymbol{H}) = \boldsymbol{G}(\boldsymbol{F} \cdot \boldsymbol{H}) - \boldsymbol{H}(\boldsymbol{F} \cdot \boldsymbol{G}) \tag{229}$$

This vector relation shows that

$$\begin{aligned} \boldsymbol{A}_{\mu}^{(2)} \times (\boldsymbol{A}^{\mu(1)} \times \boldsymbol{A}^{\nu(2)}) &= \boldsymbol{A}^{\mu(1)}(\boldsymbol{A}_{\mu}^{(2)} \cdot \boldsymbol{A}^{\nu(2)}) - \boldsymbol{A}^{\nu(2)}(\boldsymbol{A}_{\mu}^{(2)} \cdot \boldsymbol{A}^{\mu(1)}) \\ &= -A^{(0)2}\boldsymbol{A}^{\nu(2)} \end{aligned} \tag{230}$$

Using

$$g = \frac{\kappa}{A^{(0)}} = \frac{\omega}{cA^{(0)}} \tag{231}$$

it is found that

$$\Delta \boldsymbol{J}^{\nu(2)} = -\frac{\varepsilon}{c}\omega^2 \boldsymbol{A}^{\nu(2)} \tag{232}$$

On the one-electron level, the 4-current can be written in terms of the energy momentum:

$$\Delta \boldsymbol{J}^{\nu(2)} = \frac{e}{mcV}\boldsymbol{p}^{\nu(2)} \tag{233}$$

defined through the minimal prescription. From Eqs. (232) and (233), we obtain

$$\varepsilon = \frac{\chi}{c^2 V} = -\frac{e^2}{m\omega^2 V} \tag{234}$$

where χ is the one-electron susceptibility.

This result is self-consistent with the demonstration [15] that the IFE can be described through χ by using the Hamilton–Jacobi equation for one electron in the classical electromagnetic field, but the O(3) derivation is far simpler. The current $\Delta \boldsymbol{J}^{(2)}$ is due to the field-induced transverse electronic linear momentum [20].

Consider now the development of Eq. (218). From Eq. (219)

$$\partial_{\mu} \boldsymbol{H}^{\mu\nu(3)*} = \mathbf{0} \tag{235}$$

and so

$$\boldsymbol{J}^{\nu(3)*} = -ig\boldsymbol{A}_{\mu}^{(1)} \times \boldsymbol{H}^{\mu\nu(2)} \tag{236}$$

Equation (235) follows from the theoretical and experimental finding that [11–20]

$$\frac{\partial \boldsymbol{B}^{(3)}}{\partial t} = \nabla \times \boldsymbol{B}^{(3)} = \mathbf{0} \tag{237}$$

in the vacuum. In Eq. (236), $\boldsymbol{J}^{\nu(3)*}$ is induced self-consistently in the IFE as follows.

Use the constitutive relation

$$\boldsymbol{H}^{\mu\nu(2)} = \varepsilon \boldsymbol{G}^{\mu\nu(2)} \tag{238}$$

and the definition

$$\boldsymbol{G}^{\mu\nu(2)} = c(\partial^{\mu} \boldsymbol{A}^{\nu(2)} - \partial^{\nu} \boldsymbol{A}^{\mu(2)} - ig\boldsymbol{A}^{\mu(3)} \times \boldsymbol{A}^{\nu(1)}) \tag{239}$$

with

$$\boldsymbol{A}_{\mu}^{(1)} \times (\boldsymbol{A}^{\mu(3)} \times \boldsymbol{A}^{\nu(1)}) = 0 \tag{240}$$

Set $\nu = 3$ in Eq. (236) to obtain

$$\boldsymbol{J}^{3(3)*} = 2ig\varepsilon \boldsymbol{A}^{(1)} \times \boldsymbol{B}^{(2)} \tag{241}$$

which is the current induced by the nonlinear cross-product $\boldsymbol{A}^{(1)} \times \boldsymbol{B}^{(2)}$. Using

$$\boldsymbol{B}^{(2)} = \nabla \times \boldsymbol{A}^{(2)} \tag{242}$$

this current is equal to that of the orbital IFE [36]

$$\begin{aligned} \boldsymbol{J}^{3(3)*} &= ig\varepsilon c\kappa \boldsymbol{A}^{(1)} \times \boldsymbol{A}^{(2)} \\ &= -\frac{e^2}{m\omega V} \boldsymbol{B}^{(3)} \end{aligned} \tag{243}$$

and so $\boldsymbol{J}^{3(3)*}$ is the magnetization current due to $\boldsymbol{B}^{(3)}$ for one electron. There is no longitudinal source current in Eq. (218) because the source current of circularly polarized radiation is necessarily transverse, the charge in the source goes around in a circle whose plane is perpendicular to the (3) axis and the source does not move forward along the (3) axis. There is therefore no current in the (3) axis, that is, no source current in the (3) axis as argued.

The technique of RFR is simply the resonance equivalent of the IFE as argued already, but is potentially of major utility. The techniques of nuclear magnetic resonance (NMR), electron spin resonance (ESR), and magnetic resonance imaging (MRI), are widely used in contemporary analytical science and medicine, and all rely on the principle of fermion resonance induced between states of the Pauli spinor. The resonance pattern is distinct for each sample, and in MRI, an image can be built up. Optical methods have been used to enhance the subject considerably [59–65] using laser frequencies. In conventional ESR and NMR, the resonance is induced by a circularly polarized radio frequency (RF) or microwave frequency coil, and the population of the energy states of the Pauli matrices of electron or proton are separated by a very tiny amount by a powerful and homogeneous magnet, usually a superconducting magnet. The resolving power of these techniques is limited by the magnetic flux density of the magnet. This limitation can be removed by replacing the magnet with a circularly polarized electromagnetic field, resulting in RFR. In theory, the latter technique has a much greater resolving capability than does NMR or ESR and can be developed into an MRI technique based on the same principle, the induction of resonance between the states of the Pauli matrix by a circularly polarized RF field. The multi-million-dollar superconducting magnet of a conventional ESR or NMR spectrometer could be replaced in principle by an ordinary RF field.

This result emerges self-consistently at all levels of physics, from the classical nonrelativistic to the quantum electrodynamic. On the nonrelativistic classical level, the technique of RFR is due to the interaction of $\boldsymbol{B}^{(3)}$ with the Pauli matrix. One way of demonstrating this result, which has been observed empirically [37], is to extend the minimal prescription to complex $\boldsymbol{A}$, starting [66] with the Newtonian kinetic energy of the classical electron

$$H_{\mathrm{KE}} = \frac{1}{2m}\boldsymbol{p}\cdot\boldsymbol{p} \tag{244}$$

where $\boldsymbol{p}$ is its linear momentum and m is its mass. The electron interacts with the classical electromagnetic field through the O(3) covariant derivative written in momentum space, in other words, with the minimal prescription with complex $\boldsymbol{A}$, with $\boldsymbol{A}^{(1)} = \boldsymbol{A}^{(2)*}$. The interaction kinetic energy is therefore the real part of:

$$H_{\mathrm{KE}} = \frac{1}{2m}(\boldsymbol{p} - e\boldsymbol{A}^{(1)})\cdot(\boldsymbol{p} - e\boldsymbol{A}^{(2)}) \tag{245}$$

where $\boldsymbol{A}^{(1)}$ and $\boldsymbol{A}^{(2)}$ are complex conjugate transverse plane waves for simplicity of argument. The energy in Eq. (245) can be written out as

$$H_{\mathrm{KE}} = \frac{1}{2m}\boldsymbol{p}\cdot\boldsymbol{p} - \frac{e}{2m}\mathrm{Re}\,(\boldsymbol{A}^{(1)}\cdot\boldsymbol{p}) - \frac{e}{2m}\mathrm{Re}\,(\boldsymbol{p}\cdot\boldsymbol{A}^{(2)}) + \frac{e^2}{2m}\boldsymbol{A}^{(1)}\cdot\boldsymbol{A}^{(2)} \tag{246}$$

a well-known result of numerous textbooks [67]. The only difference is one of notation. In the textbooks, $\boldsymbol{A} = \boldsymbol{A}^{(1)}$ and $\boldsymbol{A}^* = \boldsymbol{A}^{(2)}$. In order to derive the RFR term, we use Pauli matrices as a basis for three-dimensional space following Sakurai [68] in his Eq. (3.18). The interaction between the classical electron and the classical electromagnetic field in this basis is described on the classical level by

$$H_{\mathrm{KE}} = -\frac{1}{2m}\sigma\cdot(\boldsymbol{p} - e\boldsymbol{A}^{(1)})\sigma\cdot(\boldsymbol{p} - e\boldsymbol{A}^{(2)}) \tag{247}$$

and consists of four terms: (1) the magnetic dipole term

$$H_1 = -\frac{e}{2m}\boldsymbol{p}\cdot(\boldsymbol{A}^{(1)} + \boldsymbol{A}^{(2)}) = \frac{e}{2m}\boldsymbol{m}_0\cdot\mathrm{Re}\,\boldsymbol{B} \tag{248}$$

where $\boldsymbol{m}_0$ is the magnetic dipole moment of the electron or proton and Re $\boldsymbol{B}$ is the real part of the magnetic component of the electromagnetic field, (2) the spin–flip term

$$H_2 = -i\frac{e}{2m}\boldsymbol{\sigma}\cdot\boldsymbol{p}\times(\boldsymbol{A}^{(2)} - \boldsymbol{A}^{(1)}) \tag{249}$$

which, for an electron or proton moving in the Z axis, can be expressed as

$$H_2 = -e\frac{A^{(0)}}{\sqrt{2}}p_Z\boldsymbol{\sigma}_Z\cdot(\boldsymbol{j}\cos\phi + \boldsymbol{i}\sin\phi) \tag{250}$$

where

$$\phi = \omega t - \kappa Z = \omega\left(t - \frac{Z}{c}\right) \tag{251}$$

[it can be seen that if $\phi = 0$, the Pauli matrix (or "spin") points in the Y axis; when $\phi = \pi/2$, in the X axis; when $\phi = \pi$, in the $-Y$ axis; when $\phi = 3\pi/2$, in the $-X$ axis; and when $\phi = 2\pi$, back in the Y axis].

Thirdly, the polarizability term which appears in the textbooks [67], is given by

$$H_3 = \frac{e^2}{2m}\boldsymbol{A}^{(1)}\cdot\boldsymbol{A}^{(2)} = \frac{e^2}{2m}A^{(0)2} \tag{252}$$

and is the basis [69] of susceptibility theory, and (4) the RFR term, which is missing from the textbooks, is given by the real-valued expression

$$H_4 = i\frac{e^2}{2m}\boldsymbol{\sigma}\cdot\boldsymbol{A}^{(1)}\times\boldsymbol{A}^{(2)} = -\frac{e^2}{2m}A^{(0)2}\boldsymbol{\sigma}\cdot\boldsymbol{k} \tag{253}$$

All four terms have been observed empirically. Terms 1–3 are well known, and term 4 has been observed as a magnetization in europium ion doped glasses by van der Ziel et al. [37] as argued already. The RFR term therefore emerges self-consistently with three other well-known and well-observed terms from what is effectively the O(3) covariant derivative.

This analysis of the classical non-relativistic level can be confirmed by writing the four Stokes parameters [70] in terms of potentials in free space:

$$\begin{aligned} S_0 &= A_X^{(1)}A_X^{(2)} + A_Y^{(1)}A_Y^{(2)} \\ S_1 &= A_X^{(1)}A_X^{(2)} - A_Y^{(1)}A_Y^{(2)} \\ S_2 &= -(A_X^{(1)}A_Y^{(2)} + A_Y^{(1)}A_X^{(2)}) \\ S_3 &= -i(A_X^{(1)}A_Y^{(2)} - A_Y^{(1)}A_X^{(2)}) \end{aligned} \tag{254}$$

For elliptically polarized electromagnetic radiation

$$S_0^2 = S_1^2 + S_2^2 + S_3^2 = S_3^2 \tag{255}$$

and for circularly polarized radiation

$$S_0 = \pm S_3 \tag{256}$$

Therefore, the existence of $\boldsymbol{A}^{(1)}\cdot\boldsymbol{A}^{(2)}$, which is proportional to S_0 and to field intensity, implies the existence of $\pm\, i\boldsymbol{A}^{(1)}\times\boldsymbol{A}^{(2)}$, which is an observable proportional to S_3. If the light intensity tensor [70] is defined as

$$\rho_{\alpha\beta} = \frac{A_\alpha^{(1)}A_\beta^{(2)}}{A^{(0)2}} \tag{257}$$

then from Eqs. (254) and (256), in circular polarization:

$$\rho_{\alpha\beta} = \frac{1}{2A^{(0)2}}\begin{bmatrix} S_0 & iS_3 \\ -iS_3 & S_0 \end{bmatrix} \tag{258}$$

Now define the Pauli matrices [6,68]

$$\sigma_X \equiv \begin{bmatrix} 0 & 1 \\ 1 & 0 \end{bmatrix}; \sigma_Y \equiv \begin{bmatrix} 1 & 0 \\ 0 & -1 \end{bmatrix}; \sigma_Z \equiv \begin{bmatrix} 0 & i \\ -i & 0 \end{bmatrix} \tag{259}$$

which are interrelated by the following cyclic relation:

$$\left[\frac{\sigma_X}{2}, \frac{\sigma_Y}{2}\right] = i\frac{\sigma_Z}{2} \tag{260}$$

The intensity tensor becomes

$$\rho_{\alpha\beta} = \frac{1}{2A^{(0)2}}(S_0 - i\boldsymbol{\sigma}_Z \cdot \boldsymbol{A}^{(1)} \times \boldsymbol{A}^{(2)}) \tag{261}$$

showing that the RFR term occurs in the fundamental definition of this tensor for circularly polarized radiation. The RFR term is as fundamental as the intensity itself, through Eq. (256).

For practical purposes, the critically important feature of the RFR term is its dependence for a given beam intensity on the inverse of frequency squared of the beam. This means that the spectral resolution [15] in RFR has the same dependence. This critically important feature is shown straightforwardly from the O(3) relations

$$\boldsymbol{B}^{(1)} = \nabla \times \boldsymbol{A}^{(1)}; \qquad \boldsymbol{B}^{(2)} = \nabla \times \boldsymbol{A}^{(2)} \tag{262}$$

so from Eq. (188), the magnetic transverse plane waves are

$$\boldsymbol{B}^{(1)} = \boldsymbol{B}^{(2)*} = \frac{B^{(0)}}{\sqrt{2}}(i\boldsymbol{i} + \boldsymbol{j})e^{i(\omega t - \kappa Z)} \tag{263}$$

and the electric transverse plane waves are

$$\boldsymbol{E}^{(1)} = \boldsymbol{E}^{(2)*} = \frac{E^{(0)}}{\sqrt{2}}(\boldsymbol{i} - i\boldsymbol{j})e^{i(\omega t - \kappa Z)} \tag{264}$$

an analysis that results in the relation between conjugate products [15]

$$\boldsymbol{A}^{(1)} \times \boldsymbol{A}^{(2)} = \frac{c^2}{\omega^2}\boldsymbol{B}^{(1)} \times \boldsymbol{B}^{(2)} = \frac{1}{\omega^2}\boldsymbol{E}^{(1)} \times \boldsymbol{E}^{(2)} \tag{265}$$

Expressing $\boldsymbol{B}^{(1)} \cdot \boldsymbol{B}^{(2)}$ in terms of beam power density (I in W/m^2) results in

$$\boldsymbol{B}^{(1)} \times \boldsymbol{B}^{(2)} = i\frac{\mu_0}{c}I\boldsymbol{e}^{(3)*} \tag{266}$$

where $\boldsymbol{\mu}_0$ is the vacuum permeability in SI units.

The basis of the RFR technique is that a probe photon at a resonance angular frequency ω_{res} can be absorbed under the resonance condition

$$\hbar\omega_{\mathrm{res}} = \frac{e^2c^2B^{(0)2}}{2m\omega^2}(1-(-1)) \tag{267}$$

defined by the transition from the negative to the positive states of the Pauli matrix σ_Z. This is precisely analogous to the basic mechanism of ESR and NMR and is a spectral absorption. The RFR resonance frequency is therefore

$$f_{\mathrm{res}} = \frac{\omega_{\mathrm{res}}}{2\pi} = \left(\frac{e^2\mu_0 c}{2\pi\hbar m}\right)\frac{I}{\omega^2} \tag{268}$$

and is inversely proportional to the square of the angular frequency ω_{res} of the circularly polarized pump electromagnetic field replacing the superconducting magnet of ESR, NMR, and MRI [69].

For ^{1}H proton resonance, the result (268) is adjusted empirically for the different experimentally observed g factors of the electron (2.002) and proton (5.5857). A more complete theory must rest on the internal structure of the proton or other nuclei. The basic theory of RFR is straightforward, however, and a term emerges with three other well-known terms. In principle, RFR can investigate nuclear properties using microwave or RF generators instead of multi-million superconducting magnets.

For proton resonance therefore, the RFR equation [15] is

$$\omega_{\mathrm{res}} = \left(\frac{5.5857e^2\mu_0 c}{2.002\hbar m}\right)\frac{I}{\omega^2} = 1.532\times 10^{25}\,\frac{I}{\omega^2} \tag{269}$$

and some data from this equation are shown in Table I, where it is seen that RFR proton resonances can be far higher than those in conventional NMR. The

TABLE I
RFR Frequencies from Eq. (27) for the Proton for $I = 10$ W/cm^2

Pump Frequency	Resonance Frequency
5000 cm^{-1} (visible)	0.28 Hz
500 cm^{-1} (infrared)	28.0 Hz
1.8 GHz	1.8 GHz (autoresonance)
1.0 GHz (microwave)	6.18 GHz
0.1 GHz (RF)	20.6 cm^{-1} (far infrared)
10.0 MHz (RF)	2,060 cm^{-1} (infrared)
1.0 MHz (RF)	206,000 cm^{-1} (ultraviolet)

concomitant resolution in RFR is also far higher than in NMR, and as will be shown, the RFR technique has its own spectral fingerprint or chemical shift pattern. The spinup–spindown population difference in RFR is also orders of magnitude greater [15] than in NMR, and because of this, the homogeneity of the pump electromagnetic field is not critical theoretically. This is another advantage of the RFR technique. Any remaining objection to the existence of RFR is removed by the empirical fact that the term (253) has been observed experimentally as a magnetization [37]. The only remaining experimental challenge is to induce resonance between the states of σ in term (253).

If RFR is applied to the electron, the same overall advantage is obtained; the equivalent of Eq. (269) is

$$\omega_{\text{res}} = 1.007 \times 10^{28} \frac{I}{\omega^2} \tag{270}$$

These conclusions can be obtained on the nonrelativistic level, and it is possible in theory to practice proton and electron spin resonance without permanent magnets, at much higher resolution, without the need for very high homogeneity, and with a novel chemical shift pattern, or spectral fingerprint, determined by a site-specific molecular property tensor, to be described later in this section.

On the classical relativistic level, the starting point is the Einstein equation

$$p^{\mu} p_{\mu} = m^2 c^2 \tag{271}$$

where p^{μ} and p_{μ} are energy/momentum 4-vectors. In order to demonstrate RFR, Eq. (271) is rewritten in the basis (260) using the gamma matrices [68]

$$\gamma^{\mu} p_{\mu} \gamma^{\mu} p_{\mu} = m^2 c^2 \tag{272}$$

and the classical electromagnetic field is introduced through the O(3) minimal prescription:

$$\gamma^{\mu}(p_{\mu} - eA_{\mu}^{(1)})\gamma^{\mu}(p_{\mu} - eA_{\mu}^{(2)}) = m^2 c^2 \tag{273}$$

In the compact Feynman slash notation [68], Eq. (272) becomes

$$\not{p}\not{p} = m^2 c^2 \tag{274}$$

and Eq. (273) becomes

$$(\not{p} - e\not{A}^{(1)})(\not{p} - e\not{A}^{(2)}) = m^2 c^2 \tag{275}$$

This is the classical relativistic expression for the interaction of an electron or proton with the classical electromagnetic field. The quantized version of Eq. (275) is the van der Waerden equation [1] as described by Sakurai [68] in his Eq. (3.24). The RFR term in relativistic classical physics is contained within the term $e^2 \not{A}^{(1)} \not{A}^{(2)}$, a result that can be demonstrated by expanding this term as follows

$$\begin{aligned} e^2 \not{A}^{(1)} \not{A}^{(2)} &= e^2 \gamma^\mu A_\mu^{(1)} \gamma^\mu A_\mu^{(2)} \\ &= e^2 (\gamma^0 A_0^{(1)} - \boldsymbol{\gamma} \cdot \boldsymbol{A}^{(1)})(\gamma^0 A_0^{(2)} - \boldsymbol{\gamma} \cdot \boldsymbol{A}^{(2)}) \end{aligned} \tag{276}$$

Using the well-known relation between the gamma and Pauli matrices [68]

$$\begin{aligned} (\boldsymbol{\gamma} \cdot \boldsymbol{p})(\boldsymbol{\gamma} \cdot \boldsymbol{p}) &= \begin{bmatrix} \mathbf{0} & \boldsymbol{\sigma} \\ -\boldsymbol{\sigma} & \mathbf{0} \end{bmatrix} \cdot \begin{bmatrix} \boldsymbol{p} & \mathbf{0} \\ \mathbf{0} & \boldsymbol{p} \end{bmatrix} \begin{bmatrix} \mathbf{0} & \boldsymbol{\sigma} \\ -\boldsymbol{\sigma} & \mathbf{0} \end{bmatrix} \cdot \begin{bmatrix} \boldsymbol{p} & \mathbf{0} \\ \mathbf{0} & \boldsymbol{p} \end{bmatrix} \\ &= \begin{bmatrix} (\boldsymbol{\sigma} \cdot \boldsymbol{p})(\boldsymbol{\sigma} \cdot \boldsymbol{p}) & \mathbf{0} \\ \mathbf{0} & (\boldsymbol{\sigma} \cdot \boldsymbol{p})(\boldsymbol{\sigma} \cdot \boldsymbol{p}) \end{bmatrix} \end{aligned} \tag{277}$$

it is found that

$$e^2 \not{A}^{(1)} \not{A}^{(2)} = e^2 (A_0^{(1)} A_0^{(2)} - \boldsymbol{A}^{(1)} \cdot \boldsymbol{A}^{(2)} - i \boldsymbol{\sigma} \cdot \boldsymbol{A}^{(1)} \times \boldsymbol{A}^{(2)}) \tag{278}$$

an expression that includes the RFR term

$$T_{\mathrm{RFR}} = -i e^2 \boldsymbol{\sigma} \cdot \boldsymbol{A}^{(1)} \times \boldsymbol{A}^{(2)} \tag{279}$$

On the nonrelativistic quantum level, both the time-independent and time-dependent Schrödinger equations can be used to demonstrate the existence of RFR. As shown by Sakurai [68], the time-independent Schrödinger–Pauli equation can be used to demonstrate ordinary ESR and NMR in the nonrelativistic quantum limit. This method is adopted here to demonstrate RFR in nonrelativistic quantum mechanics with the time-independent Schrödinger–Pauli equation [68]:

$$\hat{H}\psi = En\,\psi \tag{280}$$

where the Hamiltonian operator is

$$\hat{H} = \frac{1}{2m} (\boldsymbol{\sigma} \cdot \boldsymbol{p})(\boldsymbol{\sigma} \cdot \boldsymbol{p}) + V_0 \tag{281}$$

Here, V_0 is the potential energy, which, however, does not affect the RFR term. This method is first checked for its self-consistency using a real-valued potential function $\boldsymbol{A}$ corresponding to a static magnetic field, then the same equation is used to demonstrate the existence of the RFR term.

In a static magnetic field, the minimal prescription shows that the time-independent Schrödinger–Pauli equation of a fermion in a classical field is

$$\hat{H} \rightarrow \frac{1}{2m}(\boldsymbol{\sigma} \cdot (\boldsymbol{p} + e\boldsymbol{A}))(\boldsymbol{\sigma} \cdot (\boldsymbol{p} + e\boldsymbol{A})) + V \tag{282}$$

The usual ESR or NMR term is obtained from

$$\begin{aligned} \hat{H}\psi &= i\frac{e}{2m}(\boldsymbol{\sigma} \cdot \boldsymbol{p} \times \boldsymbol{A} + \boldsymbol{\sigma} \cdot \boldsymbol{A} \times \boldsymbol{p})\psi + \cdots \\ &= \frac{e\hbar}{2m}\boldsymbol{\sigma} \cdot (\nabla \times (\boldsymbol{A}\psi) + \boldsymbol{A} \times \nabla\psi) + \cdots \\ &= \frac{e\hbar}{2m}\boldsymbol{\sigma} \cdot ((\nabla \times \boldsymbol{A})\psi + (\nabla\psi) \times \boldsymbol{A} + \boldsymbol{A} \times (\nabla\psi)) + \cdots \\ &= \frac{e\hbar}{2m}\boldsymbol{\sigma} \cdot \boldsymbol{B}\psi + \cdots \end{aligned} \tag{283}$$

and is the famous "half-integral spin" first derived by Dirac in relativistic quantum mechanics. However, it also exists in nonrelativistic quantum mechanics as just shown [68], but is a purely quantum term with no classical equivalent because it depends on the operator relation:

$$\boldsymbol{p} \rightarrow -i\hbar\nabla \tag{284}$$

This is the spin Zeeman effect and in perturbation theory [69] gives the nonzero ground-state energy:

$$En = \frac{e\hbar}{2m}\langle 0|\boldsymbol{\sigma} \cdot \boldsymbol{B}|0\rangle \neq 0 \tag{285}$$

It is the basis for all ESR and NMR.

To obtain the RFR term on this level, the same method is used for complex-valued $\boldsymbol{A}$. This gives an extra classical term, or expectation value, which can be written as

$$En = \frac{ie^2}{2m}\boldsymbol{\sigma} \cdot \boldsymbol{A}^{(1)} \times \boldsymbol{A}^{(2)} \tag{286}$$

Perturbation theory gives the ground-state term

$$En = \frac{ie^2}{2m}\langle 0|\boldsymbol{\sigma} \cdot \boldsymbol{A}^{(1)} \times \boldsymbol{A}^{(2)}|0\rangle \tag{287}$$

which is again classical and real-valued. It has the inverse square frequency dependence described already and exists on the nonrelativistic quantum level according to the correspondence principle. Therefore the RFR term is unlike the ESR or NMR terms in that the RFR term is classical while the other two are quantum.

The time-dependent Schrödinger equations

$$H\Psi = i\hbar \frac{\partial \Psi}{\partial t} \tag{288}$$

$$H = H^{(0)} + H^{1}(t) \tag{289}$$

$$\Psi(t) = \Psi_{n} e^{-iEnt} \tag{290}$$

can also be applied to the RFR phenomenon. A two-level system can be considered to consist of the fermion in its spinup and spindown states (states of the Pauli matrix). The unperturbed two-level system has energies E_1 and E_2 and eigenfunctions ψ_1 and ψ_2. These are solutions of [69]:

$$H^{(0)}\psi_n = E_n \psi_n \tag{291}$$

In the presence of a time-dependent perturbation $H^{(1)}(t)$, the state of the system is described by a linear combination of basis functions:

$$\Psi(t) = a_1(t)\Psi_1(t) + a_2(t)\Psi_2(t) \tag{292}$$

and the system evolves under the influence of the perturbation, so a_1 and a_2 are also time-dependent. If it starts as state 1, it may evolve to state 2. The probability at any instant that the system is in state 2 is $a_2(t)a_2^*(t)$, and the probability that it remains in state 1 is.

$$a_1(t)a_1^*(t) = 1 - a_2(t)a_2^*(t) \tag{293}$$

Therefore

$$\begin{aligned} H\Psi &= a_1 H^{(0)}\Psi_1 + a_1 H^{(1)}(t)\Psi_1 + a_2 H^{(0)}\Psi_2 + a_2 H^{(1)}(t)\Psi_2 \\ &= i\hbar \frac{\partial}{\partial t}(a_1\Psi_1 + a_2\Psi_2) \\ &= i\hbar a_1 \frac{\partial \Psi_1}{\partial t} + i\hbar \frac{\partial a_1}{\partial t}\Psi_1 + i\hbar a_2 \frac{\partial \Psi_2}{\partial t} + i\hbar \frac{\partial a_2}{\partial t}\Psi_2 \end{aligned} \tag{294}$$

Each basis function satisfies

$$H^{(0)}\Psi_n = i\hbar \frac{\partial \Psi_n}{\partial t} \tag{295}$$

and therefore

$$a_1 H^{(1)}(t)\Psi_1 + a_2 H^{(1)}(t)\Psi_2 = i\hbar\dot{a}_1\Psi_1 + i\hbar\dot{a}_2\Psi_2 \tag{296}$$

This equation is

$$a_1 H^{(1)}(t)\psi_1 e^{-iE_1 t/\hbar} + a_2 H^{(1)}(t)\psi_2 e^{-iE_2 t/\hbar} = i\hbar\dot{a}_1\psi_1 e^{-iE_1 t/\hbar} + i\hbar\dot{a}_2\psi_2 e^{-iE_2 t/\hbar} \tag{297}$$

and can be multiplied through by ψ_1^* and integrated over all space. Since ψ_1 and ψ_2 are orthonormal

$$a_1 H_{11}^{(1)}(t)e^{-iE_1 t/\hbar} + a_2 H_{12}^{(1)}(t)e^{-iE_2 t/\hbar} = i\hbar\dot{a}_1 e^{-iE_1 t/\hbar} \tag{298}$$

Similarly, multiply through by ψ_2^*:

$$a_1 H_{21}^{(1)}(t)e^{-iE_1 t/\hbar} + a_2 H_{22}^{(1)}(t)e^{-iE_2 t/\hbar} = i\hbar\dot{a}_2 e^{-iE_2 t/\hbar} \tag{299}$$

Here

$$H_{ij}^{(1)}(t) \equiv \int \psi_i^* H^{(1)}(t)\psi_j \, d\tau \tag{300}$$

and ψ_1 and ψ_2 are time-dependent parts of the wavefunction of states 1 and 2 of the unperturbed fermion. Thus

$$H_{11}^{(1)}(t) \equiv \int \psi_1^* H^{(1)}(t)\psi_1 \, d\tau \equiv \langle 1|H^{(1)}(t)|1\rangle \tag{301}$$

and so on.

At this point, the RFR Hamiltonian is inputted:

$$H^{(1)}(t) = i\frac{e^2}{2m}\boldsymbol{\sigma}\cdot\boldsymbol{A}^{(1)} \times \boldsymbol{A}^{(2)} \tag{302}$$

so the existence of $H_{11}^{(1)}(t)$ and $H_{12}^{(1)}(t)$ and so on depends on the properties of σ between fermion states.

Define

$$\boldsymbol{S} \equiv \frac{1}{2}\hbar\boldsymbol{\sigma} \tag{303}$$

and

$$\alpha \equiv \left|\frac{1}{2}, \frac{1}{2}\right\rangle \equiv \text{state 1}$$
$$\beta \equiv \left|\frac{1}{2}, -\frac{1}{2}\right\rangle \equiv \text{state 2} \tag{304}$$

then

$$S_Z\alpha = \frac{1}{2}\hbar\alpha; \qquad S_Z\beta = -\frac{1}{2}\hbar\beta \tag{305}$$

and

$$\langle\alpha|S_Z|\alpha|\rangle = \frac{1}{2}\hbar = \frac{1}{2}\hbar\int\alpha^*\,\alpha\,d\tau$$
$$\langle\alpha|S_Z|\beta\rangle = 0 = -\frac{1}{2}\hbar\int\alpha^*\,\beta\,d\tau \tag{306}$$

Now define

$$\boldsymbol{B}^{(3)*} \equiv -i\frac{e}{\hbar}\boldsymbol{A}^{(1)} \times \boldsymbol{A}^{(2)} \tag{307}$$

and

$$H^{(1)}(t) = -\frac{e}{m}\boldsymbol{S}\cdot\boldsymbol{B}^{(3)} = -\frac{e}{m}S_Z B_Z^{(3)} \tag{308}$$

So Eqs. (252) and (253) become

$$a_1 H_{11}^{(1)}(t) = i\hbar\dot{a}_1 \tag{309}$$
$$a_2 H_{22}^{(1)}(t) = i\hbar\dot{a}_2 \tag{310}$$

because

$$H_{11}^{(1)}(t) = -\frac{e\hbar}{2m}B_Z^{(3)}; \qquad H_{12}^{(1)}(t) = 0$$
$$H_{22}^{(2)}(t) = \frac{e\hbar}{2m}B_Z^{(3)}; \qquad H_{21}^{(1)}(t) = 0 \tag{311}$$

Equations (309) and (310) are decoupled differential equations of the form

$$\dot{a}_1 = i\frac{eB_Z^{(3)}}{2m}a_1; \qquad \dot{a}_2 = -i\frac{eB_Z^{(3)}}{2m}a_2 \tag{312}$$

where

$$B_Z^{(3)} \equiv \frac{e}{\hbar} A^{(0)2} \tag{313}$$

with the constraint:

$$a_1 a_1^* + a_2 a_2^* = 1 \tag{314}$$

A particular solution of Eqs. (312) and (313) is

$$a_1 = \frac{1}{\sqrt{2}} \exp\left(i \frac{etB_Z^{(3)}}{2m} \right); \quad a_2 = \frac{1}{\sqrt{2}} \exp\left(-i \frac{etB_Z^{(3)}}{2m} \right) \tag{315}$$

The perturbed wave function is therefore:

$$\Psi = \frac{\Psi_1}{\sqrt{2}} \exp\left(i \frac{etB_Z^{(3)}}{2m} \right) + \frac{\Psi_2}{\sqrt{2}} \exp\left(-i \frac{etB_Z^{(3)}}{2m} \right) \tag{316}$$

and

$$\begin{aligned} p_1 &= a_1 a_1^* = 0.5 \\ p_2 &= a_2 a_2^* = 0.5 \end{aligned} \tag{317}$$

The probability of finding the system in one state or the other remains constant at 50%, and:

$$\Psi = \frac{\Psi}{\sqrt{2}} \exp(i\omega_{\text{res}} t) + \frac{\Psi}{\sqrt{2}} \exp(-i\omega_{\text{res}} t) \tag{318}$$

where

$$\omega_{\text{res}} = \frac{eB_Z^{(3)}}{2m} \tag{319}$$

is the radiatively induced resonance frequency defined by

$$\hbar \omega_{\text{res}} = H^{(1)}(t) \tag{320}$$

The final result is:

$$\Psi = \frac{\Psi_1}{\sqrt{2}} e^{i\omega_{\text{res}} t} + \frac{\Psi_2}{\sqrt{2}} e^{-i\omega_{\text{res}} t} \tag{321}$$

where

$$H\Psi = i\hbar \frac{\partial \Psi}{\partial t} \tag{322}$$

which is a combination of states with energies $\pm\hbar\omega_{\text{res}}$. The RFR term prepares or dresses the fermion in a combination of α and β spin states analogously with ESR or NMR.

On the relativistic quantum level, the Einstein equation becomes the van der Waerden equation [1,68] with the usual operator rules

$$\begin{aligned} p^{\mu} &\rightarrow i\hbar\partial^{\mu} \\ p_{\mu} &\rightarrow i\hbar\partial_{\mu} \end{aligned} \tag{323}$$

to give

$$(i\gamma^{\mu}\partial_{\mu})(i\gamma^{\mu}\partial_{\mu})\psi_W = \frac{m^2c^2}{\hbar^2}\psi_W \tag{324}$$

where ψ_W is a two component wave function as described by Sakurai [68] in his Eq. (3.24). The classical electromagnetic field is introduced into eq. (324) using O(3) covariant derivatives to give the term $e^2 \not{A}^{(1)} \not{A}^{(2)}$ on the quantum relativistic level. The Dirac equation is obtained from the van der Waerden equation [68] using standard methods, and the two equations are equivalent. The RFR term was indeed first derived [15] using the Dirac equation.

On the level of quantum electrodynamics [17], a classical expression such as

$$H = \frac{e^2}{2m}(\boldsymbol{\sigma} \cdot \boldsymbol{A}^{(1)})(\boldsymbol{\sigma} \cdot \boldsymbol{A}^{(2)}) \tag{325}$$

becomes the interaction Hamiltonian

$$H = \frac{e^2}{4m\hbar\varepsilon_0 V}\sum_k \left(\frac{I}{\omega_k} a_k^+ a_k + \sum_q \frac{\sigma^{(3)}}{\omega_q}(a_q^+ a_{k-q} - a_q a_{k-q}^+)\right) \tag{326}$$

describing the exchange of a photon that results in the change of the spin of the electron. This process is equivalent [17] to the absorption of a photon in the atomic transition $i \rightarrow j$ and the absorption of a photon in the atomic transition $j \rightarrow i$.

The free Hamiltonian term quadratic in $\boldsymbol{B}^{(3)}$ must also be considered and is

$$H_1 = \frac{e}{2\omega_q\varepsilon_0 V}\sum_{k,k',q}(a_{k+q}^+ a_k a_{k'-q}^+ a_{k'}) \tag{327}$$

This term appears only in O(3) quantum electrodynamics and describes the interaction between four photons [17]: the absorption of photons with modes

$k + q$ and $k' - q$ and the emission of photons with modes k and k'. This is a physical process where two photons interact and mutually exchange momenta, and is a process that is observable only in O(3) quantum electrodynamics. The effect has been observed empirically by Tam and Happer [71] in two interacting circularly polarized lasers and was explained using the concept of long range spins by Naik and Pradhan [72]. If the direction of the rotation of the polarization is the same, the two beams attract and vice versa. In O(3) quantum electrodynamics [17], the effect is a form of self-focusing or photon bunching that would result if the spins of the photons were aligned in the same direction, as observed empirically [71]. This result also suggests that O(3) quantum electrodynamics could account for light-squeezing effects and also photon anti-bunching if the photon spins were opposite.

The O(3) quantum electrodynamic equivalent of the RFR effect has been numerically analyzed by Crowell [17] using the Hamiltonian (327). Numerically, it is possible to consider only a finite number of photon modes, and the difference in energy between these modes is set equal to the difference between the two spin states of the fermion. More complex situations were also analyzed [17]. Crowell discovered a variety of effects numerically, including modified Rabi flopping, which has an inverse frequency dependence similar to that observed in the solid state in reciprocal noise [73]. The latter is also explained by Crowell [17] using a non-Abelian model. A variety of other effects of RFR on the quantum electrodynamical level was also reported numerically [17]. The overall result is that the occurrence, classically, of the $\boldsymbol{B}^{(3)}$ field means that there is a quantum electrodynamical Hamiltonian generated by the classical term proportional to $\frac{1}{2}B^{(3)2}$. This induces transitional behavior because it contributes to the dynamics of probability amplitudes [17]. The Hamiltonian is a quartic potential where the value of $\boldsymbol{B}^{(3)}$ determines the value of the potential. The latter has two minima: one where $\boldsymbol{B}^{(3)} = 0$ and the other for a finite value of the $\boldsymbol{B}^{(3)}$ field, corresponding to states that are invariants of the Lagrangian but not of the vacuum.

Another potentially useful feature of RFR is that its site specificity is different from that of NMR or ESR, because RFR relies on a different molecular property tensor [74]. In a precursor to RFR, called optical NMR (ONMR) [59–65], site specificity has been demonstrated at a spatial resolution corresponding to quantum dots, a dramatic demonstration of the enhancement possible with the use of circularly polarized lasers or circularly polarized microwave fields such as in RFR.

The calculation of the chemical shift in RFR is straightforward [74] and relies on a calculation of the second-order perturbation energy (SI units)

$$En = \sum_n \frac{\langle 0|H|n\rangle\langle n|H|0\rangle}{\hbar\omega_{0n}} \tag{328}$$

with the perturbation Hamiltonian

$$H = \frac{1}{2m}(\boldsymbol{p} + e(\boldsymbol{A} + \boldsymbol{A}_N))^2 + V \tag{329}$$

where

$$\boldsymbol{A}_N = \frac{\mu_0}{4\pi r^3}\boldsymbol{m}_N \times \boldsymbol{r} \tag{330}$$

is the vector potential [69] due to the nuclear dipole moment $\boldsymbol{m}_N$. The perturbation term relevant to the RFR chemical shift is the one photon off-resonance population term [74], which is by far the dominant chemical shift term (where c.c. = complex conjugates):

$$En = i\frac{e^3}{m^2\hbar\omega_{0n}}\sum_n \langle 0|\boldsymbol{p}\cdot\boldsymbol{A}|n\rangle\langle n|\boldsymbol{A}_N\cdot\boldsymbol{A}^*|0\rangle + \text{c.c.} \tag{331}$$

The transition electric dipole moment is defined by [74]

$$\langle 0|\mu|n\rangle = \frac{e}{m\omega_{0n}}\langle 0|\boldsymbol{p}|n\rangle \tag{332}$$

and the vector relations:

$$\begin{aligned} &i(\boldsymbol{\mu} \times (\boldsymbol{m}_N \times \boldsymbol{r}))\cdot(\boldsymbol{A}^{(1)} \times \boldsymbol{A}^{(2)}) \\ &\quad = i(\boldsymbol{\mu}\cdot\boldsymbol{A})((\boldsymbol{m}_N \times \boldsymbol{r})\cdot\boldsymbol{A}^{(2)}) - i(\boldsymbol{\mu}\cdot\boldsymbol{A}^{(2)})((\boldsymbol{m}_N \times \boldsymbol{r})\cdot\boldsymbol{A}) \end{aligned} \tag{333}$$

and

$$\boldsymbol{\mu} \times (\boldsymbol{\mu}_N \times \boldsymbol{r}) = (\boldsymbol{\mu}\cdot\boldsymbol{r})\boldsymbol{m}_N - (\boldsymbol{\mu}\cdot\boldsymbol{m}_N)\boldsymbol{r} \tag{334}$$

demonstrate that Eq. (331) may be written as

$$E = \zeta\left(i\frac{e^2}{2m}\boldsymbol{\sigma}\cdot\boldsymbol{A}^{(1)} \times \boldsymbol{A}^{(2)}\right) \tag{335}$$

where

$$\zeta = \frac{g_N e\mu_0}{8\pi m}\sum_n \langle \text{o}|\boldsymbol{\mu}|n\rangle\langle n|\frac{\boldsymbol{r}}{r^3}|0\rangle \tag{336}$$

Here, $\boldsymbol{m}_N = g_N(e/4m)\hbar\boldsymbol{\sigma}$ and Eq. (335) defines the RFR chemical shift factor or shielding constant. This depends on the novel molecular property tensor in

Eq. (336), which is not the tensor that defines the well-known NMR chemical shift through the Lamb shift formula of NMR [69]. The order of magnitude of ζ is about 10^{-6}, roughly the same as in NMR. The complete RFR spectrum from the protons in atoms and molecules is therefore

$$E_{\text{int}} = i\frac{e^2}{2m}(1+\zeta)\boldsymbol{\sigma}\cdot\boldsymbol{A}^{(1)}\times\boldsymbol{A}^{(2)} \tag{337}$$

and is site-specific because of the site specificity of ζ.

The experimental or empirical demonstration of RFR is a logical consequence of the detection of a term proportional to $\boldsymbol{\sigma}\cdot\boldsymbol{A}^{(1)}\times\boldsymbol{A}^{(2)}$ by van der Ziel et al. [37], and some experimental details are suggested here. It would be necessary to work initially on the interaction of a fermion beam with an electromagnetic beam. All levels of one fermion theory given in this section could then be tested under conditions that most closely approximate the theory. A successful demonstration of RFR would require careful engineering in the matter of beam interaction. The IFE has been demonstrated at 3.0 GHz by Deschamps et al. [75], and this experiment provides clues as to how to go about detecting RFR. It seems that the simplest demonstration is autoresonance, where the circularly polarized pump frequency (ω) is adjusted to be the same as the RFR frequency (ω_{res}):

$$\omega_{\text{res}} = \omega \tag{338}$$

Under this condition, the pump beam is absorbed at resonance because the pump frequency matches the resonance frequency exactly. Equation (270) simplifies to

$$\omega_{\text{res}}^3 = 1.007\times10^{28} I \tag{339}$$

Therefore, we can tune ω_{res} for a given I, or vice versa, using interacting fermion and electromagnetic beams. Since autoresonance must appear in the gigahertz range if the pump frequency is in this microwave range, the setup in Ref. 75 can be used as a starting point for the RFR design. Essentially, the magnetization 75 must be converted into a resonance. In Ref. 75, a pulsed microwave signal at 3.0 GHz was detected from a klystron delivering megawatts of power over 12 μs with a repetition rate of 10 Hz. The TE_{11} mode was circularly polarized inside a circular waveguide of 7.5 cm diameter. A plasma was created by the very intense microwave pulse. To detect RFR experimentally, the same standard of engineering would have to be reached with an electromagnetic beam interacting with an electron beam, rather than a plasma, which contains positive ions [15]. To detect resonance, the intensity of the microwave radiation would be much lower, and governed by the autoresonance equation (339). As in the design used by Deschamps et al. [75], the section of the waveguide surrounding the tube would perhaps be made of nylon coated with a micrometer-range layer of copper. The incoming electron beam would have to be guided carefully into the circular waveguide used to circularly polarize the microwave radiation. The engineering design for RFR probably has to be at least as accurate as in the experiment [75] in which magnetization was detected in the IFE at 3.0 GHz in a plasma. Cross-referencing with the detection of the term $\boldsymbol{\sigma} \cdot \boldsymbol{A}^{(1)} \times \boldsymbol{A}^{(2)}$ in Ref. 37, at least part of the signal detected by Deschamps et al. must be due to the RFR term, which is the interaction of $\boldsymbol{B}^{(3)}$ with the Pauli spinor. Contemporary IFE experiments [76] in plasma routinely detect this term and so routinely detect the $\boldsymbol{B}^{(3)}$ field. Equation (339) predicts that the resonance occurs at 3.0 GHz if I is tuned to 0.0665 W/cm^2 for an electron beam. For a circular waveguide of 7.0 cm diameter, this requires only 2.94 W of power.

The preceding estimate is based on one-fermion theory, so the observed resonance frequency in a fermion beam may be different as a result of fermion–fermion interaction. Therefore, it is strongly advisable that I be tunable over a wide range to search for the actual resonance pattern. The same experiment can then be repeated in a proton, atomic or molecular beam and the RFR effect should be I/ω^2-dependent with a pattern of resonance determined by the novel chemical shift factor ζ. Spin–spin interaction between fermions would split the spectrum as in ordinary NMR, but the RFR fingerprint would be unique.

It is to be emphasized finally that the RFR technique is simply the resonance equivalent of a magnetization term proportional to $\boldsymbol{\sigma} \cdot \boldsymbol{A}^{(1)} \times \boldsymbol{A}^{(2)}$ that has now been observed on numerous occasions [76] in the IFE in paramagnetic materials and plasma. The experimental challenge is to convert this magnetization to resonance.

VIII. CORRECTIONS TO QUANTUM ELECTRODYNAMICS IN O(3) ELECTRODYNAMICS

As discussed by Crowell [17], quantized electric and magnetic fields exist in a vacuum that is composed of virtual photons that are the result of the Heisenberg uncertainty fluctuations in the electric and magnetic fields. These fluctuations can be considered as first-order terms, and second-order terms involve fluctuations with electrons and positrons. These virtual pairs [17] are randomly distributed in the vacuum, but an electric field will preferentially align, or polarize, the virtual charge separation. Therefore a photon, with its oscillating electric field, will be associated with these virtual pairs of electrons and positrons that are polarized with the photon electric field. In the formal language of quantum electrodynamics, this is represented by Feynman diagrams [6,17].

The magnetic field is oriented perpendicular to the plane inscribed by a completely polarized electron–positron pair [17]. The virtual electron–positron is accompanied by a virtual electromagnetic field, and as discussed by Crowell [17], the charges of the virtual pair will separate under the influence of the photon electric field. The magnetic field lines of the virtual electron–positron pair will preferentially align with the magnetic field of the photon. Therefore quantum theory is the action of the vacuum on particles and fields, so there are terms such as $\boldsymbol{E}^{(1,2)} + \delta\boldsymbol{E}^{(1,2)}$ and $\boldsymbol{B}^{(1,2)} + \delta\boldsymbol{B}^{(1,2)}$ where the variational terms are quantum fluctuations. Now, following the argument by Crowell [17], consider the differential form $\boldsymbol{F} = \mathrm{d}\boldsymbol{A}$, which can be written in spacetime as

$$\boldsymbol{F} = F_{\mu\nu} dx^{\mu} \wedge dx^{\nu} \tag{340}$$

The Yang–Mills functional [17] is defined by the integration of the wedge product $\boldsymbol{F} \wedge^* \boldsymbol{F}$, where * denotes the Hodge dual-star operator

$$k = \frac{1}{8\pi^2} \int_{(m,g)} F_{\mu\nu} F_{\alpha\beta}\, dx^{\mu} \wedge dx^{\nu} \wedge dx^{\alpha} \wedge dx^{\beta} \tag{341}$$

and where k is the instanton number. The electric and magnetic fields on the manifold of three dimensions are

$$E_i = \varepsilon_{0ji} F^{0j}; \qquad B_i = \varepsilon_{kji} F^{kj} \tag{342}$$

and the Yang–Mills functional is

$$k = \frac{1}{16\pi} \int ([E_i, B_j] + [\delta E_i, \delta B_j])\, d^4x \tag{343}$$

leading to the equal time commutator [17]

$$[\delta E_i^a(\boldsymbol{r}, t), \delta B_j^b(\boldsymbol{r}', t)] = \hbar \delta_{ij} \delta^{ab} \delta(\boldsymbol{r} - \boldsymbol{r}') \delta(t - t') \tag{344}$$

where the O(3) indices are included. Quantum-mechanically, the electric and magnetic fields are conjugate variables, and the uncertainty relationship is dictated by the fluctuations in these fields in the vacuum.

These field fluctuations in the vacuum will interact with the photon's electric and magnetic fields. The fluctuation in the interaction energy due to the magnetic field is given by [17]

$$\delta \boldsymbol{E} = \int \delta(\boldsymbol{j} \cdot \boldsymbol{A}) = \int \boldsymbol{H} \cdot \delta \boldsymbol{B}\, d^3r \tag{345}$$

and can be estimated from the quantized flux $2\pi\hbar/e$. This term is responsible for the Lamb shift in the energy levels of atoms such as the hydrogen atom. The magnetic field fluctuation is defined as the magnetic flux quanta multiplied by the small area enclosed by the electron-positron pair, an area that is determined by the coordinate fluctuations of the electron and positron, and that can be estimated by using the energy fluctuation $\delta E = \delta mc^2$, the uncertainty relation between the energy and the time $\delta E \delta t = \hbar$ and the uncertainty in the position $\delta x = c\delta t$.

The magnetic field fluctuation is approximately 5.6×10^4 T over a range of about 10^{-15} m, and lasts for about 10^{-23} s. Fluctuations on this scale occur at about the classical radius of the electron.

O(3) electrodynamics predicts the existence of the $\boldsymbol{B}^{(3)}$ field, which must also have an effect on the stochastic motion of an electron on a fine scale [17]. There exists in theory [17] the commutator

$$[\delta E_i^{(3)}(\boldsymbol{r}, t), \delta B_j^{(3)}(\boldsymbol{r}', t)] = \hbar \delta_{ij} \delta(\boldsymbol{r} - \boldsymbol{r}') \delta(t - t') \tag{346}$$

and the uncertainty fluctuations:

$$\delta \boldsymbol{B}^{(3)} = \frac{e}{\hbar} (\delta \boldsymbol{A}^{(1)} \times \boldsymbol{A}^{(2)} + \boldsymbol{A}^{(1)} \times \delta \boldsymbol{A}^{(2)}) \tag{347}$$

The magnetic vector potentials will have the magnitude $|\boldsymbol{B}^{(3)}|/k$, so the magnitude of the $\boldsymbol{B}^{(3)}$ fluctuation is expected to be [17]

$$|\delta \boldsymbol{B}^{(3)}| = \frac{2e}{\hbar k^2} (|\delta \boldsymbol{B}||\boldsymbol{B}|) \tag{348}$$

The fluctuation in the ordinary magnetic field in this expression is

$$\delta \boldsymbol{B} = \frac{\pi}{2} \frac{(\delta m)^2}{e\hbar} \tag{349}$$

which is about 5.6×10^4 T. The magnetic field associated with the photon, without quantum fluctuations, is about 3×10^{-14} T, so the fluctuation in $\boldsymbol{B}^{(3)}$ is approximately 6×10^{-7} T. These result from virtual electron–positron pairs and are expected to be 10 orders of magnitude smaller than the standard magnetic field, giving measurable contributions to quantum electrodynamics in the 10-GeV range [17].

Crowell [17] argues that the vacuum contribution to the virtual $\boldsymbol{B}^{(3)}$ field is a very small effect, about a millionth of the Lamb shift.

The nonrelativistic estimate of the contribution of $\boldsymbol{B}^{(3)}$ to the Lamb shift was first carried out by Crowell [17] as follows. The interaction of the radiation field with the electron is given by

$$H = \frac{e}{c} \int d^3 r \boldsymbol{j}(\boldsymbol{r}) \cdot \boldsymbol{A}(r) \tag{350}$$

The Ampère law is next used with a covariant definition of the curl operator

$$\nabla \rightarrow D\times = \nabla\times + i\frac{e}{\hbar}\sum_i A^i\times \tag{351}$$

implying

$$\begin{aligned} \boldsymbol{j}(\boldsymbol{r})\cdot\boldsymbol{A}(\boldsymbol{r}) &= \boldsymbol{D}(\boldsymbol{r})\times\boldsymbol{H}(\boldsymbol{r})\cdot\boldsymbol{A}(\boldsymbol{r}) \\ &= \boldsymbol{H}(\boldsymbol{r})\cdot\boldsymbol{D}\times\boldsymbol{A}(\boldsymbol{r}) + \boldsymbol{D}\cdot\boldsymbol{H}(\boldsymbol{r})\times\boldsymbol{A}(\boldsymbol{r}) \end{aligned} \tag{352}$$

The last term is a boundary operator and is discarded, leaving a $\boldsymbol{B}^{(3)}$ contribution

$$H = -i\frac{e^2}{\hbar c}\int d^3r\boldsymbol{H}\cdot\boldsymbol{A}^{(1)}\times\boldsymbol{A}^{(2)} \tag{353}$$

which leads to the Lamb shift due to $\boldsymbol{B}^{(3)}$. The interaction Hamiltonian (353) will induce the spontaneous emission of a photon with wave number $\omega = ck$ and an atomic state transition $|n\rangle \rightarrow |n'\rangle$, which gives the second-order perturbation shift in energy

$$\Delta E_n = \sum_{n'}\sum_{k,\varepsilon}\left(\frac{|\langle n', k, \varepsilon|H_{\text{int}}|n, 0\rangle|^2}{E_n - E_{n'} - ck}\right) \tag{354}$$

where ϵ is the polarization state of the emitted photon. First, consider the term $\boldsymbol{B} = \nabla\times\boldsymbol{A}$ with $\boldsymbol{A} = \boldsymbol{A}\,\epsilon$. The matrix elements of the interaction Hamiltonian are

$$\langle n', k, \varepsilon|H_{\text{int}}|n, 0\rangle = \frac{e^4}{\mu^2c^2}A^3\langle\boldsymbol{p}\times\boldsymbol{\varepsilon}\cdot\boldsymbol{\varepsilon}\times\boldsymbol{\varepsilon}^*\rangle \tag{355}$$

and if the sum over the photon numbers goes to the continuum, the energy shift is

$$\Delta E_n = -\frac{e^4}{\mu^2c^2}A^3\int\frac{d^3k}{k^3}\sum_{n',\varepsilon}\frac{|\langle n', k|\boldsymbol{p}\cdot\boldsymbol{\varepsilon}^*|n, 0\rangle|^2}{E_n - E_{n'} - ck'} \tag{356}$$

where the factor $(2\pi\hbar)^{-3}$ is absorbed into $\boldsymbol{A}^{(3)}$. Now Crowell [17] sums over the polarization states and puts the integral in spherical coordinate form:

$$\Delta E_n = -\frac{e^4}{\mu^2c^2}A^3\int_0^\infty\frac{dk}{k}\sum_{n'}\frac{|\langle n'|\boldsymbol{p}\cdot\boldsymbol{\varepsilon}^*|n\rangle|^2}{E_n - E_{n'} - ck} \tag{357}$$

The integration of this result leads to

$$\Delta E_n = -\frac{e^4}{\mu^2c^2}A^3\sum_{n'}\frac{|\langle n'|\boldsymbol{p}\cdot\boldsymbol{\varepsilon}^*|n\rangle|^2}{E_n - E_{n'}}\lim_{k\to 0}\ln\left(1 + \frac{E'_n - E_n}{\hbar kc}\right) \tag{358}$$

which is divergent. This divergence is dealt with by recognizing that the probability of emitting a photon depends on the electron current as a function of wave number, so that the dipole approximation becomes

$$|\langle n', k\varepsilon|\boldsymbol{p}|n, 0\rangle| = |\langle n'|\boldsymbol{p}|n, 0\rangle|^2 |j(k)|^2 \tag{359}$$

where $j(k)$ is a current for each wave number k divided by the total current, a ratio that reflects the percentage of photons that are emitted with a given k. For a finite number of photons, this will be a Poisson distribution. If the sample size of photons is very large, but if the number of photons emitted is far less, then $|j(k)| \sim k$, and the following result is obtained

$$\Delta E_n = -\frac{e^4}{\mu^2 c^2} A^3 \sum_{n'} \frac{|\langle n'|\boldsymbol{p} \cdot \boldsymbol{\varepsilon}^*|n\rangle|^2}{E_n - E_{n'}} \ln\left(1 - \frac{\hbar kc}{E_n - E'_n}\right)\Bigg|_0^\infty \tag{360}$$

an integral that is logarithmically divergent in the ultraviolet range [17].

In U(1) quantum electrodynamics, the ultraviolet divergence is removed [17] by countering it with a similar term. For the free electron, there is the infinite term

$$\Delta E_{e^-} = \frac{2e^2}{3\pi m^2 c^2} \sum_q \langle q|\boldsymbol{p}|p\rangle^2 \int_0^\infty dk \tag{361}$$

leading to the mass renormalization of the electron from the energy shift:

$$E_{e^-} = E^0_{e^-} + \Delta E_{e^-} = \frac{1}{p}\langle|\boldsymbol{p}|\rangle + \frac{2e^2}{3\pi m^2 c^2}\langle|\boldsymbol{p}|\rangle^2 \int_0^\infty dk \tag{362}$$

An analogous process in O(3) quantum electrodynamics involves, following Crowell [17], the coupling of the electron with a nonlinear photon coupling corresponding to the energy shift:

$$\begin{aligned} \Delta E^{B^{(3)}}_{e^-} &= \sum_{k,n} |\langle n, k, \varepsilon|p^2|A|^2 A|0, 0\rangle|^2 \\ &= -\frac{8\pi}{3}\frac{\hbar^2 e^4}{m^2 c^2 \mu^4} A^3 \sum_{n'} |\langle n'|p|n\rangle|^2 \int_0^\infty dk \end{aligned} \tag{363}$$

This correction is added to the energy shift due to the $\boldsymbol{B}^{(3)}$ field to give

$$\Delta E^{B^{(3)}}_{e^-} = -\frac{8\pi}{3}\frac{\hbar^2 e^4}{m^2 c^2 \mu^4}\frac{1}{\hbar c} A^3 \sum_{n'} |\langle n'|p|n\rangle|^2 \int_0^\infty dk \frac{(E_{n'} - E_n)}{E_{n'} - E_n - \hbar kc} \tag{364}$$

which is logarithmically divergent, a divergence that is countered by the fact that the amplitudes drop off sharply for processes with frequencies $\hbar\omega > 2mc^2$, where m is the mass of the virtual electron and positron. The integral (364) can be cut off at this value, giving the final result:

$$\Delta E_{e^-}^{B^{(3)}} = -\frac{8\pi}{3}\frac{\hbar^2 e^4}{m^2 c^2 \mu^4}\alpha A^3 \sum_{n'} |\langle n'|p|n\rangle|^2 \ln\left(\frac{2mc^2}{E_{n'} - E_n - \hbar kc}\right) \tag{365}$$

The calculation of the Lamb shift due to $\boldsymbol{B}^{(3)}$ is completed by using the equations

$$H|n\rangle = E_n|n\rangle \tag{366}$$

and

$$\sum_{n'} \frac{|\langle n'|p|n\rangle|^2}{(E'_{n'} - E_n)} = \langle n'|\frac{\boldsymbol{p}}{(H_0 - E_n)} \cdot \boldsymbol{p}|n\rangle \tag{367}$$

The momentum operator acts on $(H_0 - E_n)^{-1}$ as

$$\frac{\boldsymbol{p}}{(H_0 - E_n)} = -\frac{\boldsymbol{p}}{(H_0 - E_n)^2} H_0 \tag{368}$$

and the action of the two momentum operators on the free Hamiltonian is

$$\frac{\boldsymbol{p}}{(H_0 - E_n)} \cdot \boldsymbol{p} = [\boldsymbol{p} \cdot [H_0, \boldsymbol{p}]] \tag{369}$$

In the Lamb shift, the Coulomb potential between proton and electron contributes to the commutator in the hydrogen atom, and the commutator with the free Hamiltonian becomes $(\hbar^2 e^2/2)\nabla^2(1/r)$, which gives a delta function that is evaluated in the matrix element when written out by completeness as an integral over space:

$$\frac{e^2\hbar^2}{2}\langle n|\nabla^2 \frac{1}{r}|n\rangle = \frac{e^2\hbar^2}{2}\int d^3 r \psi^*(\boldsymbol{r}) 2\pi\delta(\boldsymbol{r})\psi(\boldsymbol{r}) \tag{370}$$

For an atom in the s state, we have $|\psi|^2 = [1/\pi(na_0)^3]$, where n is the principle atomic number and a_0 is the Bohr radius. The Lamb shift due to $\boldsymbol{B}^{(3)}$ is therefore

$$\text{Lamb}(\boldsymbol{B}^{(3)}) = \frac{1}{3\pi^2}\left(\frac{e^2}{a_0}\right)\alpha^5 \ln\left(\frac{2mc^2}{E_m - E_n}\right) \tag{371}$$

which is 5.33×10^{-5} of the standard Lamb shift. This answer is about five times the quantum fluctuation estimate made already.

On the relativistic level in O(3) quantum electrodynamics [O(3) QED], the Lagrangian density is

$$\mathscr{L} = -\frac{1}{4} F^a_{\mu\nu} F^{a\mu\nu} \tag{372}$$

with the gauge covariant field:

$$F^a_{\mu\nu} = \partial_\nu A^a_\mu - \partial_\mu A^a_\nu + ig\varepsilon^{abc}[A^b_\nu, A^c_\mu] \tag{373}$$

Variational calculus with this Lagrangian density leads [17] to the field equation:

$$\partial_\mu F^{a\mu\nu} + ig\varepsilon^{abc} A^b_\mu F^{c\mu\nu} = 0 \tag{374}$$

with electric and magnetic components:

$$E^a_i = F^a_{i0} = -\dot{A}^a_i - \nabla_i A^a_0 + ig\varepsilon^{abc} A^b_0 A^c_i \tag{375}$$

In O(3) QED, the components of the vector potentials are expanded [17] in a Fourier series of

$$\varepsilon^k_{ij} B^a_k = \nabla_i A^a_j - \nabla_j A^a_i + ig\varepsilon^{abc} A^b_i A^c_j \tag{376}$$

modes, with creation and annihilation operators that act on the Fock space of states, with box normalization within a quantization volume V that has periodic boundary conditions, thus giving:

$$A^a_i(\boldsymbol{r}, t) = \sum_k \frac{1}{(2\omega V)^{1/2}} (e_i a^a(k) e^{i\boldsymbol{k}\bullet\boldsymbol{r}} + e_i a^{a^+}(k) e^{-i\boldsymbol{k}\bullet\boldsymbol{r}}) \tag{377}$$

The electric and magnetic components within O(3) QED are then

$$E^a_i = \sum_k \frac{1}{(2\omega V)^{1/2}} \left(\frac{|k|}{c} e_i a^a(k) e^{i\boldsymbol{k}\bullet\boldsymbol{r}} + \frac{|k|}{c} e_i a^{a^+}(k) e^{-i\boldsymbol{k}\bullet\boldsymbol{r}} \right) \tag{378}$$

$$\begin{aligned} \varepsilon^k_{ij} B^a_k = {} & \sum_k \frac{1}{(2\omega V)^{1/2}} (k_{[je_i]} a^a(k) e^{i\boldsymbol{k}\bullet\boldsymbol{r}} + k_{[je_i]} e_i a^{a^+}(k) e^{-i\boldsymbol{k}\bullet\boldsymbol{r}}) \\ & + ig\varepsilon^{abc} \sum_{kk'} e_{[je_i]} (a^b(k) e^{i\boldsymbol{k}\bullet\boldsymbol{r}} + a^{b^+}(k) e^{-i\boldsymbol{k}\bullet\boldsymbol{r}})(a^c(k') e^{i\boldsymbol{k}'\bullet\boldsymbol{r}} + a^{c^+}(k') e^{-i\boldsymbol{k}\bullet\boldsymbol{r}}) \end{aligned} \tag{379}$$

and the Hamiltonian for this non-Abelian field theory [17] contains novel quartic terms.

If $A_i^{(3)}$ is phase-free, as discussed in Section III, and in Ref. 15, there are no longitudinal electric field components. This also occurs if $A_i^{(3)}$ is zero [17]. The $\boldsymbol{B}^{(3)}$ field is then a Fourier sum over modes with operators $a_{k-q}^{+}a_q$ and is perpendicular to the plane defined by $\boldsymbol{A}^{(1)}$ and $\boldsymbol{A}^{(2)}$. The four-dimensional dual to this term is defined on a time-like surface, following Crowell [17], which can be interpreted as $\boldsymbol{E}^{(3)}$ under dyad vector duality in three dimensions. The $\boldsymbol{E}^{(3)}$ field vanishes because of the nonexistence of the raising and lowering operators $a^{(3)}, a^{(3)+}$. The $\boldsymbol{B}^{(3)}$ is nonzero because of the occurrence of raising and lowering operators in the expansion of $\boldsymbol{A}^{(1)}$ and $\boldsymbol{A}^{(2)}$. These facts imply that $\boldsymbol{B}^{(3)}$ is phaseless and longitudinal, but they do not necessarily represent a breakdown of duality because [15] $c\boldsymbol{B}^{(3)}$ can be dual to an imaginary valued $i\boldsymbol{E}^{(3)}$.

The effect of a local gauge transformation (Sction II) on the classical $\boldsymbol{B}^{(3)}$ field is described as

$$\boldsymbol{B}^{(3)'} = ig\boldsymbol{A}^{(1)} \times \boldsymbol{A}^{(2)} g^{-1} \tag{380}$$

where the group element g is an algebraic generator $g = e^{iX}$. So in $\hbar = c = 1$ units the effect on $\boldsymbol{B}^{(3)}$ is generated as

$$dA' = g(dA + A \wedge A)g^{-1} \tag{381}$$

where g is the group element for the O(3) theory. In the case of quantum field theory, a gauge transformation

$$A_\mu^a \rightarrow A_\mu^a + \delta A_\mu^a \tag{382}$$

is associated [17] with a unitary transform of the fermion field:

$$\psi \rightarrow \psi + \delta\psi \tag{383}$$

In quantum field theory, the gauge field is determined by its Lagrangian density, and the fermion field, by the Dirac Lagrangian density:

$$\mathscr{L}_D = -\bar{\psi}(\gamma^\mu \partial_\mu + m)\psi \tag{384}$$

In order to describe the interaction between the gauge and fermion fields, the following equation is used:

$$\partial_\mu F^{a\mu\nu} + ig\varepsilon^{abc} A_\mu^b F^{c\mu\nu} = j^\nu \tag{385}$$

Here

$$j^{\nu} = \frac{\partial \mathscr{L}}{\partial A_{\nu}} \tag{386}$$

and the addition of an interaction Lagrangian density $\mathscr{L}_i = j^{\nu} A_{\nu}$ is implied. The current term is determined by the Dirac field and is

$$j^{\nu} = \bar{\psi}\gamma^{\nu}\psi \tag{387}$$

Mass renormalization requires [15] that an additional term $\bar{\psi}\gamma^{\nu}\psi\delta m$ be added where δm is the difference between the physical and bare masses [77].

The total Lagrangian is then

$$\mathscr{L} = \mathscr{L}_G + \mathscr{L}_D + \mathscr{L}_i \tag{388}$$

and describes the interaction between the fermions and the gauge field. The Dirac field is the electron field and the gauge field is the non-Abelian electromagnetic field. The theory describes the interaction between quantized electrons and quantized photons on the O(3) level. Because it is a gauge theory, it conveys momentum from one electron to another by the virtual creation and destruction of a vector boson (the photon). There is no creation of any averaged momentum from the virtual quantum fluctuation [17].

In order to upgrade these well-known methods [6,17] of U(1) quantum field theory to involve the classical $\boldsymbol{B}^{(3)}$ field, the following prescription is used:

$$A_{\mu} \rightarrow t^a A_{\mu}^a \tag{389}$$

Here t^a is a group structure constant defined by

$$[t^a, t^b] = 2\varepsilon^{abc} t^c \tag{390}$$

The amplitude contribution from the $\boldsymbol{B}^{(3)}$ field occurs in a second-order process using the sum over all possible fluctuations of $\boldsymbol{B}^{(3)}$ in the virtual photon that causes electron–electron interaction. The amplitude due to $\boldsymbol{B}^{(3)}$ has an ultraviolet divergence [17] described by Crowell. This may be removed by regularization techniques.

This type of process is missing from U(1) quantum field theory [6]; the $\boldsymbol{B}^{(3)}$ field produces quantum vortices [17] that interact with electrons and other charged particles. The vortices are quantized states and exist as fluctuations in the QED vacuum, fluctuations that are associated, not with an $\boldsymbol{E}^{(3)}$ field, but with the $\boldsymbol{E}^{(1)} = \boldsymbol{E}^{(2)*}$ fields:

$$\delta \boldsymbol{B}^{(3)} = i\frac{e}{\hbar}\frac{1}{\omega^2}(\delta \boldsymbol{E}^{(1)} \times \boldsymbol{E}^{(2)} + \boldsymbol{E}^{(1)} \times \delta \boldsymbol{E}^{(2)}) \tag{391}$$

Therefore quantum fluctuations in $\boldsymbol{B}^{(3)}$ are accompanied by fluctuations in the transverse electric field. The ultraviolet divergence is probably unimportant [17] because of the ω^{-2} dependence of the fluctuation. The infrared divergence is also damped statistically. The divergences in U(1) electrodynamics [6] can exist as a subset of O(3) electrodynamics and can be absorbed into integrals that involve photon loop processes associated with quantum fluctuations in $\boldsymbol{B}^{(3)}$.

Crowell [17] has argued that O(3) QED is fully renormalizable. Renormalization is necessary as in any quantum field theory because the potential and propagator become divergent as the electrons approach each other. The Heisenberg uncertainty principle $\Delta p \Delta x \geq \hbar$ means that the momentum exchanged by the electrons becomes divergent [17]. The vacuum is filled with virtual quanta, as argued by Crowell [17], with enormously high momentum fluctuations: virtual quanta that may interact with systems to contribute divergences in the short wavelength limit, the ultraviolet divergences. These divergences affect the self-energy of the electron, vacuum polarization, and vertex functions [6,15,17].

In O(3) QED, there is an additional effect from the effective photon bunching or photon interaction that emerges essentially from the photon loop generated from the $\boldsymbol{A}^{(1)}$ on one photon interacting with the $\boldsymbol{A}^{(2)}$ on the other photon. The loop is associated with quanta of the $\boldsymbol{B}^{(3)}$ field with intensity $e/\hbar$ as in Eq. (347). It will be argued, following Crowell [17], that these novel fluctuations are fully renormalizable. The virtual fluctuation of a $\boldsymbol{B}^{(3)}$ field does not lead to an ultraviolet divergence, and so O(3) QED is renormalizable by dimensional regularization.

The renormalization problem generated by O(3) is similar to the interaction of the free electron with the vacuum through the Dirac equation [6,15,17] in $c = 1, h = 1$ units:

$$(\gamma^{\mu}(\partial_{\mu} - ieA_{\mu}) - m)\psi = 0 \tag{392}$$

If there is no electromagnetic field present, the quantized vector potential fluctuates according to

$$A_{\mu} = \langle A_{\mu} \rangle + \delta A_{\mu} \tag{393}$$

and the fluctuation is present in the vacuum. This phenomenon manifests itself through the zero point energy of the harmonic oscillator expansion of the fields [17]; the electron will interact with the virtual photons, an interaction which is expanded in terms of the order $\alpha = (e^2/\hbar c)$.The divergence [17] in the first term of this series is countered by a mass term, introducing a difference between the mass and the bare mass of the electron. Similar methods can be used straightforwardly [17] to show that the loop fluctuations of photon, correlated to the virtual quanta of the $\boldsymbol{B}^{(3)}$ field, can be calculated to be finite without divergence. The end result of this standard but complicated calculation [17] is that O(3) QED is free of intractable ultraviolet divergences. The Lamb shift calculation given already shows that O(3) QED is free of intractable infrared divergences.

In Section I, it was argued that O(3) electrodynamics on the classical level emerges from a vacuum configuration that can be described with an O(3) symmetry gauge group. On the QED level, this concept is developed by considering higher-order terms in the Hamiltonian

$$H = \frac{1}{2m}(\boldsymbol{p} - e\boldsymbol{A})^2 \tag{394}$$

and evolution operator $U = e^{-iH_0 t}$ [17], where:

$$U = e^{-iH_0 t} e^{A^{(1)} \bullet A^{(2)}} \tag{395}$$

Here, H_0 is the Hamiltonian without the quadratic term. The vector potentials are expanded as

$$A^{(1)} = \frac{A^{(0)}}{\sqrt{2}}(e_X + ie_Y)(a_k e^{i\boldsymbol{k}\bullet\boldsymbol{r} - i\omega t} - a_k^+ e^{-i\boldsymbol{k}\bullet\boldsymbol{r} + i\omega t}) \tag{396}$$

giving

$$\boldsymbol{A}^{(1)} \bullet \boldsymbol{A}^{(2)} = A^{(0)2}\left(a^+ a + \frac{1}{2} - \frac{1}{2}(a^{+2} e^{-2i(\boldsymbol{k}\bullet\boldsymbol{r} - \omega t)} + a^2 e^{2i(\boldsymbol{k}\bullet\boldsymbol{r} - \omega t)})\right) \tag{397}$$

The first two terms on the right-hand side [17] are precisely those obtained from the standard harmonic oscillator Hamiltonian (H_{em}) for the electromagnetic field. The evolution operator can then be written as

$$U = e^{-i(H_0 + H_{\text{em}})t} e^{(Za^{+2} + Z^* a^2)} \tag{398}$$

where $Z = te^{-2i(\boldsymbol{\kappa}\bullet\boldsymbol{r} - \omega t)}$.

The operator

$$S(Z) = \exp(Za^{+2} + Z^* a^2) \tag{399}$$

is a squeezed-state operator [17] that involves symmetries that are not precisely defined by the Hamiltonian. The quantized $\boldsymbol{B}^{(3)}$ field may correspond to such symmetries of the vacuum, coming full circle with Section I. The reason is that the $\boldsymbol{B}^{(3)}$ field is generated by writing Eq. (394) in the basis of the Pauli matrices, as discussed in Section VII.

The absence of an $\boldsymbol{E}^{(3)}$ field does not affect Lorentz symmetry, because in free space, the field equations of both O(3) electrodynamics are Lorentz-invariant, so their solutions are also Lorentz-invariant. This conclusion follows from the Jacobi identity (30), which is an identity for all group symmetries. The right-hand side is zero, and so the left-hand side is zero and invariant under the general Lorentz transformation [6], consisting of boosts, rotations, and space-time translations. It follows that the $\boldsymbol{B}^{(3)}$ field in free space Lorentz-invariant, and also that the definition (38) is invariant. The $\boldsymbol{E}^{(3)}$ field is zero and is also invariant; thus, $\boldsymbol{B}^{(3)}$ is the same for all observers and $\boldsymbol{E}^{(3)}$ is zero for all observers.

To prove the invariance of the B cyclic theorem [11–20], it is necessary only to prove the invariance of the free-space Maxwell–Heaviside equations:

$$\partial_\mu \tilde{G}^{\mu\nu(i)} = 0; \qquad i = 1, 2, 3 \tag{400}$$

Consider, for example, a Lorentz boost in the Z direction using Jackson's notation [5], and start with the 4-derivative

$$\partial'_\mu = \begin{bmatrix} 1 & 0 & 0 & 0 \\ 0 & 1 & 0 & 0 \\ 0 & 0 & \gamma & i\gamma\beta \\ 0 & 0 & -i\gamma\beta & \gamma \end{bmatrix} \begin{bmatrix} \frac{\partial}{\partial X} \\ \frac{\partial}{\partial Y} \\ \frac{\partial}{\partial Z} \\ \frac{i}{c}\frac{\partial}{\partial t} \end{bmatrix} = \begin{bmatrix} \frac{\partial}{\partial X} \\ \frac{\partial}{\partial Y} \\ \gamma\frac{\partial}{\partial Z} - \frac{\gamma\beta}{c}\frac{\partial}{\partial t} \\ -i\left(\gamma\beta\frac{\partial}{\partial Z} - \frac{\gamma}{c}\frac{\partial}{\partial t}\right) \end{bmatrix} \tag{401}$$

where

$$\gamma = \left(1 - \frac{v^2}{c^2}\right)^{-1/2}; \qquad \beta = \frac{v}{c} \tag{402}$$

Using the same Z boost

$$\begin{aligned} E'_X &= \gamma(E_X - \beta B_Y) \qquad & B'_X &= B_X + \beta E_Y \\ E'_Y &= \gamma(E_Y + \beta B_X) \qquad & B'_Y &= B_Y - \beta E_X \\ E'_Z &= E_Z \qquad & B'_Z &= B_Z \end{aligned} \tag{403}$$

so

$$\begin{aligned} (\nabla \cdot \boldsymbol{E})' &= \nabla' \cdot \boldsymbol{E}' = \gamma \nabla \cdot \boldsymbol{E} = 0 = \nabla \cdot \boldsymbol{E} \\ (\nabla \cdot \boldsymbol{B})' &= \nabla' \cdot \boldsymbol{B}' = \gamma \nabla \cdot \boldsymbol{B} = 0 = \nabla \cdot \boldsymbol{B} \end{aligned} \tag{404}$$

Considering the $\boldsymbol{i}$ component of the Faraday law in frame K:

$$\frac{\partial E_Y}{\partial Z} - \frac{\partial E_Z}{\partial Y} + \frac{\partial B_X}{\partial t} = 0 \tag{405}$$

the same component in frame K' is

$$\gamma\left(\gamma\frac{\partial}{\partial Z}-\frac{\gamma\beta}{c}\frac{\partial}{\partial t}\right)(E_Y+\beta B_X)-\gamma\frac{\partial E_Z}{\partial Y}+\gamma\left(-\gamma\beta\frac{\partial}{\partial Z}+\frac{\gamma}{c}\frac{\partial}{\partial t}\right)(B_X+\beta E_Y)=0 \tag{406}$$

On the U(1) level, we can consider E_Y and B_X to be plane waves and $E_Z = 0$. The following result is obtained in frame K':

$$\gamma^2\left(\frac{\partial E_Y}{\partial Z}+\frac{1}{c}\frac{\partial B_X}{\partial t}\right)-\frac{\gamma^2\beta^2}{c}\left(\frac{1}{c}\frac{\partial B_X}{\partial t}+\frac{\partial E_Y}{\partial Z}\right)=0 \tag{407}$$

This is true for all γ and β because

$$\frac{\partial E_Y}{\partial Z}+\frac{1}{c}\frac{\partial B_X}{\partial t}=0 \qquad \text{(Gaussian units)} \tag{408}$$

The result is obtained that Faraday's law of induction is invariant under a Z boost. Similarly, it can be shown to be invariant under the general Lorentz transformation, and all solutions are invariant. In general, on the U(1) level

$$(\partial_\mu \tilde{F}^{\mu\nu})' = \partial_\mu \tilde{F}^{\mu\nu} \equiv 0 \tag{409}$$

$$(\partial_\mu F^{\mu\nu})' = \partial_\mu \tilde{F}^{\mu\nu} \equiv 0 \tag{410}$$

It follows that the transverse field $\boldsymbol{B}^{(1)} = \boldsymbol{B}^{(2)*}$ is Lorentz-invariant in free space, and so is the B cyclic theorem:

$$\boldsymbol{B}^{(1)} \times \boldsymbol{B}^{(2)} = iB^{(0)}\boldsymbol{B}^{(3)*} \quad \text{in cyclic permutation} \tag{411}$$

The general principle being followed is that, if equations of motion are the same in any Lorentz frame, that is, to any observer, then so are the solutions.

The invariance of the definition of $\boldsymbol{B}^{(3)}$ can again be illustrated on the simplest level by considering Lorentz boosts in the Z, X and Y directions of the $\boldsymbol{B}^{(3)}$ field:

$$B_Z^{(3)'} = B_Z^{(3)} \tag{412}$$

$$B_Z^{(3)'} = \gamma B_Z^{(3)} + \gamma\beta E_X^{(3)} = \gamma B_Z^{(3)} \tag{413}$$

$$B_Z^{(3)'} = \gamma B_Z^{(3)} - \gamma\beta E_Y^{(3)} = \gamma B_Z^{(3)} \tag{414}$$

In Jackson's notation, a Z boost of $\boldsymbol{A}^{(1)}$, for example, leaves it unchanged:

$$A^{(1)} = \begin{bmatrix} 1 & 0 & 0 & 0 \\ 0 & 1 & 0 & 0 \\ 0 & 0 & \gamma & i\gamma\beta \\ 0 & 0 & -i\gamma\beta & \gamma \end{bmatrix} \begin{bmatrix} A_X^{(1)} \\ A_X^{(2)} \\ 0 \\ 0 \end{bmatrix} = \boldsymbol{A}^{(1)} \tag{415}$$

and since $\boldsymbol{A}^{(2)}$ is the complex conjugate of $\boldsymbol{A}^{(1)}$, a Z boost in free space results in

$$(\boldsymbol{B}^{(3)*} = -ig\boldsymbol{A}^{(1)} \times \boldsymbol{A}^{(2)})' \tag{416}$$

and leaves $\boldsymbol{B}^{(3)}$ invariant. The effect of a Y boost on $\boldsymbol{A}^{(1)}$ is as follows:

$$\boldsymbol{A}^{(1)} = \begin{bmatrix} 1 & 0 & 0 & 0 \\ 0 & \gamma & 0 & i\gamma\beta \\ 0 & 0 & 1 & 0 \\ 0 & -i\gamma\beta & 0 & \gamma \end{bmatrix} \begin{bmatrix} A_X^{(1)} \\ A_Y^{(1)} \\ 0 \\ 0 \end{bmatrix} = \begin{bmatrix} A_X^{(1)} \\ \gamma A_Y^{(1)} \\ 0 \\ -i\gamma\beta A_Y^{(1)} \end{bmatrix} \tag{417}$$

and using

$$B_Z^{(3)*} = -ig\varepsilon_{(1)(2)(3)}A_X^{(1)}A_Y^{(2)} \tag{418}$$

it is found that

$$\gamma\boldsymbol{B}^{(3)} = -i\gamma g\boldsymbol{A}^{(1)} \times \boldsymbol{A}^{(2)} \tag{419}$$

and the definition of $\boldsymbol{B}^{(3)}$ is again invariant. Using $B^{(0)} = \boldsymbol{\kappa}A^{(0)}$ [11–20] converts Eq. (416) into the B cyclic theorem, and both are self-consistently invariant. Therefore $\boldsymbol{B}^{(3)}$ is a fundamental field [11–20].

The $\boldsymbol{E}^{(3)}$ field is zero in frame K, and a Z boost means [from Eq. (403)] that it is zero in frame K'. This is consistent with the fact that $\boldsymbol{E}^{(3)}$ is a solution of an invariant equation, the Jacobi identity (30) of O(3) electrodynamics. Finally, we can consider two further illustrative example boosts of $\boldsymbol{E}^{(3)}$ in the X and Y directions, which both produce the following result:

$$E_Z^{(3)'} = \gamma E_Z^{(3)} \tag{420}$$

Therefore if $\boldsymbol{E}^{(3)}$ is null in frame K, it is null in frame K'. There is a symmetry between the Lorentz transforms of $\boldsymbol{B}^{(3)}$ and the hypothetical $\boldsymbol{E}^{(3)}$:

$$\begin{aligned} X: &\quad B_Z^{(3)'} = \gamma B_Z^{(3)}; \qquad E_Z^{(3)'} = \gamma E_Z^{(3)} \\ Y: &\quad B_Z^{(3)'} = \gamma B_Z^{(3)}; \qquad E_Z^{(3)'} = \gamma E_Z^{(3)} \\ Z: &\quad B_Z^{(3)'} = B_Z^{(3)}; \qquad E_Z^{(3)'} = E_Z^{(3)} \end{aligned} \tag{421}$$

This is self-consistent with the fact that $\boldsymbol{B}^{(3)}$ may be regarded [11–20] as dual to $[-i\boldsymbol{E}^{(3)}/c]$, so that $\boldsymbol{B}^{(3)2} + \boldsymbol{E}^{(3)2}$ contributes to a nonzero Lagrangian and so that $\boldsymbol{B}^{(3)}$ is a real physical field.

These are mathematically valid results, but physically, the Lorentz transform of $\boldsymbol{B}^{(3)}$ and the null $\boldsymbol{E}^{(3)}$ are governed by the equation

$$D_\mu \tilde{\boldsymbol{G}}^{\mu\nu} = \boldsymbol{0}^\nu \tag{422}$$

where:

$$\boldsymbol{0}^\nu = 0^{\nu(1)}\boldsymbol{e}^{(1)} + 0^{\nu(2)}\boldsymbol{e}^{(2)} + 0^{\nu(3)}\boldsymbol{e}^{(3)} \tag{423}$$

is a null 12-vector, whose components are null 4-vectors. The general Lorentz transform of the null 4-vector is given by

$$0'^{\mu} = \Lambda^\mu_\nu 0^\mu = 0^\mu \tag{424}$$

and a null 4-vector is a null 4-vector in all Lorentz frames. This means that the left-hand side of Eq. (422) is null in all Lorentz frames and is Lorentz-invariant. Therefore its field solutions are also all Lorentz invariant, including, of course, $\boldsymbol{B}^{(3)}$ and $\boldsymbol{E}^{(3)}$. This is self-consistent with the fact that Eq. (422) is equivalent to the Jacobi identity (30) for the group O(3). Finally, when there is field–matter interaction, all field components are Lorentz covariant, and no longer invariant, on both the U(1) and O(3) levels.

In conclusion, the homogeneous field equation of O(3) electrodynamics is Lorentz-invariant, and all its classical solutions must be also Lorentz-invariant. The same result is obtained therefore in QED.

IX. NOETHER CHARGES AND CURRENTS OF O(3) ELECTRODYNAMICS IN THE VACUUM

The first example of a vacuum current was introduced by Maxwell in order to make the equations of electrostatics and magnetostatics self-consistent. The second examples were introduced in 1979 [7] by Lehnert, and O(3) electrodynamics offers four vacuum charges and currents of topological origin as discussed already. Maxwell was led to the displacement current because the received view at the time was self-inconsistent [5]. The received view consisted of four equations

$$\nabla \cdot \boldsymbol{D} = \rho; \qquad \nabla \cdot \boldsymbol{B} = 0; \qquad \nabla \times \boldsymbol{H} = \boldsymbol{J}; \qquad \nabla \times \boldsymbol{E} + \frac{\partial \boldsymbol{B}}{\partial t} = \boldsymbol{0} \tag{425}$$

together with the continuity equation:

$$\nabla \cdot \boldsymbol{J} + \frac{\partial \rho}{\partial t} = 0 \tag{426}$$

Maxwell used the continuity equation in the Coulomb law to give

$$\nabla \cdot \left(\boldsymbol{J} + \frac{\partial \boldsymbol{D}}{\partial t} \right) = 0 \tag{427}$$

and replaced $\boldsymbol{J}$ by $\boldsymbol{J} + (\partial \boldsymbol{D}/\partial t)$. The final result is the Ampère–Maxwell law

$$\nabla \times \boldsymbol{H} = \boldsymbol{J} + \frac{\partial \boldsymbol{D}}{\partial t} \tag{428}$$

which produced electromagnetic waves and is, of course, a standard part of U(1) electrodynamics. The latter asserts, in the received view [5] currently prevailing, that in the vacuum, there is a displacement current

$$\boldsymbol{J}_D = \varepsilon_0 \frac{\partial \boldsymbol{E}}{\partial t} \tag{429}$$

using the vacuum constitutive equation $\boldsymbol{D} = \epsilon_0 \boldsymbol{E}$. The existence of Maxwell's vacuum displacement current is all-important for the theory of electromagnetic radiation. The displacement current originates in the continuity equation, which is a conservation law, similar to the laws of conservation of energy and momentum summarized in Noether's theorem [6]. The Maxwell displacement current can therefore be referred to as a "Noether current."

More than a century later, Lehnert [7] introduced and developed [7–10] the concept of vacuum charge on the classical level, and showed [7–10] that this concept leads to advantages over the Maxwell–Heaviside equations in the description of empirical data, for example, the problem of an interface with a vacuum [7–10,15]. The introduction of a vacuum charge leads to axisymmetric vacuum solutions akin to the $\boldsymbol{B}^{(3)}$ vacuum component of O(3) electrodynamics [10,15], and also leads to the Proca equation and the concept of photon mass.

The latter is therefore related to the concept of the $\boldsymbol{B}^{(3)}$ field through the Lehnert equations, which in the vacuum are

$$\nabla\cdot\boldsymbol{D}=\rho_{\text{vac}};\qquad \nabla\cdot\boldsymbol{B}=0;\qquad \nabla\times\boldsymbol{H}=\boldsymbol{J}_{\text{vac}}+\frac{\partial\boldsymbol{D}}{\partial t};\qquad \nabla\times\boldsymbol{E}+\frac{\partial\boldsymbol{B}}{\partial t}=\boldsymbol{0} \tag{430}$$

It can be seen that these are U(1) equations, but with the addition of the vacuum charge density ρ_{vac} and the vacuum current density $\boldsymbol{J}_{\text{vac}}$. On the O(3) level, the Lehnert charge density becomes

$$\underset{\text{in cyclic permutation}}{\rho_{\text{vac}}^{(1)*}=ig(\boldsymbol{A}^{(2)}\cdot\boldsymbol{D}^{(3)}-\boldsymbol{D}^{(2)}\cdot\boldsymbol{A}^{(3)})} \tag{431}$$

and the Lehnert current density becomes

$$\underset{\text{in cyclic permutation}}{\boldsymbol{J}_{\text{vac}}^{(1)*}=-ig(cA_0^{(2)}\boldsymbol{D}^{(3)}-cA_0^{(3)}\boldsymbol{D}^{(2)}+\boldsymbol{A}^{(2)}\times\boldsymbol{H}^{(3)}-\boldsymbol{A}^{(3)}\times\boldsymbol{H}^{(2)})} \tag{432}$$

and O(3) electrodynamics self-consistently produces longitudinal solutions in the vacuum typified by the phaseless $\boldsymbol{B}^{(3)}$ component. However, the magnetic charge and current allowed for by O(3) electrodynamics do not appear in the Lehnert equations (430).

The Lehnert equations are consistent [10] with the continuity equation (428) of U(1) electrodynamics. Using the vacuum continuity equation in Lehnert's vacuum Coulomb law, we find

$$\begin{aligned}\boldsymbol{J}\rightarrow\boldsymbol{J}+\frac{\partial\boldsymbol{D}}{\partial t}&\equiv\boldsymbol{J}_1\\ \nabla\times\boldsymbol{H}&=\boldsymbol{J}_1+\frac{\partial\boldsymbol{D}}{\partial t}\\ \nabla\cdot\boldsymbol{J}_1+\frac{\partial\rho_1}{\partial t}&=0\end{aligned} \tag{433}$$

Repeating this procedure gives

$$\begin{aligned}\boldsymbol{J}_1&=\boldsymbol{J}+\frac{\partial D}{\partial t}\\ &\vdots\\ \boldsymbol{J}_n&=\boldsymbol{J}+n\frac{\partial\boldsymbol{D}}{\partial t};\qquad n\rightarrow\infty\\ \rho_n&=\pm\int\nabla\cdot\boldsymbol{J}_n\,dt;\qquad n\rightarrow\infty\end{aligned} \tag{434}$$

and theoretically, there are two infinitely large densities in the vacuum given by

$$\begin{aligned} \rho_n &= \int \nabla \cdot \boldsymbol{J}_n \, dt; \qquad n \to \infty \\ \rho_n &= -\int \nabla \cdot \boldsymbol{J}_n \, dt; \qquad n \to \infty \end{aligned} \tag{435}$$

because charge density can either be negative or positive. In this process, $\boldsymbol{B}$ and $\boldsymbol{E}$ are unchanged, so the vector and scalar potentials defined by

$$\boldsymbol{B} = \nabla \times \boldsymbol{A}; \qquad \boldsymbol{E} = -\frac{\partial \boldsymbol{A}}{\partial t} - \nabla \phi \tag{436}$$

remain unchanged.

Therefore the vacuum potential energy difference is given by

$$\Delta V = \pm \int \boldsymbol{J}_n \cdot \boldsymbol{A} \, d^3x \tag{437}$$

and the rate of doing work is

$$\frac{\partial W}{\partial t} = \pm \int \boldsymbol{J}_n \cdot \boldsymbol{E} \, d^3x \tag{438}$$

In thermodynamic equilibrium, the net result is zero in both cases, but locally, there may be a non-zero rate of doing work by these vacuum charges and currents on a device, creating thermal or mechanical energy. This process is unknown in the received view but conserves energy and is consistent with Noether's theorem [6].

The existence of charge density and current density in the vacuum is not consistent with the Maxwell–Heaviside equations, but leads to a description of empirical data [10,15] superior to that of the received view. Vacuum charge and current density on the classical level are therefore postulates on the same philosophical level as the existence of displacement current in the vacuum. The latter emerges from the continuity equation (426) as argued already. If a postulate leads to an improved description of empirical data, then the postulate is valid in natural philosophy, irrespective of the received view. The role of the coefficient g on the O(3) level may be discussed in a similar philosophical vein. As argued already, the existence of g is a direct consequence of the gauge principle, and it exists in the classical vacuum (or free space), on both the U(1) and O(3) levels, in the respective covariant derivatives. It follows that $e/\hbar$ exists in the vacuum in the Maxwell–Heaviside point of view itself, if this be regarded as a U(1) Yang–Mills gauge theory as is the current practice [6]. If $e/\hbar$ exists in the vacuum on the classical level, then charge density may exist in the vacuum as argued by Lehnert, and so current density may also exist. The Lehnert equations were derived from U(1) gauge theory in Section IV. The existence of $e/\hbar$ in the vacuum on the O(3) level is therefore conceptually no different from its existence in the vacuum on the U(1) level.

As argued in Section III, the form of the received Maxwell–Heaviside equations in free space or classical vacuum is obtained for finite g. The factor g is a direct consequence of gauge theory [6] and is in general, a proportionality constant without which there is no gauge theory, and without which special relativity is violated. The coefficient g is present for all gauge groups in the vacuum, including U(1). The superiority of the O(3) gauge group over the U(1) gauge group in electrodynamics in no way depends on the introduction of g in O(3): g is also present in U(1). The gauge principle and special relativity therefore force the conclusion that e is itself topological in origin, and is not localized on the electron, a conclusion first reached by Frenkel [15]. The bosons (photons) obtained from a quantization of electrodynamics in any gauge group are not charged bosons, as discussed in Section VIII. The physical nature of g may be roughly summarized by noting the fact that g is a coupling constant that is a property of neither the source (electron) nor the field. As demonstrated in Section VIII, the classical O(3) electrodynamics may be extended without conceptual difficulty to quantum electrodynamics on both the nonrelativistic and relativistic levels. Similarly, the constant g exists in the vacuum in U(1) electrodynamics as a consequence of the gauge principle and special relativity, and U(1) electrodynamics quantizes to quantum electrodynamics without charged photons.

In field theory, electric charge [6] is a symmetry of action, because it is a conserved quantity. This requirement leads to the consideration of a complex scalar field ϕ. The simplest possibility [U(1)] is that ϕ have two components, but in general it may have more than two as in the internal space of O(3) electrodynamics which consists of the complex basis ((1),(2),(3)). The first two indices denote complex conjugate pairs, and the third is real-valued. These indices superimposed on the 4-vector A_μ give a 12-vector. In U(1) theory, the indices (1) and (2) are superimposed on the 4-vector A_μ in free space, so A_μ in U(1) electrodynamics in free space is considered as transverse, that is, determined by (1) and (2) only. These considerations lead to the conclusion that charge is not a point localized on an electron; rather, it is a symmetry of action dictated ultimately by the Noether theorem [6].

By way of introduction to the Noether currents and charges that exist in O(3) electrodynamics, the inhomogeneous field of Eq. (32) can be considered in the vacuum (source-free space) and split into two particular solutions:

$$\partial_\mu \boldsymbol{G}^{\mu\nu} = \boldsymbol{0} \tag{439}$$

$$\boldsymbol{J}^\nu = g\varepsilon_0 \boldsymbol{A}_\mu \times \boldsymbol{G}^{\mu\nu} \tag{440}$$

The first of these has been discussed in Section IV. The second is a vacuum charge–current 12-vector in SI units. On the O(3) level, it is a physical charge–current that gives rise to the energy

$$En^{(3)} = -\int \boldsymbol{J}^{\nu} \cdot \boldsymbol{A}_{\nu}\, dV \tag{441}$$

where V is the radiation volume. The energy term can be developed as follows

$$\begin{aligned} En^{(3)} &= -\frac{1}{\mu_0}\int g\boldsymbol{A}_{\mu} \times \boldsymbol{G}^{\mu\nu} \cdot \boldsymbol{A}_{\nu}\, dV \\ &= \frac{g}{\mu_0}\int \boldsymbol{G}^{\mu\nu} \cdot \boldsymbol{A}_{\mu} \times \boldsymbol{A}_{\nu}\, dV \\ &= \frac{g^2}{\mu_0}\int \boldsymbol{A}^{\mu} \times \boldsymbol{A}^{\nu} \cdot \boldsymbol{A}_{\mu} \times \boldsymbol{A}_{\nu}\, dV \\ &= \frac{1}{\mu_0}\int \boldsymbol{B}^{(3)} \cdot \boldsymbol{B}^{(3)}\, dV \end{aligned} \tag{442}$$

and is the energy due to the $\boldsymbol{B}^{(3)}$ component of O(3) electrodynamics. This is a concise way of demonstrating that the Noether charge–currents of O(3) electrodynamics give energy that in principle can be utilized for working devices. In analogy, the Maxwell displacement current of the vacuum gives rise to the electromagnetic field, which carries energy. The same principle is involved on the U(1) and O(3) levels, and the ultimate source of the energy is the topology of the vacuum, which manifests itself through the gauge principle and group theory (Section I). If g were zero in Eq. (440), there would be no energy due to $\boldsymbol{B}^{(3)}$, revealing the latter's topological origin. This energy can be thought of as originating in a covariant derivative with O(3) symmetry, and a covariant derivative is necessitated by special relativity and topology. So in this sense, the energy due to $\boldsymbol{B}^{(3)}$ can be thought of as energy from the vacuum, manifesting itself as part of the electromagnetic field. It is probable that devices can be constructed to take advantage of this property of the vacuum and convert energy of this nature efficiently into usable form.

The principle of taking energy from the vacuum is the gauge principle, and this is illustrated as follows on the U(1) level. The U(1) gauge equations in the vacuum are [6]

$$(\partial_{\mu} + igA_{\mu})\tilde{G}^{\mu\nu} \equiv 0 \tag{443}$$

$$(\partial_{\mu} + igA_{\mu})G^{\mu\nu} = 0 \tag{444}$$

where the vacuum 4-current is defined as

$$J^{\nu} = -ig\varepsilon_0 A_{\mu} G^{\mu\nu} \tag{445}$$

If we set the index $\mu = 0$ in Eq. (445), for example, the following relations are obtained:

$$J_X = ig\varepsilon_0 E_X \qquad J_Y = ig\varepsilon_0 E_Y \tag{446}$$

The average energy from this vacuum current can be defined as

$$En = c \int J^{\nu *} A_\nu \, dV \tag{447}$$

which is

$$En = -i\kappa c\varepsilon_0 \int (E_X^* A_X + E_Y^* A_Y)\, dV = \varepsilon_0 \int \kappa c E^{(0)} A^{(0)}\, dV \tag{448}$$

Using

$$E^{(0)} = \kappa c A^{(0)} \tag{449}$$

Eq. (448) becomes the familiar U(1) electromagnetic field energy:

$$En = \varepsilon_0 \int E^{(0)2}\, dV = \frac{1}{2} \int \left(\varepsilon_0 E^{(0)2} + \frac{1}{\mu_0} B^{(0)2} \right) dV \tag{450}$$

The same result is obtained from Eq. (443) using the same proportionality factor $g = \kappa / A^{(0)}$. Note carefully that without the gauge term $igA\mu$, this energy would vanish, and so the energy is due to the vacuum configuration and topology, in this case assumed to be described by the U(1) group.

Similarly, the magnitude of the linear momentum of the electromagnetic field can be obtained by using the proportionality $g = e/\hbar$ in either Eqs. (443) or (444), giving

$$\left(\partial_\mu + ie\frac{A_\mu}{\hbar} \right) \tilde{G}^{\mu\nu} \equiv 0 \qquad \left(\partial_\mu + ie\frac{A_\mu}{\hbar} \right) G^{\mu\nu} = 0 \tag{451}$$

Using the standard operator transformation of quantum mechanics

$$p_\mu = -i\hbar \partial_\mu \tag{452}$$

Eqs. (451) both become

$$\begin{aligned} \partial_\mu \tilde{G}^{\mu\nu} &\equiv 0 \\ \partial_\mu G^{\mu\nu} &= 0 \end{aligned} \tag{453}$$

and so we retrieve the familiar Maxwell–Heaviside equations in the vacuum. The momentum is obtained from the equivalence

$$\frac{e}{\hbar} = \frac{\kappa}{A^{(0)}} \tag{454}$$

giving the magnitude of the linear momentum as

$$p = \hbar\kappa = eA^{(0)} \tag{455}$$

which is again a topological or vacuum property. Using $En = \hbar\omega$, the energy is given from Eq. (455) by

$$En = ecA^{(0)} \tag{456}$$

and is again topological in origin; that is, it originates from energy inherent in a vacuum configuration described by the non-singly connected group U(1).

The principle behind this derivation is the gauge principle, and so is the same for all gauge groups. The equivalence (456) was first demonstrated on the O(3) level [15], but evidently exists for all gauge group symmetries. The gauge principle in electrodynamics therefore leads to the energy and momentum of the photon and classical field. The 4-current J_μ appears in both Eqs. (443) and (444) and is self-dual, a result that is echoed in the self-duality of the vacuum field equations:

$$\partial_\mu \tilde{G}^{\mu\nu} = \partial_\mu G^{\mu\nu} \tag{457}$$

Another advantage of this principle is that the coupling constant g is always present implicitly in the calculation, meaning that the energy and momentum have a cause, or source. This source is not the charge on the electron, but rather the structure or configuration of the vacuum itself, obtained as a direct result of the gauge principle taken to its logical conclusion.

If the procedure is repeated for the rate of doing work by the vacuum 4-current J^ν

$$\frac{dW}{dt} = c \int \boldsymbol{J}^* \cdot \boldsymbol{E}\, dV \tag{458}$$

it is found that

$$\frac{dW}{dt} = c \int (J_X^* E_X + J_Y^* E_Y)\, dV \tag{459}$$

which is zero if $\boldsymbol{E}$ is a transverse plane wave. This result means that the energy corresponding to J^ν is conserved in the vacuum because the rate of doing work is energy per unit time. Therefore the field momentum is also conserved in the vacuum. And therefore J^ν is a Noether current in the vacuum.

On the O(3) level, several new sources of energy from the vacuum emerge as follows. First, define the charge and potential 12-vectors:

$$\boldsymbol{J}^{\mu(i)} \equiv \left(\rho, \frac{\boldsymbol{J}^{(i)}}{c}\right) \tag{460}$$

$$\boldsymbol{A}^{\mu(i)} \equiv (\phi, c\boldsymbol{A}^{(i)}) \tag{461}$$

so that the energy from a vacuum configuration considered to have O(3) gauge group symmetry is

$$En = -\int (\boldsymbol{J}^{\nu(1)} \cdot \boldsymbol{A}_\nu^{(2)} + \boldsymbol{J}^{\nu(2)} \cdot \boldsymbol{A}_\nu^{(1)} + \boldsymbol{J}^{\nu(3)} \cdot \boldsymbol{A}_\nu^{(3)})\, dV \tag{462}$$

(En is used here to denote energy, not to be confuse with E as an electrical field). The 12-vector is a spinor in which the Greek indices in covariant contravariant notation are $0, 1, 2,$ and 3 and the numerical index (i) runs from 1 to 3, representing the circular basis ((1),(2),(3)). For example [11–20]), $A^{\mu(1)}$ is the 4-vector, $(\phi^{(1)}, c\boldsymbol{A}^{(1)})$, $A^{\mu(2)}$ is the 4-vector, $(\phi^{(2)}, c\boldsymbol{A}^{2})$, and $A^{\mu(3)}$ is the 4-vector $(\phi^{(3)}, c\boldsymbol{A}^{(3)})$. Each of the three 4-vectors has four components, making a 12-vector. This must not be confused with a vector of 12 components. The field 12-vector is defined as

$$\boldsymbol{G}_{\mu\nu} \equiv G_{\mu\nu}^{(1)}\boldsymbol{e}^{(1)} + G_{\mu\nu}^{(2)}\boldsymbol{e}^{(2)} + G_{\mu\nu}^{(3)}\boldsymbol{e}^{(3)} \tag{463}$$

where each component in indices (1), (2), and (3) have the structure:

$$G^{\mu\nu} \equiv \begin{bmatrix} 0 & -\frac{E^1}{c} & -\frac{E^2}{c} & -\frac{E^3}{c} \\ \frac{E^1}{c} & 0 & -B^3 & B^2 \\ \frac{E^2}{c} & B^3 & 0 & -B^1 \\ \frac{E^3}{c} & -B^2 & B^1 & 0 \end{bmatrix}; \qquad G_{\mu\nu} \equiv \begin{bmatrix} 0 & \frac{E_1}{c} & \frac{E_2}{c} & \frac{E_3}{c} \\ -\frac{E_1}{c} & 0 & -B_3 & B_2 \\ -\frac{E_2}{c} & B_3 & 0 & -B_1 \\ -\frac{E_3}{c} & -B_2 & B_1 & 0 \end{bmatrix} \tag{464}$$

The field equations in the vacuum are (31) and (32), and there are two possible vacuum charge current 12-vectors:

$$\boldsymbol{J}_H^{\nu} = -g\varepsilon_0 \boldsymbol{A}_{\mu} \times \tilde{\boldsymbol{G}}^{\mu\nu} \tag{465}$$

$$\boldsymbol{J}^{\nu} = -g\varepsilon_0 \boldsymbol{A}_{\mu} \times \boldsymbol{G}^{\mu\nu} \tag{466}$$

which, from Eq. (462), are sources of energy from energy inherent in a vacuum configuration as a direct result of the gauge principle. These two 12-vectors provide several more sources of energy, a result that can be illustrated with Eq. (466) by developing it as follows in the ((1),(2),(3)) basis:

$$\begin{aligned} \boldsymbol{J}^{(1)*} &= -ig\varepsilon_0 \boldsymbol{A}_{\mu}^{(2)} \times \boldsymbol{G}^{\mu\nu(3)} \\ \boldsymbol{J}^{(2)*} &= -ig\varepsilon_0 \boldsymbol{A}_{\mu}^{(3)} \times \boldsymbol{G}^{\mu\nu(1)} \\ \boldsymbol{J}^{(3)*} &= -ig\varepsilon_0 \boldsymbol{A}_{\mu}^{(1)} \times \boldsymbol{G}^{\mu\nu(2)} \end{aligned} \tag{467}$$

This result follows because of the negative sign in Eqs. (465) and (466). Equation (462) for the energy is therefore

$$\begin{aligned} En = -ig\varepsilon_0 \bigg(\int \boldsymbol{A}_{\mu}^{(3)} \times \boldsymbol{G}^{\mu\nu(1)} \cdot \boldsymbol{A}_{\nu}^{(2)}\, dV + \int \boldsymbol{A}_{\mu}^{(2)} \times \boldsymbol{G}^{\mu\nu(2)} \cdot \boldsymbol{A}_{\nu}^{(3)} dV \\ + \int \boldsymbol{A}_{\mu}^{(1)} \times \boldsymbol{G}^{\mu\nu(2)} \cdot \boldsymbol{A}_{\nu}^{(3)}\, dV \bigg) \end{aligned} \tag{468}$$

Now use the 3-vector identity:

$$\boldsymbol{F} \cdot \boldsymbol{G} \times \boldsymbol{H} = \boldsymbol{G} \cdot \boldsymbol{H} \times \boldsymbol{F} \tag{469}$$

to obtain

$$\begin{aligned} En = -ig\varepsilon_0 \bigg(\int \boldsymbol{G}^{\mu\nu(1)} \cdot \boldsymbol{A}_{\nu}^{(2)} \times \boldsymbol{A}_{\mu}^{(3)}\, dV + \int \boldsymbol{G}^{\mu\nu(3)} \cdot \boldsymbol{A}_{\nu}^{(1)} \times \boldsymbol{A}_{\mu}^{(2)} \boldsymbol{dV} \\ + \int \boldsymbol{G}^{\mu\nu(2)} \cdot \boldsymbol{A}_{\nu}^{(3)} \times A_{\mu}^{(1)}\, dV \bigg) \end{aligned} \tag{470}$$

The definition (461) implies that we can write

$$c^2 \boldsymbol{B}_{\nu\mu}^{(1)*} = c^2 \boldsymbol{B}_{\nu\mu}^{(2)} \equiv -ig\boldsymbol{A}_{\nu}^{(2)} \times \boldsymbol{A}_{\mu}^{(3)} \tag{471}$$

The energy terms in Eq. (470) can therefore be developed as follows:

1.

$$
\begin{aligned}
En_1 &= \varepsilon_0 c^2 \int \boldsymbol{G}^{\mu\nu(1)} \cdot \boldsymbol{B}^{(2)}_{\nu\mu}\, dV \\
&= \frac{1}{\mu_0} \int \boldsymbol{B}^{\mu\nu(1)} \cdot B^{(2)}_{\nu\mu}\, dV \\
&= \frac{1}{\mu_0} \int \boldsymbol{B}^{3\nu(1)} \cdot \boldsymbol{B}^{(2)}_{\nu 3}\, dV \\
&= \frac{1}{\mu_0} \int B^{31(1)} B^{(2)}_{13} + B^{32(1)} B^{(2)}_{23}\, dV \\
&= \frac{1}{\mu_0} \int -B^{2(1)} B^{(2)}_{2} - B^{1(1)} B^{(2)}_{1}\, dV \\
&= \frac{1}{\mu_0} \int B^{(1)}_{Y} B^{(2)}_{Y} + B^{(1)}_{X} B^{(2)}_{X}\, dV
\end{aligned} \tag{472}
$$

2.

$$
\begin{aligned}
En_2 &= \varepsilon_0 c^2 \int \boldsymbol{G}^{\mu\nu(3)} \cdot \boldsymbol{B}^{(3)}_{\nu\mu}\, dV \\
&= \varepsilon_0 c^2 \int G^{12(3)} B^{(3)}_{21} + G^{21(3)} B^{(3)}_{12}\, dV \\
&= \frac{1}{\mu_0} \int B^{(3)}_{Z} B^{(3)}_{Z} + B^{(3)}_{Z} B^{(3)}_{Z}\, dV
\end{aligned} \tag{473}
$$

3.

$$
\begin{aligned}
En_3 &= \varepsilon_0 c^2 \int \boldsymbol{G}^{\mu\nu(2)} \cdot \boldsymbol{B}^{(1)}_{\nu\mu}\, dV \\
&= \varepsilon_0 c^2 \int \boldsymbol{G}^{\mu 3(2)} \cdot \boldsymbol{B}^{(1)}_{3\mu}\, dV \\
&= \frac{1}{\mu_0} \int B^{(2)}_{Y} B^{(1)}_{Y} + B^{(2)}_{X} B^{(1)}_{X}\, dV
\end{aligned} \tag{474}
$$

These derivations are given in full detail to show that the O(3) gauge principle leads to several more terms than in U(1), where the same gauge principle leads to Eq. (450).

The overall result for the vacuum energy in U(1) is

$$
En = \frac{1}{\mu_0} \int B^{(0)2}\, dV \tag{475}
$$

and the corresponding result in O(3) is

$$En = En_1 + En_2 + En_3 \tag{476}$$

If we adopt a gauge group of higher symmetry than O(3), there will be more terms and so on, and this is a general principle. Electromagnetic charge current and electromagnetic energy depend on the configuration of the vacuum, and ultimately on the topology of the vacuum as represented in the language of gauge and group theory (Section I). Charge current is a property of the vacuum, and charge is not localized to a point as in the conventional view. On both U(1) and O(3) levels, the field equations can be expressed in terms solely of potentials that, in the language of general relativity, are connections. The constant e becomes a scaling factor and both $g = \frac{e}{\hbar} = \frac{\kappa}{A^{(0)}}$ and all field potentials are consequences of the gauge principle for all gauge groups, including U(1).

We can begin to think of the electromagnetic field in the same terms as the gravitational field, and the former is not an entity superimposed on the vacuum irrespective of the vacuum structure. This conclusion is reminiscent of Faraday's concept, as adopted by Maxwell [4], of charge as being the result of the field. In gauge theory, g is a property of neither electron nor field, but a property of the structure of the vacuum itself. The energy and charge current also come from the vacuum. These concepts are further developed in Section XII. Finally, the energy momentum of the field on the O(3) level is a 12-vector:

$$\boldsymbol{p}_\mu \equiv p_\mu^{(1)} \boldsymbol{e}^{(1)} + p_\mu^{(2)} \boldsymbol{e}^{(2)} + p_\mu^{(3)} \boldsymbol{e}^{(3)} \tag{477}$$

giving a new view of field momentum. This view is quite different from the problematic [4] view of electromagnetic energy proposed by Poynting.

Electromagnetic theory in the vacuum at the O(3) level begins to look like the theory of gravitation, the electromagnetic field can be replaced by physical potential differences, and these are primary. Analogously, mass in general relativity is a curvature of spacetime, and the gravitational field is the coordinate system itself. On the O(3) level, the potentials are connection coefficients, and charge is the result of topology expressed through gauge theory and group theory. It has been shown that the topology of the vacuum can produce energy, and that charge–current emanates from the same source. If the potential is a connection, then the field can be expressed in terms of the potential and therefore wholly in terms of the connection, and therefore in terms of topology. The view presented here of the field particle dualism of de Broglie is that all particles are pseudo particles and the vacuum electromagnetic field is the topology of the vacuum itself. This point of view rejects action at a distance, as did Newton himself. It is clear that particles result from the gauge principle, for example, photons and quarks, as the result of quantization of the potential.

The potential is again primary in canonical quantization, and it has been shown in Section IX that quantization of O(3) electrodynamics does not lead to charged photons.

X. SCALAR INTERFEROMETRY AND CANONICAL QUANTIZATION FROM WHITTAKER'S POTENTIALS

Whittaker's early work [27,28] is the precursor [4] to twistor theory and is well developed. Whittaker showed that a scalar potential satisfying the Laplace and d'Alembert equations is structured in the vacuum, and can be expanded in terms of plane waves. This means that in the vacuum, there are both propagating and standing waves, and electromagnetic waves are not necessarily transverse. In this section, a straightforward application of Whittaker's work is reviewed, leading to the feasibility of interferometry between scalar potentials in the vacuum, and to a trouble-free method of canonical quantization.

Whittaker [27,28] derived equations defining the electromagnetic field in the vacuum in terms of functions $\boldsymbol{f}$ and $\boldsymbol{g}$ with the units of magnetic flux directed longitudinally in the axis of propagation (Z)

$$[\boldsymbol{f} = F\boldsymbol{k}; \qquad \boldsymbol{g} = G\boldsymbol{k} \tag{478}$$

and defined all field components in terms of $\boldsymbol{f}$ and $\boldsymbol{g}$. The electric and magnetic field vectors in the vacuum, in SI units, are defined by

$$\boldsymbol{E} = c\nabla \times (\nabla \times \boldsymbol{f}) + \nabla \times \dot{\boldsymbol{g}} \tag{479a}$$

$$\boldsymbol{B} = \frac{1}{c}\nabla \times \dot{\boldsymbol{f}} - \nabla \times (\nabla \times \boldsymbol{g}) \tag{479b}$$

If we use the Stratton potential defined by

$$B^{\mu} \equiv (cP, \boldsymbol{S}) \tag{480}$$

where

$$\boldsymbol{E} = -\nabla \times \boldsymbol{S}; \qquad \boldsymbol{B} = -\frac{\partial \boldsymbol{S}}{\partial t} - \nabla P \tag{481}$$

and the 4-potential defined by

$$A^{\mu} \equiv (\phi, c\boldsymbol{A}) \tag{482}$$

where

$$\boldsymbol{B} = -\nabla \times \boldsymbol{A}; \qquad \boldsymbol{E} = -\frac{\partial \boldsymbol{A}}{\partial t} - \nabla\phi \tag{483}$$

it is deduced that

$$\boldsymbol{A} = -\nabla \times \boldsymbol{g} + \frac{1}{c}\dot{\boldsymbol{f}} \tag{484}$$

$$\boldsymbol{S} = -c\nabla \times \boldsymbol{f} - \dot{\boldsymbol{g}} \tag{485}$$

in the vacuum. So, in general, the Maxwell potential $\boldsymbol{A}$ and the Stratton potential $\boldsymbol{S}$ both have longitudinal components in the vacuum. Both $\boldsymbol{A}$ and $\boldsymbol{S}$ are generated from the more fundamental $\boldsymbol{f}$ and $\boldsymbol{g}$, and their longitudinal components in the vacuum are

$$A_Z = \frac{1}{c}\dot{F}; \qquad S_Z = -\dot{G} \tag{486}$$

The longitudinal magnetic and electric field components are [27,28]:

$$B_Z = \frac{\partial^2 G}{\partial X^2} + \frac{\partial^2 G}{\partial Y^2}; \qquad E_Z = \frac{\partial^2 F}{\partial Z^2} - \frac{1}{c^2}\frac{\partial^2 F}{\partial t^2} \tag{487}$$

It is now known that these equations correspond to twistor contour integral solutions for a particle with zero rest mass, and lead to an O(3) symmetry gauge group for electromagnetism in the vacuum because the Whittaker solution is a spinor formalism. Electrodynamics on the O(3) level is also a spinor, and ultimately a twistor, formalism. Using the Penrose transform [4], the full significance of the Whittaker solution becomes apparent. Later in this section, the $\boldsymbol{B}^{(3)}$ field is expressed in terms of $\boldsymbol{f}$ and $\boldsymbol{g}$, which are therefore physical. It is this property that leads to the possibility of interferometry between scalar potentials. In the received view [U(1) level], the scalar potential in the vacuum is zero or unphysical, and so the received view loses a great deal of information. The work of Whittaker therefore anticipates much of contemporary non-Abelian gauge theory applied to electrodynamics in the vacuum. In the original equations of J. C. Maxwell [78], Faraday's electrotonic state is a physical *vector potential*, a term that was introduced by Maxwell himself [79]. It is the later interpretation of Maxwell's original intent by Heaviside [80] that relegates the U(1) vector potential to a mathematical subsidiary with no physical meaning. Several refutations of Heaviside's opinion have been given in this chapter already. It is also incompatible with electromagnetism as a twistor theory, where Maxwell's original intent is realized, and vector potentials are physical on the classical level. To be precise, vector and scalar potential differences can be measured experimentally on the classical level.

Without loss of generality, it can be assumed that plane waves can be used for the transverse parts of $\boldsymbol{S}$ and $\boldsymbol{A}$, resulting in

$$\boldsymbol{S} = ic\boldsymbol{A} \tag{488}$$

We obtain, self-consistently

$$f = ig; \quad \dot{f} = i\dot{g} \tag{489}$$

The following scalar magnetic flux gives transverse plane waves for $\boldsymbol{A}$ and $\boldsymbol{S}$

$$G = \frac{A^{(0)}}{\sqrt{2}}(X - iY)e^{i(\omega t - \kappa Z)} \tag{490}$$

so that

$$\begin{aligned} \boldsymbol{A} &= -\nabla \times \boldsymbol{g} = \frac{A^{(0)}}{\sqrt{2}}(i\boldsymbol{i} + \boldsymbol{j})e^{i(\omega t - \kappa Z)} \\ \boldsymbol{B} &= -\nabla \times \boldsymbol{A} = \frac{B^{(0)}}{\sqrt{2}}(i\boldsymbol{i} + \boldsymbol{j})e^{i(\omega t - \kappa Z)} \end{aligned} \tag{491}$$

Importantly, there also exists a longitudinal propagating part of the vector potential

$$\boldsymbol{A}_L = \frac{i}{c}\dot{G}\boldsymbol{k} = -\kappa\frac{A^{(0)}}{\sqrt{2}}(X - iY)e^{i(\omega t - \kappa Z)}\boldsymbol{k} \tag{492}$$

that is not present in the received view [6]. For example, $\boldsymbol{A}_L$ is zero in the radiation and Coulomb gauges, and is considered in the received view to be unphysical in the Lorenz gauge [6]. The longitudinal vector potential gives rise to the transverse magnetic plane wave

$$\boldsymbol{B} = \nabla \times \boldsymbol{A}_L = -\frac{B^{(0)}}{\sqrt{2}}(\boldsymbol{i} + i\boldsymbol{j})e^{i(\omega t - \kappa Z)} \tag{493}$$

and to the electric field:

$$\boldsymbol{E}_L = -\frac{\partial \boldsymbol{A}_L}{\partial t} - \nabla\phi = i\frac{\kappa^2}{c}\frac{A^{(0)}}{\sqrt{2}}(X - iY)e^{i(\omega t - \kappa Z)}\boldsymbol{k} - \nabla\phi \tag{494}$$

In general, therefore, there is a longitudinal propagating component of the electric field in the vacuum. However, in the plane-wave approximation used here, there occurs the relation

$$\nabla\phi = \nabla \times \boldsymbol{S} - \frac{\partial \boldsymbol{A}}{\partial t} \tag{495}$$

and the longitudinal part of $\nabla\phi$ is

$$(\nabla\phi)_L = -\frac{\partial \boldsymbol{A}}{\partial t} \tag{496}$$

so the net longitudinal propagating electric field vanishes. Similarly, the longitudinal magnetic field is

$$\boldsymbol{B}_L = -ic\frac{\partial \boldsymbol{A}_L}{\partial t} - \nabla P = \omega^2 \frac{A^{(0)}}{\sqrt{2}}(X - iY)e^{i(\omega t - \kappa Z)} - \nabla P \tag{497}$$

and using

$$\nabla P = \nabla \times \boldsymbol{A} + \frac{\partial \boldsymbol{S}}{\partial t} \tag{498}$$

the longitudinal part of ∇P is

$$(\nabla P)_L = \frac{\partial \boldsymbol{S}}{\partial t} \tag{499}$$

and the longitudinal magnetic field vanishes. These results are consistent with Whittaker's

$$\begin{aligned} E_Z &= \frac{\partial^2 F}{\partial Z^2} - \frac{1}{c^2}\frac{\partial^2 F}{\partial t^2} = 0 \\ B_Z &= \frac{\partial^2 G}{\partial X^2} + \frac{\partial^2 G}{\partial Y^2} = 0 \end{aligned} \tag{500}$$

when F and G correspond to plane waves. The presence of a longitudinal vector potential and longitudinal $\boldsymbol{f}$ and $\boldsymbol{g}$ potentials in Whittaker's theory demonstrate that it is not a U(1) theory of electromagnetism. On the simplest level, Whittaker's theory defines the $\boldsymbol{B}^{(3)}$ field as

$$\boldsymbol{B}^{(3)*} = -i\frac{\kappa}{A^{(0)}}(\nabla \times \boldsymbol{g}) \times (\nabla \times \boldsymbol{g}^*) \tag{501}$$

so $\boldsymbol{g}$ is a physical and measurable quantity, a result that is consistent with Whittaker's own result that G and F can be expanded in terms of plane waves and are structured and physical quantities, and with the fact that Whittaker reduces the U(1) equations in the vacuum to two d'Alembert equations

$$\Box F = \Box G = 0 \tag{502}$$

which are Lorentz- and gauge-invariant. Canonical quantization can therefore proceed through consideration of F and G, giving the photon straightforwardly as demonstrated later in this section. This type of canonical quantization is free of the difficulties associated with canonical quantization [6] in the Coulomb and Lorenz gauges.

In the plane-wave approximation, all electromagnetic effects are derived from the structured time-like potential difference

$$\phi = \dot{F} = i\dot{G} = -\omega \frac{A^{(0)}}{\sqrt{2}}(X - iY)e^{i(\omega t - \kappa Z)} \tag{503}$$

which is thereby a physical observable in effects such as those observed reproducibly and repeatedly by Priore and others [81–85]. These effects have no explanation in the received view, but may be highly beneficial if properly developed. The entities known as electric and magnetic fields are double differentials of ϕ in the plane-wave approximation in the vacuum, a result that is consistent with the ontology developed in Section IX, that the topology of the vacuum is primary, and that potential differences are the result of the vacuum topology. Whittaker uses the usual Lorenz condition, and it is easily verified that

$$\nabla \cdot \boldsymbol{A}_L + \frac{1}{c^2}\frac{\partial \phi_L}{\partial t} = 0 \tag{504}$$

If gauge freedom is lost, however, the Lorenz condition is no longer valid, and a far more comprehensive view of the electromagnetic entity would be obtained by solving the O(3) equations numerically. On the O(3) level, there is no gauge freedom, and no Lorenz condition.

As discussed by Frauendiener and Tsun in Ref. 4, gauge field theory is a form of twistor theory, and as discussed in Section IX, the covariant derivative must always be used in a gauge field theory, even on the U(1) level. The covariant derivative must be used in curved spacetime, and in gauge theory when used with ordinary flat spacetime. These authors also point out that the phase on the U(1) level has no physical significance: it can be redefined by an arbitrary rotation at any point in spacetime. The role of the covariant derivative, or connection, is to compare phases at two neighboring points [4]. This property leads directly to the conclusion that electromagnetism in the vacuum is not a U(1) theory, but a Yang–Mills theory of higher symmetry. We have seen, for example, that the U(1) covariant derivative does not describe the Sagnac effect, whereas O(3) theory describes it accurately because of the nontrivial self interaction [4] resulting in the $\boldsymbol{B}^{(3)}$ field concept. Since O(3) electrodynamics is a Yang–Mills theory, it is also a spinor theory and also a twistor theory [4], which takes us full circle to the fact that Whittaker's theory is a twistor theory.

Potential differences are primary in gauge theory, because they define both the covariant derivative and the field tensor. In Whittaker's theory [27,28], potentials can exist without the presence of fields, but the converse is not true. This conclusion can be demonstrated as follows. Equation (479b) is invariant under

$$\boldsymbol{g} \rightarrow \boldsymbol{g} + \nabla \boldsymbol{a}; \qquad \nabla \times \boldsymbol{g} \rightarrow \nabla \times \boldsymbol{g} + \nabla \boldsymbol{b} \tag{505a}$$

where $\boldsymbol{a}$ and $\boldsymbol{b}$ are arbitrary. This invariance implies that:

$$\nabla \times \boldsymbol{g} \rightarrow \nabla \times \boldsymbol{g} \tag{505b}$$

The transverse part of the vector potential is therefore invariant under the transformations (505), because of the definition

$$\boldsymbol{A}_T = -\nabla \times \boldsymbol{g} \tag{506}$$

and this is a clear sign of the fact that Whittaker's theory contains something contrary to the received view that the transverse $\boldsymbol{A}_T$ is always unphysical. The gauge invariance of $\boldsymbol{A}_T$ does not occur at the U(1) level, but on the O(3) level, the vector potential is gauge covariant and physical, as in the Sagnac effect with rotating platform.

The magnetic fluxes F and G obey the Klein–Gordon equation for a massless particle in the vacuum:

$$\Box F = \Box G = 0 \tag{507}$$

and if we apply Eq. (505), we obtain

$$\Box(\nabla a) = \Box(\nabla c) = 0 \tag{508}$$

indicating that $\boldsymbol{a}$ and $\boldsymbol{c}$ are not arbitrary. Therefore $\boldsymbol{f}$ and $\boldsymbol{g}$ are physical and observable, $\boldsymbol{A}_T$ is physical and observable, and the transverse part of $\boldsymbol{A}^\mu$ is physical. These conclusions refute U(1) electrodynamics.

This result is consistent with Whittaker's main conclusion [27,28], that the scalar potential ϕ is structured and physical in the vacuum, leading to the possibility of interferometry between different scalar potentials, without the presence of fields. To reinforce this conclusion, we can differentiate Eq. (484)

$$\dot{\boldsymbol{A}} = -\nabla \times \dot{\boldsymbol{g}} + \frac{1}{c}\ddot{f} \tag{509}$$

and use the Lorenz condition (also used by Whittaker)

$$\nabla \cdot \boldsymbol{A} + \frac{1}{c^2}\frac{\partial \phi}{\partial t} = 0 \tag{510}$$

to give the following expression for the scalar potential:

$$\dot{\phi} = c^2\nabla\cdot(\nabla\times\boldsymbol{g}) - c\nabla\cdot\dot{\boldsymbol{f}} \tag{511}$$

This results in the following expression for the potential 4-vector

$$\begin{aligned} A^{\mu} &= (\phi, c\boldsymbol{A}) \\ &= \left(c^2\int\nabla\cdot(\nabla\times\boldsymbol{g})\,dt - c\nabla\cdot\boldsymbol{f}, -c\nabla\times\boldsymbol{g} + \dot{\boldsymbol{f}}\right) \\ &= \left(-c^2\int\nabla\cdot\boldsymbol{A}\,dt - c\nabla\cdot\boldsymbol{f}, -c\nabla\times\boldsymbol{g} + \dot{\boldsymbol{f}}\right) \\ &= (\phi - c\nabla\cdot\boldsymbol{f}, c\boldsymbol{A} + \dot{f}) \\ &= (\phi_T, c\boldsymbol{A}_T) + (\phi_L, c\boldsymbol{A}_L) \end{aligned} \tag{512}$$

where it is split into its transverse and longitudinal components in the vacuum. The longitudinal component is

$$A_L^{\mu} = (\phi_L, c\boldsymbol{A}_L) = (-c\nabla\cdot\boldsymbol{f}, \dot{\boldsymbol{f}}) \tag{513}$$

and is physical because $\boldsymbol{f}$ is physical. On canonical quantization, therefore, there exist physical longitudinal photons and time-like photons. By definition

$$A_L^{\mu} = \left(-c\frac{\partial F}{\partial Z}, \frac{\partial F}{\partial t}\boldsymbol{k}\right) \tag{514}$$

and in the special case where the transverse A_T^{μ} consists of plane waves, $F = iG$ and

$$A_L^{\mu} = -\frac{A^{(0)}}{\sqrt{2}}\omega(X - iY)e^{i(\omega t - \kappa Z)}(1, \boldsymbol{k}) \tag{515}$$

The vacuum longitudinal potential is light-like

$$A_{L\mu}A_L^{\mu} = 0 \tag{516}$$

and may be written as

$$A_L^{\mu} = (\phi_L, c\phi_L\boldsymbol{k}) \tag{517}$$

The potential ϕ_L obeys the massless Klein–Gordon equation

$$\Box\phi_L = 0 \tag{518}$$

and it is well known that canonical quantization of this equation is straightforward [6]. This result is consistent with Whittaker's main result that ϕ_L is physical and made up of a sum of plane waves and standing waves in the vacuum [27,28].

The Lagrangian for Eq. (518) is well known [6] to be

$$\mathscr{L} = \frac{1}{2} g^{\kappa\lambda} (\partial_\kappa \phi_L)(\partial_\lambda \phi_L^*) \tag{519}$$

from which is obtained the energy momentum tensor

$$\theta_\nu^\mu = \partial \frac{\mathscr{L}}{\partial(\partial_\mu \phi_L)} \partial_\nu \phi_L^* - \delta_\nu^\mu \mathscr{L} \tag{520}$$

and the Hamiltonian

$$H_L = \int \theta^{00} \, d^3x \tag{521}$$

In SI units, the Hamiltonian is the positive definite

$$H_L = \frac{1}{\mu_0 R^2} \int (\partial_0 \phi_L^* \partial_0 \phi_L + \nabla \phi_L^* \cdot \nabla \phi_L) \, dV \tag{522}$$

where the beam radius is $R^2 = X^2 + Y^2$. Using the relations

$$\begin{aligned}
\phi_L &= -\frac{A^{(0)}}{\sqrt{2}} \omega (X - iY) e^{i(\omega t - \kappa Z)} \\
\partial_0 \phi_L &= -\frac{1}{c} i \omega^2 (X - iY) e^{i(\omega t - \kappa Z)} \frac{A^{(0)}}{\sqrt{2}} \\
\nabla \phi_L^* &= i \frac{A^{(0)}}{\sqrt{2}} \kappa \omega (X + iY) e^{-i(\omega t - \kappa Z)} \\
\nabla \phi_L &= -i \frac{A^{(0)}}{\sqrt{2}} \kappa \omega (X - iY) e^{i(\omega t - \kappa Z)}
\end{aligned} \tag{523}$$

the Hamiltonian reduces to

$$H_L = \frac{1}{\mu_0} \int B^{(0)2} \, dV \tag{524}$$

which is identical with Eq. (450) of Section IX. This result proves that ϕ_L is physical because the result (524) is a physical vacuum electromagnetic energy.

Whittaker theory refutes U(1) theory in several ways, so it may be more appropriate to describe the result (524) as a component at the O(3) level:

$$H_L = \frac{1}{\mu_0}\int \boldsymbol{B}^{(3)} \cdot \boldsymbol{B}^{(3)} dV \tag{525}$$

It may also be argued as follows that $\boldsymbol{f}$ and $\boldsymbol{g}$ are physical. If an attempt is made to apply the usual U(1) gauge transform rule to A_L^μ

$$A_L^\mu \to A_L^\mu + \partial^\mu \chi; \quad A_L \to A_L - \nabla\chi; \quad \phi \to \phi + \frac{\partial\chi}{\partial t} \tag{526}$$

it follows that

$$\dot{\boldsymbol{f}} \to \dot{\boldsymbol{f}} - c\nabla\chi; \qquad \nabla \cdot \boldsymbol{f} \to \nabla \cdot \boldsymbol{f} - \frac{1}{c}\frac{\partial\chi}{\partial t} \tag{527}$$

and

$$\int \nabla\chi \, dt = \frac{1}{c^2}\int \frac{d\chi}{dt} dZ \tag{528}$$

It follows from Eq. (528) that the quantity χ is not random, contrary to the U(1) rule that χ must be random. Euation (528) implies solutions of the type

$$\chi = \chi_0 e^{i(\omega t - \kappa Z)} \tag{529}$$

so that

$$A_L^\mu \to A_L^\mu + A_L^{\mu'} \tag{530}$$

where

$$A_L^{\mu'} = i\omega\chi_0 e^{i\theta}(1, \boldsymbol{k}) = (\phi_L', cA_L') \tag{531}$$

and

$$\Box A_L^{\mu'} = 0; \qquad \phi_L' = 0 \tag{532}$$

The net result is

$$\begin{aligned} \phi_L &\to \phi_L + i\omega\chi_0 e^{i\Phi} \\ \Phi &= \omega t - \kappa Z \end{aligned} \tag{533}$$

If, for example

$$i\chi_0 \equiv -\frac{A^{(0)}}{\sqrt{2}}(X - iY) \tag{534}$$

then

$$\phi_L \rightarrow 2\phi_L; \qquad F \rightarrow 2F; \qquad G \rightarrow 2G \tag{535}$$

and a field such as [27,28]

$$E_X = \frac{\partial^2 F}{\partial X \partial Z} + \frac{1}{c}\frac{\partial^2 G}{\partial Y \partial t} \tag{536}$$

doubles in magnitude. The field is not invariant, contrary to the requirements of U(1) theory. The only possibility is that $\chi = 0$, and that is physical and observable.

Physical potentials are present in Whittaker's theory without fields. This is demonstrated as follows in the special case of a plane wave for the transverse parts of $\boldsymbol{E}$ and $\boldsymbol{B}$. In this special case

$$\boldsymbol{f} = i\boldsymbol{g} \tag{537}$$

and from Eqs. (479a) and (479b)

$$\boldsymbol{E} = ic\nabla \times (\nabla \times \boldsymbol{g}) + a\nabla \times \dot{\boldsymbol{g}} \tag{538a}$$

$$\boldsymbol{B} = \frac{i}{c}\nabla \times \dot{\boldsymbol{g}} - \nabla \times (\nabla \times \boldsymbol{g}) \tag{538b}$$

Under the condition

$$\nabla \times (\nabla \times \boldsymbol{g}) = \frac{i}{c}\frac{\partial}{\partial t}(\nabla \times \boldsymbol{g}) \tag{539}$$

all the components of $\boldsymbol{E}$ and $\boldsymbol{B}$ vanish. The condition (539) is satisfied by

$$\nabla \times \boldsymbol{A}_T = \frac{i}{c}\frac{\partial \boldsymbol{A}_T}{\partial t} \tag{540}$$

whose solution is

$$\boldsymbol{A}_T = \frac{A^{(0)}}{\sqrt{2}}(i\boldsymbol{i} + \boldsymbol{j})e^{-i(\omega t - \kappa Z)} \tag{541}$$

The overall result is

$$\begin{aligned}
\boldsymbol{E} &= \boldsymbol{B} = \boldsymbol{0} \\
\boldsymbol{A}_L &= -\kappa \frac{A^{(0)}}{\sqrt{2}}(X - iY)e^{-i(\omega t - \kappa Z)}\boldsymbol{k} \\
\phi_L &= -\omega \frac{A^{(0)}}{\sqrt{2}}(X - iY)e^{-i(\omega t - \kappa Z)} \\
G &= \frac{F}{i} = \frac{A^{(0)}}{\sqrt{2}}(X - iY)e^{-i(\omega t - \kappa Z)}
\end{aligned} \tag{542}$$

so there can be both transverse and longitudinal physical potentials, or connections. Electromagnetism can be described entirely without fields, and in terms of the vacuum topology.

Whittaker also argued [27,28] that longitudinal standing waves occur in the vacuum. These can be illustrated by the choice of flux

$$G = \frac{A^{(0)}}{\sqrt{2}}(X - iY)(e^{i(\omega t - \kappa Z)} + e^{-i(\omega t - \kappa Z)}) \tag{543}$$

a choice that obeys the d'Alembert equation:

$$\square G = 0 \tag{544}$$

The real part of Eq. (543) is

$$\mathrm{Re}(G) = \frac{2}{\sqrt{2}}A^{(0)}(X \cos \omega t \cos \kappa Z + Y \cos \omega t \sin \kappa Z) \tag{545}$$

which is a standing wave in the vacuum, directed along the propagation axis. Such waves do not exist in the received U(1) theory. The magnetic flux

$$\boldsymbol{g} = \frac{2}{\sqrt{2}}A^{(0)}(X \cos \omega t \cos \kappa Z + Y \cos \omega t \sin \kappa Z)\boldsymbol{k} \tag{546}$$

is a solution to the vibrating-string problem, and the idea that electromagnetism must be described in the vacuum by transverse plane waves of $\boldsymbol{E}$ and $\boldsymbol{B}$ is clearly erroneous. Fluxes of the type (546) give rise to scalar potential interferometry where there are no detectible fields.

It has been shown that the electromagnetic field in Whittaker's view originates in the vacuum, and in the plane wave approximation, in the equation

$$\phi_L = \dot{F} = i\dot{G} = -\omega \frac{A^{(0)}}{\sqrt{2}}(X - iY)e^{i(\omega t - \kappa Z)} \tag{547}$$

under conditions of circular polarization. The scalar potential ϕ_L is time-like, physical, and structured, and it propagates. An experimental design can be used to test experimentally whether $\boldsymbol{f}$ and $\boldsymbol{g}$ are physical. The principle of the design is very simple. Two dipole antennae are set up in close proximity so that the vector potentials from each antenna cancel:

$$\boldsymbol{A}_1 = -i\frac{\kappa e^{i\kappa r}}{4\pi c\varepsilon_0 r}\boldsymbol{p}_1 \tag{548}$$

Here $\boldsymbol{p}_1$ and $\boldsymbol{p}_2$ are the dipole moments of each antenna and r is the magnitude of the radius vector

$$\boldsymbol{A}_2 = i\frac{\kappa e^{i\kappa r}}{4\pi c\varepsilon_0 \boldsymbol{r}}\boldsymbol{p}_2 \tag{549}$$

in spherical coordinates [86]. It follows that

$$E = \mathbf{0}; \qquad \boldsymbol{B} = \mathbf{0}; \qquad \boldsymbol{A} = \boldsymbol{A}_1 + A_2 = \mathbf{0} \tag{550}$$

so there are no vector potentials or fields radiated into the vacuum by this antenna arrangement. Whittaker's $\boldsymbol{f}$ and $\boldsymbol{g}$ magnetic flux vectors are defined as follows by this arrangement:

$$\boldsymbol{g}_1 = -\boldsymbol{g}_2; \qquad \boldsymbol{f}_1 = -\boldsymbol{f}_2 \tag{551}$$

However, the scalar magnitudes of $\boldsymbol{g}$ and $\boldsymbol{f}$ from both antennas (G and F) are the same, because the scalar magnitude of a vector is the square root of the vector squared. Thus the following quantity is radiated into the vacuum:

$$2G = \frac{2}{\sqrt{2}}A^{(0)}(X - iY)e^{i(\omega t - \kappa Z)} \tag{552}$$

and the scalar potential

$$\phi_L \equiv 2\dot{G} \tag{553}$$

is also present in the vacuum. On canonical quantization, this scalar potential gives an ensemble of massless photons from the Klein–Gordon equation. This property will be proved later in this section. These are physical time-like photons each with the Planck energy $\hbar\omega$. The energy from these photons is therefore Eq. (524), and is phase-free. For a large number of frequencies, the photons are distributed according to the Planck distribution for blackbody radiation [69], which is radiated heat detectible by a bolometer. There are no vector potentials or fields present, so the heat is due entirely to the physical F and G. In the received view, such photons are unphysical and no heat should be detected. An improvement on this design, due to Labounsky [87], is shown in Fig. 2, which illustrates how fieldless G waves can be generated.

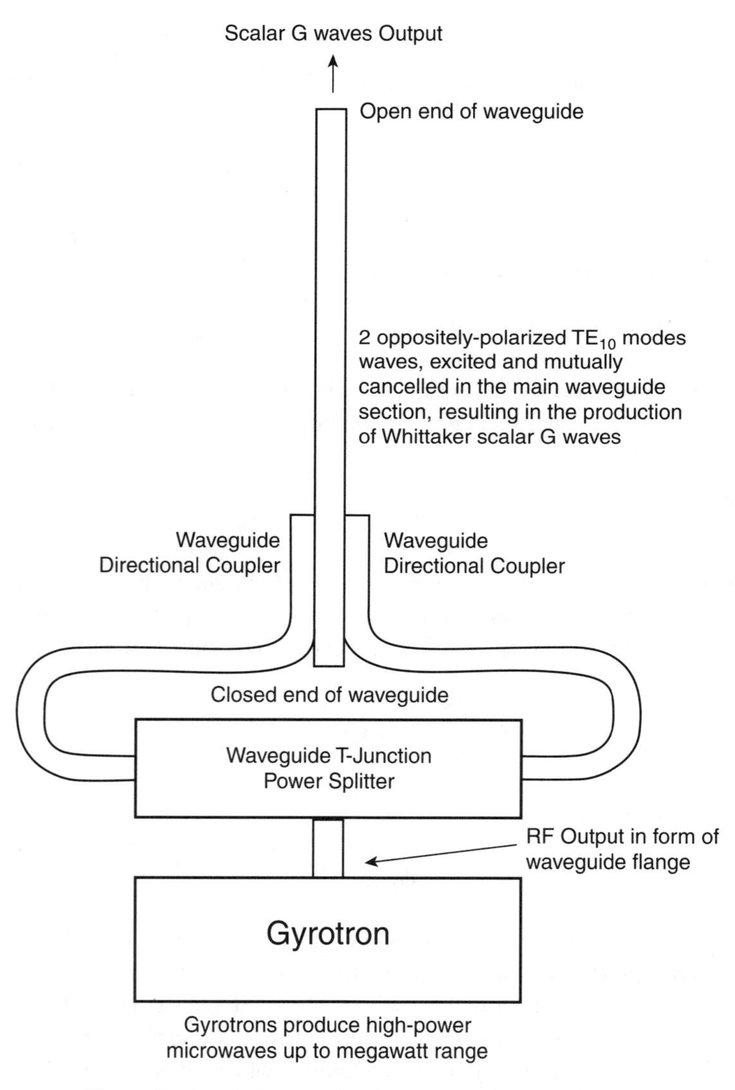

Figure 2. Practical conception for a source of scalar G waves.

Scalar interferometry is possible in this view if F and G are physical in the vacuum. When two scalar beams of the type

$$G_1 = \frac{A^{(0)}}{\sqrt{2}}(X - iY)e^{i(\omega t - \kappa Z_1)}$$
$$G_2 = \frac{A^{(0)}}{\sqrt{2}}(X - iY)e^{i(\omega t - \kappa Z_2)} \tag{554}$$

interfere, an interferogram is generated, as usual, and their combined energy density in the zone of interference is

$$\frac{En}{V} = \frac{cI}{R^2\omega^2}(1 + \cos(\kappa(Z_1 - Z_2))) \tag{555}$$

where I is the combined power density of the two beams in watts per square meter. Here, $Z_1 - Z_2$ is the path difference as usual, that is, the difference in distance traversed by each beam from source (the design in Fig. 2) to interference zone. If we now define

$$G_3 \equiv \frac{1}{G^{(0)}}(G_1 + G_2)(G_1^* + G_2^*) \tag{556}$$

then

$$\Box G_3 = B \neq 0 \tag{557}$$

and a fluctuating magnetic flux density B appears in the zone of interference even though no field is radiated by either source. The presence of a magnetic field indicates the presence of an electric field. There are magnetic and electric fields in the zone of interference but none outside. Equation (557) is a gauge-invariant construct, and the $\boldsymbol{E}$ and $\boldsymbol{B}$ fields in the zone of interference are real and physical, and so interact with matter in the zone of interference. The energy density within this zone is also gauge-invariant and physical

$$\frac{En}{V} = \frac{B^{(0)2}}{\mu_0} = \frac{GG^*}{R^4\mu_0} \tag{558}$$

where R^2 is the beam area, assumed to be the same for each beam. The lateral extent of the radiated beams from the device in Fig. 2 is constrained by the inverse fourth-power dependence on R.

It has been proved that F and G of Whittaker are physical and gauge-invariant, and it follows, as shown next, that there exist physical time-like and longitudinal photons. These have an independent existence and appear from canonical quantization of the classical, physical, and time-like scalar potential difference in vacuo [Eq. (547)]. Canonical quantization follows straightforwardly from the massless Klein–Gordon equation:

$$\Box\phi_L = 0 \tag{559}$$

The potential ϕ_L is treated as usual [6] as an operator subject to the commutator relation of quantum mechanics. This procedure gives the positive definite Hamiltonian (521) and vacuum energy (524) self-consistently. The scalar potential ϕ_L is Fourier expanded as

$$\phi_L = \int \frac{\partial^3\kappa}{(2\pi)^3 2\omega_\kappa}(a(\kappa)e^{-i\omega Z} + a^+(\kappa)e^{i\omega Z}) \tag{560}$$

a procedure that is self-consistent with Whittaker's original demonstration [27,28] that ϕ_L can be expanded in a Fourier series in the argument denoted by Whittaker in his general solution for ϕ_L. Equation (560) has frequencies $\omega_\kappa = \kappa c$ generated by the Fourier expansion. So many different photons emerge, each corresponding to a different frequency; quantization results in an ensemble [6] of physical time-like photons, each of energy $\hbar\omega$. This is consistent with the Planck quantization of energy momentum

$$p^\mu = \hbar\kappa^\mu \tag{561}$$

where the time-like component has energy $\hbar\omega$.

The coefficients a and a^+ in the expansion (560) are operators defined by the commutators [6]:

$$\begin{aligned} [a(\kappa), a(\kappa')] &= [a^+(\kappa), a^+(\kappa)] = 0 \\ [a(\kappa), a^+(\kappa')] &= (2\pi)^3 2\omega_\kappa \delta^3(\kappa - \kappa') \end{aligned} \tag{562}$$

The operator:

$$N(\kappa) = a^+(\kappa)a(\kappa) \tag{563}$$

represents the number of particles with energy $\hbar\omega$ and longitudinal momentum $\hbar\kappa$. The Hamiltonian, after quantization, takes the form

$$H = \frac{1}{2}P^2(\kappa) + \frac{\omega_\kappa^2}{2}Q^2(\kappa) \tag{564}$$

where

$$P(\kappa) = \left(\frac{\omega_\kappa}{2}\right)^{1/2}(a(\kappa) + a^+(\kappa))$$
$$Q(\kappa) = \frac{1}{(2\omega_\kappa)^{1/2}}(a(\kappa) - a^+(\kappa)) \tag{565}$$

and ϕ_L, after quantization, is an infinite sum of oscillators, that is, an ensemble of time-like photons with energy $\hbar\omega$. The operators a and a^+ respectively are therefore the annihilation and creation operators for the quanta of ϕ_L and the energy of the quantized ϕ_L is rigorously positive. The photons obtained after this type of quantization obey Bose–Einstein statistics [6], and any number of particles (photons) can exist in the same quantization state. These photons are spin zero and massless and, because they are spin zero, are not absorbed by an atom or molecule, in contrast to physical space-like photons carrying angular momentum. The received view of canonical quantization asserts [6] that these photons are unphysical. Paradoxically, the received view also asserts that the vector (561) is physical. This paradox is seen in the Compton and photoelectric effects as argued already in Section III. There are insurmountable difficulties [6] in the received methods of canonical quantization. In the radiation gauge, for example, the scalar and longitudinal parts of the 4-vector A^μ are missing, so A^μ is not fully covariant at the outset. In the Lorenz gauge, there are several difficulties well summarized by Ryder [6]. For example, there is an indefinite metric where the Lorenz condition has to be used and then discarded, then a gauge fixing term has to be used, and the final result is paradoxical in that an admixture of time-like and longitudinal photons are physical [6], but each component is unphysical. The procedure of canonical quantization in the Lorenz gauge gives photons with spin, and these are asserted to be physical transverse photons.

It is far simpler to introduce spin into the assumed massless photon by following the little group method of Wigner [6], that is, by examining the most general type of Lorentz transform possible for a particle without mass. This produces the normalized helicities -1 and 1 through parity considerations. These correspond, in the received view, to physical right and left circularly polarized photons. If the photon is massive, as implied by O(3) electrodynamics, there occurs in addition the helicity zero, corresponding to a physical longitudinal space-like photon without spin and corresponding to a physical O(3) symmetry little group. [The little group of the massless photon is the unphysical [6,15] E(2), another paradox of the received view.] As argued already, there also occurs a time-like photon that is a scalar and that is purely energetic in nature.

These various considerations point toward the O(3) definition of the energy-momentum 4-vector:

$$\boldsymbol{p}^\mu = p^{\mu(1)}\boldsymbol{e}^{(1)} + p^{\mu(2)}\boldsymbol{e}^{(2)} + p^{\mu(3)}\boldsymbol{e}^{(3)} \tag{566}$$

There are therefore three energy-momentum 4-vectors present:

$$p^{\mu(3)} = (En, c\boldsymbol{p}^{(3)}); \qquad p^{\mu(2)} = (En, c\boldsymbol{p}^{(2)}); \qquad p^{\mu(1)} = (En, c\boldsymbol{p}^{(1)}) \tag{567}$$

Energy is a scalar and so does not carry an internal gauge index. There are three momenta; $\boldsymbol{p}^{(3)}$ is longitudinal, and $\boldsymbol{p}^{(1)}$ and $\boldsymbol{p}^{(2)}$ are circularly polarized conjugates. Applying Planck quantization gives immediately a time-like photon $\hbar\omega$ without spin, a longitudinal photon $\hbar\boldsymbol{k}^{(3)}$ without spin and with energy $\hbar\omega$, and right and left circularly polarized photons $\hbar\boldsymbol{k}^{(1),(2)}$, each of energy $\hbar\omega$.

Therefore Whittaker's theory points toward the existence of O(3) electrodynamics. This conclusion is reinforced by the fact that Eqs. (479a) and (479b) are invariant under the duality transform:

$$\begin{aligned} \boldsymbol{f} &\to -\boldsymbol{g} \\ \boldsymbol{g} &\to \boldsymbol{f} \end{aligned} \tag{568}$$

and Eqs. (484) and (485) can be written as

$$\nabla \times \boldsymbol{f} + \frac{1}{c}\frac{\partial \boldsymbol{g}}{\partial t} = -\frac{1}{c}\boldsymbol{S} \tag{569}$$

$$\nabla \times \boldsymbol{g} - \frac{1}{c}\frac{\partial \boldsymbol{f}}{\partial t} = -\boldsymbol{A} \tag{570}$$

These equations are invariant under the transform:

$$\boldsymbol{f} \to -\boldsymbol{g}; \qquad \boldsymbol{g} \to \boldsymbol{f}; \qquad \boldsymbol{S} \to -c\boldsymbol{A} \tag{571}$$

Special relativity then dictates that there exists the set of equations

$$\begin{aligned} \nabla \cdot \boldsymbol{g} &= \frac{\phi}{c} \\ \nabla \times \boldsymbol{g} &= \frac{1}{c}\frac{\partial \boldsymbol{f}}{\partial t} - \boldsymbol{A} \\ \nabla \cdot \boldsymbol{f} &= -P \\ \nabla \times \boldsymbol{f} + \frac{1}{c}\frac{\partial \boldsymbol{g}}{\partial t} &= -\frac{1}{c}\boldsymbol{S} \end{aligned} \tag{572}$$

which can be written as

$$\begin{aligned} \partial_\mu \tilde{g}^{\mu\nu} &= A^\nu \\ \partial_\mu g^{\mu\nu} &= -S^\nu \end{aligned} \tag{573}$$

where

$$\tilde{g}^{\mu\nu} = \begin{bmatrix} 0 & 0 & 0 & -g^3 \\ 0 & 0 & f^3 & 0 \\ 0 & -f^3 & 0 & 0 \\ g^3 & 0 & 0 & 0 \end{bmatrix} \tag{574}$$

$$g^{\mu\nu} = \begin{bmatrix} 0 & 0 & 0 & -f^3 \\ 0 & 0 & -g^3 & 0 \\ 0 & g^3 & 0 & 0 \\ f^3 & 0 & 0 & 0 \end{bmatrix} \tag{575}$$

Equations (573) have overall O(3) symmetry, and have the same structure as the Maxwell–Heaviside equations with magnetic charge and current [3,4]. From Eqs. (573), we obtain the wave equation

$$\Box\tilde{g}^{\mu\nu} = \frac{1}{2}F^{\mu\nu} = 0 \tag{576}$$

which is consistent with Whittaker's starting point:

$$\Box G = \Box F = 0 \tag{577}$$

The received view asserts that A^{μ} is always random, but in this section, several counter arguments have been given. Several more counterarguments appear throughout this chapter and elsewhere in the literature [3].

XI. PREPARING FOR COMPUTATION

In this section, the field equations (31) and (32) are considered in free space and reduced to a form suitable for computation to give the most general solutions for the vector potentials in the vacuum in O(3) electrodynamics. This procedure shows that Eqs. (86) and (87) are true in general, and are not just particular solutions. On the O(3) level, therefore, there exist no topological monopoles or magnetic charges. This is consistent with empirical data—no magnetic monopoles of any kind have been observed in nature.

If consideration is restricted to the vacuum, the field equations (86) and (90) apply. The Jacobi identity (86) is first considered and written in the following form [6]:

$$D_{\lambda}\boldsymbol{G}_{\mu\nu} + D_{\mu}\boldsymbol{G}_{\nu\lambda} + D_{\nu}\boldsymbol{G}_{\lambda\mu} \equiv \mathbf{0} \tag{578}$$

This reduces in general to the form

$$\partial_\lambda \boldsymbol{G}_{\mu\nu} + \partial_\mu \boldsymbol{G}_{\nu\lambda} + \partial_\nu \boldsymbol{G}_{\lambda\mu} \equiv \boldsymbol{0} \tag{579}$$

because

$$\boldsymbol{A}_\mu \times \tilde{\boldsymbol{G}}^{\mu\nu} \equiv \boldsymbol{0} \tag{580}$$

is identically zero. The proof of this latter result proceeds by using the definitions

$$\begin{aligned}
\boldsymbol{G}_{\mu\nu}^{(1)*} &= \partial_\mu \boldsymbol{A}_\nu^{(1)*} - \partial_\nu \boldsymbol{A}_\mu^{(1)*} - ig\boldsymbol{A}_\mu^{(2)} \times \boldsymbol{A}_\nu^{(3)} \\
\boldsymbol{G}_{\mu\nu}^{(2)*} &= \partial_\mu \boldsymbol{A}_\nu^{(2)*} - \partial_\nu \boldsymbol{A}_\mu^{(2)*} - ig\boldsymbol{A}_\mu^{(3)} \times \boldsymbol{A}_\nu^{(1)} \\
\boldsymbol{G}_{\mu\nu}^{(3)*} &= \partial_\mu \boldsymbol{A}_\nu^{(3)*} - \partial_\nu \boldsymbol{A}_\mu^{(3)*} - ig\boldsymbol{A}_\mu^{(1)} \times \boldsymbol{A}_\nu^{(2)}
\end{aligned} \tag{581}$$

and Jacobi identities such as:

$$\boldsymbol{A}_\lambda^{(2)} \times (\boldsymbol{A}_\mu^{(1)} \times \boldsymbol{A}_\nu^{(2)}) + \boldsymbol{A}_\mu^{(2)} \times (\boldsymbol{A}_\nu^{(1)} \times \boldsymbol{A}_\lambda^{(2)}) + \boldsymbol{A}_\nu^{(2)} \times (\boldsymbol{A}_\lambda^{(1)} \times \boldsymbol{A}_\mu^{(2)}) \equiv \boldsymbol{0} \tag{582}$$

The terms

$$\begin{aligned}
&\boldsymbol{A}_\lambda^{(1)} \times (\partial_\mu \boldsymbol{A}_\nu^{(2)} - \partial_\nu \boldsymbol{A}_\mu^{(2)}) \equiv \boldsymbol{0} \\
&\boldsymbol{A}_\lambda^{(2)} \times (\partial_\mu \boldsymbol{A}_\nu^{(3)} - \partial_\nu \boldsymbol{A}_\mu^{(3)}) \equiv \boldsymbol{0} \\
&\boldsymbol{A}_\lambda^{(3)} \times (\partial_\mu \boldsymbol{A}_\nu^{(1)} - \partial_\nu \boldsymbol{A}_\mu^{(1)}) \equiv \boldsymbol{0} \\
&\qquad \cdots
\end{aligned} \tag{583}$$

vanish individually as follows:

$$\begin{aligned}
\boldsymbol{A}_\lambda^{(1)} \times (\partial_\mu \boldsymbol{A}_\nu^{(2)} - \partial_\nu \boldsymbol{A}_\mu^{(2)}) &= \varepsilon_{(1)(2)(3)} A_\lambda^{(1)} \partial_\mu A_\nu^{(2)} - \varepsilon_{(1)(2)(3)} A_\lambda^{(1)} \partial_\nu A_\mu^{(2)} \\
&= A_\lambda^{(1)} F_{\mu\nu}^{(2)} - A_\lambda^{(2)} F_{\mu\nu}^{(1)} \\
&= A_X B_X (\boldsymbol{e}^{(2)} \cdot \boldsymbol{e}^{(1)} - \boldsymbol{e}^{(1)} \boldsymbol{e}^{(2)}) \equiv 0 \\
&\cdots
\end{aligned} \tag{584}$$

Equation (584) implies that the topological magnetic charge–current

$$\boldsymbol{J}_m^\nu \propto \boldsymbol{A}_\mu \times \tilde{\boldsymbol{G}}^{\mu\nu} \equiv \boldsymbol{0} \tag{585}$$

vanishes in the vacuum, while $\boldsymbol{B}^{(3)}$ is nonzero in the vacuum, a result that is consistent with empirical data, which show the existence of $\boldsymbol{B}^{(3)}$ and the nonexistence of a magnetic monopole and magnetic current.

The computational problem reduces therefore to a numerical solution of three differential equations:

$$\partial_\lambda \boldsymbol{G}^{(1)}_{\mu\nu} + \partial_\mu \boldsymbol{G}^{(1)}_{\nu\lambda} + \partial_\nu \boldsymbol{G}^{(1)}_{\lambda\mu} \equiv \mathbf{0} \tag{586}$$

$$\partial_\lambda \boldsymbol{G}^{(2)}_{\mu\nu} + \partial_\mu \boldsymbol{G}^{(2)}_{\nu\lambda} + \partial_\nu \boldsymbol{G}^{(2)}_{\lambda\mu} \equiv \mathbf{0} \tag{587}$$

$$\partial_\lambda \boldsymbol{G}^{(3)}_{\mu\nu} + \partial_\mu \boldsymbol{G}^{(3)}_{\nu\lambda} + \partial_\nu \boldsymbol{G}^{(3)}_{\lambda\mu} \equiv \mathbf{0} \tag{588}$$

using the definitions (581). There are three equations in three unknowns, so the problem can be solved for given boundary conditions.

The work of Whittaker described in the previous section can be summarized by the potential

$$A^{(3)}_\mu \equiv (A^{(3)}_0, c\boldsymbol{A}^{(3)}) \tag{589}$$

where the magnitude of $\boldsymbol{A}^{(3)}$ is $A^{(3)}_0/c$. The O(3) theory allows $A^{(3)}_0$ and $\boldsymbol{A}^{(3)}$ to be structured, constant or zero. The $\boldsymbol{B}^{(3)}$ field exists in all three cases. If, however, $\boldsymbol{A}^{(3)}$ is zero, so is $A^{(3)}_0$ and there is no scalar potential. The conclusion reached is that there can be an infinite number of components of the 4-vector $A^{(3)}_\mu$ for a given phaseless $\boldsymbol{B}^{(3)}$. In other words, the scalar potential can be expanded in a Fourier series, or some other suitable series that includes the terms $\boldsymbol{A}^{(3)} = 0$ and $\boldsymbol{A}^{(3)} =$ constant.

The disappearance of the magnetic charge–current (585) means that the topological terms on the right-hand sides of Eqs. (95)–(100) vanish identically in the vacuum. The only topological charges and currents present are therefore those introduced by Lehnert [7–10]. There is no empirical evidence for the existence of an $\boldsymbol{E}^{(3)}$ field, so we are left with

$$\nabla \times \boldsymbol{E}^{(1)} + \frac{\partial \boldsymbol{B}^{(1)}}{\partial t} \equiv \mathbf{0} \tag{590}$$

$$\nabla \times \boldsymbol{E}^{(2)} + \frac{\partial \boldsymbol{B}^{(2)}}{\partial t} \equiv \mathbf{0} \tag{591}$$

$$\frac{\partial \boldsymbol{B}^{(3)}}{\partial t} \equiv \mathbf{0} \tag{592}$$

as first proposed some time ago [11–20]. Equation (592) has been verified empirically by Raja et al. [88] and Compton et al. [89]. The most general type of solution must be found, however, by solving Eqs. (586)–(588) numerically, so that potentials are primary and fields are derived from potentials. The mathematical structure of O(3) Yang–Mills theory applied to electrodynamics allows for $\boldsymbol{A}^{(3)} = 0$ as one of many possible solutions. However, if $\boldsymbol{A}^{(3)} = 0$, then the scalar potential is also zero, while the $\boldsymbol{B}^{(3)}$ field remains nonzero.

The vanishing of the topological magnetic current in Eqs. (98)–(100) leads to two components of the B cyclic theorem as follows. In Eq. (98)

$$A_0^{(2)} = 0; \qquad \boldsymbol{E}^{(3)} = \boldsymbol{0} \tag{593}$$

and so

$$-cA_0^{(3)}\boldsymbol{B}^{(1)} = \boldsymbol{E}^{(1)} \times \boldsymbol{A}^{(3)} \tag{594}$$

$$\boldsymbol{B}^{(1)} = \frac{1}{c}\boldsymbol{k} \times \boldsymbol{E}^{(1)} \tag{595}$$

for any $A_0^{(3)} = cA^{(3)}$. This result is self-consistent with the left-hand side of Eq. (98), because Eq. (595) is a solution of Eq. (596):

$$\nabla \times \boldsymbol{E}^{(1)} + \frac{\partial \boldsymbol{B}^{(1)}}{\partial t} \equiv \boldsymbol{0} \tag{596}$$

The B cyclic component emerges as follows:

$$\begin{aligned} \boldsymbol{B}^{(1)} \times \boldsymbol{B}^{(2)} = \frac{1}{c}(\boldsymbol{k} \times \boldsymbol{E}^{(1)}) \times \boldsymbol{B}^{(2)} = iB^{(0)}\boldsymbol{B}^{(3)*} \\ \cdots \end{aligned} \tag{597}$$

Therefore all is self-consistent.

These calculations show that $\boldsymbol{B}^{(3)}$ is not dependent on the existence of a vacuum magnetic monopole [11–20]. Therefore the explanation of phenomena based on $\boldsymbol{B}^{(3)}$ is not dependent on a topological magnetic charge or monopole. The fundamental reason for this is that $\boldsymbol{B}^{(3)}$ is defined in terms of quantities that are not dependent on a magnetic monopole, namely, g, $\boldsymbol{A}^{(1)}$, and $\boldsymbol{A}^{(2)}$. Furthermore, the structure of O(3) Yang–Mills theory forces us to conclude that $\boldsymbol{E}^{(3)}$ is zero through the structure of Eqs. (98)–(100) [11–20]. The existence of a phaseless $\boldsymbol{E}^{(3)}$ has never been observed empirically. Action at a distance in electrodynamics is obviously denied by the fact that we are working with a gauge theory, and there is no convincing evidence for superluminal phenomena in electrodynamics. It should also be clear that $\boldsymbol{B}^{(3)}$ is not a static magnetic field; rather, it is a radiated field, propagating with the third Stokes parameter.

The three equations (586)–(588) can be written in condensed form

$$\partial_\mu \tilde{\boldsymbol{G}}^{\mu\nu(i)} = \boldsymbol{0}; \qquad i = 1, 2, 3 \tag{598}$$

which is self-dual to another set of three simultaneous equations suitable for computation and derivable from Eq. (90):

$$D_\mu \boldsymbol{H}^{\mu\nu(i)} = \mathbf{0}; \qquad i = 1, 2, 3 \tag{599}$$

where $\boldsymbol{G}^{\mu\nu}$ of Eq. (90) has been replaced by $\boldsymbol{H}^{\mu\nu}$ for greater clarity and to indicate the presence of vacuum polarization Therefore

$$\partial_\mu \tilde{\boldsymbol{G}}^{\mu\nu(i)} = D_\mu \boldsymbol{H}^{\mu\nu(i)} = \mathbf{0} \tag{600}$$

represents the O(3) wave equation, which has a much richer structure than its U(1) counterpart, and many more solutions. The charge current 12-vector in vacuo, Eq. (91), is nonzero. This can be demonstrated by writing it out in component form:

$$\begin{aligned} \boldsymbol{J}^{\nu(1)*} &= -ig\boldsymbol{A}_\mu^{(2)} \times \boldsymbol{H}^{\mu\nu(3)} \\ \boldsymbol{J}^{\nu(2)*} &= -ig\boldsymbol{A}_\mu^{(3)} \times \boldsymbol{H}^{\mu\nu(1)} \\ \boldsymbol{J}^{\nu(3)*} &= -ig\boldsymbol{A}_\mu^{(1)} \times \boldsymbol{H}^{\mu\nu(2)} \end{aligned} \tag{601}$$

Terms such as

$$\boldsymbol{J}^{\nu(2)} = -ig\boldsymbol{A}_\mu^{(2)} \times (\partial^\mu \boldsymbol{A}^{\nu(3)} - \partial^\nu \boldsymbol{A}^{\mu(3)} - ig\boldsymbol{A}^{\mu(1)} \times \boldsymbol{A}^{\nu(2)}) \tag{602}$$

are obtained. The first part can be expanded as

$$\begin{aligned} \boldsymbol{A}_\mu^{(2)} \times \partial^\mu \boldsymbol{A}^{\nu(3)} - \boldsymbol{A}_\nu^{(2)} \times \partial^\nu \boldsymbol{A}^{\mu(3)} &= \varepsilon_{(2)(3)(1)} A_\mu^{(2)} \partial^\mu A^{\nu(3)} - \varepsilon_{(2)(3)(1)} A_\mu^{(2)} \partial^\nu A^{\mu(3)} \\ &= -A_\mu^{(2)} F^{\mu\nu(3)} - A_\mu^{(3)} F^{\mu\nu(2)} \end{aligned} \tag{603}$$

which is nonzero in general. The second part can be expanded as

$$\begin{aligned} \boldsymbol{A}_\mu^{(2)} \times (\boldsymbol{A}^{\mu(1)} \times \boldsymbol{A}^{\nu(2)}) &= \boldsymbol{A}^{\mu(1)} (\boldsymbol{A}_\mu^{(2)} \cdot \boldsymbol{A}^{\nu(2)}) - \boldsymbol{A}^{\nu(2)} (\boldsymbol{A}_\mu^{(2)} \cdot \boldsymbol{A}^{\mu(1)}) \\ &= -\boldsymbol{A}^{\nu(2)} (\boldsymbol{A}_\mu^{(2)} \cdot \boldsymbol{A}^{\mu(1)}) \end{aligned} \tag{604}$$

which is also nonzero in general.

Therefore we reach the important overall conclusion that the structure of the O(3) equations is a development into O(3) symmetry of the Lehnert field equations [7–10], which are written in U(1) form. The Lehnert field equations have been extensively developed and tested empirically and theoretically [7–10].

The O(3) Coulomb and Ampère–Maxwell laws in the vacuum are therefore written in terms of displacement and magnetic field strength, and are as follows. The Coulomb Law in the vacuum is

$$\begin{aligned}\nabla \cdot \boldsymbol{D}^{(1)*} &= ig(\boldsymbol{D}^{(2)} \cdot \boldsymbol{D}^{(3)} - \boldsymbol{D}^{(2)} \cdot \boldsymbol{A}^{(3)}) \\ \nabla \cdot \boldsymbol{D}^{(2)*} &= ig(\boldsymbol{A}^{(3)} \cdot \boldsymbol{D}^{(1)} - \boldsymbol{D}^{(3)} \cdot \boldsymbol{A}^{(1)}) \\ \nabla \cdot \boldsymbol{D}^{(3)*} &= ig(\boldsymbol{A}^{(1)} \cdot \boldsymbol{D}^{(2)} - \boldsymbol{D}^{(1)} \cdot \boldsymbol{A}^{(2)})\end{aligned} \tag{605}$$

and the Ampère–Maxwell law in the vacuum is

$$\begin{aligned}\nabla \times \boldsymbol{H}^{(1)*} - \frac{\partial \boldsymbol{D}^{(1)*}}{\partial t} &= -ig(cA_0^{(2)}\boldsymbol{D}^{(3)} - cA_0^{(3)}\boldsymbol{D}^{(2)} + \boldsymbol{A}^{(2)} \times \boldsymbol{H}^{(3)} - \boldsymbol{A}^{(3)} \times \boldsymbol{H}^{(2)}) \\ \nabla \times \boldsymbol{H}^{(2)*} - \frac{\partial \boldsymbol{D}^{(2)*}}{\partial t} &= -ig(cA_0^{(3)}\boldsymbol{D}^{(1)} - cA_0^{(1)}\boldsymbol{D}^{(3)} + \boldsymbol{A}^{(3)} \times \boldsymbol{H}^{(1)} - \boldsymbol{A}^{(1)} \times H^{(3)}) \\ \nabla \times \boldsymbol{H}^{(3)*} - \frac{\partial \boldsymbol{D}^{(3)*}}{\partial t} &= -ig(cA_0^{(1)}\boldsymbol{D}^{(2)} - cA_0^{(2)}\boldsymbol{D}^{(1)} + \boldsymbol{A}^{(1)} \times \boldsymbol{H}^{(2)} - \boldsymbol{A}^{(2)} \times H^{(1)})\end{aligned} \tag{606}$$

The displacement $\boldsymbol{D}^{(3)}$ for example can be developed as

$$\boldsymbol{D}^{(3)} = \varepsilon_0 \boldsymbol{E}^{(3)} + \boldsymbol{P}^{(3)} \tag{607}$$

and since $\boldsymbol{E}^{(3)}$ is zero, we obtain

$$\boldsymbol{D}^{(3)} = \boldsymbol{P}^{(3)}$$

indicating the presence of classical vacuum polarization $\boldsymbol{P}^{(3)}$ due to the topology of the vacuum as represented by a gauge field theory with an assumed O(3) gauge group symmetry. Therefore the energy inherent in the vacuum is obtained entirely from the electric charge current (91), as discussed in Section IV. The magnetic charge-current (585) vanishes, and so there is no energy inherent in the vacuum from the magnetic charge–current for an internal O(3) gauge group symmetry. On the O(3) level, there can therefore be classical vacuum polarization, whose analog in quantum electrodynamics is the photon self-energy [6].

The constitutive equations in the vacuum in O(3) electrodynamics are not the same as those of U(1) electrodynamics, and in general

$$\boldsymbol{D}^{(i)} = \varepsilon_0 \boldsymbol{E}^{(i)} + \boldsymbol{P}^{(i)}; \qquad i = 1, 2, 3 \tag{609}$$

where $\boldsymbol{P}^{(i)}$ are vacuum polarizations.

To summarize, there are three equations [Eqs. (586)–(588)] in three unknowns (indices of the vector potential appropriate to $\tilde{G}^{\mu\nu}$) and another three equations [Eq. (599)] in three unknowns (indices of the vector potential appropriate to $\boldsymbol{H}^{\mu\nu}$ in the vacuum). Simple vacuum constitutive relations such as

$$\boldsymbol{D} = \varepsilon_0 \boldsymbol{E}; \qquad \boldsymbol{H} = \frac{1}{\mu_0}\boldsymbol{B} \tag{610}$$

of U(1) electrodynamics no longer apply, because of the existence of classical vacuum polarization. The latter also occurs in the Lehnert equations [7–10], which are known to give axisymmetric solutions similar to $\boldsymbol{B}^{(3)}$, to indicate photon mass, and to be superior in ability to the Maxwell–Heaviside equations.

To put the O(3) equations into the form of the Lehnert equations, we use the definitions

$$\begin{aligned}
\boldsymbol{D} &= \boldsymbol{D}^{(1)} + \boldsymbol{D}^{(2)} + \boldsymbol{D}^{(3)} \\
\boldsymbol{H} &= \boldsymbol{H}^{(1)} + \boldsymbol{H}^{(2)} + \boldsymbol{H}^{(3)} \\
\boldsymbol{E} &= \boldsymbol{E}^{(1)} + \boldsymbol{E}^{(2)} + \boldsymbol{E}^{(3)} \\
\boldsymbol{B} &= \boldsymbol{B}^{(1)} + \boldsymbol{B}^{(2)} + \boldsymbol{B}^{(3)} \\
&\cdots
\end{aligned} \tag{611}$$

to obtain

$$\begin{aligned}
\nabla \cdot \boldsymbol{B} &= 0 \\
\nabla \times \boldsymbol{E} + \frac{\partial \boldsymbol{B}}{\partial t} &= \boldsymbol{0} \\
\nabla \cdot \boldsymbol{B} &= \rho_{\text{vac}} \\
\nabla \times \boldsymbol{H} - \frac{\partial \boldsymbol{D}}{\partial t} &= \boldsymbol{J}_{\text{vac}}
\end{aligned} \tag{612}$$

which are mathematically identical to the Lehnert equations. The O(3) gauge theory, however, shows that the origin of the vacuum charge and current postulated phenomenologically by Lehnert [7–10] is the topology of the vacuum described by an O(3) gauge group. The O(3) theory also shows, self-consistently, that there is a vacuum polarization, so that the simple constitutive relations (610) used by Lehnert do not hold. The O(3) gauge theory also reveals that the presence of the $\boldsymbol{B}^{(3)*}$ component, through its definition, is proportional to the conjugate product of potentials, $\boldsymbol{A}^{(1)} \times \boldsymbol{A}^{(2)}$. However, the mathematical form of the O(3) equations (612) is identical with that of the Lehnert equations.

Formally, the O(3) equations are written most generally as

$$
\begin{aligned}
\nabla \cdot \boldsymbol{H} &= \rho_{m,\text{vac}} \\
\nabla \times \boldsymbol{D} + \frac{\partial \boldsymbol{H}}{\partial t} &= \boldsymbol{J}_{m,\text{vac}} \\
\nabla \cdot \boldsymbol{D} &= \rho_{\text{vac}} \\
\nabla \times \boldsymbol{H} - \frac{\partial \boldsymbol{D}}{\partial t} &= \boldsymbol{J}_{\text{vac}}
\end{aligned}
\tag{613}
$$

which are identical in mathematical structure with the Harmuth equations [21,22] and Barrett equations [3,4]. However, in O(3) electrodynamics, there is no magnetic monopole or magnetic current as argued already. The structure (612) in the vacuum is identical with the structure of the Maxwell–Heaviside equations as used for field-matter interaction.

The complete computational problem in the vacuum is therefore as follows:

1. Use eqs. (586) to (588) to obtain $A_\mu^{(1)}, A_\mu^{(2)}, A_\mu^{(3)}; \mu = 0, \ldots, 3$ with the simplifying definitions

$$
\begin{aligned}
&A_0^{(i)} = A_3^{(i)} = 0; \qquad i = 1,2 \\
&A_\mu^{(3)} = (A_0^{(3)}, c\boldsymbol{A}^{(3)}) = (A_0^{(3)}, cA_3^{(3)}\boldsymbol{k}) \\
&A_0^{(1)} = A_0^{(2)} = 0; \\
&A_1^{(3)} = A_2^{(3)} = 0
\end{aligned}
\tag{614}
$$

2. Use Eqs. (101)–(103) to obtain $\boldsymbol{D}^{(1)}$, $\boldsymbol{D}^{(2)}$, and $\boldsymbol{D}^{(3)}$.
3. Use Eqs. (104)–(106) to obtain $\boldsymbol{H}^{(1)}$, $\boldsymbol{H}^{(2)}$, and $\boldsymbol{H}^{(3)}$.
4. The complete displacement and magnetic field strength vectors in the vacuum are then

$$
\begin{aligned}
\boldsymbol{D} &= \boldsymbol{D}^{(1)} + \boldsymbol{D}^{(2)} + \boldsymbol{D}^{(3)} \\
&= D_X\boldsymbol{i} + D_Y\boldsymbol{j} + D_Z\boldsymbol{k}
\end{aligned}
\tag{615}
$$

$$
\boldsymbol{H} = \boldsymbol{H}^{(1)} + \boldsymbol{H}^{(2)} + \boldsymbol{H}^{(3)} \tag{616}
$$

5. Use

$$
\begin{aligned}
\boldsymbol{B}^{(1)} &= \nabla \times \boldsymbol{A}^{(1)} \\
\boldsymbol{B}^{(2)} &= \nabla \times \boldsymbol{A}^{(2)} \\
\boldsymbol{B}^{(3)} &= -ig\boldsymbol{A}^{(1)} \times \boldsymbol{A}^{(2)}
\end{aligned}
\tag{617}
$$

to obtain $\boldsymbol{B}^{(1)}$, $\boldsymbol{B}^{(2)}$, and $\boldsymbol{B}^{(3)}$.

6. Use Eqs. (590)–(592) to obtain $\boldsymbol{E}^{(1)}$ and $\boldsymbol{E}^{(2)}$.
7. Simplify the code with

$$\begin{aligned} \boldsymbol{B}^{(1)} &= \boldsymbol{B}^{(2)*} \\ \boldsymbol{E}^{(1)} &= \boldsymbol{E}^{(2)*} \\ \boldsymbol{A}^{(1)} &= \boldsymbol{A}^{(2)*} \\ &\ldots \end{aligned} \tag{618}$$

8. Finally, find $\boldsymbol{P}^{(1)}$, $\boldsymbol{P}^{(2)}$, and $\boldsymbol{P}^{(3)}$ and, if they exist, $\boldsymbol{M}^{(1)}$, $\boldsymbol{M}^{(2)}$, and $\boldsymbol{M}^{(3)}$ in the vacuum.

The computational problem for the vacuum involves the definition of vacuum boundary conditions, which, for example, may be a volume of radiation or a beam radius. The computational method assumes no Lorenz condition, and gives a vast number of solutions. Having obtained these solutions, we can next check whether the non-Abelian Stokes theorem (153) is obeyed numerically. Essentially, everything is obtained from potentials in the vacuum, and everything is expressible in terms of these potentials, including the charge and the current. In evaluating the coupling constant $\kappa/A^{(0)}$, the denominator is the magnitude of $\boldsymbol{A}^{(1)} = \boldsymbol{A}^{(2)*}$, defined by

$$A^{(0)} = \left(\boldsymbol{A}^{(1)} \cdot \boldsymbol{A}^{(2)}\right)^{1/2} \tag{619}$$

This is then a computational solution of a classical problem in the vacuum. If g is defined as $\kappa/A^{(0)}$, then e is never used.

In field–matter interaction, the fields $\boldsymbol{B}$ and $\boldsymbol{E}$ remain unchanged. The fields $\boldsymbol{D}$ and $\boldsymbol{H}$ change because $\boldsymbol{P}$ and $\boldsymbol{M}$ change. Equations (612) have precisely the same structure as Eqs. (9-7) of Panofsky and Phillips [86] with the following identifications:

$$\rho_{\text{vac}} \equiv \rho_{\text{true}}; \qquad \boldsymbol{J}_{\text{vac}} \equiv \boldsymbol{J}_{\text{true}} \tag{620}$$

The ρ_{true} and $\boldsymbol{J}_{\text{true}}$ of Ref. 86 are therefore identified as being due to the topology of the vacuum, a topology that gives rise to potential energy inherent in the vacuum. The potential energy appears in O(3) electrodynamics through the connections $\boldsymbol{A}_{\mu}^{(i)}$, and so the connections are regarded as physical entities. Fields, currents, and charges are obtained from the potentials, or more precisely, potential energy differences that are dictated by the topology of the vacuum itself. On the classical level, $g = \kappa/A^{(0)}$ so the constant e does not appear in the vacuum. As demonstrated already in this review, the equivalent of the Poynting theorem can be obtained by considering the energy inherent in the vacuum, on both the U(1) and on the O(3) levels.

In dealing with Eqs. (612), the vacuum is treated as if it were a material, and the equations are solved with stipulated boundary conditions and constitutive relations. The ontology behind Eqs. (612) is that charge–current is the result of spacetime. Similarly, in general relativity, matter is the result of spacetime. A complete theory would obviate the need for constitutive relations and be based on grand unified field theory with an O(3) electromagnetic sector. Equations (612) deal only with the electromagnetic sector on a classical level and still utilize the concept of field as a matter of convenience. So we still write in terms of field–matter interaction, although the ontology dictates that field–matter interaction is dictated solely by the topology of spacetime.

The computational problem in the vacuum has to be solved first, to obtain the vacuum polarizations. To simulate the interaction with matter, the polarization changes in the medium must be modeled using constitutive relations, and boundary conditions defined according to the problem being solved. Integral forms of Eqs. (612) may be useful, and integral forms must be obtained through the non-Abelian Stokes theorem using O(3) covariant derivatives. For example, the integral form of Eqs. (590)–(592) is

$$\oint \boldsymbol{E}^{(1)} \cdot d\boldsymbol{r} + \frac{\partial}{\partial t} \oint \boldsymbol{B}^{(1)} \cdot d\boldsymbol{Ar} = 0 \tag{621}$$

$$\oint \boldsymbol{E}^{(2)} \cdot d\boldsymbol{r} + \frac{\partial}{\partial t} \oint \boldsymbol{B}^{(2)} \cdot d\boldsymbol{Ar} = 0 \tag{622}$$

$$\frac{\partial}{\partial t} \oint \boldsymbol{B}^{(3)} \cdot d\boldsymbol{Ar} = 0 \tag{623}$$

and the integral form of $\nabla \cdot \boldsymbol{B}^{(i)} = 0; i = 1, 2, 3$ is

$$\oint \boldsymbol{B}^{(i)} \cdot d\boldsymbol{r} = 0; \qquad i = 1, 2, 3 \tag{624}$$

A simple example of a computational problem on the U(1) level is the numerical solution of the equation

$$\nabla(\nabla \cdot \boldsymbol{A}) - \nabla^2 \boldsymbol{A} + \frac{1}{c^2}\frac{\partial}{\partial t}\nabla\phi + \frac{1}{c^2}\frac{\partial^2 \boldsymbol{A}}{\partial t^2} = \boldsymbol{0} \tag{625}$$

which is equivalent to solving the following equations simultaneously:

$$\begin{aligned} \nabla \times \boldsymbol{E} + \frac{\partial \boldsymbol{B}}{\partial t} &= \boldsymbol{0} \\ \nabla \times \boldsymbol{B} - \frac{1}{c^2}\frac{\partial \boldsymbol{E}}{\partial t} &= \boldsymbol{0} \end{aligned} \tag{626}$$

In the received opinion [5], these are the vacuum Faraday law and Ampère–Maxwell law, respectively. The vacuum charges and currents are missing in the received opinion. Nevertheless, solving Eq. (625) numerically is a useful computational problem with boundary conditions stipulated in the vacuum. The potentials and fields are related as usual by

$$\begin{aligned} \mathbf{B} &= \nabla \times \mathbf{A} \\ \mathbf{E} &= -\frac{\partial \mathbf{A}}{\partial t} - \nabla \phi \end{aligned} \tag{627}$$

In the received view, it is customary to simplify the problem of solving Eq. (625) with the Lorenz condition

$$\nabla \cdot \mathbf{A} + \frac{1}{c^2}\frac{\partial \phi}{\partial t} = 0 \tag{628}$$

to give the d'Alembert equation in vacuo

$$\nabla^2 \mathbf{A} - \frac{1}{c^2}\frac{\partial^2 \mathbf{A}}{\partial t^2} = \mathbf{0} \tag{629}$$

an equation that has analytical solutions such as plane waves. The Lorenz condition (628) is asserted to be the result of gauge freedom. The computational problem therefore consists in solving Eq. (625) with and without Eq. (628) for different boundary conditions.

Regardless of whether the Lorenz gauge is used, the equation $\square \chi = 0$ is obtained. So χ is not random after being assumed to be random (a reduction to absurdity) proof of the self-consistency of the U(1) gauge ansatz. Ludwig V. Lorenz introduced the idea of the Lorenz gauge or condition (often misattributed to Henrik Anton Lorentz) in 1867, so we can write the structured scalar potential as $\phi = \phi_0 e^{i\omega(t)}$, where (t) is the retarded time. So in this sense, we can have pure time-like potentials (something that apparently was discussed between Bearden and Wigner) in the context of a pure time-like photon. Whittaker's work depends on the Lorenz condition on the U(1) level.

Plane waves have infinite lateral extent and, for this reason, cannot be simulated on a computer because of floating-point overflow. If the lateral extent is constrained, as in Problem 6.11 of Jackson [5], longitudinal solutions appear in the vacuum, even on the U(1) level without vacuum charges and currents. This property can be simulated on the computer using boundary conditions, for example, a cylindrical beam of light. It can be seen from a comparison of Eqs. (625) and (629) that if the Lorenz condition is not used, there is no increase in the number of variables. Therefore Eq. (625) is one equation in two unknowns, ϕ and $\boldsymbol{A}$. If we use the Lorenz condition

$$\nabla \cdot \boldsymbol{A} + \frac{1}{c^2}\frac{\partial \phi}{\partial t} = 0 \tag{630}$$

we still have one equation in two unknowns. Making use of the vacuum Coulomb and Gauss laws in the received view

$$\begin{aligned} \nabla \cdot \boldsymbol{E} &= 0 \\ \nabla \cdot \boldsymbol{B} &= 0 \end{aligned} \tag{631}$$

we obtain two more equations:

$$\nabla \cdot \left(\frac{\partial \boldsymbol{A}}{\partial t} + \nabla \phi \right) = 0 \tag{632}$$

$$\nabla \cdot \nabla \times \boldsymbol{A} = 0 \tag{633}$$

So there are three equations, (625), (632), and (633), in two unknowns $\boldsymbol{A}$ and ϕ. These are enough to solve for the components of $\boldsymbol{A}$ and for ϕ for any boundary condition. For any physical boundary condition, there will be longitudinal as well as transverse components of $\boldsymbol{A}$ in the vacuum, and ϕ will in general be phase-dependent and structured. This computational exercise shows that the Lorenz condition is arbitrary and, if it is discarded, the values of $\boldsymbol{A}$ and ϕ from Eqs. (625), (632), and (633) change.

Under the U(1) gauge transform

$$A^\mu \to A^\mu + \partial^\mu \chi; \quad A^\mu \equiv (\phi, c\boldsymbol{A}); \quad \left(\text{i.e. } \phi \to \phi + \frac{1}{c}\frac{\partial \chi}{\partial t}; \quad \boldsymbol{A} \to \boldsymbol{A} - \frac{1}{c}\nabla\chi \right) \tag{634}$$

we see that $\boldsymbol{E}$ and $\boldsymbol{B}$ do not change:

$$\begin{aligned} \boldsymbol{E} &\to \boldsymbol{E} + \frac{1}{c}\frac{\partial}{\partial t}\nabla\chi - \frac{1}{c}\nabla\frac{\partial \chi}{\partial t} = \boldsymbol{E} \\ \boldsymbol{B} &\to \boldsymbol{B} - \frac{1}{c}\nabla \times (\nabla\chi) = \boldsymbol{B} \end{aligned} \tag{635}$$

and Eqs. (625), (632), and (633) do not change. This means that for any given boundary condition, we can find the solutions

$$\phi' \equiv \phi + \frac{1}{c}\frac{\partial \chi}{\partial t} \tag{636}$$

$$\boldsymbol{A}' \equiv \boldsymbol{A} - \frac{1}{c}\nabla\chi \tag{637}$$

from Eqs. (625), (632), and (633) numerically. The solutions ϕ' and $\boldsymbol{A}'$, however, are *not* arbitrary for a given boundary condition, indicating another self-inconsistency in U(1) gauge theory (Section II). Furthermore, under the same gauge transform (634), Eq. (625) indicates that χ must obey the equation

$$\Box\chi = 0 \tag{638}$$

whose general solution has been given by Whittaker [27] and is *not* arbitrary. If we arbitrarily decouple Eq. (625) into

$$\begin{gathered} \Box \boldsymbol{A} = 0 \\ \nabla \cdot \boldsymbol{A} + \frac{1}{c^2}\frac{\partial \phi}{\partial t} = 0 \end{gathered} \tag{639}$$

then Eq. (638) is obtained again, indicating that the Lorenz condition and d'Alembert equation in vacuo are arbitrary constructs, that is, particular solutions of Eq. (625). The Lorenz condition has no physical meaning, nor does the vacuum d'Alembert equation. The function χ is not arbitrary, contrary to the U(1) gauge transform ansatz, Eq. (634). In other words, the gauge transformed ϕ' and $\boldsymbol{A}'$ are not arbitrary, as they are solutions of two differential equations, (625) and (632), in two unknowns, ϕ' and $\boldsymbol{A}'$, for a given boundary condition. We conclude that ϕ' and $\boldsymbol{A}'$ are physical, not arbitrary, thus refuting Heaviside's point of view and supporting that of Maxwell and Faraday. For a self-consistent picture of electrodynamics, we have to go to the O(3) level, as discussed earlier in this section.

The same conclusion regarding the Lorenz gauge is reached by Jackson [5], who shows that:

$$\partial_\mu A^{\mu'} = \partial_\mu A^\mu + \Box\chi \tag{640}$$

However, Jackson follows the received opinion and forces

$$\Box\chi = -\partial_\mu A^\mu \tag{641}$$

through the arbitrary assumption:

$$\partial_\mu A^{\mu'} = ? \quad 0 \tag{642}$$

The latter merely reinforces the conclusion that χ is not arbitrary.

By discarding the Lorenz condition, a vacuum current $\boldsymbol{J}_{\text{vac}}$ is introduced. The vacuum current $\boldsymbol{J}_{\text{vac}}$ is conceptually similar to the one introduced by Lehnert and Roy [10]. Relativity then indicates the presence of a vacuum charge, so the field equations in vacuo become identical with those of Panofsky and Phillips [86] and those of O(3) electrodynamics [11–20] i.e., [Eqs. (612)]. Phipps [90] has also derived the same structure and describes it as "neo-Hertzian." There is therefore a remarkable degree of agreement in the literature that the structure of the Heaviside–Maxwell equations in vacuo is such that the overall symmetry is O(3). This conclusion is consistent with the fact that there is no Lorenz condition on the O(3) level, necessitating numerical solution as described earlier in this section.

The source of Eq. (625), however, is the set of vacuum Maxwell–Heaviside equations

$$\begin{aligned} \nabla\cdot\boldsymbol{B} &= 0 \\ \nabla\cdot\boldsymbol{E} &= 0 \\ \nabla\times\boldsymbol{E}+\frac{\partial\boldsymbol{B}}{\partial t} &= \mathbf{0} \\ \nabla\times\boldsymbol{B}-\frac{1}{c^2}\frac{\partial\boldsymbol{E}}{\partial t} &= \mathbf{0} \end{aligned} \tag{643}$$

and to identify Eqs. (612) with Eqs. (643), it is necessary to write the vacuum displacement as

$$\boldsymbol{D}=\varepsilon_0\boldsymbol{E}+\boldsymbol{P} \tag{644}$$

and to introduce the vacuum polarization. This result is self-consistent with our constitutive equations (607)–(609) on the O(3) level. The vacuum polarization gives rise to a polarization current

$$\boldsymbol{J}_P\equiv-\frac{\partial\boldsymbol{P}}{\partial t} \tag{645}$$

and exists if and only if we discard the Lorenz condition. It therefore becomes clear that use of the Lorenz condition prohibits the evolution of U(1) into O(3) electrodynamics and arbitrarily asserts a zero vacuum polarization. The existence of vacuum charge and currents means the existence of vacuum energy, as argued already. The experimental challenge is how to tap this energy, which is theoretically infinite, that is, extends throughout the universe.

The vacuum charge density and current density are

$$\begin{aligned} \rho_{\text{vac}} &= \frac{1}{\mu_0}\Box\phi \\ \boldsymbol{J}_{\text{vac}} &= \frac{1}{\mu_0}\Box\boldsymbol{A} \end{aligned} \tag{646}$$

and so it becomes clear that Whittaker's theory [27] is restricted severely by his adoption of the Lorenz condition. The received view is similarly restricted. The new paradigm introduced here is that the vacuum itself is the source of charge–current, including, of course, Maxwell's displacement current. The latter has nothing to do with charged electrons, and similarly, the Noether currents of O(3) electrodynamics have nothing to do with charged electrons. The received view asserts that the Maxwell displacement current is the origin of the electromagnetic field, which carries energy and momentum; the new paradigm asserts that the vacuum itself is the source of energy and momentum through the intermediary of entities labeled charge, current, and field. The topology of the vacuum is described by physical $\boldsymbol{A}$ and ϕ , which, in turn, originate in the gauge principle and group theory. We have argued that the notion of unphysical $\boldsymbol{A}$ and ϕ is untenable. It is this idea that leads to the Lorenz condition, which is, in turn, untenable.

Therefore electric and magnetic fields do not emanate from a point charge, as in the received view; both charge and field are outcomes of the topology of the vacuum. In the new paradigm, the energy that is said to be transmitted by the electromagnetic field in the received opinion is inherent in the vacuum structure; all is determined by the nature of the connection in gauge theory, and by the physical nature of the potential, which is more precisely described as potential energy difference. An intense electromagnetic field in the received view corresponds in the new paradigm to a warping of space-time by the gauge connection inherent in the covariant derivative. On the classical level, the proportionality constant g is $\boldsymbol{\kappa}/A^{(0)}$, and $e/\hbar$ is not necessary. Curvature or warping of spacetime determines the process of radiation and of detection of radiation. Causality implies that the cause precedes the effect in time. This new view of electromagnetism as being essentially the vacuum itself is similar to general relativity. The major implication is that the vacuum carries an unknown amount of electromagnetic energy; the electromagnetic field is far stronger than the gravitational field, so the amount of electromagnetic energy in the vacuum is commensurably greater.

> The vast paradox inherent in the concept of field is vividly summarized by Koestler [91, p. 502ff.]: a steel cable of a thickness equaling the diameter of the earth would not be strong enough to hold the earth in its orbit. Yet the gravitational force which holds the earth in its orbit is transmitted from the sun across 93 million miles of space without any material medium to carry that force. The paradox is further illustrated by Newton's own words, which I have quoted before, but which bear repeating: It is inconceivable, that inanimate brute matter should, without the mediation of something else, which is not material, operate upon, and affect other matter without mutual contact, $\cdots$ And this is one reason why I desired you would not ascribe innate gravity to me. That gravity should be innate, inherent, and essential to matter, so that one body may act upon another, at a

distance through a vacuum, without the mediation of anything else, by and through which their action and force may be conveyed from one to another, is to me so great an absurdity, that I believe no man who has in philosophical matters a competent faculty of thinking, can ever fall into it. Gravity must be caused by an agent acting constantly according to certain laws; but whether this agent be material or immaterial, I have left to the consideration of my readers.

The paradox is compounded greatly in electrodynamics, where, in the received view, the field is superimposed on spacetime. In the new view, both the gravitational and electromagnetic fields are the results of topology, or vacuum structure. The enormous amount of energy inherent in the vacuum is metaphorically apparent in Koestler's steel cable. The electromagnetic energy from the same source is orders of magnitude greater. Thus a few simple computational trials are needed.

XII. SU(2) × SU(2) ELECTROWEAK THEORY WITH AN O(3) ELECTROMAGNETIC SECTOR

It has been demonstrated conclusively that classical electrodynamics is not a U(1) gauge theory; therefore, the continued use of a U(1) sector in unified field theory is misleading. In this section, a first attempt is made to unify the electromagnetic and weak fields with an O(3) electromagnetic sector. The theory has SU(2) × SU(2) symmetry instead of the usual U(1) × SU(2) symmetry. The change in symmetry has several ramifications, including the appearance of a novel massive boson that has been detected empirically [92]. The use of an O(3) electromagnetic sector will also have ramifications in grand unified field theory, a paradigm shift that extends throughout field and particle physics and challenges the standard model at a fundamental level. In the new view of grand unified field theory, all four fields are manifestations of non-Abelian gauge theory. If we go a step further and drop the word "field," then all physics becomes a manifestation of vacuum topology.

The extension of U(1) × SU(2) electroweak theory to SU(2) × SU(2) electroweak theory succeeds in describing the empirically measured masses of the weakly interacting vector bosons, and predicts a novel massive boson that was been detected in 1999 [92]. The SU(2) × SU(2) theory is developed initially with one Higgs field for both parts of the twisted bundle [93], and is further developed later in this section.

The physical vacuum is assumed to be defined by the Higgs mechanism, and the SU(2) × SU(2) covariant derivative is

$$D_\mu = \partial_\mu + ig'\boldsymbol{\sigma} \cdot \boldsymbol{A}_\mu + ig\boldsymbol{\tau} \cdot \boldsymbol{b}_\mu \tag{647}$$

where $\boldsymbol{\sigma}$ and $\boldsymbol{\tau}$ are the generators for the two SU(2) gauge fields represented as Pauli matrices, and where $\boldsymbol{A}$ and $\boldsymbol{b}$ are the gauge connections defined on the two

SU(2) principal bundles. There is an additional Lagrangian for the ϕ^4 scalar field [93]:

$$\mathscr{L}_\phi = \frac{1}{2}|D_\mu(\phi)^2| - \frac{1}{2}\mu^2|\phi|^2 + \frac{1}{4}|\lambda|(|\phi|^2)^2 \tag{648}$$

The expectation value for the scalar field is then

$$\langle\phi_0\rangle = \left(0, \frac{v}{\sqrt{2}}\right) \tag{649}$$

for $v = (-\mu^2/\lambda)^{1/2}$. The generators for the theory on the broken vacuum are

$$\begin{aligned}
\langle\phi_0\rangle_{\sigma_1} &= \left(\frac{v}{\sqrt{2}}, 0\right) \\
\langle\phi_0\rangle_{\sigma_2} &= \left(i\frac{v}{\sqrt{2}}, 0\right) \\
\langle\phi_0\rangle_{\sigma_3} &= \left(0, -\frac{v}{\sqrt{2}}\right)
\end{aligned} \tag{650}$$

These are the same for the other SU(2) sector of the theory. The hypercharge formula of Nishijima, if applied directly, would lead to an electric charge

$$\begin{aligned}
Q(\phi_0) &= \frac{1}{2}\langle\phi_0\rangle(\boldsymbol{\sigma}_3 + \boldsymbol{\tau}_1) \\
&= \left(0, -\frac{v}{\sqrt{2}}\right) + \left(0, \frac{v}{\sqrt{2}}\right)
\end{aligned} \tag{651}$$

implying two unphysical oppositely charged photons. The equation for the hypercharge must therefore be modified to

$$Q(\phi_0) = \frac{1}{2}\langle\phi_0\rangle(\boldsymbol{n}_2 \cdot \boldsymbol{\tau}_3 + \boldsymbol{n}_1 \cdot \boldsymbol{\sigma}_1) = 0 \tag{652}$$

where $\boldsymbol{n}_1$ and $\boldsymbol{n}_2$ are unit vectors on the doublet defined by the two eigenstates of the vacuum. This projection on to σ_1 and τ_3 is required because we are using a single Higgs field on both bundles on both SU(2) connections. This requirement can be relaxed as discussed later in this section. At this stage of the development, the generators of the theory have a broken symmetry on the physical vacuum. Therefore, the photon is defined according to the σ_1 generator in one SU(2) sector of the theory, while the charged neutral current of the weak interaction is defined on the τ^3 generator.

The fundamental Lagrangian contains the electro-weak Lagrangians and the ϕ^4 scalar field:

$$\mathscr{L} = -\frac{1}{4}F^a_{\mu\nu}F^{a\mu\nu} - \frac{1}{4}G^a_{\mu\nu}G^{a\mu\nu} + |D_\mu\phi|^2 - \frac{1}{2}\mu^2|\phi|^2 + \frac{1}{4}\lambda(|\phi|^2)^2 \qquad (653)$$

where $G^a_{\mu\nu}$ and $F^a_{\mu\nu}$ are elements of the field strength tensors for the two SU(2) principal bundles. In order to develop the theory further, it would be necessary to include the Dirac and Yukawa Lagrangians that couple the Higgs field to the leptons and quarks. The ϕ^4 field may be developed as a small displacement in the vacuum energy:

$$\phi' = \phi + \langle\phi_0\rangle \approx \frac{(v + \xi + i\chi)}{\sqrt{2}} \qquad (654)$$

The fields ξ and χ are orthogonal components in the complex phase plane for the oscillations due to the small displacement of the scalar field, which is thereby characterized completely. The scalar field Lagrangian becomes

$$\mathscr{L}_\phi = \frac{1}{2}(\partial_\mu\xi\,\partial^\mu\xi - 2\mu^2\xi^2) \quad + \frac{1}{2}v^2\left(g'\boldsymbol{A}_\mu + g\boldsymbol{b}_\mu + \left(\frac{1}{gv} + \frac{1}{g'v}\right)\partial_\mu\chi\right)$$
$$\times \left(g'\boldsymbol{A}_\mu + g\boldsymbol{b}_\mu + \left(\frac{1}{gv} + \frac{1}{g'v}\right)\partial_\mu\chi\right) \qquad (655)$$

where Lie algebraic indices are implied. The Higgs field is described by the harmonic oscillator equation where the field has the mass $M_H \approx 1.0$ TeV/c^2.

On the physical vacuum the gauge fields are:

$$g'\boldsymbol{A}_\mu + g\boldsymbol{b}_\mu \to g'\boldsymbol{A}'_\mu + g\boldsymbol{b}'_\mu \qquad (656)$$

which corresponds to a phase rotation induced by the transition of the vacuum to the physical vacuum. The Lagrangian is now decomposed into components by expanding about the minimum of the scalar potential

$$\mathscr{L}_\phi = \frac{1}{2}(\partial_\mu\xi\,\partial^\mu\xi - 2\mu^2\xi^2) + \frac{1}{8}v^2(g'|\boldsymbol{b}^{(3)}|^3 + g'^2(|\boldsymbol{W}^+|^2 + |\boldsymbol{W}^-|^2)$$
$$+ g^2|\boldsymbol{A}^{(1)}|^2 + g^2|\boldsymbol{A}^{(3)} + i\boldsymbol{A}^{(2)}|^2) \qquad (657)$$

where the charged weak fields are identified as

$$W^\pm_\mu = \frac{1}{\sqrt{2}}(\boldsymbol{b}^{(1)}_\mu \pm i\boldsymbol{b}^{(2)}_\mu) \qquad (658)$$

with mass $g\mathrm{v}/2$. The other parts of the Lagrangian define the fields:

$$A_{\mu} = \frac{(g\boldsymbol{A}_{\mu}^{(3)} + g'\boldsymbol{b}_{\mu}^{(3)} - g\boldsymbol{A}_{\mu}^{(1)})}{(g^2 + g'^2)^{1/2}} \tag{659a}$$

$$Z_{\mu}^{0} = \frac{(g'\boldsymbol{A}_{\mu}^{(3)} + g\boldsymbol{b}_{\mu}^{(3)} - g'\boldsymbol{A}_{\mu}^{(1)})}{(g^2 + g'^2)^{1/2}} \tag{659b}$$

On scales larger than unification, the requirement $A_{\mu}^{(3)} = 0$ is needed [94] because otherwise Z_0 would have a mass greater than empirically measured, or there would be an additional massive boson along with the Z_0 neutral boson. A more complete discussion of $A_{\mu}^{(3)}$ is given later in this chapter. The additional massive boson predicted by the theory has been observed empirically [92]. The considerations thus far lead to the standard result that the mass of the photon vanishes, and that the mass of the Z_0 particle is

$$\begin{aligned} M_{Z_0} &= \frac{v}{2}(g^2 + g'^2)^{1/2} \\ &= M_W\left(1 + \left(\frac{g'}{g}\right)^2\right)^{1/2} \end{aligned} \tag{660}$$

The weak angles are defined trigonometrically by the terms $g/(g^2 + g'^2)$ and $g'/(g^2 + g'^2)$. This means that the field strength tensor satisfies

$$\begin{aligned} F_{\mu\nu}^{(3)} &= \partial_{\nu}A_{\mu}^{(3)} - \partial_{\mu}A_{\nu}^{(3)} - ig[A_{\nu}^{(1)}, A_{\mu}^{(2)}] \\ &= -ig[A_{\nu}^{(1)}, A_{\mu}^{(2)}] \end{aligned} \tag{661}$$

and that the $\boldsymbol{B}^{(3)}$ field is defined, in this notation, by

$$B_j^{(3)} = \varepsilon_j^{\mu\nu} F_{\mu\nu}^{(3)} = -ig\boldsymbol{A}^{(1)} \times \boldsymbol{A}^{(2)} \tag{662}$$

The $\boldsymbol{E}^{(3)}$ field, however, is zero, as we have seen, so that the Lagrangian is satisfactorily nonzero. The $\boldsymbol{E}^{(3)}$ field vanishes by definition [Eqs. (581)]. Specifically [11–20]

$$G^{03(3)*} = \partial^0 A^{3(3)*} - \partial^3 A^{0(3)*} - ig(A^{0(1)}A^{3(2)} - A^{3(2)}A^{0(1)}) \equiv 0 \tag{663}$$

a result that is consistent with the B cyclic theorem and with the fact that there are no magnetic monopoles or currents in O(3) electrodynamics. The $\boldsymbol{E}^{(3)}$ field also vanishes if $\boldsymbol{A}^{(3)}$ is a constant, or is structured. Therefore an SU(2) $\times$ SU(2) electroweak theory can be constructed that self-consistently describes the empirically observed Z_0, $W^{\pm}$ bosons, and the $\boldsymbol{B}^{(3)}$ field in the electromagnetic sector. The theory of electromagnetism on the physical vacuum that emerges is

$$\begin{aligned}\mathscr{L} = & -\frac{1}{4}F^{\mu\nu}F_{\mu\nu} - \frac{1}{4}G^{a\mu\nu}G^{a}_{\mu\nu} - \frac{1}{2}B^{(3)2} \\ & + M_0|Z_0|^2 + M_W|W^{\pm}|^2 + \frac{1}{2}(|\partial\xi|^2 - \partial^2_{\mu}|\xi|^2) \\ & + \text{Dirac Lagrangian} + \text{Yukawa/Fermi/Higgs}\end{aligned} \tag{664}$$

where $F_{\mu\nu}$ and $G^{a}_{\mu\nu}$are the field tensor components for standard electromagnetism and the weak interaction, and the cyclic magnetic fields define the Lagrangian in the third term. The occurrence of the massive Z_0 and $W^{\pm}$ particles breaks the gauge symmetry of the SU(2) weak interactions.

The longitudinal field $\boldsymbol{B}^{(3)}$ therefore results from the breaking of gauge invariance. There is no $\boldsymbol{E}^{(3)}$ field by definition [Eq. (663)]. Under the gauge transform

$$A^{(1)} \rightarrow UA^{(1)}U^{-1} + U\partial U^{-1} \tag{665}$$

the $\boldsymbol{B}^{(3)}$ field is invariant [11–20]:

$$B^{(3)i} = \varepsilon^{ijk}U[A^{(1)}_j, A^{(2)}_k]U^{-1} \tag{666}$$

The condition $\boldsymbol{A}^3_{\mu} = \boldsymbol{0}$ is, however, restrictive, and can be removed by the inclusion in the theory of massive fermions. This makes the SU(2) $\times$ SU(2) theory consistent with the fact that $\boldsymbol{A}^{(3)}$ is phase-dependent and structured from Eqs. (586)–(588) and with the fact that there can be many solutions for $\boldsymbol{A}^{(3)}_{\mu}$ in the vacuum. The condition is therefore a first step in the development of SU(2) $\times$ SU(2) theory. If the condition $\boldsymbol{A}^{(3)}_{\mu} = \boldsymbol{0}$ is relaxed, the currents will contain vector and axial components that obey SU(2) $\times$ $SU(2)_C$ algebra, and on the physical vacuum, fields acquire masses that violate the current conservation of the axial vector current.

The theory so far is incomplete, however, because it has two SU(2) algebras that both act on the same Fermi spinor fields, and only one Higgs mechanism is used to compute the vacuum expectations for both fields. To improve the theory, consider that each SU(2) acts on separate spinor field doublets and that there are two Higgs fields that compute separate physical vacua for each SU(2) sector independently. The Higgs fields will give 2 $\times$ 2 vacuum diagonal expectations. If two entries in each of these matrices are equal, the resulting massive fermions in each of the two spinor doublets are identical. If the spin in one doublet assumes a very large mass, then at low energies, the doublet will appear as a singlet and the gauge theory that acts on it will be O(3), with the algebra of singlets:

$$\boldsymbol{e}_i = \varepsilon_{ijk}\left[\boldsymbol{e}_j, \boldsymbol{e}_k\right] \tag{667}$$

The theory on the physical vacuum will involve transformations on a singlet according to a broken O(3) gauge theory, and transformations on a doublet according to a broken SU(2) gauge theory. The broken O(3) theory signals the existence of a very massive $\boldsymbol{A}^{(3)}$ boson, which has been observed empirically [92], and massless $\boldsymbol{A}^{(1)}$ and $\boldsymbol{A}^{(2)}$ bosons. This broken O(3) gauge theory reduces to electromagnetism with the cyclicity condition. The broken SU(2) theory reflects the occurrence, as usual, of a massive charged and neutral weak bosons. The theory can be taken further by embedding it into an SU(4) gauge theory where the gauge potentials are described by 4×4 traceless Hermitian matrices and the Dirac spinor has 16-components. The neutrality of the photon is then given by a sum over charges, a sum that vanishes because the theory is traceless. The Higgs field is described by a 4×4 matrix of entries.

By invoking the condition $A_\mu^{(3)} = 0$ in the above development, what is meant is that the *transverse* components of $A_\mu^{(3)}$ are zero. This is always the case in pure electromagnetism, because (3) is the longitudinal index. The longitudinal

$$A_\mu^{(3)} \equiv (\phi, c\boldsymbol{A}) \tag{668}$$

is evidently nonzero from the arguments of Section XI. In general, in electroweak theory, however, the indices (1), (2), and (3) denote isospin, and not the circular complex space ((1),(2),(3)). So if we take $A_\mu^{(3)}$ to denote a 4-vector with isospin index (3), it may have a transverse component that is nonzero. This would mean that the current for this gauge boson is not highly conserved with a very large mass so that the interaction scale is far smaller than that for the electromagnetic field.

If we take (1), (2), and (3) to denote isospin indices, we have in general

$$\begin{aligned} A_\mu^{\prime(1)} &= \frac{(gA_\mu^{(3)} + g'b_\mu^{(3)} - gA_\mu^{(1)})}{(g^2 + g'^2)^{1/2}} \\ Z_\mu^0 &= \frac{(gb_\mu^{(3)} + g'A_\mu^{(3)})}{(g^2 + g'^2)^{1/2}} \\ \omega_\mu^{(3)} &= \frac{g'}{(g^2 + g'^2)^{1/2}} A_\mu^{(3)} \end{aligned} \tag{669}$$

The $\omega_\mu^{(3)}$ connection has a chiral component that seems to imply that $\boldsymbol{B}^{(3)}$ has a chiral component, or is mixed with the chiral component of the other SU(2) chiral field of the electroweak theory. This is what happens to SU(2) electromagnetism at very high energies. It becomes very similar in formal structure to the theory of weak interactions and has implications for the theory of leptons. The electromagnetic interaction acts on a doublet that can be treated as an element of a Fermi doublet of charged leptons and their neutrinos in the SU(2) theory of the weak interaction.

Let ψ be a doublet that describes an electron according to the (1) field and the (3) field, where the indices (1) and (3) are isospin indices in general. The free-particle Dirac Lagrangian is ($c = 1$; $\hbar = 1$)

$$\begin{aligned} \mathscr{L} &= \bar{\psi}(i\gamma^\mu D_\mu - m)\psi = \bar{\psi}(i\gamma^\mu \partial_\mu - m)\psi - gA_\mu^b \bar{\psi}\gamma^\mu \sigma_b \psi \\ &= \mathscr{L}_{\text{free}} + A_\mu^b J_b^\mu \end{aligned} \tag{670}$$

where $\bar{\psi} = \psi^+\gamma_4$. We decompose the current J_μ^b into vector and chiral components

$$J_\mu^b = \psi^+\gamma_4\gamma_\mu(1+\gamma_5)\sigma^{(3)}\psi = V_\mu^b + \chi_\mu^b \tag{671}$$

a procedure that is analogous to the current algebra for weak and electromagnetic interactions between fermions. There are two vector current operators

$$V_\mu^a = \frac{i}{2}\bar{\psi}\gamma_\mu\sigma^a\psi \tag{672}$$

and two axial current operators

$$\chi_\mu^b = \frac{i}{2}\bar{\psi}\gamma_\mu\gamma_5\tau^b\psi \tag{673}$$

where $\gamma_5 = i\gamma_1\gamma_2\gamma_3\gamma_4$ and where τ^b are Pauli matrices. These define an algebra of equal time commutators:

$$\begin{aligned} [V_4^a, V_\mu^b] &= it^{abc}V_\mu^c \\ [V_4^a, \chi_\mu^b] &= -it^{abc}\chi_\mu^c \end{aligned} \tag{674}$$

The index $\boldsymbol{\mu} = 4$ implies the following algebra:

$$\begin{aligned} [V_4^a, V_4^b] &= it^{abc}V_4^c \\ [V_4^a, \chi_4^b] &= -it^{abc}\chi_4^c \end{aligned} \tag{675}$$

The definition

$$Q_{\pm}^{a} = \frac{1}{2}(V_4^a \pm \chi_4^a) \tag{676}$$

gives the algebra

$$\begin{aligned} [Q_+^a, Q_+^b] &= it^{abc} Q_+^c \\ [Q_-^a, Q_-^b] &= it^{abc} Q_-^c \\ [Q_+^a, Q_-^b] &= 0 \end{aligned} \tag{677}$$

which defines the SU(2) × SU(2) algebra. The parity operator P acts as follows:

$$\begin{aligned} PV_4^b P^+ &= V_4^b \\ \boldsymbol{P}\chi_4^b \boldsymbol{P}^+ &= -\chi_4^b \end{aligned} \tag{678}$$

and one SU(2) group differs from the other. The total group is therefore the chiral group SU(2) × SU(2)$_P$.

On the physical vacuum, the above theory becomes a vector gauge theory where the indices (1), (2), and (3) are now defined in the complex circular basis ((1),(2),(3)) described by

$$\begin{aligned} \boldsymbol{e}^{(1)} \times \boldsymbol{e}^{(2)} &= i\boldsymbol{e}^{(3)*} \\ &\cdots \end{aligned} \tag{679}$$

On the physical vacuum, therefore, there are no transverse components of $A_\mu^{(3)}$, and its longitudinal components are structured as in Section XI. On the physical vacuum, there is a mixture of vector and chiral gauge components within both the electromagnetic and weak-field sectors. This means that any transverse component of $\boldsymbol{A}^{(3)}$ will vanish identically at low energies, and any transverse component of $\boldsymbol{A}^{(3)}$ can exist only if (3) is regarded as an isospin index. If so, any transverse $\boldsymbol{A}^{(3)}$ will be massive and short-ranged and will quantize to the massive boson detected in Ref. 92. Clearly, a transverse component of $\boldsymbol{A}^{(3)}$ in the pure electromagnetic sector vanishes by definition, and can exist only as a result of the mixing of the electromagnetic and weak field, and then only if (3) is generalized to an isospin index from a purely spatial index $(3) = k$.

If there exists a very high energy massive $\boldsymbol{A}^{(3)}$, as the data in Ref. 92 appear to indicate, there exists the nonconserved current

$$\partial^\mu \boldsymbol{J}_\mu^{(3)} = im_\psi \psi^+ \gamma_4 \gamma_5 \sigma^{(3)} \psi \tag{680}$$

where inhomogeneous terms correspond to quark–antiquark and lepton–antilepton pairs that are formed from the decay of these particles. This breaks the chiral symmetry of the theory. The action of this current on the physical vacuum is such that when projected on a massive eigenstate for any 3-photon with transverse modes, for instance

$$\langle 0|\partial^{\mu} J_{\mu}^{(3)}|X_b\rangle = \frac{m^2}{(\omega(k)\omega(k'))^{1/2}} \langle X_{k'} \mid X_b\rangle e^{ikx} \tag{681}$$

the mass of the chiral bosons will vanish, while the mass of the chiral 3-boson will be m. Therefore $\boldsymbol{A}^{(3)}$ is a separate chiral gauge field that obeys axial vector field that does not obey axial vector conservation and occurs only at short ranges. Therefore $\boldsymbol{A}^{(3)}$ must not be confused with a transverse component of the low-energy electromagnetic $A_{\mu}^{(3)}$, which is zero by definition. Furthermore, the condition $\boldsymbol{A}^{(3)} = \mathbf{0}$ must not be taken to imply that the scalar and longitudinal vector parts of $\boldsymbol{A}^{(3)}$ are zero.

Therefore the electroweak theory is chiral at high energies, but is vector and chiral in separate sectors on the physical vacuum of low energies. The high-energy chiral field combines with the other chiral field in the twisted bundle to produce a vector field plus a broken chiral field at low energy. There are independent fields that are decoupled on the physical vacuum at low energies.

Consider two fermion fields, ψ and χ, each consisting of the two component right- and left-handed fields $R_{\psi}, L_{\phi}\, _i R_{\chi}, L_{\chi}$. These Fermi doublets have the masses m_1 and m_2. The two gauge potentials $\boldsymbol{A}_{\mu}$ and $\boldsymbol{B}_{\mu}$ interact respectively with the ψ and χ fields. In general, these Fermi fields are degeneracies that split into the multiplet of known fermions, so that there are four possible masses for these fields in the physical vacuum. The masses originate in Yukawa couplings with the Higgs field on the physical vacuum, which give Lagrangian terms of the form $Y_{\phi}R_{\psi}^{+}\phi L_{\chi} + \text{H.C.}$ and $Y_{\eta}L_{\psi}^{+}\eta R_{\chi} + \text{H.C.}$ where there are two component ϕ^4 fields for the Higgs mechanism. (H.C. = higher contributions). These components assume the minimal expectation values $\langle\phi_0\rangle$ and $\langle\eta_0\rangle$ on the physical vacuum with the Lagrangian:

$$\begin{aligned}\mathscr{L} &= \bar{\psi}(i\gamma^{\mu}(\partial_{\mu} + igA_{\mu}) - m_1)\psi \\ &\quad + \bar{\chi}(i\gamma^{\mu}(\partial_{\mu} + igB_{\mu}) - m_2)\chi - Y_{\phi}R_{\psi}^{+}L_{\chi} + \text{H.C.} \\ &\quad - Y_{\eta}L_{\psi}^{+}\eta R_{\chi} + \text{H.C.}\end{aligned} \tag{682}$$

that can be further broken into the left and right two component spinors

$$\begin{aligned}\mathscr{L} = {} & R_{\psi}^{+}i\sigma^{\mu}(\partial_{\mu} + igA_{\mu})R_{\psi} + L_{\psi}^{+}i\sigma^{\mu}(\partial_{\mu} + igA_{\mu})L_{\psi} \\ & + R_{\chi}^{+}i\sigma^{\mu}(\partial_{\mu} + igB_{\mu})R_{\chi} + L_{\chi}^{+}\sigma^{\mu}(\partial_{\mu} + igB_{\mu})L_{\chi} \\ & - m_1R_{\psi}^{+}L_{\psi} - m_1L_{\psi}^{+}R_{\psi} - m_2R_{\chi}^{+}L_{\chi} - m_2L_{\chi}^{+}R_{\chi} \\ & - Y_{\phi}R_{\psi}^{+}\phi L_{\chi} + Y_{\phi}^{*}L_{\chi}^{+}\phi^{*}R_{\psi} - Y_{\eta}L_{\psi}^{+}\eta R_{\chi} + Y_{\eta}^{*}R_{\chi}^{+}\eta^{*}L_{\psi}\end{aligned} \tag{683}$$

The gauge potentials A_μ and B_μ are 2×2 Hermitian traceless matrices, and the Higgs fields ϕ and χ are also 2×2 matrices. These expectations are real-valued, and the nonzero contributions of the Higgs field on the physical vacuum are given by the diagonal matrix entries [95]:

$$\langle\phi\rangle = \begin{bmatrix} \langle\phi^{(1)}\rangle & 0 \\ 0 & \langle\phi^{(2)}\rangle \end{bmatrix}; \qquad \langle\chi\rangle = \begin{bmatrix} \langle\chi^{(1)}\rangle & 0 \\ 0 & \langle\chi^{(2)}\rangle \end{bmatrix} \tag{684}$$

The values of the vacuum expectations are such that, at high energy, the left-handed fields R_χ and the right-handed doublet field L_ψ couple to the SU(2) vector boson field B_μ, while at low energy, the theory is one with a left-handed SU(2) doublet R_ψ that interacts with the right-handed doublet L_χ through the massive gauge fields A_μ. The mass terms from the Yukawa coupling Lagrangians will give

$$m' = Y_\eta\langle\chi^{(1)}\rangle \gg m'' = Y_\eta\langle\chi^{(2)}\rangle \gg m''' = Y_\phi\langle\phi^{(1)}\rangle \gg m''' = Y_\phi\langle\phi^{(2)}\rangle \tag{685}$$

If the SU(2) theory for B_μ potentials are right-handed chiral and the SU(2) theory for A_μ potentials are left-handed chiral, a chiral theory at high energies can become a vector theory at low energies.

This is a broken gauge theory at low energy, which can be expressed as in Eq. (686) as a gauge theory accompanied by a broken gauge symmetry. Assume a simple Lagrangian that couples the left-handed fields ψ_l to the right-handed boson A_μ and the right-handed fields ψ_r to the left-handed boson B_μ:

$$\begin{aligned} \mathscr{L} = {} & \bar{\psi}_l(i\gamma^\mu(\partial_\mu + igA_\mu) - m_1)\psi_l + \bar{\psi}_r(i\gamma^\mu(\partial_\mu + igB_\mu) - m_2)\psi_r \\ & - Y_\phi\psi_l^+\phi\psi_r - Y_\phi\psi_r^+\psi_l \end{aligned} \tag{686}$$

If the coupling constant Y_ϕ is comparable with the coupling constant g, then the Fermi expectation energies of the fermions occur at the mean expectation value for the Higgs field $\langle\phi_0\rangle$. In this case, the vacuum expectation value is proportional to the identity matrix, meaning that the masses acquired by the right chiral plus left chiral gauge bosons $A_\mu + B_\mu$ are zero, while the right chiral minus left chiral gauge bosons $A_\mu - B_\mu$ acquire masses approximately equal to $Y_\phi\langle\phi_0\rangle$. The theory at low energies is one with an unbroken vector gauge theory plus a broken chiral gauge theory [95]. The additive charges $A^{(1)(2)}, B^{(1),(2)}$ of the two chiral fields are opposite so that of the resulting vector gauge bosons are chargeless. Therefore gauge theories can change their vector and chiral character, and so also can the doublets of the theory. In so doing, this will give rise to the doublets of leptons and quarks plus doublets of very massive fermions that should be observable in the multi-TeV range.

The two parts of the twisted bundle are copies of SU(2) with a doublet fermion structure. One of the fermions has a very large mass, $m' = Y_\eta \langle \chi^{(1)} \rangle$, which is assumed to be unstable and not observed at low energies. So one sector of the twisted bundle is left with the same Abelian structure, but with a singlet fermion, meaning that the SU(2) gauge theory becomes defined by the algebra over the basis elements

$$[\hat{e}_i, \hat{e}_j] = i\varepsilon_{ijk}\hat{e}_k \tag{687}$$

To calculate the photon masses, define the Higgs field by a small expansion around the vacuum expectations

$$\begin{aligned} \eta^{(1)} &= \xi^{(1)} + \langle \eta_0^{(1)} \rangle \\ \eta^{(2)} &= \xi^{(2)} + \langle \eta_0^{(2)} \rangle \end{aligned} \tag{688}$$

The contraction of the generators $\sigma^{(1)}$ and $\sigma^{(2)}$ with the Higgs field matrix and right and left fields gives

$$\boldsymbol{\sigma}^{(1)} \cdot \eta R + \boldsymbol{\sigma}^{(2)} \cdot \eta L = 0 \tag{689}$$

so that the charges of the $\boldsymbol{A}^{(1)}$ and $\boldsymbol{A}^{(2)}$ fields are zero. On the low-energy vacuum, these fields can be thought of as massless fields composed of two gauge bosons, with masses $(m' + m'')^{1/2} \gg M_Z$ and with opposite charges. These electrically charged fields can be thought of as $\boldsymbol{A}^{\pm} = \boldsymbol{A}^{(1)} \pm \boldsymbol{A}^{(2)}$, giving rise to particles that cancel each other and massless vector photon gauge fields. The $\boldsymbol{A}^{(3)}$ field has an unstable mass that decays into particle pairs.

Therefore the more massive Higgs field acts to give the gauge theory SU(2) × O(3), where the first gauge group acts on singlets. On a lower energy scale, or longer timescale, $\boldsymbol{A}^{(3)}$ has decayed and vanished. The second gauge group is then represented by $O(3)_P$, a notation that implies "partial group." The latter describes Maxwell's equations, and the $\boldsymbol{B}^{(3)}$ field is defined through $-ig\boldsymbol{A}^{(1)} \times \boldsymbol{A}^{(2)}$. Evidently, in this scale, the isospin indices become identified with the space indices (1), (2), and (3) of the circular basis.

The second Higgs field acts in such a way that if the vacuum expectation value is zero, $\langle \phi^{(2)} \rangle = 0$, then the symmetry breaking mechanism effectively collapses to the Higgs mechanism of the standard SU(2) × U(1) electroweak theory. The result is a vector electromagnetic gauge theory $O(3)_P$ and a broken chiral SU(2) weak interaction theory. The mass of the vector boson sector is in the $\boldsymbol{A}^{(3)}$ boson plus the $W^{\pm}$ and Z^0 particles.

The two SU(2) theories can be represented as the block diagonals of the SU(4) gauge theory. The Lagrangian density for the system is then

$$\mathscr{L} = \bar{\psi}(i\gamma^{\mu}(\partial_{\mu} + igA_{\mu}) - m_1)\psi - Y\bar{\psi}\phi\psi \tag{690}$$

and the gauge potentials A_μ now have 4×4 traceless representations. The scalar field theory that describes the vacuum will satisfy field equations that involve all 16 components of the gauge potential. By selectively coupling these fields to the fermions, it might be possible to construct a theory that recovers a low energy theory that is the standard model with the $O(3)_P$ gauge theory for electromagnetism. We arrive at the important conclusion that the electroweak theory can be constructed with an O(3) electrodynamic sector to provide additional physical details at high energy.

The prediction of a heavy boson $\boldsymbol{A}^{(3)}$ has received preliminary empirical support [92,96] from an anomaly in Z decay widths that points toward the existence of Z bosons with a mass of 812 GeV^{+339}_{-152} [92,96] within the SO(1) grand unified field model, and a Higgs mechanism of 145 GeV^{+103}_{-61}. This suggests that a new massive neutral boson has been detected. Analysis of the hadronic peak cross sections obtained at LEP [96] implies a small amount of missing invisible width in Z decays. The effective number of massless neutrinos is 2.985 ± 0.008, which is below the prediction of 3 by the standard model of electroweak interactions. The weak charge Q_W in atomic parity violation can be interpreted as a measurement of the S parameter. This indicates a new $Q_W = -72.06 \pm 0.44$, which is found to be above the standard model prediction, an effect interpreted as being due to the occurrence of the Z' particle, which is referred to hereinafter as the Z_γ particle.

SO(10) has the six roots $\alpha^i, i = 1, \ldots, 6$. The angle between the connected roots are all 120°, where the roots α^3, α^4 are connected to each other and two other roots. The Dynkin diagram is

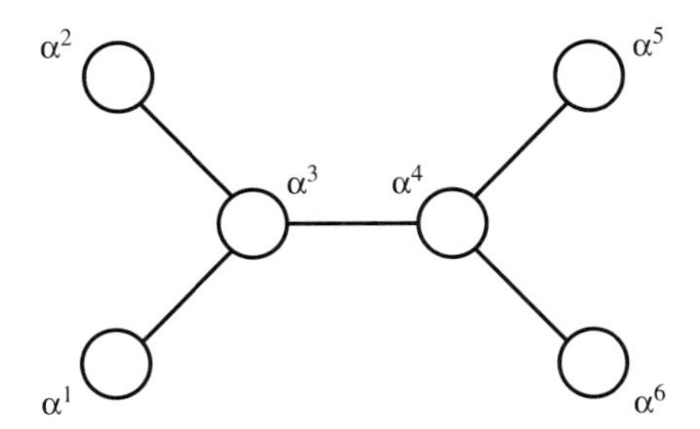

The decomposition of SO(10) to $SU(5) \times U(1)$ is performed by removing the circles representing the roots $\alpha^{1,2,5,6}$ connected by a single branch. The remaining connected graph describes the SU(5) group, and the isolated circle is the U(1) group. However, by removing either of the circles $\alpha^{3,4}$ connected by three branches forces SO(10) to decompose into $SU(2) \times SU(2) \times SU(4)$. Here, we have an SU(2) and a mirror SU(2) that describe opposite-handed chiral gauge fields, plus an SU(4) gauge field. The chiral fields are precisely the sort of electroweak structure proposed in this section and elsewhere [17,94]. Since

SU(4) can be represented by a 4, that is, $3 \oplus 1$ and $\bar{4}$ as $\bar{3} \oplus \bar{1}$, SU(4) can be decomposed into SU(3) × U(1). The neutrino short fall is furthermore a signature of the opposite chiralities of the two "mirrored" gauge fields [17,94].

The mechanism SU(2) × SU(2) → SU(2) × O(3) discussed in this section predicts the occurrence of a massive $\boldsymbol{A}^{(3)}$, so it is possible that the LEP data could corroborate the work outlined in this section, with an extended electromagnetic sector. Quantum chromodynamics (QCD) and the standard model of the electroweak theory are understood empirically. There is reasonable empirical corroboration in the TeV range and ideas about quantum gravity at 10^{19} GeV, but nothing in between. The LEP data therefore give some confidence that O(3) electrodynamics is a valid theory, and the data suggest that at high energy, electrodynamics and the weak interactions are dual-field theories in the TeV range of energy, which is expected to be accessible to the CERN heavy hadron collider.

The LEP data could be the first indication that the universe is dual according to the Olive Montonen construct [97], which asserts that coupling constants have inverse relationships. One field is weak, and the other is strong at high energy. The experimental finding [96] of the massive $\boldsymbol{A}^{(3)}$ might bring a basic change in the foundations of physics. For example, it may be conjectured that there is a dual field theory to the SU(3) nuclear interaction of QCD with a chiral SU(2) × SU(2) electroweak theory, implying the existence of an additional weak field in nature. The problem with such a program is that supergravity and superstring theories imply that, at very high energies, the universe is one of 10 or 11 dimensions [98]. The minimal grand unified field theory is the SU(5) theory that breaks into SU(3) × SU(2) × U(1) at lower energy. This is a gauge theory in six dimensions that fits into the Calabi–Yau construction of compactified manifolds. These spaces leave the four-dimensional spacetime left over and uncompactified from the 10 dimensions at high energy. A Calabi–Yau manifold of seven dimensions would accommodate an SU(3) × SU(2) × SU(2) bundle. The-low energy SU(2) × SU(2) electroweak theory would then suggest a superstring theory of 11 dimensions, which appears to preclude any SU(3) field dual to QCD because this would demand a Calabi–Yau space that can subsume an SU(3) × SU(3) × SU(2) × SU(2) bundle of 10 dimensions, and a supergravity theory of 14 dimensions.

The theory of gravitation, however, need not involve four dimensions; information [99] may exist on a two dimensional surface, such as the event horizon of a black hole. If the symmetries relevant to gravitation involve the evolution of a two-dimensional surface, then an SU(2) × SU(2) × SU(3) gauge theory plus gravity would be 11-dimensional, and duality between the two surfaces that construct spacetime would reduce this to nine dimensions. However, the issue of duality with nuclear interactions would still increase the dimensionality required to 12 or 14, and supergravity requires a total space of 11 dimensions. Strings exist at 10 dimensions.

If the nature of spacetime involves the interference of dual wave fronts of two dimensions, then there are two wave fronts, each of two dimensions, that constructively and destructively interfere, but that are determined by the same symmetry space. Gravitation can be described by the set of diffeomorphisms of a two-dimensional surface and SU(2) $\times$ SU(2) $\times$ SU(3) plus gravity involving a space of nine dimensions. The additional dimensions to spacetime are purely virtual in nature. A field dual to QCD would require a large space of 12 dimensions, and an additional constraint is required in order for this theory to satisfy current models of supergravity.

Gravitation is described by the Lie group SO(3,1) $\sim$ SL(2,C)/Z_2. It can be seen that the relevant symmetries are contained in the SL(2,C) component of two dimensions, and the Lie group has a hyperbolic metric structure. The Euclidean group for gravity is SO(4) $\sim$ (SU(2) $\times$ SU(2))/Z_2. In effect, these two groups are related by a rotation $t \longrightarrow it$, which might suggest that the electroweak interaction and gravitation can be regarded as two states of a single symmetry that may manifest itself by the action of a U(1) rotation on the Cartan center of SU(2), $\sigma'_{(3)} = e^{i\theta}\sigma^{(3)}$. At low energy, the circle associated with this rotation is reduced to a point and the direction of the angle θ determines the coupling constant for the electroweak and gravitational fields, implying a superstring theory in 11 dimensions.

If there is a field dual to the SU(3) QCD field, and if the theory is similar in form to the electroweak unification scheme outlined in this section, there may be a right–left chiral SU(3) bundle that, at low energy, combines into a right $-$ left chiral and right $+$ left chiral field. This result would indicate that QCD is a vector theory but associated with another field that is chiral or that has a broken chirality. Since QCD is the strongest force in the universe with $g = 1$, its putative dual field is one with a very weak coupling constant. For example, there may be slight chiral couplings between quarks. This would, in turn, imply the discovery of chirality with gluons, usually regarded as vector bosons.

In the absence of data, it seems best to proceed on the assumption that gauge theory at low energy is SU(2) $\times$ SU(2) $\times$ SU(3) and that the inclusion of gravity gives a space of 11 dimensions at high energy, fitting in with supergravity models. These thoughts [7,94] indicate the major impact on physics of the $\boldsymbol{B}^{(3)}$ field.

XIII. RELATIVISTIC HELICITY

In this section, we extend consideration from the Lorentz to the Poincaré group within the structure of O(3) electrodynamics, by introducing the generator of spacetime translations along the axis of propagation in the normalized (unit 12-vector) form:

$$\varepsilon_\mu \equiv \varepsilon_\mu^{(1)}\boldsymbol{e}^{(1)} + \varepsilon_\mu^{(2)}\boldsymbol{e}^{(2)} + \varepsilon_\mu^{(3)}\boldsymbol{e}^{(3)} \tag{691}$$

The relativistic helicity is then the product

$$\tilde{G}_{\nu} \equiv \tilde{G}^{(1)}_{\mu\nu}\varepsilon^{\mu(2)} + \tilde{G}^{(2)}_{\mu\nu}\varepsilon^{\mu(1)} + \tilde{G}^{(3)}_{\mu\nu}\varepsilon^{\mu(3)} \tag{692}$$

which, for $Z = (3)$ axis propagation, is the Pauli–Lubanski pseudo vector (PL vector):

$$\tilde{G}_{\nu} = \tilde{G}^{(3)}_{\mu\nu}\varepsilon^{\mu(3)} = \frac{1}{2}\varepsilon_{\mu\nu\sigma\rho}G^{\sigma\rho(3)}\varepsilon^{\mu(3)} \tag{693}$$

Evidently, this vanishes on the U(1) level, a basic paradox, because the photon has helicity after quantization. By using the Poincaré group, a fundamental geometric proof can be given for the existence of $\boldsymbol{B}^{(3)}$ in the vacuum, and helicity defined entirely through $\boldsymbol{B}^{(3)}$. This proof proceeds by constructing the PL vector from the geometric 3-manifold in 4-space, a 3-manifold that is in general a tensor of rank 3 in four dimensions, antisymmetric in all 3 indices. The PL vector is dual to this 3-tensor and has the same magnitude. The 3-tensor $S^{\mu\nu\sigma}$ is in general the following product:

$$S^{\nu\sigma\mu} \equiv \tilde{G}^{\nu\sigma}\varepsilon^{\mu} \tag{694}$$

This approach is therefore based in rigorous and general geometric tensor theory. The PL vector dual to $S^{\nu\sigma\mu}$ turns out to be the light-like invariant:

$$\tilde{B}^{\mu} = (B^{(3)}, 0, 0, B^{(3)}) \tag{695}$$

In the Lorentz group, this concept is missing, and in the Poincaré group, the relativistic helicity vanishes if $\boldsymbol{B}^{(3)}$ is not zero. Therefore $\boldsymbol{B}^{(3)}$can be regarded as the fundamental field component representing spin in the classical electromagnetic field. If $\boldsymbol{B}^{(3)}$ were zero, the PL vector would be a null vector, meaning that the space part of the equivalent hypersurface element is null. This result is a paradox, because a physical beam of light must always have a finite cross section or area perpendicular to the propagation axis of the beam, the Z or (3) axis. So if $\boldsymbol{B}^{(3)}$ vanishes, reduction to absurdity occurs, and the beam of light vanishes. This result, in turn, is self-consistent with the fact that if $\boldsymbol{B}^{(3)}$ were zero in the B cyclic theorem, $\boldsymbol{B}^{(1)}$ and $\boldsymbol{B}^{(2)}$ would also vanish, and electromagnetism would vanish.

The unit 12-vector ε_{μ} acts essentially as a normalized spacetime translation on the classical level. The concept of spacetime translation operator was introduced by Wigner, thus extending [100] the Lorentz group to the Poincaré group. The PL vector is essential for a self-consistent description of particle spin.

The dual pseudotensor of any antisymmetric tensor in 4-space arises from the integral over a two-dimensional surface in 4-space [101], in which the infinitesimal element of surface is given by the antisymmetric tensor:

$$df^{\mu\nu} = dx^{\mu}dx^{\nu'} - dx^{\nu}dx^{\mu'} \tag{696}$$

The components of this tensor are projections of the element of area on the coordinate planes. In 3-space, it is always possible to define an axial pseudovector element $d\tilde{f}_i$, dual to the antisymmetric tensor df_{jk}:

$$d\tilde{f}_i \equiv \frac{1}{2}\varepsilon_{ijk}df_{jk} \tag{697}$$

The pseudovector element $d\tilde{f}_i$ represents the same surface element as df_{jk}, and, geometrically, is a pseudovector normal to the surface element and equal in magnitude to the area of the element. In 4-space, such a pseudovector cannot be constructed from an antisymmetric tensor such as $df_{\mu\nu}$. However, the dual pseudotensor can be defined by [10]:

$$d\tilde{f}^{\mu\nu} \equiv \frac{1}{2}\varepsilon^{\mu\nu\sigma\rho}df_{\sigma\rho} \tag{698}$$

where $\varepsilon^{\mu\nu\sigma\rho}$ is the totally symmetric unit pseudotensor in four dimensions, with the property

$$\varepsilon^{0123} = -\varepsilon_{0123} = 1 \tag{699}$$

in cyclic permutation of indices. In geometric terms, $d\tilde{f}^{\mu\nu}$ is an element of surface equal and normal to the element $df_{\sigma\rho}$. All segments in it [101] are orthogonal to all segments in $df_{\sigma\rho}$, leading to the following result:

$$d\tilde{f}^{\mu\nu}df_{\mu\nu} = 0 \tag{700}$$

In general, therefore, an antisymmetric 4-tensor is an element of surface in 4-space. There are three of these elements of surface in the 12-vector $\tilde{\boldsymbol{G}}^{\mu\nu}$.

Equation (700) means that $\tilde{\boldsymbol{G}}^{\mu\nu}$ is orthogonal to $G_{\mu\nu}$ in free space

$$\tilde{G}^{\mu\nu}G_{\mu\nu} = 0 \tag{701}$$

where

$$\tilde{G}^{\mu\nu} = \frac{1}{2}\varepsilon^{\mu\nu\sigma\rho}G_{\sigma\rho} \tag{702}$$

$$G_{\mu\nu} = \frac{1}{2}\varepsilon_{\mu\nu\sigma\rho}\tilde{G}^{\sigma\rho} \tag{703}$$

In contravariant covariant notation, the field tensors are defined by [101]

$$G_{\sigma\rho} = \begin{bmatrix} 0 & \frac{E_1}{c} & \frac{E_2}{c} & \frac{E_3}{c} \\ \frac{-E1}{c} & 0 & -B_3 & B_2 \\ \frac{-E_2}{c} & B_3 & 0 & -B_1 \\ \frac{-E_3}{c} & -B_2 & B_1 & 0 \end{bmatrix}; \quad G^{\sigma\rho} = \begin{bmatrix} 0 & \frac{-E^1}{c} & \frac{-E^2}{c} & \frac{-E^3}{c} \\ \frac{E_1}{c} & 0 & -B^3 & B^2 \\ \frac{E^2}{c} & B^3 & 0 & -B^1 \\ \frac{E^3}{c} & -B^2 & B^1 & 0 \end{bmatrix} \tag{704}$$

and the dual tensors by

$$\tilde{G}^{\mu\nu} = \begin{bmatrix} 0 & -B^1 & -B^2 & -B^3 \\ B^1 & 0 & \frac{E^3}{c} & \frac{-E^2}{c} \\ B^2 & \frac{-E^2}{c} & 0 & \frac{E^1}{c} \\ B^3 & \frac{E^2}{c} & \frac{-E^1}{c} & 0 \end{bmatrix}; \quad \tilde{G}_{\mu\nu} = \begin{bmatrix} 0 & B_1 & B_2 & B_3 \\ -B_1 & 0 & \frac{E_3}{c} & \frac{-E_2}{c} \\ -B_2 & \frac{-E_3}{c} & 0 & \frac{E_1}{c} \\ -B_3 & \frac{E_2}{c} & \frac{-E_1}{c} & 0 \end{bmatrix} \tag{705}$$

It follows that

$$\tilde{G}^{\mu\nu(3)} G^{(3)}_{\mu\nu} = 0 = 4\boldsymbol{B}^{(3)} \cdot \boldsymbol{E}^{(3)} \tag{706}$$

and that $\boldsymbol{B}^{(3)}$ is zero. This result is self-consistent with earlier arguments and with the fact that the light like products of PL vectors are null:

$$B^{\mu} B^{*}_{\mu} = E^{\mu} E^{*}_{\mu} = 0 \tag{707}$$

The only nonzero components of the PL vectors $\tilde{B}^{\mu}$ and $\tilde{B}_{\mu}$ are the longitudinal and time-like components. It follows that since $\boldsymbol{B}^{(3)}$ is null, its magnitude is zero, and so $\tilde{E}^{\mu}$ and $\tilde{E}_{\mu}$ are null. This result is, in turn, consistent with the fact that the PL vector is a pseudovector, whereas $\tilde{E}^{\mu}$ is a null vector whose dual is null.

The dual axial vector in 4-space is constructed geometrically from the integral over a hypersurface, or manifold, a rank 3-tensor in 4-space antisymmetric in all three indices [101]. In three-dimensional space, the volume of the parallelepiped spanned by three vectors is equal to the determinant of the third rank formed from the components of the vectors. In four dimensions, the projections can be defined analogously of the volume of the parallelepiped (i.e., areas of the hypersurface) spanned by three vector elements: dx^{μ}, dx'^{μ} and dx''^{μ}. They are given by the determinant

$$dS^{\mu\nu\sigma} = \begin{vmatrix} dx^{\mu} & dx'^{\mu} & dx''^{\mu} \\ dx^{\nu} & dx'^{\nu} & dx''^{\nu} \\ dx^{\sigma} & dx'^{\sigma} & dx''^{\sigma} \end{vmatrix} \tag{708}$$

which forms a tensor of rank 3, antisymmetric in all three indices. The axial 4-vector element dS^{μ} dual to the tensor element $dS^{\mu\nu\sigma}$ is the element of integration over a hypersurface in four dimensions:

$$\begin{aligned} d\tilde{S}^{\mu} &= -\frac{1}{6}\varepsilon^{\mu\nu\sigma\rho} dS_{\nu\sigma\rho} \\ dS_{\nu\sigma\rho} &= \varepsilon_{\mu\nu\sigma\rho} d\tilde{S}^{\mu} \end{aligned} \tag{709}$$

so that $d\tilde{S}^0 = dS^{123}$, $d\tilde{S}^1 = dS^{023}$, and so on. The S^0 component of S^μ is therefore equivalent to the S^{123} component of $S^{\nu\sigma\rho}$, normal to it and equal to it in magnitude. The PL vector is an example of a 4-vector dual to the 3-manifold in 4-space. This is a rigorous geometric result, and if the PL vector were null, it would represent a null hypersurface in four dimensions. This, as follows, is a rigorous geometric proof of the fact that $\boldsymbol{B}^{(3)}$ is nonzero within the Poincaré group. The dual-vector S^μ is a 4-vector equal in magnitude to the area of the hypersurface to which it is dual, and is normal to this hypersurface. It is therefore perpendicular to all lines drawn in the hypersurface. In particular, the element $dS^0 = dXdYdZ$ is an element of three-dimensional volume, dV, the projection of the hypersurface on to the hyperplane $x^0 =$ constant.

In classical electromagnetic theory, the PL vector is defined through the 3-manifold

$$S^{\mu\nu\sigma} \equiv \begin{vmatrix} \partial^\mu & A^\mu & \varepsilon^\mu \\ \partial^\nu & A^\nu & \varepsilon^\nu \\ \partial^\sigma & A^\sigma & \varepsilon^\sigma \end{vmatrix} \tag{710}$$

defining the fully antisymmetric rank 3-tensor

$$S^{\nu\sigma\mu} = (\partial^\nu A^\sigma - \partial^\sigma A^\nu)\varepsilon^\mu + \cdots \tag{711}$$

which consists of three terms, the first of which is the product of ϵ^μ with the antisymmetric tensor $G^{\nu\sigma}$, a component in internal gauge space of Eq. (22). This product gives the PL vector through

$$\tilde{S}^\mu = \frac{1}{2}\varepsilon^{\mu\nu\sigma\rho}S_{\nu\sigma\rho} \tag{712}$$

The second two terms of the sum (711) can be eliminated using a combination of the free-photon minimal prescription and the quantum hypothesis

$$\partial_\mu = -i\frac{e}{\hbar}A_\mu \tag{713}$$

and the manifold defined in Eq. (711) reduces precisely to

$$S^{\nu\sigma\mu} = G^{\nu\sigma}\varepsilon^\mu \tag{714}$$

It is now possible to adopt the standard definition [6] of the PL vector to the problem at hand to give

$$\tilde{G}^\mu = \frac{1}{2}\varepsilon^{\mu\nu\sigma\rho}G_{\sigma\rho}\varepsilon_\nu \tag{715}$$

where

$$G_{\sigma\rho} = \partial_\sigma A_\rho - \partial_\rho A_\sigma \tag{716}$$

In Eq. (715), $\tilde{G}^\mu$ is dual to the third rank $G_{\sigma\rho}\varepsilon_\nu$ in four dimensions and normal to it with the same magnitude. In the received view, there is nothing normal to the purely transverse $G_{\sigma\rho}$ on the U(1) level, and therefore $\tilde{G}^\mu$ cannot be consistently dual with $G_{\sigma\rho}\varepsilon_\nu$. This result is inconsistent with the four-dimensional algebra of the Poincaré group. If we adopt the notation $\tilde{G}_\nu \equiv \tilde{B}_\nu$, we obtain

$$\tilde{B}_\nu = \tilde{\boldsymbol{G}}_{\mu\nu} \cdot \varepsilon^\mu \tag{717}$$

and the complete PL vector in consequence is

$$\begin{aligned} \tilde{B}_\nu &= \tilde{G}^{(3)}_{\mu\nu}\varepsilon^{\mu(3)} \\ &= \frac{1}{2}\varepsilon_{\mu\nu\sigma\rho}G^{\sigma\rho(3)}\varepsilon^{\mu(3)} \\ &= (B^{(3)}, 0, 0, -B^{(3)}) \end{aligned} \tag{718}$$

Similarly

$$\tilde{B}^\nu = (B^{(3)}, \boldsymbol{B}^{(3)}) \tag{719}$$

which is orthogonal to $\tilde{B}_\nu$.

The PL vector was originally constructed for particles from the generators of the Poincaré group. The PL vector corresponding to the photon's angular momentum corresponds in free space and in $c = 1$ units to

$$\tilde{J}^\mu = (J^{(3)}, 0, 0, J^{(3)}) \tag{720}$$

and the light-like momentum in $c = 1$ units is

$$p^\mu = (p^{(3)}, 0, 0, p^{(3)}) \tag{721}$$

If the mass of the photon is identically zero, its normalized helicity takes the values $+1$ and -1 because $\tilde{J}^\mu$ is proportional to p^μ [6]. The 0 component, which usually appears for a boson, is not considered but reappears if the photon has identically nonzero mass. In this case, the Wigner little group becomes O(3). The $\boldsymbol{B}^{(3)}$ field corresponds to $\tilde{\boldsymbol{J}}^{(3)}$ for the photon with a tiny but nonzero mass because, as argued earlier, the structure of the O(3) field equations is identical with that of the Lehnert equations [Eqs. (612)], which imply photon mass. Therefore p^μ and $\tilde{J}^\mu$ in the laboratory are infinitesimally different from light-like, but on an astronomical scale, they may become substantially different from light-like [11–20].

A complete consideration of relativistic helicity in the electromagnetic field therefore requires consideration of the Poincaré group. It is not sufficient to consider the Lorentz group. The vector dual to the antisymmetric field tensor introduced by Lorentz, Poincaré and Einstein could not have been defined prior to the introduction of the Pauli–Lubanski vector and Wigner's work of 1939 [100]. This work characterized all particles in terms of two Casimir invariants: one for mass and one for spin. The photon and electromagnetic field are linked by quantization, so the Wigner method must also be applied to the field. When this is done, as in this section, the relativistic helicity in O(3) electrodynamics is defined entirely by $\boldsymbol{B}^{(3)}$. U(1) electrodynamics can be described in terms of the Lorentz group, in which relativistic helicity is incompletely defined. A full understanding of $\boldsymbol{B}^{(3)}$ therefore requires the Poincaré group [11–20]. Furthermore, Noether's theorem is reduced to energy-momentum conservation only with the use of the spacetime translation generator, which within a factor $\hbar$, is the energy-momentum 4-vector itself. In the received view of the classical field [5], energy momentum is defined only through transverse components, whereas in O(3) electrodynamics, it is straightforwardly defined through $\boldsymbol{A}^{(3)}$, which is purely longitudinal at low energies.

The nature of the dual vector $(\tilde{B}^{\mu})$ can be deduced without using any equation of motion, but the dual 4-vector is a fundamental geometric property in the four dimensions of spacetime. The complete description of the electromagnetic field in O(3) electrodynamics must therefore involve boosts, rotations, and spacetime translations, meaning that $\tilde{B}^{\mu}$ is a fundamental geometric property of spacetime. The unit 4-vector ε_{μ} is orthogonal to the unit 4-vector $\tilde{B}^{\mu}$:

$$\varepsilon_{\mu}\tilde{B}^{\mu} \equiv 0 \tag{722}$$

and this is a fundamental property of the Poincaré group. The Casimir invariants of the electromagnetic field are therefore

$$\begin{aligned} \varepsilon_{\mu}\varepsilon^{\mu} &\equiv 0 \\ \tilde{B}_{\mu}\tilde{B}^{\mu} &\equiv 0 \\ \varepsilon_{\mu}\tilde{B}^{\mu} &\equiv 0 \end{aligned} \tag{723}$$

The homogeneous O(3) equations in the vacuum are obtained by considering the helicities:

$$\begin{aligned} \varepsilon_{\mu}^{(3)}\tilde{G}^{\mu\nu(3)} &= (B^{(3)}, 0, 0, B^{(3)}) \\ \varepsilon_{\mu}^{(3)}\tilde{G}^{\mu\nu(1)} &= 0 \\ \varepsilon_{\mu}^{(3)}\tilde{G}^{\mu\nu(2)} &= 0 \end{aligned} \tag{724}$$

The first of these gives the vector $\tilde{B}^{\mu}$, and the second and third give terms such as

$$-B_X^{(1)} + \frac{E_Y^{(1)}}{c} = 0 \tag{725}$$

The three relativistic helicities (724) therefore give Eqs. (590)–(592) with the addition of the following equation:

$$\nabla \cdot \boldsymbol{B}^{(3)} = 0 \tag{726}$$

In arriving at this conclusion, we have used antisymmetric tensor definitions such as

$$\tilde{G}^{\mu\nu(1)} \equiv \begin{bmatrix} 0 & -B_X^{(1)} & -B_Y^{(1)} & 0 \\ B_X^{(1)} & 0 & 0 & \frac{-E_Y^{(1)}}{c} \\ B_Y^{(1)} & 0 & 0 & \frac{E_X^{(1)}}{c} \\ 0 & \frac{E_Y^{(1)}}{c} & \frac{-E_X^{(1)}}{c} & \end{bmatrix} \tag{727}$$

By considering the conserved quantity $\tilde{B}^{\mu(3)}$, we arrive at

$$\partial_\mu \tilde{B}^{\mu(3)} = 0 \tag{728}$$

a solution of which is

$$\frac{\partial \boldsymbol{B}^{(3)}}{\partial t} = \boldsymbol{0}; \qquad \nabla \cdot \boldsymbol{B}^{(3)} = 0 \tag{729}$$

The overall structure of the O(3) equations in the vacuum is therefore

$$\partial_\mu \tilde{G}^{\mu\nu} \equiv 0 \tag{730}$$

This is the same structure as the homogenous Maxwell–Heaviside equations in the vacuum, which can therefore be obtained by a consideration of relativistic helicity.

We have seen that the overall structure of the inhomogeneous O(3) equations in the vacuum is [Eqs. (612)]

$$\partial_\mu H^{\mu\nu} = J^{\nu}_{\text{vac}} \tag{731}$$

where the vacuum charge density is defined by

$$\rho_{\text{vac}} = ig(\boldsymbol{A}^{(2)} \cdot \boldsymbol{D}^{(3)} - \boldsymbol{D}^{(2)} \cdot \boldsymbol{A}^{(3)} + \boldsymbol{A}^{(3)} \cdot \boldsymbol{D}^{(1)} - \boldsymbol{D}^{(3)} \cdot \boldsymbol{A}^{(1)} + \boldsymbol{A}^{(1)} \cdot \boldsymbol{D}^{(2)} - \boldsymbol{D}^{(1)} \cdot \boldsymbol{A}^{(2)}) \tag{732}$$

and the vacuum current density by

$$J_{\text{vac}} = -ig(cA_0^{(2)}\boldsymbol{D}^{(3)} - cA_0^{(3)}\boldsymbol{D}^{(2)} + \boldsymbol{A}^{(2)} \times \boldsymbol{H}^{(3)} - \boldsymbol{A}^{(3)} \times \boldsymbol{H}^{(2)} + cA_0^{(3)}\boldsymbol{D}^{(1)} - cA_0^{(1)}\boldsymbol{D}^{(3)} + \boldsymbol{A}^{(3)} \times \boldsymbol{H}^{(1)} - \boldsymbol{A}^{(1)} \times \boldsymbol{H}^{(3)} + cA_0^{(1)}\boldsymbol{D}^{(2)} - cA_0^{(2)}\boldsymbol{D}^{(1)} + \boldsymbol{A}^{(1)} \times \boldsymbol{H}^{(1)} - \boldsymbol{A}^{(2)} \times \boldsymbol{H}^{(1)}) \tag{733}$$

Therefore, the vacuum charge and current densities of Panofsky and Phillips [86], or of Lehnert and Roy [10], are given a topological meaning in O(3) electrodynamics. In this condensed notation, the vacuum O(3) field tensor is given by

$$H^{\mu\nu} = \begin{bmatrix} 0 & -D^1 & -D^2 & -D^3 \\ D^1 & 0 & \frac{-H^3}{c} & \frac{H^2}{c} \\ D^2 & \frac{H^3}{c} & 0 & \frac{-H^1}{c} \\ D^3 & \frac{-H^2}{c} & \frac{H^1}{c} & 0 \end{bmatrix} \tag{734}$$

and the 4-current by

$$J^\nu = \left(\rho, \frac{\boldsymbol{J}}{c}\right) \tag{735}$$

The equations of O(3) electrodynamics can therefore be written in condensed form as Eqs. (730) and (731) in the vacuum. These equations can be written as a single conservation law under all conditions (vacuum and field–matter interaction):

$$\partial_\mu \tilde{G}^\mu = \partial_\mu H^\mu = 0 \tag{736}$$

$$\tilde{G}^\mu \equiv \tilde{G}^{\mu\nu}; \qquad H^\mu \equiv H^{\mu\nu}\tilde{\varepsilon}_\nu \tag{736a}$$

In general, define the unit generators

$$\varepsilon^\mu = \left(1, \frac{v}{c}, -\frac{v}{c}, \frac{v}{c}\right) \tag{737a}$$

$$\tilde{\varepsilon}^\mu = \left(\frac{v}{c}, 1, 1, 1\right) \tag{737b}$$

where v is linear velocity and c the speed of light. Equation (737a) defines a unit energy-momentum 4-vector orthogonal to the unit energy momentum 4-vector in Eq. (737b). The existence of such generators signals that the electromagnetic field in general has a rotation–translation character, so forward momentum is always accompanied simultaneously by a transverse momentum. Thus $\epsilon_\mu \tilde{\epsilon}^\mu = 0$, that is, ϵ_μ is orthogonal to $\tilde{\epsilon}_\mu$. This feature develops Eq. (736) into two field equations. In the vacuum, $v = c$, and these field equations become Eqs. (730) and (731) with vacuum charge and current defined by Eqs. (732) and (733), respectively. In field–matter interaction, $v < c$ in the charge–current 4-vector of Eq. (735). If $\boldsymbol{B}^{(3)}$ is zero, the vacuum electromagnetic field is lost. Because of its simultaneous rotation and translation, the electromagnetic field has left- and right-handed circular polarization and is chiral. The Pauli–Lubanski construct can be either a pseudovector or vector.

We first consider the conservation law

$$\partial_\mu \tilde{G}^\mu = 0 \tag{738}$$

where ($c = 1$ units)

$$\tilde{G}^\mu = \tilde{G}^{\mu\nu}\varepsilon_\nu = \left(-\frac{v}{c}B^1 + \frac{v}{c}B^2 - \frac{v}{c}B^3, B^1 - \frac{v}{c}E^2, B^2 + \frac{v}{c}E^1, B^3 + \frac{v}{c}E^2 + \frac{v}{c}E^1\right) \tag{739}$$

giving the conservation equation:

$$\begin{aligned}\frac{v}{c}(-\partial_0 B^1 + \partial_0 B^2 - \partial_0 B^3) + \partial_1\left(B^1 - \frac{v}{c}E^2\right)\\ + \partial_2\left(B^2 + \frac{v}{c}E^1\right) + \partial_3\left(B^3 + \frac{v}{c}E^2 + \frac{v}{c}E^1\right) = 0\end{aligned} \tag{740}$$

In vector form, this becomes (in SI units)

$$\boldsymbol{v}\boldsymbol{\cdot}\left(\frac{\partial \boldsymbol{B}}{\partial t} + \nabla \times \boldsymbol{E}\right) = c^2 \nabla \boldsymbol{\cdot} \boldsymbol{B} \tag{741}$$

which is a balance of the Faraday law of induction and the Gauss law for all v, including $v = c$. This result is true for all v, and therefore under all conditions, and is precisely equivalent to the result (730), the condensed form of Eqs. (95)–(100) of O(3) electrodynamics. Apparently, magnetic monopole was never observed and the Faraday law was never violated. This is consistent with O(3) electrodynamics as argued already.

Next, we consider the conservation law:

$$\partial_\mu H^\mu = 0 \tag{742}$$

where ($c = 1$ units)

$$H^\mu = H^{\mu\nu}\tilde{\varepsilon}_\nu = \left(D^1 + D^2 + D^3, \frac{v}{c}D^1 + H^3 - H^2, \frac{v}{c}D^2 - H^3 + H^1, \frac{v}{c}D^3 + H^2 - H^1\right)$$

Using Eq. (742)

$$\partial_0(D^1 + D^2 + D^3) + \partial_1\left(\frac{v}{c}D^1 + H^3 - H^2\right) + \partial_2\left(\frac{v}{c}D^2 - H^3 + H^1\right) + \partial_3\left(\frac{v}{c}D^3 + H^2 - H^1\right) = 0 \tag{744}$$

which in vector form is (in SI units):

$$\nabla \times \mathrm{H} - \frac{\partial \boldsymbol{D}}{\partial t} = v(\nabla \cdot \boldsymbol{D}) \tag{745}$$

and is a combination of the Ampère–Maxwell law:

$$\nabla \times \mathrm{H} - \frac{\partial \boldsymbol{D}}{\partial t} = \boldsymbol{J} = v(\nabla \cdot \boldsymbol{D}) = v\rho \tag{746}$$

and the Coulomb law:

$$\nabla \cdot \boldsymbol{D} = \rho \tag{747}$$

Equation (745) can be written as

$$\nabla \times \boldsymbol{H} = \left(\frac{\partial}{\partial t} + \mathbf{v}\nabla \cdot\right)\boldsymbol{D} = \frac{\partial \boldsymbol{D}}{\partial t} \tag{748}$$

where

$$\frac{\partial}{\partial t} \equiv \frac{\partial}{\partial t} + \mathbf{v}\nabla \cdot \tag{749}$$

is the convective derivative. The charge–current 4-vector in general is

$$J^{\mu} \equiv \left(\rho, \frac{\boldsymbol{J}}{c}\right) = \left(\rho, \frac{v}{c}\rho\right) \tag{750}$$

and in the vacuum is

$$J^{\mu}_{\text{vac}} \equiv \left(\rho_{\text{vac}}, \frac{1}{c}J_{\text{vac}}\right); \quad v = c \tag{751}$$

Therefore, charge density and current density in the vacuum and in matter take the same form, [see Eqs. (732) and (733)]. This is a general result of assuming an O(3) vacuum configuration as in Section I. Equations (736) are a form of Noether's theorem and charge/current enters the scene as the result of conservation and topology. Similarly, mass is curvature of the gravitational field.

In the vacuum

$$v = c; \quad \varepsilon^{\mu} = (1, 0, 0, 1) \tag{752}$$

and conservation of the PL pseudovector gives the continuity equation

$$\nabla \times \boldsymbol{D}^{(3)} = \nabla \times \boldsymbol{P}^{(3)} = -\frac{\partial \boldsymbol{B}^{(3)}}{\partial t} - c\nabla \cdot \boldsymbol{B}^{(3)}\boldsymbol{e}^{(3)} = \mathbf{0} \tag{753}$$

which is a post-Noether-invariant. We have used the vacuum relation:

$$\boldsymbol{D}^{(3)} = \varepsilon_0 \boldsymbol{E}^{(3)} + \boldsymbol{P}^{(3)} = \boldsymbol{P}^{(3)} \tag{754}$$

The vacuum polarization component $\boldsymbol{P}^{(3)}$ is equal to the vacuum displacement $\boldsymbol{D}^{(3)}$ and aligned along one axis, so its curl vanishes. If $\boldsymbol{B}^{(3)}$ were zero, then for a light-like $\epsilon^{\mu}, \tilde{G}^{\mu}$ would be null and the electromagnetic field would vanish a reduction to absurdity proof of the existence of $\boldsymbol{B}^{(3)}$ if we adopt the Poincaré group. The adoption of the latter group leads to the post-Noether-invariant equations (736), which break out into the field equations of O(3) electrodynamics. Since U(1) is an O(3) symmetry with one null axis (the Z axis), U(1) is in a sense a sub symmetry of O(3), and this property leads to the fact that O(3) equations can be expressed in the form of U(1) equations without self-contradiction. The following diagram, which outlines the rules for connecting U(1) and O(3), may help the reader understand how this process occurs.

Rules for Connecting U(1) and O(3)

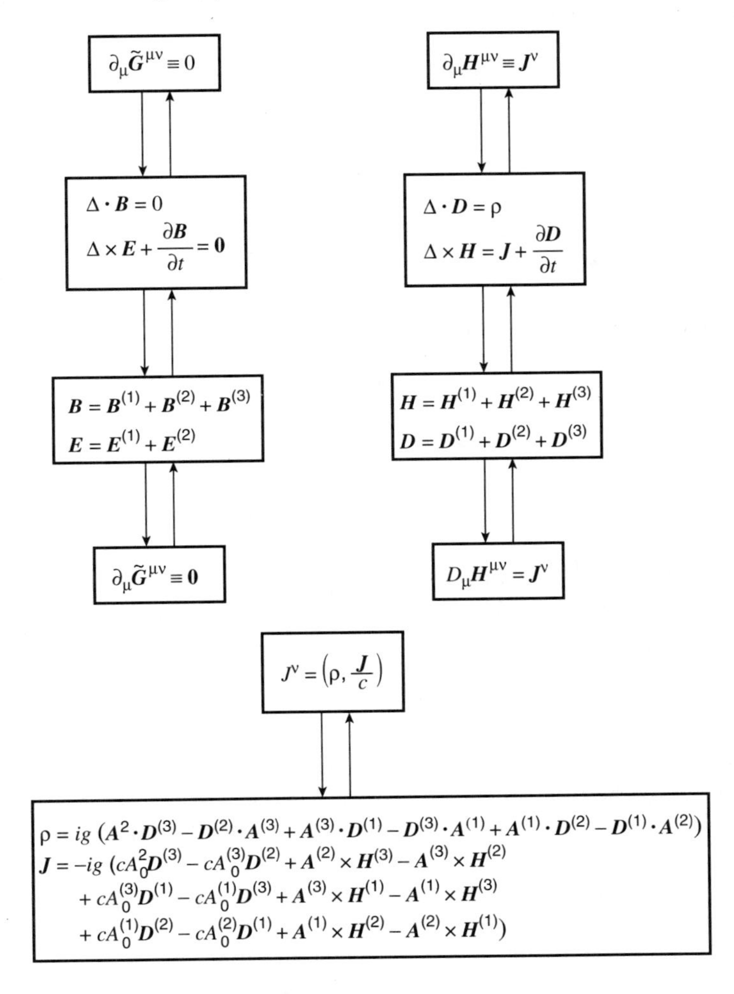

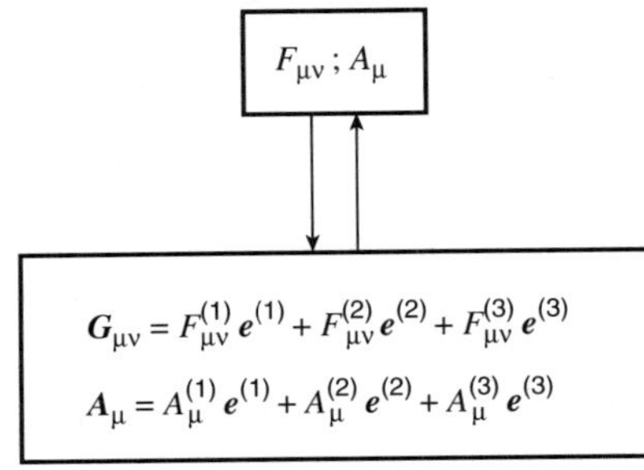

In the vacuum limit, we also obtain the following equation for the vacuum displacement $\boldsymbol{D}^{(3)}$ and vacuum polarization $\boldsymbol{P}^{(3)}$:

$$\frac{\partial \boldsymbol{D}^{(3)}}{\partial t} + c\frac{\partial \boldsymbol{D}^{(3)}}{\partial Z} = \boldsymbol{0} \tag{755}$$

Now use

$$\nabla \times \mathrm{H} = \left(\frac{\partial \boldsymbol{D}}{\partial t} + v(\nabla \cdot \boldsymbol{D})\right) \tag{756}$$

in the limit $v \longrightarrow c$, and take the (3) component to find that:

$$\nabla \times \boldsymbol{H}^{(3)} = \boldsymbol{0} \tag{757}$$

which gives

$$\nabla \times (\boldsymbol{A}^{(1)} \times \boldsymbol{A}^{(2)}) = \boldsymbol{0} \tag{758}$$

a result that is consistent with the definition of $\boldsymbol{B}^{(3)}$ in the vacuum, Eq. (38), because the curl of $\boldsymbol{A}^{(1)} \times \boldsymbol{A}^{(2)}$ is zero. The 3-component of Eq. (741) is simply

$$\nabla \cdot \boldsymbol{B}^{(3)} = \frac{\partial \boldsymbol{B}^{(3)}}{\partial t} = \boldsymbol{0} \tag{759}$$

because $\boldsymbol{E}^{(3)}$ is zero as proved already. The fact that $\boldsymbol{E}^{(3)}$ is zero is a direct consequence of the Jacobi identities (86) or (578). The same identities imply that there is no magnetic monopole or magnetic current in O(3) electrodynamics under any circumstances. The $\boldsymbol{B}^{(3)}$ component is topological in origin, and does not originate in a magnetic monopole *as a material particle*. These theoretical results are consistent with empirical data [11–20], which imply the presence of $\boldsymbol{B}^{(3)}$ and the absence of a magnetic monopole in nature.

In the Poincaré group, therefore, the fundamental spin of the electromagnetic field is represented ineluctably by the PL vector:

$$\tilde{B}^{\mu} = (B^{(3)}, 0, 0, B^{(3)}) \tag{760}$$

The integral of $\tilde{B}^{\mu}$ over a hypersurface in four-dimensions is always zero, a result of the ordinary Stokes theorem in four dimensions:

$$\oint \tilde{B}_{\mu}\, dx^{\mu} = \frac{1}{2}\int (\partial_{\mu}\tilde{B}_{\nu} - \partial_{\nu}\tilde{B}_{\mu})\partial\sigma^{\mu\nu} = 0 \tag{761}$$

The equivalent result in 3-space dimensions has been given by Evans and Jeffers [102]:

$$\oint \boldsymbol{B}^{(3)} \cdot d\boldsymbol{r} = 0 \tag{762}$$

and is simply a consequence of the fact that $\boldsymbol{B}^{(3)}$ is irrotational by definition. Therefore we obtain from Eq. (761) the results

$$\partial_\mu \tilde{B}_\nu = \partial_\nu \tilde{B}_\mu = 0 \tag{763}$$

and

$$\partial_\mu H_\nu = \partial_\nu H_\mu = J_{\mu\nu} = J_{\nu\mu} \tag{764}$$

These are alternative forms of the Lehnert or Panofsky–Phillips equations (612), which can be expanded out into the O(3) equations (95)–(106) using the rules in the above flowchart shown above [after text that followss Eq. (754)]. Conservation of helicity therefore requires the charge current tensor to be symmetric. Similarly, conservation of angular momentum requires the energy-momentum tensor to be symmetric in dynamics [6]. Therefore conservation of helicity generates the field equations and new conservation laws based on topology. Charge current itself is the result of topology as discussed by Ryder. [6, p. 93].

The Lie algebra of the PL vector within the Poincaré group is not well known and is given here for convenience. The PL vector is defined by

$$\tilde{W}_\mu = \tilde{J}_{\mu\nu} P^\nu \tag{765}$$

where

$$\tilde{J}_{\mu\nu} = \begin{bmatrix} 0 & J_1 & J_2 & J_3 \\ -J_1 & 0 & K_3 & -K_2 \\ -J_2 & -K_3 & 0 & K_1 \\ -J_3 & K_2 & -K_1 & 0 \end{bmatrix} \tag{766}$$

is a matrix of Poincaré group generators: the boost (K) and rotation (J) generators [6,11–20]. Here, P^ν is the generator of spacetime translation, which is missing from the Lorentz group. Therefore the PL vectors written out in full are

$$\begin{aligned} \tilde{W}_0 &= -J_1 P^1 + J_2 P^2 + J_3 P^3 \\ \tilde{W}_1 &= -J_1 P^0 + K_3 P^2 - K_2 P^3 \\ \tilde{W}_2 &= -J_2 P^0 - K_3 P^1 + K_1 P^3 \\ \tilde{W}_3 &= -J_1 P^0 + K_2 P^1 - K_1 P^2 \end{aligned} \tag{767}$$

These are linear operator relations implying the property

$$[P^0, \tilde{W}^\mu] = 0 \tag{768}$$

showing that the Hamiltonian operator $H = P^0$ [6,11–20] commutes with the complete vector $\tilde{W}^\mu$ under all conditions. Equation (768) implies

$$\partial_\mu \tilde{W}^\mu = 0 \tag{769}$$

as in Eq. (736). Relativistic helicity has no 4-divergence. From Eqs. (767), we obtain the closed Lie algebra

$$\begin{aligned}
[\tilde{W}^1, \tilde{W}^2] &= i(P^0\tilde{W}^3 + P^3\tilde{W}^0) \\
[\tilde{W}^2, \tilde{W}^3] &= i(P^0\tilde{W}^1 + P^1\tilde{W}^0) \\
[\tilde{W}^3, \tilde{W}^1] &= i(P^0\tilde{W}^2 - P^2\tilde{W}^0) \\
[\tilde{W}^0, \tilde{W}^1] &= i(P^3\tilde{W}^2 - P^2\tilde{W}^3) \\
[\tilde{W}^0, \tilde{W}^2] &= i(P^1\tilde{W}^3 - P^3\tilde{W}^1) \\
[\tilde{W}^0, \tilde{W}^3] &= i(P^2\tilde{W}^1 + P^1\tilde{W}^2)
\end{aligned} \tag{770}$$

and Jacobi identities such as

$$[\tilde{W}^1, [\tilde{W}^2, \tilde{W}^3]] + [\tilde{W}^2, [\tilde{W}^3, \tilde{W}^1]] + [\tilde{W}^3, [\tilde{W}^1, \tilde{W}^2]] = 0 \tag{771}$$

checking that $\tilde{W}^\mu$ is a valid generator of the Poincaré group. The Casimir invariants $P_\mu P^\mu$ and $\tilde{W}_\mu \tilde{W}^\mu$ are the two fundamental invariants of the Poincaré group.

In electromagnetic theory, we replace $\tilde{W}^\mu$ by $\tilde{G}^\mu$ the relativistic helicity of the field. Therefore, Eq. (770) forms a fundamental Lie algebra of classical electrodynamics within the Poincaré group. From first principles of the Lie algebra of the Poincaré group, the field $\boldsymbol{B}^{(3)}$ is nonzero.

If a light beam is considered propagating at c in Z, we obtain from Eqs. (770) the Lie algebra of the E(2) Euclidean group [6,11–20], which is a mathematical group with no physical meaning:

$$[\tilde{W}^1, \tilde{W}^2] = 0 \tag{772a}$$

$$[\tilde{W}^2, \tilde{W}^3] = iP^0\tilde{W}^1 \tag{772b}$$

$$[\tilde{W}^3, \tilde{W}^1] = iP^0\tilde{W}^2 \tag{772c}$$

compared with the O(3) Lie algebra

$$[\tilde{W}^1, \tilde{W}^2] = iP^0\tilde{W}^3 \tag{772d}$$

$$[\tilde{W}^2, \tilde{W}^3] = iP^0\tilde{W}^1 \tag{772e}$$

$$[\tilde{W}^3, \tilde{W}^1] = iP^0\tilde{W}^2 \tag{772f}$$

and similarly for $\tilde{G}^\mu$. The E(2) group is the Wigner little group for a particle whose mass is identically zero, and so such a particle does not exist in nature. This proves that the photon and neutrino both have identically nonzero mass. The Wigner little group for a particle with mass is the physical O(3) group. In terms of field components, Eq. (772b) gives (in $c = 1$ units)

$$[B^2 - E^1, B^3] = iB^{(0)}B^1 \tag{773}$$

which is satisfied by

$$\begin{aligned} [B^2, B^3] &= iB^{(0)}B^1 \\ [B^3, E^1] &= iB^{(0)}E^2 \end{aligned} \tag{774}$$

The first of these equations is an equation of the B cyclic theorem, which therefore emerges from the symmetry of the Poincaré group in free space. Similarly, Eq. (772c) gives:

$$[B^3, B^1 + E^2] = iB^{(0)}(B^2 - E^1) \tag{775}$$

which is satisfied by

$$\begin{aligned} [B^3, B^1] &= iB^{(0)}B^2 \\ [B^3, E^2] &= -iB^{(0)}E^1 \end{aligned} \tag{776}$$

The first of this pair of equations is another of the B cyclic equations. Finally, Eq. (772a) gives

$$[B^1 + E^2, B^2 - E^1] = 0 \tag{777}$$

which is satisfied by

$$\begin{aligned} [B^1, B^2] &= [E^1, E^2] \\ [E^2, B^2] &\equiv [E^2, B^2] \end{aligned} \tag{778}$$

where the first of this pair give the third and final equation of the B cyclic theorem.

The structure of the O(3) equations in condensed form [i.e., Eqs. (612)] emerges from the symmetry of the Poincaré group. Consider, for example, the three equations:

$$\begin{aligned}[P_2, J_3] &= iP_1 \\ [P_3, J_2] &= -iP_1 \\ [P_0, K_1] &= iP_1\end{aligned} \tag{779}$$

By definition, the generator of space-time translation is

$$P \equiv i\partial_\mu \tag{780}$$

so Eq. (779) becomes

$$([\partial_2, J_3] - [\partial_3, J_2] - [\partial_0, K_1])\psi = P_1\psi \tag{781}$$

where ψ is an eigenfunction. Equation (781) can be written as

$$(\partial_2 J_3 - \partial_3 J_2 - \partial_0 K_1 - (J_3\partial_2 - J_2\partial_3 - K_1\partial_0))\psi = 0 \tag{782}$$

which is a relation between operators on ψ. Now use

$$\begin{aligned}J_3\psi &= j_3\psi \\ J_2\psi &= j_2\psi \\ J_1\psi &= j_1\psi\end{aligned} \tag{783}$$

where lowercase letters denote eigenvalues. We have

$$\begin{aligned}\partial_2(j_3\psi) &= (\partial_2 j_3)\psi + j_3(\partial_2\psi) \\ \partial_3(j_2\psi) &= (\partial_3 j_2)\psi + j_2(\partial_3\psi) \\ \partial_0(k_1\psi) &= (\partial_0 k_1)\psi + k_1(\partial_0\psi)\end{aligned} \tag{784}$$

Assume that

$$J_3(\partial_2\psi) + J_2(\partial_3\psi) + K_1(\partial_0\psi) = j_3(\partial_2\psi) + j_2(\partial_3\psi) + k_1(\partial_0\psi) \tag{785}$$

an equation that is compatible with:

$$(\partial_2 + \partial_3 + \partial_0)\psi = \text{constant } \psi \tag{786}$$

Equations (781)–(786) give the eigenvalue relation

$$\partial_2 j_3 - \partial_3 j_2 - \partial_0 k_1 = P_1 \tag{787}$$

which is one component of

$$\nabla \times \boldsymbol{J} - \frac{1}{c}\frac{\partial \boldsymbol{K}}{\partial t} = \boldsymbol{P} \tag{788}$$

If we write

$$\psi \equiv e^{i\phi}\psi_0 \tag{789}$$

where ϕ is a phase factor, then

$$J_3\psi = J_3(e^{i\phi}\psi_0) = j_3^{(0)}e^{i\phi}\psi_0 \equiv j_3\psi \tag{790}$$

and so on. Therefore the eigenvalues appearing in Eq. (788) are phase-dependent in general. It is clear that the structure of Eq. (788) is the same as one of Eqs. (612). The complete set of operator relations leading to this equation is

$$\begin{aligned} ([\partial_1, J_2] - [\partial_2, J_1] - [\partial_0, K_3])\psi &= P_3\psi \\ ([\partial_2, J_3] - [\partial_3, J_2] - [\partial_0, K_1])\psi &= P_1\psi \\ ([\partial_3, J_1] - [\partial_1, J_3] - [\partial_0, K_2])\psi &= P_2\psi \end{aligned} \tag{791}$$

Similarly, the Lie algebra

$$([\partial_2, K_3] - [\partial_3, K_2] + [\partial_0, J_3])\psi = 0 \tag{792}$$

and so on leads to the eigenvalue relation

$$\nabla \times \boldsymbol{k} + \frac{1}{c}\frac{\partial \boldsymbol{j}}{\partial t} = \boldsymbol{0} \tag{793}$$

as another of Eqs. (612).

The Lie algebra

$$([\partial_1, J_1] + [\partial_2, J_2] + [\partial_3, J_3])\psi = 0 \tag{794}$$

gives

$$((\partial_1 J_1 - J_1\partial_1) + (\partial_2 J_2 - J_2\partial_2) + (\partial_3 J_3 - J_3\partial_3))\psi = 0 \tag{795}$$

Using

$$\begin{aligned} J_1\psi &= j_1\psi \\ \partial_1(j_1\psi) &= j_1(\partial_1\psi) + (\partial_1 j_1)\psi \end{aligned} \tag{796}$$

and assuming

$$J_1(\partial_1\psi) + J_2(\partial_2\psi) + J_3(\partial_3\psi) = j_1(\partial_1\psi) + j_2(\partial_2\psi) + j_3(\partial_3\psi) \tag{797}$$

leads to

$$\begin{aligned} \partial_1 j_1 + \partial_2 j_2 + \partial_3 j_3 &= 0 \\ \nabla \cdot \boldsymbol{j} &= 0 \end{aligned} \tag{798}$$

Therefore the complete set of equations (612) emerges in the form

$$\begin{aligned} \nabla \cdot \boldsymbol{k} &= 3p_0 \\ \nabla \cdot \boldsymbol{j} &= 0 \\ \nabla \times \boldsymbol{k} + \frac{1}{c}\frac{\partial \boldsymbol{j}}{\partial t} &= \boldsymbol{0} \\ \nabla \times \boldsymbol{j} - \frac{1}{c}\frac{\partial \boldsymbol{k}}{\partial t} &= \boldsymbol{p} \end{aligned} \tag{799}$$

simply by considering the symmetry of the Poincaré group. The vacuum charge–current is therefore intrinsic to the structure of the Poincaré group, but not of the Lorentz group, in which $\boldsymbol{p}$ is undefined. Structure (799) exists under all conditions because the Poincaré group applies under all conditions. Therefore O(3) electrodynamics emerges self-consistently from the symmetry of the Poincaré group, without a magnetic monopole or magnetic current *as material entities*, but with vacuum charge and current. This is a powerful result of symmetry.

Consideration of the symmetry of the Poincaré group also shows that the B cyclic theorem is independent of Lorentz boosts in any direction, and also reveals the physical meaning of the E(2) little group of Wigner. This group is unphysical for a photon without mass, but is physical for a photon with mass. This proves that Poincaré symmetry leads to a photon with identically nonzero mass. The proof is as follows. Consider in the particle interpretation the PL vector

$$W^{\mu} = -\frac{1}{2}\varepsilon^{\lambda\mu\nu\rho}P_{\mu}J_{\nu\rho} \tag{800}$$

Barut [102] shows that this PL vector obeys the cyclic conditions:

$$[W^{\lambda}, W^{\mu}] = -i\varepsilon^{\mu\nu\sigma\rho}P_{\sigma}W_{\rho} \tag{801}$$

For a particle (including the photon) with mass, the spacetime translation operator P^{μ} in the rest frame is

$$P^{\mu} = (P^0, 0, 0, 0) \tag{802}$$

and in the light-like condition

$$P^{\mu} = (P^0, 0, 0, P^0) \tag{803}$$

In the rest frame, Eq. (801) becomes [15]

$$\begin{aligned} [J_1, J_2] &= iJ_3 \\ [J_2, J_3] &= iJ_1 \\ [J_3, J_1] &= iJ_2 \end{aligned} \tag{804}$$

which is the Lie algebra of the rotation generators of the Lorentz group [6]. In the light-like condition, Eq. (801) becomes

$$\begin{aligned} [J_X + K_Y, J_Y - K_X] &= i(J_Z - J_Z) \\ [J_Y - K_X, J_Z] &= i(K_Y + J_X) \\ [K_Y + J_X, J_Z] &= i(K_X - J_Y) \end{aligned} \tag{805}$$

which has the symmetry of the E(2) group. Equation (805) can be written as

$$\begin{aligned} [J_X, J_Y] + [K_X, K_Y] &= iJ_Z - iJ_Z \\ [J_Y, J_Z] + [J_Z, K_X] &= iJ_X + iK_Y \\ -[J_Z, J_X] + [K_Y, J_Z] &= -iJ_Y + iK_X \end{aligned} \tag{806}$$

If we assume that the Lie algebra (804) is independent of Lorentz boosts in any direction, we obtain the Lie algebra:

$$\begin{aligned} [K_X, K_Y] &= -iJ_Z \\ [J_Z, K_X] &= iK_Y \\ [K_Y, J_Z] &= iK_X \end{aligned} \tag{807}$$

This is a Lie algebra of the Poincaré group [15] and of the Lorentz group [6], and is therefore self-consistently independent of spacetime translation. Therefore the meaning of the E(2) little group of Wigner is that it is a combination of the Lie algebra (804), which is independent of Lorentz boosts and spacetime translations; and of the Lie algebra (807), which is independent of spacetime translations. Note that the relation

$$[K_X, K_Y] = -iJ_Z \tag{808}$$

is the Thomas precession [6].

In the field interpretation [11–20], the Lie algebra (804) becomes [15]

$$[\hat{B}^{(1)}, \hat{B}^{(2)}] = -iB^{(0)}\hat{B}^{(3)} \\ \cdots \tag{809}$$

in the basis ((1),(2),(3)), which in vector notation is the B cyclic theorem:

$$\boldsymbol{B}^{(1)} \times \boldsymbol{B}^{(2)} = iB^{(0)}\boldsymbol{B}^{(3)*} \\ \cdots \tag{810}$$

The latter is therefore independent of Lorentz boosts of any kind, and independent of spacetime translations of any kind. As demonstrated previously in this chapter, this result can be arrived at independently and self-consistently by considering the following definition:

$$\boldsymbol{B}^{(3)*} \equiv -ig\boldsymbol{A}^{(1)} \times \boldsymbol{A}^{(2)} \tag{811}$$

The B cyclic theorem is therefore a Lie algebra independent of boosts and spacetime translations and is the same in the rest mass and light-like conditions for the photon. This result leads to the Lie algebra (807) for a particle with mass. The E(2) group becomes physical if the photon with mass is boosted to the speed of light, or, more precisely, infinitesimally close to the speed of light.

This symmetry analysis of the generators of the Poincaré group also shows in the field interpretation that the E(2) group contains the $\boldsymbol{B}^{(3)}$ field (corresponding in the particle interpretation to the J_Z generator) but does not contain the $\boldsymbol{E}^{(3)}$ field, corresponding in the particle interpretation to the K_Z generator. The Poincaré group also gives the structure of the O(3) equations of motion, Eqs. (799). In the field interpretation, the P^{μ} generator of the particle interpretation corresponds to charge–current. Therefore charge is analogous with energy and current with linear momentum. The magnetic field is analogous with the rotation generator, and the electric field is analogous with the boost generator. The Poincaré group Lie algebra produces the O(3) equations (799), and not the Maxwell–Heaviside equations. Our analysis throughout this chapter is therefore shown to be entirely self-consistent on the O(3) level, while there are many self-inconsistencies on the U(1) level. The normalized helicity of the photon with mass is -1, 0, 1, as for any boson with mass. In the rest frame, there is no helicity, because there is no forward momentum for a particle in its own rest frame. In the light-like condition (i.e., infinitesimally near the light-like condition), the three helicities are the space parts of the PL vector in that state:

$$\begin{aligned} W_1 &= J_1P_0 + K_2P_3 \\ W_2 &= J_2P_0 - K_1P_3 \\ W_3 &= J_3P_0 \end{aligned} \tag{812}$$

The time-like part of the PL vector is

$$W_0 = -J_3 P_3 \tag{813}$$

It can be seen that the PL vector is not proportional to P^{μ} in the light-like condition, thus removing another paradox [6] of the concept of massless photon.

In the U(1) gauge the vacuum field equations are:

$$\begin{aligned}(\partial^{\nu} + igA^{\nu})\tilde{F}_{\mu\nu} &= 0\\(\partial^{\nu} + igA^{\nu})F_{\mu\nu} &= 0\end{aligned} \tag{814}$$

and become the Maxwell equations if and only if

$$\begin{aligned}A^{\nu}\tilde{F}_{\mu\nu} &= 0\\A^{\nu}F_{\mu\nu} &= 0'\end{aligned} \tag{815}$$

which in vector notation correspond to

$$\begin{aligned}\boldsymbol{A}\cdot\boldsymbol{B} &= 0\\\boldsymbol{A}\times\boldsymbol{E} &= \mathbf{0}\\\boldsymbol{A}\cdot\boldsymbol{E} &= 0\\\boldsymbol{A}\times\boldsymbol{B} &= \mathbf{0}\end{aligned} \tag{816}$$

Therefore $\boldsymbol{A}\cdot\boldsymbol{B} = 0$ in the U(1) gauge in the vacuum. Unfortunately, the helicity in the U(1) gauge is defined by [103]

$$h \equiv \int \boldsymbol{A}\cdot\boldsymbol{B}\, dV \tag{817}$$

which is the linking number of field lines. This is zero because $\boldsymbol{A}\cdot\boldsymbol{B} = 0$, and helicity cannot be defined in the vacuum in the U(1) gauge. It is necessary to go to the O(3) level and to define helicity by

$$h_{O(3)} \equiv \int \boldsymbol{A}^{(1)}\cdot\boldsymbol{B}^{(2)}\, dV \tag{818}$$

It is only on this level that the link between helicity and topological quantization [103] can be understood properly. The O(3) group, like the U(1) group, is multiply connected. The group space of U(1) is a circle [6, p. 105]. As explained earlier in this review, this is not simply connected because a path that goes twice around a circle cannot be continuously deformed while staying on a circle into one that goes around once. The group space of SU(2) is S^3 [6, p. 411]. Every closed curve S^1 on S^3 may be shrunk to a point. The group O(3) is not simply connected but doubly connected, [6, p. 412]. Therefore the Aharonov–Bohm effect is possible only in O(3), as described in early sections of this review. We have the relation SO(3) = SU(2)/Z2. There are only two types of closed path S^1 in the group space of O(3): homotopic to a point and line [6]; therefore it is doubly connected. The topological theory of classical electromagnetism proposed by Ranada [103] thus can be extended systematically to the O(3) level. On the U(1) level used by Ranada, the electromagnetic knot is locally equivalent to the Maxwell–Heaviside equations. The electromagnetic knot is a field defined by the condition that their force lines are closed curves, and any pair of magnetic or electric lines is a link [103]. The linking lines are two integers that are interpreted as the Hopf indices of two applications from the sphere S^3 to the sphere S^2 at any instant. In the vacuum, the knots are such that $n_m = n_e$. Since $\boldsymbol{A} \cdot \boldsymbol{B}$ is identically zero in the U(1) gauge (Maxwell–Heaviside theory), this elegant theory needs to be upgraded to the O(3) level.

XIV. GAUGE FREEDOM AND THE LAGRANGIAN

We have just seen that the symmetry of the Poincaré group leads to vacuum charge and current as proposed by Panofsky and Phillips [86], Lehnert and Roy [10], and others. We must therefore seek a Lagrangian that gives the structure of the O(3) equations, a structure that, in condensed form, is identical with the Panofsky–Phillips and Lehnert–Roy equations. The Lagrangian leading to the Maxwell–Heaviside equations is deficient. It must also be explained why photon mass can enter gauge theory without making ten Lagrangian gauge not invariant. The problem with the Proca equation is that it removes gauge freedom, but at the expense of rendering the Lagrangian gauge noninvariant [6]. The original Proca equation is not therefore an entirely satisfactory approach to photon mass. The origin of photon mass (m_0) in O(3) electrodynamics is therefore topological, because the origin of charge–current is topological. The topology is expressed through gauge theory and group theory as discussed in Section I. On the U(1) level in the received view, a Lagrangian that does not contain a photon mass term is needed, Euler Lagrange equations have to be constructed, and constraints are needed to reduce the number of field variables so that there are no undetermined multipliers.

This program is not consistent with the Proca equation on the U(1) level. If the Proca equation

$$\partial_\mu H^{\mu\nu} - J^\nu = -\varepsilon_0 \frac{m_0^2 c^2}{\hbar^2} A^\nu \tag{819}$$

is used by ansatz, then it follows, by taking its divergence [6,15], that

$$m_0^2 \partial_\nu A^\nu = 0 \tag{820}$$

and if m_0 is not zero, the Lorenz condition is always obtained

$$\partial_\nu A^\nu = 0 \tag{821}$$

and the d'Alembert equation becomes

$$-\Box A_\mu = \frac{m_0^2 c^4}{\hbar^2} A_\mu \tag{822}$$

A condition is imposed on one of the four components of A_μ so that there are only three free components. However, the Lagrangian leading to the Proca equation is not gauge invariant due to the presence of a mass term [15]

$$\mathscr{L}_{m_0} = \frac{m_0^2}{2} A_\mu A^\mu \tag{823}$$

and the Proca equation always leads to the Lorenz condition, which is arbitrary and self-inconsistent. These disadvantages offset the advantages of the Proca equation; for example, it allows a three dimensional particle interpretation of the photon and it can be quantized without difficulty.

In U(1) gauge theory, the Lagrangian in general [6] contains the mass term (823), but in order to obtain the inhomogeneous Maxwell equations, this is discarded. This procedure is outlined, for example, on pp. 89ff. of Ref. 6. The U(1) Lagrangian in general is, in reduced units

$$\mathscr{L} = D_\mu \phi D_\mu \phi^* - m^2 \phi^* \phi - \frac{1}{4} H^{\mu\nu} H_{\mu\nu} \tag{824}$$

where ϕ is a scalar complex field and $F_{\mu\nu}$ is the electromagnetic field tensor. The Euler–Lagrange equation in the U(1) gauge is

$$\frac{\partial \mathscr{L}}{\partial A_\mu} - \partial_\nu \left(\frac{\partial \mathscr{L}}{\partial(\partial_\nu A_\mu)} \right) = 0 \tag{825}$$

and Eqs. (824) and (825) give

$$\partial_\mu H^{\mu\nu} = J^\nu = -ig(\phi^* D^\nu \phi - \phi D^\nu \phi^*) \tag{826}$$

The photon mass term in the Lagrangian

$$\mathscr{L} = -\frac{1}{4}H_{\mu\nu}H^{\mu\nu} + \frac{1}{2}m_0^2 A_\mu A^\mu \tag{827}$$

leading to the Proca equation in the received view [6] is not invariant under the gauge transformation

$$A^\mu \rightarrow A^\mu + \partial^\mu \chi \tag{828}$$

and is discarded in order to obtain the inhomogeneous Maxwell–Heaviside equation (826). The constant g appears in this theory as a coupling constant; it couples the ϕ and A^μ electromagnetic fields.

Therefore the fact that $\partial^\mu \chi$ is arbitrary in U(1) theory compels that theory to assert that photon mass is zero. This is an unphysical result based on the Lorentz group. When we come to consider the Poincaré group, as in section XIII, we find that the Wigner little group for a particle with identically zero mass is E(2), and this is unphysical. Since $\partial^\mu \chi$ in the U(1) gauge transform is entirely arbitrary, it is also unphysical. On the U(1) level, the Euler–Lagrange equation (825) seems to contain four unknowns, the four components of A^μ, and the field tensor $H^{\mu\nu}$ seems to contain six unknowns. This situation is simply the result of the term $H^{\mu\nu}$ in the initial Lagrangian (824) from which Eq. (826) is obtained. However, the fundamental field tensor is defined by the 4-curl:

$$F_{\mu\nu} = \partial_\mu A_\nu - \partial_\nu A_\mu \tag{829}$$

and the six components of the field are interrelated automatically by a constraint. The field tensor therefore contains only the four unknowns of A^μ by definition, and this definition is the constraint. The physical nature of the potential has been reviewed by Barrett [3,4].

It is well known that the Proca equation [6], Eq. (809), for a massive photon is not gauge-invariant because the Lagrangian (827) corresponding to it is not gauge-invariant. In SI units, this Lagrangian is

$$\mathscr{L} = -\frac{\varepsilon_0}{4}V_R H_{\mu\nu}H^{\mu\nu} + \frac{\varepsilon_0 V_R}{2}\frac{m_0^2 c^4}{\hbar^2}A_\mu A^\mu \tag{830}$$

where V_R is the radiation volume, ϵ_0 is the permittivity in vacuo, $H^{\mu\nu}$ is the field tensor, and m_0 is the mass of the photon. It is customary to adopt reduced units, so the Lagrangian becomes [6] Eq. (827), with:

$$\varepsilon_0 V = 1; \quad \frac{c^4}{\hbar^2} = 1 \tag{831}$$

The term $m_0^2 A_\mu A^\mu$ is not gauge-invariant under a local U(1) transform of A^μ. This problem can be circumvented by adopting the notion of the vacuum as the ground state of a scalar field ϕ:

$$\frac{\partial V}{\partial \phi} = 0 \tag{832}$$

where V is potential energy. This definition of the vacuum depends on the spontaneous symmetry breaking [6] of the Lagrangian:

$$\begin{aligned}\mathscr{L} &= (\partial_\mu \phi)(\partial^\mu \phi^*) - m^2 \phi^* \phi - \lambda(\phi^* \phi)^2 \\ &= (\partial_\mu \phi)(\partial^\mu \phi^*) - V(\phi, \phi^*)\end{aligned} \tag{833}$$

where λ is the self-interaction parameter, and assuming that $\mathscr{L}$ is invariant under the local transformation

$$\phi \rightarrow e^{i\Lambda(x^\mu)} \phi \tag{834}$$

the vacuum is the ground state

$$\frac{\partial V}{\partial \phi} = 0 = m^2 \phi^* + 2\lambda \phi^* (\phi^* \phi) \tag{835}$$

and the parameter m is allowed to become negative. This is the basis of the Higgs mechanism of introducing mass. If $m < 0$, there is a minimum at

$$a^2 \equiv |\phi|^2 = -\frac{m^2}{2\lambda}; \quad |\phi| = a \tag{836}$$

from the equation defining the vacuum [Eq. (835)]. In reduced units, spontaneous symmetry breaking of this type leads to the Lagrangian (824) and to the inhomogeneous field equation (826).

The charge–current density

$$J^\mu = -ig(\phi^* D^\mu \phi - \phi D^\mu \phi^*) \tag{837}$$

is a vacuum current because g exists in the vacuum and Eq. (837) is obtained from the definition of the vacuum, Eq. (835), as the ground state of the scalar field ϕ. The fundamental field $F_{\mu\nu}$ is completely defined in terms of the commutator of covariant derivatives:

$$F_{\mu\nu} \equiv \frac{i}{g}[D_\mu, D_\nu] \tag{838}$$

The Lagrangian (824) can be rewritten using Eq. (836) as [6]

$$\mathscr{L} = -\frac{1}{4}H_{\mu\nu}H^{\mu\nu} + \frac{1}{2}g^2a^2A_\mu A^\mu + \frac{1}{2}(\partial_\mu\phi_1)^2 + \frac{1}{2}(\partial_\mu\phi_2)^2 - 2\lambda a^2\phi_1^2 + \sqrt{2}gaA^\mu\partial_\mu\phi_2 + \cdots \tag{839}$$

The two Lagrangians (824) and (839) contain the same physical information, but in the form (839), the mass of the photon appears as the term $\frac{1}{2}g^2a^2A_\mu A^\mu$ in these reduced units. In SI units, the mass of the photon is

$$m_0\frac{c}{\hbar} = ga = g|\phi| \tag{840}$$

and using

$$g = \frac{\kappa}{|\phi|} \tag{841}$$

we recover the de Broglie guidance theorem [15]:

$$m_0c^2 = \hbar\omega \tag{842}$$

from the Higgs mechanism. The Proca equation is recovered in gauge-invariant form from the Lagrangian (839) if it is assumed that ϕ_2 vanishes as the result of spontaneous symmetry breaking. Using the Euler–Lagrange equation

$$\frac{\partial\mathscr{L}}{\partial A_\mu} - \partial_\nu\left(\frac{\partial\mathscr{L}}{\partial(\partial_\nu A_\mu)}\right) = 0 \tag{843}$$

the gauge-invariant Proca equation is as follows, in SI units:

$$\partial_\mu H^{\mu\nu} = J^\nu = -\varepsilon_0 g^2|\phi|^2\frac{c^2}{\hbar^2}A^\nu \tag{844}$$

The Lagrangian (824), which is the same as the Lagrangian (839), gives the inhomogeneous equation (826) using the same Euler–Lagrange equation (843). Therefore the photon mass can be identified with the vacuum charge–current density as follows (in SI units):

$$J^\nu = -\varepsilon_0 g^2|\phi|^2\frac{c^2}{\hbar^2}A^\nu = -ig\varepsilon_0\frac{c^2}{\hbar^2}(\phi^*D^\mu\phi - \phi D^\mu\phi^*) \tag{845}$$

This result, in turn, shows that the O(3) equations in their condensed form, Eq. (612), indicate the existence of photon mass. This is precisely the result obtained by Lehnert and Roy [10]. Canonical quantization of the gauge-invariant Proca equation proceeds without any problem to give the photon as a boson with helicities $-1, 0, 1$. This procedure is described in Ref. 6. In summary, it has been shown that the vacuum charge–current density and photon mass are the result of the Higgs mechanism.

Photon mass is shown to be self-consistent with O(3) electrodynamics by considering the O(3) Lagrangian [6] in reduced units:

$$\mathscr{L} = \frac{1}{2} D_\mu \phi_i)(D^\mu \phi_i) - \frac{m^2}{2} \phi_i \phi_i - \lambda(\phi_i \phi_i)^2 - \frac{1}{4} H^i_{\mu\nu} H^{i\mu\nu} \tag{846}$$

where i is the internal gauge index and D^μ is the covariant derivative of O(3) electrodynamics. The latter gives the usual results

$$\begin{aligned} D_\mu \phi_i &= \partial_\mu \phi_i + g\varepsilon_{ijk} A^j_\mu \phi_k \\ G^i_{\mu\nu} &= \partial_\mu A^i_\nu - \partial_\nu A^i_\mu + g\varepsilon^{ijk} A^j_\mu A^k_\nu \end{aligned} \tag{847}$$

and the potential V has a minimum [6] at

$$|\phi_0| = a = \left(-\frac{m^2}{4\lambda}\right)^{1/2} \tag{848}$$

where

$$\phi_0 = a\boldsymbol{e}_3 \equiv a\boldsymbol{e}^{(3)} \tag{849}$$

The O(3) Lagrangian becomes

$$\begin{aligned} \mathscr{L} = {} & \frac{1}{2}((\partial_\mu \phi_1)^2 + (\partial_\mu \phi_2)^2 + (\partial_\mu(\phi_3 - a))^2) + ag((\partial_\mu \phi_1) A^\mu_2 - (\partial_\mu \phi_2) A^\mu_1) \\ & + \frac{a^2 g^2}{2}((A^1_\mu)^2 + (A^2_\mu)^2) - \frac{1}{4} H^i_{\mu\nu} H^{i\mu\nu} - 4a^2 \lambda \chi^2 \end{aligned} \tag{850}$$

and contains the photon mass term

$$\mathscr{L}_m = \frac{a^2 g^2}{2}((A^1_\mu)^2 + (A^2_\mu)^2) \tag{851}$$

in gauge-invariant form. The photon mass in O(3) electrodynamics is therefore given again by Eq. (840). If it is assumed that

$$g = \frac{\kappa}{|\phi_0|} \tag{852}$$

the de Broglie guidance theorem (842) is again recovered self-consistently.

The Lagrangian (850) shows that O(3) electrodynamics is consistent with the Proca equation. The inhomogeneous field equation (32) of O(3) electrodynamics is a form of the Proca equation where the photon mass is identified with a vacuum charge-current density. To see this, rewrite the Lagrangian (850) in vector form as follows:

$$\mathscr{L} = D_\mu \boldsymbol{\phi} \cdot D^\mu \boldsymbol{\phi} - m^2 \boldsymbol{\phi} \cdot \boldsymbol{\phi} - \frac{1}{4} \boldsymbol{H}_{\mu\nu} \cdot \boldsymbol{H}^{\mu\nu} \tag{853}$$

The inhomogeneous O(3) field equation (32) is obtained through the Euler–Lagrange equation:

$$\frac{\partial \mathscr{L}}{\partial (A^i_\mu)} = \partial_\nu \left(\frac{\partial \mathscr{L}}{\partial (\partial_\nu A^i_\mu)} \right) \tag{854}$$

which gives Eq. (32) with the current term (in SI units):

$$D_\mu \boldsymbol{H}^{\mu\nu} = \boldsymbol{J}^\nu = \frac{\varepsilon_0 c^2}{\hbar^2} g (D^\nu \boldsymbol{\phi}) \times \boldsymbol{\phi} \tag{855}$$

In analogy with Eq. (845), the photon mass is defined in SI units by

$$D_\mu \boldsymbol{H}^{\mu\nu} = \boldsymbol{J}^\nu = -\varepsilon_0 g^2 |\phi_0|^2 \frac{c^2}{\hbar^2} \boldsymbol{A}^\nu \tag{856}$$

The individual terms of the charge current density ($\boldsymbol{J}^\nu$) in the vacuum are Noether currents of the type (101)–(106) and we have the following identifications under all conditions:

$$\begin{aligned}
\rho^{(1)} &= ig(\boldsymbol{A}^{(2)} \cdot \boldsymbol{D}^{(3)} - \boldsymbol{D}^{(2)} \cdot \boldsymbol{A}^{(3)}) \\
\rho^{(2)} &= ig(\boldsymbol{A}^{(3)} \cdot \boldsymbol{D}^{(1)} - \boldsymbol{D}^{(3)} \cdot \boldsymbol{A}^{(1)}) \\
\rho^{(3)} &= ig(\boldsymbol{A}^{(1)} \cdot \boldsymbol{D}^{(2)} - \boldsymbol{D}^{(1)} \cdot \boldsymbol{A}^{(2)}) \\
\boldsymbol{J}^{(1)*} &= -ig(cA_0^{(2)} \boldsymbol{D}^{(3)} - cA_0^{(3)} \boldsymbol{D}^{(2)} + \boldsymbol{A}^{(2)} \times \boldsymbol{H}^{(3)} - \boldsymbol{A}^{(3)} \times \boldsymbol{H}^{(2)}) \\
\boldsymbol{J}^{(2)*} &= -ig(cA_0^{(3)} \boldsymbol{D}^{(1)} - cA_0^{(1)} \boldsymbol{D}^{(3)} + \boldsymbol{A}^{(3)} \times \boldsymbol{H}^{(1)} - \boldsymbol{A}^{(1)} \times \boldsymbol{H}^{(3)}) \\
\boldsymbol{J}^{(3)*} &= -ig(cA_0^{(1)} \boldsymbol{D}^{(2)} - cA_0^{(2)} \boldsymbol{D}^{(1)} + \boldsymbol{A}^{(1)} \times \boldsymbol{H}^{(2)} - \boldsymbol{A}^{(2)} \times \boldsymbol{H}^{(1)})
\end{aligned} \tag{857}$$

The photon with mass has three degrees of freedom, so the O(3) procedure is again self-consistent. The key advantage of the O(3) procedure is that it produces a Proca equation that does not indicate the necessity for the Lorenz condition.

The U(1) Proca equation (819) implies that the Lorenz condition always holds, because Eq. (819) leads to

$$\partial_\nu A^\nu = 0 \tag{858}$$

The O(3) Proca equation (856) does not have this artificial constraint on the potentials, which are regarded as physical in this chapter. This overall conclusion is self-consistent with the inference by Barrett [104] that the Aharonov–Bohm effect is self-consistent only in O(3) electrodynamics, where the potentials are, accordingly, physical.

Having derived the Proca equation in gauge-invariant form on the U(1) and O(3) levels, canonical quantization can be attempted. Defining the photon mass in reduced units as

$$m_0 \equiv g|\phi|, \quad (c = 1, \hbar = 1) \tag{859}$$

canonical quantization of the Proca equation is similar to that of the Klein–Gordon equation discussed in section X. The difference is that the Klein–Gordon equation produces a massless photon. With the definition $m_0 = g|\phi|$, the canonical momentum from the gauge-invariant Lagrangian (827) is

$$\pi^\mu = \frac{\partial \mathscr{L}}{\partial \dot{A}_\mu} = \partial^\mu A^0 - \dot{A}^\mu \tag{860}$$

from which [6] it follows that

$$\pi^i = -\dot{A}^i; \quad \pi^0 = 0 \tag{861}$$

So on this U(1) level, the scalar photon represented by A^0 is set to zero and the Lorenz condition always applies, meaning no gauge freedom. This is self-inconsistent because the original Lagrangian from which Eq. (827) is obtained is a U(1) Lagrangian with gauge freedom. If so, the Lorenz condition cannot always apply. Leaving these problems aside for the sake of argument, the commutation relations fundamental to the method of canonical quantization become [6]

$$[A^i(\boldsymbol{x}, t), \pi_j(\boldsymbol{x}', t)] = i\delta^i_j \delta^3(\boldsymbol{x} - \boldsymbol{x}') \tag{862}$$

$$[\dot{A}_i(\boldsymbol{x}, t), A_j(\boldsymbol{x}', t)] = i g_{ij} \delta^3(\boldsymbol{x} - \boldsymbol{x}') \tag{863}$$

and the field can be expanded in the Fourier series

$$A_\mu(k) = \int \frac{\partial^3 k}{(2\pi)^3 2k_0} \sum_{\lambda=1}^{3} \varepsilon_\mu^{(\lambda)}(k)(a^{(\lambda)}(k)e^{-ikx} + a^{(\lambda)+}e^{ikx}) \tag{864}$$

implying

$$[a^{(\lambda)}(k), a^{(\lambda')+}(k')] = \delta_{\lambda\lambda'}\partial k^0 (2\pi)^3 \delta^3(\boldsymbol{k} - \boldsymbol{k}') \tag{865}$$

and a Hamiltonian:

$$H = \int \frac{\partial^3 k}{(2\pi)^3 2k_0} k_0 \sum_{\lambda=1}^{3} a^{(\lambda)+}(k) a^{(\lambda)}(k) \tag{866}$$

This gives a straightforward interpretation of the photon with mass as a particle, but this interpretation is self-inconsistent on the U(1) level, as argued.

Self-consistent quantization of the photon with mass can occur using the Higgs mechanism. Symmetry breaking of a U(1) theory gives one massive photon, A_μ; and symmetry breaking on the O(3) level gives one massive photon, A_μ^1, and one massive photon, A_μ^2. On the U(1) level, the time-like component of the photon is canceled by the scalar field, leaving three polarization states for the space-like part of the photon. On the O(3) level, symmetry breaking leads to one massive scalar field and two massive vector fields. The massive scalar field can be interpreted as a physical time-like photon with mass. This massive scalar field appears in the term $-4a^2\lambda\chi^2$ in the Lagrangian (850), where $\chi = \phi_3 - a$. It is also possible to define an effective physical longitudinal photon whose amplitude is the same as that of the physical scalar photon. This should not be confused with the superheavy photon that emerges from electroweak theory with an O(3) electromagnetic sector and observed as described in Section XII. In summary, physical time-like and longitudinal photons are missing from symmetry breaking of a U(1) theory, but are present after symmetry breaking of an O(3) theory. It can be seen from Eq. (826) that electric charge current density is defined by the scalar field ϕ, and the basic requirement for charge to exist from Noether's theorem [6] is that ϕ be complex. It is therefore possible to build up electromagnetic theory from topological considerations, in particular the complex scalar field ϕ, whose ground state is the vacuum.

From the foregoing, it becomes clear that fields and potentials are freely intermingled in the symmetry-broken Lagrangians of the Higgs mechanism. To close this section, we address the question of whether potentials are physical (Faraday and Maxwell) or mathematical (Heaviside) using the non-Abelian Stokes theorem for any gauge symmetry:

$$\oint D_\mu \, dx^\mu = -\frac{1}{2}\int [D_\mu, D_\nu] \, d\sigma^{\mu\nu} \tag{867}$$

On the U(1) level, this becomes

$$\oint A_\mu \, dx^\mu = -\frac{1}{2}\int F_{\mu\nu} \, d\sigma^{\mu\nu} \tag{868}$$

or in vector notation

$$\oint \boldsymbol{A} \cdot d\boldsymbol{r} = \int \boldsymbol{B} \cdot d\boldsymbol{Ar} = \int \nabla \times \boldsymbol{A} \cdot d\boldsymbol{Ar} \tag{869}$$

The gauge transformation rule on the U(1) level is

$$\boldsymbol{A} \rightarrow \boldsymbol{A} - \nabla\chi \tag{870}$$

and when applied to Eq. (869), it is found that

$$\oint \nabla\chi \cdot d\boldsymbol{r} = 0 \tag{871}$$

which is self-consistent with

$$\nabla \times (\nabla\chi) = \boldsymbol{0} \tag{872}$$

The Dirac phase factor

$$\exp\left(ig \oint A_\mu \, dx^\mu \right) = \exp\left(-i\frac{g}{2} \int F_{\mu\nu} \, d\sigma^{\mu\nu} \right) \tag{873}$$

is therefore gauge-invariant [3,4] and fully describes the electromagnetic phase factor on the U(1) level.

On the O(3) level, a gauge transformation applied to the theorem (867) produces

$$\oint \left(SA_\mu S^{-1} - \frac{i}{g}(\partial_\mu S)S^{-1} \right) dx^\mu = -\frac{1}{2} \int SG_{\mu\nu}S^{-1} \, d\sigma^{\mu\nu} \tag{874}$$

where

$$S = \exp\left(iM^a \Lambda^a(x^\mu)\right); \quad A^\mu = M^a A^a_\mu \tag{875}$$

Here, M^a are *physical* rotation generators of the O(3) group and Λ^a are *physical* angles [11–20]. The gauge transform produces

$$A^{(i)}_\mu \rightarrow A^{(i)}_\mu + \frac{1}{g}\partial_\mu \Lambda^{(i)}(x^\mu); \qquad i = 1, 2, 3 \tag{876}$$

so that the potential components of:

$$\boldsymbol{A}_\mu = A^{(1)}_\mu \boldsymbol{e}^{(1)} + A^{(2)}_\mu \boldsymbol{e}^{(2)} + A^{(3)}_\mu \boldsymbol{e}^{(3)} \tag{877}$$

are also *physical*. The gauge transform (874) also produces the result

$$\oint \partial_\mu \Lambda^a(x^\mu)\, dx^\mu = 0 \tag{878}$$

which means that

$$\partial_\nu \partial_\mu \Lambda^a = \partial_\mu \partial_\nu \Lambda^a \tag{879}$$

This result, however, is an identity of Minkowski spacetime itself, namely, $\partial_\nu\partial_\mu$ operating on a function of x^μ produces the same result as $\partial_\nu\partial_\mu$ operating on a function of x^μ. Equation (879) does not mean that Λ^a can take any value. We reach the important conclusion that the vector identity (872) of U(1) is a property of three-dimensional space itself and can always be interpreted as such. Therefore even on the U(1) level, Eq. (872) does not mean that χ can take any value. Even on the U(1) level, therefore, potentials can be interpreted physically, as was the intent of Faraday and Maxwell. On the O(3) level, potentials are always physical.

XV. BELTRAMI ELECTRODYNAMICS AND NONZERO $\boldsymbol{B}^{(3)}$

In this final section, it is shown that the three magnetic field components of electromagnetic radiation in O(3) electrodynamics are Beltrami vector fields, illustrating the fact that conventional Maxwell–Heaviside electrodynamics are incomplete. Therefore Beltrami electrodynamics can be regarded as foundational, structuring the vacuum fields of nature, and extending the point of view of Heaviside, who reduced the original Maxwell equations to their presently accepted textbook form. In this section, transverse plane waves are shown to be solenoidal, complex lamellar, and Beltrami, and to obey the Beltrami equation, of which $\boldsymbol{B}^{(3)}$ is an identically nonzero solution. In the Beltrami electrodynamics, therefore, the existence of the transverse $\boldsymbol{B}^{(1)} = \boldsymbol{B}^{(2)*}$ implies that of $\boldsymbol{B}^{(3)}$, as in O(3) electrodynamics.

As argued by Reed [4], the Beltrami vector field originated in hydrodynamics and is force-free. It is one of the three basic types of field: solenoidal, complex lamellar, and Beltrami. These vector fields originated in hydrodynamics and describe the properties of the velocity field, flux or streamline, $\boldsymbol{v}$, and the vorticity $\nabla \times \boldsymbol{v}$. The Beltrami field is also a Magnus force free fluid flow and is expressed in hydrodynamics as

$$\boldsymbol{v} \times (\nabla \times \boldsymbol{v}) = \boldsymbol{0} \tag{880}$$

The solenoidal vector field is:

$$\nabla \cdot \boldsymbol{v} = 0 \tag{881}$$

and the complex lamellar vector field is

$$\boldsymbol{v}(\nabla \times \boldsymbol{v}) = \mathbf{0} \tag{882}$$

The Beltrami condition can also be represented [4] as:

$$\nabla \times \boldsymbol{v} = k\boldsymbol{v} \tag{883}$$

where

$$k = \frac{1}{v^2}\boldsymbol{v} \cdot \nabla \times \boldsymbol{v} \tag{884}$$

for real-valued v.

Beltrami fields have been advanced [4] as theoretical models for astrophysical phenomena such as solar flares and spiral galaxies, plasma vortex filaments arising from plasma focus experiments, and superconductivity. Beltrami electrodynamic fields probably have major potential significance to theoretical and empirical science. In plasma vortex filaments, for example, energy anomalies arise that cannot be described with the Maxwell–Heaviside equations. The three magnetic components of O(3) electrodynamics are Beltrami fields as well as being complex lamellar and solenoidal fields. The component $\boldsymbol{B}^{(3)}$ is identically nonzero in Beltrami electrodynamics if $\boldsymbol{B}^{(1)} = \boldsymbol{B}^{(2)*}$ is so. In the Beltrami electrodynamics, $\boldsymbol{B}^{(3)}$ is a particular solution of the general solution given by Chandrasekhar and Kendall [4] of the Beltrami equation:

$$\nabla \times \boldsymbol{B} = k\boldsymbol{B} \tag{885}$$

This argument shows again that Maxwell–Heaviside electrodynamics is incomplete, because $\boldsymbol{B}^{(3)}$ is zero. General solutions are given in this section of the Beltrami equation, which is an equation of O(3) electrodynamics. Therefore these solutions are also general solutions of O(3) electrodynamics in the vacuum.

The three components of the B cyclic theorem (411) are solenoidal, complex lamellar, and Beltrami. This is a remarkable property of Beltrami electrodynamics when recognized as O(3) electrodynamics for the special case when $\boldsymbol{B}^{(1)} = \boldsymbol{B}^{(2)*}$ are plane waves. Specifically

$$\begin{aligned}
&\nabla \cdot \boldsymbol{B}^{(1)} = 0; \quad \boldsymbol{B}^{(1)} \cdot \nabla \times \boldsymbol{B}^{(1)} = 0; \quad \boldsymbol{B}^{(1)} \times (\nabla \times \boldsymbol{B}^{(1)}) = \mathbf{0} \\
&\nabla \cdot \boldsymbol{E}^{(1)} = 0; \quad \boldsymbol{E}^{(1)} \cdot \nabla \times \boldsymbol{E}^{(1)} = 0; \quad \boldsymbol{E}^{(1)} \times (\nabla \times \boldsymbol{E}^{(1)}) = \mathbf{0} \\
&\nabla \cdot \boldsymbol{A}^{(1)} = 0; \quad \boldsymbol{A}^{(1)} \cdot \nabla \times \boldsymbol{A}^{(1)} = 0; \quad \boldsymbol{A}^{(1)} \times (\nabla \times \boldsymbol{A}^{(1)}) = \mathbf{0}
\end{aligned} \tag{886}$$

and also for indices (2) and (3). Multiplying the Beltrami equation:

$$\nabla \times \boldsymbol{B}^{(1)} = k\boldsymbol{B}^{(1)} \tag{887}$$

on both sides by $\boldsymbol{B}^{(2)}$, it is seen that

$$\boldsymbol{B}^{(2)} \cdot \nabla \times \boldsymbol{B}^{(1)} = k\boldsymbol{B}^{(1)} \cdot \boldsymbol{B}^{(2)} \tag{888}$$

so the constant k is not necessarily zero when dealing with complex fields. To prove that k can be different from zero, consider the complex transverse magnetic plane wave

$$\boldsymbol{A}^{(1)} = \frac{A^{(0)}}{\sqrt{2}}(i\boldsymbol{i} + \boldsymbol{j})e^{i(\omega t - \kappa Z)} \tag{889}$$

which obeys the B cyclic theorem (411). From Eqs. (883) and (884)

$$k = \frac{1}{A^{(0)2}}\boldsymbol{A}^* \cdot \nabla \times \boldsymbol{A} = \frac{\boldsymbol{A}^* \cdot \boldsymbol{B}}{A^{(0)2}} = \kappa \tag{890a}$$

$$\nabla \times \boldsymbol{A}^{(1)} = \kappa A^{(1)} \tag{890b}$$

and all three components—(1), (2) and (3)—are solutions of the same Beltrami equation. Similarly, if we define the complete magnetic field vector by

$$\boldsymbol{B} \equiv \boldsymbol{B}^{(1)} + \boldsymbol{B}^{(2)} + \boldsymbol{B}^{(3)} \tag{891}$$

the complete vector $\boldsymbol{B}$ obeys Eq. (885).

On the U(1) level, if we start with the free-space Maxwell–Heaviside equations

$$\nabla \times \boldsymbol{E} + \frac{\partial \boldsymbol{B}}{\partial t} = \boldsymbol{0}; \quad \nabla \times \boldsymbol{B} - \frac{1}{c^2}\frac{\partial \boldsymbol{E}}{\partial t} = \boldsymbol{0} \tag{892}$$

it follows that

$$\nabla \times \boldsymbol{B} = k\boldsymbol{B} \tag{893a}$$

$$\nabla \times \boldsymbol{E} = k\boldsymbol{E} \tag{893b}$$

$$\nabla \times \boldsymbol{A} = k\boldsymbol{A} \tag{893c}$$

where $\boldsymbol{B} = \nabla \times \boldsymbol{A}$ as usual, and where $k = \pm\boldsymbol{\kappa}$. Here, k is a pseudo scalar that changes sign between left and right circularly polarized radiation. The Beltrami equation for $\boldsymbol{B}^{(3)}$ is

$$\nabla \times \boldsymbol{B}^{(3)} = k\boldsymbol{B}^{(3)} \tag{894}$$

where $k = 0$. It follows that all components of transverse plane waves are described by Beltrami equations in vacuo. For left-handed plane waves

$$\begin{aligned}
\boldsymbol{E}_L^{(1)} &= \frac{E^{(0)}}{\sqrt{2}}(\boldsymbol{i} - i\boldsymbol{j})e^{-i(\omega t - \kappa Z)} = \boldsymbol{E}_L^{(2)*} \\
\boldsymbol{B}_L^{(1)} &= \frac{B^{(0)}}{\sqrt{2}}(i\boldsymbol{i} + \boldsymbol{j})e^{-i(\omega t - \kappa Z)} = \boldsymbol{B}_L^{(2)*} \\
\boldsymbol{A}_L^{(1)} &= \frac{A^{(0)}}{\sqrt{2}}(i\boldsymbol{i} + \boldsymbol{j})e^{-i(\omega t - \kappa Z)} = \boldsymbol{A}_L^{(2)*}
\end{aligned} \tag{895}$$

For right-handed transverse plane waves

$$\begin{aligned}
\boldsymbol{E}_R^{(1)} &= \frac{E^{(0)}}{\sqrt{2}}(\boldsymbol{i} + i\boldsymbol{j})e^{-i(\omega t - \kappa Z)} = \boldsymbol{E}_R^{(2)*} \\
\boldsymbol{B}_R^{(1)} &= \frac{B^{(0)}}{\sqrt{2}}(-i\boldsymbol{i} + \boldsymbol{j})e^{-i(\omega t - \kappa Z)} = \boldsymbol{B}_R^{(2)*} \\
\boldsymbol{A}_R^{(1)} &= \frac{A^{(0)}}{\sqrt{2}}(-i\boldsymbol{i} + \boldsymbol{j})e^{-i(\omega t - \kappa Z)} = \boldsymbol{A}_R^{(2)*}
\end{aligned} \tag{896}$$

and for the longitudinal $\boldsymbol{B}^{(3)}$*field*

$$\boldsymbol{B}_L^{(3)} = -\boldsymbol{B}_R^{(3)} = B^{(0)}\boldsymbol{k} \tag{897}$$

Therefore

$$\begin{aligned}
\nabla \times \boldsymbol{B}_L^{(1)} &= -\kappa \boldsymbol{B}_L^{(1)}; &\quad \nabla \times \boldsymbol{B}_R^{(1)} &= \kappa \boldsymbol{B}_R^{(1)} \\
\nabla \times \boldsymbol{E}_L^{(1)} &= -\kappa \boldsymbol{E}_L^{(1)}; &\quad \nabla \times \boldsymbol{E}_R^{(1)} &= \kappa \boldsymbol{E}_R^{(1)} \\
\nabla \times \boldsymbol{A}_L^{(1)} &= -\kappa \boldsymbol{A}_L^{(1)}; &\quad \nabla \times \boldsymbol{A}_R^{(1)} &= \kappa \boldsymbol{A}_R^{(1)}
\end{aligned} \tag{898}$$

and similarly for index (2). For the longitudinal index (3)

$$\nabla \times \boldsymbol{B}_R^{(3)} = \nabla \times \boldsymbol{B}_L^{(3)} = \boldsymbol{0} \tag{899}$$

and all components are described by Beltrami equations in vacuo. Since $\boldsymbol{E}$ and $\boldsymbol{B}$ are the fundamental fields of electrodynamics, these equations are valid under all conditions. In particular, Eq. (893c) for the potential is not gauge-invariant under the transform:

$$\boldsymbol{A} \rightarrow \boldsymbol{A} - \nabla \chi \tag{900}$$

revealing that in Beltrami electrodynamics, $\boldsymbol{A}$ is physical. This result again supports Maxwell's postulate of a physical vector potential and does not support Heaviside's postulate of an unphysical vector potential. Equation (893c) is self-consistent, however, on the O(3) level, where potentials are physical. The covariant form of Eq. (893c) is

$$F_{\mu\nu} = \kappa A_{\mu\nu} \tag{901}$$

so the field tensor is directly proportional to an axial potential 4-tensor. This suggests that the vector potential can be polar or axial in nature. The solutions of Eq. (901) are also solutions of the d'Alembert equation in vacuo. In this view, the field tensor is directly proportional to the axial potential tensor $A_{\mu\nu}$, and so gauge freedom is lost because, if $F_{\mu\nu}$ is gauge-invariant, so is $A_{\mu\nu}$. This result is another internal inconsistency of the Maxwell–Heaviside point of view.

The Faraday law of induction does not distinguish between left and right circular polarization, that is, the structure of the equation is the same for R and L:

$$\begin{aligned} \nabla \times \boldsymbol{E}_L^{(1)} &= -\frac{\partial \boldsymbol{B}_L^{(1)}}{\partial t} \\ \nabla \times \boldsymbol{E}_R^{(1)} &= -\frac{\partial \boldsymbol{B}_R^{(1)}}{\partial t} \end{aligned} \tag{902}$$

On the other hand, the corresponding Beltrami equations are distinct:

$$\begin{aligned} \nabla \times \boldsymbol{E}_L^{(1)} &= -\kappa \boldsymbol{E}_L^{(1)} \\ \nabla \times \boldsymbol{E}_R^{(1)} &= \kappa \boldsymbol{E}_R^{(1)} \end{aligned} \tag{903}$$

The handedness, or chirality, inherent in foundational electrodynamics at the U(1) level manifests itself clearly in the Beltrami form (903). The chiral nature of the field is inherent in left- and right-handed circular polarization, and the distinction between axial and polar vector is lost. This result is seen in Eq. (901), where $A_{\mu\nu}$ is a tensor form that contains axial and polar components of the potential. This is precisely analogous with the fact that the field tensor $F_{\mu\nu}$ contains polar (electric) and axial (magnetic) components intermixed. Therefore, in propagating electromagnetic radiation, there is no distinction between polar and axial. In the received view, however, it is almost always asserted that $\boldsymbol{E}$ and $\boldsymbol{A}$ are polar vectors and that $\boldsymbol{B}$ is an axial vector.

The $\boldsymbol{B}^{(3)}$ component [which is nonzero only on the O(3) level] is a solution of the Beltrami equation (885) with $k = 0$. Therefore, in Beltrami electrodynamics, $\boldsymbol{B}^{(3)}$ is a solenoidal, irrotational, complex lamellar and Beltrami field in the vacuum, and is also a propagating field. The $\boldsymbol{B}^{(3)}$ component in Beltrami electrodynamics is part of the general solution of the solenoidal Beltrami equation given in Ref. 4, and is identically nonzero in the vacuum. This statement is equivalent to saying that electrodynamics is an O(3) Yang–Mills theory in the vacuum. The general solution in cylindrical components of Eq. (885) is

$$\boldsymbol{B} = \sum_{m,n} B_{mn}\boldsymbol{b}^{mn}(r, \theta, z) \tag{904}$$

where m is a nonnegative integer and where $\boldsymbol{b}^{mn}$ depends on ϕ and Z through $\phi = m\theta + nZ$. The expressions for the modes depend on linear combinations of Bessel and Neumann functions, J_m and N_m, similar to the solutions of the Helmholtz equation [5]. When the domain of solution involves the axis $r = 0$, and solutions are restricted to axisymmetric wave equations, then

$$\frac{1}{r}\frac{\partial}{\partial r}\left(r\frac{\partial\psi}{\partial r}\right) = -k^2\psi \tag{905}$$

The solution of this equation is [4]

$$\psi = C\,J_0(kr) \tag{906}$$

where C is any constant, and the solution specializes to:

$$\boldsymbol{B} = B_0(0, J_1(kr), J_0(kr)) \tag{907}$$

for the mode $m = n = 0; a = (0, 0, 1)$. Therefore the unit vector $a = (0, 0, 1)$ designates the Z axis. The solution for the $\boldsymbol{B}^{(3)}$ component is

$$\boldsymbol{B}^{(3)} = B_0(0, J_1(0), J_0(0)) \tag{908}$$

and depends on the Bessel functions $J_1(0)$ and $J_0(0)$. Therefore

$$\begin{aligned} \boldsymbol{B}^{(3)} &= \boldsymbol{B}(k = 0, m = 0, n = 0) \\ &= B_0(0, 0, 1) = B^{(0)}\boldsymbol{k} \end{aligned} \tag{909}$$

and $\boldsymbol{B}^{(3)}$ is an identically nonzero, phaseless function directed in the Z axis. This result is self-consistent with that of O(3) electrodynamics.

In conducting media, the wave number $\boldsymbol{\kappa}$ becomes complex [5], and by separating real and imaginary parts, we can obtain the Beltrami equations:

$$\nabla \times \boldsymbol{A} = k\boldsymbol{A}; \quad k \equiv i\kappa'' \tag{910}$$

$$\nabla \times B = kB; \quad k \equiv i\kappa'' \tag{911}$$

Taking the curl of Eq. (910) gives

$$\nabla \times (\nabla \times \boldsymbol{A}) = k\nabla \times \boldsymbol{A} = k^2\boldsymbol{A} \tag{912}$$

which can be rewritten as

$$\nabla^2\boldsymbol{A} = \kappa''^2\boldsymbol{A} \tag{913}$$

using the vector identity

$$\nabla \times (\nabla \times \boldsymbol{A}) = \nabla \times (\nabla \cdot \boldsymbol{A}) - \nabla^2\boldsymbol{A} \tag{914}$$

The covariant form of Eq. (914) is

$$\Box A^\mu = -\kappa''^2 A^\mu \tag{915}$$

If we assume

$$\kappa'' = \frac{m_0 c}{\hbar} \tag{916}$$

Eq. (915) becomes the Proca equation, and Eq. (916), the de Broglie guidance theorem.

Similarly, Eq. (911) becomes the equation of the Meissner effect in superconductivity:

$$\nabla^2\boldsymbol{B} = \kappa''^2\boldsymbol{B} \tag{917}$$

Finally, using

$$\nabla \times \boldsymbol{B} = k\boldsymbol{B} = k\nabla \times \boldsymbol{A} = k^2\boldsymbol{A} \tag{918}$$

we obtain the London equation:

$$\boldsymbol{J} = \nabla \times \boldsymbol{B} = -\kappa''^2\boldsymbol{A} \tag{919}$$

It is seen that the acquisition of mass by the photon is the result of an equation of superconductivity, and this is, of course, the basis of spontaneous symmetry breaking and the Higgs mechanism (Section XIV). Beltrami equations account for all these phenomena, and are foundational in nature. Note that the London equation (919) is not gauge-invariant on the U(1) level because a physical gauge-invariant current is proportional to the vector potential, which, in the received view, is gauge-noninvariant. This is another flaw of U(1) electrodynamics in the received opinion. The electric field from the London equation is zero because the current $\boldsymbol{J}$ is time-independent:

$$\boldsymbol{E} = -\frac{\partial \boldsymbol{A}}{\partial t} = \boldsymbol{0} \tag{920}$$

By Ohm's law, the resistance of the conducting medium vanishes, and the medium becomes a superconductor. The Higgs mechanism and spontaneous symmetry breaking were derived using the properties of superconductors.

TECHNICAL APPENDIX A: THE NON-ABELIAN STOKES THEOREM

The non-Abelian Stokes theorem is a relation between covariant derivatives for any gauge group symmetry:

$$\oint D_\mu \, dx^\mu = -\frac{1}{2}\int [D_\mu, D_\nu]\, d\sigma^{\mu\nu} \tag{A.1}$$

This expression can be expanded as

$$\oint (\partial_\mu - igA_\mu) dx^\mu = -\frac{1}{2}\int \left[\partial_\mu - igA_\mu,\ \partial_\nu - igA_\nu\right] d\sigma^{\mu\nu} \tag{A.2}$$

The terms

$$\oint \partial_\mu \, dx^\mu = \left[\partial_\mu,\ \partial_\nu\right] = 0 \tag{A.3}$$

are zero because by symmetry

$$\partial_\nu \partial_\mu = \partial_\mu \partial v \tag{A.4}$$

so

$$\oint \partial_\mu \, dx^\mu = -\frac{1}{2}\int [\partial_\mu,\ \partial v]\, d\sigma^{\mu\nu} = 0 \tag{A.5}$$

The half-commutators are evaluated as follows

$$[A_\mu, \partial_\nu] = -\partial_\nu A_\mu; \qquad [\partial_\mu, A_\nu] = \partial_\mu A_\nu \tag{A.6}$$

giving the non-Abelian Stokes theorem

$$\oint A_\mu \, dx^\mu = -\frac{1}{2}\int G_{\mu\nu}\, d\sigma^{\mu\nu} \tag{A.7}$$

where the field tensor for any gauge group is

$$G_{\mu\nu} \equiv \partial_\mu A_\nu - \partial_\nu A_\mu - ig[A_\mu, A_\nu] \tag{A.8}$$

On the U(1) level, the 4-potential is

$$A_\mu = (\phi, c\boldsymbol{A}) \tag{A.9}$$

and the field tensor is

$$F_{\mu\nu} = \begin{bmatrix} 0 & \frac{E_1}{c} & \frac{E_2}{c} & \frac{E_3}{c} \\ \frac{-E_1}{c} & 0 & B_3 & -B_2 \\ \frac{-E_2}{c} & -B_3 & 0 & B_1 \\ \frac{-E_3}{c} & B_2 & -B_1 & 0 \end{bmatrix} \tag{A.10}$$

Summing over repeated indices gives the time-like relation

$$\oint \phi \, dt = \frac{1}{2c^2}\left(\int E_X \, d\sigma^{01} + \int E_Y \, d\sigma^{02}\right) \tag{A.11}$$

where the SI units on either side are those of electric field strength multiplied by area. Summing over space indices gives

$$\oint A_1 \, dx^1 + A_2 \, dx^2 + A_3 \, dx^3 = -\frac{1}{2}\int F_{ij} \, d\sigma^{ij} \tag{A.12}$$

which can be rewritten as

$$\begin{aligned} \oint A_1 \, dx^1 &= -\frac{1}{2}\int F_{23} \, d\sigma^{23} + F_{32} \, d\sigma^{32} = -\int B_1 \, d\sigma^{23} \\ \oint A_2 \, dx^2 &= -\frac{1}{2}\int F_{31} \, d\sigma^{31} + F_{13} \, d\sigma^{13} = -\int B_2 \, d\sigma^{31} \\ \oint A_3 \, dx^3 &= -\frac{1}{2}\int F_{12} \, d\sigma^{12} + F_{21} \, d\sigma^{21} = -\int B_3 \, d\sigma^{12} \end{aligned} \tag{A.13}$$

In Cartesian coordinates, this is

$$\begin{aligned} \oint A_X \, dX &= \int B_x \, d\sigma^{YZ} \\ \oint A_Y \, dY &= \int B_Y \, d\sigma^{ZX} \\ \oint A_Z \, dZ &= \int B_Z \, d\sigma^{XY} \end{aligned} \tag{A.14}$$

or in condensed notation

$$\oint \boldsymbol{A} \cdot d\boldsymbol{r} = \int \boldsymbol{B} \cdot d\boldsymbol{Ar} \tag{A.15}$$

This is the Stokes theorem as usually found in textbooks. For plane waves, $\boldsymbol{A}$ is always perpendicular to the path, so in free space

$$\oint \boldsymbol{A} \cdot d\boldsymbol{r} = 0 \equiv \oint A_Z dZ \Rightarrow \nabla \times \boldsymbol{A}_Z = \boldsymbol{0} \tag{A.16}$$

On the O(3) level, there is a nonzero commutator and an additional term

$$\oint A_3^{(3)}\, dx^3 = -i\frac{g}{2}\left(\int [A_1^{(1)}, A_2^{(2)}]\, d\sigma^{12} + \int [A_2^{(1)}, A_1^{(2)}]\, d\sigma^{21}\right) \tag{A.17}$$

in the basis ((1),(2),(3)) defined by

$$\begin{gathered} \boldsymbol{e}^{(1)} \times \boldsymbol{e}^{(2)} = i\boldsymbol{e}^{(3)*} \\ \cdots \end{gathered} \tag{A.18}$$

In Cartesian form, Eq. (A.17) becomes

$$\oint A_Z^{(3)}\, dZ = -ig \int [A_X^{(1)}, A_Y^{(2)}]\, dAr = \int B_Z^{(3)}\, dAr \tag{A.19}$$

and explains the Sagnac effect as in the text. There are time-like relations such as

$$\oint A_0\, dx^0 = -\frac{1}{2}\int \partial_0 A_\nu - \partial_\nu A_0 - ig[A_0, A_\nu]\, d\sigma^{0\nu} \tag{A.20}$$

which define the scalar potential in O(3) electrodynamics to be nonzero and structured.

TECHNICAL APPENDIX B: 4-VECTOR MAXWELL–HEAVISIDE EQUATIONS

In this second technical appendix, it is shown that the Maxwell–Heaviside equations can be written in terms of a field 4-vector $G^\mu = (0, c\boldsymbol{B} + i\boldsymbol{E})$ rather than as a tensor. Under Lorentz transformation, G^μ transforms as a 4-vector. This shows that the field in electromagnetic theory is not uniquely defined as a 4-tensor. The Maxwell–Heaviside equations can be written in terms of the 4-vectors:

$$G^\mu = (0, c\boldsymbol{B} + i\boldsymbol{E}) \tag{B.1}$$

and

$$H^{\mu} = (0, \boldsymbol{H} + ic\boldsymbol{D}) \tag{B.2}$$

as

$$\begin{aligned} \partial_{\mu}G^{\mu} &= 0 \\ &[\partial_i, G_j] + i[\partial_0, G_k] = 0 \\ \partial_{\mu}H^{\mu} &= i\rho c \\ &[\partial_i, H_j] + i[\partial_0, H_k] = J_k \end{aligned} \tag{B.3}$$

Under Lorentz transformation:

$$\begin{aligned} G_{\mu}G^{\mu} &= G'_{\mu}G^{\mu'} \\ H_{\mu}H^{\mu} &= H'_{\mu}H^{\mu'} \end{aligned} \tag{B.4}$$

Using the fact that ρ and $\boldsymbol{J}$ themselves form the components of a 4-vector, the Maxwell–Heaviside equations for field matter interaction can be combined into one relation between 4-vectors:

$$(-i\partial_{\mu}H^{\mu}, ([\partial_i, H_j] + i[\partial_0, H_k])) = c\left(\rho, \frac{1}{c}J_k\right) \tag{B.5}$$

The free-space equivalent is

$$(\partial_{\mu}G^{\mu}, [\partial_i, G_i] + i[\partial_0, G_k]) = 0 \tag{B.6}$$

A Lorentz boost in the Z direction of the vector G^{μ} produces

$$\begin{aligned} cB'_X + iE'_X &= cB_X + iE_X \\ cB'_Y + iE'_Y &= cB_Y + iE_Y \\ cB'_Z + iE'_Z &= \gamma(cB'_Z + iE'_Z) \\ cB'_0 + iE'_0 &= -\gamma\beta(cB'_Z + iE'_Z) \end{aligned} \tag{B.7}$$

but a Lorentz transform in the Z direction applied to $F^{\mu\nu}$ produces

$$\begin{aligned} cB'_X &= \gamma(cB_X + \beta E_Y) \\ cB'_Y &= \gamma(cB_Y - \beta E_X) \\ cB'_Z &= cB_Z \\ B'_0 &= 0 \end{aligned} \tag{B8}$$

The results (B.7) and (B.8) are different, even though both describe a boost of the same vector equations, the Maxwell–Heaviside equations:

$$\nabla \cdot \boldsymbol{B} = 0$$
$$\nabla \times E + \frac{\partial \boldsymbol{B}}{\partial t} = \boldsymbol{0} \tag{B.9}$$

The only common factor is that the charge–current 4-tensor transforms in the same way. The vector representation develops a time-like component under Lorentz transformation, while the tensor representation does not. However, the underlying equations in both cases are the Maxwell–Heaviside equations, which transform covariantly in both cases and obviously in the same way for both vector and tensor representations.

If we define the vectors

$$\boldsymbol{a} \equiv \frac{1}{2}(c\boldsymbol{B} + i\boldsymbol{E})$$
$$\boldsymbol{b} \equiv \frac{1}{2}(c\boldsymbol{B} - i\boldsymbol{E}) \tag{B.10}$$

then

$$[a_X, a_Y] = ia_Z \cdots$$
$$[b_X, b_Y] = ib_Z \cdots \tag{B.11}$$
$$[a_i, a_i] = 0 \qquad (i,j = X, Y, Z)$$

and $\boldsymbol{a}$ and $\boldsymbol{b}$ both generate a group SU(2). The Lorentz group is then SU(2) $\otimes$ SU(2) and transforms in a well-defined way labeled by two angular momenta $(\boldsymbol{j}, \boldsymbol{j}')$, the first corresponding to $\boldsymbol{a}$ and the second to $\boldsymbol{b}$. Thus $\boldsymbol{a}$ and $\boldsymbol{b}$ are generators of the Lorentz group. The vector G^μ also transforms as a rest frame Pauli–Lubanski vector, suggesting that the vector representation is suitable for intrinsic photon spin, and the tensor representation for orbital angular momentum. This is also suggested by O(3) electrodynamics where the fundamental intrinsic spin of the field is $\boldsymbol{B}^{(3)}$.

TECHNICAL APPENDIX C: ON THE ABSENCE OF MAGNETIC MONOPOLES AND CURRENTS IN O(3) ELECTRODYNAMICS

The non-Abelian Stokes theorem

$$\oint D_\mu \, dx^\mu + \frac{1}{2}\int [D_\mu, D_\nu] \, d\sigma^{\mu\nu} = 0 \tag{C.1}$$

is the integral form of the Jacobi identity

$$\sum_{\sigma,\mu,\nu} [D_\sigma, [D_\mu, D_\nu]] = 0 \tag{C.2}$$

which is an identity between spacetime translation generators of the Poincaré group. Since

$$D_\mu = \partial_\mu - igA_\mu \tag{C.3}$$

for any gauge group symmetry, it follows that the identity (C.2) holds for the different components of D_μ. In an O(3) gauge, group symmetry identity (C.2) can be written as the field equation (31) of the text, so it follows that

$$\partial_\mu \tilde{\boldsymbol{G}}^{\mu\nu} \equiv \boldsymbol{0} \tag{C.4}$$

$$A_\mu \times \tilde{\boldsymbol{G}}^{\mu\nu} \equiv \boldsymbol{0} \tag{C.5}$$

Equation (C.5) means that there are no magnetic charge or current densities in O(3) electrodynamics.

It follows that

$$\boldsymbol{A}^{(2)} \cdot \boldsymbol{B}^{(3)} - \boldsymbol{B}^{(2)} \cdot \boldsymbol{A}^{(3)} = \boldsymbol{0} \tag{C.6}$$

$$\boldsymbol{A}^{(3)} \cdot \boldsymbol{B}^{(1)} - \boldsymbol{B}^{(3)} \cdot \boldsymbol{A}^{(1)} = \boldsymbol{0} \tag{C.7}$$

$$\boldsymbol{A}^{(1)} \cdot \boldsymbol{B}^{(2)} - \boldsymbol{B}^{(1)} \cdot \boldsymbol{A}^{(2)} = \boldsymbol{0} \tag{C.8}$$

The third equation is always true if

$$\boldsymbol{B}^{(1)} = \nabla \times \boldsymbol{A}^{(1)}; \qquad B^{(2)} = \nabla \times A^{(2)} \tag{C.9}$$

because of the vector identity

$$\nabla \cdot (\boldsymbol{F} \times \boldsymbol{G}) = \boldsymbol{G} \cdot (\nabla \times \boldsymbol{F}) - \boldsymbol{F} \cdot (\nabla \times \boldsymbol{G}) \tag{C.10}$$

and the first two equations are always true because (3) is always orthogonal to (1) and (2).

It also follows that

$$cA_0^{(3)} \boldsymbol{B}^{(2)} - cA_0^{(2)} \boldsymbol{B}^{(3)} + \boldsymbol{A}^{(2)} \times \boldsymbol{E}^{(3)} - \boldsymbol{A}^{(3)} \times \boldsymbol{E}^{(2)} = \boldsymbol{0} \tag{C.11}$$

$$cA_0^{(1)} \boldsymbol{B}^{(3)} - cA_0^{(3)} \boldsymbol{B}^{(1)} + \boldsymbol{A}^{(3)} \times \boldsymbol{E}^{(1)} - \boldsymbol{A}^{(1)} \times \boldsymbol{E}^{(3)} = \boldsymbol{0} \tag{C.12}$$

$$cA_0^{(2)} \boldsymbol{B}^{(1)} - cA_0^{(1)} \boldsymbol{B}^{(2)} + \boldsymbol{A}^{(1)} \times \boldsymbol{E}^{(2)} - \boldsymbol{A}^{(2)} \times \boldsymbol{E}^{(1)} = \boldsymbol{0} \tag{C.13}$$

and using

$$A_0^{(2)} = A_0^{(1)} \equiv 0; \qquad |\boldsymbol{A}^{(3)}| = A_0^{(3)} \tag{C.14}$$

Eqs. (C.11) and (C.12) give

$$c\boldsymbol{B}^{(1)} = \boldsymbol{k} \times \boldsymbol{E}^{(1)} - \boldsymbol{A}^{(1)} \times \frac{\boldsymbol{E}^{(3)}}{cA_0^{(3)}} \tag{C.15}$$

$$c\boldsymbol{B}^{(2)} = \boldsymbol{k} \times \boldsymbol{E}^{(2)} - \boldsymbol{A}^{(2)} \times \frac{\boldsymbol{E}^{(3)}}{cA_0^{(3)}} \tag{C.16}$$

However, we know from Eq. (C.4) that

$$\nabla \times E^{(1)} + \frac{\partial \boldsymbol{B}^{(1)}}{\partial t} = \boldsymbol{0} \tag{C.17}$$

$$\nabla \times E^{(2)} + \frac{\partial \boldsymbol{B}^{(2)}}{\partial t} = \boldsymbol{0} \tag{C.18}$$

so

$$c\boldsymbol{B}^{(1)} = \boldsymbol{k} \times \boldsymbol{E}^{(1)} \tag{C.19}$$

$$c\boldsymbol{B}^{(2)} = \boldsymbol{k} \times \boldsymbol{E}^{(2)} \tag{C.20}$$

and $\boldsymbol{E}^{(3)}$ is identically zero because $A_0^{(3)}$, $\boldsymbol{A}^{(1)}$, and $\boldsymbol{A}^{(2)}$are nonzero. It follows that

$$\frac{\partial \boldsymbol{B}^{(3)}}{\partial t} = \boldsymbol{0} \tag{C.21}$$

and there is no Faraday induction due to $\boldsymbol{B}^{(3)}$. Equation (C.13) gives

$$\boldsymbol{A}^{(1)} \times \boldsymbol{E}^{(2)} = \boldsymbol{A}^{(2)} \times \boldsymbol{E}^{(1)} \tag{C.22}$$

which is self-consistent with Eqs. (C.9), (C.17), and (C.18).

The B cyclic theorem follows from

$$c\boldsymbol{B}^{(1)} \times \boldsymbol{B}^{(2)} = c\boldsymbol{B}^{(1)} \times (\boldsymbol{k} \times \boldsymbol{E}^{(2)}) \tag{C.23}$$

which becomes

$$\boldsymbol{B}^{(1)} \times \boldsymbol{B}^{(2)} = iB^{(0)}\boldsymbol{B}^{(3)*} \tag{C.24}$$

using the vector identity

$$\boldsymbol{F} \times (\boldsymbol{G} \times \boldsymbol{H}) = \boldsymbol{G}(\boldsymbol{F} \cdot \boldsymbol{H}) - \boldsymbol{H}(\boldsymbol{F} \cdot \boldsymbol{G}) \tag{C.25}$$

Similarly

$$c\boldsymbol{B}^{(1)} \times \boldsymbol{B}^{(3)} = (\boldsymbol{k} \times \boldsymbol{E}^{(1)}) \times \boldsymbol{B}^{(3)} \tag{C.26}$$

becomes

$$\boldsymbol{B}^{(3)} \times \boldsymbol{B}^{(1)} = iB^{(0)}\boldsymbol{B}^{(2)*} \tag{C.27}$$

using

$$\boldsymbol{E}^{(1)} = -ic\boldsymbol{B}^{(1)} \tag{C.28}$$

and we obtain the Poincaré invariant B cyclic theorem because $\boldsymbol{E}^{(3)}$ is zero, and because there are no magnetic charge and current desities:

$$\begin{aligned} \boldsymbol{B}^{(1)} \times \boldsymbol{B}^{(2)} &= iB^{(0)}\boldsymbol{B}^{(3)*} \\ \boldsymbol{B}^{(2)} \times \boldsymbol{B}^{(3)} &= iB^{(0)}\boldsymbol{B}^{(1)*} \\ \boldsymbol{B}^{(3)} \times \boldsymbol{B}^{(1)} &= iB^{(0)}\boldsymbol{B}^{(2)*} \end{aligned} \tag{C.29}$$

References

1. B. L. van der Waerden, *Group Theory and Quantum Mechanics*, Springer-Verlag, Berlin, 1974.
2. L. O'Raigheartaigh, *Rep. Prog. Phys.* **42**, 159 (1979).
3. T. W. Barrett, in A. Lakhtakia (Ed.), *Essays on the Formal Aspects of Electromagnetic Theory*, World Scientific, Singapore, 1993.
4. T. W. Barret, in T. W. Barrett and D.M. Grimes (Eds.), *Advanced Electrodynamics*, World Scientific, Singapore, 1995.
5. J. D. Jackson, *Classical Electrodynamics*, Wiley, New York, 1962.
6. L. H. Ryder, *Quantum Field Theory*, 2nd ed., Cambridge Univ. Press, Cambridge, UK, 1987.
7. B. Lehnert, TRITA-EPP-79-13, *Electron and Plasma Physics*, Royal Institute of Technology, Stockholm, Sweden, 1979.
8. B. Lehnert, *Optik* **99**, 113 (1995).
9. B. Lehnert, *Physica Scripta* **19**, 204 (1996).
10. B. Lehnert and S. Roy, *Extended Electromagnetic Theory*, World Scientific, Singapore, 1998.
11. M. W. Evans, *Physica B* **182**, 227 (1992).
12. M. W. Evans, *Found. Phys.* **24**, 1519 (1994).
13. M. W. Evans, *Physica A* **214**, 605 (1995).

14. M. W. Evans and S. Kielich (Eds.), *Modern Non-Linear Optics*, Vol. 85 in I. Prigogine and S.A. Rice (Series Eds.), *Advances in Chemical Physics*, Wiley, New York, 1992, 1993, 1997, Part 2.
15. M. W. Evans, J. P. Vigier, S. Roy, and S. Jeffers, *The Enigmatic Photon*, Kluwer, Dordrecht, 1994–1999 (in five vols.).
16. M. W. Evans and A. A. Hasanein, *The Photomagneton in Quantum Field Theory*, World Scientific, Singapore, 1994.
17. M. W. Evans and L. B. Crowell, *Classical and Quantum Electrodynamics and the* $\boldsymbol{B}^{(3)}$ *Field*, World Scientific, Singapore, 1999.
18. M. W. Evans et al., AIAS group paper, *Found. Phys. Lett.* 12, 187 (1999); M. W. Evans and L. B. Crowell, *Found. Phys. Lett.* (in press, two papers); M. W. Evans et al., AIAS group paper, *Found. Phys. Lett.* (in press, two papers); M. W. Evans et al., a collection of ~60 AIAS group papers, *J. New Energy* (in press); U.S. Dept. Energy website http://www.ott.doe.gov/electromagnetic/.
19. M.W. Evans et al., AIAS group paper, *Optik* (in press).
20. M. W. Evans et al., AIAS group paper, *Physica Scripta* (in press).
21. H. F. Harmuth, *Information Theory Applied to Space-time Physics*, World Scientific, Singapore, 1993.
22. H. F. Harmuth and M. G. M. Hussain, *Propagation of Electromagnetic Signals*, World Scientific, Singapore, 1994.
23. M. W. Evans and J. P. Vigier, *The Enigmatic Photon*, Kluwer, Dordrecht, 1994, Vol. 1.
24. M. W. Evans et al., AIAS group paper, *Optik*, submitted for publication.
25. M. W. Evans et al., AIAS group paper, *Physica Scripta* (in press).
26. T. W. Barrett, *Apeiron*, special issue on electrodynamics (2000).
27. E. T. Whittaker, *Math. Ann.* **57**, 333 (1903).
28. E. T. Whittaker, *Proc. Lond. Math. Soc.* **1**, 367 (1904).
29. G. Nash and S. Sen, *Topology and Geometry for Physicists*, Academic, London, 1983.
30. Y. Choquet-Brulat, C. deWitt-Morette, and M. Dillard-Bleink, *Analysis, Manifolds and Physics*, North Holland, Amsterdam, 1982.
31. W. Dreschler and M. E. Mayer, *Fiber Bundle Techniques in Gauge Theory*, Springer-Verlag, Berlin, 1977.
32. D. Blecker, *Gauge Theory and Variational Principles*, Addison-Wesley, Reading, MA, 1981.
33. T. T. Wu and C. N. Yang, *Phys. Rev. D* **12**, 3845 (1975).
34. M. W. Evans et al. AIAS group paper, *Found. Phys. Lett.* (in press).
35. M. W. Evans, G. J. Evans, W. T. Coffey, and P. Grigolini, *Molecular Dynamics*, Wiley, New York, 1982.
36. B. Talin, V. P. Kaftandjan, and L. Klein, *Phys. Rev. A* **11**, 648 (1975).
37. P. S. Pershan, J. P. van der Ziel, and L. D. Malmstrom, *Phys. Rev.* **143**, 574 (1966).
38. L. D. Landau and E. M. Lifshitz, *The Classical Theory of Fields*, 4th ed., Pergamon, Oxford, 1975.
39. P. W. Atkins, *Molecular Quantum Mechanics*, 2nd ed., Oxford Univ. Press, Oxford, 1983.
40. F. Rohrlich, *Ann. Phys.* **13**, 93 (1961).
41. P. A. M. Dirac, *Directions in Physics*, Wiley, New York, 1978.
42. A. van der Merwe and A. Garrucio (Eds.), *Waves and Particles in Light and Matter*, Plenum, New York, 1994.

43. M. W. Evans et al., AIAS group paper, *Physica Scripta* (in press); *Found. Phys. Lett.* (in press).
44. T. W. Barrett, *Apeiron*, double issue on new electrodynamics (2000).
45. T. W. Barrett, *Ann. Fond. L. de Broglie* **14**, 37 (1989).
46. M. Mészáros, President of the Alpha Foundation, Budapest, Hungary, personal communication.
47. E. J. Post, *Rev. Mod. Phys.* **39**, 475 (1967).
48. P. Fleming, in F. Selleri (Ed.), *Open Questions in Relativistic Physics*, Apeiron, Montreal, 1998.
49. H. R. Bilger, G. E. Stedman, W. Schreiber, and M. Schneider, *IEE Trans.* **44IM**(2), 468 (1995).
50. M. G. Sagnac, *J. Phys.* **4**, 177 (1914).
51. F. Hasselbach and M. Nicklaus, *Phys. Rev. A* **48**, 243 (1993).
52. A. G. Kelly, in F. Selleri (Ed.), *Open Questions in Relativistic Physics*, Apeiron, Montreal, 1998.
53. M. von Laue, *Ann. Phys.* **62**, 448 (1920).
54. G. B. Pegram, *Phys. Rev.* **10**, 591 (1917).
55. M. W. Evans, G. J. Evans, W. T. Coffey, and P. Grigolini, *Molecular Dynamics*, Wiley, New York, 1982.
56. P. S. Pershan, *Phys. Rev.* **130**, 919 (1963).
57. G. L. J. A. Rikken, *Opt. Lett.* **20**, 846 (1995); M. W. Evans, *Found. Phys. Lett.* **9**, 61 (1996).
58. S. Wozniak, M. W. Evans, and G. Wagniere, *Mol. Phys.* **75**, 81 (1992).
59. M. W. Evans, *J. Mol. Spectrosc.* **143**, 327 (1990); *Chem. Phys.* **1571**, 1 (1991); *J. Phys. Chem.* **96**, 2256 (1991); *J. Mol. Liq.* **49**, 77 (1991); *Int. J. Mod. Phys. B* **5**, 1963 (1991); M. W. Evans, S. Wozniak, and G. Wagniere, *Physica B* **1873**, 357 (1991); **175**, 416 (1991).
60. D. Gammon, S. W. Brown, E. S. Snow, T. A. Kennedy, D. S. Katzer, and D. Park, *Science* **277**, 85 (1997).
61. S. W. Brown, T. A. Kennedy, E. R. Glaser, and D. S. Katzer, *Phys. Rev. D* **30**, 1411 (1997).
62. S. W. Brown, T. A. Kennedy and D. Gammon, *Solid State NMR* **11**, 49 (1998).
63. S. W. Brown, T. A. Kennedy, D. Gammon, and E. S. Snow, *Phys. Rev. B* **54**, R17 339 (1996).
64. E. R. Glaser, T. A. Kennedy, A. E. Wickenden, D. D. Koleske, and J. A. Freitas Jr., *Mat. Res. Soc. Symp.* **449**, 543 (1997).
65. E. R. Glaser, T. A. Kennedy, K. Doverspike, L. B. Rowland, D. K. Gaskill, J. A. Freitas Jr., M. Asif Khan, D. T. Olson, J. N. Kuznia, and D. K. Wickenden, *Phys. Rev. B* **51**, 13326 (1995).
66. J. B. Marion and S.T. Thornton, *Classical Dynamics* Harcourt Brace Jovanovich New York, 1988.
67. E. R. Pike and S. Sarkav, *The Quantum Theory of Radiation*, Oxford Univ. Press, Oxford, 1995, example 3.4.
68. J. J. Sakurai, *Advanced Quantum Mechanics*, 11th printing, Addison-Wesley, New York, 1967, Chap. 3.
69. P. W. Atkins, *Molecular Quantum Mechanics*, 2nd ed., Oxford Univ. Press, Oxford, 1983.
70. L. D. Barron, *Molecular Light Scattering and Optical Activity*, Cambridge Univ. Press, Cambridge, UK, 1982.
71. A. C. Tam and W. Happer, *Phys. Rev. Lett.* **38**, 278 (1977).
72. P. C. Naik and T. Pradhan, *J. Phys. A* **14**, 2795 (1981).
73. R. V. Sole, S. C. Manrubia, M. Benton, and P. Bak, *Nature* **388**, 644 (1997).
74. R. A. Harris and I. Tinoco, *J. Chem. Phys.* **101**, 9289 (1994).
75. J. Deschamps, M. Fitaire and M. Lagoutte, *Phys. Rev. Lett.* **25**, 1330 (1970); *Rev. Appl. Phys.* **7**, 155 (1972).

76. G. H. Wagnière, *Linear and Non-Linear Optical Properties of Molecules*, VCH, Basel, 1993.
77. J. D. Bjorken and S. D. Drell, *Relativistic Quantum Mechanics*, McGraw-Hill, New York, 1964.
78. J. C. Maxwell, *Trans. Cambridge Phil. Soc.* **10**, 27 (1856).
79. A. M. Bork, (article on Maxwell and the vector potential) *Isis* **58**, 210 (1967).
80. Reviewed in Ref. 3.
81. A. Priore, *Brevet d'invention*, P.V. No. 899414 and 1342772, Oct. 7, 1963.
82. A. Priore, U.S. Patent 3,280,816 (1966).
83. E. Perisse, thesis, *Effects of Electromagnetic Waves and Magnetic Fields on Cancer and Experimental Trypanosomiasis* Thèse d'Etat, Univ. Bordeaux, 1984.
84. M. R. Courrier, Permanent Secretary of the French Academy of Sciences, a discussion on the Priore device, April 26, 1977.
85. A. Priore, *Healing of Acute and Chronic Experimental Trypanosomiasis by the Combined Action of Magnetic Fields and Modulated Electromagnetic Waves*, thesis, (in French), Univ. Bordeaux, 1973; (transl. by A. Beaulieu into English).
86. W. K. H. Panofsky and M. Phillips, *Classical Electricity and Magnetism*, Addison-Wesley, Reading, MA, 1962).
87. A. Labounsky, *Principal Engineer*, Boeing Company, communication to AIAS group (1999).
88. M. Y. A. Raja, W. N. Sisk, M. Yousaf, and S. Allen, *Appl. Phys. B* **64**, 79 (1997); M. W. Evans, *Found. Phys. Lett.* **10**, 255 (1997).
89. R. A. Compton, Oak Ridge National Labs., personal communications.
90. T. E. Phipps, Jr., in M. W. Evans (Ed.), *The New Electrodynamics*, a special issue of *Apeiron* (1999/2000).
91. A. Koestler, *The Sleepwalkers*, Hutchinson, London, Danube Edition, 1968.
92. LEP and SLD Collaborations, CERN Report CERN-EP/9915, Feb. 8, 1999.
93. P. Ramond, *Field Theory*, Princeton Univ. Press, Prisceton, NJ, 1982.
94. M. W. Evans and L. B. Crowell, *Found. Phys. Lett.* **12**, 12,373 (1999).
95. S. Weinberg, *Phys. Rev. Lett.* **19**, 1264 (1967).
96. Website: http://xxx.lanl.gov/abs/hep-ph/9910315.
97. C. Montonen and S. Olive, *Nucl. Phys.* **110B**, 237 (1976).
98. M. B. Green, J. H. Schwarz, and E. Witten, *Superstring Theory*, Cambridge, Univ. Press, Cambridge, UK, 1987.
99. J. D. Berkenstein, *Phys. Rev. D* **23D**, 287 (1974).
100. E. P. Wigner, *Ann. Math.* **40**, 149 (1939).
101. L. D. Landau and E. M. Lifshitz, *The Classical Theory of Fields*, 4th ed., Pergamon, Oxford, 1975.
102. M. W. Evans and S. Jeffers, *Found. Phys. Lett.* **9**, 587 (1996).
103. A. F. Ranada, *J. Phys. A* **25**, 1621 (1992).
104. T. W. Barrett, in M. W. Evans (Ed.), *Apeiron The New Electrodynamics*, a special issue of (2000).

SYMMETRY AND EXACT SOLUTIONS OF THE MAXWELL AND $SU(2)$ YANG–MILLS EQUATIONS

R. Z. ZHDANOV*

Department of Theoretical Physics II, Complutense University, Madrid, Spain

V. I. LAHNO

State Pedagogical University, Poltava, Ukraine

CONTENTS

*On leave of absence from the Institute of Mathematics of NAN of Ukraine, 3 Tereshchenkivska Street, 01601 Kyiv-4, Ukraine

Modern Nonlinear Optics, Part 2, Second Edition, Advances in Chemical Physics, Volume 119,
Edited by Myron W. Evans. Series Editors I. Prigogine and Stuart A. Rice.
ISBN 0-471-38931-5

I. INTRODUCTION

The famous paper [1] written by Yang and Mills is a milestone of modern quantum physics, where the role played by the equations introduced in the paper (now called the $SU(2)$ *Yang–Mills equations*) can be compared only to that of the Klein–Gordon–Fock, Schrödinger, Maxwell, and Dirac equations. However, the real importance of the Yang–Mills equations was first understood only in the late 1960s, when the concept of the gauge fields as being responsible for all four fundamental physical interactions (gravitational, electromagnetic, weak, and strong interactions) became widespread.

The simplest example of gauge theory in $(1+3)$-dimensional space is the system of the Maxwell equations for the 4-component vector potential of the electromagnetic field, whose gauge group is the single-parameter group $U(1)$. The simplest example of the non-Abelian gauge group is group $SU(2)$. This group is realized as the symmetry group admitted by the Yang–Mills equations describing the triplet of the gauge fields (called in the sequel the *Yang–Mills field*) $\mathbf{A}_\mu(x) = (A^a_\mu(x), a = 1, 2, 3)$, where $\mu = 1, 2, 3, 4$, $x = (x_1, x_2, x_3, x_4)$ for the case of the four-dimensional Euclid space and $\mu = 0, 1, 2, 3$, $x = (x_0 = t, x_1, x_2, x_3) = (t, \mathbf{x})$ for the case of the Minkowski space. The matrix vector field $A_\mu = A_\mu(x)$ is defined as follows:

$$A_\mu = e\frac{\sigma_a}{2i}A^a_\mu$$

Here σ_a, $(a = 1, 2, 3)$ are the Pauli matrices

$$\sigma_1 = \begin{pmatrix} 0 & 1 \\ 1 & 0 \end{pmatrix}, \quad \sigma_2 = \begin{pmatrix} 0 & -i \\ i & 0 \end{pmatrix}, \quad \sigma_3 = \begin{pmatrix} 1 & 0 \\ 0 & -1 \end{pmatrix}$$

and e is the real constant called the *gauge coupling constant.*

Using the matrix gauge potentials, one constructs the matrix-valued field

$$F_{\mu\nu} \equiv \partial^\mu A_\nu - \partial^\nu A_\mu + [A_\mu,\ A_\nu], \quad \mu, \nu = 0, 1, 2, 3$$

Writing these expressions componentwise yields

$$F_{\mu\nu} \equiv e\frac{\sigma_a}{2i}F^a_{\mu\nu}, \quad F^a_{\mu\nu} = \partial^\mu A^a_\nu - \partial^\nu A^a_\mu + ef^a_{bc}A^b_\mu A^c_\nu$$

where $\mu, \nu = 0, 1, 2, 3$, $a = 1, 2, 3$, and the symbols f^a_{bc} $(a, b, c = 1, 2, 3)$ stand for the structure constants determining the Lie algebra of the gauge group [note that for the case of the group $SU(2)$, $f^{\ a}_{\ bc} = \varepsilon_{abc}$, where ε_{abc} is the antisymmetric tensor with $\varepsilon_{123} = 1$, $a, b, c = 1, 2, 3$].

Hereafter we use the following designations:

$$\partial_\mu = \partial_{x_\mu} = \frac{\partial}{\partial x_\mu}$$

Furthermore, lowering and raising the indices μ, ν is performed with the help of the metric tensor of the space of the variables x_μ, and summation over the repeated indices is carried out.

The $SU(2)$ Yang–Mills equations are obtained from the Lagrangian

$$\mathscr{L} = -\frac{1}{4} F_{\mu\nu} F^{\mu\nu}$$

and are of the form

$$\partial_\mu F_{\mu\nu} + [A^\mu, F_{\mu\nu}] = [D_\mu, F_{\mu\nu}] = 0 \tag{1}$$

where $D_\mu = \partial_\mu + A^\mu$ is the covariant derivative, $\mu, \nu = 0, 1, 2, 3$.

One of the most popular and exciting parts of the general theory of the Yang–Mills equations is that devoted to constructing their exact analytical solutions. There is a vast literature devoted solely to constructing and analyzing exact solutions of (1) (see the review by Actor [2] and the monograph by Radjaraman [3] for an extensive list of references). Most of the results are obtained for the case of the Yang–Mills equations in Euclidean space. The principal reason for this is the fact that Eq. (1) in Euclidean space have the monopole and instanton solutions [3,4], which admit numerous physical interpretations and have highly nontrivial geometric and algebraic properties. Note that these and some other classes of exact solutions of system (1) can also be obtained by solving the so-called self-dual Yang–Mills equations

$$F_{\mu\nu} = *F_{\mu\nu}, \quad \mu, \nu = 0, 1, 2, 3. \tag{2}$$

Evidently, each solution of (2) satisfies (1), while the reverse assertion does not hold.

Provided we consider the Euclidean case, $*F_{\mu\nu} = \frac{1}{2}\varepsilon_{\mu\nu\lambda\rho} F_{\lambda\rho}$, $(\mu, \nu, \lambda, \rho = 1, 2, 3, 4)$, where $\varepsilon_{\mu\nu\lambda\rho}$ is the completely antisymmetric tensor, and equations (2) form the system of four real first-order partial differential equations.

It was the self-duality property of the instanton solutions of (1) in Euclidean space that had enabled the use of ansatz, suggested by t'Hooft [5], Corrigan and

Fairlie [6], Wilczek [7], and Witten [8], in order to construct these solutions. Furthermore, the well-known monopole solution by Prasad and Sommerfield [9], as well as the solutions obtainable via the Atiyah–Hitchin–Drinfeld–Manin method [10] exploit explicitly the self-duality condition.

One more important property of the self-dual Yang–Mills equations is that they are equivalent to the compatibility conditions of some overdetermined system of linear partial differential equations [11,12]. In other words, the self-dual Yang–Mills equations admit the Lax representation and, in this sense, are integrable. For this very reason it is possible to reduce Eq. (2) to the widely studied solitonic equations, such as the Euler–Arnold, Burgers, and Devy–Stuardson equations [13,14] and Liouville and sine–Gordon equations [15] by use of the symmetry reduction method.

For the case, when the Yang–Mills field is defined in the Minkowski space, we have $*F_{\mu\nu} = \frac{i}{2}\,\varepsilon_{\mu\nu\lambda\rho}\,F^{\lambda\rho}, \;\; (\mu, \nu, \lambda, \rho = 0, 1, 2, 3)$. Consequently, Eq. (2) form the system of complex first-order differential equations. In view of this fact, exploitation of the abovementioned methods and results for study of the $SU(2)$ Yang–Mills equations (1) in the Minkowski space yields complex-valued solutions. That is why the abovementioned methods for solving Eq. (1) fail to be efficient for the case of the Minkowski space. Consequently, there is a need for developing new methods that do not rely on the self-duality condition. This problem has been addressed by one of the creators of the inverse scattering technique, V. E. Zaharov, who wrote, in the foreword to the Russian translation of the monograph by Calogero and Degasperis [16], that a number of important problems of nonlinear mathematical physics (including Yang–Mills equations in Minkowski space) still await new, more efficient solutions.

On the other hand, it is known [17] (see also Ref. 18) that Eq. (1) have rich symmetry. Specifically, their maximal (in the Lie sense) symmetry group is the group $G \otimes SU(2)$, where G is

- The conformal group $C(1,3)$, if the Yang–Mills equations are defined in Minkowski space
- The conformal group $C(4)$, if the Yang–Mills equations are defined in Euclidean space
- The conformal group $C(2,2)$, if the Yang–Mills equations are defined in pseudo-Euclidean space having the metric tensor with the signature $(-,\ -,\ +,\ +)$

Note that the maximal symmetry groups admitted by the self-dual Yang–Mills equations (2) coincide with the symmetry groups of the corresponding equations (1).

The rich symmetry of Eqs. (1) and (2) enables efficient exploitation of the symmetry reduction routine for the sake of dimensional reduction of

Yang–Mills equations either to ordinary differential equations integrable by quadratures or to integrable solitonic equations in two or three independent variables [19–21]. In particular, some subgroups of the generalized Poincaré group $P(2,2)$ which is the subgroup of the conformal group $C(2,2)$, were used in order to reduce the self-dual Yang–Mills equations, defined in the pseudo-Euclidean space having the metric tensor with the signature $(-,-,+,+)$, to a number of known integrable systems, such as the Ernst, cubic Schrödinger, and Euler–Calogero–Moser equations (see Ref. 22 and references cited therein). Legaré et al. have carried out systematic investigation of the problem of symmetry reduction of system (2) in Euclidean space by subgroups of the Euclid group $E(4) \in C(4)$ [23,24]. What is more, some of the known analytical solutions of Eq. (1) in Euclidean space (viz. the non-self-dual meron solution, obtained by de Alfaro et al. [25], and the instanton solution, constructed by Belavin, et al. [26]), can also be obtained within the framework of the symmetry reduction approach [21].

To the best of our knowledge, the first paper devoted to symmetry reduction of the $SU(2)$ Yang–Mills equations in Minkowski space has been published by Fushchych and Shtelen [27] (see also Ref. 21). They use two conformally invariant ansatzes in order to perform reduction of Eqs. (1) to systems of ordinary differential equations. Integrating the latter yields several exact solutions of Yang–Mills equations (1).

Let us note that the full solution of the problem of symmetry reduction of fundamental equations of relativistic physics, whose symmetry groups are subgroups of the conformal group $C(1,3)$, has been obtained for the scalar wave equation only (for further details, see Refs. 21 and 28–30). This fact is explained by the extreme cumbersomity of the calculations needed to perform a systematic symmetry reduction of systems of partial differential equations by all inequivalent subgroups of the conformal group $C(1,3)$. The complete solution of the problem of symmetry reduction to systems of ordinary differential equations has been obtained for the conformally invariant nonlinear spinor equations [31–33], which generalize the Dirac equation for an electron. We have carried out symmetry reduction of the Yang–Mills equations (1) and (2) by subgroups of the Poincaré group and have constructed a number of their exact solutions [34–39].

The principal aim of the present chapter is twofold. First, we will review the already known ideas, methods, and results centered around the solution techniques that are based on the symmetry reduction method for the Yang–Mills equations (1) in Minkowski space. Second, we will describe the general reduction routine, developed by us in the 1990s, which enables the unified treatment of both the classical and nonclassical symmetry reduction approaches for an arbitrary relativistically invariant system of partial differential equations. As a byproduct, this approach yields exhaustive solution of the problem of symmetry reduction of the vacuum Maxwell equations

$$\begin{aligned} \operatorname{rot}\mathbf{E} &= -\frac{\partial \mathbf{H}}{\partial t}, \quad \operatorname{div}\mathbf{H} = 0 \\ \operatorname{rot}\mathbf{H} &= \frac{\partial \mathbf{E}}{\partial t}, \quad \operatorname{div}\mathbf{E} = 0 \end{aligned} \tag{3}$$

The history of the study of symmetry properties of Eq. (3) goes back to the beginning of the twentieth century. Invariance properties of Maxwell equations have been studied by Lorentz [40] and Poincaré [41,42]. They have proved that Eq. (3) are invariant with respect to the transformation group named by the Poincaré suggestion the Lorentz group. Furthermore, Larmor [43] and Rainich [44] have found that equations (3) are invariant with respect the single-parameter transformation group

$$\mathbf{E} \to \mathbf{E}\cos\theta + \mathbf{H}\sin\theta, \quad \mathbf{H} \to \mathbf{H}\cos\theta - \mathbf{E}\sin\theta \tag{4}$$

now known as the Heaviside–Larmor–Rainich group. Later, Bateman [45] and Cunningham [46] showed that Maxwell equations are invariant with respect to the conformal group.

Much later, Ibragimov [47] proved that the group $C(1,3) \otimes H$, where $C(1,3)$ is the group of conformal transformations of Minkowski space and H is the Heaviside–Larmor–Rainich group (4), is maximal in Lie's sense invariance group of equations (3). Note that this result coincides with that obtained earlier without explicit use of the infinitesimal Lie algorithm [19,20]. Further progress in the study of symmetries of the Maxwell equations became possible when Fushchych and Nikitin suggested a non-Lie approach to investigating symmetry properties of linear systems of partial differential equations [48].

The present review is based mainly on our publications [33,35–39,49–53]. In Section II we give a detailed description of the general reduction routine for an arbitrary relativistically invariant systems of partial differential equations. The results of Section II are used in Section III to solve the problem of symmetry reduction of Yang–Mills equations (1) by subgroups of the Poincaré group $P(1,3)$ and to construct their exact (non-Abelian) solutions. In Section IV we review the techniques for nonclassical reductions of the $SU(2)$ Yang–Mills equations, which are based on their conditional symmetry. These techniques enable us to obtain the principally new classes of exact solutions of (1), which are not derivable within the framework of the standard symmetry reduction technique. In Section V we give an overview of the known invariant solutions of the Maxwell equations and construct multiparameter families of new ones.

II. CONFORMALLY INVARIANT ANSATZES FOR AN ARBITRARY VECTOR FIELD

In this section we describe the general approach to constructing conformally invariant ansatzes applicable to any (linear or nonlinear) system of partial differential equations, on whose solution set a linear covariant representation of the conformal group $C(1,3)$ is realized. Since the majority of the equations of the relativistic physics, including the Klein–Gordon–Fock, Maxwell, massless Dirac, and Yang–Mills equations, respect this requirement, they can be handled within the framework of this approach.

Note that all our subsequent considerations are local and the functions involved are supposed to be as many times continuously differentiable, as this is necessary for performing the corresponding mathematical operations.

A. The Linear Form of Invariant Ansatzes

Consider the system of partial differential equations (which we denote as S)

$$S : \underset{A}{F}(\mathbf{x}, \mathbf{u}, \underset{1}{\mathbf{u}}, \dots, \underset{r}{\mathbf{u}}) = 0, \quad A = 1, \dots, m \tag{5}$$

defined on the open subset $M \subset X \times U \simeq \mathbf{R}^p \times \mathbf{R}^q$ of the space of p independent and q dependent variables. In (5) we use the notations $\mathbf{x} = (x_1, \dots, x_p) \in X$, $\mathbf{u} = (u^1, \dots, u^q) \in U$, $\underset{l}{\mathbf{u}} = \{\frac{\partial^l u^k}{\partial x_1^{\alpha_1} \partial x_2^{\alpha_2} \dots \partial x_p^{\alpha_p}}, 0 \le \alpha_i \le l, \sum_{i=1}^{p} \alpha_i = l, k = 1, \dots, q\}$, $l = 1, 2, \dots, r$, and F_A, which are sufficiently smooth functions of the given arguments.

Let G be a local transformation group that acts on M and is the symmetry group of system (5). Next, let the basis operators of the Lie algebra g of the group G be of the form

$$X_a = \xi_a^i(\mathbf{x}, \mathbf{u})\partial_{x_i} + \eta_j^a(\mathbf{x}, \mathbf{u})\partial_{u^j}, \ a = 1, \dots, n \tag{6}$$

where ξ_a^i, η_j^a are arbitrary smooth functions on M, $\partial_{u^j} = \frac{\partial}{\partial u^j}$, $i = 1, \dots, p$, $j = 1, \dots, q$. By definition, operators (6) satisfy the commutation relations

$$[X_a, X_b] \equiv X_a X_b - X_b X_a = C_{ab}^c X_c, \ a, b, c = 1, \dots, n$$

where C_{ab}^c are the structure constants, which determine uniquely the type of the Lie algebra g.

We say that a solution $\mathbf{u} = \mathbf{f}(\mathbf{x})$ $(\mathbf{f} = (f^1, \dots, f^q))$ of system (6) is G-invariant if the manifold $\mathbf{u} - \mathbf{f}(\mathbf{x}) = 0$ is invariant with respect to the action of the group G. This means that for an arbitrary $\gamma \in G$, the functions $\mathbf{f}$ and $\gamma(\mathbf{f})$ coincide in the intersection of the domains, where they are defined. More precisely, we can define a G-invariant solution of system (5) as the solution

$\mathbf{u} = \mathbf{f}(\mathbf{x})$, whose graph $\Gamma_{\mathbf{f}} = \{(\mathbf{x}, \mathbf{f}(\mathbf{x}))\} \subset M$ is locally G-invariant subset of the set M.

If G is the symmetry group of system (5), then, under some additional assumption of regularity of the action of the group G, we can find all its G-invariant solutions by solving the reduced system of differential equations S/G. Note that by construction the system S/G has fewer independent variables; that is, the dimension of the initial system is reduced (hence this procedure is called the *symmetry reduction method*).

In the sequel, we will restrict our considerations to the case of the projective action of the group G in M. This means that all the transformations γ from G are of the form

$$(\bar{\mathbf{x}}, \bar{\mathbf{u}}) = \gamma((\mathbf{x}, \mathbf{u})) = (\Psi_\gamma(\mathbf{x}), \Phi_\gamma(\mathbf{x}, \mathbf{u})).$$

In other words, the transformation law for the independent variables $\mathbf{x}$ does not involve the dependent variables [for the Lie algebra g of the group G, this implies that in formulas (2) $\xi_a^i = \xi_a^i(\mathbf{x})$]. This defines the projective action of the group G $\bar{\mathbf{x}} = \gamma(\mathbf{x}) = \Psi_\gamma(\mathbf{x})$ in an arbitrary subset Ω of the set X.

In what follows, we will suppose that the action of the group G in M and its projective actions in Ω are regular and the orbits of these actions have the same dimension s. This dimension is called the rank of the group G (or, alternatively, the rank of the Lie algebra g). Note that the condition $\operatorname{rank} G = s$ is equivalent to the requirement that the relation

$$\operatorname{rank} \| \xi_a^i(\mathbf{x}_0) \| = \operatorname{rank} \| \xi_a^i(\mathbf{x}_0), \eta_j^a(\mathbf{x}_0, \mathbf{u}_0) \| = s \tag{7}$$

holds in an arbitrary point $(\mathbf{x}_0, \mathbf{u}_0) \in M$ [19]. We will also suppose that $s < p$ (the case $s = p$ is trivial, and furthermore, G-invariant functions do not exist under $s > p$).

If these assumptions hold, then there are $p - s$ fuctionally independent invariants $y^1 = \omega^1(\mathbf{x}), y^2 = \omega^2(\mathbf{x}), \ldots, y^{p-s} = \omega^{p-s}(\mathbf{x})$ (the first set of invariants) of the group G acting projectively in Ω, and each of them is the invariant of the group G acting in M. Furthermore, there are q functionally independent invariants $v^1 = g^1(\mathbf{x}, \mathbf{u}),\ v^2 = g^2(\mathbf{x}, \mathbf{u}), \ldots, v^q = g^q(\mathbf{x}, \mathbf{u})$ of the group G acting in M (the second set of invariants) [19,20]. Using the shorthand notation, we represent the full set of invariants of the group G in the following way:

$$\mathbf{y} = \mathbf{w}(\mathbf{x}), \quad \mathbf{v} = \mathbf{g}(\mathbf{x}, \mathbf{u}) \tag{8}$$

Owing to the validity of the relation

$$\operatorname{rank} \left\| \frac{\partial g^j}{\partial u^i} \right\| = q, \quad i, j = 1, \ldots, q$$

we can solve locally the second system of equations from (8) with respect to $\mathbf{u}$

$$\mathbf{u} = \tilde{\mathbf{r}}(\mathbf{x}, \mathbf{v}) \tag{9}$$

Using the relation

$$\operatorname{rank}\left\|\frac{\partial\omega^j}{\partial x_i}\right\| = p - s, \qquad j = 1,\ldots,p-s, \quad i = 1,\ldots,p$$

we choose $p - s$ independent variables $\tilde{\mathbf{x}} = (\tilde{x}_1,\ldots,\tilde{x}_{p-s})$ so that

$$\operatorname{rank}\left\|\frac{\partial\omega^j}{\partial\tilde{x}_i}\right\| = p - s, \quad i,j = 1,\ldots,p-s$$

We call these variables *principal*. The remaining s independent variables $\hat{\mathbf{x}} = (\hat{x}_1,\ldots,\hat{x}_s)$ are called *parametric* (they enter all the subsequent formulas as parameters).

Now we can solve the first system from (8) with respect to the principal variables

$$\tilde{\mathbf{x}} = \mathbf{z}(\hat{\mathbf{x}},\mathbf{y}) \tag{10}$$

Inserting (10) into (9), we get the equality

$$\mathbf{u} = \tilde{\mathbf{r}}(\hat{\mathbf{x}},\mathbf{z},\mathbf{v})$$

or

$$\mathbf{u} = \mathbf{r}(\hat{\mathbf{x}},\mathbf{y},\mathbf{v}) \tag{11}$$

Note that in (9)–(11), $\tilde{\mathbf{r}} = (\tilde{r}^1, \ldots, \tilde{r}^q), \mathbf{r} = (r^1,\ldots,r^q), \mathbf{z} = (z^1,\ldots,z^{p-s})$. The so constructed G-invariant function (11) is called the *ansatz*. Inserting ansatz (11) into system (5) yields the system of partial differential equations for the functions $\mathbf{v}$ of the variables $\mathbf{y}$, which do not explicitly involve the parametric variables [19]. These equations form the reduced (or factor) system S/G having the fewer number of independent variables $y^1,\ldots,y^{p-s}$, as compared with the initial system (5). Now, if we are given a solution $\mathbf{v} = \mathbf{h}(\mathbf{y})$ of the reduced system, then inserting it into (11) yields a G-invariant solution of system (5).

Summing up, we formulate the algorithm of symmetry reduction and construction of invariant solutions of systems of partial differential equations, that admit nontrivial Lie symmetry.

1. Using the infinitesimal Lie method, we compute the maximal symmetry group G admitted by the equation under study.
2. We fix the symmetry degree s of the invariant solutions to be constructed and find the optimal system of subgroups of the group G having the rank s. We can do this because the subgroup classification problem reduces to classifying inequivalent subalgebras of the rank s of the Lie algebra g of

the group G. This classification is performed within the action of the inner automorphism group of the algebra g.

3. For each of the so obtained subgroups we construct the full set of functionally independent invariants, which yields the invariant ansatz.
4. Inserting the abovementioned ansatz into the system of partial differential equations under study reduces it to the one having $p - s$ independent variables.
5. We investigate the reduced system and construct its exact solutions. Each of them corresponds to the invariant solution of the initial system.

Symmetry properties of the overwhelming majority of physically significant differential equations [including the Maxwell and $SU(2)$ Yang–Mills equations] are well known. The most important symmetry groups are those isomorphic to the Euclid, Galileo, and Poincaré groups and their natural extensions (the Schrödinger and conformal groups). This fact motivated Patera et al. to investigate the subgroup structure of these fundamental groups [54]. They have suggested the general method for classifying continuous subgroups of Lie groups and illustrated its efficiency by rederiving the known classification of inequivalent subgroups of the Poincaré group $P(1,3)$. Exploiting this method has enabled investigators to fully describe the continuous subgroups of a number of important symmetry groups arising in theoretical and mathematical physics, including the Euclid, Galileo, Poincaré, Schrödinger, and conformal groups (see, e.g., Ref. 30 and references cited therein).

Thus, to completely solve the problem of symmetry reduction within the framework of the formulated algorithm above, we need to be able to perform steps 3–5 listed above. However, solving these problems for a system of partial differential equations requires enormous amount of computations; moreover, these computations cannot be fully automatized with the aid of symbolic computation routines. On the other hand, it is possible to simplify drastically the computations, if one notes that for the majority of physically important realizations of the Euclid, Galileo, and Poincaré groups and their extensions, the corresponding invariant solutions admit linear representation. It was this very idea that enabled us to construct broad classes of invariant solutions of a number of nonlinear spinor equations [31–33].

In the paragraphs that follow, we will concentrate on the case of the 15-parameter conformal group $C(1,3)$, admitted both by the Maxwell and $SU(2)$ Yang–Mills equations. We emphasize that the same reasoning applies directly to the case of the 11-parameter Schrödinger group $Sch(1,3)$, which is the analog of the conformal group in nonrelativistic physics. The group $C(1,3)$ acts in the open domain $M \subset \mathbf{R}^{1,3} \times \mathbf{R}^q$ of the four-dimensional Minkowski spacetime of the independent variables $x_0, \mathbf{x} = (x_1, x_2, x_3)$ and of the q-dimensional space of dependent variables $\mathbf{u} = \mathbf{u}(x_0, \mathbf{x}),\ \mathbf{u} = (u^1, u^2, \ldots, u^q)$.

The Lie algebra $c(1,3)$ of the conformal group $C(1,3)$ is spanned by the generators of the translation P_μ ($\mu = 0,1,2,3$), rotation J_{ab} ($a, b = 1,2,3$, $a < b$), Lorentz rotation J_{0a} ($a = 1,2,3$), dilation D, and conformal K_μ ($\mu = 0,1,2,3$) transformations. The basis elements of $c(1,3)$ satisfy the following commutation relations:

$$[P_\mu, P_\nu] = 0, \quad [P_\mu, J_{\alpha\beta}] = g_{\mu\alpha}P_\beta - g_{\mu\beta}P_\alpha$$

$$[J_{\mu\nu}, J_{\alpha\beta}] = g_{\mu\beta}J_{\nu\alpha} + g_{\nu\alpha}J_{\mu\beta} - g_{\mu\alpha}J_{\nu\beta} - g_{\nu\beta}J_{\mu\alpha} \tag{12}$$

$$[P_\mu, D] = P_\mu, \quad [J_{\mu\nu}, D] = 0 \tag{13}$$

$$[K_\mu, J_{\alpha\beta}] = g_{\mu\alpha}K_\beta - g_{\mu\beta}K_\alpha, \ [D, K_\mu] = K_\mu$$

$$[K_\mu, K_\nu] = 0, \ [P_\mu, K_\nu] = 2(g_{\mu\nu}D - J_{\mu\nu}) \tag{14}$$

Here $\mu, \nu, \alpha, \beta = 0,1,2,3$ and $g_{\mu\nu}$ is the metric tensor of the Minkowski spacetime $\mathbf{R}^{1,3}$:

$$g_{\mu\nu} = \begin{cases} 1, & \mu = \nu = 0 \\ -1, & \mu = \nu = 1,2,3 \\ 0, & \mu \neq \nu \end{cases}$$

The group $C(1,3)$ contains the following important subgroups:

1. The Poincaré group $P(1,3)$, whose Lie algebra $p(1,3)$ is spanned by the operators $P_\mu, J_{\mu\nu}$ ($\mu, \nu = 0,1,2,3$) satisfying commutation relations (12);
2. the extended Poincaré group $\tilde{P}(1,3)$, whose Lie algebra $\tilde{p}(1,3)$ is spanned by the operators $P_\mu, J_{\mu\nu}, D$ ($\mu, \nu = 0,1,2,3$) satisfying commutation relations (12) and (13)

Analysis of the symmetry groups of the equations of relativistic physics shows that for the majority of them the generators of the Poincaré, extended Poincaré, and conformal groups can be represented in the following form [19,21,33,48]:

$$\begin{aligned}
P_\mu &= \partial_{x_\mu} \\
J_{\mu\nu} &= x^\mu\partial_{x_\nu} - x^\nu\partial_{x_\mu} - (S_{\mu\nu}\mathbf{u} \cdot \partial_{\mathbf{u}}) \\
D &= x_\mu\partial_{x_\mu} - k(E\mathbf{u} \cdot \partial_{\mathbf{u}}) \\
K_0 &= 2x_0D - (x_\nu x^\nu)\partial_{x_0} - 2x_a(S_{0a}\mathbf{u} \cdot \partial_{\mathbf{u}}) \\
K_1 &= -2x_1D - (x_\nu x^\nu)\partial_{x_1} + 2x_0(S_{01}\mathbf{u} \cdot \partial_{\mathbf{u}}) \\
&\quad - 2x_2(S_{12}\mathbf{u} \cdot \partial_{\mathbf{u}}) - 2x_3(S_{13}\mathbf{u} \cdot \partial_{\mathbf{u}}) \\
K_2 &= -2x_2D - (x_\nu x^\nu)\partial_{x_2} + 2x_0(S_{02}\mathbf{u} \cdot \partial_{\mathbf{u}}) \\
&\quad + 2x_1(S_{12}\mathbf{u} \cdot \partial_{\mathbf{u}}) - 2x_3(S_{23}\mathbf{u} \cdot \partial_{\mathbf{u}}) \\
K_3 &= -2x_3D - (x_\nu x^\nu)\partial_{x_3} + 2x_0(S_{03}\mathbf{u} \cdot \partial_{\mathbf{u}}) \\
&\quad + 2x_1(S_{13}\mathbf{u} \cdot \partial_{\mathbf{u}}) + 2x_2(S_{23}\mathbf{u} \cdot \partial_{\mathbf{u}})
\end{aligned} \tag{15}$$

In these formulas, $S_{\mu\nu}$ are constant $q \times q$ matrices, which realize a representation of the Lie algebra $o(1,3)$ of the pseudoorthogonal group $O(1,3)$ and satisfy the commutation relations

$$[S_{\mu\nu}, S_{\alpha\beta}] = g_{\mu\beta}S_{\nu\alpha} + g_{\nu\alpha}S_{\mu\beta} - g_{\mu\alpha}S_{\nu\beta} - g_{\nu\beta}S_{\mu\alpha} \tag{16}$$

$\mu, \nu, \alpha, \beta = 0, 1, 2, 3$; $g_{\mu\nu}$ is the metric tensor of Minkowski space $\mathbf{R}^{1,3}$, E is the unit $q \times q$ matrix, $\mathbf{u} = (u^1, u^2, \ldots, u^q)^T$, $\partial_{\mathbf{u}} = (\partial_{u^1}, \partial_{u^2}, \ldots, \partial_{u^q})^T$, and the symbol $(* \cdot *)$ stands for the scalar product in the vector space $\mathbf{R}^q$. We remind the reader that the repeated indices imply summation over the corresponding interval and raising and lowering the indices is carried out with the help of the metric $g_{\mu\nu}$. Also, k is some fixed real number called the conformal degree of the group $C(1,3)$.

It follows from relations (15) that the basis elements of the Lie algebra $c(1,3)$ have the form (6), where the functions ξ_a^i depend on $\mathbf{x} \in X = \mathbf{R}^p$ only and the functions η_j^a are linear in $\mathbf{u}$. We will prove that owing to these properties of the basis elements of $c(1,3)$, the ansatzes invariant under subalgebras of the algebra (15) admit linear representation.

Let a local transformation group G act projectively in M, and let $g = \langle X_1, \ldots, X_n \rangle$ be its Lie algebra spanned by the infinitesimal operators of the form

$$X_a = \xi_a^i(\mathbf{x})\partial_{x_i} + \rho_{jk}^a(\mathbf{x})u^k\partial_{u^j} \tag{17}$$

where $a = 1, \ldots, n,\ i = 1, \ldots, p, j, k = 1, \ldots, q$.

According to the discussion above, the group G has the two types of invariants. The first set of invariants is formed by $p - s$ (where s is the rank of the group G) functionally independent invariants

$$\mathbf{w} = \mathbf{w}(\mathbf{x}), \quad \mathbf{w} = (\omega^1, \ldots, \omega^{p-s}) \tag{18}$$

The second set is formed by q invariants

$$\mathbf{h} = \mathbf{h}(\mathbf{x}, \mathbf{u}), \quad \mathbf{h} = (h^1, \ldots, h^q) \tag{19}$$

The functions $\mathbf{w}$ and $\mathbf{h}$ are invariants of the group G if and only if they are, respectively, solutions of the following systems of partial differential equations:

$$\xi_a^i(\mathbf{x})\frac{\partial\omega^b}{\partial x_i} = 0 \tag{20}$$

$$\xi_a^i(\mathbf{x})\frac{\partial h^l}{\partial x_i} + \rho_{jk}^a(\mathbf{x})u^k\frac{\partial h^l}{\partial u^j} = 0 \tag{21}$$

where the indices take the following values: $a = 1, \ldots, n$, $b = 1, \ldots, p - s$, $i = 1, \ldots, p$, $j, k, l = 1, \ldots, q$.

Generically, a G-invariant ansatz has the form (11), where $\mathbf{v} \equiv \mathbf{h}$. However, provided the infinitesimal operators of the group G are of the form (17), G-invariant ansatz for the vector field $\mathbf{u}$ can be represented in the linear form [33]

$$\mathbf{u} = \Lambda(\mathbf{x})\mathbf{h}(\mathbf{w}) \tag{22}$$

where $\Lambda(\mathbf{x})$ is some $q \times q$ matrix nonsingular in $\Omega \subset M$, $\mathbf{u} = (u^1, \ldots, u^q)^T$, $\mathbf{h} = (h^1, \ldots, h^q)^T$.

The matrix $\Lambda(\mathbf{x})$ from (22) is obtained by integrating the system of partial differential equations to be derived below.

Lemma 1. *Let a G-invariant ansatz be of the form* (22). *Then there is $q \times q$ matrix $H(\mathbf{x}) = \Lambda^{-1}(\mathbf{x})$ nonsingular in Ω satisfying the matrix partial differential equation*

$$\xi_a^i(\mathbf{x}) \frac{\partial H(\mathbf{x})}{\partial x_i} + H(\mathbf{x})\Gamma_a(\mathbf{x}) = 0 \tag{23}$$

where $\Gamma_a(\mathbf{x})$ is the $q \times q$ matrices, whose (i,j)th entry reads as $\rho_{ij}^a(\mathbf{x})$, $i, j = 1, \ldots, q$.

Proof. Provided a G-invariant ansatz is of the form (22), the relation

$$\mathbf{h} = H(\mathbf{x})\mathbf{u}$$

with $H(\mathbf{x}) = \Lambda^{-1}(\mathbf{x})$ holds. So, the second set of invariants (19) of the group G consists of the functions, which are linear in u^j and, consequently, can be represented in the form

$$h^b = h_{bl}(\mathbf{x})u^l, \quad b, l = 1, \ldots, q$$

The function h^b is the invariant of the group G, if and only if it satisfies Eq. (21)

$$\xi_a^i(\mathbf{x}) \frac{\partial h_{bl}(\mathbf{x})}{\partial x_i} u^l + \rho_{jl}^a(\mathbf{x}) u^l h_{bj}(\mathbf{x}) = 0$$

Splitting this relation by u^l ensures that the system of partial differential equations

$$\xi_a^i(\mathbf{x}) \frac{\partial h_{bl}(\mathbf{x})}{\partial x_i} + h_{bj}(\mathbf{x})\rho_{jl}^a(\mathbf{x}) = 0 \tag{24}$$

holds for all the values of b, l. The indices in (24) take the following values: $a = 1, \dots, n, i = 1, \dots, p, b, j, l = 1, \dots, q$.

It is readily seen that the second term of the left-hand side of Eq. (24) is the (b, l)th entry of the matrix $H(\mathbf{x})\Gamma_a(\mathbf{x})$, $(a = 1, \dots, n)$. Hence it follows that the matrix $H(\mathbf{x})$ satisfies Eq. (13). The lemma is proved.

The forms of the matrices Γ_a for the basis operators of the algebra $c(1,3)$ are as follows:

- Matrices Γ_a, $a = 1, 2, 3, 4$ corresponding to the operators P_μ, $(\mu = 0, 1, 2, 3)$ are zero $q \times q$ matrices.
- Matrices Γ_a, $a = 1, \dots, 6$ corresponding to the operators $J_{\mu\nu}$, $(\mu, \nu = 0, 1, 2, 3)$ are equal to $-S_{\mu\nu}$, where $S_{\mu\nu}$ are constant $q \times q$ matrices realizing a representation of the algebra $o(1,3)$ and satisfying commutation relations (16).
- Matrix Γ_1 corresponding to the dilation operator D reads as $-kE$, where k is the conformal degree of the algebra $c(1,3)$ and E is the unit $q \times q$ matrix.
- Matrices Γ_a, $a = 1, 2, 3, 4$ corresponding to the operators K_μ, $(\mu = 0, 1, 2, 3)$ are given by the following formulas:

$$\begin{aligned}
\Gamma_1 &= -2x_0 kE - 2x_1 S_{01} - 2x_2 S_{02} - 2x_3 S_{03} \\
\Gamma_2 &= 2x_1 kE + 2x_0 S_{01} - 2x_2 S_{12} - 2x_3 S_{13} \\
\Gamma_3 &= 2x_2 kE + 2x_0 S_{02} + 2x_1 S_{12} - 2x_3 S_{23} \\
\Gamma_4 &= 2x_3 kE + 2x_0 S_{03} + 2x_1 S_{13} + 2x_2 S_{23}
\end{aligned}$$

With the explicit forms of the matrices Γ_a in hand, we can determine the structure of the matrices $H = \Lambda^{-1}$ for ansatz (22) invariant under a subalgebra g of the conformal algebra $c(1,3)$.

If $g \in p(1,3) = \langle P_\mu, J_{\mu\nu} | \mu, \nu = 0, 1, 2, 3 \rangle$, then the corresponding matrices Γ_a are linear combinations of the matrices $S_{\mu\nu}$. Hence it follows that the matrix H can be sought in the form

$$H = \tilde{H} = \prod_{\mu<\nu} \exp(\theta_{\mu\nu} S_{\mu\nu}) \tag{25}$$

where $\theta_{\mu\nu} = \theta_{\mu\nu}(x_0, \mathbf{x})$ are arbitrary smooth functions defined in $\tilde{\Omega} \subset \mathbf{R}^{1,3}$.

Next, if g is a subalgebra of the conformal algebra $c(1,3)$ with a nonzero projection on the vector space spanned by the operators D, K_0, K_1, K_2, K_3, then the corresponding matrices Γ_a are linear combinations of the matrices E and $S_{\mu\nu}$. That is why the matrix H should be sought in the more general form

$$H = \exp(\theta E)\tilde{H} \tag{26}$$

where $\theta = \theta(x_0, \mathbf{x})$ is an arbitrary smooth function defined in $\tilde{\Omega}$ and $\tilde{H}$ is the matrix given in (25).

B. Subalgebras of the Conformal Algebra $c(1,3)$ of Rank 3

Now we turn to the problem of constructing conformally invariant ansatzes that reduce systems of partial differential equations invariant under the group $C(1,3)$ to systems of ordinary differential equations.

As a second step of the algorithm of symmetry reduction formulated above, we have to describe the optimal system of subalgebras of the algebra $c(1,3)$ of the rank $s = 3$. Indeed, the initial system has $p = 4$ independent variables. It has to be reduced to a system of differential equations in $4 - s = 1$ independent variables, so that $s = 3$.

Classification of inequivalent subalgebras of the algebras $p(1,3)$, $\tilde{p}(1,3)$, $c(1,3)$ within actions of different automorphism groups [including the groups $P(1,3)$, $\tilde{P}(1,3)$ and $C(1,3)$] is already available [30]. Since we will concentrate on conformally invariant systems, it is natural to restrict our disscussion to the classification of subalgebras of $c(1,3)$ that are inequivalent within the action of the conformal group $C(1,3)$.

In order to get the full lists of the subalgebras in question we have to check that relation (7) with $s = 3$ holds for each element of the lists of inequivalent subalgebras of the algebras $p(1,3), \tilde{p}(1,3), c(1,3)$ given elsewhere [30]. Evidently, we can restrict our considerations to subalgebras having the dimension not less than 3.

Let $c(1,3)$ be the conformal algebra having the basis operators (15) and $c^{(1)}(1,3)$ be the conformal algebra spanned by the operators

$$\begin{aligned} P^{(1)}_{\mu} &= \partial_{x_\mu}, \quad J^{(1)}_{\mu\nu} = x^{\mu}\partial_{x_\nu} - x^{\nu}\partial_{x_\mu}, \quad D^{(1)} = x_{\mu}\partial_{x_\mu} \\ K^{(1)} &= 2x^{\mu}D^{(1)} - (x_{\nu}x^{\nu})\partial_{x_\mu} \end{aligned} \tag{27}$$

where $\mu, \nu = 0, 1, 2, 3$.

Note that the conformal group $C(1,3)$ generated by the infinitesimal operators (27) acts in the space of independent variables $\mathbf{R}^{1,3}$ only. That is why the basis operators of the algebra $c^{(1)}(1,3)$ act in the space of dependent variables $\mathbf{R}^q$ as zero operators.

Lemma 2. *Let L be a subalgebra of the algebra $c(1,3)$ of the rank s and let $s^{(1)}$ be the rank of the projection of L on $c^{(1)}(1,3)$. Then, from the equality $s = s^{(1)}$, it follows that* $\dim L = s$.

Proof. Suppose that the reverse assertion holds, namely, that $\dim L \neq s$. As $\dim L \geq s$, it follows that $\dim L > s$. Choose the basis elements $X_1, \ldots, X_m$ of the algebra L so that

- The rank of the matrix M, whose entries are projections of the operators $X_1, \ldots, X_m$ on $c^{(1)}(1,3)$, is equal to s.
- The linear space spanned by the operators $X_1, \ldots, X_m$ contains $L \cap \langle P_0, P_1, P_2, P_3 \rangle$.

We denote as $S_0^{(1)}$ the point $(x_0^0, \mathbf{x}^0) \in \tilde{\Omega}$ in which the rank of the matrix M equals to s. Let the vector fields X_i be equal to $X_1^0, \ldots, X_s^0, X_{s+1}^0, \ldots X_m^0$ in $S_0^{(1)}$. Then there are constants $\alpha_1, \ldots, \alpha_s$, such that the vector field $\alpha_1 X_1^0 + \cdots + \alpha_s X_s^0 + X_{s+1}^0$ restricted to the space of dependent variables $U = \mathbf{R}^q$ is a nonzero operator. Indeed, if this operator vanishes identically on $\mathbf{R}^q$ for any choice of $\alpha_1, \ldots, \alpha_s$, then the vector fields $X_1^0, \ldots, X_s^0, X_{s+1}^0$ belong to the vector space $\langle P_0, P_1, P_2, P_3 \rangle$, and this fact contradicts to the assumptions that $\dim L > s$, rank $L = s$. Consequently, the matrix formed by the coefficients of the vector fields $X^1, \ldots, X_s, \alpha_1 X_1 + \cdots + \alpha_s X_s + X_{s+1}$ has a nonzero minor of the order $s+1$ in some point $(x_0^0, \mathbf{x}^0, \mathbf{u}^0)$ (the first four coordinates are same as those of the point S_0). This contradicts the assumption that rank $L > s$. Hence we conclude that $\dim L = s$. The lemma is proved.

It follows from Lemma 2 that the validity of the relation (7) with $s = 3$ should be ascertained only for the three-dimensional subalgebras of the algebras $p(1,3), \tilde{p}(1,3), c(1,3)$ given elsewhere [30]. Moreover, we need to check the first condition from (7) only.

Consider the subalgebras of the algebra $p(1,3)$, whose basis operators are of the form (15). Among the three-dimensional subalgebras of the algebra $p(1,3)$ listed elsewhere [30], there are only five subalgebras $\langle G_1, P_0 + P_3, > P_1 \rangle$, $\langle J_{12}, P_1, P_2 \rangle$, $\langle J_{03}, P_0, P_3 \rangle$, $\langle J_{12}, J_{13}, J_{23} \rangle$, $\langle J_{01}, J_{02}, J_{12} \rangle$, that do not respect the first condition, (7). These subalgebras give rise to the so-called partially invariant solutions [19]. Partially invariant solutions cannot be handled in a generic way; they should always be considered within the context of a specific system of partial differential equation to be reduced. We will not consider the partially invariant solutions further. The remaining inequivalent subalgebras are listed in the assertion below.

Assertion 1. *The list of subalgebras of the algebra $p(1,3)$ of the rank 3, defined within the action of the inner automorphism group of the algebra $c(1,3)$, is exhausted by the following subalgebras:*

$$
\begin{array}{ll}
L_1 = \langle P_0, P_1, P_2 \rangle; & L_2 = \langle P_1, P_2, P_3 \rangle \\
L_3 = \langle M, P_1, P_2 \rangle; & L_4 = \langle J_{03} + \alpha J_{12}, P_1, P_2 \rangle \\
L_5 = \langle J_{03}, M, P_1 \rangle; & L_6 = \langle J_{03} + P_1, P_0, P_3 \rangle \\
L_7 = \langle J_{03} + P_1, M, P_2 \rangle; & L_8 = \langle J_{12} + \alpha J_{03}, P_0, P_3 \rangle \\
L_9 = \langle J_{12} + P_0, P_1, P_2 \rangle; & L_{10}^j = \langle J_{12} + (-1)^j P_3, P_1, P_2 \rangle \\
L_{11}^j = \langle J_{12} + (-1)^j 2T, P_1, P_2 \rangle; & L_{12} = \langle G_1, M, P_2 + \alpha P_1 \rangle
\end{array}
$$

$$L^j_{13} = \langle G_1 + (-1)^j P_2, M, P_1\rangle; \quad L_{14} = \langle G_1 + 2T, M, P_2\rangle$$
$$L_{15} = \langle G_1 + 2T, M, P_1 + \alpha P_2\rangle; \quad L_{16} = \langle J_{12}, J_{03}, M\rangle$$
$$L^j_{17} = \langle G^j_1, G^j_2, M\rangle; \quad L_{18} = \langle J_{03}, G_1, P_2\rangle$$
$$L_{19} = \langle G_1, J_{03}, M\rangle; \quad L_{20} = \langle G_1, J_{03} + P_2, M\rangle$$
$$L_{21} = \langle G_1, J_{03} + P_1 + \alpha P_2, M\rangle; \quad L_{22} = \langle G_1, G_2, J_{03} + \alpha J_{12}\rangle$$

Here $\alpha \in \mathbf{R}$; $M = P_0 + P_3$, $T = \frac{1}{2}(P_0 - P_3)$, $G_a = J_{0a} - J_{a3}$, $(a = 1, 2)$; $G^j_1 = G_1 + (-1)^j P_2$, $G^j_2 = G_2 - (-1)^j P_1 + \alpha P_2$; $j = 1, 2$.

In the same way, we handle the three-dimensional subalgebras of the algebras $\tilde{p}(1,3)$ and $c(1,3)$. We have skipped from the list of subalgebras of the algebra $\tilde{p}(1,3)$ those conjugate to subalgebras of $p(1,3)$. Furthermore, we have skipped from the list of subalgebras of the conformal algebra those conjugate to subalgebras of the algebra $\tilde{p}(1,3)$. The results obtained are presented in the two assertions below.

Assertion 2. *The list of subalgebras of the algebra $\tilde{p}(1,3)$ of the rank 3, defined within the action of the inner automorphism group of the algebra $c(1,3)$, is exhausted by the subalgebras given in Assertion* 1 *and by the following subalgebras*:

$$F_1 = \langle D, P_0, P_3\rangle; \quad F_2 = \langle J_{12} + \alpha D, P_0, P_3\rangle$$
$$F_3 = \langle J_{12}, D, P_0\rangle; \quad F_4 = \langle J_{12}, D, P_3\rangle$$
$$F_5 = \langle J_{03} + \alpha D, P_0, P_3\rangle; \quad F_6 = \langle J_{03} + \alpha D, P_1, P_2\rangle$$
$$F_7 = \langle J_{03} + \alpha D, M, P_1\rangle \ (a \neq 0)$$
$$F_8 = \langle J_{03} + D + (-1)^j 2T, P_1, P_2\rangle$$
$$F_9 = \langle J_{03} + D + (-1)^j 2T, M, P_1\rangle; \quad F_{10} = \langle J_{03}, D, P_1\rangle$$
$$F_{11} = \langle J_{03}, D, M\rangle; \quad F_{12} = \langle J_{12} + \alpha J_{03} + \beta D, P_0, P_3\rangle \ (\alpha \neq 0)$$
$$F_{13} = \langle J_{12} + \alpha J_{03} + \beta D, P_1, P_2\rangle \ (\alpha \neq 0)$$
$$F_{14} = \langle J_{12} + \alpha(J_{03} + D + 2T), P_1, P_2\rangle \ (\alpha \neq 0)$$
$$F_{15} = \langle J_{12} + \alpha J_{03}, D, M\rangle \ (\alpha \neq 0)$$
$$F_{16} = \langle J_{03} + \alpha D, J_{12} + \beta D, M\rangle \ (0 \leq |\alpha| \leq 1, \ \beta \geq 0, |\alpha| + |\beta| \neq 0)$$
$$F_{17} = \langle J_{03} + D + (-1)^j 2T, J_{12} + 2\alpha T, M\rangle \ (\alpha \in R)$$
$$F_{18} = \langle J_{03} + D, J_{12} + (-1)^j 2T, M\rangle; \quad F_{19} = \langle J_{03}, J_{12}, D\rangle$$
$$F_{20} = \langle G_1, J_{03} + \alpha D, P_2\rangle \ (0 < |\alpha| \leq 1)$$
$$F_{21} = \langle J_{03} + D, G_1 + (-1)^j P_2, M\rangle$$
$$F_{22} = \langle J_{03} - D + (-1)^j M, G_1, P_2\rangle$$
$$F_{23} = \langle J_{03} + 2D, G_1 + (-1)^j 2T, M\rangle$$
$$F_{24} = \langle J_{03} + 2D, G_1 + (-1)^j 2T, P_2\rangle$$

Here $M = P_0 + P_3,\ G_1 = J_{01} - J_{13},\ T = \frac{1}{2}(P_0 - P_3)$, *the parameters* α, β *are positive (if otherwise is not indicated)*; $j = 1, 2$.

Assertion 3. *The list of subalgebras of the algebra* $c(1,3)$ *of the rank* 3, *defined within the action of the inner automorphism group of the algebra* $c(1,3)$, *is exhausted by the subalgebras of the algebras* $p(1,3), \tilde{p}(1,3)$ *given in Assertions* 1 *and* 2 *and by the following subalgebras:*

$$
\begin{aligned}
C_1 &= \langle S + T + J_{12}, G_1 + P_2, M \rangle \\
C_2 &= \langle S + T + J_{12} + G_1 + P_2, G_2 - P_1, M \rangle \\
C_3 &= \langle J_{12}, S + T, M \rangle; \quad C_4 = \langle S + T, Z, M \rangle \\
C_5 &= \langle S + T + \alpha J_{12}, Z, M \rangle \ (\alpha \neq 0) \\
C_6 &= \langle S + T + J_{12} + \alpha Z, G_1 + P_2, M \rangle \ (\alpha \neq 0) \\
C_7 &= \langle S + T + J_{12}, Z, G_1 + P_2 \rangle \\
C_8 &= \langle S + T + \beta Z, J_{12} + \alpha Z, M \rangle \quad (\alpha, \beta \in R, |\alpha| + |\beta| \neq 0) \\
C_9 &= \langle J_{12}, S + T, Z \rangle; \quad C_{10} = \langle D - J_{03}, S, T \rangle \\
C_{11} &= \langle P_2 + K_2 + \sqrt{3}(P_1 + K_1) + K_0 - P_0, \\
&\qquad - D + J_{02} - \sqrt{3}J_{01}, P_0 + K_0 - 2(K_2 - P_2) \rangle \\
C_{12} &= \langle P_0 + K_0 \rangle \oplus \langle J_{12}, K_3 - P_3 \rangle \\
C_{13} &= \langle 2J_{12} + K_3 - P_3, 2J_{13} - K_2 + P_2, 2J_{23} + K_1 - P_1 \rangle \\
C_{14} &= \langle P_1 + K_1 + 2J_{03}, P_2 + K_2 + K_0 - P_0, 2J_{12} + K_3 - P_3 \rangle
\end{aligned}
$$

where $M = P_0 + P_3,\ G_{0a} = J_{0a} - J_{a3}\ (a = 1, 2),\ Z = J_{03} + D, S = \frac{1}{2}(K_0 + K_3)$, $T = \frac{1}{2}(P_0 - P_3)$.

Remark 1. While classifying subalgebras of the extended Poincaré algebra $\tilde{p}(1,3)$, the discrete equivalence transformations Φ_1, Φ_2, Φ_3, that leave the algebra $\tilde{p}(1,3)$ invariant, were exploited elsewhere [30]. The result of the action of these groups on the operators of the algebra $\tilde{p}(1,3)$ is given in Table I. That is why we have completed the list of subalgebras of the algebras $p(1,3), \tilde{p}(1,3)$ obtained earlier [30] by the subalgebras obtainable by acting on these subalgebras with the discrete transformation groups Φ_1, Φ_2, Φ_3.

C. Construction of Conformally Invariant Ansatzes

Now we turn to constructing $C(1,3)$-invariant ansatzes that reduce conformally invariant systems of partial differential equations to systems of ordinary differential equations. To this end, we use the lists of subalgebras of the algebra $c(1,3)$ given in Assertions 1–3. Note that all the subsequent computations are

TABLE I
Effect of Equivalence Transformations on Extended Poincaré Algebra

	Action on $\tilde{p}(1,3)$		
Operators	Φ_1	Φ_2	Φ_3
P_0	$-P_0$	P_0	$-P_0$
P_1	$-P_1$	$-P_1$	P_1
$P_a\,(a=2,3)$	$-P_a$	P_a	$-P_a$
J_{03}	J_{03}	J_{03}	J_{03}
J_{12}	J_{12}	$-J_{12}$	$-J_{12}$
G_1	G_1	$-G_1$	$-G_1$
G_2	G_2	G_2	G_2
M	$-M$	M	$-M$
T	$-T$	T	$-T$
D	D	D	D

performed under supposition that the basis operators of $c(1,3)$ are of the form (15).

As shown in Section II.A, the ansatzes in question can be searched for in linear form (22) and matrices $H = \Lambda^{-1}$, in the form (26). According to Lemma 1, the matrix H has to satisfy equations (23), whose coefficients are defined uniquely by the choice of a subalgebra of the conformal algebra of the rank 3. Thus the problem of complete description of conformally invariant ansatzes reduces to solving system of partial differential equations (20), (23) for each subalgebra of the conformal algebra, which requires a vast amount of computation. The calculations are simplified if we take into account the general structure of the subalgebras listed in Assertions 1–3.

For further convenience, we will use the following basis of the algebra $o(1,3)$: $S_{03}, S_{12}, H_a, \tilde{H}_a$ $(a=1,2)$, where $H_a = S_{0a} - S_{a3}$, $\tilde{H}_a = S_{0a} + S_{a3}$ $(a=1,2)$. It is not difficult to check that these matrices satisfy the commutation relations

$$
\begin{aligned}
&[S_{03}, S_{12}] = [H_1, H_2] = [\tilde{H}_1, \tilde{H}_2] = 0 \\
&[H_a, S_{03}] = H_a\, [\tilde{H}_a, S_{03}] = -\tilde{H}_a \quad (a=1,2) \\
&[H_1, S_{12}] = -H_2\, [H_2, S_{12}] = H_1 \\
&[\tilde{H}_1, S_{12}] = -\tilde{H}_2\ [\tilde{H}_2, S_{12}] = \tilde{H}_1 \\
&[H_1, \tilde{H}_1] = [H_2, \tilde{H}_2] = -2S_{03} \\
&[\tilde{H}_2, H_1] = [H_2, \tilde{H}_1] = 2S_{12}
\end{aligned} \tag{28}
$$

In particular, relations (28) imply that the matrices H_1, H_2, S_{12}, S_{03} and $\tilde{H}_1, \tilde{H}_2, S_{12}, S_{03}$ realize two matrix representations of the Euclid algebra $\tilde{e}(2)$ (here the matrix S_{03} is identified with the dilation generator and the matrices H_1, H_2 and $\tilde{H}_1, \tilde{H}_2$ are identified with the translation generators). Furthermore, as E is the unit matrix, it commutes with all the basis elements of $o(1,3)$, namely

$$[E, S_{12}] = [E, S_{03}] = [E, H_a] = [E, \tilde{H}_a] = 0 \tag{29}$$

where $a = 1, 2$.

Analyzing the structure of the basis elements of the subalgebras of the conformal algebra given in Assertions 1–3, we see that the corresponding matrices Γ_a are most conveniently represented in terms of the matrices S_{03}, S_{12}, H_a, $\tilde{H}_a$ $(a = 1, 2)$. Hence we conclude that the matrix $H = H(x_0, \mathbf{x}) = \Lambda^{-1}(x_0, \mathbf{x})$ can be searched for in the form

$$\begin{aligned} H = {} & \exp\{(-\ln\theta)E\}\exp(\theta_0 S_{03})\exp(-\theta_3 S_{12})\exp(-2\theta_1 H_1) \\ & \times \exp(-2\theta_2 H_2)\exp(-2\theta_4 \tilde{H}_1)\exp(-2\theta_5 \tilde{H}_2) \end{aligned} \tag{30}$$

where $\theta = \theta(x_0, \mathbf{x})$, $\theta_0 = \theta_0(x_0, \mathbf{x}), \theta_m = \theta_m(x_0, \mathbf{x})$ $(m = 1, 2, \ldots, 5)$ are arbitrary smooth functions defined in an open domain $\tilde{\Omega} \subset \mathbf{R}^{1,3}$ of the Minkowski space of the independent variables $x_0, \mathbf{x} = (x_1, x_2, x_3)$.

Let $L = \langle X_a | a = 1, 2, 3 \rangle$ be a subalgebra of the algebra $c(1,3)$ of rank 3. By assumption, the basis operators of L can be written in the following form

$$X_a = \xi_a^\mu(x_0, \mathbf{x})\partial_{x_\mu} + (\tilde{\Gamma}_a \mathbf{u} \cdot \partial_{\mathbf{u}}), \quad (a = 1, 2, 3) \tag{31}$$

and

$$\tilde{\Gamma}_a = f^a E + f_0^a S_{03} + f_1^a H_1 + f_2^a H_2 + f_3^a S_{12} + f_4^a \tilde{H}_1 + f_5^a \tilde{H}_2, \quad (a = 1, 2, 3) \tag{32}$$

where $f^a = f^a(x_0, \mathbf{x})$, $f_0^a = f_0^a(x_0, \mathbf{x}), f_m^a = f_m^a(x_0, \mathbf{x})$ $(m = 1, \ldots, 5)$ are some fixed smooth functions. In particular, if the operator X_a is a linear combination of the translation generators, then $\Gamma_a = 0$, and therefore, $f^a = f_0^a = f_m^a = 0$ in (32).

Owing to Lemma 1, in order to construct ansatz (22) invariant under the subalgebra L, we have to solve systems (20), (23), which in the case under consideration read as

$$\xi_a^\mu \frac{\partial \omega}{\partial x_\mu} = 0 \tag{33}$$

$$\xi_a^\mu \frac{\partial H}{\partial x_\mu} + H\tilde{\Gamma}_a = 0 \tag{34}$$

where $a = 1, 2, 3$, $\mu = 0, 1, 2, 3$. The functions $\xi_a^\mu = \xi_a^\mu(x_0, \mathbf{x})$ and the variable matrices $\tilde{\Gamma}_a = \tilde{\Gamma}(x_0, \mathbf{x})$ are the coefficients of the basis operators of the subalgebra L [note that $\tilde{\Gamma}_a$ is of the form (32)]. Matrix function H (30) and the scalar function $\omega = \omega(x_0, \mathbf{x})$ are to be determined while integrating (33) and (34).

Next, we prove a technical assertion to be used in the sequel for simplifying the form of system (34).

Lemma 3. *Let H be of the form* (30). *Then the following identity holds true:*

$$
\begin{aligned}
\xi_a^\mu \frac{\partial H}{\partial x_\mu} = H\Big\{ & -\theta^{-1}\xi_a^\mu \frac{\partial \theta}{\partial x_\mu} E + \xi_a^\mu \frac{\partial \theta_0}{\partial x_\mu}[(1 + 8\theta_1\theta_4 + 8\theta_2\theta_5)S_{03} \\
& + 8(\theta_1\theta_5 - \theta_2\theta_4)S_{12} + 2\theta_1 H_1 + 2\theta_2 H_2 - 2(\theta_4 + 4\theta_1\theta_4^2 + 8\theta_2\theta_4\theta_5 \\
& - 4\theta_1\theta_5^2)\tilde{H}_1 - 2(\theta_5 + 4\theta_2\theta_5^2 + 8\theta_1\theta_4\theta_5 - 4\theta_2\theta_4^2)\tilde{H}_2] \\
& - \xi_a^\mu \frac{\partial \theta_3}{\partial x_\mu}[8(\theta_2\theta_4 - \theta_1\theta_5)S_{03} + (1 + 8\theta_1\theta_4 + 8\theta_2\theta_5)S_{12} \\
& + 2\theta_2 H_1 - 2\theta_1 H_2 + 2(\theta_5 + 4\theta_2\theta_5^2 - 4\theta_2\theta_4^2 + 8\theta_1\theta_4\theta_5)\tilde{H}_1 \\
& - 2(\theta_4 + 4\theta_1\theta_4^2 - 4\theta_1\theta_5^2 + 8\theta_2\theta_4\theta_5)\tilde{H}_2] \\
& - 2\xi_a^\mu \frac{\partial \theta_1}{\partial x_\mu}[4\theta_4 S_{03} + 4\theta_5 S_{12} + H_1 + 4(\theta_5^2 - \theta_4^2)\tilde{H}_1 - 8\theta_4\theta_5\tilde{H}_2] \\
& - 2\xi_a^\mu \frac{\partial \theta_2}{\partial x_\mu}[4\theta_5 S_{03} - 4\theta_4 S_{12} + H_2 - 8\theta_4\theta_5\tilde{H}_1 + 4(\theta_4^2 - \theta_5^2)\tilde{H}_2] \\
& - 2\xi_a^\mu \frac{\partial \theta_4}{\partial x_\mu}\tilde{H}_1 - 2\xi_a^\mu \frac{\partial \theta_5}{\partial x_\mu}\tilde{H}_2]\Big\},
\end{aligned}
$$

where $a = 1, 2, 3,\ \mu = 0, 1, 2, 3.$

Proof. Acting by the linear differential operator $\xi_a^\mu \partial_{x_\mu}$ on matrix H (30) yields an equality whose right-hand side can be decomposed into the sum of seven terms having the same structure:

$$\xi_a^\mu \frac{\partial H}{\partial x_\mu} = \sum_{i=1}^{7} D_i \tag{35}$$

As each term D_i is handled in the same way, we give the calculation details for one of them, say, for

$$D_4 = \exp\{(-\ln\theta)E\} \prod_{i=1}^{3} \Lambda_i \Big(-2\xi_a^\mu \frac{\partial \theta_1}{\partial x_\mu} H_1\Big) \prod_{j=4}^{6} \Lambda_j \tag{36}$$

Note that in (36) we use the following designations:

$$\begin{aligned} \Lambda_1 &= \exp(\theta_0 S_{03}), & \Lambda_2 &= \exp(-\theta_3 S_{12}) \\ \Lambda_3 &= \exp(-2\theta_1 H_1), & \Lambda_4 &= \exp(-2\theta_3 H_2) \\ \Lambda_5 &= \exp(-2\theta_4 \tilde{H}_1), & \Lambda_6 &= \exp(-2\theta_5 \tilde{H}_2) \end{aligned} \tag{37}$$

Having multiplied the right-hand side of (36) by the matrix HH^{-1} on the left, we arrive at the equality

$$D_4 = H\left(-2\xi_a^\mu \frac{\partial\theta_1}{\partial x_\mu}\right)\Lambda_6^{-1}\Lambda_5^{-1}\Lambda_4^{-1}H_1\Lambda_4\Lambda_5\Lambda_6 \tag{38}$$

where the matrices $\Lambda_4, \Lambda_5, \Lambda_6$ are given in (37).

To simplify the right-hand side of (38), we exploit the Campbell–Hausdorff formula

$$\exp(\tau A)B\exp(-\tau A) = \sum_{n=0}^{\infty}\frac{\tau^n}{n!}\{A,B\}^n$$
$$\{A,B\}^n = [A,\{A,B\}^{n-1}], \quad \{A,B\}^0 = B$$

which holds for arbitrary square matrices A, B.

With account of commutation relations (28) and (29), we get

$$\Lambda_4^{-1}H_1\Lambda_1 = \exp(2\theta_2 H_2)H\exp(-2\theta_2 H_2) = H_1$$

whence

$$\begin{aligned} \Lambda_5^{-1}\Lambda_4^{-1}H\Lambda_4\Lambda_5 = \Lambda_5^{-1}H_1\Lambda_5 &= \exp(2\theta_4\tilde{H}_1)H_1\exp(-2\theta_4\tilde{H}_1) \\ &= H_1 + 4\theta_4 S_{03} - 4\theta_4^2\tilde{H}_1 \end{aligned}$$

Consequently

$$\begin{aligned} \Lambda_6^{-1}\Lambda_5^{-1}\Lambda_4^{-1}H_1\Lambda_4\Lambda_5\Lambda_6 &= \exp(2\theta_5\tilde{H}_2)(H_1 + 4\theta_4 S_{03} - 4\theta_2^2\tilde{H}_1)\exp(-2\theta_5\tilde{H}_2) \\ &= H_1 + 4\theta_4 S_{03} + 4\theta_5 S_{12} + 4(\theta_5^2 - \theta_4^2)\tilde{H}_1 - 8\theta_4\theta_5\tilde{H}_2 \end{aligned}$$

Finally, we have

$$D_4 = H\left(-2\xi_a^\mu\frac{\partial\theta_2}{\partial x_\mu}\right)[H_1 + 4\theta_4 S_{03} + 4\theta_5 S_{12} + 4(\theta_5^2 - \theta_4^2)\tilde{H}_1 - 8\theta_4\theta_5\tilde{H}_2]$$

The same reasoning, when applied to the remaining terms of the right-hand side of the equality (35), completes the proof of the lemma.

Assertion 4. *System* (34) *is equivalent to the system of partial differential equations for the functions* $\theta, \theta_0, \theta_m$ $(m = 1, 2, \ldots, 5)$.

$$
\begin{aligned}
\xi_a^\mu \frac{\partial \theta}{\partial x_\mu} &= f^a \theta \\
\xi_a^\mu \frac{\partial \theta_0}{\partial x_\mu} &= 4(\theta_4 f_1^a + \theta_5 f_2^a) - f_0^a \\
\xi_a^\mu \frac{\partial \theta_1}{\partial x_\mu} &= 4(\theta_1\theta_4 + \theta_2\theta_5) f_1^a + 4(\theta_1\theta_5 - \theta_2\theta_4) f_2^a - \theta_1 f_0^a - \theta_2 f_3^a + \frac{1}{2} f_1^a \\
\xi_a^\mu \frac{\partial \theta_2}{\partial x_\mu} &= 4(\theta_2\theta_4 - \theta_1\theta_5) f_1^a + 4(\theta_2\theta_5 + \theta_1\theta_4) f_2^a - \theta_2 f_0^a + \theta_1 f_3^a + \frac{1}{2} f_2^a \\
\xi_a^\mu \frac{\partial \theta_3}{\partial x_\mu} &= 4(\theta_4 f_2^a - \theta_5 f_1^a) + f_3^a \\
\xi_a^\mu \frac{\partial \theta_4}{\partial x_\mu} &= \theta_4 f_0^a - 2(\theta_4^2 - \theta_5^2) f_1^a - 4\theta_4\theta_5 f_2^a - \theta_5 f_3^a + \frac{1}{2} f_4^a \\
\xi_a^\mu \frac{\partial \theta_5}{\partial x_\mu} &= \theta_5 f_0^a - 4\theta_4\theta_5 f_1^a + 2(\theta_4^2 - \theta_5^2) f_2^a + \theta_4 f_3^a + \frac{1}{2} f_5^a
\end{aligned}
\tag{39}
$$

In (39) $\mu = 0, 1, 2, 3;$ $a = 1, 2, 3$. *The coefficients of linear differential operators* $\xi_a^\mu \partial_{x_\mu}$ *and the functions* f^a, f_0^a, f_m^a $(m = 1, 2, \ldots, 5)$ *are defined by the coefficients of the basis operators of the subalgebra L of the algebra* $c(1,3)$ *of the rank* 3.

Proof. Inserting the expression for $\xi_a^\mu \frac{\partial H}{\partial x_\mu}$ given in Lemma 3 into the left-hand side of (34) and multiplying the equation thus obtained by the inverse of the nonsingular matrix H we arrive at the system of matrix equations, whose left-hand sides are the linear combinations of the linearly independent matrices $E, S_{01}, S_{12}, H_a, \tilde{H}_a$ $(a = 1, 2)$. Splitting the system obtained by these matrices, and taking into account the forms of the matrices $\tilde{\Gamma}_a$, and performing some simplifications yield system of Eqs. (39). The assertion is proved.

Summarizing we conclude that the problem of constructing conformally invariant ansatzes reduces to finding the fundamental solution of the system of linear partial differential equations (33) and particular solutions of first-order systems of nonlinear partial differential equations (39).

The next subsections are devoted to constructing the ansatzes invariant under the subalgebras of the Poincaré, extended Poincaré, and conformal algebras given in Assertions 1–3. The solution procedure is based on the above derived identities and, essentially, on Assertion 4.

1. $P(1,3)$-*Invariant Ansatzes*

Subalgebras listed in Assertion 1 give rise to $P(1,3)$ (Poincaré)-invariant ansatzes. Analysis of the structure of these subalgebras shows that we can put $\theta = 1, \theta_4 = \theta_5 = 0$ in formula (30) for the matrix H. Moreover, the form of the basis elements of these subalgebras imply that in formulas (32) and (38) $f^a = f_4^a = f_5^a = 0$, for all the values of $a = 1, 2, 3$. Therefore system (39) for the matrix H takes the form of 12 first-order partial differential equations for the functions $\theta_0, \theta_1, \theta_2, \theta_3$

$$
\begin{aligned}
&\xi_a^\mu \frac{\partial \theta_0}{\partial x_\mu} = -f_0^a, \quad \xi_a^\mu \frac{\partial \theta_3}{\partial x_\mu} = f_3^a \\
&\xi_a^\mu \frac{\partial \theta_1}{\partial x_\mu} = -\theta_1 f_0^a - \theta_2 f_3^a + \frac{1}{2} f_1^a \\
&\xi_a^\mu \frac{\partial \theta_2}{\partial x_\mu} = -\theta_2 f_0^a + \theta_1 f_3^a + \frac{1}{2} f_2^a
\end{aligned}
\tag{40}
$$

where $\mu = 0, 1, 2, 3; a = 1, 2, 3$.

We integrate system (33), (40) for the case of the subalgebra $L_{22} = \langle G_1, G_2, J_{03} + \alpha J_{12} \rangle$ ($\alpha \in \mathbf{R}$) (all other cases are handled in a similar way).

System (33) for finding the function $\omega = \omega(x_0, \mathbf{x})$ reads as

$$
\begin{aligned}
G_1^{(1)} \omega &= [(x_0 - x_3)\partial_{x_1} + x_1(\partial_{x_0} + \partial_{x_3})]\omega = 0 \\
G_2^{(1)} \omega &= [(x_0 - x_3)\partial_{x_2} + x_2(\partial_{x_0} + \partial_{x_3})]\omega = 0 \\
(J_{03}^{(1)} + \alpha J_{12}^{(1)})\omega &= [x_0 \partial_{x_3} + x_3 \partial_{x_0} + \alpha(x_2 \partial_{x_1} - x_1 \partial_{x_2})]\omega = 0, \quad \alpha \in \mathbf{R}
\end{aligned}
\tag{41}
$$

Performing the change of variables

$$
\begin{aligned}
&y_0 = (x_0 + x_3)(x_0 - x_3), \quad y_1 = \sqrt{x_1^2 + x_2^2} \\
&y_2 = \arctan \frac{x_2}{x_1}, \quad y_3 = x_0 - x_3
\end{aligned}
$$

reduces system (41) to the form

$$
\begin{aligned}
y_1 \frac{\partial \omega}{\partial y_1} + 2y_1^2 \frac{\partial \omega}{\partial y_0} - \tan y_2 \frac{\partial \omega}{\partial y_2} &= 0 \\
y_1 \frac{\partial \omega}{\partial y_1} + 2y_1^2 \frac{\partial \omega}{\partial y_0} - (\tan y_2)^{-1} \frac{\partial \omega}{\partial y_2} &= 0 \\
y_3 \frac{\partial \omega}{\partial y_3} + \alpha \frac{\partial \omega}{\partial y_2} &= 0
\end{aligned}
$$

The fundamental solution of this system reads as $\omega = y_0 - y_1^2$. Returning to the initial variables, we get the fundamental solution of system (41), $\omega = x_\mu x^\mu = x_0^2 - x_1^2 - x_2^2 - x_3^2$.

Next, taking into account the forms of the basis elements of the subalgebra L_{22}, we get the expressions for the functions f_μ^a ($\mu = 0, 1, 2, 3;\ a = 1, 2, 3$)

$$\begin{aligned} G_1&: \quad f_0^1 = f_2^1 = f_3^1 = 0, \quad f_1^1 = -1 \\ G_2&: \quad f_0^2 = f_1^2 = f_3^2 = 0, \quad f_2^2 = -1 \\ J_{03} + \alpha J_{12}&: \quad f_0^3 = -1, \quad f_1^3 = f_2^3 = 0, \quad f_3^3 = -\alpha \qquad (\alpha \in \mathbf{R}) \end{aligned}$$

So system (40) takes the form

$$\begin{aligned} &G_1^{(1)}\theta_0 = G_1^{(1)}\theta_2 = G_1^{(1)}\theta_3 = 0, \quad G_1^{(1)}\theta_1 = -\frac{1}{2} \\ &G_2^{(1)}\theta_0 = G_2^{(1)}\theta_1 = G_2^{(1)}\theta_3 = 0, \quad G_2^{(1)}\theta_2 = -\frac{1}{2} \\ &(J_{03}^{(1)} + \alpha J_{12}^{(1)})\theta_0 = 1, \quad (J_{03}^{(1)} + \alpha J_{12}^{(1)})\theta_3 = -\alpha \\ &(J_{03}^{(1)} + \alpha J_{12}^{(1)})\theta_1 = \theta_1 + \alpha\theta_2, \quad (J_{03}^{(1)} + \alpha J_{12}^{(1)})\theta_2 = \theta_2 - \alpha\theta_1 \end{aligned} \tag{42}$$

As we have already mentioned, to construct the matrix H it suffices to find particular solutions of system (42). The system for determination of the function θ_0 reads as

$$\begin{aligned} &(x_0 - x_3)\frac{\partial\theta_0}{\partial x_a} + x_a\left(\frac{\partial\theta_0}{\partial x_0} + \frac{\partial\theta_0}{\partial x_3}\right) = 0, \quad (a = 1, 2) \\ &x_0\frac{\partial\theta_0}{\partial x_3} + x_3\frac{\partial\theta_0}{\partial x_0} + \alpha\left(x_2\frac{\partial\theta_0}{\partial x_1} - x_1\frac{\partial\theta_0}{\partial x_2}\right) = 1 \end{aligned} \tag{43}$$

We look for its particular solution of the form $\theta_0 = f(x_0 - x_3)$. On direct check, we become convinced of the fact that this function satisfies the first two equations of system (43) and that, the third one reduces to the ordinary differential equation

$$-\xi\frac{df}{d\xi} = 1, \quad \xi = x_0 - x_3$$

whose solution reads as $f = -\ln|\xi|$.

Thus we can choose $\theta_0 = -\ln|x_0 - x_3|$. The first two equations for the function θ_3 coincide with those from (43), and the third equation

$$x_0\frac{\partial\theta_3}{\partial x_3} + x_3\frac{\partial\theta_3}{\partial x_0} + \alpha\left(x_2\frac{\partial\theta_3}{\partial x_1} - x_1\frac{\partial\theta_3}{\partial x_2}\right) = -\alpha$$

differs from the third equation from system (43) by the constant $-\alpha$ in the right-hand side. Thus we easily get the final form of the particular solution of (43) $\theta_3 = \alpha \ln |x_0 - x_3|$.

According to (42) the system for finding the functions θ_1, θ_2 has the form

$$\begin{aligned}
(x_0 - x_3)\frac{\partial\theta_1}{\partial x_1} + x_1\left(\frac{\partial\theta_1}{\partial x_0} + \frac{\partial\theta_1}{\partial x_3}\right) &= -\frac{1}{2}\\
(x_0 - x_3)\frac{\partial\theta_2}{\partial x_1} + x_1\left(\frac{\partial\theta_2}{\partial x_0} + \frac{\partial\theta_2}{\partial x_3}\right) &= 0\\
(x_0 - x_3)\frac{\partial\theta_1}{\partial x_2} + x_2\left(\frac{\partial\theta_1}{\partial x_0} + \frac{\partial\theta_2}{\partial x_3}\right) &= 0\\
(x_0 - x_3)\frac{\partial\theta_2}{\partial x_2} + x_2\left(\frac{\partial\theta_2}{\partial x_0} + \frac{\partial\theta_2}{\partial x_3}\right) &= -\frac{1}{2}\\
x_0\frac{\partial\theta_1}{\partial x_3} + x_3\frac{\partial\theta_1}{\partial x_0} - \alpha\left(x_1\frac{\partial\theta_1}{\partial x_2} - x_2\frac{\partial\theta_1}{\partial x_1}\right) &= \theta_1 + \alpha\theta_2\\
x_0\frac{\partial\theta_2}{\partial x_3} + x_3\frac{\partial\theta_2}{\partial x_0} - \alpha\left(x_1\frac{\partial\theta_2}{\partial x_2} - x_2\frac{\partial\theta_2}{\partial x_1}\right) &= \theta_2 - \alpha\theta_1
\end{aligned} \tag{44}$$

We seek for its solutions of the form

$$\theta_1 = g(\xi, x_1), \quad \theta_2 = h(\xi, x_2), \quad \xi = x_0 - x_3 \tag{45}$$

Inserting functions (45) into system (44) reduces it to the form

$$\begin{aligned}
&\xi\frac{\partial g}{\partial x_1} = -\frac{1}{2}, \quad \xi\frac{\partial h}{\partial x_2} = -\frac{1}{2}\\
&-\xi\frac{\partial g}{\partial \xi} + \alpha x_2\frac{\partial g}{\partial x_1} = g + \alpha h\\
&-\xi\frac{\partial h}{\partial \xi} - \alpha x_1\frac{\partial h}{\partial x_2} = h - \alpha g
\end{aligned}$$

By direct check we verify that the functions $g = -x_1(2\xi)^{-1}$, $h = -x_2(2\xi)^{-1}$ satisfy this system so that we can choose

$$\theta_1 = -\frac{1}{2}x_1(x_0 - x_3)^{-1}, \quad \theta_2 = -\frac{1}{2}x_2(x_0 - x_3)^{-1}$$

Performing the same calculations for the remaining subalgebras listed in Assertion 1, we arrive at the following statement.

Assertion 5. *Each subalgebra L_j $(j = 1, 2, \ldots, 22)$ from the list given in Assertion 1 yields invariant ansatz (22) with*

$$\Lambda^{-1} = H = \exp(\theta_0 S_{03}) \exp(-\theta_3 S_{12}) \exp(-2\theta_1 H_1) \exp(-2\theta_2 H_2)$$

In addition, the functions $\theta_\mu = \theta_\mu(x_0, \mathbf{x})$ $(\mu = 0, 1, 2, 3)$, $\omega = \omega(x_0, \mathbf{x})$ are given by one of the corresponding formulas below:

L_1: $\theta_\mu = 0, \quad (\mu = 0, 1, 2, 3), \quad \omega = x_3$

L_2: $\theta_\mu = 0, \quad (\mu = 0, 1, 2, 3), \quad \omega = x_0$

L_3: $\theta_\mu = 0, \quad (\mu = 0, 1, 2, 3), \quad \omega = \xi$

L_4: $\theta_0 = -\ln|\xi|, \quad \theta_1 = \theta_2 = 0, \quad \theta_3 = \alpha \ln|\xi|, \quad \omega = \xi \cdot \eta$

L_5: $\theta_0 = -\ln|\xi|, \quad \theta_1 = \theta_2 = \theta_3 = 0, \quad \omega = x_2$

L_6: $\theta_0 = x_1, \quad \theta_1 = \theta_2 = \theta_3 = 0, \quad \omega = x_2$

L_7: $\theta_0 = x_1, \quad \theta_1 = \theta_2 = \theta_3 = 0, \quad \omega = x_1 + \ln|\xi|$

L_8: $\theta_0 = \alpha \arctan x_1 x_2^{-1}, \quad \theta_1 = \theta_2 = 0,$

$\theta_3 = -\arctan x_1 x_2^{-1}, \quad \omega = x_1^2 + x_2^2;$

L_9: $\theta_0 = \theta_1 = \theta_2 = 0, \quad \theta_3 = -x_0, \quad \omega = x_3$

L_{10}: $\theta_0 = \theta_1 = \theta_2 = 0, \quad \theta_3 = -(-1)^i x_3, \quad \omega = x_0$

L_{11}: $\theta_0 = \theta_1 = \theta_3 = 0, \quad \theta_2 = -\dfrac{(-1)^i}{2}\xi, \quad \omega = \eta$

L_{12}: $\theta_0 = \theta_2 = \theta_3 = 0, \quad \theta_1 = -\dfrac{1}{2}(x_1 - \alpha x_2)\xi^{-1}, \quad \omega = \xi$

L_{13}: $\theta_0 = \theta_2 = \theta_3 = 0, \quad \theta_1 = -\dfrac{(-1)^i}{2} x_2, \quad \omega = \xi$

L_{14}: $\theta_0 = \theta_2 = \theta_3 = 0, \quad \theta_1 = -\dfrac{1}{4}\xi, \quad \omega = \xi^2 - 4x_1$

L_{15}: $\theta_0 = \theta_2 = \theta_3 = 0, \quad \theta_1 = -\dfrac{1}{4}\xi, \quad \omega = \alpha\xi^2 - 4(\alpha x_1 - x_2)$

L_{16}: $\theta_0 = -\ln|\xi|, \quad \theta_1 = \theta_2 = 0, \quad \theta_3 = -\arctan x_1 x_2^{-1}, \quad \omega = x_1^2 + x_2^2$

L_{17}: $\theta_0 = \theta_3 = 0, \quad \theta_1 = -\dfrac{1}{2}[(-1)^i x_2 + (\alpha + \xi)x_1][1 + (\alpha + \xi)\xi]^{-1},$

$\theta_2 = \dfrac{1}{2}[(-1)^i x_1 - x_2\xi][1 + (\alpha + \xi)\xi]^{-1}, \quad \omega = \xi$

L_{18}: $\theta_0 = -\ln|\xi|, \quad \theta_1 = -\dfrac{1}{2}x_1\xi^{-1}, \quad \theta_2 = \theta_3 = 0, \quad \omega = \xi\eta - x_1^2$

L_{19}: $\theta_0 = -\ln|\xi|, \quad \theta_1 = -\dfrac{1}{2}x_1\xi^{-1}, \quad \theta_2 = \theta_3 = 0, \quad \omega = x_2$

L_{20}: $\theta_0 = -\ln|\xi|, \quad \theta_1 = -\dfrac{1}{2}x_1\xi^{-1}, \quad \theta_2 = \theta_3 = 0, \quad \omega = \ln|\xi| + x_2$

$$L_{21}:\quad \theta_0 = -\ln|\xi|,\quad \theta_1 = -\frac{1}{2}(x_1 + \ln|\xi|)\xi^{-1}, \theta_2 = \theta_3 = 0,$$
$$\omega = \alpha \ln|\xi| + x_2$$

$$L_{22}:\quad \theta_0 = -\ln|\xi|,\quad \theta_1 = -\frac{1}{2}x_1\xi^{-1},\quad \theta_2 = -\frac{1}{2}x_2\xi^{-1}, \theta_3 = \alpha \ln|\xi|,$$
$$\omega = x_\mu x^\mu \ (\mu = 0, 1, 2, 3)$$

Here $i = 1, 2;\ \alpha \in \mathbf{R};\ \xi = x_0 - x_3,\ \eta = x_0 + x_3$

2. $\tilde{P}(1,3)$-*Invariant Ansatzes*

Generically, the list of $\tilde{P}(1,3)$-invariant ansatzes is exhausted by $P(1,3)$-invariant ansatzes given in Assertion 5 and by ansatzes invariant with respect to the subalgebras F_j $(j = 1, 2, \ldots, 24)$ listed in Assertion 2. For this reason, to construct all inequivalent $\tilde{P}(1,3)$-invariant ansatzes, it suffices to consider the cases of the subalgebras F_j $(j = 1, 2, \ldots 24)$ only.

A preliminary analysis of these algebras shows that for the algebras F_j with j taking the values $2, 3, 4, 12, 13, \ldots, 19$, we can choose $\theta_1 = \theta_2 = \theta_4 = \theta_5 = 0$ in (30) and, in addition, we can put $f_1^a = f_2^a = f_4^a = f_5^a = 0$ $(a = 1, 2, 3)$ in (39). As a consequence, system (39) for the subalgebras in question reads as

$$\xi_a^\mu \frac{\partial\theta}{\partial x_\mu} = f^a\theta,\quad \xi_a^\mu \frac{\partial\theta_0}{\partial x_\mu} = -f_0^a,\quad \xi_a^\mu \frac{\partial\theta_3}{\partial x_\mu} = f_3^a,$$

where $\mu = 0, 1, 2, 3;\ a = 1, 2, 3$.

For the remaining subalgebras from the list in Assertion 2, the following equalities hold, $\theta_b = 0, f_b^a = 0$ $(b = 2, 3, 4, 5;\ a = 1, 2, 3)$, and system (39) takes the form

$$\xi_a^\mu \frac{\partial\theta}{\partial x_\mu} = f^a\theta,\quad \xi_a^\mu \frac{\partial\theta_0}{\partial x_\mu} = -f_0^a,\quad \xi_a^\mu \frac{\partial\theta_1}{\partial x_\mu} = -\theta_1 f_0^a + \frac{1}{2}f_1^a$$

where $\mu = 0, 1, 2, 3;\ a = 1, 2, 3$.

Summarizing, we conclude that the problem of construction of $\tilde{P}(1,3)$-invariant ansatzes reduces to finding solutions of linear systems of first-order partial differential equations that are integrated by rather standard methods of the general theory of partial differential equations.

We omit the cumbersome intermediate calculations, which are very similar to those performed in the previous subsection, and give the final result.

Assertion 6. *Each subalgebra* F_j $(j = 1, 2, \ldots, 24)$ *from the list given in Assertion 2 yields invariant ansatz* (22) *with*

$$\Lambda^{-1} = H = \exp\{(-\ln\theta)E\}\exp(\theta_0 S_{03})\exp(-\theta_3 S_{12})\exp(-2\theta_1 H_1),\ \theta_1\theta_3 = 0$$

Moreover, the functions $\theta = \theta(x_0, \mathbf{x})$, $\theta_0 = \theta_0(x_0, \mathbf{x})$, $\theta_1 = \theta_1(x_1, \mathbf{x})$, $\theta_3 = \theta_3(x_3, \mathbf{x})$, $\omega = \omega(x_0, \mathbf{x})$ *are given by one of the following corresponding formulas*

F_1: $\theta = |x_1|^{-k}, \quad \theta_0 = \theta_1 = \theta_3 = 0, \quad \omega = x_2 x_1^{-1}$

F_2: $\theta = (x_1^2 + x_2^2)^{-(k/2)}, \quad \theta_0 = \theta_1 = 0, \quad \theta_3 = \arctan x_2 x_1^{-1}$

$\omega = \ln(x_1^2 + x_2^2) + 2\alpha \arctan x_2 x_1^{-1}, \quad \alpha > 0$

F_3: $\theta = |x_3|^{-k}, \quad \theta_0 = \theta_1 = 0, \quad \theta_3 = \arctan x_2 x_1^{-1}, \quad \omega = (x_1^2 + x_2^2) x_3^{-2}$

F_4: $\theta = |x_0|^{-k}, \quad \theta_0 = \theta_1 = 0, \quad \theta_3 = \arctan x_2 x_1^{-1}, \quad \omega = (x_1^2 + x_2^2) x_0^{-2}$

F_5: $\theta = |x_1|^{-k}, \quad \theta_0 = \alpha^{-1} \ln |x_1|, \quad \theta_1 = \theta_3 = 0, \quad \omega = x_2 x_1^{-1}, \quad \alpha > 0$

F_6: $\theta = |\xi\eta|^{-(k/2)}, \quad \theta_0 = \frac{1}{2} \ln |\eta\xi^{-1}|, \quad \theta_1 = \theta_3 = 0$

$\omega = (1 - \alpha) \ln |\eta| + (1 + \alpha) \ln |\xi|, \quad \alpha > 0$

F_7: $\theta = |x_2|^{-k}, \quad \theta_0 = \alpha^{-1} \ln |x_2|, \quad \theta_1 = \theta_3 = 0, \quad \omega = |\xi|^{\alpha} |x_2|^{1-\alpha}, \quad \alpha > 0$

F_8: $\theta = |\eta|^{-(k/2)}, \quad \theta_0 = \frac{1}{2} \ln |\eta|, \quad \theta_1 = \theta_3 = 0$

$\omega = \xi - (-1)^j \ln |\eta|, \quad j = 1, 2$

F_9: $\theta = |x_2|^{-k}, \quad \theta_0 = \ln |x_2|, \quad \theta_1 = \theta_3 = 0$

$\omega = \xi - 2(-1)^j \ln |x_2|, \quad j = 1, 2$

F_{10}: $\theta = |x_2|^{-k}, \quad \theta_0 = \ln |\eta x_2^{-1}|, \quad \theta_1 = \theta_3 = 0, \quad \omega = \xi\eta x_2^{-2}$

F_{11}: $\theta = |x_2|^{-1}, \quad \theta_0 = -\ln |\xi x_2^{-1}|, \quad \theta_1 = \theta_3 = 0, \quad \omega = x_2 x_1^{-1}$

F_{12}: $\theta = (x_1^2 + x_2^2)^{-(k/2)}, \quad \theta_0 = -\alpha \arctan x_2 x_1^{-1}, \quad \theta_1 = 0$

$\theta_3 = \arctan x_2 x_1^{-1}, \omega = \ln (x_1^2 + x_2^2) + 2\beta \arctan x_2 x_1^{-1}$

$\alpha \neq 0, \quad \beta > 0$

F_{13}: $\theta = |\xi\eta|^{-k/2}, \quad \theta_0 = -\frac{1}{2} \ln |\eta\xi^{-1}|, \quad \theta_1 = 0$

$\theta_3 = -\frac{1}{2\alpha} \ln |\eta\xi^{-1}|, \quad \omega = (\alpha - \beta) \ln |\eta| + (\alpha + \beta) \ln |\xi|$

$\alpha \neq 0, \quad \beta > 0$

F_{14}: $\theta = |\eta|^{-(k/2)}, \theta_0 = \frac{1}{2} \ln |\eta|, \quad \theta_1 = 0, \quad \theta_3 = -\frac{1}{2} \ln |\eta|, \quad \omega = \xi - \ln |\eta|$

F_{15}: $\theta = (x_1^2 + x_2^2)^{-(k/2)}, \quad \theta_0 = -\alpha \arctan x_2 x_1^{-1}, \quad \theta_1 = 0$

$\theta_3 = \arctan x_2 x_1^{-1}, \quad \omega = \ln (x_1^2 + x_2^2) \xi^{-2} + 2\alpha \arctan x_2 x_1^{-1}, \quad \alpha \neq 0$

F_{16}: $\theta = (x_1^2 + x_2^2)^{-(k/2)}, \quad \theta_0 = \dfrac{1}{2} \ln (x_1^2 + x_2^2)\xi^{-2}, \quad \theta_1 = 0$

$\theta_3 = \arctan x_2 x_1^{-1}, \quad \omega = \ln (x_1^2 + x_2^2)^{1-\alpha}\xi^{2\alpha} + 2\beta \arctan x_2 x_1^{-1}$

$0 \le |\alpha| \le 1, \quad \beta \ge 0, |\alpha| + |\beta| \ne 0$

F_{17}: $\theta = (x_1^2 + x_2^2)^{-(k/2)}, \theta_0 = \dfrac{1}{2} \ln (x_1^2 + x_2^2)$

$\theta_1 = 0, \quad \theta_3 = \arctan x_2 x_1^{-1}$

$\omega = \xi - (-1)^j \ln (x_1^2 + x_2^2) + 2\alpha \arctan x_2 x_1^{-1}, \quad \alpha \in R, \quad j = 1, 2$

F_{18}: $\theta = (x_1^2 + x_2^2)^{-(k/2)}, \quad \theta_0 = \dfrac{1}{2} \ln (x_1^2 + x_2^2), \quad \theta_1 = 0$

$\theta_3 = \arctan x_2 x_1^{-1}, \quad \omega = \xi + 2(-1)^j \arctan x_2 x_1^{-1}, \quad j = 1, 2$

F_{19}: $\theta = (x_1^2 + x_2^2)^{-(k/2)}, \quad \theta_0 = -\dfrac{1}{2} \ln |\xi\eta^{-1}|, \quad \theta_1 = 0$

$\theta_3 \arctan x_2 x_1^{-1}, \quad \omega = (x_1^2 + x_2^2)(\xi\eta)^{-1}$

F_{20}: $\theta = |\xi\eta - x_1^2|^{-(k/2)}, \quad \theta_0 = \dfrac{1}{2\alpha} \ln |\xi\eta - x_1^2|, \quad \theta_1 = -\dfrac{1}{2} x_1 \xi^{-1}$

$\theta_3 = 0, \quad \omega = |\xi|^{2\alpha}|\xi\eta - x_1^2|^{1-\alpha}, \quad 0 \le |\alpha| \le 1$

F_{21}: $\theta = |x_1 - (-1)^j \xi x_2|^{-k}, \quad \theta_0 = \ln |x_1 - (-1)^j \xi x_2|, \quad \theta_1 = -\dfrac{(-1)^j}{2} x_2$

$\theta_3 = 0, \quad \omega = \xi, = 1, 2$

F_{22}: $\theta = |\xi|^{-k/2}, \quad \theta_0 = -\dfrac{1}{2} \ln |\xi|, \quad \theta_1 = -\dfrac{1}{2} x_1 \xi^{-1}$

$\theta_3 = 0, \quad \omega = \eta - x_1^2 \xi^{-1} + (-1)^j \ln |\xi|, \quad j = 1, 2$

F_{23}: $\theta = |x_2|^{-k}, \quad \theta_0 = \dfrac{1}{2} \ln |x_2|, \quad \theta_1 = -\dfrac{(-1)^j}{4} \xi^{-1}$

$\theta_3 = 0, \quad \omega = (\xi^2 - 4(-1)^j x_1) x_2^{-1}, \quad j = 1, 2$

F_{24}: $\theta = |\xi^2 - 4(-1)^j x_1|^{-k}, \quad \theta_0 = \dfrac{1}{2} \ln |\xi^2 - 4(-1)^j |x_1|, \quad \theta_1 = -\dfrac{(-1)^j}{4} \xi$

$\theta_3 = 0, \quad \omega = \left(\eta - (-1)^j x_1 \xi + \dfrac{1}{6}\xi^3\right)^2 (\xi^2 - 4(-1)^j x_1)^{-3}, \quad j = 1, 2$

Here k is an arbitrarily fixed constant [*the conformal degree of the algebra* $c(1, 3)$], $\xi = x_0 - x_3, \;\; \eta = x_0 + x_3$.

3. $C(1, 3)$-*Invariant Ansatzes*

To obtain the full description of conformally invariant ansatzes it suffices to consider the subalgebras C_j, $(j = 1, 2, \ldots, 14)$ listed in Assertion 3.

The preliminary analysis of these subalgebras shows that we can put $\theta_4 = \theta_5 = f_4^a = f_5^a = 0$ $(a = 1, 2, 3)$ for the subalgebras C_j $(j = 1, 2, \ldots, 10)$. As a result, system (39) corresponding to these subalgebras takes the following form:

$$\xi_a^{\mu} \frac{\partial \theta}{\partial x_{\mu}} = f^a \theta, \quad \xi_a^{\mu} \frac{\partial \theta_0}{\partial x_{\mu}} = -f_0^a, \quad \xi_a^{\mu} \frac{\partial \theta_3}{\partial x_{\mu}} = f_3^a$$

$$\xi_a^{\mu} \frac{\partial \theta_1}{\partial x_{\mu}} = -\theta_1 f_0^a - \theta_2 f_3^a + \frac{1}{2} f_1^a, \quad \xi_a^{\mu} \frac{\partial \theta_2}{\partial x_{\mu}} = -\theta_2 f_0^a + \theta_1 f_3^a + \frac{1}{2} f_2^a$$

where $a = 1, 2, 3$.

Thus the problem of constructing ansatzes invariant under the subalgebras $C_j\,(j = 1, 2, \ldots, 10)$ is again reduced to solving linear first-order partial differential equations. However, for the remaining subalgebras C_j $(j = 11, 12, 13, 14)$, system (39) is not linear. It has been solved for the case of the spinor field elsewhere [33]. The obtained expressions for the functions are so cumbersome that they prove to be useless within the context of symmetry reduction of the conformally invariant nonlinear Dirac equation. For this reason, we do not give here the ansatzes corresponding to the subalgebras C_j $(j = 11, 12, 13, 14)$.

Assertion 7. *Each subalgebra C_j $(j = 1, 2, \ldots, 10)$ from the list in Assertion 3 yields invariant ansatz* (22) *with*

$$\Lambda^{-1} = H = \exp\{(-\ln\theta)E\}\exp(\theta_0 S_{03})\exp(-\theta_3 S_{12})\exp(-2\theta_1 H_1) \times \exp(-2\theta_2 H_2).$$

Also, the functions $\theta = \theta(x_0, \mathbf{x})$, $\theta_{\mu} = \theta_{\mu}(x_0, \mathbf{x})$ $(\mu = 0, 1, 2, 3)$, $\omega = \omega(x_0, \mathbf{x})$ are given by one of the corresponding formulas below.

$$C_1: \quad \theta = (1 + \xi^2)^{-(k/2)}, \quad \theta_0 = -\frac{1}{2}\ln(1 + \xi^2)$$

$$\theta_1 = -\frac{1}{2}(x_2 + x_1\xi)(1 + \xi^2)^{-1}, \quad \theta_2 = \frac{1}{2}(x_1 - \xi x_2)(1 + \xi^2)^{-1}$$

$$\theta_3 = -\arctan\xi, \quad \omega = (x_1 - x_2\xi)(1 + \xi^2)^{-1}$$

$$C_2: \quad \theta = (1 + \xi^2)^{-(k/2)}, \quad \theta_0 = -\frac{1}{2}\ln(1 + \xi^2)$$

$$\theta_1 = -\frac{1}{2}(x_2 + x_1\xi)(1 + \xi^2)^{-1}, \quad \theta_2 = \frac{1}{2}(x_1 - x_2\xi)(1 + \xi^2)^{-1}$$

$$\theta_3 = -\arctan\xi, \quad \omega = (x_2 + x_1\xi)(1 + \xi^2)^{-1} - \arctan\xi$$

C_3: $\theta = (1+\xi^2)^{-k/2}, \quad \theta_0 = -\frac{1}{2}\ln(1+\xi^2)$

$\theta_1 = -\frac{1}{2}x_1\xi(1+\xi^2)^{-1}, \quad \theta_2 = -\frac{1}{2}x_2\xi(1+\xi^2)^{-1}$

$\theta_3 = \arctan x_2x_1^{-1}, \quad \omega = (1+\xi^2)(x_1^2+x_2^2)^{-1}$

C_4: $\theta = |x_1|^{-k}, \quad \theta_0 = \ln|x_1| - \ln(1+\xi^2)$

$\theta_1 = -\frac{1}{2}x_1\xi(1+\xi^2)^{-1}, \quad \theta_2 = -\frac{1}{2}x_2\xi(1+\xi^2)^{-1}$

$\theta_3 = 0, \quad \omega = x_2x_1^{-1}$

C_5: $\theta = ((x_1^2+x_2^2)(1+\xi^2))^{-k/2}, \quad \theta_0 = \frac{1}{2}\ln(x_1^2+x_2^2)(1+\xi^2)^{-1}$

$\theta_1 = -\frac{1}{2}x_1\xi(1+\xi^2)^{-1}, \quad \theta_2 = -\frac{1}{2}x_2\xi(1+\xi^2)^{-1}$

$\theta_3 = \arctan x_2x_1^{-1}, \quad \omega = \arctan x_2x_1^{-1} + \alpha\arctan\xi, \quad \alpha \neq 0$

C_6: $\theta = [(x_1-x_2\xi)^2(1+\xi^2)^{-1}]^{-(k/2)}, \quad \theta_0 = \frac{1}{2}\ln[(x_1-x_2\xi)^2(1+\xi^2)^{-3}]$

$\theta_1 = -\frac{1}{2}(x_2+x_1\xi)(1+\xi^2)^{-1}, \quad \theta_2 = \frac{1}{2}(x_1-x_2\xi)(1+\xi^2)^{-1}$

$\theta_3 = -\arctan\xi, \quad \omega = \alpha\arctan\xi - \ln[(x_1-x_2\xi)(1+\xi^2)^{-1}], \quad \alpha \neq 0$

C_7: $\theta = [(x_1-x_2\xi)^2(1+\xi^2)^{-1}]^{-(k/2)}, \quad \theta_0 = \frac{1}{2}\ln[(x_1-x_2\xi)^2(1+\xi^2)^{-3}]$

$\theta_1 = -\frac{1}{2}(x_2+x_1\xi)(1+\xi^2)^{-1}, \quad \theta_2 = \frac{1}{2}(x_1-x_2\xi)(1+\xi^2)^{-1}$

$\theta_3 = -\arctan\xi$

$\omega = [\eta(1+\xi^2)^2 - 2x_1(x_2+x_1\xi) - \xi(x_1^2\xi^2 - x_2^2)][x_1-\xi x_2]^{-2} - \xi$

C_8: $\theta = (x_1^2+x_2^2)^{-(k/2)}, \quad \theta_0 = \frac{1}{2}\ln[(x_1^2+x_2^2)(1+\xi^2)^{-2}]$

$\theta_1 = -\frac{1}{2}x_1\xi(1+\xi^2)^{-1}, \quad \theta_2 = -\frac{1}{2}x_2\xi(1+\xi^2)^{-1}$

$\theta_3 = \arctan x_2x_1^{-1}$

$\omega = \ln(x_1^2+x_2^2)(1+\xi^2)^{-1} + 2\alpha\arctan x_2x_1^{-1} - 2\beta\arctan\xi$

$\alpha, \beta \in R, |\alpha| + |\beta| \neq 0$

C_9: $\theta = (x_1^2+x_2^2)^{-(k/2)}, \quad \theta_0 = \frac{1}{2}\ln(x_1^2+x_2^2) - \ln(1+\xi^2)$

$\theta_1 = -\frac{1}{2}x_1\xi(1+\xi^2)^{-1}, \quad \theta_2 = -\frac{1}{2}x_2\xi(1+\xi^2)^{-1}$

$\theta_3 = \arctan x_2x_1^{-1}, \quad \omega = \eta(1+\xi^2)(x_1^2+x_2^2)^{-1} - \xi$

$$C_{10}:\quad \theta = (x_1^2 + x_2^2)^{-(k/2)},\quad \theta_0 = -\frac{1}{2}\ln(x_1^2 + x_2^2)$$
$$\theta_1 = -\frac{1}{2}x_1\eta(x_1^2 + x_2^2)^{-1},\quad \theta_2 = -\frac{1}{2}x_2\eta(x_1^2 + x_2^2)^{-1}$$
$$\theta_3 = 0,\quad \omega = x_2 x_1^{-1}$$

Here k is an arbitrarily fixed constant [*the conformal degree of the algebra* $c(1,3)$], $\xi = x_0 - x_3$, $\eta = x_0 + x_3$.

III. EXACT SOLUTIONS OF THE YANG–MILLS EQUATIONS

In this section we apply the technique described above in order to perform in-depth analysis of the problems of symmetry reduction and construction of exact invariant solutions of the $SU(2)$ Yang–Mills equations in the (1+3)-dimensional Minkowski space of independent variables. Since the general method to be used relies heavily on symmetry properties of the equations under study, we will briefly review the group-theoretic properties of the $SU(2)$ Yang–Mills equations.

A. Symmetry Properties of the Yang–Mills Equations

The classical Yang–Mills equations of $SU(2)$ gauge theory in the Minkowski spacetime $\mathbf{R}^{1,3}$ form the system of nonlinear second-order partial differential equations of the form

$$\begin{aligned}\partial_\nu\partial^\nu\mathbf{A}_\mu - \partial^\mu\partial_\nu\mathbf{A}_\nu + e[(\partial_\nu\mathbf{A}_\nu)\times\mathbf{A}_\mu - 2(\partial_\nu\mathbf{A}_\mu)\times\mathbf{A}_\nu \\ + (\partial^\mu\mathbf{A}_\nu)\times\mathbf{A}^\nu] + e^2\mathbf{A}_\nu\times(\mathbf{A}^\nu\times\mathbf{A}_\mu) = 0\end{aligned} \tag{46}$$

Hereafter in this section, the indices $\mu, \nu, \alpha, \beta, \gamma, \delta, \sigma$ take the values $0, 1, 2, 3$; $\partial_\mu = \partial_{x_\mu} = \frac{\partial}{\partial x_\mu}$; rising and lowering the indices is performed with the use of the metric tensor $g_{\mu\nu}$ of Minkowski space and the summation convention over the repeated indices is used. Furthermore, $\mathbf{A}_\mu = \mathbf{A}_\mu(x_0, \mathbf{x}) = (A_\mu^1(x_0, \mathbf{x}), A_\mu^2(x_0, \mathbf{x}), A_\mu^3(x_0, \mathbf{x}))^{\mathbf{T}}$ is the vector potential of the Yang–Mills field (for brevity it is called the Yang–Mills field hereafter) and e is the gauge coupling constant.

The maximal symmetry group admitted by Eqs. (46) is the group $C(1,3)\otimes SU(2)$ [17], where $C(1,3)$ is the 15-parameter conformal group generated by the following vector fields

$$\begin{aligned}P_\mu &= \partial_{x_\mu} \\ J_{\mu\nu} &= x^\mu\partial_{x_\nu} - x^\nu\partial_{x_\mu} + A^{a\mu}\partial_{A_\nu^a} - A^{a\nu}\partial_{A_\mu^a} \\ D &= x_\mu\partial_{x_\mu} - A_\mu^a\partial_{A_\mu^a} \\ K_\mu &= 2x^\mu D - (x_\nu x^\nu)\partial_{x_\mu} + 2A^{a\mu}x_\nu\partial_{A_\nu^a} - 2A_\nu^a x^\nu\partial_{A_\mu^a}\end{aligned} \tag{47}$$

and $SU(2)$ is the infinite-parameter unitary gauge transformation group having the generator

$$Q = (\varepsilon_{abc}A^b_\mu\omega^c(x_0,\mathbf{x}) + e^{-1}\partial_{x_\mu}\omega^a(x_0,\mathbf{x}))\partial_{A^a_\mu} \tag{48}$$

In formulas (47) and (48), $\partial_{A^a_\mu} = \frac{\partial}{\partial A^a_\mu}$, $\omega^c(x_0,\mathbf{x})$ stand for arbitrary real functions, $a,b,c = 1,2,3$ and ε_{abc} is the antisymmetric third-order tensor with $\varepsilon_{123} = 1$.

It is not difficult to check that vector fields (47) can be rewritten in the form (15) if we put

$$S_{01} = \begin{pmatrix} 0 & -I & 0 & 0 \\ -I & 0 & 0 & 0 \\ 0 & 0 & 0 & 0 \\ 0 & 0 & 0 & 0 \end{pmatrix}, \quad S_{02} = \begin{pmatrix} 0 & 0 & -I & 0 \\ 0 & 0 & 0 & 0 \\ -I & 0 & 0 & 0 \\ 0 & 0 & 0 & 0 \end{pmatrix}$$
$$S_{03} = \begin{pmatrix} 0 & 0 & 0 & -I \\ 0 & 0 & 0 & 0 \\ 0 & 0 & 0 & 0 \\ -I & 0 & 0 & 0 \end{pmatrix}, \quad S_{12} = \begin{pmatrix} 0 & 0 & 0 & 0 \\ 0 & 0 & -I & 0 \\ 0 & I & 0 & 0 \\ 0 & 0 & 0 & 0 \end{pmatrix} \tag{49}$$
$$S_{13} = \begin{pmatrix} 0 & 0 & 0 & 0 \\ 0 & 0 & 0 & -I \\ 0 & 0 & 0 & 0 \\ 0 & I & 0 & 0 \end{pmatrix}, \quad S_{23} = \begin{pmatrix} 0 & 0 & 0 & 0 \\ 0 & 0 & 0 & 0 \\ 0 & 0 & 0 & -I \\ 0 & 0 & I & 0 \end{pmatrix}$$

where 0 and I are the zero and unit 3×3 matrices, correspondingly. Next, we choose the matrix E to be the 12×12 unit matrix and the conformal degree k of the algebra $c(1,3)$ to be equal to 1.

One important application of the symmetry admitted by Yang–Mills equations is a possibility of getting new exact solutions with the help of the solution generation formulas. This method is based on the fact that the symmetry group maps the set of solutions of an equation admitting this group into itself. We give the corresponding formulae without proof [20,21,33].

Assertion 8. *Let*

$$\bar{x}_i = f_i(\mathbf{x},\mathbf{u},\tau), \quad i = 1,2,\ldots,p$$
$$\bar{u}_j = g_j(\mathbf{x},\mathbf{u},\tau), \quad j = 1,2,\ldots,q$$

where $\tau = (\tau_1,\tau_2,\ldots,\tau_r)$, *be the r-parameter invariance group admitted by a system of partial differential equations and* $U_j(\mathbf{x})$, $j = 1,2,\ldots,q$ *be a particular solution of the latter. Then the q-component function* $\mathbf{u}(\mathbf{x}) = (u^1(x), \ldots, u^q(x))$, *defined in implicit way by the formulas*

$$U_j(\mathbf{f}(\mathbf{x}, \mathbf{u}, \tau)) = g_j(\mathbf{x}, \mathbf{u}, \tau)$$

with $\mathbf{f} = (f_1, \ldots, f_p),\ j = 1, 2, \ldots, q$, *is also a solution of the system in question.*

To take advantage of Assertion 8, we need the following formulas for the final transformations generated by the basis operators (47) and (48) of the Lie algebra of the group $C(1,3) \otimes SU(2)$ [2,21]:

1. The translation group (the generator is $X = \tau_\mu P_\mu$)

$$\bar{x}_\mu = x_\mu + \tau_\mu, \quad \bar{A}^d_\mu = A^d_\mu;$$

2. The Lorentz group $O(1,3)$
   a. The rotation group (the generator is $X = \tau J_{ab}$)

$$\begin{aligned}
\bar{x}_0 &= x_0, \quad \bar{x}_c = x_c, \quad c \neq a, \quad c \neq b \\
\bar{x}_a &= x_a \cos\tau + x_b \sin\tau \\
\bar{x}_b &= x_b \cos\tau - x_a \sin\tau \\
\bar{A}^d_0 &= A^d_0, \quad \bar{A}^d_c = A^d_c, \quad c \neq a,\ c \neq b \\
\bar{A}^d_a &= A^d_a \cos\tau + A^d_b \sin\tau \\
\bar{A}^d_b &= A^d_b \cos\tau - A^d_a \sin\tau
\end{aligned}$$

   b. The Lorentz transformations (the generator is $X = \tau J_{0a}$)

$$\begin{aligned}
\bar{x}_0 &= x_0 \cosh\tau + x_a \sinh\tau \\
\bar{x}_a &= x_a \cosh\tau + x_0 \sinh\tau \\
\bar{A}^d_0 &= A^d_0 \cosh\tau + A^d_a \sinh\tau \\
\bar{A}^d_a &= A^d_a \cosh\tau + A^d_0 \sinh\tau \\
\bar{x}_b &= x_b, \quad \bar{A}^d_b = A^d_b, \quad b \neq a;
\end{aligned}$$

3. The scale transformation group (the generator is $X = \tau D$)

$$\bar{x}_\mu = x_\mu e^\tau, \quad \bar{A}^d_\mu = A^d_\mu e^{-\tau}$$

4. The group of conformal transformations (the generation is $X = \tau_\mu K^\mu$)

$$\begin{aligned}
\bar{x}_\mu &= (x_\mu - \tau_\mu x_\nu x^\nu)\sigma^{-1}(x_0, \mathbf{x}) \\
\bar{A}^d_\mu &= [g_{\mu\nu}\sigma(x_0, \mathbf{x}) + 2(x_\mu\tau_\nu - x_\nu\tau_\mu \\
&\quad + 2\tau_\alpha x^\alpha \tau_\mu x_\nu - x_\alpha x^\alpha \tau_\mu \tau_\nu - \tau_\alpha \tau^\alpha x_\mu x_\nu]A^{d\nu}
\end{aligned}$$

5. The gauge transformation group (the generator is $X = Q$)

$$\begin{aligned}\bar{x}_\mu &= x_\mu \\ \bar{A}^d_\mu &= A^d_\mu \cos\omega + \varepsilon_{dbc} A^b_\mu n^c \sin\omega + 2n^d n^b A^b_\mu \sin^2\frac{\omega}{2} \\ &\quad + e^{-1}\left[\frac{1}{2} n^d \partial_{x_\mu}\omega + \frac{1}{2}(\partial_{x_\mu} n^d)\sin\omega + \varepsilon_{dbc}(\partial_{x_\mu} n^b) n^c\right]\end{aligned}$$

In these formulas $\sigma(x_0, \mathbf{x}) = 1 - \tau_\alpha\tau^\alpha + (\tau_\alpha\tau^\alpha)(x_\beta x^\beta)$, $n^a = n^a(x_0, \mathbf{x})$ are the components of the unit vector given by the relations $\omega^a(x_0, \mathbf{x}) = \omega(x_0, \mathbf{x}) n^a(x_0, \mathbf{x})$ with $a, b, c, d = 1, 2, 3$.

Using Assertion 8, it is not difficult to derive the following formulas for generating solutions of the Yang–Mills equations by the transformation groups enumerated above [33]:

1. The translation group

$$A^a_\mu(x) = u^a_\mu(x + \tau)$$

2. The Lorentz group

$$\begin{aligned}A^d_\mu(x) &= a_\mu u^d_0(ax, bx, cx, dx) + b_\mu u^d_1(ax, bx, cx, dx) \\ &\quad + c_\mu u^d_2(ax, bx, cx, dx) + d_\mu u^d_3(ax, bx, cx, dx)\end{aligned}$$

3. The scale transformation group

$$A^d_\mu(x) = e^\tau u^d_\mu(xe^\tau)$$

4. The group of conformal transformations

$$\begin{aligned}A^d_\mu(x) &= [g_{\mu\nu}\sigma^{-1}(x) + 2\sigma^{-2}(x)(x_\mu\tau_\nu - x_\nu\tau_\mu + 2\tau_\alpha x^\alpha \tau_\mu x_\nu \\ &\quad - x_\alpha x^\alpha \tau_\mu\tau_\nu - \tau_\alpha\tau^\alpha x_\mu x_\nu)] u^{d\nu}((x - \tau(x_\alpha x^\alpha))\sigma^{-1}(x))\end{aligned}$$

5. The gauge transformation group

$$\begin{aligned}A^d_\mu(x) &= u^d_\mu \cos\omega + \varepsilon_{dbc} u^b_\mu n^c \sin\omega + 2n^d n^b u^b_\mu \sin^2\frac{\omega}{2} \\ &\quad + e^{-1}\left[\frac{1}{2} n^d \partial_{x_\mu}\omega + \frac{1}{2}(\partial_{x_\mu} n^d)\sin\omega + \varepsilon_{dbc}(\partial_{x_\mu} n^b) n^c\right]\end{aligned}$$

Here $u^d_\mu(x)$ is an arbitrary particular solution of the Yang–Mills equations; $x = (x_0, \mathbf{x})$; τ, τ_μ are arbitrary parameters; $a_\mu, b_\mu, c_\mu, d_\mu$ are arbitrary constants satisfying the relations

$$\begin{aligned} a_\mu a^\mu &= -b_\mu b^\mu = -c_\mu c^\mu = -d_\mu d^\mu = 1 \\ a_\mu b^\mu &= a_\mu c^\mu = a_\mu d^\mu = b_\mu c^\mu = b_\mu d^\mu = c_\mu d^\mu = 0 \end{aligned} \tag{50}$$

In addition, we use the following notations:

$$x + \tau = \{x_\mu + \tau_\mu, \mu = 0, 1, 2, 3\}$$
$$ax = a_\mu x^\mu, \quad bx = b_\mu x^\mu, \quad cx = c_\mu x^\mu, \quad dx = d_\mu x^\mu$$

Thus, using the solution generation formulae enables extending a single solution of the Yang–Mills equations to a multiparameter family of exact solutions.

Let us also discuss briefly the discrete symmetries of equations (46). It is straightforward to check that the Yang–Mills equations admit the following groups of discrete transformations:

$$\begin{aligned} \Psi_1:\quad & \bar{x}_\mu = -x_\mu, \quad \bar{\mathbf{A}}_\mu = -\mathbf{A}_\mu \\ \Psi_2:\quad & \bar{x}_0 = -x_0, \quad \bar{x}_1 = -x_1, \quad \bar{x}_2 = x_2, \quad \bar{x}_3 = x_3 \\ & \bar{\mathbf{A}}_0 = \mathbf{A}_0, \quad \bar{\mathbf{A}}_1 = -\mathbf{A}_1, \quad \bar{\mathbf{A}}_2 = \mathbf{A}_2, \quad \bar{\mathbf{A}}_3 = \mathbf{A}_3 \\ \Psi_3:\quad & \bar{x}_0 = -x_0, \quad \bar{x}_1 = x_1, \bar{x}_2 = -x_2, \ \bar{x}_3 = -x_3 \\ & \bar{\mathbf{A}}_0 = -\mathbf{A}_0, \quad \bar{\mathbf{A}}_1 = \mathbf{A}_1, \quad \bar{\mathbf{A}}_2 = \mathbf{A}_2, \quad \bar{\mathbf{A}}_3 = -\mathbf{A}_3 \end{aligned}$$

Action of these transformation groups on the basis elements (47) of the symmetry algebra admitted by Eqs. (46) is described in Table II, where $G_m = J_{0m} - J_{m3}$ $(m = 1, 2)$, $M = P_0 + P_3$, $T = \frac{1}{2}(P_0 - P_3)$.

While classifying the subalgebras of the algebras $p(1, 3)$ and $\tilde{p}(1, 3)$ of rank 3 we have exploited the discrete symmetries Φ_a given in Table II. Comparing Tables I and II, we see that the actions of the discrete symmetries Φ_a and Ψ_a on the operators $P_\mu, J_{\mu\nu}, D$ give identical results, namely

$$\Phi_a P_\mu = \Psi_a P_\mu, \quad \Phi_a J_{\mu\nu} = \Psi_a J_{\mu\nu}, \quad \Phi_a D = \Psi_a D$$

for all $a = 1, 2, 3$. This fact makes it possible to use the discrete symmetries in order to simplify the forms of the basis operators of subalgebras of the algebras $p(1, 3)$ $\tilde{p}(1, 3)$.

B. Ansatzes for the Yang–Mills Field

Conformally invariant ansatzes for the Yang–Mills field, that reduce equations (46) to systems of ordinary differential equations, can be represented in the linear form

$$\mathbf{A}_\mu(x_0, \mathbf{x}) = \Lambda_{\mu\nu}\mathbf{B}_\nu(\omega) \tag{51}$$

TABLE II
Discrete Symmetries of Eq. (46)

Generators	Action of Ψ_a		
	Ψ_1	Ψ_2	Ψ_3
P_0	$-P_0$	P_0	$-P_0$
P_1	$-P_1$	$-P_1$	P_1
P_k $(k = 2, 3)$	$-P_k$	P_k	$-P_k$
J_{03}	J_{03}	J_{03}	J_{03}
J_{12}	J_{12}	$-J_{12}$	$-J_{12}$
G_1	G_1	$-G_1$	$-G_1$
G_2	G_2	G_2	G_2
M	$-M$	M	$-M$
T	$-T$	T	$-T$
D	D	D	D
K_0	$-K_0$	K_0	$-K_0$
K_1	$-K_1$	$-K_1$	K_1
K_m $(m = 2, 3)$	$-K_m$	K_m	$-K_m$

where $\Lambda_{\mu\nu} = \Lambda_{\mu\nu}(x_0, \mathbf{x})$ are some fixed nonsingular 3×3 matrices and $\mathbf{B}_\nu(\omega) = (B_\nu^1(\omega),\ B_\nu^2(\omega), B_\nu^3(\omega))^T$ are new unknown vector functions of the new independent variable $\omega = \omega(x_0, \mathbf{x})$. In the following text, we will denote the 12×12 matrix having the matrix entries $\Lambda_{\mu\nu}$ as Λ.

Because of space limitations, we restrict our discussion to the ansatzes invariant under the subalgebras of the Poincaré algebra. For further details on extended Poincaré algebra, see Ref. 39.

The structure of the matrix Λ for the case of arbitrary vector field is described in Assertion 5. Adapting the formula for Λ to the case in hand, we have

$$\Lambda = \exp(2\theta_1 H_1)\exp(2\theta_2 H_2)\exp(-\theta_0 S_{03})\exp(\theta_3 S_{12})$$

where $\theta_\mu = \theta_\mu(x_0, \mathbf{x})$ are some real-valued functions, and $H_1 = S_{01} - S_{13}$, $H_2 = S_{02} - S_{23}$, and $S_{\mu\nu}$ are matrices (49), which realize the matrix representation of the Lie algebra $o(1,3)$ of the Lorentz group $O(1,3)$.

Computing the exponents with the help of the Campbell–Hausdorff formula yields

$$\Lambda = \begin{pmatrix} [\cosh\theta_0 + \Phi] & -2[\Psi_1] & 2[\Psi_2] & [\sinh\theta_0 - \Phi] \\ [-2\theta_1 e^{-\theta_0}] & [\cos\theta_3] & [-\sin\theta_3] & [2\theta_1 e^{-\theta_0}] \\ [-2\theta_2 e^{-\theta_0}] & [\sin\theta_3] & [\cos\theta_3] & [2\theta_2 e^{-\theta_0}] \\ [\sinh\theta_0 + \Phi] & -2[\Psi_1] & 2[\Psi_2] & [\cosh\theta_0 + \Phi] \end{pmatrix}$$

where $\Phi = 2(\theta_1^2 + \theta_2^2)e^{-\theta_0}$, $\Psi_1 = \theta_1\cos\theta_3 + \theta_2\sin\theta_3$, $\Psi_2 = \theta_1\sin\theta_3 - \theta_2 \cos\theta_3$ and the symbol $[f]$ stands for fI, where I is the unit 3×3 matrix.

Inserting the expression obtained for the matrix Λ into (51) yields the final form of the Poincaré-invariant ansatz for the Yang–Mills field

$$\begin{aligned}
\mathbf{A}_0 &= \cosh\theta_0\mathbf{B}_0 + \sinh\theta_0\mathbf{B}_3 - 2(\theta_1\cos\theta_3 + \theta_2\sin\theta_3)\mathbf{B}_1 \\
&\quad + 2(\theta_1\sin\theta_3 - \theta_2\cos\theta_3)\mathbf{B}_2 + 2(\theta_1^2 + \theta_2^2)e^{-\theta_0}(\mathbf{B}_0 - \mathbf{B}_3) \\
\mathbf{A}_1 &= \cos\theta_3\mathbf{B}_1 - \sin\theta_3\mathbf{B}_2 - 2\theta_1 e^{-\theta_0}(\mathbf{B}_0 - \mathbf{B}_3) \\
\mathbf{A}_2 &= \sin\theta_3\mathbf{B}_1 + \cos\theta_3\mathbf{B}_2 - 2\theta_2 e^{-\theta_0}(\mathbf{B}_0 - \mathbf{B}_3) \\
\mathbf{A}_3 &= \sinh\theta_0\mathbf{B}_0 + \cosh\theta_0\mathbf{B}_3 - 2(\theta_1\cos\theta_3 + \theta_2\sin\theta_3)\mathbf{B}_1 \\
&\quad + 2(\theta_1\sin\theta_3 - \theta_2\cos\theta_3)\mathbf{B}_2 + 2(\theta_1^2 + \theta_2^2)e^{-\theta_0}(\mathbf{B}_0 - \mathbf{B}_3)
\end{aligned} \tag{52}$$

where $\mathbf{B}_\mu = \mathbf{B}_\mu(\omega)$ and the forms of the functions θ_μ, ω are given in Assertion 5.

Inserting (52) into (46) yields a system of ordinary differential equations for the functions $\mathbf{B}_\mu(\omega)$. If we succeed in constructing its general or particular solution, then substituting it into (52) gives an exact solution of the Yang–Mills equations (46). However, the so-constructed solution will have an unpleasant feature of being asymmetric in the variables x_μ, while Eqs. (46) are symmetric in these.

To get exact solutions that are symmetric in all the variables, we exploit the formulas for generating solutions by Lorentz transformations (see the previous subsection) and thus come to the following general form of the Poincaré-invariant ansatz:

$$\mathbf{A}_\mu(x) = a_{\mu\nu}(x)\mathbf{B}^\nu(\omega) \tag{53}$$

where

$$\begin{aligned}
a_{\mu\nu}(x) &= (a_\mu a_\nu - d_\mu d_\nu)\cosh\theta_0 + (d_\mu a_\nu - d_\nu a_\mu)\sinh\theta_0 \\
&\quad + 2(a_\mu + d_\mu)[(\theta_1\cos\theta_3 + \theta_2\sin\theta_3)b_\nu + (\theta_2\cos\theta_3 - \theta_1\sin\theta_3)c_\nu \\
&\quad + (\theta_1^2 + \theta_2^2)e^{-\theta_0}(a_\nu + d_\nu)] + (b_\mu c_\nu - b_\nu c_\mu)\sin\theta_3 \\
&\quad - (c_\mu c_\nu + b_\mu b_\nu)\cos\theta_3 - 2e^{-\theta_0}(\theta_1 b_\mu + \theta_2 c_\mu)(a_\nu + d_\nu)
\end{aligned} \tag{54}$$

Here $\mu, \nu = 0, 1, 2, 3$; $x = (x_0, \mathbf{x})$ and $a_\mu, b_\mu, c_\mu, d_\mu$ are arbitrary parameters that satisfy relations (50). Thus, we have represented Poincaré-invariant ansatzes (52) in the explicitly covariant form.

Before giving the corresponding forms of the functions θ_μ, ω for the above mentioned ansatz, we remind the reader that using the discrete symmetries Φ_a $(a = 1, 2, 3)$ enables us to simplify the forms of the subalgebras of the algebra $p(1, 3)$. Specifically, at the expense of these symmetries, we can put $j = 2$ in the subalgebras L_i^j $(i = 10, 11, 13, 17)$. Consequently, for the corresponding ansatzes we have $(-1)^j = 1$. With this remark the forms of the functions θ_μ, ω, defining ansatzes (53), (54) invariant with respect to subalgebras from Assertion 1, read as

$$L_1: \quad \theta_\mu = 0, \quad \omega = dx$$

$$L_2: \quad \theta_\mu = 0, \quad \omega = ax$$

$$L_3: \quad \theta_\mu = 0, \quad \omega = kx$$

$$L_4: \quad \theta_0 = -\ln|kx|, \quad \theta_1 = \theta_2 = 0, \quad \theta_3 = \alpha \ln|kx|, \quad \omega = (ax)^2 - (dx)^2$$

$$L_5: \quad \theta_0 = -\ln|kx|, \quad \theta_1 = \theta_2 = \theta_3 = 0, \quad \omega = cx$$

$$L_6: \quad \theta_0 = -bx, \quad \theta_1 = \theta_2 = \theta_3 = 0, \ \omega = cx$$

$$L_7: \quad \theta_0 = -bx, \quad \theta_1 = \theta_2 = \theta_3 = 0, \quad \omega = bx - \ln|kx|$$

$$L_8: \quad \theta_0 = \alpha \arctan(bx(cx)^{-1}), \quad \theta_1 = \theta_2 = 0$$
$$\theta_3 = -\arctan(bx(cx)^{-1}), \quad \omega = (bx)^2 + (cx)^2$$

$$L_9: \quad \theta_0 = \theta_1 = \theta_2 = 0, \quad \theta_3 = -ax, \quad \omega = dx$$

$$L_{10}: \quad \theta_0 = \theta_1 = \theta_2 = 0, \quad \theta_3 = dx, \quad \omega = ax$$

$$L_{11}: \quad \theta_0 = \theta_1 = \theta_3 = 0, \quad \theta_2 = -\frac{1}{2}kx, \quad \omega = ax - dx$$

$$L_{12}: \quad \theta_0 = 0, \quad \theta_1 = \frac{1}{2}(bx - \alpha cx)(kx)^{-1}, \quad \theta_2 = \theta_3 = 0, \quad \omega = kx$$

$$L_{13}: \quad \theta_0 = \theta_2 = \theta_3 = 0, \quad \theta_1 = \frac{1}{2}cx, \quad \omega = kx$$

$$L_{14}: \quad \theta_0 = \theta_2 = \theta_3 = 0, \quad \theta_1 = -\frac{1}{4}kx, \quad \omega = 4bx + (kx)^2$$

$$L_{15}: \quad \theta_0 = \theta_2 = \theta_3 = 0, \quad \theta_1 = -\frac{1}{4}kx, \quad \omega = 4(\alpha bx - cx) + \alpha(kx)^2$$

$$L_{16}: \quad \theta_0 = -\ln|kx|, \quad \theta_1 = \theta_2 = 0, \quad \theta_3 = -\arctan(bx(cx)^{-1})$$
$$\omega = (bx)^2 + (cx)^2$$

$$L_{17}: \quad \theta_0 = \theta_3 = 0, \quad \theta_1 = \frac{1}{2}(cx + (\alpha + kx)bx)(1 + kx(\alpha + kx))^{-1}$$
$$\theta_2 = -\frac{1}{2}(bx - cx \cdot kx)(1 + kx(\alpha + kx))^{-1}, \quad \omega = kx$$

$$L_{18}: \quad \theta_0 = -\ln|kx|, \quad \theta_1 = \frac{1}{2}bx(kx)^{-1}$$
$$\theta_2 = \theta_3 = 0, \quad \omega = (ax)^2 - (bx)^2 - (dx)^2$$

$$L_{19}: \quad \theta_0 = -\ln|kx|, \quad \theta_1 = \frac{1}{2}bx(kx)^{-1}, \quad \theta_2 = \theta_3 = 0, \quad \omega = cx$$

$$L_{20}: \quad \theta_0 = -\ln|kx|, \quad \theta_1 = \frac{1}{2}bx(kx)^{-1}$$
$$\theta_2 = \theta_3 = 0, \quad \omega = \ln|kx| - cx$$
$$L_{21}: \quad \theta_0 = -\ln|kx|, \quad \theta_1 = \frac{1}{2}(bx - \ln|kx|)(kx)^{-1}$$
$$\theta_2 = \theta_3 = 0, \quad \omega = \alpha\ln|kx| - cx$$
$$L_{22}: \quad \theta_0 = -\ln|kx|, \quad \theta_1 = -\frac{1}{2}bx(kx)^{-1}, \quad \theta_2 = -\frac{1}{2}cx(kx)^{-1}$$
$$\theta_3 = \alpha\ln|kx|, \quad \omega = (ax)^2 - (bx)^2 - (cx)^2 - (dx)^2$$

As earlier, we use the shorthand notations for the scalar product in Minkowski space:

$$ax = a_\mu x^\mu, \quad bx = b_\mu x^\mu, \quad cx = c_\mu x^\mu, \quad dx = d_\mu x^\mu$$

and also $kx = ax + dx$.

C. Symmetry Reduction of the Yang–Mills Equations

Ansatzes (53)–(55) are given in explicitly covariant form. This fact enables us to perform symmetry reduction of Eqs. (46) in a unified way. First, we give without derivation three important identities for the tensor $a_{\mu\nu}$ [35]:

$$a^\gamma_\mu a_{\gamma\nu} = g_{\mu\nu} \tag{56}$$

$$\begin{aligned} a^\gamma_\mu \frac{\partial a_{\gamma\nu}}{\partial x_\delta} &= -(a_\mu d_\nu - a_\nu d_\mu)\frac{\partial\theta_0}{\partial x_\delta} + (b_\mu c_\nu - c_\mu b_\nu)\frac{\partial\theta_3}{\partial x_\delta} \\ &\quad + 2e^{-\theta_0}[(k_\mu b_\nu - k_\nu b_\mu)\cos\theta_3 - (k_\mu c_\nu - k_\nu c_\mu)\sin\theta_3]\frac{\partial\theta_1}{\partial x_\delta} \\ &\quad + 2e^{-\theta_0}[(k_\mu b_\nu - k_\nu b_\mu)\sin\theta_3 + (k_\mu c_\nu - k_\nu c_\mu)\cos\theta_3]\frac{\partial\theta_2}{\partial x_\delta} \end{aligned} \tag{57}$$

$$\begin{aligned} a^\gamma_\mu \Box a_{\gamma\nu} &= (a_\mu a_\nu - d_\mu d_\nu)\frac{\partial\theta_0}{\partial x_\gamma}\frac{\partial\theta_0}{\partial x^\gamma} - (a_\mu d_\nu - a_\nu d_\mu)\Box\theta_0 \\ &\quad + 2e^{-\theta_0}k_\mu b_\nu\bigg[(\Box\theta_1)\cos\theta_3 + (\Box\theta_2)\sin\theta_3 - 2\frac{\partial\theta_1}{\partial x_\gamma}\frac{\partial\theta_3}{\partial x^\gamma}\sin\theta_3 \\ &\quad + 2\frac{\partial\theta_2}{\partial x_\gamma}\frac{\partial\theta_3}{\partial x^\gamma}\cos\theta_3\bigg] + 2e^{-\theta_0}k_\mu c_\nu\bigg[(\Box\theta_2)\cos\theta_3 - (\Box\theta_1)\sin\theta_3 \\ &\quad - 2\frac{\partial\theta_1}{\partial x_\gamma}\frac{\partial\theta_3}{\partial x^\gamma}\cos\theta_3 - 2\frac{\partial\theta_2}{\partial x_\gamma}\frac{\partial\theta_3}{\partial x^\gamma}\sin\theta_3\bigg] + 4e^{-2\theta_0}k_\mu k_\nu \\ &\quad \times\left(\frac{\partial\theta_1}{\partial x_\gamma}\frac{\partial\theta_1}{\partial x^\gamma} + \frac{\partial\theta_2}{\partial x_\gamma}\frac{\partial\theta_2}{\partial x^\gamma}\right) + (b_\mu b_\nu + c_\mu c_\nu)\frac{\partial\theta_3}{\partial x_\gamma}\frac{\partial\theta_3}{\partial x^\gamma} \\ &\quad + (b_\mu c_\nu - c_\mu b_\nu)\Box\theta_3 \end{aligned} \tag{58}$$

Hereafter, we denote the derivatives of the functions in one variable ω by the dots over the symbols of the functions, for example

$$\frac{df}{d\omega}=\dot{f}, \quad \frac{d^2f}{d\omega^2}=\ddot{f}$$

Assertion 9. *Let ansatz (53) reduce system (46) to a system of second-order ordinary differential equations. Then the reduced system is necessarily of the form*

$$\begin{aligned} k_{\mu\gamma}\ddot{\mathbf{B}}^{\gamma} + l_{\mu\gamma}\dot{\mathbf{B}}^{\gamma} + m_{\mu\gamma}\mathbf{B}^{\gamma} + eg_{\mu\nu\gamma}\dot{\mathbf{B}}^{\nu} \times \mathbf{B}^{\gamma} \\ + eh_{\mu\nu\gamma}\mathbf{B}^{\nu} \times \mathbf{B}^{\gamma} + e^2\mathbf{B}_{\gamma} \times (\mathbf{B}^{\gamma} \times \mathbf{B}_{\mu}) = 0 \end{aligned} \tag{59}$$

and its coefficients are given by the relations

$$\begin{aligned} k_{\mu\gamma} &= g_{\mu\gamma}F_1 - G_{\mu}G_{\gamma}, \quad l_{\mu\gamma} = g_{\mu\gamma}F_2 + 2S_{\mu\gamma} - G_{\mu}H_{\gamma} - G_{\mu}\dot{G}_{\gamma} \\ m_{\mu\gamma} &= R_{\mu\gamma} - G_{\mu}\dot{H}_{\gamma}, \quad g_{\mu\nu\gamma} = g_{\mu\gamma}G_{\nu} + g_{\nu\gamma}G_{\mu} - 2g_{\mu\nu}G_{\gamma} \\ h_{\mu\nu\gamma} &= \frac{1}{2}(g_{\mu\gamma}H_{\nu} - g_{\mu\nu}H_{\gamma}) - T_{\mu\nu\gamma} \end{aligned} \tag{60}$$

where $F_1, F_2, G_{\mu}, H_{\mu}, S_{\mu\nu}, R_{\mu\nu}, T_{\mu\nu\gamma}$ represent the following functions of ω:

$$\begin{aligned} F_1 &= \frac{\partial\omega}{\partial x_{\mu}}\frac{\partial\omega}{\partial x^{\mu}}, \quad F_2 = \Box\omega, \quad G_{\mu} = a_{\gamma\mu}\frac{\partial\omega}{\partial x_{\gamma}} \\ H_{\mu} &= \frac{\partial a_{\gamma\mu}}{\partial x_{\gamma}}, \quad S_{\mu\nu} = a_{\mu}^{\gamma}\frac{\partial a_{\gamma\nu}}{\partial x_{\delta}}\frac{\partial\omega}{\partial x^{\delta}}, \quad R_{\mu\nu} = a_{\mu}^{\gamma}\Box a_{\gamma\nu} \\ T_{\mu\nu\gamma} &= a_{\mu}^{\delta}\frac{\partial a_{\delta\nu}}{\partial x_{\sigma}}a_{\sigma\gamma} + a_{\nu}^{\delta}\frac{\partial a_{\delta\gamma}}{\partial x_{\sigma}}a_{\sigma\mu} + a_{\gamma}^{\delta}\frac{\partial a_{\delta\mu}}{\partial x_{\sigma}}a_{\sigma\nu} \end{aligned} \tag{61}$$

Proof. Inserting ansatz (53) into Eq. (46) and performing some simplifications yield the following identities:

$$\begin{aligned} \Box\mathbf{A}_{\mu} - \partial^{\mu}(\partial_{\nu}\mathbf{A}_{\nu}) &= \left(\Box a_{\mu\gamma} - \frac{\partial^2 a_{\nu\gamma}}{\partial x^{\mu}\partial x_{\nu}}\right)\mathbf{B}^{\gamma} \\ &+ \left(2\frac{\partial a_{\mu\gamma}}{\partial x_{\nu}}\frac{\partial\omega}{\partial x^{\nu}} + a_{\mu\gamma}\Box\omega - \frac{\partial a_{\nu\gamma}}{\partial x_{\nu}}\frac{\partial\omega}{\partial x^{\mu}} - \frac{\partial a_{\nu\gamma}}{\partial x^{\mu}}\frac{\partial\omega}{\partial x_{\nu}}\right. \\ &\left. - a_{\nu\gamma}\frac{\partial^2\omega}{\partial x_{\nu}\partial x^{\mu}}\right)\dot{\mathbf{B}}^{\gamma} + \left(a_{\mu\gamma}\frac{\partial\omega}{\partial x_{\nu}}\frac{\partial\omega}{\partial x^{\nu}} - a_{\nu\gamma}\frac{\partial\omega}{\partial x_{\nu}}\frac{\partial\omega}{\partial x^{\mu}}\right)\ddot{\mathbf{B}}^{\gamma} \end{aligned} \tag{62}$$

$$\begin{aligned} &(\partial_{\nu}\mathbf{A}_{\nu}) \times \mathbf{A}_{\mu} - 2(\partial_{\nu}\mathbf{A}_{\mu}) \times \mathbf{A}_{\nu} + (\partial^{\mu}\mathbf{A}_{\nu}) \times \mathbf{A}^{\nu} \\ &= \left(a_{\mu\gamma}\frac{\partial a_{\nu\alpha}}{\partial x_{\nu}} - 2\frac{\partial a_{\mu\alpha}}{\partial x_{\nu}}a_{\nu\gamma} + \frac{\partial a_{\nu\alpha}}{\partial x^{\mu}}a_{\gamma}^{\nu}\right)\mathbf{B}^{\alpha} \times \mathbf{B}^{\gamma} \\ &+ \left(a_{\mu\gamma}a_{\nu\alpha}\frac{\partial\omega}{\partial x_{\nu}} - 2a_{\mu\alpha}a_{\nu\gamma}\frac{\partial\omega}{\partial x_{\nu}} + a_{\nu\alpha}a_{\gamma}^{\nu}\frac{\partial\omega}{\partial x^{\mu}}\right)\dot{\mathbf{B}}^{\alpha} \times \mathbf{B}^{\gamma} \end{aligned} \tag{63}$$

$$\mathbf{A}_{\nu} \times (\mathbf{A}^{\nu} \times \mathbf{A}_{\mu}) = a_{\nu\beta}a_{\alpha}^{\nu}a_{\mu\gamma}\mathbf{B}^{\beta} \times (\mathbf{B}^{\alpha} \times \mathbf{B}^{\gamma}) \tag{64}$$

Here $\alpha, \beta = 0, 1, 2, 3$.

Convoluting the left- and right-hand sides of these expressions with a^{μ}_{δ} and taking into account (3.11) yield

$$\begin{aligned} a^{\mu}_{\delta}\mathbf{A}_{\nu} \times (\mathbf{A}^{\nu} \times \mathbf{A}_{\mu}) &= a^{\mu}_{\delta} a_{\nu\beta} a^{\nu}_{\alpha} a_{\mu\gamma} \mathbf{B}^{\beta} \times (\mathbf{B}^{\alpha} \times \mathbf{B}^{\gamma}) \\ &= g_{\beta\alpha} g_{\delta\gamma} \mathbf{B}^{\beta} \times (\mathbf{B}^{\alpha} \times \mathbf{B}^{\gamma}) = \mathbf{B}_{\alpha} \times (\mathbf{B}^{\alpha} \times \mathbf{B}_{\delta}) \end{aligned}$$

Consequently, convoluting (62) and (63) with a^{μ}_{δ}, we get the equalities such that their right-hand sides are linear combinations of $\mathbf{B}^{\gamma}, \dot{\mathbf{B}}^{\gamma}, \ddot{\mathbf{B}}^{\gamma}$, $\mathbf{B}^{\alpha} \times \mathbf{B}^{\gamma}, \dot{\mathbf{B}}^{\alpha} \times \mathbf{B}^{\gamma}$. Furthermore, the coefficients of these combinations are the functions of ω only. Consider first equality (62). The coefficients of $\mathbf{B}^{\gamma}, \dot{\mathbf{B}}^{\gamma}, \ddot{\mathbf{B}}^{\gamma}$ read as

$$\mathbf{B}^{\gamma}\colon \quad a^{\mu}_{\delta} \Box a_{\mu\gamma} - a^{\mu}_{\delta} \frac{\partial^2 a_{\nu\gamma}}{\partial x^{\mu} \partial x_{\nu}} = F_{\delta\gamma}(\omega) \tag{65}$$

$$\begin{aligned} \dot{\mathbf{B}}^{\gamma}\colon \quad & 2a^{\mu}_{\delta} \frac{\partial a_{\mu\gamma}}{\partial x_{\nu}} \frac{\partial \omega}{\partial x^{\nu}} + g_{\delta\gamma} \Box \omega - a^{\mu}_{\delta} \frac{\partial a_{\nu\gamma}}{\partial x_{\nu}} \frac{\partial \omega}{\partial x^{\mu}} \\ & - a^{\mu}_{\delta} \frac{\partial a_{\nu\gamma}}{\partial x^{\mu}} \frac{\partial \omega}{\partial x_{\nu}} - a^{\mu}_{\delta} a_{\nu\gamma} \frac{\partial^2 \omega}{\partial x_{\nu} \partial x^{\mu}} = G_{\delta\gamma}(\omega) \end{aligned} \tag{66}$$

$$\ddot{\mathbf{B}}^{\gamma}\colon \quad g_{\delta\gamma} \frac{\partial \omega}{\partial x_{\nu}} \frac{\partial \omega}{\partial x^{\nu}} - a^{\mu}_{\delta} a_{\nu\gamma} \frac{\partial \omega}{\partial x_{\nu}} \frac{\partial \omega}{\partial x^{\mu}} = H_{\delta\gamma}(\omega) \tag{67}$$

For coefficient (67), convoluting the function $H_{\delta\gamma}(\omega)$ with the metric tensor $g^{\delta\gamma} = g_{\delta\gamma}$ yields

$$\begin{aligned} g^{\delta\gamma} H_{\delta\gamma}(\omega) &= 4\frac{\partial \omega}{\partial x_{\nu}} \frac{\partial \omega}{\partial x^{\nu}} - a_{\mu\delta} a^{\delta}_{\nu} \frac{\partial \omega}{\partial x_{\nu}} \frac{\partial \omega}{\partial x_{\mu}} = 4\frac{\partial \omega}{\partial x_{\nu}} \frac{\partial \omega}{\partial x^{\nu}} - g_{\mu\nu} \frac{\partial \omega}{\partial x_{\nu}} \frac{\partial \omega}{\partial x_{\mu}} \\ &= 4\frac{\partial \omega}{\partial x_{\nu}} \frac{\partial \omega}{\partial x^{\nu}} - \frac{\partial \omega}{\partial x_{\nu}} \frac{\partial \omega}{\partial x^{\nu}} = 3\frac{\partial \omega}{\partial x_{\nu}} \frac{\partial \omega}{\partial x^{\nu}} \end{aligned}$$

Hence $\frac{\partial \omega}{\partial x_{\nu}} \frac{\partial \omega}{\partial x^{\nu}}$ is a function of ω only, as follows:

$$\frac{\partial \omega}{\partial x_{\nu}} \frac{\partial \omega}{\partial x^{\nu}} = F_1(\omega) \tag{68}$$

Therefore

$$a^{\mu}_{\delta} a_{\nu\gamma} \frac{\partial \omega}{\partial x_{\nu}} \frac{\partial \omega}{\partial x^{\mu}} = a_{\mu\delta} \frac{\partial \omega}{\partial x_{\mu}} a_{\nu\gamma} \frac{\partial \omega}{\partial x_{\nu}} = \tilde{H}_{\delta\gamma}(\omega)$$

whence

$$a_{\mu\delta} \frac{\partial \omega}{\partial x_{\mu}} = G_{\delta}(\omega) \tag{69}$$

In view of (68), (69) we get the equality

$$H_{\delta\gamma}(\omega) = g_{\delta\gamma}F_1 - G_\delta G_\gamma$$

Thus the function $H_{\delta\gamma}(\omega)$ coincides with $k_{\mu\gamma}$ from (60).

Convoluting (66) with the metric tensor $g^{\delta\gamma}$ gives

$$g^{\delta\gamma}G_{\delta\gamma}(\omega) = 2a_\delta^\mu \frac{\partial a_\mu^\delta}{\partial x_\nu}\frac{\partial \omega}{\partial x^\nu} + 4\square\omega - a_{\mu\delta}\frac{\partial \omega}{\partial x_\mu}\frac{\partial a_\nu^\delta}{\partial x_\nu} - a_{\mu\delta}\frac{\partial a_\nu^\delta}{\partial x_\nu}\frac{\partial \omega}{\partial x_\mu} - a_{\mu\delta}a_\nu^\delta\frac{\partial^2\omega}{\partial x_\mu \partial x_\nu} \tag{70}$$

Now using (56), we ensure that the relation

$$a_\delta^\mu \frac{\partial a_\mu^\delta}{\partial x_\nu} = \frac{1}{2}\frac{\partial}{\partial x_\nu}(a_\delta^\mu a_\mu^\delta) = \frac{1}{2}\frac{\partial}{\partial x_\nu}(g_\beta^\beta) = 0 \tag{71}$$

as well as the relation

$$\frac{\partial}{\partial x_\nu}\left[a_{\mu\delta}a_\nu^\delta\frac{\partial \omega}{\partial x_\mu}\right] = \frac{\partial a_{\mu\delta}}{\partial x_\nu}a_\nu^\delta\frac{\partial \omega}{\partial x_\mu} + a_{\mu\delta}\frac{\partial a_\nu^\delta}{\partial x_\nu}\frac{\partial \omega}{\partial x_\mu} + a_{\mu\delta}a_\nu^\delta\frac{\partial^2\omega}{\partial x_\mu \partial x_\nu}$$

hold true. Because

$$\frac{\partial a_{\mu\delta}}{\partial x_\nu}a_\nu^\delta\frac{\partial \omega}{\partial x_\mu} = \frac{\partial a_\mu^\delta}{\partial x_\nu}a_{\nu\delta}\frac{\partial \omega}{\partial x_\mu}$$

we make sure that the relation

$$\frac{\partial a_{\mu\delta}}{\partial x_\nu}a_\nu^\delta\frac{\partial \omega}{\partial x_\mu} = a_{\mu\delta}\frac{\partial a_\nu^\delta}{\partial x_\mu}\frac{\partial \omega}{\partial x_\nu}$$

is valid, whence

$$a_{\mu\delta}\frac{\partial \omega}{\partial x_\mu}\frac{\partial a_\nu^\delta}{\partial x_\nu} + a_{\mu\delta}\frac{\partial a_\nu^\delta}{\partial x_\nu}\frac{\partial \omega}{\partial x_\nu} = \frac{\partial}{\partial x_\nu}\left[a_{\mu\delta}a_\nu^\delta\frac{\partial \omega}{\partial x_\mu}\right] - a_{\mu\delta}a_\nu^\delta\frac{\partial^2\omega}{\partial x_\mu \partial x_\nu} = \frac{\partial}{\partial x_\nu}\left[g_{\mu\nu}\frac{\partial \omega}{\partial x_\mu}\right] - g_{\mu\nu}\frac{\partial^2\omega}{\partial x_\mu \partial x_\nu} = 0 \tag{72}$$

Factoring in Eqs. (70)–(72), we obtain

$$g^{\delta\gamma}G_{\delta\gamma}(\omega) = 4(\square\omega) - g_{\mu\nu}\frac{\partial^2\omega}{\partial x_\mu \partial x_\nu} = 3\square\omega$$

Consequently, the relation

$$\Box\omega = F_2(\omega) \tag{73}$$

holds.

Next, making sure that the equalities

$$a_{\mu\delta}\frac{\partial a_{\nu\gamma}}{\partial x_\mu}\frac{\partial\omega}{\partial x_\nu} + a_{\mu\delta}a_{\nu\gamma}\frac{\partial^2\omega}{\partial x_\nu \partial x_\mu} = a_{\mu\delta}\frac{\partial}{\partial x_\mu}\left(a_{\nu\gamma}\frac{\partial\omega}{\partial x_\nu}\right)$$
$$= a_{\mu\delta}\frac{\partial}{\partial x_\nu}G_\gamma(\omega) = a_{\mu\delta}\dot{G}_\gamma(\omega)\frac{\partial\omega}{\partial x^\mu} = \dot{G}_\gamma(\omega)G_\delta(\omega)$$

hold and taking into account (66) yield

$$2a^\mu_\delta\frac{\partial a_{\mu\gamma}}{\partial x_\nu}\frac{\partial\omega}{\partial x^\nu} - G_\delta\frac{\partial a_{\nu\gamma}}{\partial x^\nu} = \tilde{G}_{\delta\gamma}(\omega) \tag{74}$$

Now, convoluting (74) with $g_{\mu\nu}$, we have

$$g_{\mu\nu}\tilde{G}_{\delta\gamma}(\omega) = g_{\mu\nu}G_\delta(\omega)\left(2\frac{\partial a_{\mu\nu}}{\partial x_\mu} - g_{\mu\nu}\frac{\partial a_{\nu\gamma}}{\partial\partial x_\nu}\right)$$

or, equivalently

$$\frac{\partial a_{\mu\nu}}{\partial x_\mu} = H_\nu(\omega), \quad a^\mu_\delta\frac{\partial a_{\mu\gamma}}{\partial x_\nu}\frac{\partial\omega}{\partial x^\nu} = S_{\delta\gamma}(\omega) \tag{75}$$

Combining (69), (73), and (75), we find that the coefficient of $\dot{\mathbf{B}}^\gamma$ in the reduced system (59) coincides with $l_{\mu\gamma}$ in Eq. (60).

Finally, from the relation

$$a_{\mu\delta}\frac{\partial^2 a_{\mu\gamma}}{\partial x_\mu \partial x_\nu} = a_{\mu\delta}\frac{\partial}{\partial x_\nu}\left(\frac{\partial a_{\nu\gamma}}{\partial x_\nu}\right) = a_{\mu\delta}\frac{\partial}{\partial x_\mu}(H_\gamma(\omega)) = G_\delta(\omega)\dot{H}_\gamma(\omega)$$

and (65) it follows that

$$a^\mu_\delta\Box a_{\mu\gamma} = R_{\delta\gamma}(\omega)$$

Consequently, the function in the right-hand side of (65) coincides with $m_{\mu\gamma}$ from (60).

Analysis of (63) is carried out in the same way (we do not present the corresponding calculations here). Assertion 9 is proved.

Thanks to Assertion 9, the problem of symmetry reduction of Yang–Mills equations by subalgebras of the algebra $p(1,3)$ reduces to routine substitution of the corresponding expressions for $a_{\mu\nu}, \omega$ into (61). We give below the final forms of the coefficients (60) of the reduced system of ordinary differential equations (59) for each subalgebras of the algebra $p(1,3)$:

$$
\begin{aligned}
L_1:\quad & k_{\mu\gamma} = -g_{\mu\gamma} - d_\mu d_\gamma, \quad l_{\mu\gamma} = m_{\mu\gamma} = 0\\
& g_{\mu\nu\gamma} = g_{\mu\gamma} d_\nu + g_{\nu\gamma} d_\mu - 2g_{\mu\nu} d_\gamma, \quad h_{\mu\nu\gamma} = 0\\
L_2:\quad & k_{\mu\gamma} = g_{\mu\gamma} - a_\mu a_\gamma, \quad l_{\mu\gamma} = m_{\mu\gamma} = 0\\
& g_{\mu\nu\gamma} = g_{\mu\gamma} a_\nu + g_{\nu\gamma} a_\mu - 2g_{\mu\nu} a_\gamma, \quad h_{\mu\nu\gamma} = 0\\
L_3:\quad & k_{\mu\gamma} = k_\mu k_\gamma, \quad l_{\mu\gamma} = m_{\mu\gamma} = 0\\
& g_{\mu\nu\gamma} = g_{\mu\gamma} k_\nu + g_{\nu\gamma} k_\mu - 2g_{\mu\nu} k_\gamma, \quad h_{\mu\nu\gamma} = 0\\
L_4:\quad & k_{\mu\gamma} = 4g_{\mu\gamma}\omega - a_\mu a_\gamma(\omega+1)^2 - d_\mu d_\gamma(\omega-1)^2\\
& \qquad - (a_\mu d_\gamma + a_\gamma d_\mu)(\omega^2 - 1)\\
& l_{\mu\gamma} = 4(g_{\mu\gamma} + \alpha(b_\mu c_\gamma - c_\mu b_\gamma)) - 2k_\mu(a_\gamma - d_\gamma + k_\gamma\omega)\\
& m_{\mu\gamma} = 0\\
& g_{\mu\nu\gamma} = \epsilon(g_{\mu\gamma}(a_\nu - d_\nu + k_\nu\omega) + g_{\nu\gamma}(a_\mu - d_\mu + k_\mu\omega)\\
& \qquad - 2g_{\mu\nu}(a_\gamma - d_\gamma + k_\gamma\omega))\\
& h_{\mu\nu\gamma} = \frac{\epsilon}{2}[g_{\mu\gamma}k_\nu - g_{\mu\nu}k_\gamma] + \alpha\epsilon[(b_\mu c_\nu - c_\mu b_\nu)k_\gamma\\
& \qquad + (b_\nu c_\gamma - c_\nu b_\gamma)k_\mu + (b_\gamma c_\mu - c_\gamma b_\mu)k_\nu]\\
L_5:\quad & k_{\mu\gamma} = -g_{\mu\gamma} - c_\mu c_\gamma, \quad l_{\mu\gamma} = -\epsilon c_\mu k_\gamma, \quad m_{\mu\gamma} = 0\\
& g_{\mu\nu\gamma} = g_{\mu\gamma} c_\nu + g_{\nu\gamma} c_\mu - 2g_{\mu\nu} c_\gamma\\
& h_{\mu\nu\gamma} = \frac{\epsilon}{2}(g_{\mu\gamma}k_\nu - g_{\mu\nu}k_\gamma)\\
L_6:\quad & k_{\mu\gamma} = -g_{\mu\gamma} - c_\mu c_\gamma, \quad l_{\mu\gamma} = 0\\
& m_{\mu\gamma} = -(a_\mu a_\gamma - d_\mu d_\gamma), \quad g_{\mu\nu\gamma} = g_{\mu\gamma} c_\nu + g_{\nu\gamma} c_\mu - 2g_{\mu\nu} c_\gamma\\
& h_{\mu\nu\gamma} = -[(a_\mu d_\nu - a_\nu d_\mu)b_\gamma + (a_\nu d_\gamma - a_\gamma d_\nu)b_\mu\\
& \qquad + (a_\gamma d_\mu - a_\mu d_\gamma)b_\nu]\\
L_7:\quad & k_{\mu\gamma} = -g_{\mu\gamma} - (b_\mu - \epsilon k_\mu e^\omega)(b_\gamma - \epsilon k_\gamma e^\omega)\\
& l_{\mu\gamma} = -2(a_\mu d_\gamma - a_\gamma d_\mu) + \epsilon e^\omega(b_\mu - \epsilon k_\mu e^\omega)k_\gamma\\
& m_{\mu\gamma} = -(a_\mu a_\gamma - d_\mu d_\gamma)\\
& g_{\mu\nu\gamma} = g_{\mu\gamma}(b_\nu - \epsilon k_\nu e^\omega) + g_{\nu\gamma}(b_\mu - \epsilon k_\mu e^\omega) - 2g_{\mu\nu}(b_\gamma - \epsilon k_\gamma e^\omega)\\
& h_{\mu\nu\gamma} = -[(a_\mu d_\nu - a_\nu d_\mu)b_\gamma + (a_\nu d_\gamma - a_\gamma d_\nu)b_\mu + (a_\gamma d_\mu - a_\mu d_\gamma)b_\nu]
\end{aligned}
$$

$$
\begin{aligned}
L_8:\quad & k_{\mu\gamma} = -4\omega(g_{\mu\gamma} + c_\mu c_\gamma), \quad l_{\mu\gamma} = -4(g_{\mu\gamma} + c_\mu c_\gamma) \\
& m_{\mu\gamma} = -\frac{1}{\omega}(\alpha^2(a_\mu a_\gamma - d_\mu d_\gamma) + b_\mu b_\gamma) \\
& g_{\mu\nu\gamma} = 2\sqrt{\omega}(g_{\mu\gamma} c_\nu + g_{\nu\gamma} c_\mu - 2 g_{\mu\nu} c_\gamma) \\
& h_{\mu\nu\gamma} = \frac{1}{2\sqrt{\omega}}(g_{\mu\gamma} c_\nu - g_{\mu\nu} c_\gamma) + \frac{\alpha}{\sqrt{\omega}}((a_\mu d_\nu - a_\nu d_\mu) b_\gamma \\
& \qquad + (a_\nu d_\gamma - d_\nu a_\gamma) b_\mu + (a_\gamma d_\mu - a_\mu d_\gamma) b_\nu) \\
L_9:\quad & k_{\mu\gamma} = -g_{\mu\gamma} - d_\mu d_\gamma, \quad l_{\mu\gamma} = 0 \\
& m_{\mu\gamma} = b_\mu b_\gamma + c_\mu c_\gamma \\
& g_{\mu\nu\gamma} = g_{\mu\gamma} d_\nu + g_{\nu\gamma} d_\mu - 2 g_{\mu\nu} d_\gamma \\
& h_{\mu\nu\gamma} = a_\gamma(b_\mu c_\nu - c_\mu b_\nu) + a_\mu(b_\nu c_\gamma - c_\nu b_\gamma) \\
& \qquad + a_\nu(b_\gamma c_\mu - c_\gamma b_\mu) \\
L_{10}:\quad & k_{\mu\gamma} = g_{\mu\gamma} - a_\mu a_\gamma, \quad l_{\mu\gamma} = 0 \\
& m_{\mu\gamma} = -(b_\mu b_\gamma + c_\mu c_\gamma) \\
& g_{\mu\nu\gamma} = g_{\mu\gamma} a_\nu + g_{\nu\gamma} a_\mu - 2 g_{\mu\nu} a_\gamma \\
& h_{\mu\nu\gamma} = -[d_\gamma(b_\mu c_\nu - c_\mu b_\nu) + d_\mu(b_\nu c_\gamma - c_\nu b_\gamma) \\
& \qquad + d_\nu(b_\gamma c_\mu - c_\gamma b_\mu)] \\
L_{11}:\quad & k_{\mu\gamma} = -(a_\mu - d_\mu)(a_\gamma - d_\gamma), l_{\mu\gamma} = 2(b_\mu c_\gamma - c_\mu b_\gamma) \\
& m_{\mu\gamma} = 0 \\
& g_{\mu\nu\gamma} = g_{\mu\gamma}(a_\nu - d_\nu) + g_{\nu\gamma}(a_\mu - d_\mu) - 2 g_{\mu\nu}(a_\gamma - d_\gamma) \\
& h_{\mu\nu\gamma} = \frac{1}{2}[(k_\gamma(b_\mu c_\nu - c_\mu b_\nu) + k_\mu(b_\nu c_\gamma - c_\nu b_\gamma) + k_\nu(b_\gamma c_\mu - c_\gamma b_\mu)] \\
L_{12}:\quad & k_{\mu\gamma} = -k_\mu k_\gamma, \quad l_{\mu\gamma} = -\omega^{-1} k_\mu k_\gamma, \quad m_{\mu\gamma} = -\alpha^2 \omega^{-2} k_\mu k_\gamma \\
& g_{\mu\nu\gamma} = g_{\mu\gamma} k_\nu + g_{\nu\gamma} k_\mu - 2 g_{\mu\nu} k_\gamma \\
& h_{\mu\nu\gamma} = \frac{1}{2}\omega^{-1}(g_{\mu\gamma} k_\nu - g_{\mu\nu} k_\gamma) + \alpha\omega^{-1}((k_\mu b_\nu - k_\nu b_\mu) c_\gamma \\
& \qquad + (k_\nu b_\gamma - k_\gamma b_\nu) c_\mu + (k_\gamma b_\mu - k_\mu b_\gamma) c_\nu) \qquad (76) \\
L_{13}:\quad & k_{\mu\gamma} = -k_\mu k_\gamma, \quad l_{\mu\gamma} = 0, \quad m_{\mu\gamma} = -k_\mu k_\gamma \\
& g_{\mu\nu\gamma} = g_{\mu\gamma} k_\nu + g_{\nu\gamma} k_\mu - 2 g_{\mu\nu} k_\gamma \\
& h_{\mu\nu\gamma} = -((k_\mu b_\nu - k_\nu b_\mu) c_\gamma + (k_\nu b_\gamma - k_\gamma b_\nu) c_\mu + (k_\gamma b_\mu - k_\mu b_\gamma) c_\nu) \\
L_{14}:\quad & k_{\mu\gamma} = -16(g_{\mu\gamma} + b_\mu b_\gamma), \quad l_{\mu\gamma} = m_{\mu\gamma} = h_{\mu\nu\gamma} = 0 \\
& g_{\mu\nu\gamma} = 4(g_{\mu\gamma} b_\nu + g_{\nu\gamma} b_\mu - 2 g_{\mu\nu} b_\gamma) \\
L_{15}:\quad & k_{\mu\gamma} = -16[(1+\alpha^2) g_{\mu\gamma} + (c_\mu - \alpha b_\mu)(c_\gamma - \alpha b_\gamma)] \\
& l_{\mu\gamma} = m_{\mu\gamma} = h_{\mu\nu\gamma} = 0 \\
& g_{\mu\nu\gamma} = -4[g_{\mu\gamma}(c_\nu - \alpha b_\nu) + g_{\nu\gamma}(c_\mu - \alpha b_\mu) - 2 g_{\mu\nu}(c_\gamma - \alpha b_\gamma)]
\end{aligned}
$$

$$L_{16}:\quad k_{\mu\gamma} = -4\omega(g_{\mu\gamma} + c_\mu c_\gamma), \quad l_{\mu\gamma} = -4(g_{\mu\gamma} + c_\mu c_\gamma) - 2\epsilon k_\gamma c_\mu \sqrt{\omega}$$

$$m_{\mu\gamma} = -\omega^{-1} b_\mu b_\gamma, \quad g_{\mu\nu\gamma} = 2\sqrt{\omega}(g_{\mu\gamma}c_\nu + g_{\nu\gamma}c_\mu - 2g_{\mu\nu}c_\gamma)$$

$$h_{\mu\nu\gamma} = \frac{1}{2}\left[\epsilon(g_{\mu\gamma}k_\nu - g_{\mu\nu}k_\gamma) + \frac{1}{\sqrt{\omega}}(g_{\mu\gamma}c_\nu - g_{\mu\nu}c_\gamma)\right]$$

$$L_{17}:\quad k_{\mu\gamma} = -k_\mu k_\gamma, \quad l_{\mu\gamma} = -\frac{2\omega + \alpha}{\omega(\omega + \alpha) + 1} k_\mu k_\gamma$$

$$m_{\mu\gamma} = -4k_\mu k_\gamma (1 + \omega(\alpha + \omega))^{-2}$$

$$g_{\mu\nu\gamma} = g_{\mu\gamma}k_\nu + g_{\nu\gamma}k_\mu - 2g_{\mu\nu}k_\gamma$$

$$h_{\mu\nu\gamma} = \frac{1}{2}(\alpha + 2\omega)(g_{\mu\gamma}k_\nu - g_{\mu\nu}k_\gamma)(1 + \omega(\alpha + \omega))^{-1}$$
$$- 2(1 + \omega(\omega + \alpha))^{-1}((k_\mu b_\nu - k_\nu b_\mu)c_\gamma$$
$$+ (k_\nu b_\gamma - k_\gamma b_\nu)c_\mu + (k_\gamma b_\mu - k_\mu b_\gamma)c_\nu)$$

$$L_{18}:\quad k_{\mu\gamma} = 4\omega g_{\mu\gamma} - (k_\mu \omega + a_\mu - d_\mu)(k_\gamma \omega + a_\gamma - d_\gamma)$$

$$l_{\mu\gamma} = 6g_{\mu\gamma} + 4(a_\mu d_\gamma - a_\gamma d_\mu) - 3k_\gamma(k_\mu \omega + a_\mu - d_\mu)$$

$$m_{\mu\gamma} = -k_\mu k_\gamma, \quad g_{\mu\nu\gamma} = \epsilon(g_{\mu\gamma}(k_\nu \omega + a_\nu - d_\nu)$$
$$+ g_{\nu\gamma}(k_\mu \omega + a_\mu - d_\mu) - 2g_{\mu\nu}(k_\gamma \omega + a_\gamma - d_\gamma))$$

$$h_{\mu\nu\gamma} = \epsilon(g_{\mu\gamma}k_\nu - g_{\mu\nu}k_\gamma)$$

$$L_{19}:\quad k_{\mu\gamma} = -g_{\mu\gamma} - c_\mu c_\gamma, \quad l_{\mu\gamma} = 2\epsilon k_\gamma c_\mu, \quad m_{\mu\gamma} = -k_\mu k_\gamma$$

$$g_{\mu\nu\gamma} = g_{\mu\gamma}c_\nu + g_{\nu\gamma}c_\mu - 2g_{\mu\nu}c_\gamma, \quad h_{\mu\nu\gamma} = \epsilon(g_{\mu\gamma}k_\nu - g_{\mu\nu}k_\gamma$$

$$L_{20}:\quad k_{\mu\gamma} = -g_{\mu\gamma} - (c_\mu - \epsilon k_\mu)(c_\gamma - \epsilon k_\gamma)$$

$$l_{\mu\gamma} = 2\epsilon k_\gamma c_\mu - 2k_\mu k_\gamma, \quad m_{\mu\gamma} = -k_\mu k_\gamma$$

$$g_{\mu\nu\gamma} = g_{\mu\gamma}(\epsilon k_\nu - c_\nu) + g_{\nu\gamma}(\epsilon k_\mu - c_\mu) - 2g_{\mu\nu}(\epsilon k_\gamma - c_\gamma)$$

$$h_{\mu\nu\gamma} = \epsilon(g_{\mu\gamma}k_\nu - g_{\mu\nu}k_\gamma)$$

$$L_{21}:\quad k_{\mu\gamma} = -g_{\mu\gamma} - (c_\mu - \alpha\epsilon k_\mu)(c_\gamma - \alpha\epsilon k_\gamma)$$

$$l_{\mu\gamma} = 2(\epsilon k_\gamma c_\mu - \alpha k_\mu k_\gamma), \quad m_{\mu\gamma} = -k_\mu k_\gamma$$

$$g_{\mu\nu\gamma} = -g_{\mu\gamma}(c_\nu - \alpha\epsilon k_\nu) - g_{\nu\gamma}(c_\mu - \alpha\epsilon k_\mu)$$
$$+ 2g_{\mu\nu}(c_\gamma - \alpha\epsilon k_\gamma), \quad h_{\mu\nu\gamma} = \epsilon(g_{\mu\gamma}k_\nu - g_{\mu\nu}k_\gamma)$$

$$L_{22}:\quad k_{\mu\gamma} = -4\omega g_{\mu\gamma} - (a_\mu - d_\mu + k_\mu \omega)(a_\gamma - d_\gamma + k_\gamma \omega)$$

$$l_{\mu\gamma} = 4[2g_{\mu\gamma} + \alpha(b_\mu c_\gamma - c_\mu b_\gamma) - a_\mu a_\gamma + d_\mu d_\gamma - \omega k_\mu k_\gamma]$$

$$m_{\mu\gamma} = -2k_\mu k_\gamma, \quad g_{\mu\nu\gamma} = \epsilon(g_{\mu\gamma}(a_\nu - d_\nu + k_\nu \omega)$$
$$+ g_{\nu\gamma}(a_\mu - d_\mu + k_\mu \omega) - 2g_{\mu\nu}(a_\gamma - d_\gamma + k_\gamma \omega))$$

$$h_{\mu\nu\gamma} = \frac{3\epsilon}{2}(g_{\mu\gamma}k_\nu - g_{\mu\nu}k_\gamma) - \epsilon\alpha[k_\gamma(b_\mu c_\nu - c_\mu b_\nu)$$
$$+ (k_\mu(b_\nu c_\gamma - c_\nu b_\gamma) + k_\nu(b_\gamma c_\mu - c_\gamma b_\mu)]$$

In these formulas [Eq. (76)] we have $\epsilon = 1$ for $kx > 0$ and $\epsilon = -1$ for $kx < 0$. Furthermore, α is an arbitrary parameter.

D. Exact Solutions

Clearly, efficiency of the symmetry reduction procedure is subject to our ability to integrate the reduced systems of ordinary differential equations. Since the reduced equations are nonlinear, it is not at all clear that it will be possible to construct their particular or general solutions. That it why we devote the first part of this subsection to describing our technique for integrating the reduced systems of nonlinear ordinary differential equations (further details can be found in Ref. 33).

Note that in contrast to the case of the nonlinear Dirac equation, it is not possible to construct the *general* solutions of the reduced systems (59)–(61). For this reason, we give whenever possible their *particular* solutions, obtained by reduction of systems of equations in question by the number of components of the dependent function. Let us emphasize that the miraculous efficiency of the t'Hooft ansatz [5] for the Yang–Mills equations is a consequence of the fact that it reduces the system of 12 differential equations to a single conformally invariant wave equation.

Consider system (59)–(61), which corresponds to the subalgebra L_8. We adopt the following ansatz

$$\mathbf{B}_\mu = a_\mu \mathbf{e}_1 f(\omega) + d_\mu \mathbf{e}_2 g(\omega) + b_\mu \mathbf{e}_3 h(\omega) \tag{77}$$

for the vector function $\mathbf{B}_\mu$, where $f(\omega),\ g(\omega),\ h(\omega)$ are new unknown smooth functions of ω and

$$\mathbf{e}_1 = (1,0,0)^T, \quad \mathbf{e}_2 = (0,1,0)^T, \quad \mathbf{e}_3 = (0,0,1)^T$$

Now inserting (77) into (59), where the coefficients (60) are listed in (76) for the case of the subalgebra L_8, we arrive at the system of relations

$$\begin{aligned} &a_\mu \mathbf{e}_1[-4\omega\ddot{f} - 4\dot{f} - \frac{\alpha^2}{\omega} f + \frac{2\alpha e}{\sqrt{\omega}} gh + e^2(h^2 + g^2)f] \\ &+ d_\mu \mathbf{e}_2[-4\omega\ddot{g} - 4\dot{g} - \frac{\alpha^2}{\omega} g - \frac{2\alpha e}{\sqrt{\omega}} fh + e^2(h^2 - f^2)g] \\ &+ b_\mu \mathbf{e}_3[-4\omega\ddot{h} - 4\dot{h} + \omega^{-1} h - \frac{2\alpha e}{\sqrt{\omega}} fg + e^2(g^2 - f^2)h] = 0 \end{aligned}$$

This is equivalent to the following system of three ordinary differential equations:

$$
\begin{aligned}
4\omega\ddot{f} + 4\dot{f} + \frac{\alpha^2}{\omega} f - \frac{2\alpha e}{\sqrt{\omega}} gh - e^2(h^2 + g^2)f &= 0 \\
4\omega\ddot{g} + 4\dot{g} + \frac{\alpha^2}{\omega} g + \frac{2\alpha e}{\sqrt{\omega}} fh - e^2(h^2 - f^2)g &= 0 \\
4\omega\ddot{h} + 4\dot{h} - \omega^{-1}h + \frac{2\alpha e}{\sqrt{\omega}} fg - e^2(g^2 - f^2)h &= 0
\end{aligned}
\tag{78}
$$

so that we reduce the system of 12 ordinary differential equations (59) to a system containing three equations only.

Next, choosing

$$
\mathbf{B}_\mu = k_\mu \mathbf{e}_1 f(\omega) + b_\mu \mathbf{e}_2 g(\omega) \tag{79}
$$

and inserting this expression into (59) with coefficients given by formulas (76) for the case of the subalgebra L_8 under $\alpha = 0$ yield the system of two ordinary differential equations

$$
4\omega\ddot{f} + 4\dot{f} - e^2 g^2 f = 0, \quad 4\omega\ddot{g} + 4\dot{g} - \omega^{-1}g = 0
$$

Note that the second equation of this system is linear.

In a similar way we have reduced some other systems of ordinary differential equations (59) to systems of two or three equations. Below we list the substitutions for $\mathbf{B}_\mu(\omega)$ and corresponding systems of ordinary differential equations. Numbering of the systems below reflects numbering of the corresponding subalgebras L_j of the algebra $p(1,3)$:

1. $\mathbf{B}_\mu = a_\mu \mathbf{e}_1 f(\omega) + b_\mu \mathbf{e}_2 g(\omega) + c_\mu \mathbf{e}_3 h(\omega) \quad \ddot{f} - e^2(g^2 + h^2)f = 0$
 $g + e^2(f^2 - h^2)g = 0, \quad \ddot{h} + e^2(f^2 - g^2)h = 0$

2. $\mathbf{B}_\mu = b_\mu \mathbf{e}_1 f(\omega) + c_\mu \mathbf{e}_2 g(\omega) + d_\mu \mathbf{e}_3 h(\omega), \quad \ddot{f} + e^2(g^2 + h^2)f = 0$
 $\ddot{g} + e^2(f^2 + h^2)g = 0, \quad \ddot{h} + e^2(f^2 + g^2)h = 0$

5. $\mathbf{B}_\mu = k_\mu \mathbf{e}_1 f(\omega) + b_\mu \mathbf{e}_2 g(\omega), \quad \ddot{f} - e^2 g^2 f = 0, \quad \ddot{g} = 0$

8.1. $(\alpha = 0)\ \mathbf{B}_\mu = k_\mu \mathbf{e}_1 f(\omega) + b_\mu \mathbf{e}_2 g(\omega), \quad 4\omega\ddot{f} + 4\dot{f} - e^2 g^2 f = 0$
 $4\omega\ddot{g} + 4\dot{g} - \omega^{-1}g = 0$

8.2. $\mathbf{B}_\mu = a_\mu \mathbf{e}_1 f(\omega) + d_\mu \mathbf{e}_2 g(\omega) + b_\mu \mathbf{e}_3 h(\omega)$

$$
4\omega\ddot{f} + 4\dot{f} - \frac{\alpha^2}{\omega} f - \frac{2\alpha e}{\sqrt{\omega}} gh - e^2(h^2 + g^2)f = 0
$$

$$4\omega\ddot{g} + 4\dot{g} + \frac{\alpha^2}{\omega} g + \frac{2\alpha e}{\sqrt{\omega}} fh + e^2(f^2 - h^2)g = 0$$

$$4\omega\ddot{h} + 4\dot{h} - \omega^{-1}h + \frac{2\alpha e}{\sqrt{\omega}} fg + e^2(f^2 - g^2)h = 0$$

14.1. $\mathbf{B}_\mu = a_\mu \mathbf{e}_1 f(\omega) + d_\mu \mathbf{e}_2 g(\omega) + c_\mu \mathbf{e}_3 h(\omega), \quad 16\ddot{f} - e^2(h^2 + g^2)f = 0$

$$16\ddot{g} + e^2(f^2 - h^2)g = 0, \quad 16\ddot{h} + e^2(f^2 - g^2)h = 0$$

14.2. $\mathbf{B}_\mu = k_\mu \mathbf{e}_1 f(\omega) + c_\mu \mathbf{e}_2 g(\omega) \quad 16\ddot{f} - e^2 g^2 f = 0, \quad \ddot{g} = 0$

15.1. $\mathbf{B}_\mu = a_\mu \mathbf{e}_1 f(\omega) + d_\mu \mathbf{e}_2 g(\omega) + (1 + \alpha^2)^{-(1/2)}(\alpha c_\mu + b_\mu)\mathbf{e}_3 h(\omega)$

$$16(1 + \alpha^2)\ddot{f} - e^2(h^2 + g^2)f = 0, \quad 16(1 + \alpha^2)\ddot{g} + e^2(f^2 - h^2)g = 0$$

$$16(1 + \alpha^2)\ddot{h} + e^2(f^2 - g^2)h = 0 \tag{80}$$

15.2. $\mathbf{B}_\mu = k_\mu \mathbf{e}_1 f(\omega) + (1 + \alpha^2)^{-(1/2)}(\alpha c_\mu + b_\mu)\mathbf{e}_2 g(\omega)$

$$16(1 + \alpha^2)\ddot{f} - e^2 f g^2 = 0, \quad \ddot{g} = 0$$

16. $\mathbf{B}_\mu = k_\mu \mathbf{e}_1 f(\omega) + b_\mu \mathbf{e}_2 g(\omega), \quad 4\omega\ddot{f} + 4\dot{f} - e^2 g^2 f = 0$

$$4\omega\ddot{g} + 4\dot{g} - \omega^{-1} g = 0$$

18. $\mathbf{B}_\mu = b_\mu \mathbf{e}_1 f(\omega) + c_\mu \mathbf{e}_2 g(\omega), \quad 4\omega\ddot{f} + 6\dot{f} + e^2 g^2 f = 0$

$$4\omega\ddot{g} + 6\dot{g} + e^2 f^2 g = 0$$

19. $\mathbf{B}_\mu = k_\mu \mathbf{e}_1 f(\omega) + b_\mu \mathbf{e}_2 g(\omega), \quad \ddot{f} - e^2 g^2 f = 0, \quad \ddot{g} = 0$

20. $\mathbf{B}_\mu = k_\mu \mathbf{e}_1 f(\omega) + b_\mu \mathbf{e}_2 g(\omega), \quad \ddot{f} - e^2 g^2 f = 0, \quad \ddot{g} = 0$

21. $\mathbf{B}_\mu = k_\mu \mathbf{e}_1 f(\omega) + b_\mu \mathbf{e}_2 g(\omega), \quad \ddot{f} - e^2 g^2 f = 0, \quad \ddot{g} = 0$

22. $(\alpha = 0) \quad \mathbf{B}_\mu = b_\mu \mathbf{e}_1 f(\omega) + c_\mu \mathbf{e}_2 g(\omega)$

$$4\omega\ddot{f} + 8\dot{f} + e^2 g^2 f = 0, \quad 4\omega\ddot{g} + 8\dot{g} + e^2 f^2 g = 0$$

So, combining symmetry reduction by the number of independent variables and direct reduction by the number of the components of the function to be found, we have reduced the $SU(2)$ Yang–Mills equations (46) to comparatively simple systems of ordinary differential equations (80).

As a next step, we briefly review the procedure of integration of equations (80). Choosing $f = 0, \ g = h = u(\omega)$ reduces system 1 [in Eq. (80)] to the equation

$$\ddot{u} = e^2 u^3 \tag{81}$$

which is integrated in terms of the elliptic functions. Note that this equation has solution that is expressed in terms of elementary functions:

$$u = \sqrt{2}(e\omega - C)^{-1}, \quad C \in \mathbf{R}$$

System 2 with $f = g = h = u(\omega)$ reduces to the equation

$$\ddot{u} + 2e^2u^3 = 0$$

which is also integrated in terms of the elliptic functions.

Integrating the second equation of system 5, we get

$$g = C_1\omega + C_2, \quad C_1, C_2 \in \mathbf{R}$$

Provided $C_1 \neq 0$, the constant C_2 is negligible and we may put $C_2 = 0$. With this condition, the first equation of system 5 reads as

$$\ddot{f} - e^2C_1^2\omega^2 f = 0 \tag{82}$$

The general solution of Eq. (82), which is equivalent to the Bessel equation, is given by the formula

$$f = \sqrt{\omega}Z_{1/4}\left(\frac{ie}{2}C_1\omega^2\right)$$

Here we use the designations $Z_\nu(\omega) = C_3J_\nu(\omega) + C_4Y_\nu(\omega)$, where J_ν, Y_ν are the Bessel functions and C_3, C_4 are arbitrary constants.

Given the condition $C_1 = 0$, $C_2 \neq 0$, the general solution of the first equation of system 5 reads as

$$f = C_3\cosh(C_2e\omega) + C_4\sinh(C_2e\omega)$$

where C_3, C_4 are arbitrary real constants.

Finally, if $C_1 = C_2 = 0$, then the general solution of the first equation of system 5 is given by the formula $f = C_3\omega + C_4, C_3, C_4 \in \mathbf{R}$.

Next, we integrate the second equation of system 8.1 to obtain

$$g = C_1\sqrt{\omega} + C_2(\sqrt{\omega})^{-1}$$

where C_1, C_2 are arbitrary integration constants. Inserting the function g into the first equation of system 8.1 yields the linear differential equation

$$4\omega^2\ddot{f} + 4\omega\dot{f} - e^2(C_1\omega + C_2)^2 f = 0 \tag{83}$$

For the case $C_1C_2 \neq 0$, Eq. (83) is related to the Whittaker equation. Here we consider only the case $C_1C_2 = 0$, thus getting

$$\text{(a)} \quad C_1 \neq 0, \quad C_2 = 0, \quad f = Z_0\left[\frac{ie}{2}C_1\omega\right]$$

$$\text{(b)} \quad C_1 = 0, \quad C_2 \neq 0, \quad f = C_3\omega^{eC_2/2} + C_4\omega^{-(eC_2/2)}$$

$$\text{(c)} \quad C_1 = C_2 = 0, \quad f = C_3\ln\omega + C_4$$

where C_3, C_4 are arbitrary integration constants.

Analyzing systems 14.1 and 14.2 in Eq. (80), we conclude that they reduce to systems 1 and 5, correspondingly, if we replace e by $\frac{e}{4}$. Analogously, replacing in systems 1 and 5 the parameter e by $\frac{e}{4}(1+\alpha^2)^{-(1/2)}$ yields systems 15.1 and 15.2, respectively.

Finally, system 22 with $\alpha = 0$ is reduced by the change of the dependent variable $f = g = u(\omega)$ to the Emden–Fauler equation

$$\omega\ddot{u} + 2\dot{u} + \frac{e^2}{4}u^3 = 0$$

which has the following particular solution $u = e^{-1}\omega^{-(1/2)}$.

We have not succeeded in integrating systems of ordinary differential systems 8.2 and 18 [Eq. (80)]. Furthermore, systems 19–21 coincide with system 5 and system 16 coincides with system 8.1.

Inserting the forms of the functions f, g, h obtained into (80) with the subsequent substitution of the latter expression into the corresponding ansatz (53)–(55) yields invariant solutions of the $SU(2)$ Yang–Mills equations (46). Note that solutions of systems 5, 8.1, 14.2, 15.2, 16, and 19–21, with $g = 0$, give rise to Abelian solutions of the Yang–Mills equation, namely, to solutions satisfying the additional restriction $\mathbf{A}_\mu \times \mathbf{A}_\nu = \mathbf{0}$. Such solutions are of low interest for physical applications and are not considered here. Below we give the full list of non-Abelian invariant solutions of Eqs. (46):

1. $\mathbf{A}_\mu = (\mathbf{e}_2 b_\mu + \mathbf{e}_3 c_\mu)\sqrt{2}(e\,dx - \lambda)^{-1}$

2. $\mathbf{A}_\mu = (\mathbf{e}_2 b_\mu + \mathbf{e}_3 c_\mu)\left[\lambda \mathrm{sn}\left(\frac{\sqrt{2}}{2}e\lambda dx\right)\mathrm{dn}\left(\frac{\sqrt{2}}{2}e\lambda dx\right)\right]\left[\mathrm{cn}\,\frac{\sqrt{2}}{2}e\lambda dx)\right]^{-1}$

3. $\mathbf{A}_\mu = (\mathbf{e}_2 b_\mu + \mathbf{e}_3 c_\mu)\lambda[\mathrm{cn}\,(e\lambda dx)]^{-1}$

4. $\mathbf{A}_\mu = (\mathbf{e}_1 b_\mu + \mathbf{e}_2 c_\mu + \mathbf{e}_3 c_\mu)\lambda \mathrm{cn}\,(e\lambda ax)$

5. $\mathbf{A}_\mu = \mathbf{e}_1 k_\mu |kx|^{-1}\sqrt{cx}Z_{1/4}\left[\frac{i}{2}e\lambda(cx)^2\right] + \mathbf{e}_2 b_\mu \lambda cx$

6. $\mathbf{A}_\mu = \mathbf{e}_1 k_\mu |kx|^{-1}[\lambda_1 \cosh(e\lambda cx) + \lambda_2 \sinh(e\lambda cx)] + \mathbf{e}_2 b_\mu \lambda$

7. $\mathbf{A}_\mu = \mathbf{e}_1 k_\mu Z_0\left[\frac{i}{2}e\lambda((bx)^2 + (cx)^2)\right] + \mathbf{e}_2(b_\mu cx - c_\mu bx)\lambda$

8. $\mathbf{A}_\mu = \mathbf{e}_1 k_\mu[\lambda_1((bx)^2 + (cx)^2))^{e\lambda/2} + \lambda_2((bx)^2 + (cx)^2)^{-(e\lambda/2)}$
$\quad + \mathbf{e}_2(b_\mu cx - c_\mu bx)\lambda((bx)^2 + (cx)^2)^{-1}$

9. $\mathbf{A}_\mu = \left[\mathbf{e}_2\left(\frac{1}{8}(d_\mu - k_\mu(kx)^2) + \frac{1}{2}b_\mu kx\right) + \mathbf{e}_3 c_\mu\right]\lambda\,\mathrm{sn}\left(\frac{e\sqrt{2}}{8}\lambda(4bx\right.$
$\quad \left.+ (kx)^2)\right)\mathrm{dn}\left(\frac{e\sqrt{2}}{8}\lambda(4bx + (kx)^2)\right)\left(\mathrm{cn}\left(\frac{e\sqrt{2}}{8}\lambda(4bx + (kx)^2)\right)\right)^{-1}$

$$10. \quad \mathbf{A}_\mu = \left[\mathbf{e}_2\left(\frac{1}{8}(d_\mu - k_\mu(kx)^2) + \frac{1}{2}b_\mu kx\right) + \mathbf{e}_3 c_\mu\right] \times \lambda\left[\operatorname{cn}\left(\frac{e\sqrt{2}\lambda}{8}(4bx + (kx)^2)\right)\right]^{-1}$$

$$11. \quad \mathbf{A}_\mu = \left[\mathbf{e}_2\left(\frac{1}{8}(d_\mu - k_\mu(kx)^2\right) + \frac{1}{2}b_\mu kx) + \mathbf{e}_3 c_\mu\right] \times 4\sqrt{2}(e(4bx + (kx)^2) - \lambda)^{-1}$$

$$12. \quad \mathbf{A}_\mu = \mathbf{e}_1 k_\mu \sqrt{4bx + (kx)^2} Z_{1/4}\left(\frac{ie\lambda}{8}(4bx + (kx)^2)^2\right) + \mathbf{e}_2 c_\mu \lambda(4bx + (kx)^2)$$

$$13. \quad \mathbf{A}_\mu = \mathbf{e}_1 k_\mu \left(\lambda \cosh\left(\frac{e\lambda}{4}(4bx + (kx)^2\right)\right) + \lambda_2 \sinh\left(\frac{e\lambda}{4}(4bx + (kx)^2))\right) + \mathbf{e}_2 c_\mu \lambda$$

$$\begin{aligned} 14. \quad \mathbf{A}_\mu = {} & \left\{\mathbf{e}_2\left(d_\mu - \frac{1}{8}k_\mu(kx)^2 - \frac{1}{2}b_\mu kx\right) + \mathbf{e}_3\left(\alpha c_\mu + b_\mu + \frac{1}{2}k_\mu kx\right)(1 + \alpha^2)^{-(1/2)}\right\} \\ & \times \lambda \operatorname{sn}\left[\frac{e\lambda\sqrt{2}}{8}(4(\alpha bx - cx) + \alpha(kx)^2)(1 + \alpha^2)^{-(1/2)}\right] \\ & \times \operatorname{dn}\left[\frac{e\lambda\sqrt{2}}{8}(4(\alpha bx - cx) + \alpha(kx)^2)(1 + \alpha^2)^{-(1/2)}\right] \\ & \times \left\{\operatorname{cn}\left[\frac{e\lambda\sqrt{2}}{8}((4\alpha bx - cx) + \alpha(kx)^2)(1 + \alpha^2)^{-(1/2)}\right]\right\}^{-1} \end{aligned} \tag{84}$$

$$\begin{aligned} 15. \quad \mathbf{A}_\mu = {} & \left\{\mathbf{e}_2\left(d_\mu - \frac{1}{8}k_\mu(kx)^2) - \frac{1}{2}b_\mu kx\right) + \mathbf{e}_3\left(\alpha c_\mu + b_\mu + \frac{1}{2}k_\mu kx\right)(1 + \alpha^2)^{-(1/2)}\right\} \\ & \times \left\{\operatorname{cn}\left[\frac{e\lambda}{4}(4\alpha bx - cx) + \alpha(kx)^2(1 + \alpha^2)^{-(1/2)}\right]\right\}^{-1} \end{aligned}$$

$$\begin{aligned} 16. \quad \mathbf{A}_\mu = {} & \left\{\mathbf{e}_2\left(d_\mu - \frac{1}{8}k_\mu(kx)^2 - \frac{1}{2}b_\mu kx\right) + \mathbf{e}_3\left(\alpha c_\mu + b_\mu + \frac{1}{2}k_\mu kx\right)(1 + \alpha^2)^{-(1/2)}\right\} \\ & \times 4\sqrt{2}(1 + \alpha^2)^{1/2}[e(4(\alpha bx - cx) + \alpha(kx)^2)]^{-1} \end{aligned}$$

17. $$\mathbf{A}_\mu = \mathbf{e}_1 k_\mu \left\{ \sqrt{4(\alpha bx - cx) + \alpha(kx)^2} Z_{1/4}\left(\frac{ie\lambda}{8}(4(\alpha bx - cx) + \alpha(kx)^2)^2 \right)(1+\alpha^2)^{-(1/2)} \right\} + \mathbf{e}_2\left(\alpha c_\mu + b_\mu + \frac{1}{2}k_\mu kx \right)\lambda(4(\alpha bx - cx) + \alpha(kx)^2)(1+\alpha^2)^{-(1/2)}$$

18. $$\mathbf{A}_\mu = \mathbf{e}_1 k_\mu \left\{ \operatorname{cn}\left[\frac{e\lambda}{4}(1+\alpha^2)^{-(1/2)}(4(\alpha bx - cx) + \alpha(kx)^2) \right] + \lambda_2 \sinh\left[\frac{e\lambda}{4}(1+\alpha^2)^{-(1/2)}(4(\alpha bx - cx) + \alpha(kx)^2 \right] \right\} + \mathbf{e}_2\left(\alpha c_\mu + b_\mu + \frac{1}{2}k_\mu kx \right)\lambda(1+\alpha^2)^{-(1/2)}$$

19. $$\mathbf{A}_\mu = \mathbf{e}_1 k_\mu |kx|^{-1} Z_0\left[\frac{ie\lambda}{2}((bx)^2 + (cx)^2) \right] + \mathbf{e}_2(b_\mu cx - c_\mu bx)\lambda$$

20. $$\mathbf{A}_\mu = \mathbf{e}_1 k_\mu |kx|^{-1}[\lambda_1((bx)^2 + (cx)^2)^{(e\lambda/2)} + \lambda((bx)^2 + (cx)^2)^{-(e\lambda/2)}] + \mathbf{e}_2(b_\mu cx - c_\mu bx)\lambda((bx)^2 + (cx)^2)^{-1}$$

21. $$\mathbf{A}_\mu = \mathbf{e}_1 k_\mu |kx|^{-1}\sqrt{cx}Z_{1/4}\left(\frac{ie\lambda}{2}(cx)^2 \right) + \mathbf{e}_2(b_\mu - k_\mu bx(kx)^{-1})\lambda cx$$

22. $$\mathbf{A}_\mu = \mathbf{e}_1 k_\mu |kx|^{-1}[\lambda_1 \cosh(\lambda ecx) + \lambda_2 \sinh(\lambda ecx)] + \mathbf{e}_2(b_\mu - k_\mu bx(kx)^{-1})\lambda$$

23. $$\mathbf{A}_\mu = \mathbf{e}_1 k_\mu |kx|^{-1}\sqrt{\ln|kx| - cx}Z_{1/4}\left(\frac{ie\lambda}{2}(\ln|kx| - cx)^2 \right) + \mathbf{e}_2(b_\mu - k_\mu bx(kx)^{-1})\lambda(\ln|kx| - cx)$$

24. $$\mathbf{A}_\mu = \mathbf{e}_1 k_\mu |kx|^{-1}[\lambda_1 \cosh(\lambda e(\ln|kx| - cx)) + \lambda_2 \sinh(\lambda e(\ln|kx| - cx))] + \mathbf{e}_2[b_\mu - k_\mu bx(kx)^{-1}]\lambda$$

25. $$\mathbf{A}_\mu = \mathbf{e}_1 k_\mu |kx|^{-1}\sqrt{\alpha\ln|kx| - cx}Z_{1/4}\left(\frac{ie\lambda}{2}(\alpha\ln|kx| - cx)^2 \right) + \mathbf{e}_2(b_\mu - k_\mu bx - \ln|kx|)(kx)^{-1})\lambda(\alpha\ln|kx| - cx)$$

26. $$\mathbf{A}_\mu = \mathbf{e}_1 k_\mu |kx|^{-1}[\lambda_1 \cosh(\lambda e(\alpha\ln|kx| - cx)) + \lambda_2 \sinh(\lambda e(\alpha\ln|kx| - cx))] + \mathbf{e}_2(b_\mu - k_\mu(bx - \ln|kx|^{-1})(kx)^{-1})\lambda$$

27. $\mathbf{A}_\mu = \{\mathbf{e}_1(b_\mu - k_\mu bx(kx)^{-1}) + \mathbf{e}_2(c_\mu - k_\mu cx(kx)^{-1})\}e^{-1}(x_\nu x^\nu)^{-(1/2)}$

28. $\mathbf{A}_\mu = \{\mathbf{e}_1(b_\mu - k_\mu bx(kx)^{-1}) + \mathbf{e}_2(c_\mu - k_\mu cx(kx)^{-1})\}f(x_\nu x^\nu)$

In these formulas the symbol $Z_\alpha(\omega)$ stands for the Bessel function, $\mathrm{sn}\,(\omega), \mathrm{dn}\,(\omega), \mathrm{cn}\,(\omega)$ are the Jacobi elliptic functions having the module $\frac{\sqrt{2}}{2}$; $f(x_\nu x^\nu)$ is the general solution of the ordinary differential equation

$$\omega^2\ddot{f} + 2\omega\dot{f} + \frac{e^2}{4}f^3 = 0$$

and $\lambda, \lambda_1, \lambda_2$ are arbitrary real constants.

IV. CONDITIONAL SYMMETRY AND NEW SOLUTIONS OF THE YANG–MILLS EQUATIONS

With all the wealth of exact solutions obtainable through Lie symmetries of the Yang–Mills equations, it is possible to construct solutions that cannot be derived by the symmetry reduction method. The source of these solutions is *conditional* or *nonclassical* symmetry of the Yang–Mills equations.

The first paper devoted to nonclassical symmetry of partial differential equations was published by Bluman and Cole [57]. However, the real importance of these symmetries was understood much later after the explanations given in several papers [31,32,58–61] where the method of conditional symmetries had been used in order to construct new exact solutions of a number of nonlinear partial differential equations.

The methods for dimensional reduction of partial differential equations based on their conditional symmetry can be conventionally classified into two principal groups. The first group is formed by the direct methods (the ansatz method by Fushchych and the direct method by Clarkson and Kruskal [60]), relying on a special ad hoc representation of the solution to be found in the form of the ansatz containing some arbitrary elements (functions) $f_1, f_2, \ldots, f_n$ and unknown functions $\varphi_1, \varphi_2, \ldots, \varphi_m$ with fewer dependent variables. Inserting the ansatz in question into the equation under study and requiring the relation obtained to be equivalent to a system of partial differential equations for the functions $\varphi_1, \varphi_2, \ldots, \varphi_m$ yield nonlinear determining equations for the functions $f_1, f_2, \ldots, f_n$. Solution of the latter yields a number of ansatzes reducing a given partial differential equation to an equation with fewer dependent variables. The second group of methods (the nonclassical method by Bluman and Cole, the method of conditional symmetries by Fushchych, and the method of side conditions by Olver and Rosenau [58]) may be regarded as infinitesimal ones.

They are in line with the traditional Lie approach to the reduction of partial differential equations, since they exploit symmetry properties of the equation under study in order to construct its invariant solutions. And again, any deviation from the standard Lie approach requires solving overdetermined system of nonlinear determining equations. A more profound analysis of similarities and differences between these approaches can be found elsewhere [33,56,64].

So the principal idea of the method of ansatzes, as well as of the direct method of reduction of partial differential equations is a special choice of the class of functions to which the solution to be found should belong. Within the framework of the preceding methods, a solution of system (46) is sought in the form

$$\mathbf{A}_\mu = H_\mu(x, \mathbf{B}_\nu(\omega(x))), \quad \mu = 0, 1, 2, 3$$

where H_μ are smooth functions chosen in such a way that substitution of the above expressions into the Yang–Mills equations yields a system of ordinary differential equations for new unknown vector functions $\mathbf{B}_\nu$ of one variable ω. However, when posed in this way, the problem of reduction of the Yang–Mills equations seems to be hopeless. Indeed, even if we restrict ourselves to the case of a linear dependence of the above ansatz on B_ν

$$\mathbf{A}_\mu(x) = R_{\mu\nu}(x)\mathbf{B}^\nu(\omega) \tag{85}$$

where $\mathbf{B}_\nu(\omega)$ are new unknown vector functions and $\omega = \omega(x)$ is the new independent variable, then the requirement of reduction of (46) to a system of ordinary differential equations by virtue of (85) gives rise to the system of nonlinear partial differential equations for 17 unknown functions $R_{\mu\nu}, \omega$. Moreover, the system obtained is not at all simpler than the initial Yang–Mills equations (46). Consequently, some additional information about the structure of the matrix function $R_{\mu\nu}$ should be input into ansatz (85). This can be done in various ways, but the most natural one is to use the information about the structure of solutions provided by the Lie symmetry of the equation under study.

In a previous work [33] we suggest an effective approach to study of conditional symmetry of the nonlinear Dirac equation based on its Lie symmetry. We have observed that all the Poincaré-invariant ansatzes for the Dirac field $\psi(x)$ can be represented in the unified form by introducing several arbitrary elements (functions) $u_1(x),\ u_2(x), \ldots, u_N(x)$. As a result, we get an ansatz for the field $\psi(x)$ that reduces the nonlinear Dirac equation to system of ordinary differential equations, provided functions $u_i(x)$ satisfy some compatible over-determined system of nonlinear partial differential equations. After integrating it, we have obtained a number of new ansatzes that cannot in principle be obtained within the framework of the classical Lie approach.

Here, following Ref. 49, we will show that the same idea proves to work efficiently for obtaining new (non-Lie) reductions of the Yang–Mills equations and for constructing new exact solutions of system (46).

A. Nonclassical Reductions of the Yang–Mills Equations

In the previous section we gave a complete list of $P(1,3)$-inequivalent ansatzes for the Yang–Mills field, which are invariant under the three-parameter subgroups of the Poincaré group $P(1,3)$. These ansatzes can be represented in the unified form (53), where $\mathbf{B}_\nu(\omega)$ are new unknown vector functions, $\omega = \omega(x)$ is the new independent variable, and the functions $a_{\mu\nu}(x)$ are given by (54).

In (54), $\theta_\mu(x)$ are some smooth functions, and $\theta_a = \theta_a(\xi, b_\mu x^\mu, c_\mu x^\mu)$, $a = 1,2$; $\xi = (\frac{1}{2})k_\mu x^\mu = (\frac{1}{2})(a_\mu x^\mu + d_\mu x^\mu)$; $a_\mu, b_\mu, c_\mu, d_\mu$ are arbitrary constants satisfying relations (50).

The choice of the functions $\omega(x), \theta_\mu(x)$ is determined by the requirement that substitution of ansatz (53) into the Yang–Mills equations yield a system of ordinary differential equations for the vector function $\mathbf{B}_\mu(\omega)$. By direct check, one can prove the validity of the following statement [33,49].

Assertion 10. *Ansatz* (53),(54) *reduces the Yang–Mills equations* (46) *to a system of ordinary differential equations, if and only if the functions* $\omega(x), \theta_\mu(x)$ *satisfy the following system of partial differential equations*:

$$
\begin{aligned}
&1.\quad \omega_{x_\mu}\omega_{x^\mu} = F_1(\omega)\\
&2.\quad \Box\omega = F_2(\omega)\\
&3.\quad a_{\alpha\mu}\omega_{x_\alpha} = G_\mu(\omega)\\
&4.\quad a_{\alpha\mu x_\alpha} = H_\mu(\omega)\\
&5.\quad a^\alpha_\mu a_{\alpha\nu x_\beta}\omega_{x^\beta} = Q_{\mu\nu}(\omega)\\
&6.\quad a^\alpha_\mu \Box a_{\alpha\nu} = S_{\mu\nu}(\omega)\\
&7.\quad a^\alpha_\mu a_{\alpha\nu x_\beta}a_{\beta\gamma} + a^\alpha_\nu a_{\alpha\gamma x_\beta}a_{\beta\mu} + a^\alpha_\gamma a_{\alpha\mu x_\beta}a_{\beta\nu} = T_{\mu\nu\gamma}(\omega)
\end{aligned}
\tag{86}
$$

where $F_1, F_2, G_\mu, \ldots, T_{\mu\nu\gamma}$ *are some smooth functions of* ω, $\mu, \nu, \gamma = 0,1,2,3$ *and the reduced system has the form*

$$
\begin{aligned}
&k_{\mu\gamma}\ddot{\mathbf{B}}^\gamma + l_{\mu\gamma}\dot{\mathbf{B}}^\gamma + m_{\mu\gamma}\mathbf{B}^\gamma + eq_{\mu\nu\gamma}\dot{\mathbf{B}}^\nu \times \mathbf{B}^\gamma + eh_{\mu\nu\gamma}\mathbf{B}^\nu \times \mathbf{B}^\gamma\\
&\quad + e^2\mathbf{B}_\gamma \times (\mathbf{B}^\gamma \times \mathbf{B}_\mu) = \mathbf{0}
\end{aligned}
\tag{87}
$$

where

$$
\begin{aligned}
k_{\mu\gamma} &= g_{\mu\gamma}F_1 - G_\mu G_\gamma\\
l_{\mu\gamma} &= g_{\mu\gamma}F_2 + 2Q_{\mu\gamma} - G_\mu H_\gamma - G_\mu \dot{G}_\gamma
\end{aligned}
$$

$$
\begin{aligned}
m_{\mu\gamma} &= S_{\mu\gamma} - G_\mu \dot{H}_\gamma \\
q_{\mu\nu\gamma} &= g_{\mu\gamma}G_\nu + g_{\nu\gamma}G_\mu - 2g_{\mu\nu}G_\gamma \\
h_{\mu\nu\gamma} &= \frac{1}{2}(g_{\mu\gamma}H_\nu - g_{\mu\nu}H_\gamma) - T_{\mu\nu\gamma}
\end{aligned}
\tag{88}
$$

Consequently, to describe all the ansatzes of the form (53),(54) reducing the Yang–Mills equations to a system of ordinary differential equations, one has to construct the general solution of the overdetermined system of partial differential equations (54),(86). Let us emphasize that system (54),(86) is compatible since the ansatzes for the Yang–Mills field $\mathbf{A}_\mu(x)$ invariant under the three-parameter subgroups of the Poincaré group satisfy equations (54),(86) with some specific choice of the functions $F_1, F_2, \ldots, T_{\mu\nu\gamma}$ [35].

Integration of systems of nonlinear partial differential equations (54),(86) has been performed [33,49]. Here we indicate the principal steps of the integration procedure. While integrating (54),(86), we essentially apply the fact that the general solution of system of equations 1,2 from (86) is known [62]. With already known $\omega(x)$ in hand, we proceed to integrating linear partial differential equations 3,4 from (86). Next, we insert the results obtained into the remaining equations and get the final forms of the functions $\omega(x)$, $\theta_\mu(x)$.

Before presenting the results of the integration of the system of partial differential equations (54),(86), we make the following remark. As direct check shows, the structure of ansatz (53),(54) is not altered by the change of variables

$$
\begin{aligned}
&\omega \to \omega' = T(\omega), \quad \theta_0 \to \theta_0' = \theta_0 + T_0(\omega) \\
&\theta_1 \to \theta_1' = \theta_1 + e^{\theta_0}(T_1(\omega)\cos\theta_3 + T_2(\omega)\sin\theta_3) \\
&\theta_2 \to \theta_2' = \theta_2 + e^{\theta_0}(T_2(\omega)\cos\theta_3 - T_1(\omega)\sin\theta_3) \\
&\theta_3 \to \theta_3' = \theta_3 + T_3(\omega)
\end{aligned}
\tag{89}
$$

where $T(\omega)$, $T_\mu(\omega)$ are arbitrary smooth functions. That is why solutions of system (54),(86) connected by relations (89) are considered as equivalent.

Integrating the system of partial differential equations under study within the equivalence relations above, we obtain a set of ansatzes containing those equivalent to the Poincaré-invariant ansatzes obtained in the previous section. That is why we concentrate on essentially new (non-Lie) ansatzes. It so happens that our approach gives rise to non-Lie ansatzes, provided the functions $\omega(x), \theta_\mu(x)$ within the equivalence relations (89) have the form

$$
\theta_\mu = \theta_\mu(\xi, bx, cx), \quad \omega = \omega(\xi, bx, cx) \tag{90}
$$

where, as earlier, $bx = b_\mu x^\mu, cx = c_\mu x^\mu$.

The list of inequivalent solutions of the system of partial differential equations (54),(86) satisfying (90) is completed with the following solutions:

$$
\begin{aligned}
1.\quad & \theta_0 = \theta_3 = 0, \quad \omega = \frac{1}{2}kx, \quad \theta_1 = w_0(\xi)bx + w_1(\xi)cx \\
& \theta_2 = w_2(\xi)bx + w_3(\xi)cx \\
2.\quad & \omega = bx + w_1(\xi), \quad \theta_0 = \alpha(cx + w_2(\xi)) \\
& \theta_a = -\left(\frac{1}{4}\right)\dot{w}_a(\xi), \quad a = 1, 2, \quad \theta_3 = 0 \\
3.\quad & \theta_0 = T(\xi), \quad \theta_3 = w_1(\xi), \quad \omega = bx \cos w_1 + cx \sin w_1 + w_2(\xi) \\
& \theta_1 = \left(\left(\frac{1}{4}\right)(\varepsilon e^T + \dot{T})(bx \sin w_1 - cx \cos w_1) + w_3(\xi)\right) \sin w_1 \\
& \quad + \left(\frac{1}{4}\right)(\dot{w}_1(bx \sin w_1 - cx \cos w_1) - \dot{w}_2)\cos w_1 \\
& \theta_2 = -\left(\left(\frac{1}{4}\right)(\varepsilon e^T + \dot{T})(bx \sin w_1 - cx \cos w_1) + w_3(\xi)\right) \cos w_1 \\
& \quad + \left(\frac{1}{4}\right)(\dot{w}_1(bx \sin w_1 - cx \cos w_1) - \dot{w}_2) \sin w_1 \\
4.\quad & \theta_0 = 0, \quad \theta_3 = \arctan\left([cx + w_2(\xi)][bx + w_1(\xi)]^{-1}\right) \\
& \theta_a = -\left(\frac{1}{4}\right)\dot{w}_a(\xi), \quad a = 1, 2 \\
& \omega = \left([bx + w_1(\xi)]^2 + [cx + w_2(\xi)]^2\right)^{1/2}
\end{aligned}
\tag{91}
$$

Here $\alpha \neq 0$ is an arbitrary constant, $\varepsilon = \pm 1$; w_0, w_1, w_2, w_3 are arbitrary smooth functions on $\xi = \frac{1}{2}kx$; and $T = T(\xi)$ is a solution of the nonlinear ordinary differential equation

$$(\dot{T} + \varepsilon e^T)^2 + \dot{w}_1^2 = \varkappa e^{2T}, \varkappa \in \mathbf{R} \tag{92}$$

a dot over the symbol denotes differentiation with respect to ξ.

Inserting ansatz (53), where $a_{\mu\nu}(x)$ are given by formulas (54) and (91), into the Yang–Mills equations yields systems of nonlinear ordinary differential equations of the form (87), where

$$
\begin{aligned}
1.\quad & k_{\mu\gamma} = -\frac{1}{4}k_\mu k_\gamma, \quad l_{\mu\gamma} = -(w_0 + w_3)k_\mu k_\gamma \\
& m_{\mu\gamma} = -4\,(w_0^2 + w_1^2 + w_2^2 + w_3^2)k_\mu k_\gamma - (\dot{w}_0 + \dot{w}_3)k_\mu k_\gamma
\end{aligned}
$$

$$q_{\mu\nu\gamma} = \frac{1}{2}(g_{\mu\gamma}k_\nu + g_{\nu\gamma}k_\mu - 2g_{\mu\nu}k_\gamma)$$

$$h_{\mu\nu\gamma} = (w_0 + w_3)(g_{\mu\gamma}k_\nu - g_{\mu\nu}k_\gamma) + 2(w_1 - w_2)((k_\mu b_\nu - k_\nu b_\mu)c_\gamma + (b_\mu c_\nu - b_\nu c_\mu)k_\gamma + (c_\mu k_\nu - c_\nu k_\mu)b_\gamma)$$

2. $k_{\mu\gamma} = -g_{\mu\gamma} - b_\mu b_\gamma, \quad l_{\mu\gamma} = 0, \quad m_{\mu\gamma} = -\alpha^2(a_\mu a_\gamma - d_\mu d_\gamma)$

$$q_{\mu\nu\gamma} = g_{\mu\gamma}b_\nu + g_{\nu\gamma}b_\mu - 2g_{\mu\nu}b_\gamma$$

$$h_{\mu\nu\gamma} = \alpha((a_\mu d_\nu - a_\nu d_\mu)c_\gamma + (d_\mu c_\nu - d_\nu c_\mu)a_\gamma + (c_\mu a_\nu - c_\nu a_\mu)d_\gamma)$$

3. $k_{\mu\gamma} = -g_{\mu\gamma} - b_\mu b_\gamma, \quad l_{\mu\gamma} = -\left(\frac{\varepsilon}{2}\right)b_\mu k_\gamma \qquad (93)$

$$m_{\mu\gamma} = -\left(\frac{\varkappa}{4}\right)k_\mu k_\gamma, \quad q_{\mu\nu\gamma} = g_{\mu\gamma}b_\nu + g_{\nu\gamma}b_\mu - 2g_{\mu\nu}b_\gamma$$

$$h_{\mu\nu\gamma} = \left(\frac{\varepsilon}{4}\right)(g_{\mu\gamma}k_\nu - g_{\mu\nu}k_\gamma)$$

4. $k_{\mu\gamma} = -g_{\mu\gamma} - b_\mu b_\gamma, \quad l_{\mu\gamma} = -\omega^{-1}(g_{\mu\gamma} + b_\mu b_\gamma)$

$$m_{\mu\gamma} = -\omega^{-2}c_\mu c_\gamma, \quad q_{\mu\nu\gamma} = g_{\mu\gamma}b_\nu + g_{\nu\gamma}b_\mu - 2g_{\mu\nu}b_\gamma$$

$$h_{\mu\nu\gamma} = \frac{1}{2}\omega^{-1}(g_{\mu\gamma}b_\nu - g_{\mu\nu}b_\gamma)$$

B. Exact Solutions

Systems (87) and (91) contain 12 nonlinear second-order ordinary differential equations with variable coefficients. That is why there is little hope for constructing their general solutions. Nevertheless, it is possible to obtain particular solutions of system (87), whose coefficients are given by formulas 2–4 from (91).

Consider, as an example, system of ordinary differential equations (87) with coefficients given by the formula 2 from (93). We look for its solutions of the form

$$\mathbf{B}_\mu = k_\mu \mathbf{e}_1 f(\omega) + b_\mu \mathbf{e}_2 g(\omega), \quad fg \neq 0, \tag{94}$$

where $\mathbf{e}_1 = (1, 0, 0), \ \mathbf{e}_2 = (0, 1, 0)$.

Substituting expression (94) into the abovementioned system, we get

$$\ddot{f} + (\alpha^2 - e^2 g^2)f = 0, \quad f\dot{g} + 2\dot{f}g = 0 \tag{95}$$

The second ordinary differential equation from (4.11) is easily integrated

$$g = \lambda f^{-2}, \quad \lambda \in \mathbf{R}, \quad \lambda \neq 0 \tag{96}$$

Inserting the result obtained into the first ordinary differential equation from (95) yields the Ermakov-type equation for $f(\omega)$

$$\ddot{f} + \alpha^2 f - e^2\lambda^2 f^{-3} = 0,$$

which is integrated in elementary functions [63]

$$f = (\alpha^{-2}C^2 + \alpha^{-2}(C^4 - \alpha^2 e^2\lambda^2)^{1/2}\sin 2|\alpha|\omega)^{1/2} \tag{97}$$

Here $C \neq 0$ is an arbitrary constant.

Substituting (94),(96),(97) into the corresponding ansatz for $\mathbf{A}_\mu(x)$, we get the following class of exact solutions of the Yang-Mills equations (46):

$$\begin{aligned}\mathbf{A}_\mu = {} & \mathbf{e}_1 k_\mu \exp\,(-\alpha cx - \alpha w_2)(\alpha^{-2}C^2 + \alpha^{-2}(C^4 - \alpha^2 e^2\lambda^2)^{1/2} \\ & \times \sin 2|\alpha|(bx + w_1))^{1/2} + \mathbf{e}_2\lambda(\alpha^{-2}C^2 + \alpha^{-2}(C^4 - \alpha^2 e^2\lambda^2)^{1/2} \\ & \times \sin 2|\alpha|(bx + w_1))^{-1}\left(b_\mu + \frac{1}{2}k_\mu \dot{w}_1\right)\end{aligned}$$

In a similar way we have obtained the five other classes of exact solutions of the Yang–Mills equations

$$\begin{aligned}\mathbf{A}_\mu = {} & \mathbf{e}_1 k_\mu e^{-T}(bx\cos w_1 + cx\sin w_1 + w_2)^{1/2} Z_{1/4}\left(\left(\frac{ie\lambda}{2}\right)(bx\cos w_1\right. \\ & \left. + cx\sin w_1 + w_2)^2\right) + \mathbf{e}_2\lambda\,(bx\cos w_1 + cx\sin w_1 + w_2) \\ & \times\left(c_\mu\cos w_1 - b_\mu\sin w_1 + 2k_\mu\left[\frac{1}{4}(\varepsilon e^T + \dot{T})(bx\sin w_1\right.\right. \\ & \left.\left. - cx\cos w_1) + w_3\right]\right)\end{aligned}$$

$$\begin{aligned}\mathbf{A}_\mu = {} & \mathbf{e}_1 k_\mu e^{-T}(C_1\cosh[e\lambda(bx\cos w_1 + cx\sin w_1 + w_2)] + C_2\sinh[e\lambda \\ & \times (bx\cos w_1 + cx\sin w_1 + w_2)]) + \mathbf{e}_2\lambda\left(c_\mu\cos w_1 - b_\mu\sin w_1\right. \\ & \left. + 2k_\mu\left[\frac{1}{4}(\varepsilon e^T + \dot{T})(bx\sin w_1 - cx\cos w_1) + w_3\right]\right)\end{aligned}$$

$$\begin{aligned}\mathbf{A}_\mu = {} & \mathbf{e}_1 k_\mu e^{-T}(C^2(bx\cos w_1 + cx\sin w_1 + w_2)^2 + \lambda^2 e^2 C^{-2})^{1/2} \\ & + \mathbf{e}_2\lambda(C^2(bx\cos w_1 + cx\sin w_1 + w_2)^2 + \lambda^2 e^2 C^{-2})^{-1} \\ & \times\left(b_\mu\cos w_1 + c_\mu\sin w_1 - \left(\frac{1}{2}\right)k_\mu[\dot{w}_1(bx\sin w_1\right. \\ & \left. - cx\cos w_1) - \dot{w}_2]\right)\end{aligned}$$

$$\mathbf{A}_\mu = \mathbf{e}_1 k_\mu Z_0\left(\left(\frac{ie\lambda}{2}\right)[(bx+w_1)^2+(cx+w_2)^2]\right) + \mathbf{e}_2\lambda\left(c_\mu(bx+w_1)\right.$$
$$\left. - b_\mu(cx+w_2) - \left(\frac{1}{2}\right)k_\mu[\dot{w}_1(cx+w_2) - \dot{w}_2(bx+w_1)]\right)$$
$$\mathbf{A}_\mu = \mathbf{e}_1 k_\mu (C_1[(bx+w_1)^2+(cx+w_2)^2]^{e\lambda/2} + C_2[(bx+w_1)^2$$
$$+ (cx+w_2)^2]^{-e\lambda/2}) + \mathbf{e}_2\lambda[(bx+w_1)^2+(cx+w_2)^2]^{-1}$$
$$\times\left(c_\mu(bx+w_1) - b_\mu(cx+w_2) - \left(\frac{1}{2}\right)k_\mu[\dot{w}_1(cx+w_2)\right.$$
$$\left. - \dot{w}_2(bx+w_1)]\right)$$

Here $C_1, C_2, C \neq 0, \lambda$ are arbitrary parameters; w_1, w_2, w_3 are arbitrary smooth functions on $\xi = \frac{1}{2}kx$; $T = T(\xi)$ is a solution of ordinary differential equation (92). Besides that, we use the following notations:

$$kx = k_\mu x^\mu, \quad bx = b_\mu x^\mu, \quad cx = c_\mu x^\mu$$
$$Z_s(\omega) = C_1 J_s(\omega) + C_2 Y_s(\omega)$$
$$\mathbf{e}_1 = (1,0,0), \quad \mathbf{e}_2 = (0,1,0)$$

where J_s, Y_s are the Bessel functions.

Thus, we have obtained the broad families of exact non-Abelian solutions of the Yang–Mills equations (46). We can verify by direct and rather involved computation that the solutions obtained are not self-dual, that is, that they do not satisfy the self-dual Yang–Mills equations.

C. Conditional Symmetry Formalism

Now we briefly discuss the problem of conditional symmetry interpretation of ansatzes (53), (54), and (91). Consider, as an example, the ansatz determined by formula 1 from (91). As direct computation shows, generators of a three-parameter Lie group G leaving it invariant are of the form

$$\begin{aligned} Q_1 &= k_\alpha \partial_\alpha \\ Q_2 &= b_\alpha\partial_\alpha - 2[w_0(k_\mu b_\nu - k_\nu b_\mu) + w_2(k_\mu c_\nu - k_\nu c_\mu)]\sum_{a=1}^{3} A^{a\nu}\partial_{A^{a\mu}} \\ Q_3 &= c_\alpha\partial_\alpha - 2[w_1(k_\mu b_\nu - k_\nu b_\mu) + w_3(k_\mu c_\nu - k_\nu c_\mu)]\sum_{a=1}^{3} A^{a\nu}\partial_{A^{a\mu}} \end{aligned} \tag{98}$$

Evidently, the system of partial differential equations (46) is invariant under the one-parameter group G_1 having the generator Q_1. However, it is not invariant under the one-parameter groups G_2, G_3 having the generators Q_2, Q_3. Consider, as an example, the generator Q_2. Acting by the second prolongation of the operator Q_2 (which is constructed in the standard way; see, e.g., Refs. 19 and 20) on the system of partial differential equations (46), we see that the resulting expression does not vanish on the solution set of Eqs. (46). However, if we consider the constrained Yang–Mills equations

$$\mathbf{L}_\mu = \mathbf{0}, \quad Q_a \mathbf{A}_\mu = \mathbf{0}, \quad a = 1, 2, 3$$

then we see that the system obtained is invariant under the group G_2. In the preceding formulas we use the designations

$$\begin{aligned}
\mathbf{L}_\mu &\equiv \Box \mathbf{A}_\mu - \partial^\mu \partial_\nu \mathbf{A}_\nu + e((\partial_\nu \mathbf{A}_\nu) \times \mathbf{A}_\mu - 2(\partial_\nu \mathbf{A}_\mu) \times \mathbf{A}_\nu \\
&\quad + (\partial^\mu \mathbf{A}_\nu) \times \mathbf{A}^\nu) + e^2 \mathbf{A}_\nu \times (\mathbf{A}^\nu \times \mathbf{A}_\mu) \\
Q_1 \mathbf{A}_\mu &\equiv k_\alpha \partial_\alpha \mathbf{A}_\mu \\
Q_2 \mathbf{A}_\mu &\equiv b_\alpha \partial_\alpha \mathbf{A}_\mu + 2(w_0(k_\mu b_\nu - k_\nu b_\mu) + w_2(k_\mu c_\nu - k_\nu c_\mu))\mathbf{A}^\nu \\
Q_3 \mathbf{A}_\mu &\equiv c_\alpha \partial_\alpha \mathbf{A}_\mu + 2(w_1(k_\mu b_\nu - k_\nu b_\mu) + w_3(k_\mu c_\nu - k_\nu c_\mu))\mathbf{A}^\nu
\end{aligned}$$

The same assertion holds for the Lie transformation group G_3 having the generator Q_3. Consequently, the Yang–Mills equations are conditionally-invariant with respect to the three-parameter Lie transformation group $G = G_1 \otimes G_2 \otimes G_3$. This means that solutions of the Yang–Mills equations obtained with the help of the ansatz invariant under the group with generators (98) cannot be found by means of the classical symmetry reduction procedure. We refer the reader interested in further details to two monographs [21,33].

As very cumbersome computations show, the ansatzes determined by formulas 2–4 from (91) also correspond to the conditional symmetry of Yang–Mills equations. Hence it follows, in particular, that Yang–Mills equations should be included in the long list of mathematical and theoretical physics equations possessing nontrivial conditional symmetry [21].

V. SYMMETRY REDUCTION AND EXACT SOLUTIONS OF THE MAXWELL EQUATIONS

In this section we exploit symmetry properties of the (vacuum) Maxwell equations in order to construct their exact solutions.

It is well known that the electromagnetic field for the case of the vanishing current is described by the Maxwell equations in vacuum

$$\operatorname{rot} \mathbf{E} = -\frac{\partial \mathbf{H}}{\partial x_0}, \quad \operatorname{div} \mathbf{H} = 0$$
$$\operatorname{rot} \mathbf{H} = \frac{\partial \mathbf{E}}{\partial x_0}, \quad \operatorname{div} \mathbf{E} = 0$$

for the vector fields $\mathbf{E} = \mathbf{E}(x_0, \mathbf{x})$ and $\mathbf{H} = \mathbf{H}(x_0, \mathbf{x})$ (we shall call them the *Maxwell fields*).

First, we give a brief overview of symmetry properties of Eqs. (99) following [48].

A. Symmetry of the Maxwell Equations

As we have mentioned in the introduction, the maximal symmetry group admitted by Eqs. (99) is the 16-parameter group $C(1,3) \otimes H$. This group is the direct product of the conformal group $C(1,3)$ generated by the Lie vector fields

$$\begin{aligned}
P_\mu &= \partial_{x_\mu}, \quad J_{0a} = x_0\partial_{x_a} + x_a\partial_{x_0} + \varepsilon_{abc}(E_b\partial_{H_c} - H_b\partial_{E_c}) \\
J_{ab} &= x_b\partial_{x_a} - x_a\partial_{x_b} + E_b\partial_{E_a} - E_a\partial_{E_b} + H_b\partial_{H_a} - H_a\partial_{H_b} \\
D &= x_\mu\partial_{x_\mu} - 2(E_a\partial_{E_a} + H_a\partial_{H_a}) \\
K_0 &= 2x_0 D - x_\mu x^\mu\partial_{x_0} + 2x_a\varepsilon_{abc}(E_b\partial_{H_c} - H_b\partial_{E_c}) \\
K_a &= -2x_a D - x_\mu x^\mu\partial_{x_a} - 2x_0\varepsilon_{abc}(E_b\partial_{H_c} - H_b\partial_{E_c}) \\
&\quad - 2H_a(x_b\partial_{H_b}) - 2E_a(x_b\partial_{E_b}) + 2(x_bH_b)\partial_{H_a} + 2(x_bE_b)\partial_{E_a}
\end{aligned} \tag{100}$$

and of the one-parameter Heaviside–Larmor–Rainich group H having the generator

$$Q = E_a\partial_{H_a} - H_a\partial_{E_a} \tag{101}$$

where ε_{abc} is the third-order antisymmetric tensor with $\varepsilon_{123} = 1$. In this section the indices denoted by the Latin alphabet letters a, b, c take the values $1, 2, 3$, and the ones denoted by the Greek alphabet letters take the values $0, 1, 2, 3$, and the summation convention is used.

It is readily seen from (100) and (101) that the action of the group $C(1,3) \otimes H$ in the space $\mathbf{R}^{1,3} \times \mathbf{R}^6$, where $\mathbf{R}^{1,3}$ is Minkowski space of the variables x_0, $\mathbf{x} = (x_1, x_2, x_3)$ and $\mathbf{R}^6$ is the six-dimensional space of the functions $\mathbf{E} = (E_1, E_2, E_3)$, $\mathbf{H} = (H_1, H_2, H_3)$, is projective. Furthermore, the basis generators of this group can be represented in the form (15).

The matrices $S_{\mu\nu}$ read as

$$S_{01} = \begin{pmatrix} 0 & \tilde{S}_{23} \\ -\tilde{S}_{23} & 0 \end{pmatrix}, \qquad S_{02} = \begin{pmatrix} 0 & -\tilde{S}_{13} \\ \tilde{S}_{13} & 0 \end{pmatrix}$$

$$S_{03} = \begin{pmatrix} 0 & \tilde{S}_{12} \\ -\tilde{S}_{12} & 0 \end{pmatrix}, \qquad S_{12} = \begin{pmatrix} \tilde{S}_{12} & 0 \\ 0 & \tilde{S}_{12} \end{pmatrix}$$

$$S_{13} = \begin{pmatrix} \tilde{S}_{13} & 0 \\ 0 & \tilde{S}_{13} \end{pmatrix}, \qquad S_{23} = \begin{pmatrix} \tilde{S}_{23} & 0 \\ 0 & \tilde{S}_{23} \end{pmatrix}$$

where 0 is the zero 3×3 matrix and

$$\tilde{S}_{12} = \begin{pmatrix} 0 & -1 & 0 \\ 1 & 0 & 0 \\ 0 & 0 & 0 \end{pmatrix}, \quad \tilde{S}_{13} \begin{pmatrix} 0 & 0 & -1 \\ 0 & 0 & 0 \\ 1 & 0 & 0 \end{pmatrix}, \quad \tilde{S}_{23} = \begin{pmatrix} 0 & 0 & 0 \\ 0 & 0 & -1 \\ 0 & 1 & 0 \end{pmatrix}$$

E is the unit 6×6 matrix. The matrix $-A$ corresponding to operator Q (101) is given by the formula

$$A = \begin{pmatrix} 0 & -I \\ I & 0 \end{pmatrix} \tag{103}$$

where 0 and I are zero and unit 3×3 matrices, correspondingly.

Hence, it follows that $C(1,3) \otimes H$-invariant ansatzes for the Maxwell fields, which reduce (99) to systems of ordinary differential equations, can be represented in the form (22), namely,

$$\mathbf{V} = \Lambda(x_0, \mathbf{x})\tilde{\mathbf{V}}(\omega) \tag{104}$$

with

$$\mathbf{V} = \begin{pmatrix} E_1 \\ E_2 \\ E_3 \\ H_1 \\ H_2 \\ H_3 \end{pmatrix}, \quad \tilde{\mathbf{V}} = \begin{pmatrix} \tilde{E}_1 \\ \tilde{E}_2 \\ \tilde{E}_3 \\ \tilde{H}_1 \\ \tilde{H}_2 \\ \tilde{H}_3 \end{pmatrix}$$

Here $\Lambda(x_0, \mathbf{x})$ is the 6×6 matrix, which is nonsingular in some open domain of the space $\mathbf{R}^{1,3}$ and $\tilde{E}_a = \tilde{E}_a(\omega)$, $\tilde{H}_a = \tilde{H}_a(\omega)$ are new unknown functions of the variable $\omega = \omega(x_0, \mathbf{x})$.

In addition, the Maxwell equations admit the following discrete symmetry group [48]:

$$\Psi\colon \bar{x}_\mu = -x_\mu, \quad \bar{\mathbf{E}} = -\mathbf{E}, \quad \bar{\mathbf{H}} = -\mathbf{H} \tag{105}$$

The transformation properties of operators (100), (101) with respect to the action of the group Ψ read as

$$P_\mu \to -P_\mu, \quad J_{\mu\nu} \to J_{\mu\nu}, \quad D \to D, \quad K_\mu \to -K_\mu, \quad Q \to Q$$

so that actions of discrete symmetry groups Ψ (5.7) and Φ_1 from Table I on the basis operators of the algebra $\tilde{p}(1,3)$ coincide. Therefore, we can use Assertions 5 and 6 and choose the parameter j to be equal to 2, namely, $(-1)^j = 1$.

In what follows we exploit invariance of the Maxwell equations under the conformal group $C(1,3)$ in order to construct their invariant solutions.

B. Conformally Invariant Ansatzes for the Maxwell Fields

First we will give two assertions that substantially simplify the full description of invariant solutions of the Maxwell equations.

Assertion 11. *If* $\mathbf{E} = \mathbf{E}(x_0, x_3)$, $\mathbf{H} = \mathbf{H}(x_0, x_3)$, *then it is possible to construct the general solution of equations (5.1). It has the form*

$$\begin{aligned} E_1 &= \varphi_1(\xi) + \psi_1(\eta), & H_1 &= -\varphi_2(\xi) + \psi_2(\eta) \\ E_2 &= \varphi_2(\xi) + \psi_2(\eta), & H_2 &= \varphi_1(\xi) - \psi_1(\eta) \\ E_3 &= C_1, & H_3 &= C_2 \end{aligned}$$

where $\varphi_1, \varphi_2, \psi_1, \psi_2$ *are arbitrary smooth functions*; $\xi = x_0 - x_3$, $\eta = x_0 + x_3$; $C_1, C_2 \in \mathbf{R}$.

Assertion 12. *If* $\mathbf{E} = \mathbf{E}(x_1, x_2, \xi)$, $\mathbf{H} = \mathbf{H}(x_1, x_2, \xi)$, *where* $\xi = \frac{1}{2}(x_0 - x_3)$, *then it is possible to construct the general solution of the Maxwell equations* (99). *It is given by the following formulas*:

$$\begin{aligned} E_1 &= \frac{1}{2}(R + R^* + T_1 + T_1^*), & H_1 &= \frac{1}{2}(iR - iR^* - T_2 - T_2^*) \\ E_2 &= \frac{1}{2}(iR - iR^* + T_2 + T_2^*), & H_2 &= \frac{1}{2}(R + R^* - T_1 - T_1^*) \\ E_3 &= S + S^*, & H_3 &= iS - iS^* \end{aligned}$$

where

$$\begin{aligned} T_j &= \frac{\partial^2 \theta_j}{\partial \xi^2} \quad (j = 1, 2) \\ S &= \frac{\partial \theta_1}{\partial \xi} + i\frac{\partial \theta_2}{\partial \xi} + \lambda(z) \\ R &= -2\left(\frac{\partial \theta_1}{\partial z} + i\frac{\partial \theta_2}{\partial z}\right) + \xi\frac{d\lambda}{dz} \end{aligned}$$

Here $\theta_j = \theta_j(z, \xi)$, $\lambda(z)$ *are arbitrary functions analytic by the variable* $z = x_1 + ix_2$; $j = 1, 2$; *i is the imaginary unit, namely,* $i^2 = -1$.

Proof of these assertions can be found in Refs. 50–53.

It follows from Assertions 11 and 12 that we have to exclude from further consideration those subalgebras of the conformal algebra that yield solutions of the form covered by these assertions. It is straightforward to check that we have to skip subalgebras L of rank 3 fulfilling the conditions

$$\langle P_0 + P_3 \rangle \not\subset L, \quad \langle P_0 - P_3 \rangle \not\subset L, \quad 1\langle P_0, P_3 \rangle \not\subset L, \quad \langle P_1, P_2 \rangle \not\subset L$$

Thus, to get the full description of conformally invariant solutions of the Maxwell equations, it suffices to consider the following subalgebras of the conformal algebra $c(1,3)$ (note, that we have also made use of the discrete symmetry group Ψ in order to simplify their basis elements):

$$\begin{aligned}
M_1 &= \langle J_{03}, G_1, P_2 \rangle \\
M_2 &= \langle G_1, G_2, J_{03} + \alpha J_{12} \rangle, \quad \alpha \in \mathbf{R} \\
M_3 &= \langle J_{12}, D, P_0 \rangle \\
M_4 &= \langle J_{12}, D, P_3 \rangle \\
M_5 &= \langle J_{03}, D, P_1 \rangle \\
M_6 &= \langle J_{03}, J_{12}, D \rangle \\
M_7 &= \langle G_1, J_{03} + \alpha D, P_2 \rangle \quad (0 < |\alpha| \leq 1) \\
M_8 &= \langle J_{03} - D + M, G_1, P_2 \rangle \\
M_9 &= \langle J_{03} + 2D, G_1 + 2T, P_2 \rangle \\
M_{10} &= \langle J_{12}, S + T, Z \rangle \\
M_{11} &= \langle S + T + J_{12}, Z, G_1 + P_2 \rangle \\
M_{12} &= \langle P_2 + K_2 + \sqrt{3}(P_1 + K_1) + K_0 - P_0, J_{02} - D - \sqrt{3}J_{01} \\
&\qquad P_0 + K_0 - 2(K_2 - P_2) \rangle \\
M_{13} &= \langle P_0 + K_0 \rangle \oplus \langle J_{12}, K_3 - P_3 \rangle \\
M_{14} &= \langle 2J_{12} + K_3 - P_3, 2J_{13} - K_2 + P_2, 2J_{23} + K_1 - P_1 \rangle \\
M_{15} &= \langle P_1 + K_1 + 2J_{03}, P_2 + K_2 + K_0 - P_0, 2J_{12} + K_3 - P_3 \rangle
\end{aligned}$$

Here we use the following designations:

$$\begin{aligned}
M &= P_0 + P_3, \quad G_{0j} = J_{0j} - J_{j3} \qquad (j = 1, 2) \\
Z &= J_{03} + D, \quad S = \frac{1}{2}(K_0 + K_3), \quad T = \frac{1}{2}(P_0 - P_3)
\end{aligned}$$

Below, we consider the first 10 subalgebras from the preceding list. For these subalgebras we can represent the matrix Λ from ansatz (104) as

$$\Lambda = \exp\{(\ln\theta)E\} \exp(2\theta_1 H_1) \exp(2\theta_2 H_2) \exp(-\theta_0 S_{03}) \exp(\theta_3 S_{12})$$

where the matrices $S_{\mu\nu}$ have the form (102). Thus, we have

$$\Lambda = \theta\begin{pmatrix} C & G \\ -G & C \end{pmatrix}$$

where

$$C = \begin{pmatrix} \cosh\theta_0\cos\theta_3 - r_1 & -\cosh\theta_0\sin\theta_3 + r_2 & 2\theta_1 \\ \cosh\theta_0\sin\theta_3 + r_2 & \cosh\theta_0\cos\theta_3 + r_1 & 2\theta_2 \\ -2s_1 & 2s_2 & 1 \end{pmatrix}$$

$$G = \begin{pmatrix} \sinh\theta_0\sin\theta_3 + r_2 & \sinh\theta_0\cos\theta_3 + r_1 & 2\theta_2 \\ -\sinh\theta_0\cos\theta_3 + r_1 & \sinh\theta_0\sin\theta_3 - r_2 & -2\theta_1 \\ 2s_2 & 2s_1 & 0 \end{pmatrix}$$

and

$$\begin{aligned} r_1 &= 2[(\theta_1^2 - \theta_2^2)\cos\theta_3 + 2\theta_1\theta_2\sin\theta_3]e^{-\theta_0} \\ r_2 &= 2[(\theta_1^2 - \theta_2^2)\sin\theta_3 - 2\theta_1\theta_2\cos\theta_3]e^{-\theta_0} \\ s_1 &= 2[\theta_1\cos\theta_3 + \theta_2\sin\theta_3]e^{-\theta_0} \\ s_2 &= 2[\theta_1\sin\theta_3 - \theta_2\cos\theta_3]e^{-\theta_0} \end{aligned}$$

After some algebra, we obtain the following form of the conformally invariant ansatz for the Maxwell fields:

$$\begin{aligned} E_1 &= \theta\{(\tilde{E}_1\cos\theta_3 - \tilde{E}_2\sin\theta_3)\cosh\theta_0 \\ &\quad + (\tilde{H}_1\sin\theta_3 + \tilde{H}_2\cos\theta_3)\sinh\theta_0 \\ &\quad + 2\theta_1\tilde{E}_3 + 2\theta_2\tilde{H}_3 + 4\theta_1\theta_2\Sigma_1 + 2(\theta_1^2 - \theta_2^2)\Sigma_2\} \\ E_2 &= \theta\{(\tilde{E}_2\cos\theta_3 + \tilde{E}_1\sin\theta_3)\cosh\theta_0 \\ &\quad + (\tilde{H}_2\sin\theta_3 - \tilde{H}_1\cos\theta_3)\sinh\theta_0 \\ &\quad - 2\theta_1\tilde{H}_3 + 2\theta_2\tilde{E}_3 + 4\theta_1\theta_2\Sigma_2 - 2(\theta_1^2 - \theta_2^2)\Sigma_1\} \\ E_3 &= \theta\{\tilde{E}_3 + 2\theta_1\Sigma_2 + 2\theta_2\Sigma_1\} \\ H_1 &= \theta\{(\tilde{H}_1\cos\theta_3 - \tilde{H}_2\sin\theta_3)\cosh\theta_0 \\ &\quad - (\tilde{E}_1\sin\theta_3 + \tilde{E}_2\cos\theta_3)\sinh\theta_0 \\ &\quad + 2\theta_1\tilde{H}_3 - 2\theta_2\tilde{E}_3 - 4\theta_1\theta_2\Sigma_2 + 2(\theta_1^2 - \theta_2^2)\Sigma_1\} \\ H_2 &= \theta\{(\tilde{H}_2\cos\theta_3 + \tilde{H}_1\sin\theta_3)\cosh\theta_0 \\ &\quad + (\tilde{E}_1\cos\theta_3 - \tilde{E}_2\sin\theta_3)\sinh\theta_0 \\ &\quad + 2\theta_1\tilde{E}_3 + 2\theta_2\tilde{H}_3 + 4\theta_1\theta_2\Sigma_1 + 2(\theta_1^2 - \theta_2^2)\Sigma_2\} \\ H_3 &= \theta\{\tilde{H}_3 + 2\theta_1\Sigma_1 - 2\theta_2\Sigma_2\} \end{aligned} \tag{106}$$

Here

$$\Sigma_1 = [(\tilde{H}_2 - \tilde{E}_1)\sin\theta_3 - (\tilde{E}_2 + \tilde{H}_1)\cos\theta_3]e^{-\theta_0}$$
$$\Sigma_2 = [(\tilde{E}_2 + \tilde{H}_1)\sin\theta_3 + (\tilde{H}_2 - \tilde{E}_1)\cos\theta_3]e^{-\theta_0}$$

The form of the functions θ, θ_μ, ω for each of the subalgebras M_j, $(j = 1, 2, \ldots, 10)$ is obtained from Assertions 5–7 with $k = 2$:

$$M_1:\quad \theta = 1,\quad \theta_0 = -\ln|x_0 - x_3|,\quad \theta_1 = -\frac{1}{2}x_1(x_0 - x_3)^{-1}$$
$$\theta_2 = \theta_3 = 0,\quad \omega = x_0^2 - x_1^2 - x_3^2$$

$$M_2:\quad \theta = 1,\quad \theta_0 = -\ln|x_0 - x_3|,\quad \theta_1 = -\frac{1}{2}x_1(x_0 - x_3)^{-1}$$
$$\theta_2 = -\frac{1}{2}x_2(x_0 - x_3)^{-1},\quad \theta_3 = \alpha\ln|x_0 - x_3|$$
$$\omega = x_0^2 - x_1^2 - x_2^2 - x_3^2,\qquad \alpha \in \mathbf{R}$$

$$M_3:\quad \theta = (x_3)^{-2},\quad \theta_0 = \theta_1 = \theta_2 = 0$$
$$\theta_3 = \arctan\frac{x_2}{x_1},\ \omega = (x_1^2 + x_2^2)x_3^{-2}$$

$$M_4:\quad \theta = (x_0)^{-2},\quad \theta_0 = \theta_1 = \theta_2 = 0$$
$$\theta_3 = \arctan\frac{x_2}{x_1},\quad \omega = (x_1^2 + x_2^2)x_0^{-2}$$

$$M_5:\quad \theta = (x_2)^{-2},\quad \theta_0 = \ln|(x_0 + x_3)x_2^{-1}|,\quad \theta_1 = \theta_2 = \theta_3 = 0$$
$$\omega = (x_0^2 - x_3^2)x_2^{-2}$$

$$M_6:\quad (x_1^2 + x_2^2)^{-1},\quad \theta_0 = -\frac{1}{2}\ln|(x_0 - x_3)(x_0 + x_3)^{-1}|$$
$$\theta_1 = \theta_2 = 0,\quad \theta_3 = \arctan\frac{x_2}{x_1},\quad \omega = (x_1^2 + x_2^2)(x_0^2 - x_3^2)^{-1}$$

M_7: 1.L $\alpha = -1$

$$\theta = (x_0 - x_3)^{-1},\quad \theta_0 = -\frac{1}{2}\ln|x_0 - x_3|$$
$$\theta_1 = -\frac{1}{2}x_1(x_0 - x_3)^{-1},\quad \theta_2 = \theta_3 = 0$$
$$\omega = x_0 + x_3 - x_1^2(x_0 - x_3)^{-1}$$

2. $\alpha \neq -1$

$$\theta = |x_0^2 - x_1^2 - x_3^2|^{-1},\quad \theta_0 = \frac{1}{2\alpha}\ln|x_0^2 - x_1^2 - x_3^2|$$
$$\theta_1 = -\frac{1}{2}x_1(x_0 - x_3)^{-1},\quad \theta_2 = \theta_3 = 0$$
$$\omega = 2\alpha\ln|x_0 - x_3| + (1 - \alpha)\ln|x_0^2 - x_1^2 - x_3^2|$$

$$M_8:\quad \theta = |x_0 - x_3|^{-1}, \quad \theta_0 = -\frac{1}{2}\ln|x_0 - x_3|, \quad \theta_1 = -\frac{1}{2}x_1(x_0 - x_3)^{-1}$$

$$\theta_2 = \theta_3 = 0, \quad \omega = x_0 + x_3 - x_1^2(x_0 - x_3)^{-1} + \ln|x_0 - x_3|$$

$$M_9:\quad \theta = [(x_0 - x_3)^2 - 4x_1]^{-2}, \quad \theta_0 = \frac{1}{2}\ln|(x_0 - x_3)^2 - 4x_1|$$

$$\theta_1 = -\frac{1}{4}(x_0 - x_3), \quad \theta_2 = \theta_3 = 0$$

$$\omega = \left[x_0 + x_3 - x_1(x_0 - x_3) + \frac{1}{6}(x_0 - x_3)^3\right]^2 [(x_0 - x_3)^2 - 4x_1]^{-3}$$

$$M_{10}:\quad \theta = [(x_1 - (x_0 - x_3)x_2)^2(1 + (x_0 - x_3)^2)^{-1}]^{-1}$$

$$\theta_0 = \frac{1}{2}\ln[(x_1 - (x_0 - x_3)x_2)^2(1 + (x_0 - x_3)^2)^{-3}]$$

$$\theta_1 = -\frac{1}{2}(x_2 + (x_0 - x_3)x_1)(1 + (x_0 - x_3)^2)^{-1}$$

$$\theta_2 = \frac{1}{2}(x_1 - (x_0 - x_3)x_2)(1 + (x_0 - x_3)^2)^{-1}$$

$$\begin{aligned}\theta_3 &= -\arctan(x_0 - x_3), \quad \omega = [(x_0 + x_3)(1 + (x_0 - x_3)^2)^2 \\ &- 2x_1(x_2 + (x_0 - x_3)x_1) - (x_0 - x_3)(x_1^2(x_0 - x_3)^2 - x_2^2)] \\ &\times [x_1 - (x_0 - x_3)x_2]^{-2} - x_0 + x_3\end{aligned}$$

C. Exact Solutions of the Maxwell Equations

Now we have to insert ansatzes (106) into (99). However, it is more convenient to rewrite the Maxwell equations (99) in the following equivalent form:

$$\begin{aligned}\partial_{x_1}(E_1 + H_2) + \partial_{x_2}(E_2 - H_1) &= (\partial_{x_0} - \partial_{x_3})E_3 \\ \partial_{x_1}(E_1 - H_2) + \partial_{x_2}(E_2 + H_1) &= -(\partial_{x_0} + \partial_{x_3})E_3 \\ \partial_{x_1}(E_2 - H_1) - \partial_{x_2}(E_1 + H_2) &= -(\partial_{x_0} - \partial_{x_3})H_3 \\ \partial_{x_1}(E_2 + H_1) - \partial_{x_2}(E_1 - H_2) &= -(\partial_{x_0} + \partial_{x_3})H_3 \\ (\partial_{x_0} + \partial_{x_3})(E_1 + H_2) &= \partial_{x_1}E_3 + \partial_{x_2}H_3 \\ (\partial_{x_0} - \partial_{x_3})(E_1 - H_2) &= -\partial_{x_1}E_3 + \partial_{x_2}H_3 \\ (\partial_{x_0} - \partial_{x_3})(E_2 + H_1) &= -\partial_{x_2}E_3 - \partial_{x_1}H_3 \\ (\partial_{x_0} + \partial_{x_3})(E_2 - H_1) &= \partial_{x_2}E_3 - \partial_{x_1}H_3\end{aligned} \tag{107}$$

We will give the calculation details for the case of the subalgebra M_1 only, since the remaining subalgebras are handled in a similar way. For the case in hand, ansatz (106) can be written in the form

$$\begin{aligned} E_1 + H_2 &= fe^{\theta_0} + 4\theta_1 \tilde{E}_3 - 4\theta_1^2 e^{-\theta_0} h \\ E_1 - H_2 &= he^{-\theta_0}, \quad E_2 + H_1 = \rho e^{-\theta_0} \\ E_2 - H_1 &= ge^{\theta_0} - 4\theta_1 \tilde{H}_3 + 4\theta_1^2 e^{-\theta_0} \rho \\ E_3 &= \tilde{E}_3 - 2\theta_1 he^{-\theta_0}, \quad H_3 = \tilde{H}_3 - 2\theta_1 \rho e^{-\theta_0} \end{aligned} \tag{108}$$

where $\theta_0 = -\ln|x_0 - x_3|$, $\theta_1 = -\frac{1}{2}x_1(x_0 - x_3)^{-1}$. The functions $\tilde{E}_3, \tilde{H}_3$, and

$$\begin{aligned} f = f(\omega) = \tilde{E}_1 + \tilde{H}_2, \quad g = g(\omega) = \tilde{E}_2 - \tilde{H}_1 \\ h = h(\omega) = \tilde{E}_1 - \tilde{H}_2, \quad \rho = \rho(\omega) = \tilde{E}_2 + \tilde{H}_1 \end{aligned} \tag{109}$$

are arbitrary smooth functions of the variable $\omega = x_0^2 - x_1^2 - x_3^2$.

Inserting (108) into the second and fourth equations from (107) gives equations

$$\dot{\tilde{E}}_3 = 0, \quad \dot{\tilde{H}}_3 = 0 \tag{110}$$

We remind the reader that the dot over the symbol stands for the derivative with respect to the variable ω.

Similarly, we get from the sixth and seventh equations of system (107) the following reduced equations:

$$2\omega\dot{h} + 3h = 0, \quad 2\omega\dot{\rho} + 3\rho = 0 \tag{111}$$

Next, the fifth and eighth equations give rise to ordinary differential equations of the form

$$2\dot{f} - h = 0, \quad 2\dot{g} + \rho = 0 \tag{112}$$

Finally, substituting ansatz (5.10) into the first and third equations from (107) yields

$$\begin{aligned} 4\varepsilon\theta_1[\omega\dot{h} + h + \dot{f}] = 2\xi^{-1}\tilde{E}_3 \\ 4\varepsilon\theta_1[\dot{g} - \omega\dot{\rho} - \rho] = -2\xi^{-1}\tilde{H}_3 \end{aligned} \tag{113}$$

where $\varepsilon = 1$ for $\xi = x_0 - x_3 > 0$ and $\varepsilon = -1$ for $x_0 - x_3 < 0$.

Taking into account (111)–(113), we see that $\tilde{E}_3 = 0$ and $\tilde{H}_3 = 0$.

Summing up, we conclude that the ansatz invariant with respect to the subalgebra M_1 reduces the Maxwell equations to the following system of ordinary differential equations:

$$
\begin{aligned}
2\omega\dot{h} + 3h = 0, \quad 2\omega\dot{\rho} + 3\rho = 0, \quad 2\dot{f} - h = 0 \\
2\dot{g} + \rho = 0, \qquad \tilde{E}_3 = 0, \qquad \tilde{H}_3 = 0
\end{aligned}
\tag{114}
$$

Below we give the reduced systems for the ansatzes invariant with respect to the remaining subalgebras M_2–M_{10}: [note that the functions f, g, h, ρ are of the form (109)]:

1. System (114)
2. $\dot{f} = 0,\ \tilde{E}_3 = 0,\ \dot{g} = 0,\ \tilde{H}_3 = 0,\ \omega\dot{h} + 2h + \alpha\rho = 0$

   $\omega\dot{\rho} + 2\rho - \alpha h = 0, \quad \alpha \in \mathbf{R}$
3. $2\omega(1+\omega)\ddot{\tilde{E}}_3 + (7\omega + 2)\dot{\tilde{E}}_3 + 3\tilde{E}_3 = 0$

   $f = h = -2\sqrt{\omega}(\tilde{E}_3 + (1+\omega)\dot{\tilde{E}}_3)$

   $2\omega(1+\omega)\ddot{\tilde{H}}_3 + (7\omega + 2)\dot{\tilde{H}}_3 + 3\tilde{H}_3 = 0$

   $g = -\rho = 2\sqrt{\omega}(\tilde{H}_3 + (1+\omega)\dot{\tilde{H}}_3)$
4. $2\omega(\omega - 1)\ddot{\tilde{E}}_3 + (7\omega - 2)\dot{\tilde{E}}_3 + 3\tilde{E}_3 = 0$

   $f = -h = 2\sqrt{\omega}(\tilde{E}_3 + (\omega - 1)\dot{\tilde{E}}_3)$

   $2\omega(\omega - 1)\ddot{\tilde{H}}_3 + (7\omega - 2)\dot{\tilde{H}}_3 + 3\tilde{H}_3 = 0$

   $g = \rho = -2\sqrt{\omega}(\tilde{H}_3 + (\omega - 1)\dot{\tilde{H}}_3)$
5. $2\omega(\omega - 1)\ddot{\tilde{E}}_3 + (7\omega - 2)\dot{\tilde{E}}_3 + 3\tilde{E}_3 = 0$

   $g = -\omega^{-1}\rho = 2\varepsilon[\tilde{E}_3 + (\omega - 1)\dot{\tilde{E}}_3]$

   $2\omega(\omega - 1)\ddot{\tilde{H}}_3 + (7\omega - 2)\dot{\tilde{H}}_3 + 3\tilde{H}_3 = 0$

   $f = \omega^{-1}h = 2\varepsilon[\tilde{H}_3 + (\omega - 1)\dot{\tilde{H}}_3]$

   $\varepsilon = 1$ for $(x_0 + x_3)x_2^{-1} > 0$

   $\varepsilon = -1$ for $(x_0 + x_3)x_2^{-1} < 0$
6. $(\omega - 1)\dot{\tilde{E}}_3 + \tilde{E}_3 = 0, \qquad 2\omega\dot{f} + f = -2\varepsilon_2\sqrt{|\omega|}\dot{\tilde{E}}_3$

   $2\omega\dot{h} + h = 2\varepsilon_1\sqrt{|\omega|}\dot{\tilde{E}}_3, \qquad (\omega - 1)\dot{\tilde{H}}_3 + \tilde{H}_3 = 0$

   $2\omega\dot{\rho} + \rho = 2\varepsilon_1\sqrt{|\omega|}\dot{\tilde{H}}_3 \qquad 2\omega\dot{g} + g = 2\varepsilon_2\sqrt{|\omega|}\dot{\tilde{H}}_3$

   $\varepsilon_1 = 1$ for $x_0 + x_3 > 0$

   $\varepsilon_1 = -1$ for $x_0 + x_3 < 0$

   $\varepsilon_2 = 1$ for $x_0 - x_3 > 0$

   $\varepsilon_2 = -1$ for $x_0 - x_3 < 0$

7. 1. $\dot{\tilde{E}}_3 = 0, \quad 2\dot{f} = \varepsilon h, \quad \dot{\tilde{H}}_3 = 0, \quad 2\dot{g} = -\varepsilon\rho$

$\varepsilon = 1 \quad \text{for} \quad x_0 - x_3 > 0$

$\varepsilon = -1 \quad \text{for} \quad x_0 - x_3 < 0$

2. $\tilde{E}_3 = 0, \quad 2(1+\alpha)\dot{h} - \left(1 + \frac{1}{\alpha}\right)h = 0$

$\left(\frac{1}{\alpha} - 2\right)f + 2(1-\alpha)\dot{f} = \varepsilon e^{-(1/\alpha)\omega}h$

$\tilde{H}_3 = 0, \quad 2(1+\alpha)\dot{\rho} - (1 + \frac{1}{\alpha})\rho = 0$

$\left(\frac{1}{\alpha} - 2\right)g + 2(1-\alpha)\dot{g} = -\varepsilon e^{-\frac{1}{\alpha}\omega}\rho \qquad 0 < |\alpha| \le 1, \ \alpha \ne -1$

$\varepsilon = 1 \quad \text{for} \quad x_0^2 - x_1^2 - x_3^2 > 0$

$\varepsilon = -1 \quad \text{for} \quad x_0^2 - x_1^2 - x_3^2 < 0$

8. $\dot{h} = 0, \quad \dot{\tilde{E}}_3 = 0, \quad \dot{\rho} = 0, \quad \dot{\tilde{H}}_3 = 0, \quad 2\varepsilon\dot{f} - h = 0$

$2\varepsilon\dot{g} + \rho = 0$

$\varepsilon = 1 \quad \text{for} \quad x_0 - x_3 > 0$

$\varepsilon = -1 \quad \text{for} \quad x_0 - x_3 < 0$

9. $h = 4\varepsilon f, \quad \rho = -4\varepsilon g, \quad \tilde{E}_3 = -\left(9\omega^2 + \frac{\varepsilon}{4}\right)\dot{f} - 15\omega f = 0,$

$(36\omega^2 + \varepsilon)\ddot{f} + 180\omega\dot{f} + 140f = 0, \quad \tilde{H}_3 = 15\omega g + \left(9\omega^2 + \frac{\varepsilon}{4}\right)\dot{g}$

$(36\omega^2 + \varepsilon)\ddot{g} + 180\omega\dot{g} + 140g = 0$

$\varepsilon = 1 \quad \text{for} \quad \sigma > 0$

$\varepsilon = -1 \quad \text{for} \quad \sigma < 0$

$\sigma = 4x_1 - (x_0 - x_3)^2$

10. $\dot{f} = \dot{h}, \quad h = (\omega^2+1)\dot{\tilde{E}}_3 + \omega\tilde{E}_3, \quad (\omega^2+1)\ddot{\tilde{E}}_3 + 4\omega\dot{\tilde{E}}_3 + 2\tilde{E}_3 = 0$

$\dot{g} = -\dot{\rho}, \quad \rho = (\omega^2+1)\dot{\tilde{H}}_3 + \omega\tilde{H}_3, \ (\omega^2+1)\ddot{\tilde{H}}_3 + 4\omega\dot{\tilde{H}}_3 + 2\tilde{H}_3 = 0$

These systems are linear and therefore are easily integrated (the integration details can be found in Refs. 50–53). Below we give the final result; specifically, we present the families of exact solutions of the Maxwell equations (99) invariant with respect to the subalgebras M_1–M_{10}.

M_1: $E_1 = C_2(x_0 - x_3)^{-1} - 2x_3C_1|x_0^2 - x_1^2 - x_3^2|^{-(3/2)}$

$E_2 = C_4(x_0 - x_3)^{-1} + 2x_0C_3|x_0^2 - x_1^2 - x_3^2|^{-(3/2)}$

$E_3 = 2x_1C_1|x_0^2 - x_1^2 - x_3^2|^{-(3/2)}$

$H_1 = -C_4(x_0 - x_3)^{-1} - 2x_3C_3|x_0^2 - x_1^2 - x_3^2|^{-(3/2)}$

$H_2 = C_2(x_0 - x_3)^{-1} - 2x_0C_1|x_0^2 - x_1^2 - x_3^2|^{-(3/2)}$

$$H_3 = 2x_1 C_3 |x_0^2 - x_1^2 - x_3^2|^{-(3/2)}$$

$$M_2: \quad E_1 = |\xi|^{-1}\Big\{C_1 \cos(\alpha \ln|\xi|) - C_2 \sin(\alpha \ln|\xi|) - x_1x_2[h \sin(\alpha \ln|\xi|) + \rho \cos(\alpha \ln|\xi|)] + \frac{1}{2}(\xi^2 - x_1^2 + x_2^2)[h \cos(\alpha \ln|\xi|) - \rho \sin(\alpha \ln|\xi|)]\Big\}$$

$$E_2 = |\xi|^{-1}\Big\{C_2 \cos(\alpha \ln|\xi|) + C_1 \sin(\alpha \ln|\xi|) + x_1x_2[\rho \sin(\alpha \ln|\xi|) - h \cos(\alpha \ln|\xi|)] + \frac{1}{2}(\xi^2 + x_1^2 - x_2^2)[h \sin(\alpha \ln|\xi|) + \rho \cos(\alpha \ln|\xi|)]\Big\}$$

$$E_3 = \varepsilon\{h[x_1 \cos(\alpha \ln|\xi|) + x_2 \sin(\alpha \ln|\xi|)] + \rho[x_2 \cos(\alpha \ln|\xi|) - x_1 \sin(\alpha \ln|\xi|)]\}$$

$$H_1 = |\xi|^{-1}\Big\{ - C_2 \cos(\alpha \ln|\xi|) - C_1 \sin(\alpha \ln|\xi|) - x_1x_2[\rho \sin(\alpha \ln|\xi|) - h \cos(\alpha \ln|\xi|)] + \frac{1}{2}(\xi^2 - x_1^2 + x_2^2)[h \sin(\alpha \ln|\xi|) + \rho \cos(\alpha \ln|\xi|)]\Big\}$$

$$H_2 = |\xi|^{-1}\Big\{C_1 \cos(\alpha \ln|\xi|) - C_2 \sin(\alpha \ln|\xi|) - x_1x_2[h \sin(\alpha \ln|\xi|) + \rho \cos(\alpha \ln|\xi|)] - \frac{1}{2}(\xi^2 + x_1^2 - x_2^2)[h \cos(\alpha \ln|\xi|) - \rho \sin(\alpha \ln|\xi|)]\Big\}$$

$$H_3 = \varepsilon\{h[x_1 \sin(\alpha \ln|\xi|) - x_2 \cos(\alpha \ln|\xi|)] + \rho[x_1 \cos(\alpha \ln|\xi|) + x_2 \sin(\alpha \ln|\xi|)]\}$$

where $\xi = x_0 - x_3, \; h = \omega^{-2}[C_4 \cos(\alpha \ln|\omega|) - C_3 \sin(\alpha \ln|\omega|)]$

$$\rho = \omega^{-2}[C_3 \cos(\alpha \ln|\omega|) + C_4 \sin(\alpha \ln|\omega|)], \; \omega = x_\mu x^\mu$$

$\alpha \in \mathbf{R}, \; \varepsilon = 1,$ *for* $\xi > 0$ and $\varepsilon = -1$ for $\xi < 0$

$$M_3: \quad E_a = -\frac{2C_1 x_a}{x_3(x_1^2 + x_2^2)} + x_a \sigma^{-(3/2)} A_{12}, \quad E_3 = x_3 \sigma^{-(3/2)} A_{12}$$

$$H_a = -\frac{2C_3 x_a}{x_3(x_1^2 + x_2^2)} + x_a \sigma^{-(3/2)} A_{34}, \quad H_3 = x_3 \sigma^{-(3/2)} A_{34}$$

where $A_{ij} = C_i\left(\ln\left|\frac{\sqrt{\sigma} - x_3}{\sqrt{\sigma} + x_3}\right| + 2x_3^{-1}\sqrt{\sigma}\right) + C_j$

$$\sigma = x_1^2 + x_2^2 + x_3^2, \; a = 1, 2$$

$$M_4: \quad 1. \quad E_a = \varepsilon_{ab} x_b \left\{\frac{2C_4}{x_0(x_1^2 + x_2^2)} - \sigma^{-(3/2)} A_{34}\right\}, \quad E_3 = x_0 \sigma^{-(3/2)} A_{12}$$

$$H_a = -\varepsilon_{ab} x_b \left\{\frac{2C_2}{x_0(x_1^2 + x_2^2)} - \sigma^{-(3/2)} A_{12}\right\}, H_3 = x_0 \sigma^{-(3/2)} A_{34}$$

$$\text{where } A_{ij} = C_i + C_j\left(\ln\left|\frac{\sqrt{\sigma}-x_0}{\sqrt{\sigma}+x_0}\right| + 2x_0^{-1}\sqrt{\sigma}\right)$$

$$\sigma = x_0^2 - x_1^2 - x_2^2 > 0,\ a,b = 1,2$$

2. $$E_a = -\varepsilon_{ab}x_b\left\{\frac{C_4}{x_0(x_1^2+x_2^2)} - \sigma^{-(3/2)}B_{34}\right\},\quad E_3 = x_0\sigma^{-(3/2)}B_{12}$$

$$H_a = -\varepsilon_{ab}x_b\left\{\frac{C_2}{x_0(x_1^2+x_2^2)} - \sigma^{-3/2}B_{12}\right\},\quad H_3 = x_0\sigma^{-3/2}B_{34}$$

$$\text{where } B_{ij} = C_i + C_j\left(x_0^{-1}\sqrt{\sigma} - \arctan\frac{\sqrt{\sigma}}{x_0}\right),\ \sigma = x_1^2 + x_2^2 - x_0^2 > 0$$

$a,b = 1,2$

Here $\varepsilon_{ab},\ (a,b=1,2)$ is the antisymmetric tensor of the second order with $\varepsilon_{12} = 1$.

M_5: 1. $$E_1 = \frac{2x_0C_4}{x_2(x_0^2-x_3^2)} - x_0\sigma^{-(3/2)}A_{34},\ E_2 = \frac{2x_3C_2}{x_2(x_0^2-x_3^2)} - x_3\sigma^{-(3/2)}A_{12}$$

$$H_1 = -\frac{2x_0C_2}{x_2(x_0^2-x_3^2)} + x_0\sigma^{-(3/2)}A_{12},\ H_2 = \frac{2x_3C_4}{x_2(x_0^2-x_3^2)} - x_3\sigma^{-(3/2)}A_{34}$$

$$E_3 = x_2\sigma^{-(3/2)}A_{12},\quad H_3 = x_2\sigma^{-(3/2)}A_{34}$$

$$\text{where } A_{ij} = C_i + C_j\left(2\frac{\sqrt{\sigma}}{x_2} - \ln\left|\frac{\sqrt{\sigma}-x_2}{\sqrt{\sigma}+x_2}\right|\right),\ \sigma = x_2^2 + x_3^2 - x_0^2 > 0$$

2. $$E_1 = \frac{x_0C_4}{x_2(x_0^2-x_3^2)} - x_0\sigma^{-(3/2)}B_{34},\ E_2 = \frac{x_3C_2}{x_2(x_0^2-x_3^2)} - x_3\sigma^{-(3/2)}B_{12}$$

$$H_1 = -\frac{x_0C_2}{x_2(x_0^2-x_3^2)} + x_0\sigma^{-(3/2)}B_{12},\ H_2 = \frac{x_3C_4}{x_2(x_0^2-x_3^2)} - x_3\sigma^{-(3/2)}B_{34}$$

$$E_3 = x_2\sigma^{-(3/2)}B_{12},\quad H_3 = x_2\sigma^{-(3/2)}B_{34}$$

$$\text{where } B_{ij} = C_i + C_j\left(\frac{\sqrt{\sigma}}{x_2} - \arctan\frac{\sqrt{\sigma}}{x_2}\right),\quad \sigma = x_0^2 - x_2^2 - x_3^2 > 0$$

M_6: $$E_1 = \frac{1}{2}\left[\frac{\xi(x_1C_2 - x_2C_5) + \eta(x_1C_3 - x_3C_6)}{\xi\eta(x_1^2+x_2^2)} - \frac{\varepsilon_1\xi(x_1C_1 + x_2C_4) - \varepsilon_2\eta(x_1C_1 - x_2C_4)}{\sigma(x_1^2+x_2^2)}\right]$$

$$E_2 = \frac{1}{2}\left[\frac{\xi(x_1C_5 + x_2C_2) + \eta(x_1C_6 + x_2C_3)}{\xi\eta(x_1^2+x_2^2)} + \frac{\varepsilon_1\xi(x_1C_4 - x_2C_1) + \varepsilon_2\eta(x_1C_4 + x_2C_1)}{\sigma(x_1^2+x_2^2)}\right]$$

$$H_1 = \frac{1}{2}\left[\frac{\eta(x_1C_6 + x_2C_3) - \xi(x_1C_5 + x_2C_2)}{\xi\eta(x_1^2 + x_2^2)} + \frac{\varepsilon_1\xi(x_2C_1 - x_1C_4) + \varepsilon_2\eta(x_1C_4 + x_2C_1)}{\sigma(x_1^2 + x_2^2)}\right]$$

$$H_2 = \frac{1}{2}\left[\frac{\xi(x_1C_2 - x_2C_5) - \eta(x_1C_3 - x_2C_6)}{\xi\eta(x_1^2 + x_2^2)} - \frac{\varepsilon_1\xi(x_1C_1 + x_2C_4) + \varepsilon_2\eta(x_1C_1 - x_2C_4)}{\sigma(x_1^2 + x_2^2)}\right]$$

$$E_3 = C_1\sigma^{-1}, \quad H_3 = C_4\sigma^{-1}$$

where $\sigma = x_1^2 + x_2^2 + x_3^2 - x_0^2$, $\xi = x_0 + x_3, \eta = x_0 - x_3$ and

$$\varepsilon_1 = \begin{cases} 1 & \text{if} \quad x_0 + x_3 > 0 \\ -1 & \text{if} \quad x_0 + x_3 < 0 \end{cases} \qquad \varepsilon_2 = \begin{cases} 1 & \text{if} \quad x_0 - x_3 > 0 \\ -1 & \text{if} \quad x_0 - x_3 < 0 \end{cases}$$

M_7: 1. $\alpha = -1$

$$E_1 = |\eta|^{-(3/2)}\left(C_1 + \frac{1}{4}F\right) - x_1\eta^{-2}C_2 - \frac{1}{2}\varepsilon|\eta|^{-(1/2)}f(x_1^2\eta^{-2} - 1)$$

$$E_2 = |\eta|^{-(3/2)}\left(C_3 - \frac{1}{4}G\right) + x_1\eta^{-2}C_4 + \frac{1}{2}\varepsilon|\eta|^{-(1/2)}g(x_1^2\eta^{-2} + 1)$$

$$H_1 = -|\eta|^{-(3/2)}\left(C_3 - \frac{1}{4}G\right) - x_1\eta^{-2}C_4 - \frac{1}{2}\varepsilon|\eta|^{-(1/2)}g(x_1^2\eta^{-2} - 1)$$

$$H_2 = |\eta|^{-(3/2)}\left(C_1 + \frac{1}{4}F\right) - x_1\eta^{-2}C_3 - \frac{1}{2}\varepsilon|\eta|^{-(1/2)}f(x_1^2\eta^{-2} + 1)$$

$$E_3 = \eta^{-1}C_2 + x_1|\eta|^{-(3/2)}f, \quad H_3 = \eta^{-1}C_4 + x_1|\eta|^{-(3/2)}g$$

where $f = f(\omega)$, $g = g(\omega)$, $F = F(\omega)$, $G = G(\omega)$ are arbitrary smooth functions, $\dfrac{dF}{d\omega} = f$, $\dfrac{dG}{d\omega} = g, \omega = \xi - x_1^2\eta^{-1}$, and

$$\xi = x_0 + x_3, \quad \eta = x_0 - x_3$$

$$\varepsilon = \begin{cases} 1 \text{ if} & x_0 - x_3 > 0 \\ -1 \text{ if} & x_0 - x_3 < 0 \end{cases}$$

2. $0 < |\alpha| \le 1$

$$E_1 = x_3|\sigma|^{-(3/2)}C_1 + C_2\eta^{(2\alpha-1)/(1-\alpha)}, \ E_2 = x_0|\sigma|^{-(3/2)}C_3 + C_4\eta^{(2\alpha-1)/(1-\alpha)}$$

$$E_3 = -x_1|\sigma|^{-(3/2)}C_1$$

$$H_1 = -x_3|\sigma|^{-(3/2)}C_3 - C_4\eta^{(2\alpha-1)/(1-\alpha)}, \; H_2 = x_0|\sigma|^{-(3/2)}C_1 + C_2\eta^{(2\alpha-1)/(1-\alpha)}$$

$$H_3 = x_1|\sigma|^{-(3/2)}C_3$$

If $\alpha = 1$, then $C_2 = C_4 = 0$. Here $\sigma = x_0^2 - x_1^2 - x_3^2$, $\eta = x_0 - x_3$.

M_8:
$$E_1 = -x_1\eta^{-2}C_1 + \frac{1}{4}|\eta|^{-(3/2)}C_2(\xi + 2\eta - 3x_1^2\eta^{-1} + \ln|\eta|) + |\eta|^{-(3/2)}C_3$$

$$E_2 = x_1\eta^{-2}C_4 - \frac{1}{4}|\eta|^{-(3/2)}C_5(\xi - 2\eta - 3x_1^2\eta^{-1} + \ln|\eta|) + |\eta|^{-(3/2)}C_6$$

$$H_1 = -x_1\eta^{-2}C_4 + \frac{1}{4}|\eta|^{-(3/2)}C_5(\xi + 2\eta - 3x_1^2\eta^{-1} + \ln|\eta|) - |\eta|^{-(3/2)}C_6$$

$$H_2 = -x_1\eta^{-2}C_1 + \frac{1}{4}|\eta|^{-(3/2)}C_2(\xi - 2\eta - 3x_1^2\eta^{-1} + \ln|\eta|) + |\eta|^{-(3/2)}C_3$$

$$E_3 = \eta^{-1}C_1 + x_1|\eta|^{-(3/2)}C_2, \; H_3 = \eta^{-1}C_4 + x_1|\eta|^{-(3/2)}C_5$$

where $\xi = x_0 + x_3$, $\eta = x_0 - x_3$

M_9: 1.
$$E_1 = \varphi^{-2}[A_{12}(\varphi^{1/2} - \varphi^{-(1/2)}(\eta^2 - 4) - 12\eta\omega) - \eta B_{12}]$$

$$E_2 = \varphi^{-2}[A_{34}(\varphi^{1/2} - \varphi^{-(1/2)}(\eta^2 + 4) - 12\eta\omega) - \eta B_{34}]$$

$$E_3 = \varphi^{-2}[4A_{12}(\eta\varphi^{-(1/2)} + 6\omega) + 2B_{12}]$$

$$H_1 = -\varphi^{-2}[A_{34}(\varphi^{1/2} - \varphi^{-(1/2)}(\eta^2 - 4) - 12\eta\omega) - \eta B_{34}]$$

$$H_2 = \varphi^{-2}[A_{12}(\varphi^{1/2} - \varphi^{-(1/2)}(\eta^2 + 4) - 12\eta\omega) - \eta B_{12}]$$

$$H_3 = -\varphi^{-2}[4A_{34}(\eta\varphi^{-(1/2)} + 6\omega) + 2B_{34}]$$

where
$$A_{ij} = (1 + 36\omega^2)^{-(3/2)}[C_i\sigma^{(1/3)}(4\sqrt{1 + 36\omega^2} - 72\omega) + C_j\sigma^{-(1/3)}(4\sqrt{1 + 36\omega^2} + 72\omega)]$$

$$B_{ij} = 16(1 + 36\omega^2)^{-(1/2)}(C_i\sigma^{(1/3)} - C_j\sigma^{-(1/3)})$$

$$\sigma = 6\omega + \sqrt{36\omega^2 + 1}, \; \omega = (\xi - x_1\eta + \frac{1}{6}\eta^3)\varphi^{-(3/2)}$$

$$\varphi = 4x_1 - \eta^2 > 0$$

$$\xi = x_0 + x_3, \; \eta = x_0 - x_3$$

$$
\begin{aligned}
2.\quad E_1 &= \varphi^{-2}[A_{12}(\varphi^{1/2} - \varphi^{-(1/2)}(\eta^2+4) + 42\eta\omega) - \eta B_{12}] \\
E_2 &= \varphi^{-2}[A_{34}(\varphi^{1/2} - \varphi^{-(1/2)}(\eta^2-4) - 42\eta\omega) - \eta B_{34}] \\
E_3 &= -\varphi^{-2}[4A_{12}(\eta\varphi^{-(1/2)} + 21\omega) - 2B_{12}] \\
H_1 &= \varphi^{-2}[A_{34}(\varphi^{1/2} - \varphi^{-(1/2)}(\eta^2+4) + 42\eta\omega) - \eta B_{34}] \\
H_2 &= \varphi^{-2}[A_{12}(\varphi^{1/2} - \varphi^{-(1/2)}(\eta^2-4) + 42\eta\omega) - \eta B_{12}] \\
H_3 &= \varphi^{-2}[A_{34}(\eta\varphi^{-(1/2)} + 21\omega) - 2B_{34}]
\end{aligned}
$$

$$
\begin{aligned}
\text{where } A_{ij} &= (1-36\omega^2)^{-(3/2)}\{\cos\sigma[72\omega C_j - 4C_i\sqrt{1-36\omega^2}] \\
&\quad - \sin\sigma[72\omega C_i + 4C_j\sqrt{1-36\omega^2}]\} \\
B_{ij} &= 16(1-36\omega^2)^{-(1/2)}[C_i\sin\sigma - C_j\cos\sigma], \ \sigma = \frac{1}{3}\arcsin 6\omega \\
|6\omega| &< 1, \ \varphi = \eta^2 - 4x_1 > 0, \ \omega = \left(\xi - x_1\eta + \frac{1}{6}\eta^3\right)\varphi^{-(3/2)} \\
\xi &= x_0 + x_3, \ \eta = x_0 - x_3
\end{aligned}
$$

$$
\begin{aligned}
3.\quad E_1 &= \varphi^{-2}[A_{12}(\varphi^{1/2} - \varphi^{-(1/2)}(\eta^2+4) - 12\eta\omega) - \eta B_{12}] \\
E_2 &= \varphi^{-2}[A_{34}(\varphi^{1/2} - \varphi^{-(1/2)}(\eta^2-4) - 12\eta\omega) - \eta B_{34}] \\
E_3 &= \varphi^{-2}[-4A_{12}(\eta\varphi^{-(1/2)} - 6\omega) + 2B_{12}] \\
H_1 &= -\varphi^{-2}[A_{34}(\varphi^{1/2} - \varphi^{-(1/2)}(\eta^2+4) - 12\eta\omega) - \eta B_{34}] \\
H_2 &= \varphi^{-2}[A_{12}(\varphi^{1/2} - \varphi^{-(1/2)}(\eta^2-4) - 12\eta\omega) - \eta B_{12}] \\
H_3 &= \varphi^{-2}[4A_{34}(\eta\varphi^{-(1/2)} - 6\omega) - 2B_{34}]
\end{aligned}
$$

$$
\begin{aligned}
\text{where } A_{ij} &= (36\omega^2-1)^{-(3/2)}[C_i\sigma^{1/3}(4\sqrt{36\omega^2-1} - 72\omega) \\
&\quad + C_j\sigma^{-(1/3)}(4\sqrt{36\omega^2-1} + 72\omega)] \\
B_{ij} &= 16(36\omega^2-1)^{-(3/2)}[C_i\sigma^{1/3} - C_j\sigma^{-(1/3)}] \\
\sigma &= 6\omega + \sqrt{36\omega^2-1}, \ |6\omega| > 1, \ \varphi = 4x_1 - \eta^2 > 0 \\
\omega &= \left(\xi - x_1\eta + \frac{1}{6}\eta^3\right)\varphi^{-(3/2)}, \ \xi = x_0 + x_3, \ \eta = x_0 - x_3
\end{aligned}
$$

$$
\begin{aligned}
M_{10}:\quad E_1 &= \sigma^{-1}(1+\xi^2)^{-1}\Big\{x_1C_5 - x_2C_6 - (1+\omega^2)^{-1}\Big[\xi x_1(C_1\omega + C_2) \\
&\quad + \xi x_2(C_3\omega + C_4) - \frac{1}{2}(1-\xi^2)(x_1(C_1 - \omega C_2) \\
&\quad + x_2(C_3 - \omega C_4))\Big]\Big\} + \frac{1}{2}\sigma^{-2}(1+\xi^2)(1+\omega^2)^{-1}[x_1(C_1 - \omega C_2) \\
&\quad - x_2(C_3 - \omega C_4)]
\end{aligned}
$$

$$E_2 = \sigma^{-1}(1+\xi^2)^{-1}\Big\{x_1C_6 + x_2C_5 + (1+\omega^2)^{-1}\Big[\xi x_1(C_3\omega + C_4)$$

$$- \xi x_2(C_1\omega + C_2) + \frac{1}{2}(1-\xi^2)(x_2(C_1 - \omega C_2) - x_1(C_3 - \omega C_4))\Big]\Big\}$$

$$+\frac{1}{2}\sigma^{-2}(1+\xi^2)(1+\omega^2)^{-1}[x_1(C_3 - \omega C_4) + x_2(C_1 - \omega C_2)]$$

$$E_3 = \sigma^{-1}(1+\omega^2)^{-1}[C_1(\omega+\xi) + C_2(1-\xi\omega)]$$

$$H_1 = -\sigma^{-1}(1+\xi^2)^{-1}\Big\{x_1C_6 + x_2C_5 + (1+\omega^2)^{-1}\Big[\xi x_1(C_3\omega + C_4)$$

$$- \xi x_2(C_1\omega + C_2) + \frac{1}{2}(1-\xi^2)(x_2(C_1 - \omega C_2) - x_1(C_3 - \omega C_4))\Big]\Big\}$$

$$+\frac{1}{2}\sigma^{-2}(1+\xi^2)(1+\omega^2)^{-1}[x_1(C_3 - \omega C_4) + x_2(C_1 - \omega C_2)]$$

$$H_2 = \sigma^{-1}(1+\xi^2)^{-1}\Big\{x_1C_5 - x_2C_6 - (1+\omega^2)^{-1}\Big[\xi x_1(C_1\omega + C_2)$$

$$+ \xi x_2(C_3\omega + C_4) - \frac{1}{2}(1-\xi^2)(x_1(C_1 - \omega C_2) + x_2(C_3 - \omega C_4))\Big]\Big\}$$

$$-\frac{1}{2}\sigma^{-2}(1+\xi^2)(1+\omega^2)^{-1}[x_1(C_1 - \omega C_2) - x_2(C_3 - \omega C_4)]$$

$$H_3 = \sigma^{-1}(1+\omega^2)^{-1}[C_3(\omega+\xi) + C_4(1-\xi\omega)]$$

where $\sigma = x_1^2 + x_2^2$, $\omega = \eta(1+\xi^2)\sigma^{-1} - \xi$, $\eta = x_0 + x_3$, $\xi = x_0 - x_3$

In these formulas C_j, $(j = 1, 2, \ldots, 6)$ are arbitrary real constants.

Note that the constructed Maxwell fields are, generally speaking, non-orthogonal. However, provided some additional restrictions on the parameters $C_1, \ldots, C_6$ are imposed, they become orthogonal. Consider, as an example, the last solution from the preceding list. Imposing the orthogonality condition $\mathbf{E} \cdot \mathbf{H} = 0$ yields the following restrictions on the choice of $C_1, \ldots, C_6$:

$$C_2C_6 = C_4C_5, \quad C_1C_6 = C_1C_3 + C_2C_4 + C_3C_5$$

Next, for the solution invariant under the subalgebra M_1, the orthogonality condition leads to the following set of algebraic equations to be satisfied by the parameters $C_1, \ldots, C_6$:

$$C_2C_3 = C_1C_4, \quad C_1C_3 = 0$$

VI. CONCLUDING REMARKS

The range of applications of the Lie group methods for solving systems of linear and nonlinear partial differential equations is so wide that it is simply impossible to give a detailed account of all the available techniques, even if we consider only some fixed group, such as the conformal group $C(1,3)$. However, the basic ideas and methods presented in this review chapter are easily adapted to the cases of other groups of importance for modern physics. In particular, it is straightforward to modify the general reduction method suggested here in order to make it applicable for solving equations of nonrelativistic physics, where the central role is played by the Galileo and Schrödinger groups.

Furthermore, the general method presented in this chapter applies directly to solving the full Maxwell equations with currents. It can also be used to construct exact classical solutions of Yang–Mills equations with Higgs fields and their generalizations. Generically, the method developed in this chapter can be efficiently applied to any conformally invariant wave equation, on the solution set of which a covariant representation of the conformal algebra in Eq. (15) is realized.

We do not consider here the solution techniques based on the symmetry reduction of different versions of the self-dual Yang–Mills equations to integrable models (we refer the interested reader to several papers [13–15],[22–24,65] for a detailed exposition of the results in this field available to date).

The results of exact solutions of nonlinear generalizations of the Maxwell equations are also beyond the scope of the present review. A survey of these results, as well as an extensive list of references, can be found in Fushchych et al. [21].

References

1. C. N. Yang and R. L. Mills, *Phys. Rev.* **96**, 191–195 (1954).
2. A. Actor, *Rev. Mod. Phys.* **51**(3), 461–525 (1979).
3. R. Radjaraman, *Solitons and Instantons in the Quatum Field Theory*, Mir, Moscow, 1985 (in Russian).
4. M. K. Prasad, in *Geometry Ideas in Physics*, Mir, Moscow, 1983, pp. 64–96 (in Russian).
5. G. t'Hooft, *Phys. Rev. D* **14**(12), 3432–3450 (1976).
6. E. Corrigan and D. B. Fairlie, *Phys. Lett. B* **67**(1), 69–71 (1977).
7. F. Wilczek, in *Quark Confinement and Field Theory*, Wiley, New York, 1977, pp. 211–219.
8. E. Witten, *Phys. Rev. Lett.* **38**(3), 121–124 (1977).
9. M. K. Prasad and C. M. Sommerfield, *Phys. Rev. Lett.* **35**(12), 760–762 (1975).
10. M. F. Atiyah, N. J. Hitchin, V. G. Drinfeld and Yu. I. Manin, *Phys. Lett. A* **65**(3), 185–187 (1978).
11. R. S. Ward, *Phys. Lett. A*, **61**(2), 81–82 (1977).
12. A. A. Belavin and V. E. Zakharov, *Phys. Lett. B* **73**(1), 53–57 (1978).

13. S. Chakravarty, M. J. Ablowitz, and P. A. Clarkson, *Phys. Rev. Lett.* **65**(9), 1085–1087 (1990).
14. S. Chakravarty, S. L. Kent, and E. T. Newman, *J. Math. Phys.* **36**(2), 763–772 (1995).
15. J. Tafel, *J. Math. Phys.* **34**(5), 1892–1907 (1993).
16. F. Calogero and A. Degasperis, *Spectral Transformations and Solitons. Methods for Solution and Study of Evolutionary Equations*, Mir, Moscow, 1985 (in Russian).
17. F. Schwarz, *Lett. Math. Phys.* **6**(5), 355–359 (1982).
18. B. A. Dubrovin, S. P. Novikov, and A. T. Fomenko, *Modern Geometry: Methods and Applications*, Nauka, Moscow, 1986.
19. L. V. Ovsyannikov, *Group Analysis of Differential Equations*, Nauka, Moscow, 1978 (in Russian).
20. P. Olver, *Applications of Lie Groups to Differential Equations*, Springer, New York, 1986.
21. W. I. Fushchych, W. Shtelen, and N. Serov, *Symmetry Analysis and Exact Solutions of Nonlinear Mathematical Physics*, Kluwer, Dordrecht, 1993.
22. T. A. Ivanova and A. D. Popov, *Phys. Lett. A* **205**(2–3), 158–166 (1995).
23. M. Kovalyov, M. Legaré, and L. Gagnon, *J. Math. Phys.* **34**(7), 3245–3268 (1993).
24. M. Legaré, *J. Nonlin. Math. Phys.* **3**(3–4), 266–285 (1996).
25. V. de Alfaro, S. Fubini, and G. Furlan, *Phys. Lett. B* **65**(2), 163–166 (1976).
26. A. A. Belavin, A. M. Polyakov, A. S. Schwartz, and Yu. S. Tyupkin, *Phys. Lett. B* **59**(1), 85–87 (1975).
27. W. I. Fushchych and W. M. Shtelen, *Lett. Nuovo Cimento* **38**(2), 37–40 (1983).
28. A. M. Grundland, J. Harnad, and P. Winternitz, *J. Math. Phys.* **25**(4), 791–806 (1984).
29. P. Winternitz, A. M. Grundland, and J. A. Tuszynski, *J. Math. Phys.* **28**(9), 2194–2212 (1987).
30. W. I. Fushchych, L. F. Barannik, and A. F. Barannik, *Subgroup Analysis of the Galilei and Poincaré Groups and Reduction of Nonlinear Equations*, Naukova Dumka, Kiev, 1991 (in Russian).
31. W. I. Fushchych and R. Z. Zhdanov, *Sov. J. Particles Nuclei* **19**(5), 498–516 (1988).
32. W. I. Fushchych and R. Z. Zhdanov, *Phys. Rep.* **172**(4), 123–174 (1989).
33. W. I. Fushchych and R. Z. Zhdanov, *Symmetries and Exact Solutions of Nonlinear Dirac Equations*, Mathematical Ukraina Publisher, Kiev, 1997.
34. R. Z. Zhdanov and V. I. Lahno, *Proc. Acad. Sci. Ukraine* No. 8, 26–31 (1994).
35. V. I. Lahno, R. Zhdanov, and W. I. Fushchych, *J. Nonlin. Math. Phys.* **2**(1), 51–72 (1995).
36. V. I. Lahno, *J. Nonlin. Math. Phys.* **3**(3–4), 291–295 (1996).
37. R. Z. Zhdanov, V. I. Lahno, and W. I. Fushchych, *Ukrain. Math. J.* **47**(4), 456–462 (1995).
38. V. I. Lahno and W. I. Fushchych, *Theor. Math. Phys.* **110**(3), 416–432 (1997).
39. V. I. Lahno, *Proc. Acad. Sci. Ukraine* No. 7, 21–26 (1997) (in Russian).
40. G. A. Lorentz, *Proc. Acad. Sci. (Amsterdam)* **6**, 809–830 (1904).
41. H. Poincaré, *C. R. Acad. Sci.* **140**, 1504–1506 (1905).
42. H. Poincaré, *Rendiconti del Circolo Mathem. Palermo* **21**, 129–160 (1906).
43. I. Larmor, *Collected Papers*, Clarendon Press, London, 1928.
44. G. I. Rainich, *Trans. Am. Math. Soc.* **27**, 106–109 (1925).
45. H. Bateman, *Proc. Lond. Math. Soc.* **8**, 223–264 (1909).
46. E. Cunningham, *Proc. Lond. Math. Soc.* **8**, 77–98 (1909).
47. N. Kh. Ibragimov, *Proc. Acad. Sci. USSR* **178**(3), 566–568 (1968) (in Russian).

48. W. I. Fushchych and A. G. Nikitin, *Symmetry of Maxwell Equations*, Naukova Dumka, 1983 (in Russian; English transl.: *Symmetries of Maxwell's Equations*, Reidel, Dordrecht, 1987).

49. R. Z. Zhdanov and W. I. Fushchych, *J. Phys. A: Math. Gen.* **28**(21), 6253–6263 (1995).

50. R. Z. Zhdanov, V. F. Smalij, and V. I. Lahno, *Proc. Acad. Sci. Ukraine, Ser. A* Nos. 4–7 (in Russian).

51. V. Lahno and V. F. Smalij, *Proc. Acad. Sci. Ukraine* No. 12, 49–54 (1996) (in Russian).

52. V. Lahno, *J. Nonlin. Math. Phys.* **4**(3–4), 392–400 (1997).

53. V. I. Lahno, in *Symmetry and Analytical Methods in Mathematical Physics*, Proc. Inst. Mathematics of NAN of Ukraine, **19**, Kiev, 1998, pp. 123–129.

54. J. Patera, P. Winternitz, and H. Zassenhaus, *J. Math. Phys.* **16**(8), 1597–1614 (1975).

55. W. I. Fushchych, *Ukrain. Math. J.* **39**(1), 116–123 (1987).

56. W. I. Fushchych and R. Z. Zhdanov, *Ukrain. Math. J.* **44**(7), 970–982 (1992).

57. G. Bluman and J. D. Cole, *J. Math. Mech.* **18**(11), 1025–1042 (1969).

58. P. J. Olver and P. Rosenau, *Phys. Lett.* **114A**(3), 107–112 (1986).

59. W. I. Fushchych and I. M. Tsyfra, *J. Phys. A: Math. Gen.* **20**(2), L45–L48 (1987).

60. P. Clarkson and M. D. Kruskal, *J. Math. Phys.* **30**(10), 2201–2213 (1989).

61. D. Levi and P. Winternitz, *J. Phys. A: Math. Gen.* **22**(15), 2915–2924 (1989).

62. W. I. Fushchych, R. Z. Zhdanov, and I. V. Revenko, *Ukrain. Math. J.* **43**(11), 1471–1486 (1991).

63. E. Kamke, *Differentialgleichungen. Lösungmethoden und Lösungen*, Akademische Verlagsgesellschaft, Leipzig, 1961.

64. R. Z. Zhdanov, I. M. Tsyfra, and R. O. Popovych, *J. Math. Anal. Appl.* **238**(1), 101–103 (1999).

65. J. Tafel, *J. Math. Phys.* **31**(5), 1234–1236 (1990).

CHAOS IN OPTICAL SYSTEMS

P. SZLACHETKA AND K. GRYGIEL

Nonlinear Optics Division, Institute of Physics, Adam Mickiewicz University, Poznań, Poland

CONTENTS

Modern Nonlinear Optics, Part 2, Second Edition, Advances in Chemical Physics, Volume 119, Edited by Myron W. Evans. Series Editors I. Prigogine and Stuart A. Rice.
ISBN 0-471-38931-5

I. INTRODUCTION

It was assumed that a description of evolution of deterministic systems required a solution of the equations of motion, starting from some initial conditions. Although Poincaré [1] knew that it was not always true, this opinion was common. Since the work of Lorenz [2] in 1963, unpredictability of deterministic systems described by differential nonlinear equations has been discovered in many cases. It has been established that given infinitesimally different initial conditions, the outcomes can be wildly different, even with the simplest equations of motion. This feature means the occurrence of *deterministic chaos*. The literature devoted to this multidisciplinary and rapidly developing discipline of science is huge. There are many excellent textbooks, monographs, and collections of main papers, and we mention only a few [3–8].

In this overview we focus our attention on some problems of *optical chaos*. In many optical effects and devices intrinsic instabilities occur and for over thirty years they have been extensively investigated. The literature on optical chaos is widespread and a few excellent reviews and collections of papers should be recalled [9–13].

After an overview of the main papers devoted to chaos in lasers (Section I.A) and in nonlinear optical processes (Section I.B), we present a more detailed analysis of dynamics in a process of second-harmonic generation of light (Section II) as well as in Kerr oscillators (Section III). The last case we consider particularly in the context of coupled nonlinear systems. Finally, we present a cumulant approach to the problem of quantum corrections to the classical dynamics in second-harmonic generation and Kerr processes (Section IV).

A. Chaos in Lasers

Since the discovery of lasers it has been known that a derivation of time-dependent equations governing interaction of molecules with electromagnetic cavity modes leads to the so-called spontaneous instabilities. These laser instabilities were also observed experimentally — even for the first laser built by Maiman in 1960. A random, periodic, or quasiperiodic train of spikes in a laser generation is a fundamental instability due to nonlinearity of laser equations. A comprehensive review of this specific laser-related topics was published in 1983 [14].

A major development reported in 1964 was the first numerical solution of the laser equations by Buley and Cummings [15]. They predicted the possibility of undamped chaotic oscillations far above a gain threshold in lasers. Precisely, they numerically found "almost random spikes" in systems of equations adopted to a model of a single-mode laser with a bad cavity. Thus optical chaos became a subject soon after the appearance Lorenz' paper [2].

Real development in the field of chaotic properties of laser action began over 10 years later. In 1975 Haken [16] used the model of a single-mode laser with a *homogeneously broadened line* (HBL) described by the Maxwell–Bloch equations and after some approximations showed the equivalence with an appropriate Lorenz system of equations. The model was extended to a multimode case [17]. For a modulated external field, certain laser systems are described by a driven Van der Pol oscillator, and the existence of chaos was found numerically for these systems [18]. In the case of HBL lasers, a spatial inhomogeneity of pump leading to a coupling of different modes, could give rise to an undamped spiking behavior of lasers. This instability is chaotic and was found numerically in a two-mode laser case [19]. A detuning was also incorporated in this model [20], and the exact equivalence between a bad-cavity laser with a modulated inversion and nonlinear oscillator in the Toda potential driven by an external modulation was presented a few years later [21]. The parameters in HBL lasers for which chaos is expected are highly unreal because of big loss in cavity rates. For a detailed discussion of instabilities in HBL lasers, we refer the reader to a treatise by Milonni et al. [13] and a paper by Harrison and Biswas [22].

The Haken model can be easily extended to the case of a single-mode *inhomogeneously broadened line* (IBL) laser [13]. Numerical investigation of the Maxwell–Bloch equations has been carried out for the case of a Doppler broadening and for different parameter ranges, leading to findings of period doubling and intermittency routes to chaos [23,24]. A phenomenon of metastable chaos was also observed. The Maxwell–Bloch equations with an inhomogeneously broadened line were also studied in the context of mode splitting [13,25], bad-cavity instability conditions [26], ring laser configuration [27], Hopf bifurcations [28], and a period doubling route to chaos [29].

Laser instabilities were *experimentally* investigated in many kinds of lasers (see an overview of early papers [14]), but the first experimental observation of the optical chaos was performed by Arecchi et al. [30] in 1982. They used a stabilized CO_2 laser with modulated cavity loss $\Gamma = \gamma(1 + a\cos\Omega t)$ and by changing the frequency of modulation Ω, they found a few period doubling oscillations of the output intensity, both numerically and experimentally.

A detailed analysis shows that the case of the IBL laser is more convenient in experimental investigations because the value of the threshold gain coefficient needed in a laser setup is much smaller. Some spontaneous instability for this case was first discovered experimentally by Casperson quite early [31] in a low-pressure, electric discharge HeXe laser at 3.51 μm. For a special choice of parameters the laser worked in the regime of the so-called self-pulsing instability. But the first chaotic output from an IBL laser was experimentally shown in 1982 by Weiss and King [32] in a HeNe laser (3.39 μm). A period doubling route to chaos was found. In a HeXe laser, Gioggia and Abraham [33] in 1983 reported a chaotic behavior of a generated signal and confirmed period doubling and intermittency routes to chaos. Similarly, chaotic emission was observed in a ring cavity laser [34]. For an overview of early papers devoted to IBL laser instabilities and chaos, see the study on self-pulsing and chaos in continuous-wave (cw)-excited lasers by Abraham et al. [35].

To achieve the instability of homogeneous broadened line lasers, a satisfaction of much more difficult conditions is required: large gain and the so-called bad-cavity properties. This special regime for damping constants and mode intensity is fulfilled in the far-infrared lasers [36]. In 1985 Weiss et al. [37,38] experimentally found a period doubling route to chaos in the NH_3 laser. Further experimental investigation of chaotic dynamics in such lasers was reported later [39].

The CO_2 lasers were also investigated in connection with chaotic behavior, and here we mention the most important papers in the field. The chaotic behavior associated with a transverse mode structure in a cw CO_2 laser was observed in 1985 [40]. In the CO_2 laser with elastooptically modulated cavity length, a period doubling route to chaos was also found [41].

Chaos was also investigated in *solid-state lasers*, and the important role of a pump nonuniformity leading to a chaotic lasing was pointed out [42]. A modulation of pump of a solid-state NdP_5O_{14} laser leads to period doubling route to chaos [43]. The same phenomenon was observed in the case of laser diodes with modulated currents [44,45]. Also a chaotic dynamics of outputs in Nd:YAG lasers was also discovered [46–48]. In semiconductor lasers a period doubling route to chaos was found experimentally and theoretically in 1993 [49].

An important technique of *chaos control* [50] was introduced in laser systems in 1992 by Roy et al. [51]. They adopted the so-called occasional proportional feedback method to stabilize limit cycles in a multimode Nd:YAG laser with KTP crystal (doubling the basic frequency), pumped by a diode laser. The CO_2 laser with cavity loss modulation was used to implement the control method of output signals proposed by Pyragas [52] and Bielawski et al. [53]. The experimental investigation of the control scheme based on a "washout spectral filter" has been performed in the chaotic regimes of the CO_2 laser with modulated loss [54] as well as in the CO_2 laser with intensity feedback [55]. In 1998, a control of chaos was demonstrated in Nd-doped laser with modulated loss and pump and nonfeedback methods were adopted [56]. These important methods of stabilization of chaotic systems are related to communication theory. In particular, a synchronization of lasers in chaotic regimes has many potential applications. In 1994 Roy and Thornburg proved experimentally for the first time the possibility of synchronization of chaotic lasers [57], with possible applications in digital communication [58]. The last experiments with chaotic lasers revealed a possibility of transmitting a desired message in a very fast way as well as encoding and decoding information in output lasers signals [59–61].

B. Chaos in Nonlinear Optics

Nonlinear optics is a very convenient area to investigate the phenomenon of deterministic chaos both from theoretical and experimental points of view.

The *Jaynes–Cummings* model describing an ensemble of two-level atoms in a resonant cavity with a single-mode field is a basic paradigm in quantum optics. Numerical calculations of the appropriate Maxwell–Bloch equations have revealed a chaotic behavior of the system in a semiclassical approach when no rotating wave approximation is used [62,63]. In a full quantum-mechanical approach, Graham [17] determined the eigenvalues and eigenstates of the coupled atom-field system by numerical diagonalization, and the basis for a quantum description of chaos was prepared. Later, different aspects of chaos in the Jaynes–Cummings model were investigated in a semiclassical or in a full quantum model [64–68].

A complex dynamical behavior was experimentally and numerically found in a system of spin-$\frac{1}{2}$ atoms in an optical resonator with near-resonant cw laser light and external static magnetic field [69]. Three-dimensional Bloch equations were solved, and a chaotic motions was found and compared with experiment.

Quite early optical chaos was found in *optical bistability*. In 1979 Ikeda used a ring cavity configuration for an optically bistable system with two-level absorbing atoms [70]. Ikeda constructed an iterated map of a such system and solving it, found the chaotic output of transmitted field strengths. Moreover, by changing the input light intensities, he proved a period doubling route to chaos. Later, chaos was investigated in the case of off-resonant (dispersive) bistability [71–74]. The first experimental observation of chaos in optical bistability system was made in 1981 by Gibbs et al. [75] in an optical device with electronically introduced delay time. Nakatsuka et al. [76] in 1983 observed experimentally the first chaotic generation in the phenomenon of dispersive bistability. Next, experimental and theoretical evidence of chaotic behavior of signals generated in bistable systems was checked by a few groups [77–82].

Second-harmonic generation of light is a nonlinear phenomenon in which chaotic behavior was discovered in 1983 [83] (for details, see Secction II). In the *Kerr effect* with an external time-dependent pump, a chaotic output may also occur, which was proved for the first time in 1990 by Milburn [84] (see also Section III).

Many kinds of molecular systems pumped by a strong laser light show chaotic dynamics. Indeed, in a semiclassical model of a *multiphoton excitation* on molecular vibration, chaos was discovered by Ackerhalt et al. [85] and theoretically and numerically investigated in detail [86,87]. Moreover, the equations of motion that describe a rotating molecule in a laser field can exhibit a chaotic behavior and have been applied in the classical case of a rigid-rotator approximation [87,88].

Dynamical instabilities and chaos were discovered in many light scattering processes. For example Milonni et al. in 1983 [89] found a chaotic strange attractor in *stimulated Raman scattering*. They solved numerically the classical coupled wave equations in the case of perfect phase-matching conditions. Next, a period doubling route to chaos was found, and a fractal dimension of the attractor was calculated by Nath and Ray [90]. Chaos in *stimulated Brillouin scattering* was found in 1984 by Candall and Albritton [91]. The dynamics of generated signals in stimulated scattering processes in optical fibers has also been investigated [92].

Another class of good candidates for a study of chaos in nonlinear optics are *wave-mixing processes* in which chaos appears in the propagation of laser light through passive nonlinear media [93]. A chaotic behavior was observed in three-wave mixing [94] and in four-wave mixing [95].

Experimental work and theoretical investigation show an important role of spatial chaos in optical fibers, directional couplers, and generally in all-optical switching devices [96/97].

The problem of *quantum chaos* in optics has been studied in a few areas. For a short review, see Section IV.

II. CHAOS IN SECOND-HARMONIC GENERATION OF LIGHT

A. Introduction

Nonlinear optics deals with physical systems described by Maxwell equations with an nonlinear polarization vector. One of the best known nonlinear optical processes is the second-harmonic generation (SHG) of light. In this section we consider a well-known set of equations describing generation of the second harmonic of light in a medium with second-order nonlinear susceptibility $\chi^{(2)}$. The classical approach of this section is extended to a quantum case in Section IV.

The first experimental evidence of SHG was reported by Franken et al. [98], who focused a ruby laser beam ($\lambda_L = 0.694$ nm) on a quartz crystal and analyzed the two outgoing beams by a standard method (the second-harmonic beam was observed in the UV region $2\lambda_S = 0.347$ nm). This experiment was soon followed by a theoretical analysis by Armstrong et al. [99]. Since then many articles have appeared on the subject (bibliographies are presented in Refs. 100 and 101).

To analyze the dynamics of SHG, we use time-dependent ordinary differential equations. At the beginning, Maxwell's equations governing SHG were studied, and a simple analytical time dependent solutions was found [99]. The classical case of SHG was discussed by Bloembergen [102], and the present-day state in the dynamics of SHG without damping and pumping was clarified

[103]. The same equations, albeit with damping and coherent external driving field, were studied by Drummond et al. [104] as a particular case of sub/second-harmonic generation. They proved that below a critical pump intensity, the system can reach a stable state (field of constant amplitude). However, beyond the critical intensity, the steady state is unstable. They predicted the existence of various instabilities as well as both first- and second-order phase transition-like behavior. For certain sets of parameters they found an amplitude self-modulation of the second harmonic and of the fundamental field in the cavity as well as new bifurcation solutions. Mandel and Erneux [105] constructed explicitly and analytically new time-periodic solutions and proved their stability in the vicinity of the transition points.

SHG equations were used also to analyze of deterministic chaos. Savage and Walls were the first [83] to prove the existence of chaos in the case of nonzero detuning between laser and cavity modes. They found a period-doubling route to chaos. Bistability, self-pulsing, and chaos were also studied Lugiato et al. [106]. The dynamics of SHG in the case of time-dependent external pumping was investigated by the present authors. Numerical analysis of the equation of motions was performed for the modulated pump amplitude [107] as well as for the external pump of rectangular pulses [108]. Alekseeva et al. [109] presented a detailed study of the spatial evolution of multifrequency fundamental and second-harmonic radiation and showed that the system may exhibit a spatial chaos due to multiple competing processes. Also, a hyperchaotic dynamic in SHG was numerically predicted [110,111].

B. Basic Equations

Let us consider an optical system with two modes at the frequencies ω and 2ω interacting through a nonlinear crystal with second-order susceptibility placed within a Fabry–Pérot interferometer. In a general case, both modes are damped and driven with external phase-locked driving fields. The input external fields have the frequencies ω_L and $2\omega_L$. The classical equations describing second-harmonic generation are [104,105]:

$$\begin{aligned} \frac{d\alpha_1}{dt} &= -i\Delta_1'\alpha_1 - \Gamma_1\alpha_1 + \kappa\alpha_1^*\alpha_2 + F_1 \\ \frac{d\alpha_2}{dt} &= -i\Delta_2'\alpha_2 - \Gamma_2\alpha_2 - \frac{1}{2}\kappa\alpha_1^2 + F_2 \end{aligned} \tag{1}$$

Rapid oscillations (at the frequencies $\omega, \omega_L, 2\omega, 2\omega_L$) are removed from Eq. (1) by frequency-matching conditions in the usual way. The quantities $\Delta_1' = \omega - \omega_L$ and $\Delta_2' = 2\omega - 2\omega_L$ are frequency mismatches between the cavity and external fields. Slowly varying in time, complex variables α_1 and α_2 are the electric field amplitudes of the two modes $E_1(t) = \alpha_1(t)\exp(i\omega t)$ and $E_2(t) = \alpha_2(t)\exp(2i\omega t)$

describing fundamental and second-harmonic modes, respectively. Similarly, F_1 and F_2 are proportional to the electric field amplitudes of the two external pumped modes $\mathscr{F}_1(t) = F_1(t)\exp(i\omega_L t)$ and $\mathscr{F}_2(t) = F_2(t)\exp(2i\omega_L t)$. Two constants, Γ_1 and Γ_2, are the cavity loss rates for the appropriate modes. The coupling constant κ between the two modes is proportional to a nonlinear susceptibility $\chi^{(2)}$ of the nonlinear medium. With a special choice of the spatial mode functions, we can assume that κ is real, and we exclude from our investigation of polarization effects—all fields have linear polarization in the same directions [104].

For numerical investigation, it is convenient to reduce the number of relevant parameters in Eq. (1). On substituting

$$\tau = \kappa t, \quad \Delta_{1(2)} = \frac{\Delta'_{1(2)}}{\kappa}, \quad \gamma_{1(2)} = \frac{\Gamma_{1(2)}}{\kappa}, \quad f_{1(2)}(\tau) = \frac{F_{1(2)}(\tau/\kappa)}{\kappa} \tag{2}$$

into (2), we get the following redefined set of equations:

$$\begin{aligned} \frac{d\alpha_1}{d\tau} &= -i\Delta_1\alpha_1 - \gamma_1\alpha_1 + \alpha_1^*\alpha_2 + f_1(\tau) \\ \frac{d\alpha_2}{d\tau} &= -i\Delta_2\alpha_2 - \gamma_2\alpha_2 - \frac{1}{2}\alpha_1^2 + f_2(\tau) \end{aligned} \tag{3}$$

where f_i are taken to be real. The above equations can be written in real variables. On inserting

$$\begin{aligned} \alpha_1 &= \mathrm{Re}(\alpha_1) + i\,\mathrm{Im}(\alpha_1) = y_1 + i\,y_3 \\ \alpha_2 &= \mathrm{Re}(\alpha_2) + i\,\mathrm{Im}(\alpha_2) = y_2 + i\,y_4 \end{aligned} \tag{4}$$

we obtain four equations of motion:

$$\begin{aligned} \frac{dy_1}{dt} &= \Delta_1 y_3 - \gamma_1 y_1 + y_1 y_2 + y_3 y_4 + f_1 \\ \frac{dy_2}{dt} &= \Delta_2 y_4 - \gamma_2 y_2 - \frac{1}{2}(y_1^2 - y_3^2) + f_2 \\ \frac{dy_3}{dt} &= -\Delta_1 y_1 - \gamma_1 y_3 + y_1 y_4 - y_2 y_3 \\ \frac{dy_4}{dt} &= -\Delta_2 y_2 - \gamma_2 y_4 - y_1 y_3 \end{aligned} \tag{5}$$

These four equations of motion describe the dynamics of SHG in the four-dimensional phase space $(\mathrm{Re}\,\alpha_1,\ \mathrm{Im}\,\alpha_1,\ \mathrm{Re}\,\alpha_2,\ \mathrm{Im}\,\alpha_2)$. In practice, we can observe the motion only in the reduced phase space (phase surface). For example, with the help of two-dimensional phase portraits (Re α_i, Im α_j), (Re α_i, Re α_j) and (Im α_i, Im α_j), we can qualify the kind of motion of our system, which may be periodic, quasiperiodic, or chaotic.

To identify chaotic behavior of a dynamical system, it is convenient to use the *Lyapunov exponents* [112,113]. In particular, the procedure proposed by Wolf et al. [114] is a very useful and efficient method that gives such exponents. In this method we have to linearize the set of equations (5), and next the linearized equations are solved together with the primary equations. Moreover, we solve the eigenproblem for the Jacobi matrix of the set of linearized equations in the so-called tangent space. Then, after Gram–Schmidt reorthonormalization, we obtain the set of Lyapunov exponents λ_i as eigenvalues of the long-time product Jacobi matrix. So, in this method the number of exponents, is equal to a dimension of phase space [115]. In our case we have a set $\{\lambda_1\lambda_2\lambda_3\lambda_4\}$; thus, we get a *spectrum* of Lyapunov exponents. Such a spectrum is ordered from maximal to minimal value. The quantity λ_1 is traditionally termed the *maximal Lyapunov exponent* (MLE), and its positive value points to chaotic motion. If $\lambda_1 \leq 0$, the dynamical system behaves nonchaotically (orderly).

A highly unstable system can manifest hyperchaotic behavior [116]. This means that we have two positive Lyapunov exponents in a spectrum. The phenomenon of hyperchaos have been investigated in many papers [117–120]. A route to hyperchaos was also investigated [121], and a method of controlling of hyperchaos was introduced [122].

In next three sections we present a short overview of investigations of chaotic and hyperchaotic behavior in the process of SHG.

C. Simplest Case: $\gamma_i = 0$, $\Delta_i = 0$, $f_i = 0$

In the simplest case of a free evolution without damping, pumping, and mismatch, the equations of motion (3) are solved analytically. One easily notes that the system (3) now belongs to the class of Hamiltonian systems with two constants of motion:

$$\begin{aligned} I_1 &= \alpha_1^*\alpha_1 + 2\alpha_2^*\alpha_2 \\ I_2 &= -\frac{1}{2}i\left(\alpha_1^2\alpha_2^* - \alpha_2\alpha_1^{*2}\right) \end{aligned} \tag{6}$$

They reduce the set (5) of four equations in real variables to two equations. This means that we can have only regular, periodic, or quasiperiodic behavior, never chaos. Chaos in a dynamical system governed by ordinary differential equations can arise only if the number of equations is equal to or greater than 3. We remember that we refer to the case of perfect phase matching ($\Delta k = k_1 - 2k_2 = 0$), and the well-known monotonic evolution of fundamental and

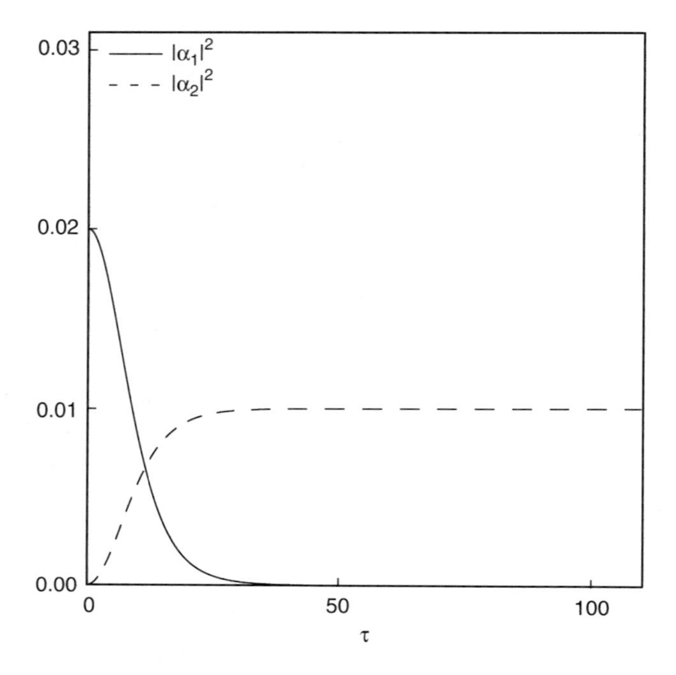

Figure 1. Monotonic behaviour of the fundamental and second-harmonic modes. Solution of Eqs. (7) for the initial conditions $\alpha_{10} = 0.1 + i\,0.1$ and $\alpha_{20} = 0$.

second-harmonic mode intensities has been found [99,102] and is shown on Fig. 1. If $\Delta k \neq 0$, we obtain three equations of motion for $\alpha_1, \alpha_2, \Delta k$ (six equations in real variables) and well known solutions show an oscillation behavior in such cases of SHG. Detailed analysies are available in the literature [99,102,103].

Let us focus on the role of initial conditions in this case of SHG. The equations of motion

$$\frac{d\alpha_1}{d\tau} = \alpha_1^* \alpha_2 \,, \qquad \frac{d\alpha_2}{d\tau} = -\frac{1}{2}\alpha_1^2 \tag{7}$$

were solved with initial conditions $\alpha_1(0) = \alpha_{10}$ and $\alpha_2(0) = \alpha_{20}$. The case of $(\alpha_{10} \neq 0, \alpha_{20} = 0)$ is often called a *second-harmonic generation process* (Fig. 1). For the case of $(\alpha_{10} \neq 0, \alpha_{20} \neq 0)$, that is, when both fields start from the nonzero initial conditions, we deal with a mixed process of sub/second-harmonic generation. Throughout this work the symbol SHG refers to both these cases. In Fig. 2 we see the evolution of the system from the initial conditions: $\alpha_{10} = 0.1 + i0.1$ and $\alpha_{20} = 0.01 + i0.01$. One can observe in Fig. 2a the periodic oscillation in intensity of both modes. However, in the phase space the motion of

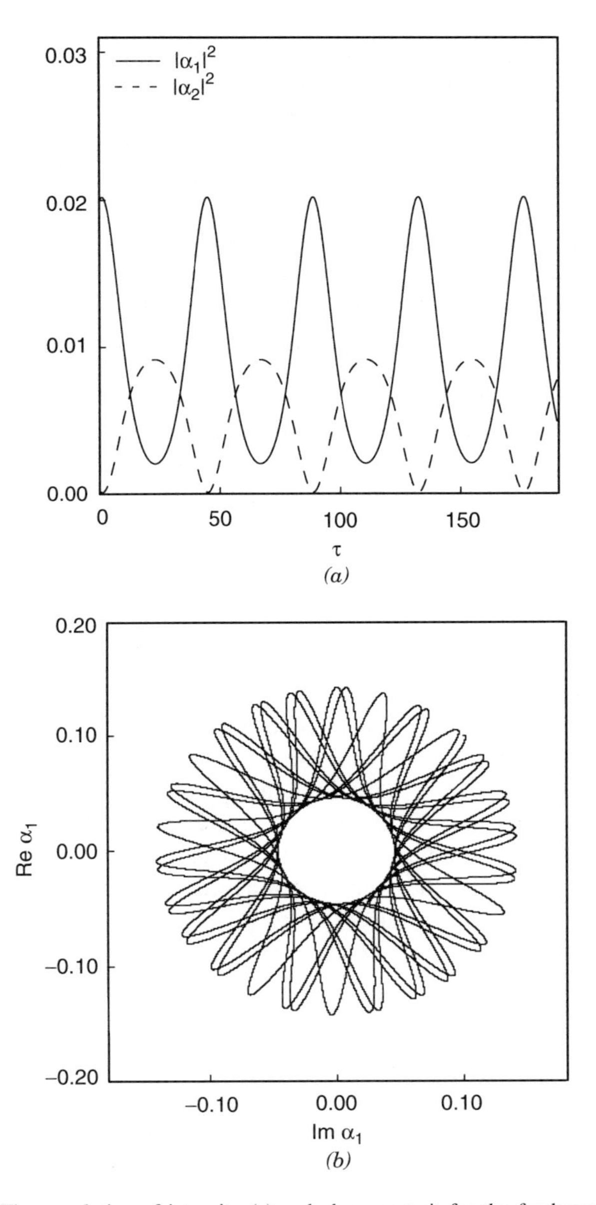

Figure 2. Time evolution of intensity (a) and phase portrait for the fundamental mode (b). Solution of Eqs. (7) for the initial conditions $\alpha_{10} = 0.1 + i0.1$ and $\alpha_{20} = 0.01 + i0.01$. Quasi-periodic behavior.

the system is quasiperiodic, as seen in Fig. 2b. The phase point draws a nonclosed path within the rosette area. The rosette becomes increasingly denser with time, and finally we get a blackened area. A similar rosette is obtained for the second-harmonic mode. For the case of $(\alpha_{10} = 0, \alpha_{20} \neq 0)$, Eq. (7) have constant solutions, and we do not observe any time changes in the SHG system. There is no subharmonic generation without an external pumping f_2.

To sum up, in nonlinear systems the influence of initial conditions on dynamics of a system is essential, therefore the three different initial conditions discribed above lead to different dynamics within the same equations of motion.

D. Coherent External Field

Another clear example of a system generating second harmonics is the one employing an external coherent pump field $f_i =$ constant without dumping $(\gamma_i = 0)$ and frequency mismatch $(\Delta_i = 0)$. The system belongs to the class of Hamiltonian systems. The function (Hamiltonian)

$$H(\tau) = if_1(\alpha_1^* - \alpha_1) + if_2(\alpha_2^* - \alpha_2) - \frac{1}{2}i(\alpha_1^2\alpha_2^* - \alpha_2\alpha_1^{*2}) \tag{8}$$

is a constant of motion for Eq. (3). Since we have only pumping, the trajectory shows an expanding nature [123].

If we now include damping (without mismatch), we get results in compliance with Ref. 104. As did Mandel and Erneux [105], we introduce the notions of good $(\gamma_1 \ll \gamma_2)$ and bad $(\gamma_1 \cong \gamma_2)$ frequency conversion limits in our discussions. We denote them as GCL and BCL, respectively. The case of a coherent pump field was also studied by Drummond et al. [104] with a nonrescaled version of Eq. (1). To get the compact results we use, in accordance with (3), the parameters $f_0 = 2$, $\tau = 10t$, and $\gamma_1 = \gamma_2 = 0.34$ (BCL) or $\gamma_1 = 0$, $\gamma_2 = 0.34$ (GCL). For the intensity of the coherent pump

$$f_1 = (2\gamma_1 + \gamma_2)\sqrt{2\gamma_2(\gamma_1 + \gamma_2)} \tag{9}$$

and $f_2 = 0$, we get a transition from monotonic solutions of (3) to a self-pulsation. As we see in Fig. 3a, after transient effects the system manifests self-pulsation and an appropriate phase portrait for the fundamental mode is presented in Fig. 3b. The limit cycle indicates a periodic motion of the system. If the pump f_1 increases some multiperiodic oscillations occur (Fig. 4). If we change the parameters of pumping f_1 and f_2, we can find [104] that this system exhibits both first- and second-order phase transition-like behavior and also has a hard mode transition. Farther numerical and analytical analysis [105] indicated a new transition involves an hysteresis cycle.

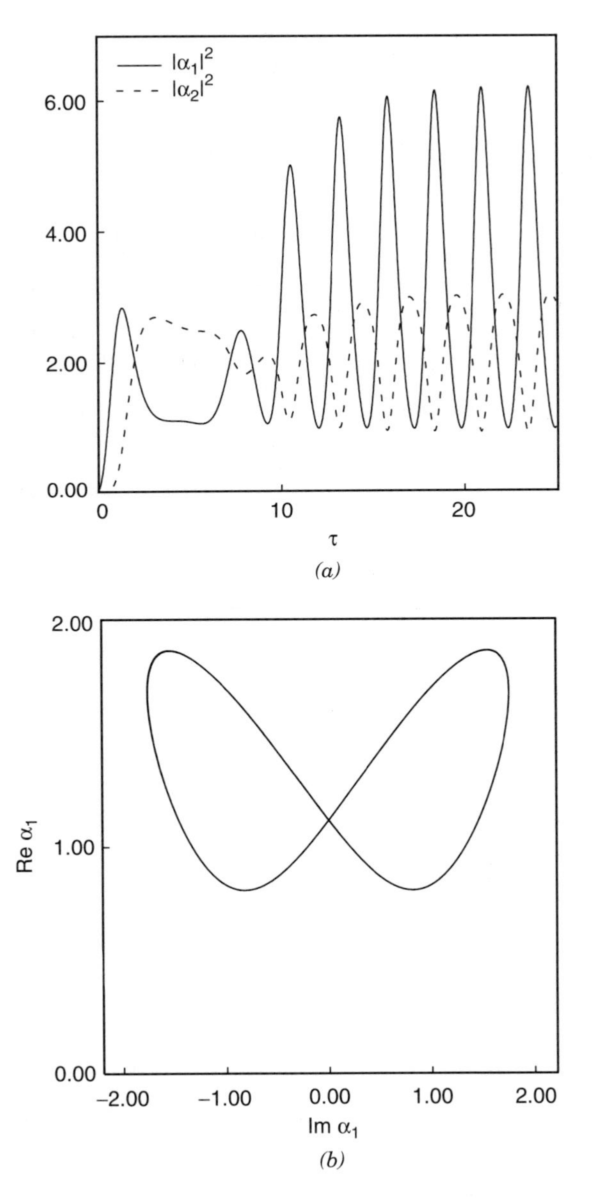

Figure 3. Time evolution of intensity (a) and phase portrait for the fundamental mode (b). Solution of Eqs. (3) for $f_1 = 2, f_2 = 0, \gamma_1 = \gamma_2 = 0.34$ (BCL) and initial conditions $\alpha_{10} = 0.1 + i\,0.1$ and $\alpha_{20} = 0$. Self-pulsation.

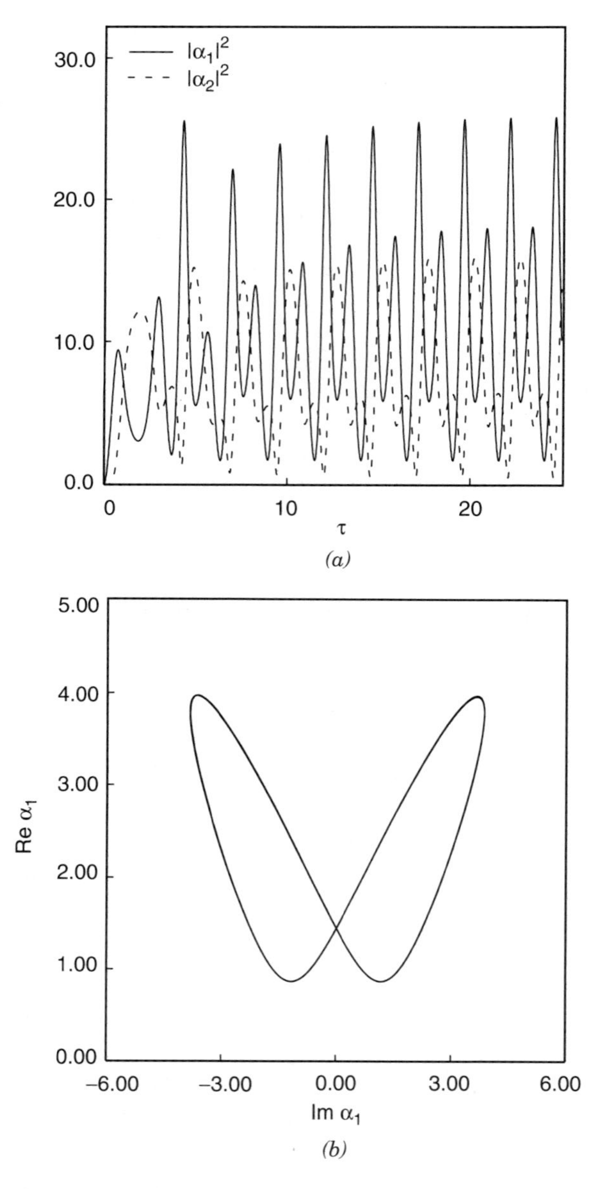

Figure 4. Multiperiodic behaviour in SHG. The same as in Fig. 3 but $f_1 = 5$.

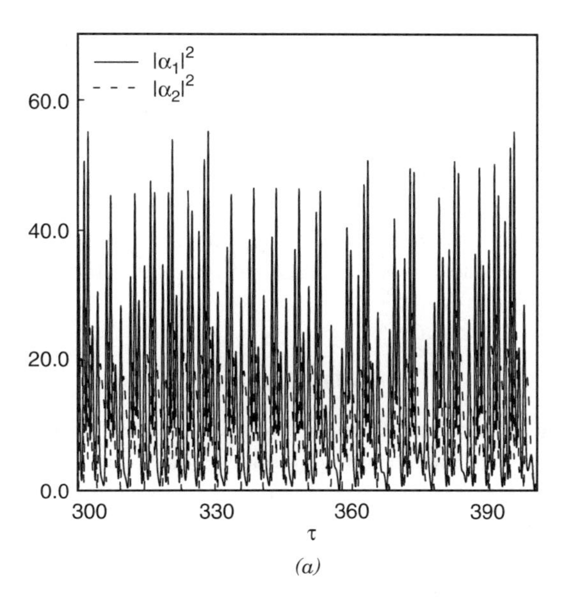

(a)

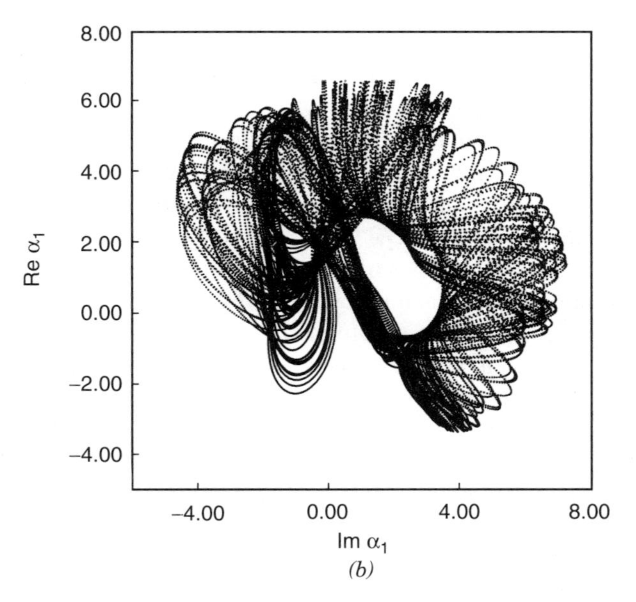

(b)

Figure 5. Time evolution of intensity (a) and phase portrait for the fundamental mode (b). Solutions of Eqs. (3) with parameters $\Delta_1 = \Delta_2 = 1, f_1 = 5.5, f_2 = 0, \gamma_1 = \gamma_2 = 0.34$ (BCL). The initial conditions are $\alpha_{10} = 0.1 + i\,0.1$ and $\alpha_{20} = 0.01 + i\,0.01$. Chaos.

We get a very similar time behavior in subharmonic generation where $f_1 = 0$ and $f_2 \neq 0$. Self-pulsation and multiperiodic evolution of intensities have been found. However, these findings are not investigated here.

The case of a frequency mismatch between laser pumps and cavity modes was investigated [83], and for the first time, chaos in SHG was found. When the pump intensity is increased, we observe a period doubling route to chaos for $\Delta_1 = \Delta_2 = 1$. Now, for $f_1 = 5.5$, Eq. (3) give aperiodic solutions and we have a chaotic evolution in intensities (Fig. 5a) and a chaotic attractor in phase plane $(\mathrm{Im}\,\alpha_1, \mathrm{Re}\,\alpha_1)$ (Fig. 5b).

E. Modulated External Field

A more complicated behavior of the system (3) is manifested if the time-dependent driving field and damping are taken into account. Let us assume that the driving amplitude has the form $f_1(\tau) = f_0(1 + \sin(\Omega\tau))$, meaning that the external pump amplitude is modulated with the frequency Ω around f_0. Moreover, $f_2 = 0$ and $\Delta_1 = \Delta_2 = 0$. It is obvious that if we now examine Eq. (3), the situation in the phase space changes sharply. In our system there are two competitive oscillations. The first belongs to the multiperiodic evolution mentioned in Section II.D, and the second is generated by the modulated external pump field. Consequently, we observe a rich variety of nonlinear oscillations in the SHG process.

The frequency of modulation Ω is now the main parameter, and we are able to switch the system of SHG between different dynamics by changing the value of Ω. To find the regions of Ω where a chaotic motion occurs, we calculate a Lyapunov spectrum versus the "knob" parameter Ω. The first Lyapunov exponent λ_1 from the spectrum is of the greatest importance; its sign determines the chaos occurrence. The maximal Lyapunov exponent λ_1 as a function of Ω is presented for GCL in Fig. 6a and for BCL in Fig. 6b. We see that for some frequencies Ω the system behaves chaotically ($\lambda_1 > 0$) but orderly ($\lambda_1 < 0$) for others. The system in the second case is much more damped than in the first case and consequently much more stable. By way of example, for $\Omega = 0.9$ the system of SHG becomes chaotic as illustrated in Fig. 7a, showing the evolution of second-harmonic and fundamental mode intensities. The phase point of the fundamental mode draws a chaotic attractor as seen in the phase portrait (Fig. 7b). However, the phase point loses its chaotic features and settles into a symmetric limit cycle if we change the frequency to $\Omega = 1.1$ as shown in Fig. 8b, while Fig. 8a shows a seven-period oscillation in intensities. To avoid transient effects, the evolution is plotted for $450 < \tau < 500$.

Let us emphasize that for other values of parameter Ω we can also observe in the phase plane intricate symmetric limit cycles [107,123], such as the five-period oscillations we get for $\Omega = 0.78$.

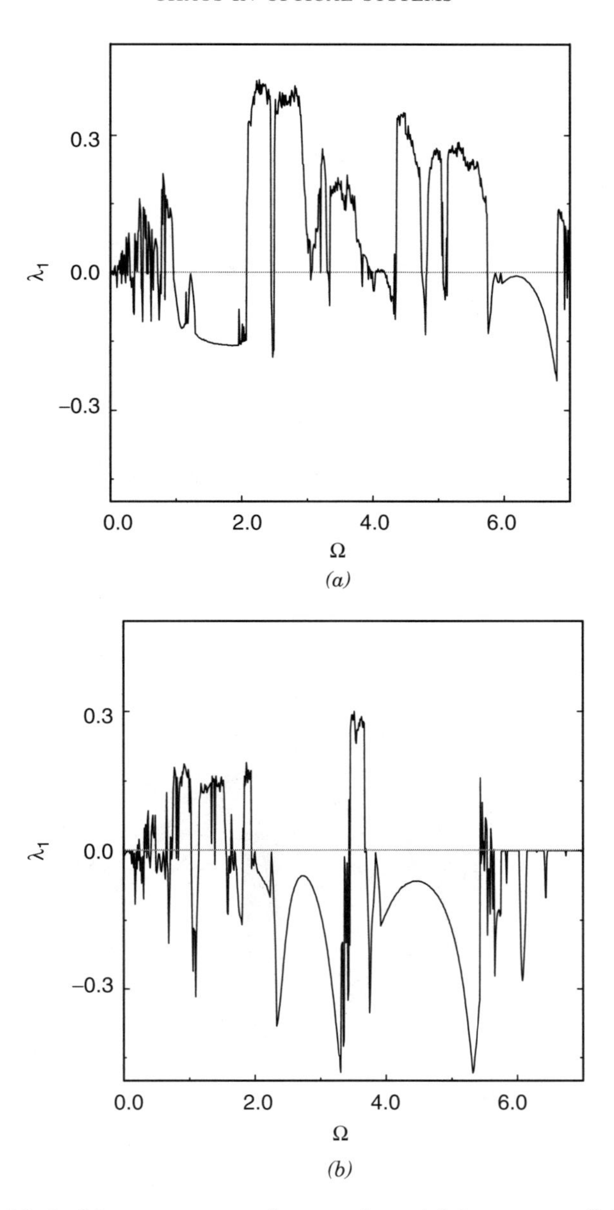

Figure 6. Maximal Lyapunov exponent λ_1 versus the modulation parameter Ω for $f_0 = 2$ and the initial conditions are $\alpha_{10} = 0.1 + i\,0.1$ and $\alpha_{20} = 0.01 + i\,0.01$. (a) GCL and (b) BCL.

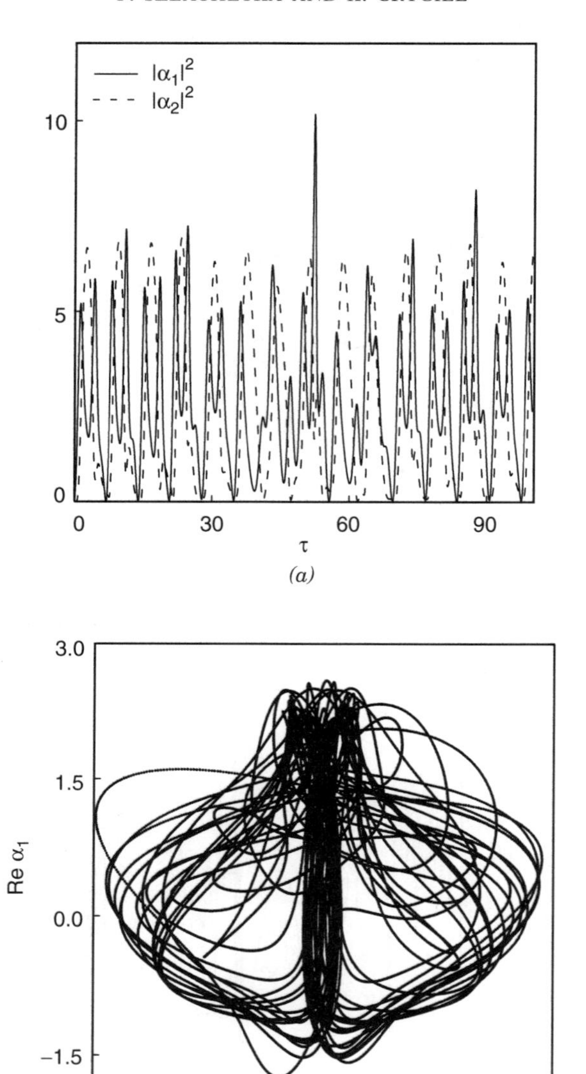

Figure 7. Time evolution of intensity (a) and phase portrait for the fundamental mode for $0 < \tau < 300$ (b). Parameters are the same as in Fig. 6b (BCL), but with $\Omega = 0.9$. Chaos.

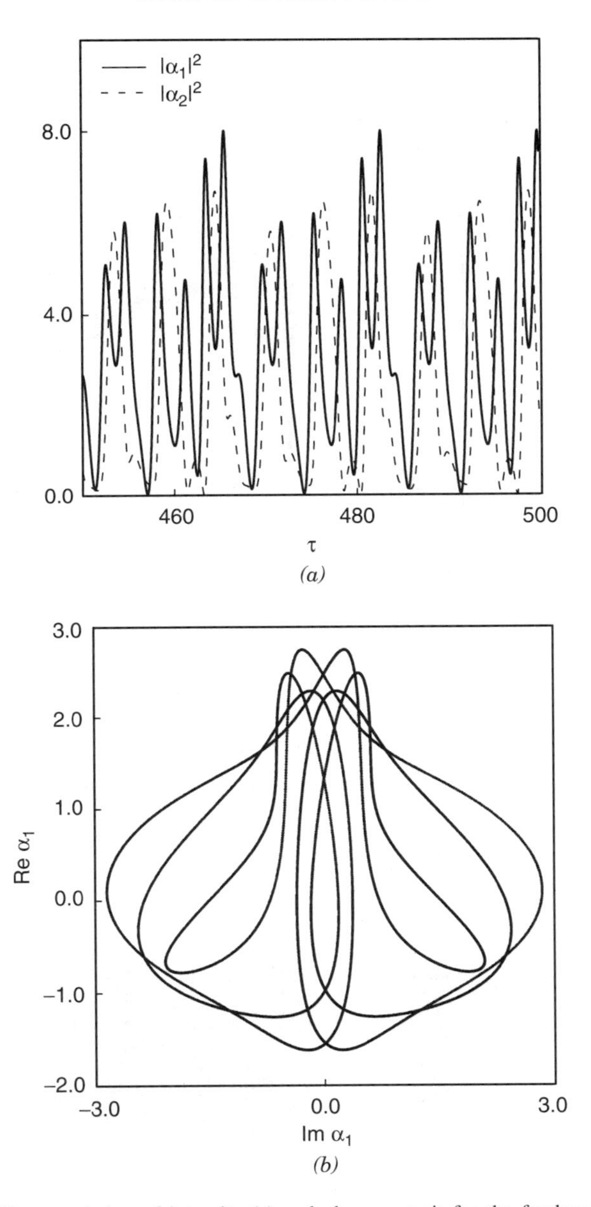

Figure 8. Time evolution of intensity (a) and phase portrait for the fundamental mode for $\tau > 450$ (b). Parameters are the same as in Fig. 6b (BCL) but with $\Omega = 1.1$. Limit cycle.

Highly unstable systems lead to two positive Lyapunov exponents that show the *hyperchaotic* behavior [116]. Now, Eq. (3) is numerically examined with damping constants $\gamma_1 = \gamma_2 = 0.01$. In Fig. 9a we see only the two largest Lyapunov exponents of all the spectrum *versus* the modulation parameter Ω. The

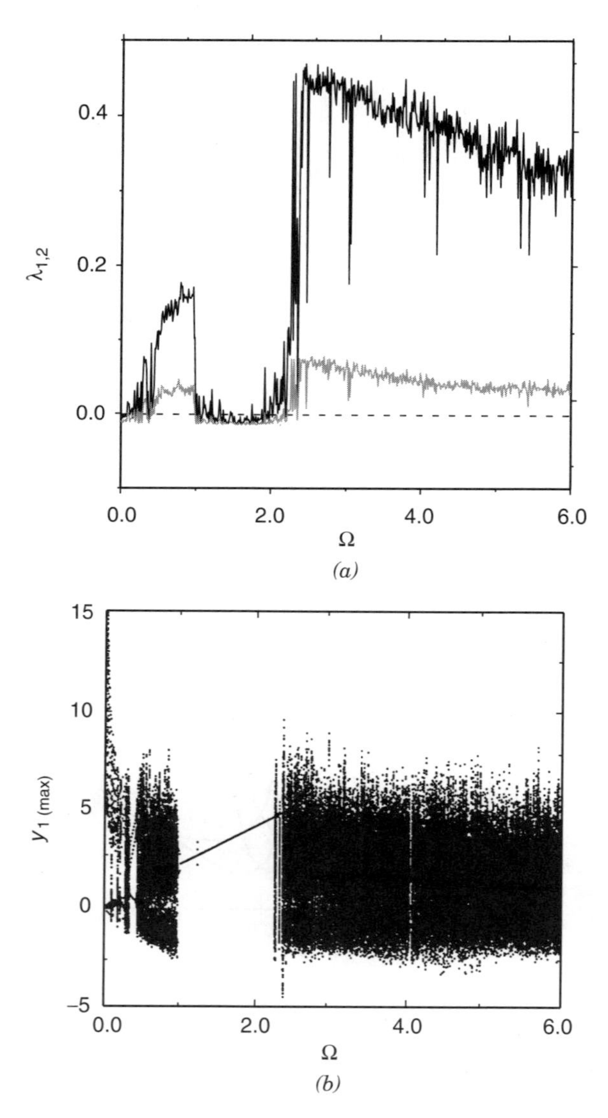

Figure 9. The two largest Lyapunov exponents (a) and the bifurcation diagram (the maxima of y_1) (b) versus the modulation parameter Ω. Parameters are $f_0 = 1, \gamma_1 = \gamma_2 = 0.01$ and the initial conditions are $\alpha_{10} = 0.1 + i0.1$ and $\alpha_{20} = 0.01 + i0.01$. Hyperchaos.

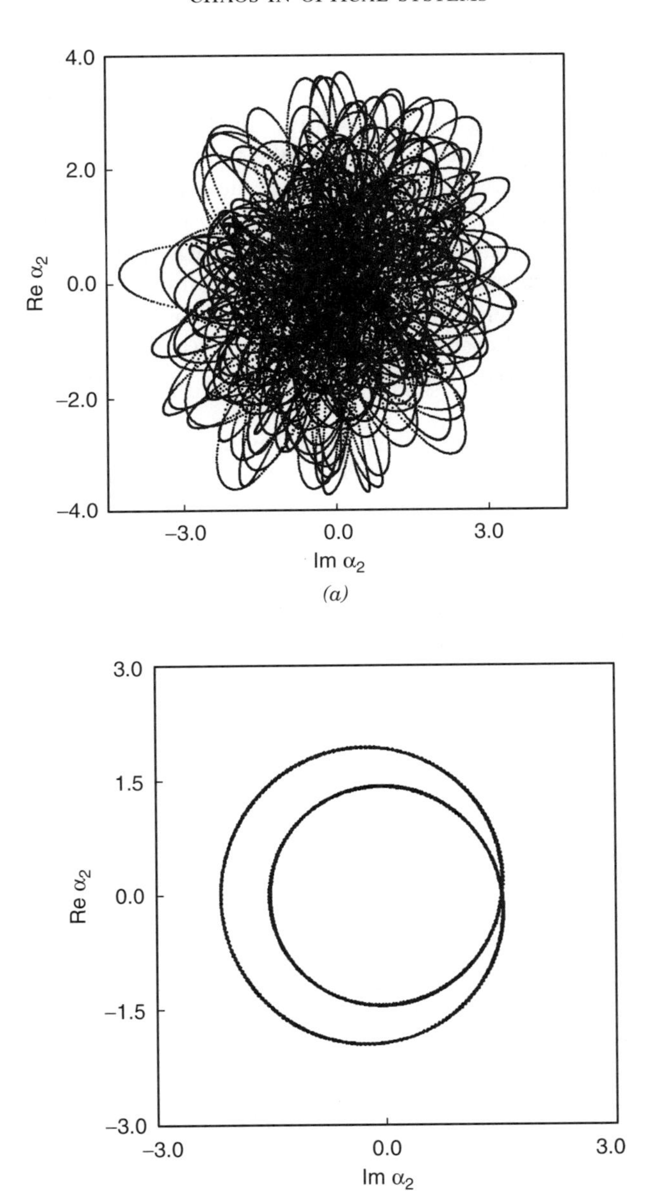

Figure 10. The phase portraits $\mathrm{Re}\,\alpha_2$ versus $\mathrm{Im}\,\alpha_2$ for $f_0 = 1, \gamma_1 = \gamma_2 = 0.01, \alpha_{10} = 0.1 + i\,0.1$, and $\alpha_{20} = 0.01 + i0.01$. The hyperchaotic trajectory for $\Omega = 0.8$ (a) and the limit cycle for $\Omega = 1.55$(b). The time is $400 < \tau < 500$.

damping is so weak that we can state that the system is hyperchaotic. There are two extensive regions of hyperchaos between $0.45 < \Omega < 0.98$ and $\Omega > 2.22$, where two Lyapunov exponents are positive. In the region $0.98 < \Omega < 2.22$ hyperchaos does not appear at all. Generally, the region $0.98 < \Omega < 2.22$ can be treated as nonchaotic apart from a few values of the parameter Ω for which only one Lyapunov exponent is positive. These regions of stability and instability are best visualized in the bifurcation diagram (Fig. 9b), where we plot the maxima of $\mathrm{Re}\,\alpha_1 = y_1$ versus the parameter of modulation Ω. It is obvious that a change in Ω switches the system among chaos, hyperchaos, or limit cycles. For $\Omega = 0.8$, we observe a hyperchaotic orbit in the phase portrait of the second-harmonic mode (Fig. 10a). The same orbit, except for $\Omega = 1.55$, becomes a limit cycle (Fig. 10b).

When the damping in the system is increased, the regions of hyperchaos disappear. Moreover, it is interesting that the region of order that we obtained in Fig. 9 is very stable despite changing damping constants, so we can chose the frequency of modulation of an external field in such a way ($1 < \Omega < 1.8$) that the system remains stable even for a relatively small damping.

F. Pulsed External Field

In this section we consider a case particularly important for experimental investigation. The external driving field $f_1(\tau)$ applied to Eq. (3) has the form of a train of pulses that are simulated by a computer. The length of the pulse is denoted by T_1, and the height of the pulse by f_0. The distance between two pulses is denoted by T_2. For $f_0 \neq 0$ and $T_2 = 0$, the train of pulses becomes a coherent driving field (Section II.D). The second driving field f_2 is assumed to be zero and $\Delta_1 = \Delta_2 = 0$. We examine the dynamical system (3) in the same way as in Section II.E. In Fig. 11 we present the maximal Lyapunov exponent λ_1 as a function of the length of the pulse T_1 (for $T_2 = 1$). As shown in Fig. 11a,b, at the beginning λ_1 is negative, implying the appearance of order in the range $0 < T_1 < 0.085$ for GCL and $0 < T_1 < 0.55$ for BCL. The fundamental ($| \alpha_1 |^2$)) and second-harmonic ($| \alpha_2 |^2$) intensities tend to oscillatory states in the course of time [108]. This is the short-pulse regime, and the appropriate evolution of both intensities is shown in Fig. 12. Here, one can easy recognize moments of time where the pulses are switched on and off. The period of sawtooth-like oscillations is equal to the repetition rate of pulses. The typical phase portrait for the short-pulse case is presented in Fig. 13. Finally, we observe a limit cycle where the phase point moves up and down only a segment of a straight line (shaded dark in Fig. 13b).

For $0.085 < T_1 < 0.5$ (GCL) and for $0.55 < T_1 < 0.97$ (BCL), the maximal Lyapunov exponents λ_1 are near zero; consequently, we obtained quasiperiodic trajectories. Typical quasiperiodic trajectories for both cases are shown in

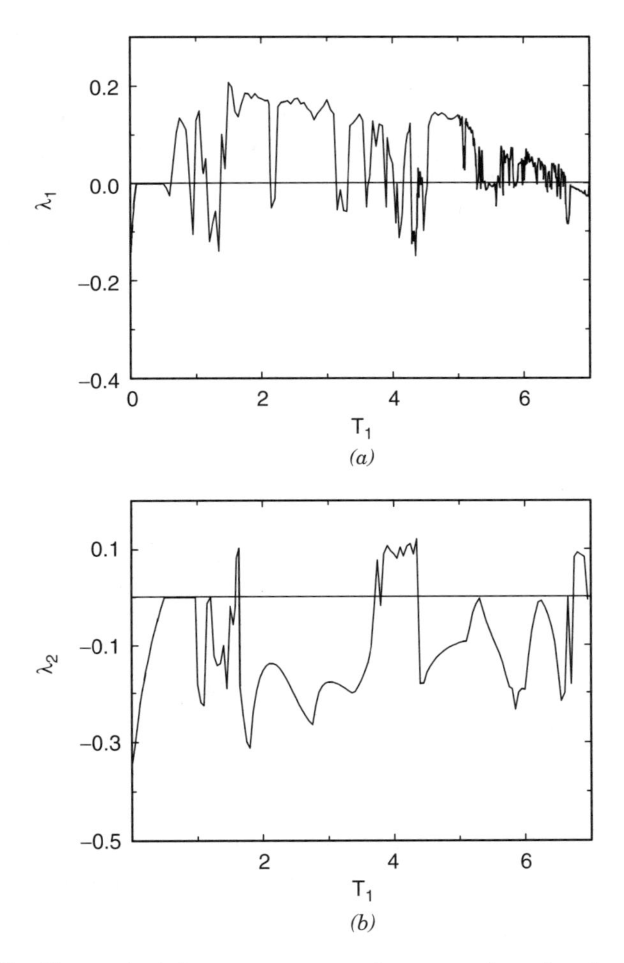

Figure 11. The maximal Lyapunov exponent λ_1 versus the pulse duration T_1, for $f_0 = 2, T_2 = 1, \alpha_{10} = 0.1 + i\,0.1, \alpha_{20} = 0$: (a) the case of GCL $\gamma_1 = 0, \gamma_2 = 0.34$; (b) the case of BCL $\gamma_1 = 0.34, \gamma_2 = 0.34$.

Fig. 14. The trajectory is a nonclosed path, and for long times we get a blackened area.

A more complicated behavior of the MLE is observed for higher values of T_1. Varying the length of the pulse T_1, we observe regions of order and chaos. By way of an example, the phase portrait $\mathrm{Re}\,\alpha_1$ versus $\mathrm{Im}\,\alpha_1$ for a chaotic attractor is shown in Fig. 15.

Within the region of order $(\lambda_1 \leq 0)$ we see intricate symmetric and nonsymmetric limit cycles in phase diagrams. For example, for $T_1 = 4.1$ we see in

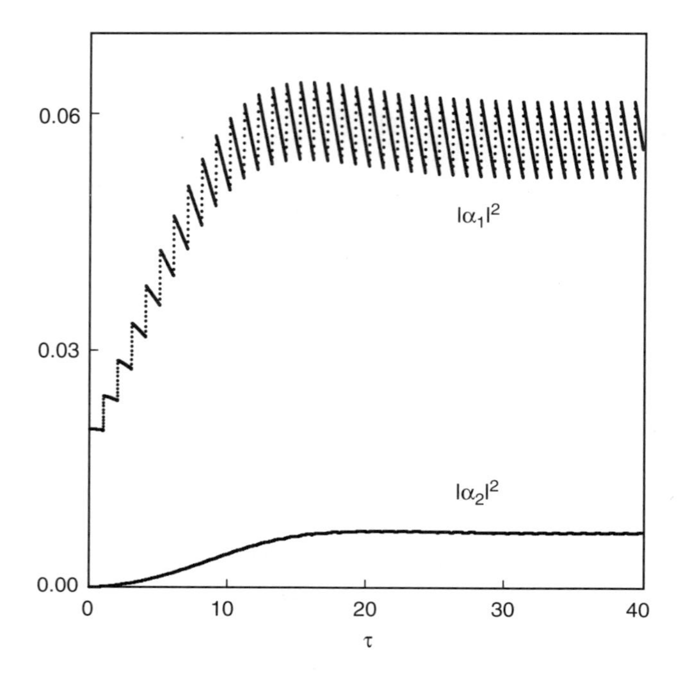

Figure 12. Intensities in the short-pulse regime for the GCL case. The parameters are the same as for Fig. 11a but $T_1 = 0.01$.

Fig. 16a symmetric limit cycles for the second-harmonic mode (GCL) and in Fig. 16b, an nonsymmetric phase portrait example for $T_1 = 0.5$ for BCL. In both cases the phase point settles down into a closed-loop trajectory, although not earlier than about $\tau > 200$. An intricate limit cycle is usually related to multiperiod oscillations. For example, the cycle in Fig. 16a corresponds to five-period oscillations of the fundamental and SHG modes intensity, and the phase portrait in Fig. 16b resembles the four-period oscillations (see Fig. 17). Generally, for $T_1 > 0.5$, we observe many different multiperiod (even 12-period) oscillations in intensity and a rich variety of phase portraits.

Some hyperchaotic behavior in SHG with pump of pulses has been shown [111]. The two largest Lyapunov exponents versus a duration of pulse T_1 are presented in Fig. 18a for the cases of BCL. There are a two regions of hyperchaos. A Typical hyperchaotic phase portrait is presented in Fig. 18b.

G. Final Remarks

Small changes in the modulated pump parameters Ω, f_o and in the pulse parameters T_1, T_2, f_o induce dramatic changes the output fields. Therefore

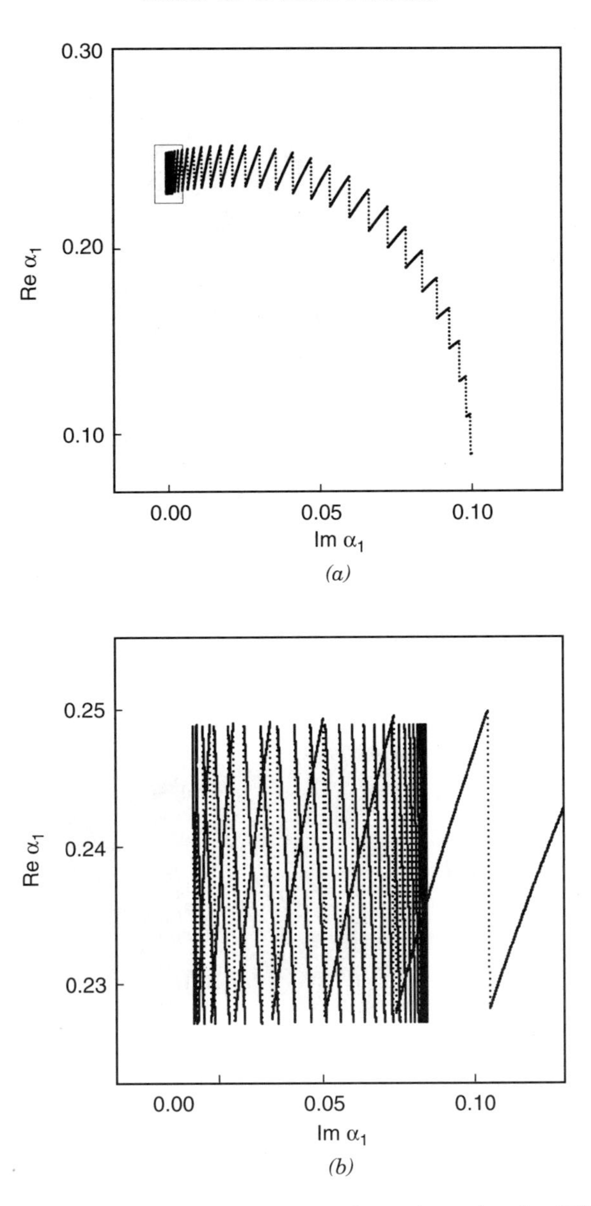

Figure 13. (a) A typical phase portrait in the short-pulse regime for GCL case; (b) an enlargement of the signed region of Fig. 13a. The parameters are the same as for Fig. 11a but $T_1 = 0.01$.

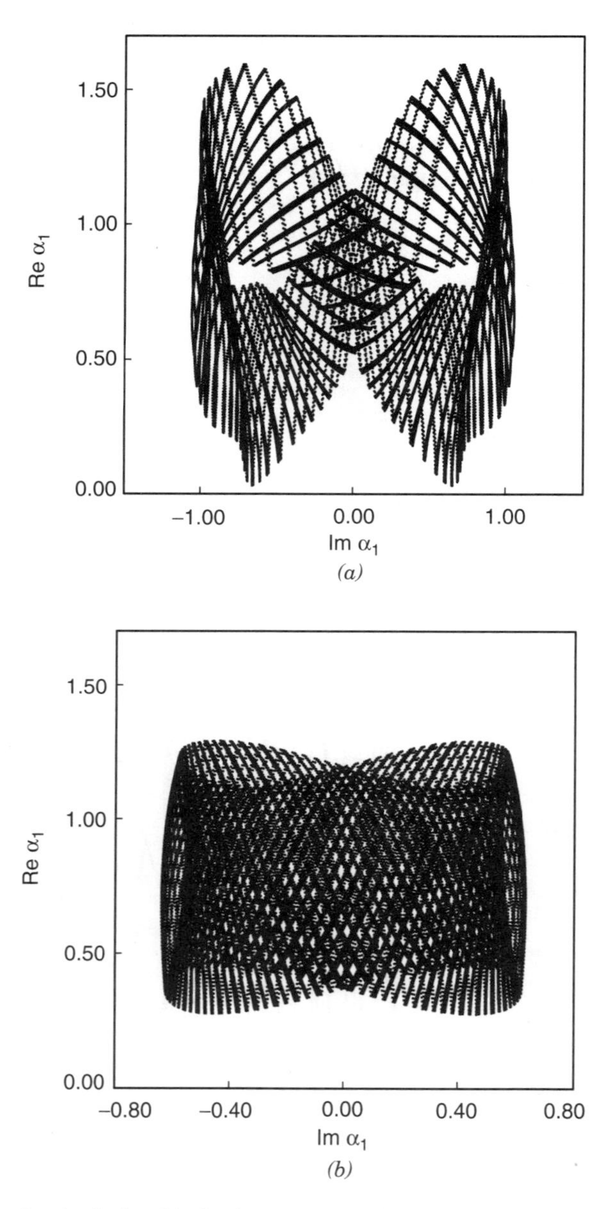

Figure 14. Quasiperiodic orbits for the parameters of Fig. 11 but (a) $T_1 = 0.5$ (GCL) and (b) $T_1 = 0.8$ (BCL). The time is $0 \leq \tau \leq 300$.

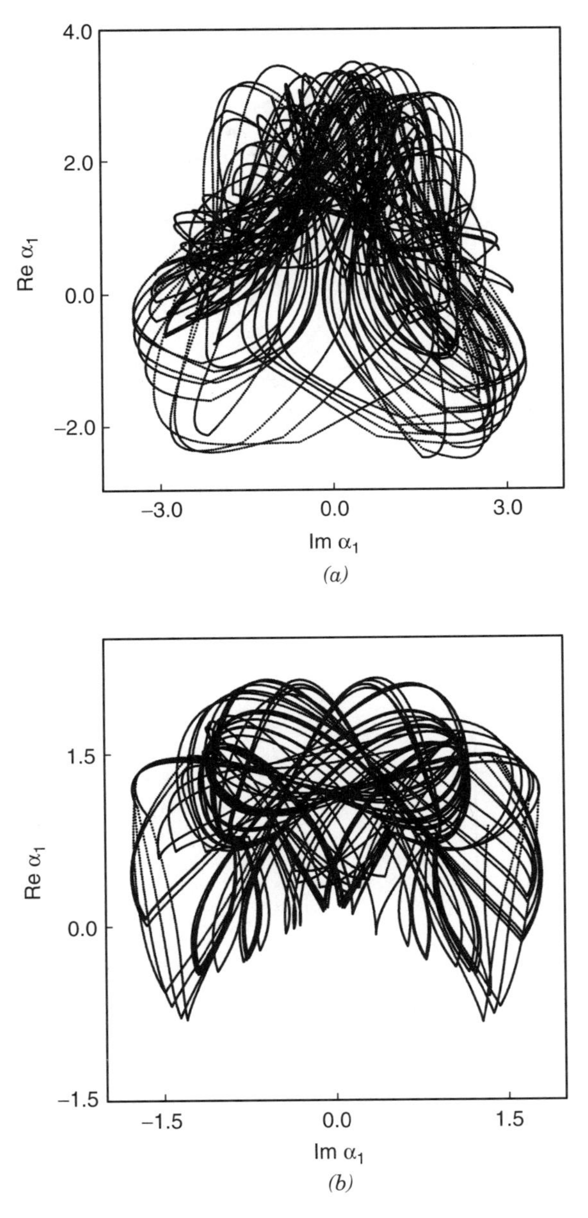

Figure 15. Chaotic attractors for the parameters of Fig. 11 but $T_1 = 0.5$ (GCL) (a) and $T_1 = 4$ (BCL) (b). The time is $0 \leq \tau \leq 300$.

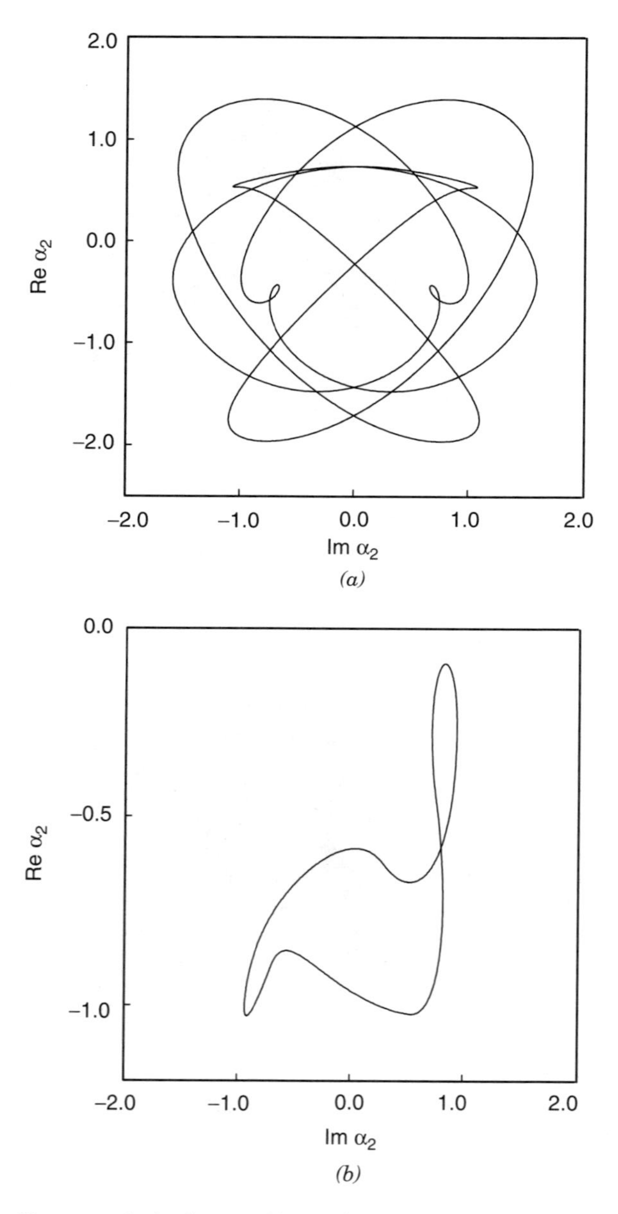

Figure 16. Phase portraits for the second-harmonic mode: (a) symmetric example for GCL, (b) nonsymetric example for BCL. The parameters are the same as for Fig. 11, and the time is $200 < \tau < 500$.

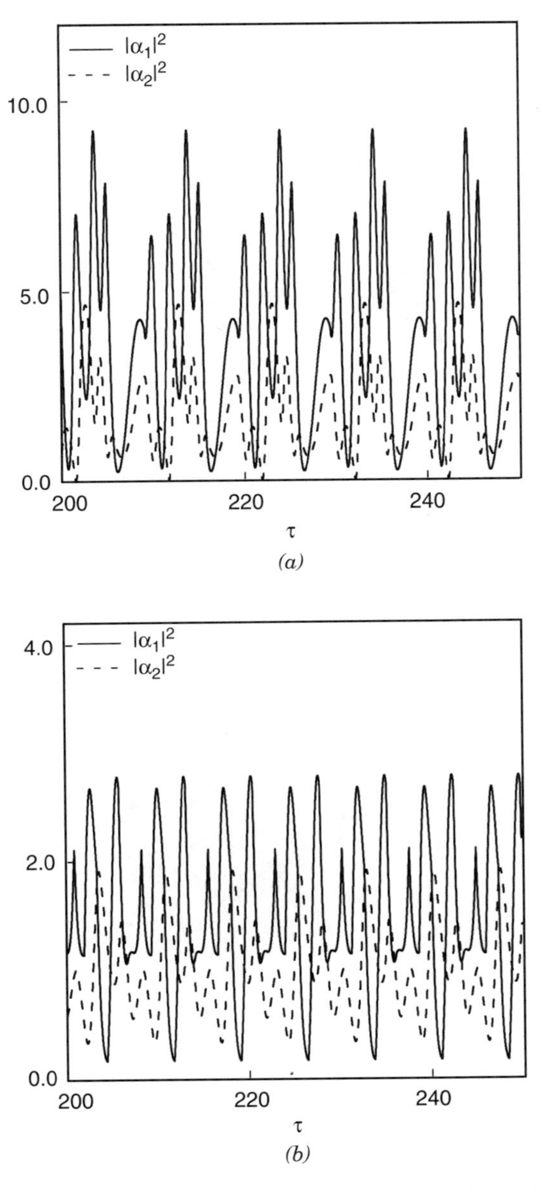

Figure 17. Evolution of the intensities related to the cases of Fig. 16: (a) five-period oscillations in GCL; (b) four-period oscillations in BCL.

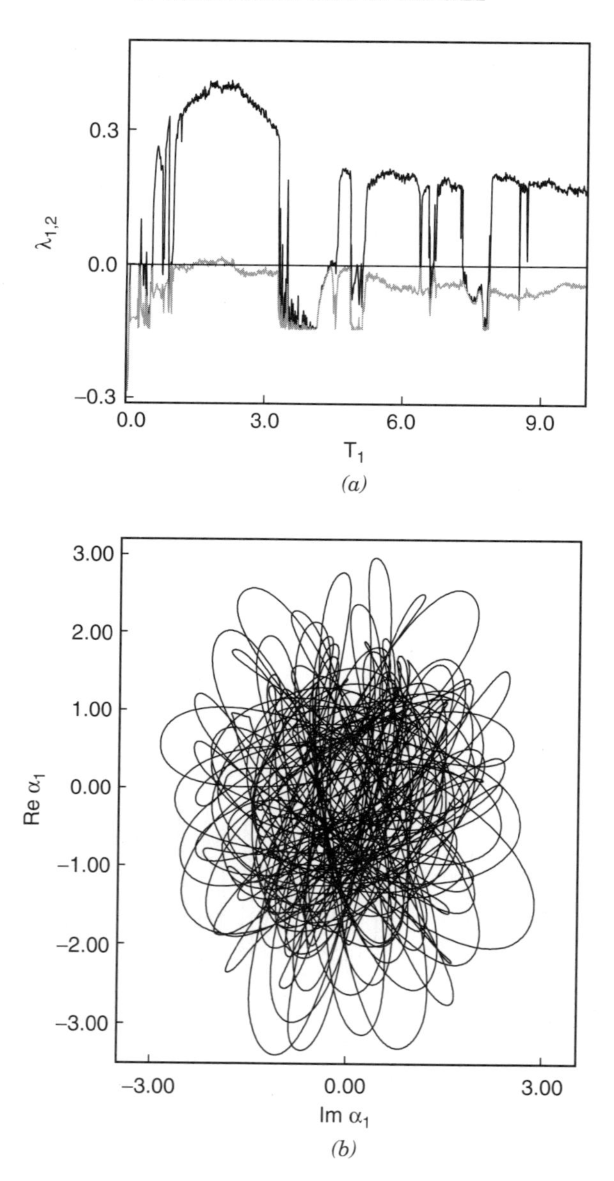

Figure 18. (a) The two largest Lyapunov exponents λ_1 and λ_2 versus the pulse duration T_1, for $f_0 = 2, \alpha_{10} = 0.1 + i\,0.1, \alpha_{20} = 0$ and for BCL; (b) typical hyperchaotic phase portrait for pulse duration $T_1 = 2.0$.

SHG can be used as a source of signals with chaotic or even hyperchaotic amplitudes that can be suddenly switched to the periodic regimes. This kind of performance can be employed for communications devices. We mention here the possibility of encoding a message within chaotic dynamics [124].

In order to relate the theory and numerical calculations to the physical parameters, we followed the estimations of Drummond et al. [104]. For a typical spherical Fabry–Pérot interferometer of length 10 cm with an appropriate crystal (e.g., KDP of length 1 cm), one can get approximate values of parameters of the SHG system. The typical damping constant for the mirror reflectivities 0.995 is $\gamma \simeq 10^6$. The coupling constant κ was estimated in the interval of 50–500 s^{-1}. These coupling constant values permit experimental verification of dynamical behavior of SHG. In preceding, sections the coupling constant κ is given by relation $\kappa = \tau/t$, where τ and t are the rescaled and real times, respectively. Therefore the parameter of modulation Ω can change between 0 and 3500 Hz (in our calculations $0 < \Omega < 7$ in arbitrary units). We also obtained the appropriate pulse repetition rate in an interval from 10^{-3} up to 10^{-2} s. This rather rough estimation allows experimental verification of our numerical analysis.

III. CHAOS IN KERR OSCILLATORS

A. Introduction

Since 1990 considerable interest has been devoted to mutually coupled dynamical systems. Different kinds of new dynamical behavior have been revealed and studied, including synchronization effects [125–128], ON-OFF intermittency [129], two-state ON-OFF intermittency [130], uncertain destination dynamics [131], or riddled basins of attractions [132]. Other interesting topics in the field of coupled nonlinear systems are generation of beats and their properties. The structure of beats has been intensely studied mainly in quantum and nonlinear optics. The intricate beats are frequently referred to as "revivals" and "collapse phenomena" [133]. The revivals and collapses, representing the structure of complicated modulations, remain quasiperiodic functions [134,135]. It is well known that beats in linear systems originate from the superposition of periodic functions with slightly different periods. The question is what are the changes in the structure of beats in a linear system if the linear system is supplemented by a nonlinear term and whether it is possible to generate chaotic beats.

One of the best known and most intensively studied optical models is an oscillator with Kerr nonlinearity. Mutually coupled Kerr oscillators can be successfully used for a study of couplers; the systems consist of a pair of coupled Kerr fibers. The first two-mode Kerr coupler was proposed by Jensen [136] and investigated in depth [136,137]. Kerr couplers affected by quantization can exhibit various quantum properties such as squeezing of vacuum fluctuations, sub-Poissonian statistics, collapses, and revivals [138,139].

In this section we consider a model of interactions between the Kerr oscillators applied by J. Fiurášek et al. [139] and Peřinová and Karská [140]. Each Kerr oscillator is externally pumped and damped. If the Kerr nonlinearity is turned off, the system is linear. This enables us to perform a simple comparison of the linear and nonlinear dynamics of the system, and we have found a specific nonlinear version of linear filtering. We study numerically the possibility of synchronization of chaotic signals generated by the Kerr oscillators by employing different feedback methods.

B. Basic Equations

The Hamilton function for a single Kerr oscillator is defined by

$$H(p,q) = \frac{p^2}{2} + \frac{\omega_0^2 q^2}{2} + \epsilon\left(\frac{p^2}{2} + \frac{\omega_0^2 q^2}{2}\right)^2 \tag{10}$$

where ϵ is the Kerr parameter. If $\epsilon = 0$, the Hamiltonian expressed here describes a simple harmonic oscillator with the natural frequency ω_0. The dynamical variables p and q denote the momentum and generalized coordinate, respectively. The Hamilton equations

$$\frac{dq}{dt} = \frac{\partial H}{\partial p} \tag{11}$$

$$\frac{dp}{dt} = -\frac{\partial H}{\partial q} \tag{12}$$

applied to the Hamiltonian (10) lead to the following coupled equations of motion:

$$\frac{dq}{dt} = p[1 + \epsilon(p^2 + \omega_0^2 q^2)] \tag{13}$$

$$\frac{dp}{dt} = -\omega_0^2 q[1 + \epsilon(p^2 + \omega_0^2 q^2)] \tag{14}$$

If the initial state of the system is determined by the initial conditions $q(0) = q_0$ and $p(0) = p_0$, the solution of the system (13)–(14) is given by

$$q(t) = q_0 \cos\omega_0[1 + \epsilon(p_0^2 + \omega_0^2 q_0^2)]t + \frac{p_0}{\omega_0}\sin\omega_0[1 + \epsilon(p_0^2 + \omega_0^2 q_0^2)]t \tag{15}$$

$$p(t) = p_0 \cos\omega_0[1 + \epsilon(p_0^2 + \omega_0^2 q_0^2)]t - q_0\omega_0 \sin\omega_0[1 + \epsilon(p_0^2 + \omega_0^2 q_0^2)]t \tag{16}$$

The system (13)–(14) has two independent constants of motion (first integrals): the Hamilton function (10) and

$$\phi(p, q; t) = -\omega_0 \left[1 + \epsilon\left(p^2 + \omega_0^2 q^2\right)\right] t + \arctan\left(\frac{\omega_0 q}{p}\right) \tag{17}$$

For $\epsilon = 0$, the quantities (10) and (17) become first integrals for the harmonic oscillator [141]. It is obvious from (15)–(16) that a trajectory in phase space (p, q) for the Kerr oscillator is analytically the same ellipse as for the harmonic oscillator

$$\frac{p^2}{p_0^2 + \omega_0^2 q_0^2} + \frac{\omega_0^2 q^2}{p_0^2 + \omega_0^2 q_0^2} = 1 \tag{18}$$

The only difference is that for the harmonic oscillator the phase point draws the ellipse with the frequency ω_0, whereas for the Kerr oscillator with the frequency, $\Omega = \omega_0[1 + \epsilon(p_0^2 + \omega_0^2 q_0^2)]$. The frequency Ω depends on the initial conditions, which is a feature typical of nonlinear conservative systems [143].

The set of equations (13)–(14) describes a conservative system. However, the effect of linear dissipation can be incorporated phenomenologically. Then, Eqs. (13)–(14) have the form

$$\frac{dq}{dt} = p[1 + \epsilon(p^2 + \omega_0^2 q^2)] - \gamma q \tag{19}$$

$$\frac{dp}{dt} = -\omega_0^2 q[1 + \epsilon(p^2 + \omega_0^2 q^2)] - \gamma p \tag{20}$$

where the terms γq and γp describe a loss mechanism, with the damping constant γ. The solution of the preceding equations is given by [142]

$$q(t) = e^{-\gamma t}\left(q_0 \cos N(t) + \frac{p_0}{\omega_0} \sin N(t)\right) \tag{21}$$

$$p(t) = e^{-\gamma t}(p_0 \cos N(t) - q_0 \omega_0 \sin N(t)) \tag{22}$$

where

$$N(t) = \omega_0 t + \frac{\epsilon\omega_0}{2\gamma}(p_0^2 + \omega_0^2 q_0^2)(1 - e^{-2\gamma t}) \tag{23}$$

If $\epsilon = 0$, the system (19)–(20) describes a damped linear oscillator governed by the equation

$$\frac{d^2 q}{dt^2} + 2\gamma \frac{dq}{dt} + (\omega_0^2 + \gamma^2) q = 0 \tag{24}$$

Generally, if Kerr systems are driven by external time-dependent forces, the equations of motion are nonintegrable and have to be studied numerically.

C. Dynamics of Linearly Coupled Kerr Oscillators

Let us consider a system of two classical oscillators with Kerr nonlinearity. Both oscillators interact with each other by way of a linear coupling; moreover, they are pumped by external time-dependent forces. The Hamiltonian for the system is given by

$$H = \sum_{i=1}^{2} [H_i + \epsilon_i H_i^2 - q_i F_i(t)] - \alpha q_1 q_2 \tag{25}$$

where the Hamiltonian $H_i = \frac{1}{2}(p_i^2 + \omega_0^2 q_i^2)$ describes a simple harmonic oscillator with the frequency ω_o. Moreover, $F_i(t) = A_i \cos\omega_i t$ is the time-dependent force, with the amplitude A_i and the frequency ω_i. The parameter of Kerr nonlinearity is denoted by ϵ_i. The interaction between the Kerr oscillators is governed by the term $\alpha q_1 q_2$, where α plays the role of an interaction parameter. The equations of motion for the system described by the Hamiltonian (25) are given by

$$\frac{dq_1}{dt} = p_1[1 + \epsilon_1(p_1^2 + \omega_0^2 q_1^2)] - \gamma_1 q_1 \tag{26}$$

$$\frac{dp_1}{dt} = -\omega_0^2 q_1[1 + \epsilon_1(p_1^2 + \omega_0^2 q_1^2)] + \alpha q_2 - \gamma_1 p_1 + A_1 \cos\omega_1 t \tag{27}$$

$$\frac{dq_2}{dt} = p_2[1 + \epsilon_2(p_2^2 + \omega_0^2 q_2^2)] - \gamma_2 q_2 \tag{28}$$

$$\frac{dp_2}{dt} = -\omega_0^2 q_2[1 + \epsilon_2(p_2^2 + \omega_0^2 q_2^2)] + \alpha q_1 - \gamma_2 p_2 + A_2 \cos\omega_2 t \tag{29}$$

where the terms $\gamma_i q_i$ and $\gamma_i p_i$ describe a loss mechanism. The loss mechanism has been incorporated phenomenologically. If the linear coupling parameter α is equal to zero, both anharmonic oscillators behave independently; that is, they do not interact with each other. Therefore, for $\alpha = 0$ the equations of motion (26)–(29) form two independent sets of equations. The equations of motion (26)–(29) give a four-dimensional nonautonomous system that can be easily autonomized [115] if we put $t = q_3$ in the functions $\cos\omega_i t$. Then, time becomes a dynamical variable and the fifth equation is given by

$$\frac{dq_3}{dt} = 1, \quad q_3(0) = 0 \tag{30}$$

In general, the system (26)–(30) is nonintegrable and its dynamics has to be studied numerically. We examined it with the help of a fourth-order Runge–Kutta method. To calculate Lyapunov exponents, we used the procedure proposed by Wolf et al. [114]. The spectrum of the autonomized system (26)–(30) is denoted by the symbols $\{\lambda_1, \lambda_2, \lambda_3, \lambda_4, \lambda_5\}$.

1. Noninteracting Oscillators

Let us first consider the case of noninteracting oscillators that takes place when the interaction parameter α in Eqs. (26)–(29) is equal to zero. Then, the system (26)–(29) consists of two independent subsystems in the dynamical variables (q_1, p_1) and (q_2, p_2). The parameters of the subsystems are $A_1 = A_2 = 200$, $\omega_0 = 1$, $\epsilon_1 = \epsilon_2 = 0.1$, $\gamma_1 = 0.05$, $\gamma_2 = 0.5$. The frequencies $\omega_{1,2}$ of the external driving forces vary in the range $0 < \omega_{1,2} < 3.2$. The autonomized spectrum of Lyapunov exponents $\{\lambda_1, \lambda_2, \lambda_3\}$ for the first oscillator *I* versus the frequency ω_1 is presented in Fig. 19a. We observe three types of spectra: $\{+, 0, -\}$, $\{0, 0, -\}$, and $\{0, -, -\}$. The first indicates a chaotic attractor; the second, a quasiperiodic orbit; and the third, a limit cycle. Therefore a change in the frequency ω_1 switches the chaotic oscillations (chaotic attractors) into nonchaotic oscillations (quasiperiodic orbits, limit cycles) and inversely. The autonomized spectrum of Lyapunov exponents for the second oscillator II versus the frequency ω_2 is shown in Fig. 19b. The difference between the two figures is essential. The chaotic regions in Fig. 19b do not appear at all because of the increase in damping in the system. The only attractors are limit cycles $\{0, -, -\}$. By way of an example, for identical frequencies $\omega_1 = \omega_2 = 0.55$, the Lyapunov spectra for the first and second oscillators are $\{0.08, 0.00, -0.23\}^{\mathrm{I}}$ and $\{0.00, -0.55, -0.90\}^{\mathrm{II}}$, respectively. The topology of the chaotic attractor in the phase space (q_1, p_1) is shown in Fig. 20a. The phase point starts from the initial conditions $q_{10} = 10$ and $p_{10} = 10$ and moves within the blackened area, which makes an attractor, after $t > 200$. In the phase plane (q_2, p_2) the phase point draws a limit cycle (Fig. 20b). The intricate structure of the limit cycle is related to multiperiodic oscillations of the system. The blackened areas at the top and bottom of the limit cycle have a periodic structure invisible in the scale of the phase portrait.

The single Kerr anharmonic oscillator has one more interesting feature. It is obvious that for $\epsilon_j = 0$ and $\gamma_j = 0$, the Kerr oscillator becomes a simple linear oscillator that in the case of a resonance $\omega_i = \omega_0$ manifests a primitive instability; in the phase space the phase point draws an expanding spiral. On adding the Kerr nonlinearity, the linear unstable system becomes highly chaotic. For example, putting $A_1 = 200$, $\omega_1 = \omega_0 = 1$, $\epsilon_1 = 0.1$ and $\gamma_1 = 0$, the spectrum of Lyapunov exponents for the first oscillator is $\{0.20, 0, -0.20\}^{\mathrm{I}}$. However, the system does not remain chaotic if we add a small damping. For example, if $\gamma_1 = 0.05$, then the spectrum of Lyapunov exponents has the form $\{0.00, -0.03, -0.12\}^{\mathrm{I}}$, which indicates a limit cycle.

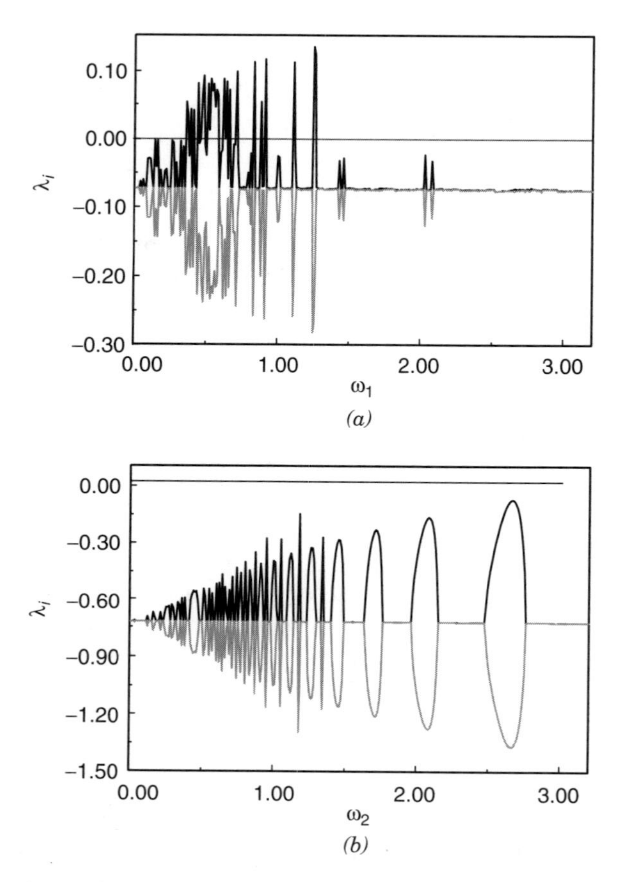

Figure 19. Spectra of Lyapunov exponents for the system (26)-(30) with $\alpha = 0$. The initial conditions are $q_{10} = 10, p_{10} = 10, q_{20} = 10$, and $p_{20} = 10$. (a) Spectrum $\{\lambda_1, \lambda_2, \lambda_3\}^{\mathrm{I}}$ for the first oscillator (I) versus the frequency ω_1 for $\omega_0 = 1, A_1 = 200$. $\gamma_1 = 0.05$, and $\epsilon_1 = 0.1$. (b) The same for the second oscillator (II) with the parameters: $\omega_0 = 1, A_2 = 200, \gamma_2 = 0.5$, and $\epsilon_2 = 0.1$.

2. *Interacting Oscillators*

If the interaction parameter α is switched on, the system of coupled oscillators (26)–(29) manifests a rich variety of spectacular behavior. Below, we concentrate on the most interesting ones. First, we answer the question as to how the attractors in Fig. 20 change when both oscillators interact with each other.

1. The Case $A_1 = A_2$, $\gamma_1 < \gamma_2$. The dynamics of the coupled oscillators is investigated for an interaction parameter α varying in the range $0 < \alpha < 1$. The

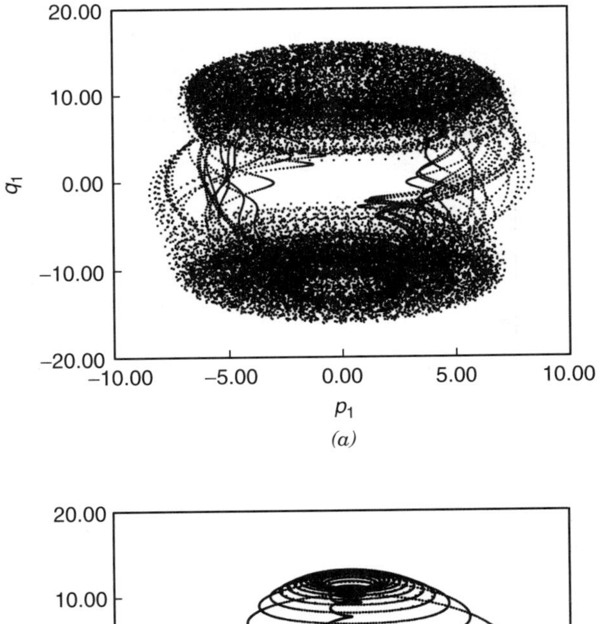

(a)

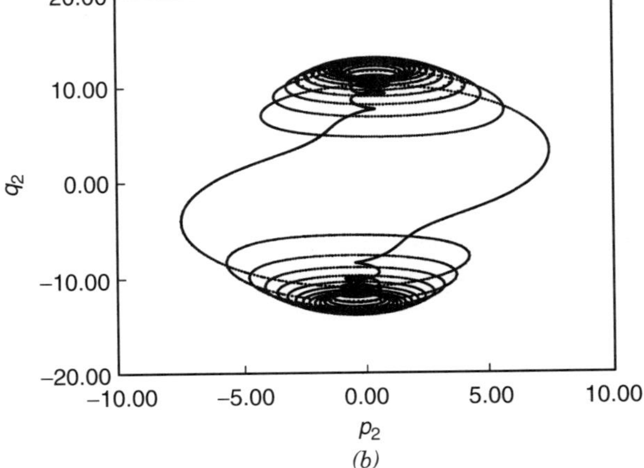

(b)

Figure 20. Phase portraits for the system (26)–(30) with $\alpha = 0$. The initial conditions are $q_{10} = 10, p_{10} = 10, q_{20} = 10$, and $p_{20} = 10$. (a) Phase portrait (q_1, p_1) of the first oscillator for $A_1 = 200, \omega_0 = 1, \epsilon_1 = 0.1, \omega_1 = 0.55$, and $\gamma_1 = 0.05$. (b) Phase portrait (q_2, p_2) of the second oscillator for $A_2 = 200, \omega_0 = 1, \epsilon_1 = 0.1, \omega_2 = 0.55$, and $\gamma_1 = 0.5$.

joint autonomized spectrum of Lyapunov exponents $\{\lambda_1, \lambda_2, \lambda_3, \lambda_4, \lambda_5\}$ versus the interaction parameter α is shown in Fig. 21. The value $\alpha = 0$ is a limit value related to the dynamics of the uncoupled oscillators. This has already been done in Section III.C.1 In the region $0 < \alpha < 0.74$ the chaotic behavior of the coupled oscillator system predominates over the nonchaotic one; thus, for most values of

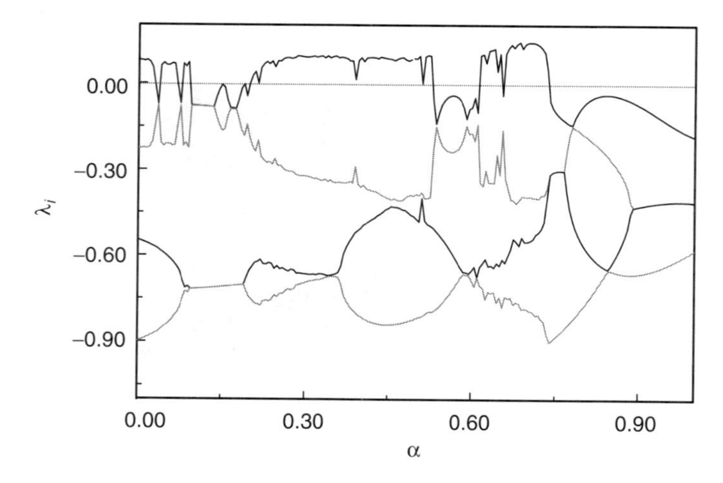

Figure 21. Spectrum of Lyapunov exponents $\{\lambda_1, \lambda_2, \lambda_3, \lambda_4, \lambda_5\}$ for the system (26)–(30) versus the interaction parameter α. The other parameters are $A_1 = A_2 = 200, \omega_0 = 1, \omega_1 = \omega_2 = 0.55$, $\epsilon_1 = \epsilon_2 = 0.1, \gamma_1 = 0.05$, and $\gamma_2 = 0.5$. The initial conditions are $q_{10} = 10, p_{10} = 10$, $q_{20} = 10$, and $p_{20} = 10$.

the parameter α, the maximal Lyapunov exponent is positive. For $0.68 < \alpha < 0.71$ we get the maximum chaos. For $\alpha > 0.74$ the system does not show chaotic behavior. Generally, the only spectra of Lyapunov exponents that appear in Fig. 21 are of types $\{+, 0, -, -, -\}$, $\{0, 0, -, -, -\}$, and $\{0, -, -, -, -\}$. These three types of spectra (for $\alpha > 0$) do not allow us to ascertain which of the two interacting oscillators is more (or less) chaotic than the other unless $\alpha = 0$. However, the dynamics of individual oscillators can be estimated with the help of the appropriate phase portraits. For example, if the interaction coupling is equal to $\alpha = 0.7$, the spectrum of Lyapunov exponents has the form $\{0.14, 0.00, -0.39, -0.55, -0.79\}$, and the appropriate phase portraits are as shown in Fig. 22. The attractors for the interacting oscillators shown in Fig. 22 are reminiscent of the attractors for noninteracting oscillators presented in Fig. 20. Let us note that the maximal Lyapunov exponent for the system of interacting oscillators, which is equal to $\lambda_1 = 0.14$, is greater than the maximal Lyapunov exponent for the uncoupled oscillators, which equals $\lambda_1 = 0.08$. Therefore, for $0.67 < \alpha < 0.72$, the coupled oscillators are more chaotic than their uncoupled version. However, as is seen from Fig. 21, this is not a rule. In the range $0.2 < \alpha < 0.5$ the values of the maximal Lyapunov exponent are of the rank ~ 0.08, which corresponds to the value for uncoupled oscillators (a measure of chaos in the coupled and uncoupled oscillators is in practice the same). Therefore, the linear coupling here is relatively small in order to

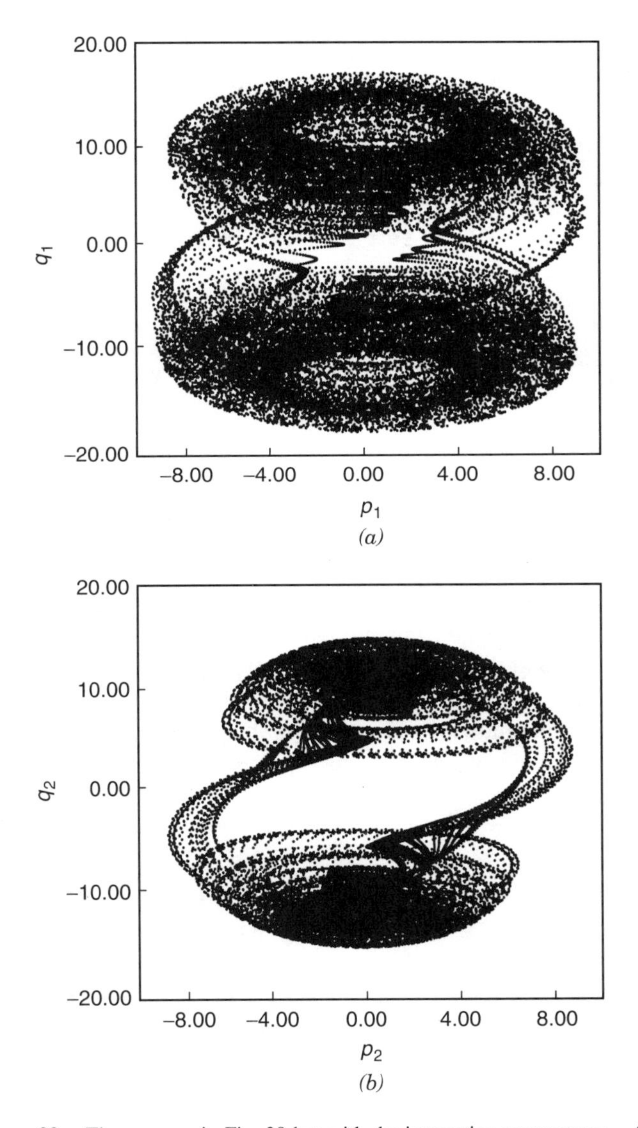

Figure 22. The same as in Fig. 20 but with the interaction parameter $\alpha = 0.7$.

additionally increase the instability of the system. Rather chaos flows from one oscillator to the other by the coupling term α.

2. The Case $A_1 = A$, $A_2 = 0$, $\gamma_1 = \gamma_2 = \gamma$. In what follows, we consider a simple version of the system (26)–(29), namely: both oscillators are equally damped ($\gamma_1 = \gamma_2 = \gamma$) and only the first oscillator is externally pumped ($A_1 = A, A_2 = 0$). Therefore, the equations of motion are

$$\frac{dq_1}{dt} = p_1[1 + \epsilon(p_1^2 + \omega_0^2 q_1^2)] - \gamma q_1 \tag{31}$$

$$\frac{dp_1}{dt} = -\omega_0^2 q_1[1 + \epsilon(p_1^2 + \omega_0^2 q_1^2)] + \alpha q_2 - \gamma p_1 + A\cos\omega t \tag{32}$$

$$\frac{dq_2}{dt} = p_2[1 + \epsilon(p_2^2 + \omega_0^2 q_2^2]) - \gamma q_2 \tag{33}$$

$$\frac{dp_2}{dt} = -\omega_0^2 q_2[1 + \epsilon(p_2^2 + \omega_0^2 q_2^2)] + \alpha q_1 - \gamma p_2 \tag{34}$$

This system in its linear version (i.e., when $\epsilon = 0$) is a dynamical filter. Suppose that the oscillators interact with each other with the interaction parameter $\alpha = 0.9$. The frequency ω of the external driving field varies in the range $0 < \omega < 4.2$. The other parameters of the system are $A = 200$, $\omega_0 = 1$, $\epsilon = 0.1$, and $\gamma = 0.05$. The autonomized spectrum of Lyapunov exponents $\{\lambda_1, \lambda_2, \lambda_3, \lambda_4, \lambda_5\}$ versus the frequency ω is presented in Fig. 23. In the range $0 < \omega < 0.2$ the system does not exhibit chaotic oscillation. Here, the maximal Lyapunov exponent $\lambda_1 = 0$ and the spectrum is of the type $\{0, -, -, -, -\}$ (limit cycles). For example, for $\omega = 0.05$ we have $\{0.00, -0.07, -0.07, -0.07, -0.07\}$, and the limit cycles are shown in Fig. 24. The blackened areas in Fig. 24 have a periodic structure invisible in the scale applied. In the range $0.21 < \omega < 3.41$,

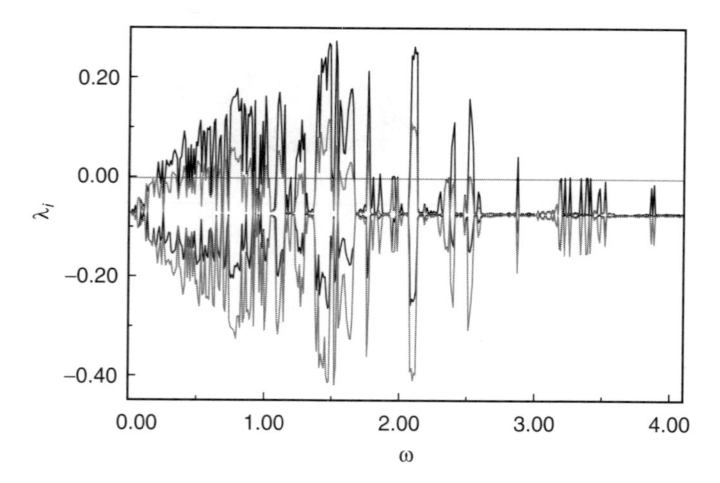

Figure 23. Spectrum of Lyapunov exponents $\{\lambda_1, \lambda_2, \lambda_3, \lambda_4, \lambda_5\}$ for the system (31–(34) versus the pump frequency ω. The other parameters are $A = 200, \omega_0 = 1, \gamma = 0.05, \epsilon = 0.1$, and $\alpha = 0.9$. The initial conditions are $q_{10} = 10, p_{10} = 10, q_{20} = 10$, and $p_{20} = 10$.

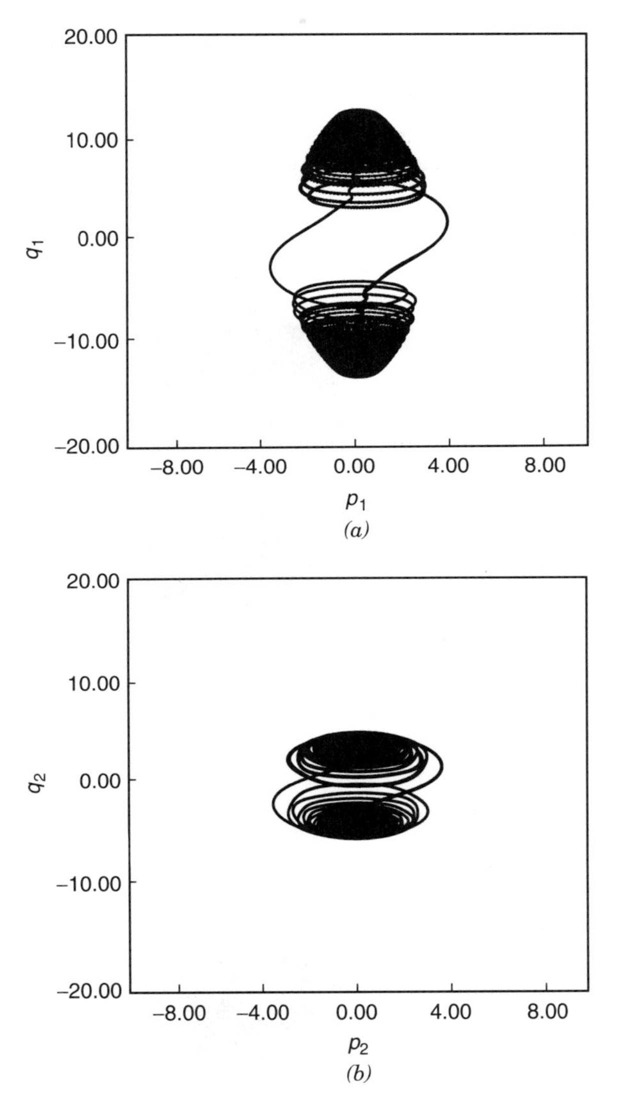

Figure 24. Phase portraits (q_1, p_1) and (q_2, p_2) for the system (31)-(34) with $\alpha = 0.9$. The other parameters are $A = 200, \omega_0 = 1, \epsilon = 0.1, \gamma = 0.05$, and $\omega = 0.05$. The initial conditions are the same as for Fig. 23. Limit cycles.

new types of spectra appear: $\{0, 0, -, -, -\}$, $\{+, 0, -, -, -\}$, and $\{+, +, 0, -, -\}$. The first indicate a quasiperiodic orbit; the second, a chaotic attractor, the third, a hyperchaotic attractor. Let us concentrate on the last and the most interesting case, with two positive Lyapunov exponents. The system reaches the highest degree of hyperchaos for $\omega = 2.1$. Then, the spectrum is $\{0.26, 0.10, 0.00, -0.25, -0.41\}$, and the behavior of the phase point is presented in the phase diagrams in Fig. 25. Here, the phase point starts from the initial state $q_{10} = q_{20} = p_{10} = p_{20} = 10$ and moves into the hyperchaotic attractor after $t > 50$. For $\omega > 3.41$ the system behaves orderly.

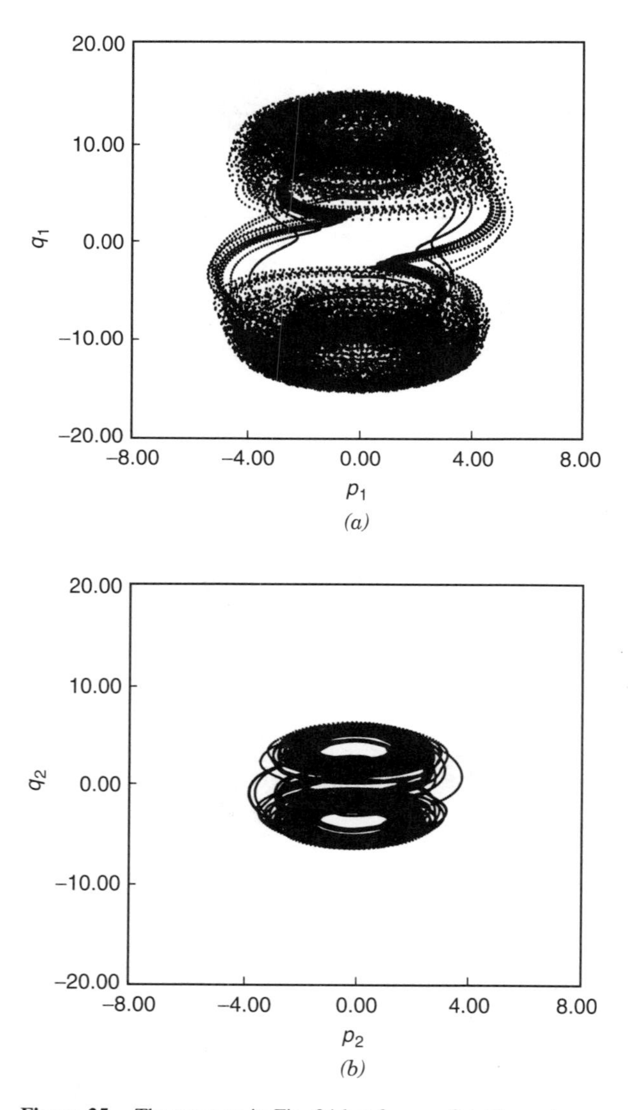

Figure 25. The same as in Fig. 24 but for $\omega = 2.1$. Hyperchaos.

The system (31)–(34) with $\epsilon = 0$ is a linear system with the normal frequencies $\Omega_1 = \sqrt{\omega_0^2 - \alpha}$ and $\Omega_2 = \sqrt{\omega_0^2 + \alpha}$. For $\omega_0 = 1$ and $\alpha = 0.9$, we have $\Omega_1 = 0.32$ and $\Omega_2 = 1.38$. It is known from linear dynamics that if $\Omega_1 > \omega > \Omega_2$, the steady-state amplitude of the first oscillator is greater than the steady-state amplitude of the second oscillator. If $\Omega_1 < \omega < \Omega_2$, we observe the inverse situation—the steady-state amplitude of the second oscillator is now greater than that of the first oscillator. This behavior is known as *dynamical filtering* of the signal $F(t) = A\cos\omega t$. The frequency range (Ω_1, Ω_2) is called the *charge-transfer band*, whereas Ω_1 and Ω_2 are the lower- and upper-band frequencies, respectively. The question is whether the filtering is, in a sense, also maintained in our nonlinear system. A detailed analysis shows that thevibrations of the first oscillator are always greater than the oscillations of the second oscillator, irrespective of the value of ω. This is also seen from the phase portraits in Figs. 24 and 25, which show that the *volume* of the attractor in the phase space (p_2, q_2) is always less than the attractors in the phase space (p_1, q_1).

The linear version $(\epsilon = 0)$ of the system (31)—(34) has one more interesting feature; namely, if $\gamma = 0$, $\omega = \omega_0$ and the following initial conditions are satisfied $(q_{10} = 0,\ p_{10} = 0,\ q_{20} = -A/\alpha\ ,\ p_{20} = 0)$, then the solutions of the linear equations of motion are $q_1(t) = 0, p_1(t) = 0, q_2(t) = (-A/\alpha)\cos\omega_0 t$ and $p_2(t) = (A\omega_0/\alpha)\sin\omega_0 t$. Therefore, the first oscillator remains in a state of rest and the second performs harmonic vibrations; such a system is frequently referred to as a *dynamical damper*. However, a nonlinear counterpart of the linear dynamical damper does not exist. For $\epsilon = 0.1$, $A = 9$, $\alpha = 0.9$, and $\omega = \omega_o = 1$, the system behaves hyperchaotically. The spectrum of Lyapunov exponents is $\{0.68, 0.04, 0.00, -0.04, -0.68\}$.

Finally, let us briefly consider the dynamical properties of the system (31)–(34) without damping, that is, when $\gamma = 0$. The other parameters are $\omega_0 = 1$, $\alpha = 0.9$, and $A = 200$. The appropriate spectrum of Lyapunov exponents $\{\lambda_1, \lambda_2, \lambda_3, \lambda_4, \lambda_5\}$ versus the frequency $0 < \omega < 2$ is presented in Fig. 26. As is seen from Fig. 26, the system is completely hyperchaotic. Here, the only type of spectrum is $\{+, +, 0, -, -\}$. This type of spectrum is a case of the symmetric spectrum (the axis symmetry is the Lyapunov exponent $\lambda_3 = 0$).

3. *Synchronization*

In chaotic motion trajectories starting from insignificantly different initial conditions diverge from each other exponentially. The question is whether we can converge chaotic signals from two identically or slightly different subsystems, both starting from different initial conditions. This behavior is possible by linking them with a common signal and synchronizing both outputs. We show that two single Kerr oscillators are a convenient system for synchronization. According to the continuous feedback method [52,61,125,127], we consider two Kerr subsystems (oscillators) where one subsystem is called the

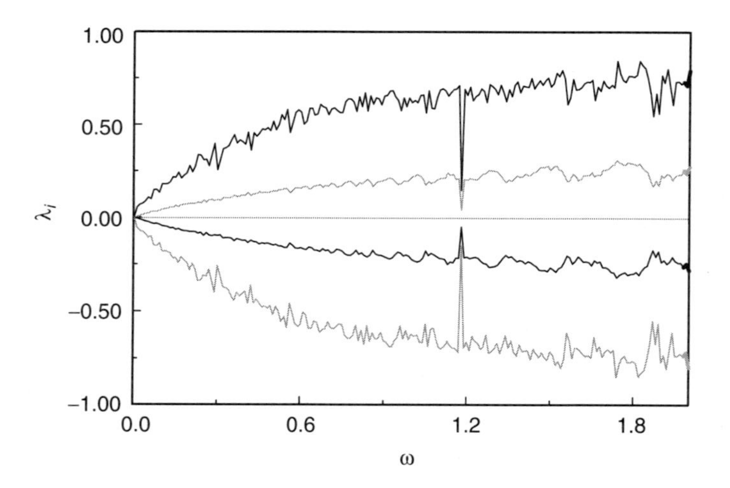

Figure 26. The same as in Fig. 23 but for $\gamma = 0$.

drive and the other the *response*. Both systems are coupled *unidirectionally* by a difference signal. The behavior of the response system depends only on the drive system, but not vice versa. The dynamics of our system is governed by the following set of equations:

$$\frac{dq_1}{dt} = p_1[+\epsilon_1(p_1^2 + \omega_0^2 q_1^2)] - \gamma_1 q_1 \tag{35}$$

$$\frac{dp_1}{dt} = -\omega_0^2 q_1[1 + \epsilon_1(p_1^2 + \omega_0^2 q_1^2)] - \gamma_1 p_1 + A_1 \cos\omega_1 t \; + S \tag{36}$$

$$\frac{dq_2}{dt} = p_2[1 + \epsilon_2(p_2^2 + \omega_0^2 q_2^2)] - \gamma_2 q_2 \tag{37}$$

$$\frac{dp_2}{dt} = -\omega_0^2 q_2[1 + \epsilon_2(p_2^2 + \omega_0^2 q_2^2)] - \gamma_2 p_2 + A_2 \cos\omega_2 t \tag{38}$$

where $S = \kappa(q_2 - q_1)$ is the difference signal and κ is the control parameter. As is seen, the second oscillator (drive) pumps a signal to the first oscillator (response) via the term S in Eq. (36) . The synchronization of chaos (for a chosen parameter κ and the initial conditions $q_{10} \neq q_{20}$ and $p_{10} \neq p_{20}$) takes place if the chaotic trajectory $q_1 = q_1(t)$ of the response oscillator jumps after some time into the chaotic trajectory $q_2 = q_2(t)$ of the drive oscillator. The time needed to uniform chaotic motions of subsystems is called a *synchronization time*.

Let us consider the dynamics of synchronization for the system (35)–(38) with the parameters $A_1 = A_2 = 200$, $\omega_0 = 1$, $\omega_1 = \omega_2 = 0.55$, $\gamma_1 = \gamma_2 = 0.05$, and $\epsilon_1 = \epsilon_2 = 0.1$. The initial conditions for the drive and response systems are $(q_{10}, p_{10}) = (10, 10)$ and $(q_{20}, p_{20}) = (5, 5)$, respectively. For $\kappa = 0$ both systems draw different chaotic orbits. Figure 27 shows the measure of synchronization $\Delta_s = q_1(t) - q_2(t)$ versus time t for $\kappa = 0.33$. The appearance of the

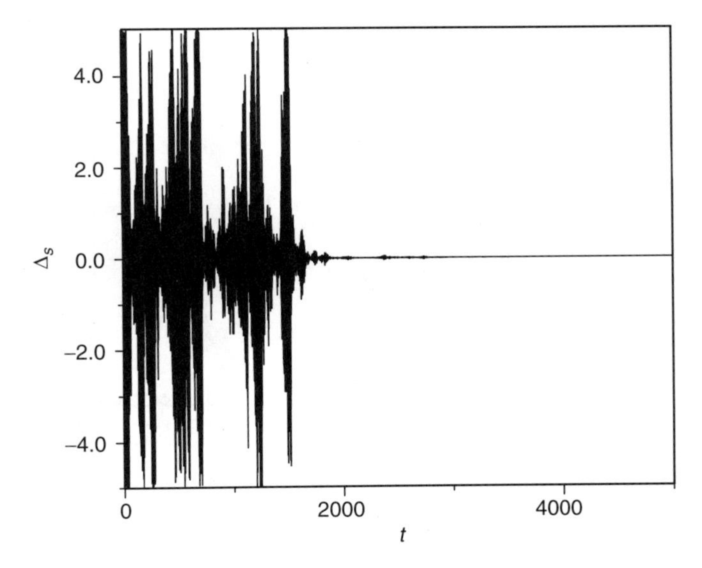

Figure 27. The time evolution of $\Delta_s = q_1 - q_2$ for $\kappa = 0.33$. The parameters and the initial conditions of the system (35)-(38) are $\omega_1 = \omega_2 = 0.55, A_1 = A_2 = 200, \omega_0 = 1, \epsilon_1 = \epsilon_2 = 0.1$, $\gamma_1 = \gamma_2 = 0.05, (p_{10}, q_{10}) = (10, 10)$, and $(p_{20}, q_{20}) = (5, 5)$.

straight line after $T_s \cong 2900$ clearly implies that both chaotic orbits have just been synchronized: $q_1(t) = q_2(t)$. The efficiency of the synchronization process depends on the parameter κ. This is illustrated in Fig. 28a, where the synchronization time T_s is presented as a function of the parameter κ. We observe four regions of synchronization: $0.28 < \kappa < 0.35$, $0.42 < \kappa < 0.54$, $0.77 < \kappa < 0.80$, and $\kappa > 1.58$. In the other regions it is not possible to achieve the synchronization effect. The synchronization time takes the minimum value $T_s \cong 200$ for $\kappa > 1.59$.

In physical terms the unidirectional synchronization means that the drive oscillator plays the role of an external source. The situation is different if one considers the problem of a *mutual synchronization* of two oscillators, which we may assume to be identical in all respect except for the initial conditions: $q_{10} \neq q_{20}$ and $p_{10} \neq p_{20}$. Let us consider the following model of mutual synchronization

$$\frac{dq_1}{dt} = p_1[1 + \epsilon_1(p_1^2 + \omega_0^2 q_1^2)] - \gamma_1 q_1 \tag{39}$$

$$\frac{dp_1}{dt} = -\omega_0^2 q_1[1 + \epsilon_1(p_1^2 + \omega_0^2 q_1^2)] - \gamma_1 p_1 + A_1 \cos\omega_1 t + S \tag{40}$$

$$\frac{dq_2}{dt} = p_2[1 + \epsilon_2(p_2^2 + \omega_0^2 q_2^2)] - \gamma_2 q_2 \tag{41}$$

$$\frac{dp_2}{dt} = -\omega_0^2 q_2[1 + \epsilon_2(p_2^2 + \omega_0^2 q_2^2)] - \gamma_2 p_2 + A_2 \cos\omega_2 t - S \tag{42}$$

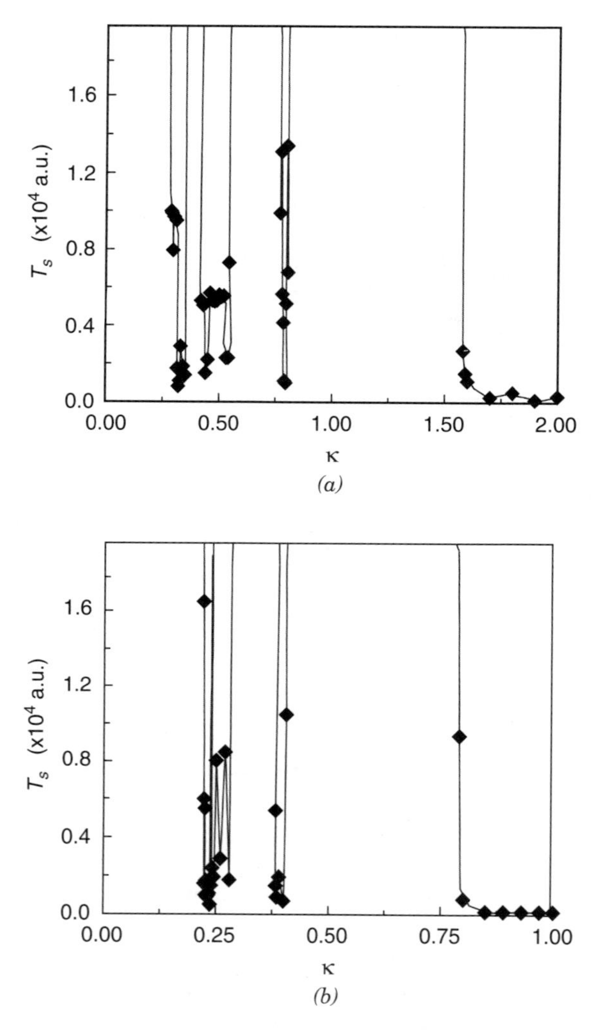

Figure 28. Synchronization time T_s versus κ. (a) for Eqs. (35)–(38); (b) for Eqs. (39)–(42). The parameters and the initial conditions are the same as for Fig. 27.

where $S = \kappa(q_2 - q_1)$. The system is similar to that governed by Eqs. (26)–(29). The equations of motion (39)–(42) can be derived from the Hamiltonian (25) if, instead of $\alpha q_1 q_2$, we put $0.5\kappa(q_1 - q_2)^2$. The values of the parameters and the initial conditions for the model of mutual synchronization are the same as for the unidirectional model. Synchronization takes place in the ranges $0.22 < \kappa < 0.27$, $0.38 < \kappa < 0.41$ and $\kappa > 0.79$, as is shown in Fig. 28b. For $\kappa > 0.80$ we obtain the minimum value of $T_s \cong 150$

It is interesting to note that the regions of unidirectional and mutual synchronization do not overlap. In both cases we have the critical value of κ (1.59, unidirectional synchronization; 0.80, mutual synchronization), after which the coupling is strong enough to maximize the process of synchronization.

4. *Chaotic Beats*

Let us now concentrate on the problem of beats generated by the system (26)–(29) without a loss mechanism ($\gamma_1 = \gamma_2 = 0$). For $\gamma_1 = \gamma_2 = 0$, and $\alpha = 0$, the dynamics of the system (26)–(29) is reduced to two noninteracting Kerr subsystems:

$$\frac{dq_j}{dt} = p_j[1 + \epsilon_j(p_j^2 + \omega_0^2 q_j^2)] \tag{43}$$

$$\frac{dp_j}{dt} = -\omega_0^2 q_j[1 + \epsilon_j(p_j^2 + \omega_0^2 q_j^2)] + A_j \cos\omega_j t\,, \quad j = 1, 2 \tag{44}$$

These Kerr oscillators, with $\epsilon_1 = \epsilon_2 = 0$, are linear subsystems that in the case of resonance ($\omega = \omega_1 = \omega_2$) exhibit a common instability—the solutions of Eqs. (43) and (44) for $t \to \infty$ grow linearly without bound. This resonance instability of our linear subsystems vanishes for $\epsilon_1 \neq 0$ and $\epsilon_2 \neq 0$. The subsystems become stable but only for small values of ϵ_1 and ϵ_2. For example, beats generated by the first oscillator for $\epsilon_1 = 10^{-9}$, $A_1 = 200$, and $\omega_0 = \omega_1 = 1$ are illustrated in Fig. 29a, and the appearing beats originate from the Kerr nonlinearity.

Beats generated by the second oscillator for $\epsilon_2 = 10^{-9}$, $A_2 = 200$, $\omega_0 = 1$, and $\omega_2 = 1.05$ are shown in Fig. 29b. The Lyapunov analysis of beats presented in Fig. 29 leads to the conclusion that the beats have a quasiperiodic nature, or, as we frequently say, they are almost periodic solutions and our system can be treated as a nearly linear system [143]. The structure of beats in the coupled system (26)–(29) is much more intricate than for the individual noninteracting subsystems (43)–(44), where the beats are quasiperiodic functions. Let us suppose that the individual noninteracting oscillators ($\alpha = 0$) behave as presented in Fig. 29 and answer the question as to how the structure of beats in both figures change when the oscillators interact with each other ($\alpha \neq 0$), that is, how the occurrence of beats in the coupled oscillators depends on the selected value of α. Numerical calculations show that the coupled system generates distinct beats if $\alpha < 0.3$. Let us now have a look at the Lyapunov analysis of beats. The autonomized spectrum of Lyapunov exponents for the system (26)–(29) versus the coupling parameter ($0 < \alpha < 0.16$) is presented in Fig. 30. As is seen, the most spectacular behavior of the system is observed for $0.01 < \alpha < 0.13$. In this range our system generates beats and behaves hyperchaotically. The magnitude of chaos depends on the value of the coupling parameter α. The highest degree

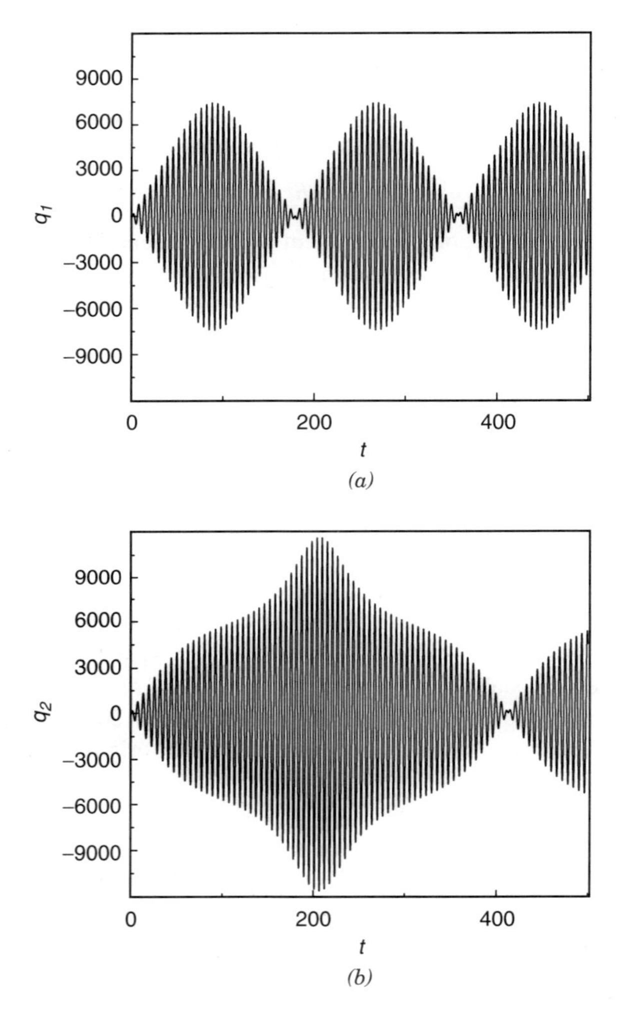

Figure 29. Evolution of q_1 and q_2 versus t for Eqs. (26)–(29) with $\alpha = 0$. The initial conditions are $q_{10} = 10, p_{10} = 10, q_{20} = 10$, and $p_{20} = 10$. The other parameters are $A_1 = A_2 = 200, \gamma_1 = \gamma_2 = 0, \epsilon_1 = \epsilon_2 = 10^{-9}$, and $\omega_0 = \omega_1 = 1$ (a), $\omega_0 = 1, \omega_2 = 1.05$ (b).

of hyperchaos is achieved at $\alpha = 0.04$, and the spectrum of the Lyapunov exponents is given by the following set $\{0.013, 0.003, 0.000, -0.003, -0.013\}$. The beats with chaotic envelopes in the q_1- component are shown in Fig. 31a. The envelope function is very sensitive to the interaction parameter α. A small change in α, for example, from $\alpha = 0.04$ to $\alpha = 0.05$ drastically changes the shape of the envelope function (Fig. 31a,b), leaving the basic frequency of oscillations almost unchanged (Fig. 31, window). As seen in Fig. 31, the

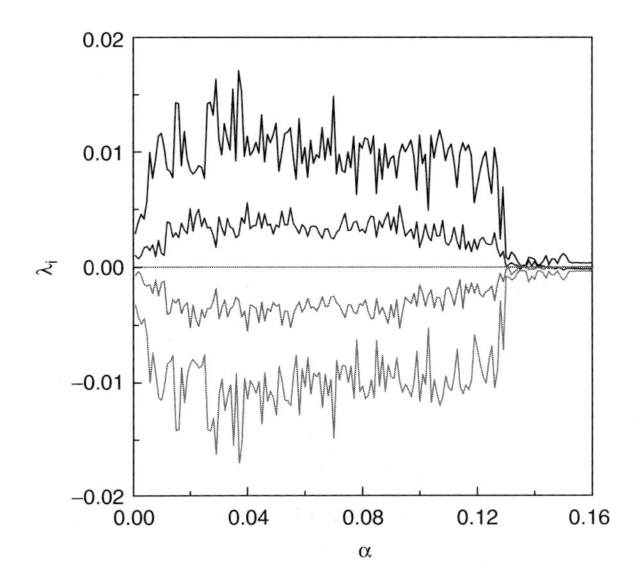

Figure 30. Spectra of Lyapunov exponents $\{\lambda_1, \lambda_2, \lambda_3, \lambda_4, \lambda_5\}$ for the system (26)–(29) versus the coupling constant α. The other parameters are $A_1 = A_2 = 200, \gamma_1 = \gamma_2 = 0, \omega_0 = 1, \omega_1 = 1, \omega_2 = 1.05$, and $\epsilon_1 = \epsilon_2 = 10^{-9}$. The initial conditions are $q_{10} = 10, q_{20} = 10, p_{10} = 10$, and $p_{20} = 10$.

envelope functions can be drawn as smooth functions, in contradistinction to the envelopes of beats generated stochastically [144,145]. For $\alpha > 0.16$ the beats lose their chaotic behavior and for $\alpha > 0.4$, the beats vanish completely.

It is interesting that envelope functions can also behave as multiperiod oscillations. This takes place if we take into account small damping. By way of an example, for the damping constant $\gamma_1 = \gamma_2 = 0.1$, the envelope function has a feature of two period doubling oscillations.

5. *Final Remarks*

The dynamics of two linear coupled Kerr oscillators strongly depends on the value of the interaction parameter α, frequency of pumping fields ω_j, and the damping constants γ_j. If the oscillators are coupled, both undergo a homogenization regarding the nature of their motion; either both are chaotic, or both are ordered, as is obvious from the phase graphs. For some parameters chaotic signals generated by the Kerr oscillators can be synchronized. Both unidirectional and mutual synchronization have been studied. The phenomenon of beats appears in linear and nonlinear systems whenever an impressed frequency is close to a natural frequency of a linear system or whenever two slightly different frequencies are impressed on a system regardless of what its natural frequencies

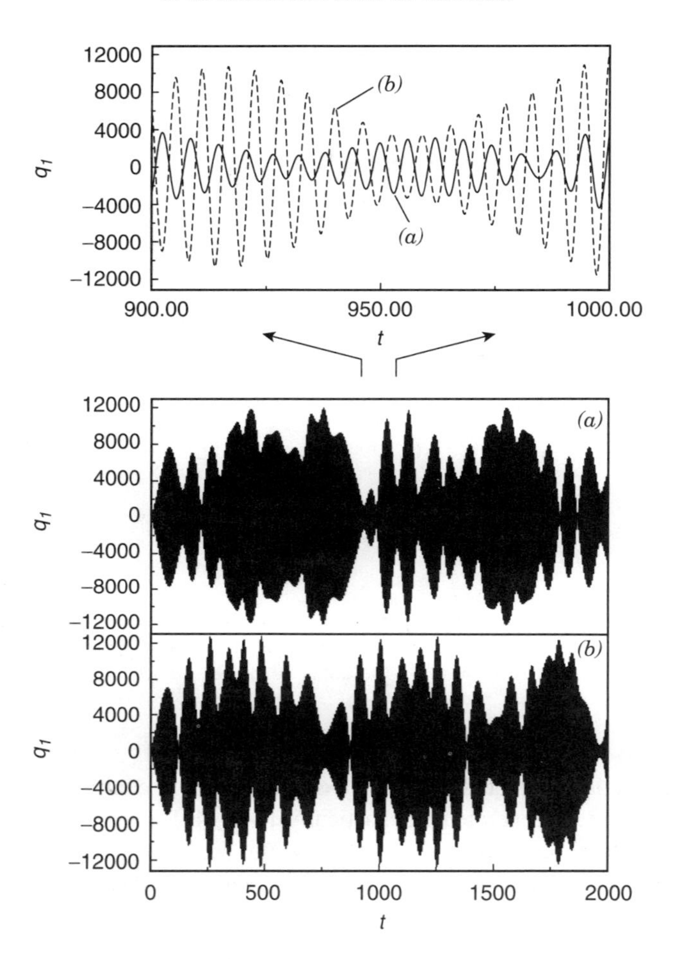

Figure 31. Evolution of $q_1(t)$ versus for Eqs. (26)-(29) from the initial conditions $q_{10} = 10, p_{10} = 10, q_{20} = 10$, and $p_{20} = 10$ for (a) $\alpha = 0.04$, (b) $\alpha = 0.05$. The other parameters are $A_1 = A_2 = 200, \gamma_1 = \gamma_2 = 0, \omega_0 = 1, \omega_1 = 1, \omega_2 = 1.05$, and $\epsilon_1 = \epsilon_2 = 10^{-9}$.

may be. For the small parameters of nonlinearity $\epsilon = 10^{-9}$, the quasiperiodic beats in uncoupled Kerr oscillators become beats with chaotic envelopes if the Kerr oscillators are linearly coupled. A small change in the interaction parameter rapidly changes the shape of the envelopes, whereas the basic frequencies of vibrations remains practically unchanged. Therefore the coupled oscillators can be used as a source of signals with chaotic envelopes and stable fundamental frequency. The appropriate materials useful for the generation of beats with chaotic envelopes could be optical systems consisting of a pair of coupled Kerr fibers [138,139,146]. Since the pioneering work by Jensen [136], twin-core nonlinear fibers (so-called couplers) have been one of the highest-priority topics of fiberoptic research. The couplers are expected to find important applications as all-optical switches [147] in photonics, for example. Another interesting problem connected with optical application to secure communication is synchronization of coupled systems [148,149].

D. Dynamics of Nonlinearly Coupled Kerr Oscillators

Let us now consider a system of two nonlinearly coupled Kerr oscillators. Now, we write the Hamiltonian (25) in the form

$$H = \sum_{i=1}^{2} [H_i + \epsilon_i H_i^2 - q_i F_i(t)] + 2\epsilon_{12} H_1 H_2 \tag{45}$$

where ϵ_{12} the intermodal coupling constant. The autonomized equations of motion for the Hamiltonian (45) have the following form:

$$\frac{dq_1}{dt} = p_1[1 + \epsilon_{11}(p_1^2 + \omega_0^2 q_1^2) + \epsilon_{12}(p_2^2 + \omega_0^2 q_2^2)] - \gamma_1 q_1 \tag{46}$$

$$\begin{aligned}\frac{dp_1}{dt} &= -\omega_1^2 q_1[1 + \epsilon_{11}(p_1^2 + \omega_0^2 q_1^2) + \epsilon_{12}(p_2^2 + \omega_0^2 q_2^2)] \\ &\quad - \gamma_1 p_1 + A_1 \cos\omega_1 t\end{aligned} \tag{47}$$

$$\frac{dq_2}{dt} = p_2[1 + \epsilon_{22}(p_2^2 + \omega_0^2 q_2^2) + \epsilon_{12}(p_1^2 + \omega_0^2 q_1^2)] - \gamma_2 q_2 \tag{48}$$

$$\begin{aligned}\frac{dp_2}{dt} &= -\omega_0^2 q_2[1 + \epsilon_{22}(p_2^2 + \omega_0^2 q_2^2) + \epsilon_{12}(p_1^2 + \omega_0^2 q_1^2)] \\ &\quad - \gamma_2 p_2 + A_2 \cos\omega_2 t\end{aligned} \tag{49}$$

$$\frac{dq_3}{dt} = 1, \qquad q_3(0) = 0 \tag{50}$$

Let us emphasize that if $A_j = 0$, the set of equations (46)–(50) is integrable and has a relatively simple analytic solution. If the initial state of the system is determined by the initial conditions $q_j(0) = q_{j0}$ i $p_j(0) = p_{j0}$, the analytic solution is given by [110]

$$q_j(t) = e^{-\gamma_j t}\left(q_{j0}\cos E_j(t) + \frac{p_{j0}}{\omega_j}\sin E_j(t)\right) \tag{51}$$

$$p_j(t) = e^{-\gamma_j t}(p_{j0}\cos E_j(t) - \omega_j q_{j0}\sin E_j(t)), \qquad j = 1,2 \tag{52}$$

where

$$E_1(t) = \omega_0 t + \frac{\epsilon_1 \omega_0}{\gamma_1} H_{10}(1 - e^{-2\gamma_1 t}) + \frac{\epsilon_{12}\omega_0}{\gamma_2} H_{20}(1 - e^{-2\gamma_2 t}) \tag{53}$$

$$E_2(t) = \omega_0 t + \frac{\epsilon_2 \omega_0}{\gamma_2} H_{20}(1 - e^{-2\gamma_2 t}) + \frac{\epsilon_{12}\omega_0}{\gamma_1} H_{10}(1 - e^{-2\gamma_1 t}) \tag{54}$$

$$H_{j0} = \frac{p_{j0}^2}{2} + \frac{\omega_0^2 q_{j0}^2}{2}, \qquad j = 1, 2 \tag{55}$$

The system (46)–(50) is examined numerically with the following parameters: $A_1 = A_2 = 200$, $\omega_0 = 1$, $\epsilon_1 = \epsilon_2 = 0.1$, $\gamma_1 = 0.05$, $\gamma_2 = 0.5$. The frequencies $\omega_{1,2}$ of the external driving forces and the cross interaction Kerr constant ϵ_{12} vary in the range $0 < \omega_{1,2} < 3$ and $0 < \epsilon_{12} < 1.5$, respectively. Therefore we study the dynamics of two nonlinearly coupled oscillators, I and II, which differ only in the value of the damping constants γ_1 and γ_2.

1. *Noninteracting Oscillators*

The case of noninteracting oscillators takes place when the coupling constant ϵ_{12} is equal to zero. Then, the systems (46)–(50) with $\epsilon_{12} = 0$ and (26)–(29) with $\alpha = 0$ are identical, and their dynamics are considered in Section III.C.1.

2. *Interacting Oscillators*

Let us now consider the behavior of the system when the Kerr coupling constant is switched on ($\epsilon_{12} \neq 0$). For brevity and clarity, we restrict our discussion to the question of how the attractors in Fig. 20 change when both oscillators interact with each other. To answer this question, let us have a look at the joint autonomized spectrum of Lyapunov exponents for the two oscillators $\{\lambda_1, \lambda_2, \lambda_3, \lambda_4, \lambda_5\}$ versus the interaction parameter $0 < \epsilon_{12} < 0.7$. The spectrum is seen in Fig. 32 and describes the dynamical properties of our oscillators in a global sense. The dynamics of individual oscillators can be glimpsed at the appropriate phase portraits. Let us now fix our attention on a detailed analysis of Fig. 32. For the limit value $\epsilon_{12} = 0$, the dynamics of the uncoupled oscillators has already been presented in Fig. 20. In the case of very weak interaction $0 < \epsilon_{12} < 0.0005$, the system of coupled oscillators manifests chaotic behavior. For $\epsilon_{12} = 0.0005$ we obtain the spectrum $\{0.06, 0.00, -0.21, -0.54, -0.89\}$. It is interesting to note that the maximal Lyapunov exponent $\lambda_1 = 0.08$ for the system of noninteracting oscillators ($\epsilon_{12} = 0$) is greater than the maximal Lyapunov exponent $\lambda_1 = 0.06$ for the coupled system with the parameter $\epsilon_{12} = 0.0005$. Therefore, in this case, the uncoupled system is more chaotic than the coupled system. A further increase in the interacting parameter ϵ_{12} leads to the disappearance of chaos. In the region $0.0005 < \epsilon_{12} < 0.15$ the oscillators behave orderly and nonchaotically. By way of example, for $\epsilon_{12} = 0.1$, all the values of

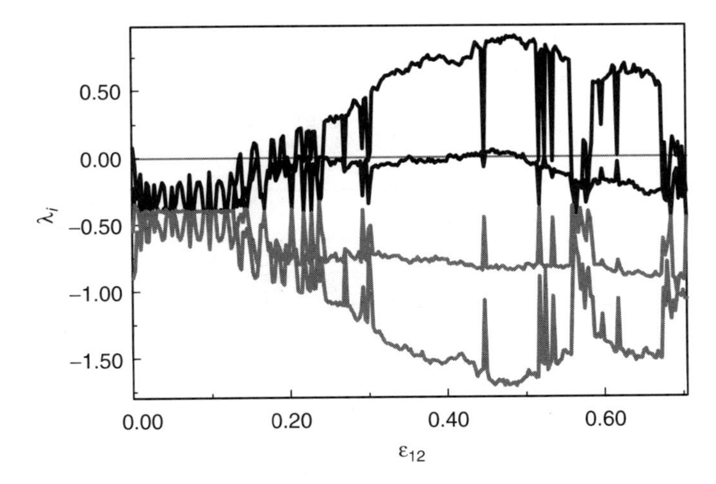

Figure 32. Spectrum of Lyapunov exponents $\{\lambda_1, \lambda_2, \lambda_3, \lambda_4, \lambda_5\}$ for the system (46)–(50) versus the Kerr coupling constant ϵ_{12}. The other parameters are $\omega_0 = 1, \omega_1 = \omega_2 = 0.55$, $A_1 = A_2 = 200$, and $\epsilon_1 = \epsilon_2 = 0.1$. The initial conditions are $q_{10} = 10, p_{10} = 10, q_{20} = 10$, and $p_{20} = 10$.

Lyapunov exponents are nonpositive: $\{0.00, -0.12, -0.26, -0.53, -0.68\}$. In this case the appropriate limit cycles are shown in Fig. 33a,b. The intricate structure of the limit cycles is reminiscent of the structure seen in Fig. 20. The blackened areas in Fig. 33 contain some pattern structure invisible in the scale used. As we see from Fig. 32, the situation changes in the region $0.15 < \epsilon_{12} < 0.43$. Chaotic behavior of the system predominates over nonchaotic behavior—for most values of the parameter ϵ_{12}, one Lyapunov exponent is positive. The most spectacular behavior of the coupled oscillators is observed in the region $0.43 < \epsilon_{12} < 0.49$. Here, two positive Lyapunov exponents in the spectrum indicate hyperchaotic behavior of the system. The highest degree of hyperchaos is achieved by the system at $\epsilon_{12} = 0.46$. The spectrum of the Lyapunov exponents is given by the set $\{0.87, 0.05, 0.00, -0.83, -1.71\}$, pointing to the existence of an hyperchaotic attractor. Its topology in the phase portraits (q_1, p_1) and (q_2, p_2) is shown in Fig. 34a,b. Precisely, in the phase portraits the system initially manifests a transient behavior but then (for $t > 500$) settles into a hyperchaotic attractor.

For $\epsilon_{12} \geq 0.49$ we observe a reduction of hyperchaos to chaos. Generally, in the region $0.49 \leq \epsilon_{12} \leq 0.75$ chaos dominates order and is maximal for the value $\epsilon_{12} = 0.63$, and the spectrum is $\{0.67, 0.00, -0.20, -0.90, -1.48\}$. Spectacular chaotic attractors appear for $\epsilon_{12} = 0.7$. Their phase portraits are presented in Fig. 35, where both attractors make impressions of spread limit cycles, as

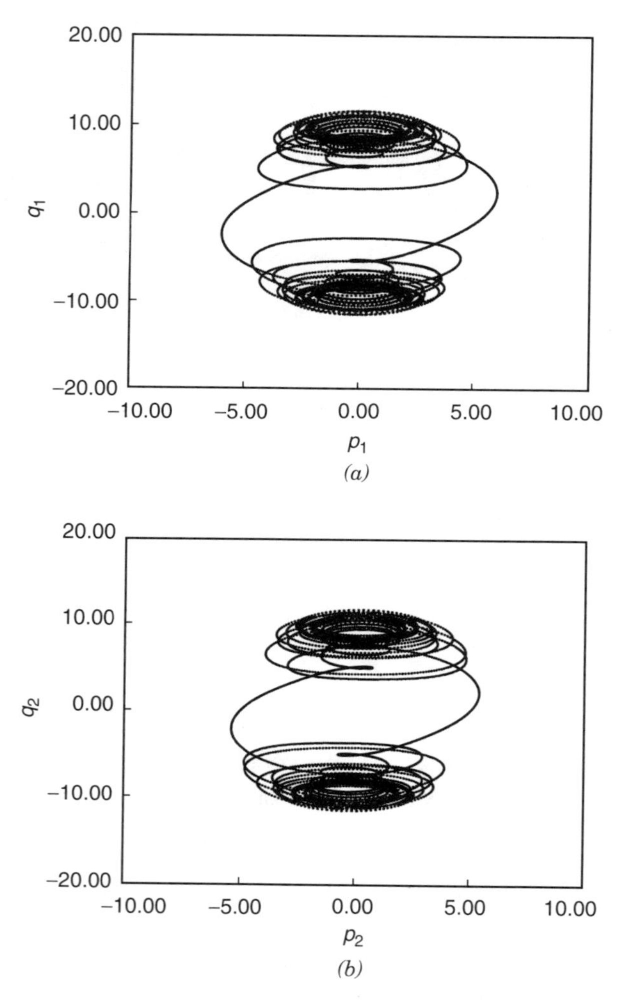

Figure 33. Phase portraits for the system (46)–(50) for $\epsilon_{12} = 0.1$ with the initial conditions $q_{10} = 10, p_{10} = 10, q_{20} = 10$, and $p_{20} = 10$: (a) phase portrait (q_1, p_1) of the first oscillator for $A_1 = 200, \omega_0 = 1, \epsilon_1 = 0.1, \omega_1 = 0.55$, and $\gamma_2 = 0.05$; (b) phase portrait (q_2, p_2) of the second oscillator for $A_2 = 200, \omega_0 = 1, \epsilon_2 = 0.1, \omega_2 = 0.55$, and $\gamma_2 = 0.5$. Order.

chaos is relatively small here. The spectrum of Lyapunov exponents is $\{0.06, 0.00, -0.31, -0.92, -1.01\}$.

3. *Final Remarks*

The emergence of order and chaos in the system of two oscillators depends on the value of the Kerr coupling constant ϵ_{12}. For the fixed parameters of damping

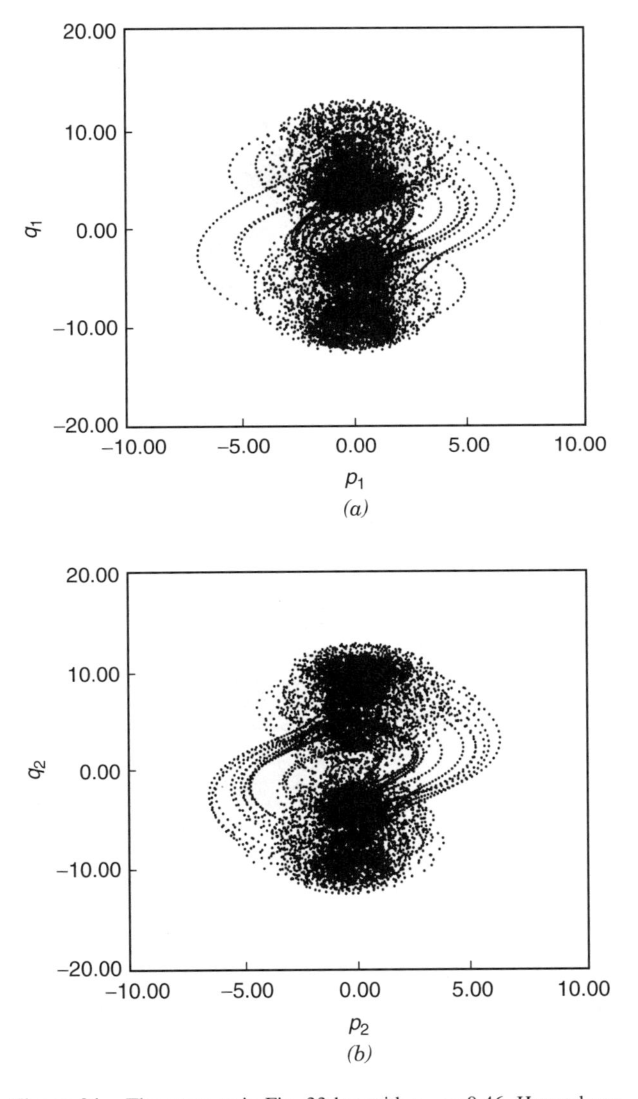

Figure 34. The same as in Fig. 33 but with $\epsilon_{12} = 0.46$. Hyperchaos.

γ_i, the sum of all exponents in the Lyapunov spectrum is not an invariant of the parameter ϵ_{12}. For the noninteracting oscillators ($\epsilon_{12} = 0$), the sum is equal to $\sum_{i=1}^{5} \lambda_i = -1.60$ and tends to the value $\sum_{i=1}^{5} \lambda_i = -2.25$ if $\epsilon_{12} \to 0.7$. Therefore we can say that the coupling term with ϵ_{12} in the equations of motion has an attribute of damping. These negative values result from nonconservation of volume in phase space (for conservative systems, the sum of Lyapunov

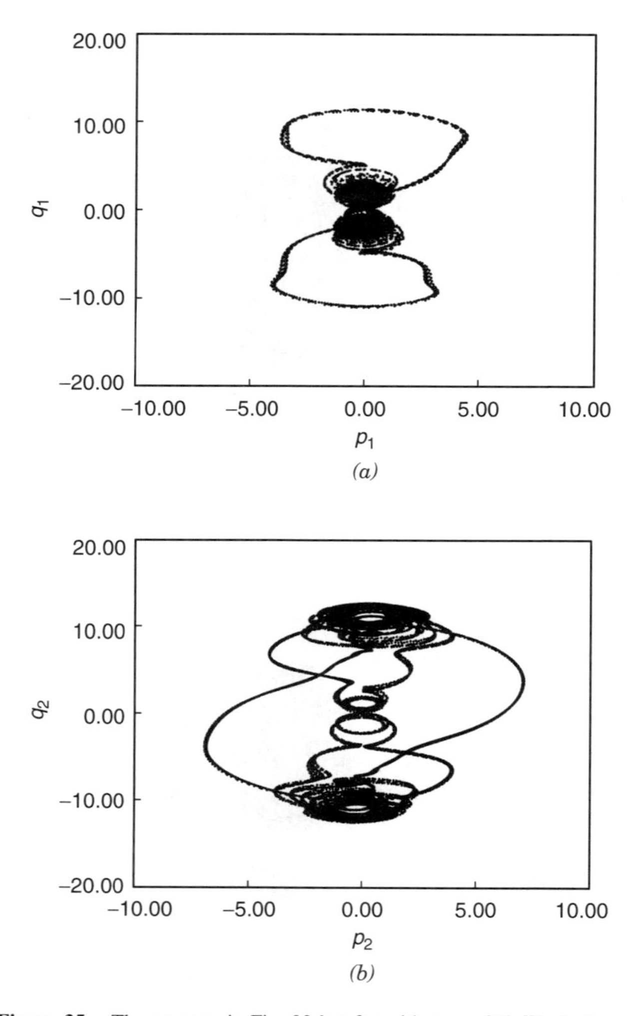

Figure 35. The same as in Fig. 33 but for with $\epsilon_{12} = 0.7$. Weak chaos.

exponents equals zero). Obviously, even if the volume of the system is suppressed, this does not mean that its length is equally suppressed in *all* directions. Some directions are stretched. In the direction of stretching we observe only an exponential separation of the trajectories, namely, chaotic or hyperchaotic behavior of the system. Finally, let us emphasize that the appropriate media for the experimental studies of chaotic behavior generated by Kerr nonlinearities could be optical fibers. The appearance of chaotic output signals generated by Kerr media means that the signals are unstable. The instability depends on the value of the coupling constant ϵ_{12}. Therefore, by changing the value of the coupling constant, we can turn the output chaotic signals into the periodic ones and vice versa. Promising materials for the implementation of nonlinear Kerr oscillators also seem to be some organic polymers [150].

IV. QUANTUM CHAOS

The modifications introduced by quantum mechanics into the dynamics of classical systems that manifest deterministic chaotic behavior are frequently referred to collectively as "quantum chaos" [4,6,13,151– 161]. It is rather conceded that quantization drastically modifies classically chaotic behavior. For example, suppression of chaos to quasiperiodicity is observed in the quantum kicked rotator, whose classical counterpart behaves chaotically [6,151,152]. In the system of a hydrogen atom in a microwave field, quantum effects suppress diffusive ionization by the mechanism of quantum localization [153,154]. Certain manifestations of chaos also become apparent in quantum optics [84, 162–167]. It seems that Wigner's formulation of quantum mechanics offers the simplest comparison between quantum and classical chaos in contradistinction to the conventional procedure. The conventional way is to study how a wavepacket initially fixed around a certain position q and momentum p follows the appropriate classical trajectory. However, this involves a disadvantage. Specifically, the wavepacket spreads in the course of time and is no longer sharply fixed around a particular position and momentum, rendering dubious the comparison with the respective classical trajectory. To avoid this spreading problem, we can make use of the so-called Wigner symbols, which are a quantum generalization of classical variables. For example, we can compare the time evolution of the Wigner symbols for the position $\hat{q}$ and momentum $\hat{p}$ operators with the classical evolution of the position q and momentum p, respectively. Generally, Wigner's formulation of quantum mechanics leads to a c-number representation of the density matrix, that is, to the quantum analog of a classical probability density in (p, q) space. In quantum optics three kinds of c-number functions are the most popular, the P representation, the Q function, and the Wigner function W [168]. All these three functions are defined in ($\alpha = p + iq$, $\alpha^* = p - iq$) space instead of in (p, q) space. This is due to the coherent state technique. The P representation is related to normal ordering of the creation $\hat{a}^+$ and annihilation $\hat{a}$ operators, the Q function is related to antinormal ordering of the operators, and the Wigner function W is related to symmetric (Weyl) ordering. The c-number approach makes it possible to treat quantum systems in a "classical way," including all their quantum features and contrasting the quantum and classical dynamics within the framework of a phase picture. The equations for the Wigner-like functions P and Q belong to the class of generalized Fokker–Planck equations whose solutions are known only for some simple optical models. The Wigner approach can also be used to study both "kicked" dynamics (i.e., a quantum map) and a continuous flow. Kicked models are easier to analyze numerically than continuous models but are more difficult to verify practically. On the other hand, continuous models seem to be mathematically more cumbersome, resembling the complexity of hydrodynamical systems. In the latter case we usually make some truncations leading to a set of ordinary differential equations. Historically, for the first time in the treatment of classical dynamical systems, a truncation method was used by Lorenz [2]. A similar truncation method can be used for generalized Fokker–Planck equations if we note that these equations generate a hierarchic and infinite set of ordinary differential equations for statistical cumulants [169–171]. The first truncation always leads to equations having the form of classical equations of motion. The second truncation plays the role of the first quantum correction, and so on. The cumulant method has also been applied to the study of some aspects of chaos in classical and quantum mechanics [173,174] and in quantum optics [165,166,171,172]. To identify chaotic behavior of a classical dynamical system, it suffices to use the maximal Lyapunov exponent. A quantum analog of the Lyapunov exponent involving the Q function has been proposed by Toda and Ikeda [175]. However, as we have already mentioned, the equation for the Q (P, W) function is mathematically cumbersome, and its analytical solution is unknown for most nonlinear systems. This poses additional difficulties when it comes to calculate the Lyapunov exponents. However, this problem can be solved indirectly and approximately by finite cumulant expansion [165], enabling us to use the classical calculation method of Lyapunov exponents for equations with statistical cumulants.

A. Chaos in a Kerr Oscillator

We write the Hamiltonian in the form

$$\hat{H} = \hat{H}_1 + \hat{H}_2 + \hat{H}_3 \tag{56}$$

where

$$\hat{H}_1 = \hbar\omega\hat{a}^\dagger\hat{a} + \frac{\hbar\chi}{2}\hat{a}^{\dagger 2}\hat{a}^2 \tag{57}$$

$$\hat{H}_2 = i\hbar F(\hat{a}^\dagger - \hat{a}) \tag{58}$$

$$\hat{H}_3 = \hbar\sum_j \Omega_j \hat{b}_j^\dagger\hat{b}_j + \hbar\sum_j (K_j\hat{b}_j\hat{a}^\dagger + K_j^*\hat{b}_j^\dagger\hat{a}) \tag{59}$$

In the single-mode Hamiltonian $\hat{H}_1$, the quantities $\hat{a}\,(\hat{a}^\dagger)$ are the photon annihilation (creation) operators, respectively; ω is the frequency of the harmonic oscillator, and χ is the anharmonicity parameter. The Hamiltonian $\hat{H}_2$ describes the interaction between the classical external driving field F and the single-mode field. The loss mechanism is described by the coupling to a heat bath governed by the reservoir Hamiltonian $\hat{H}_3$. Here, the $(\hat{b}_j)\hat{b}_j^\dagger$ are the boson annihilation (creation) operators of the reservoir. The frequencies of the reservoir modes are denoted by Ω_j. The quantities K_j are the coupling constants between the optical and reservoir modes. On eliminating the reservoir operators, we obtain the master equation for the density operator $\hat{\rho}$ in the following form:

$$\frac{\partial \hat{\rho}}{\partial t} = \frac{-i}{\hbar}[\hat{H}_1 + \hat{H}_2, \hat{\rho}] + L_{ir}[\hat{\rho}] \tag{60}$$

The irreversible term $L_{ir}[\hat{\rho}]$ describes damping and has the following form:

$$\begin{aligned} L_{ir}[\hat{\rho}] = {} & \frac{\Gamma}{2}(2\hat{a}\hat{\rho}\hat{a}^\dagger - \hat{a}^\dagger\hat{a}\hat{\rho} - \hat{\rho}\hat{a}^\dagger\hat{a}) \\ & + \Gamma\langle n\rangle(\hat{a}^\dagger\hat{\rho}\hat{a} + \hat{a}\hat{\rho}\hat{a}^\dagger - \hat{a}^\dagger\hat{a}\hat{\rho} - \hat{\rho}\hat{a}\hat{a}^\dagger) \end{aligned} \tag{61}$$

The parameter Γ is the damping constant, and $\langle n\rangle$ is the mean number of reservoir photons. The quantum theory of damping assumes that the reservoir spectrum is flat, so the mean number of reservoir oscillators $\langle n\rangle = \langle\hat{b}_j^\dagger(0)\hat{b}_j(0)\rangle = (\exp(\hbar\omega/kT) - 1)^{-1}$ in the jth mode is independent of j. Thus the reservoir oscillators form a thermal system. The case $\langle n\rangle = 0$ corresponds to vacuum fluctuations (zero-temperature heat bath). It is convenient to consider the quantum dynamics of the system (56)–(59) in the interaction picture. Then the master equation for the density operator $\hat{\rho}$ is given by

$$\begin{aligned} \frac{\partial \hat{\rho}}{\partial \tau} = {} & -i\left[\frac{1}{2}\hat{a}^{\dagger 2}\hat{a}^2 + i\mathscr{F}(\hat{a}^\dagger - \hat{a}), \hat{\rho}\right] + \frac{\gamma}{2}(2\hat{a}\hat{\rho}\hat{a}^\dagger - \hat{a}^\dagger\hat{a}\hat{\rho} - \hat{\rho}\hat{a}^\dagger\hat{a}) \\ & + \gamma\langle n\rangle(\hat{a}^\dagger\hat{\rho}\hat{a} + \hat{a}\hat{\rho}\hat{a}^\dagger - \hat{a}^\dagger\hat{a}\hat{\rho} - \hat{\rho}\hat{a}\hat{a}^\dagger) \end{aligned} \tag{62}$$

where $\tau = t\chi$ is the redefined time, $\gamma = \Gamma/\chi$, and $\mathscr{F} = F/\chi$. The term $\omega\hat{a}^\dagger\hat{a}$ does not appear in Eq.(62) as a consequence of the interaction picture.

The master equation (62) can be transformed to a c-number partial differential equation. Three kinds of equations can be derived from (62): (1) an equation for the Wigner function $\Phi_{(\mathrm{Sym})}$ related to symmetric (Weyl) ordering of the field operators $\hat{a}, \hat{a}^\dagger$, (2) an equation for the Wigner-like function $\Phi_{(A)}$ related to antinormal ordering of the operators, and (3) an equation for the Wigner-like function Φ_N related to normal ordering. The statistical properties of the Φ functions are discussed fully in the book by Peřina [168]. These are quasidistribution functions in the complex plane (α, α^*), where the quantity α is an eigenvalue of the annihilation operator $\hat{a}$, i.e. $\hat{a} \mid \alpha\rangle = \alpha \mid \alpha\rangle$. Here, $\mid \alpha\rangle$ is a coherent state.

For convenience we introduce the so-called s ordering of the field operators $\hat{a}, \hat{a}^\dagger$. Then we can write $\Phi_{(\mathrm{Sym})} = \Phi_{(0)}$, $\Phi_{(A)} = \Phi_{(-1)}$ and $\Phi_{(N)} = \Phi_{(1)}$.

From (62) we get the generalized Fokker–Planck equation for the quasidistribution $\Phi_{(s)}(\alpha^*, \alpha; \tau)$ related to the s ordering [165]:

$$\frac{\partial \Phi_{(s)}}{\partial \tau} = L_{\mathrm{class}} + L_{\mathrm{quant}} \tag{63}$$

where

$$\begin{aligned}
L_{\mathrm{class}} &= \frac{\partial}{\partial \alpha}\left[\left(\frac{1}{2}\gamma\alpha - \mathscr{F} + i\alpha|\alpha|^2\right)\Phi_{(s)}\right] \\
&\quad + \frac{\partial}{\partial \alpha^*}\left[\left(\frac{1}{2}\gamma\alpha^* - \mathscr{F} - i\alpha^*|\alpha|^2\right)\Phi_{(s)}\right] + \gamma\langle n\rangle \frac{\partial^2 \Phi_{(s)}}{\partial\alpha\,\partial\alpha^*} \\
L_{\mathrm{quant}} &= -i\Bigg[(1-s)\frac{\partial}{\partial\alpha}\alpha\Phi_{(s)} - (1-s)\frac{\partial}{\partial\alpha^*}\alpha^*\Phi_{(s)} \\
&\quad + \frac{s}{2}\frac{\partial^2}{\partial\alpha^2}\alpha^2\Phi_{(s)} - \frac{s}{2}\frac{\partial^2}{\partial\alpha^{*2}}\alpha^{*2}\Phi_{(s)} \\
&\quad + \frac{(s^2-1)}{4}\frac{\partial^3}{\partial\alpha^{*2}\,\partial\alpha}\alpha^*\Phi_{(s)} - \frac{(s^2-1)}{4}\frac{\partial^3}{\partial\alpha^2\,\partial\alpha^*}\alpha\Phi_{(s)}\Bigg] \\
&\quad + \gamma\frac{(1-s)}{2}\frac{\partial^2\Phi_{(s)}}{\partial\alpha\,\partial\alpha^*}
\end{aligned} \tag{64}$$

Let us emphasize that there is no difference among the equations for $\Phi_{(\mathrm{sym})}$, $\Phi_{(A)}$, and $\Phi_{(N)}$ as long as the system (56)–(59) is classical. This problem has been studied elsewhere [176,177]. In the classical limit the term L_{quant} in Eq.(63) vanishes and $\Phi_{(s)}$ is a classical distribution function. For $L_{\mathrm{quant}} = 0$ and $\gamma = 0$, Eq.(63) reduces to the classical Liouville equation, and for $L_{\mathrm{quant}} = 0$ and $\gamma \neq 0$, to the classical Fokker–Planck equation. So, we can say that the L_{class} term governs classical dynamics whereas the L_{quant} term adds the quantum (operator) correction. The decision as to whether chaos appears in the system (56)–(59) can be made by investigating the separation rate of two peaks of a $\Phi_{(s)}$ function initially close to each other or by the analysis of equations for the statistical moments originating in Eq. (63). Thus, instead of attempting to solve the partial differential equation (63), we deal with the problem of solving a set of ordinary differential equations for the statistical moments.

The calculation of statistical moments with the help of $\Phi_{(s)}$ is simple. For example, if we want to calculate the average number of photons $\langle a^{\dagger}a\rangle$, we use one of the three function $\Phi_{(N)}$, $\Phi_{(A)}$ or $\Phi_{(\text{sym})}$. We have

$$\langle \hat{a}^{\dagger}\hat{a}\rangle = \int \alpha^{*}\alpha\, \Phi_{(N)}(\alpha^{*},\alpha)\, d^{2}\alpha \tag{65}$$

$$\langle \hat{a}^{\dagger}\hat{a}\rangle = \int (\alpha^{*}\alpha - 1)\, \Phi_{(A)}(\alpha^{*},\alpha)\, d^{2}\alpha \tag{66}$$

$$\langle \hat{a}^{\dagger}\hat{a}\rangle = \int \left(\alpha^{*}\alpha - \frac{1}{2}\right) \Phi_{(\text{sym})}(\alpha^{*},\alpha)\, d^{2}\alpha \tag{67}$$

The value of $\langle \hat{a}^{\dagger}\hat{a}\rangle$ is always the same, but the averaging procedure differs in each case. The relations (65)–(67) are a simple consequence of the boson commutation relation $[\hat{a},\hat{a}^{\dagger}] = 1$ and the definition

$$\langle \alpha^{*}\alpha\rangle_{(s)} = \int \alpha^{*}\alpha\, \Phi_{(s)}(\alpha^{*},\alpha)\, d^{2}\alpha \tag{68}$$

where $\langle \alpha^{*}\alpha\rangle_{(N)} = \langle \hat{a}^{\dagger}\hat{a}\rangle$, $\langle \alpha^{*}\alpha\rangle_{(A)} = \langle \hat{a}\hat{a}^{\dagger}\rangle$, and $\langle \alpha^{*}\alpha\rangle_{(\text{sym})} = \frac{1}{2}\langle \hat{a}^{\dagger}\hat{a} + \hat{a}\hat{a}^{\dagger}\rangle$. It is obvious that some expectation values do not depend on ordering, for example, $\langle \hat{a}^{\dagger n}\rangle = \langle \alpha^{*n}\rangle_{(N)} = \langle \alpha^{*n}\rangle_{(A)} = \langle \alpha^{*n}\rangle_{(\text{sym})}$. The function $\Phi_{(s)}$ allows us to define the quantum cumulants. The cumulants of first order are given by

$$\langle \alpha^{*}\rangle_{(s)} = \xi^{*}, \quad \langle \alpha\rangle_{(s)} = \xi \tag{69}$$

The cumulants of second order have the forms

$$\langle \alpha^{*}\alpha\rangle_{(s)} - \langle \alpha^{*}\rangle_{(s)}\langle \alpha\rangle_{(s)} = B_{(s)} \tag{70}$$

$$\begin{aligned} \langle \alpha^{*2}\rangle_{(s)} - \langle \alpha^{*}\rangle_{(s)}^{2} &= C^{*} \\ \langle \alpha^{2}\rangle_{(s)} - \langle \alpha\rangle_{(s)}^{2} &= C \end{aligned} \tag{71}$$

It is easy to note that simple relations hold among $B_{(N)}$, $B_{(A)}$, and $B_{(\text{sym})}$, namely, $B_{(A)} = B_{(N)} + 1$ and $B_{(\text{sym})} = \frac{1}{2}(2B_{(N)} + 1)$. Thus the average number of photons can be expressed with the help of s ordering as follows: $\langle \hat{a}^{\dagger}\hat{a}\rangle = G_{(s)} + \xi^{*}\xi$, where $G_{(s)} = B_{(s)} - \frac{1-s}{2}$.

Analytical solutions of quantum Fokker–Planck equations such as Eq. (63) are known only in special cases. Thus, some special methods have been developed to obtain approximate solutions. One of them is the statistical moment method, based on the fact that the equation for the probability density generates an infinite hierarchic set of equations for the statistical moments and vice versa.

However, for numerical reasons the set of equations has to be truncated to a finite number, which means approximation. In this section we restrict ourselves to the second truncation (Gaussian approximation), namely, to the equations for ξ, C and $G_{(s)}$. We arrive at the following set of equations:

$$\frac{d\xi}{d\tau} = -\frac{1}{2}\gamma\,\xi + \mathscr{F} - i[2G_{(s)}\xi + C\xi^* + \xi^2\xi^*] \tag{72}$$

$$\frac{dC}{d\tau} = -\gamma\,C_{(s)} - i[\xi^2(1 + 2G_{(s)}) + C(1 + 4|\xi|^2)] - 6iG_{(s)}C \tag{73}$$

$$\frac{dG_{(s)}}{d\tau} = -\gamma\,G_{(s)} + i[C\xi^{*2} - C^*\xi^2] + \gamma\langle n\rangle \tag{74}$$

We examine the dynamics of this system with the initial conditions $\xi(0) = 1 + i$ and $G_{(s)}(0) = C(0) = 0$. The driving field $\mathscr{F}$ is assumed in the form of a train of rectangular computer simulated pulses. The length of the pulse is denoted by T_1, whereas T_2 is the distance between the pulses, and $\mathscr{F}_0$ is their height. Moreover, we put $\langle n\rangle = 0$, $\gamma = 0.5$, $\mathscr{F}_0 = 2$, $T_2 = 1$ and $0 < T_1 < 7.5$. The physical sense of the truncation is clear if we note that the first truncation [Eq. (63) is without s terms] gives only the classical equation for the anharmonic oscillator:

$$\frac{d\xi}{d\tau} = -\frac{1}{2}\gamma\xi + \mathscr{F}(\tau) - i\xi^2\xi^* \tag{75}$$

Thus $\langle\hat{a}^\dagger\hat{a}\rangle = |\xi|^2$ is a classical intensity. The system (75) is nonautonomous if the function $\mathscr{F}$ is explicitly time-dependent. The autonomized version of Eq.(75) is given by

$$\begin{aligned} \frac{d\xi}{d\tau} &= -\frac{1}{2}\gamma\xi + \mathscr{F}(w) - i\xi^2\xi^* \\ \frac{dw}{d\tau} &= 1, \qquad w(0) = 0 \end{aligned} \tag{76}$$

It is readily seen that the set of equations (76) consists of three equations of motion in the real variables $\mathrm{Re}\,\xi$, $\mathrm{Im}\,\xi$, w. If $\mathscr{F}(\tau) = \text{constant}$, chaos in the system does not appear since the set (76) becomes a two-dimensional autonomous system. The maximal Lyapunov exponents for the systems (75) and (72)–(74) plotted versus the pulse duration T_1 are presented in Fig. 36. We note that within the classical system (75) by fluently varying the length of the pulse T_1, we turn order into chaos and chaos into order. For $0 < T_1 < 0.84$ and $1.08 < T_1 < 7.5$, the maximal Lyapunov exponents λ_1 are negative or equal to zero and, consequently, lead to limit cycles and quasiperiodic orbits. In the points where $\lambda_1 = 0$, the system switches its periodicity. The situation changes dramatically if,

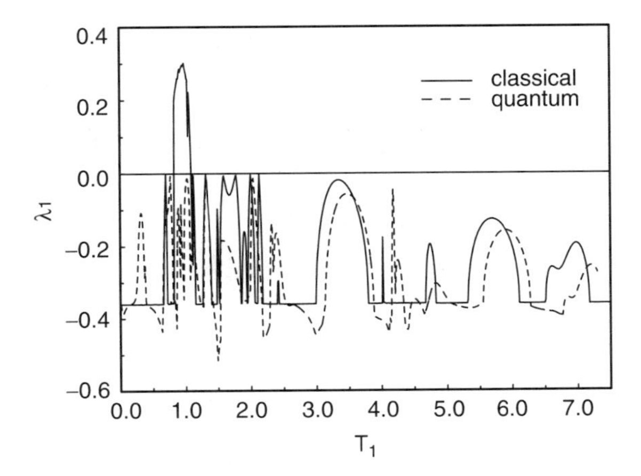

Figure 36. Maximal Lyapunov exponents for the system before (solid line) and after quantum correction (dashed line).

instead of Eq. (75), its quantum version, Eqs. (72)–(74), is taken into account. For the quantum system the maximal Lyapunov exponent is not positive, Therefore the chaotic oscillations due to quantum correction vanish (Fig. 37). The regular oscillations remain regular, but their structures change [165].

B. Chaos in Second-Harmonic Generation of Light

Let us consider a quantum optical system with two interacting modes at the frequencies ω_1 and $\omega_2 = 2\omega_1$, respectively, interacting by way of a nonlinear crystal with second-order susceptibility. Moreover, let us assume that the nonlinear crystal is placed within a Fabry–Pérot interferometer. Both modes are damped via a reservoir. The fundamental mode is driven by an external field with the frequency ω_L and amplitude F. The Hamiltonian for our system is given by [169,178]:

$$\hat{H} = \hat{H}_{\text{rev}} + \hat{H}_{\text{irrev}} \tag{77}$$

$$\begin{aligned}\hat{H}_{\text{rev}} &= \hbar\omega_1 \hat{a}_1^\dagger \hat{a}_1 + \hbar\omega_2 \hat{a}_2^\dagger \hat{a}_2 + i\hbar F(\hat{a}_1^\dagger e^{-i\omega_L t} - \hat{a}_1 e^{i\omega_L t}) \\ &\quad + i\hbar \frac{\kappa}{2}(\hat{a}_1^{\dagger 2}\hat{a}_2 - \hat{a}_1^2 \hat{a}_2^\dagger)\end{aligned} \tag{78}$$

$$\hat{H}_{\text{irrev}} = \hbar \sum_j \sum_{i=1}^{2} (\Omega_j^{(i)} \hat{b}_j^{\dagger(i)} \hat{b}_j^{(i)} + K_j^{(i)} \hat{b}_j^{(i)} \hat{a}_i^\dagger + K_j^{*(i)} \hat{b}_j^{\dagger(i)} \hat{a}_i) \tag{79}$$

where $\hat{H}_{\text{rev}}$ describes the reversible part of interaction and $\hat{H}_{\text{irrev}}$ is the irreversible part responsible for the loss mechanism. The quantities $\hat{a}_1, (\hat{a}_1^\dagger); \hat{a}_2, (\hat{a}_2^\dagger)$ are the

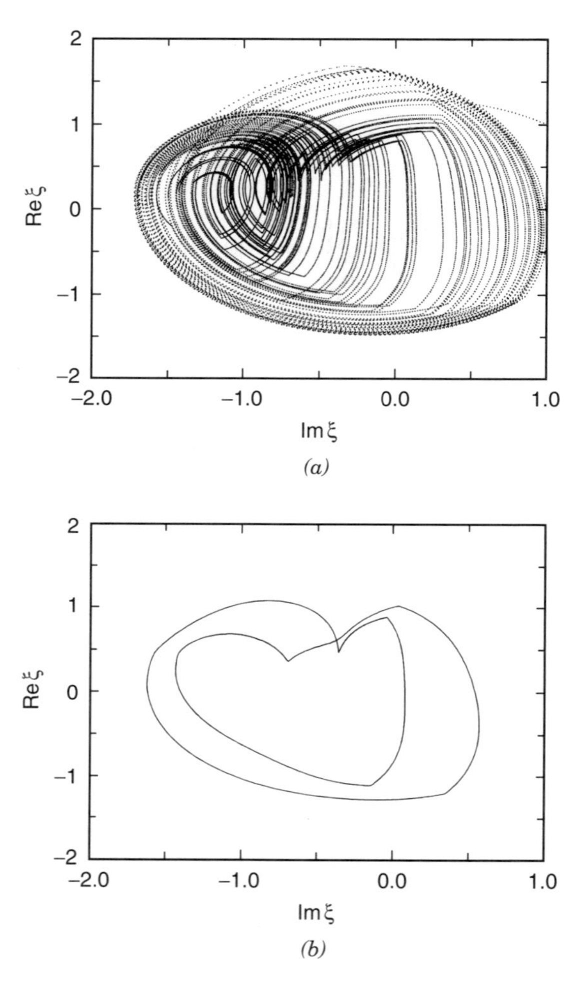

Figure 37. Phase portraits Re ξ versus Im ξ. (a) the classical case; Eq. (75) with the initial condition $\xi(0) = 1 + i$. The parameters of the pulse are $T_1 = 0.98, T_2 = 1$, and $F_0 = 2$. The damping constant is $\gamma = 0.5$, and the time is $100 < \tau < 200$. (b) The quantum system; Eqs. (72)–(74) with the initial conditions $\xi(0) = 1 + i$ and $G_{(s)}(0) = C(0) = 0$. The parameters of the pulse are $T_1 = 0.98, T_2 = 1$, and $F_0 = 2$. The damping constant is $\gamma = 0.5$, and the time $100 < \tau < 200$.

photon annihilation (creation) operators for the fundamental and second-harmonic modes, respectively. The parameter κ is taken to be real and acts as a nonlinear coupling constant between the two modes. Finally, the operators $\hat{b}_j^{\dagger(i)}, \hat{b}_j^{(i)}$ are the boson annihilation (creation) operators of the reservoir. The frequencies of the reservoir oscillations are denoted by $\Omega_j^{(i)}$ and the coupling constant between the optical and reservoir modes, by $K_j^{(i)}$. The dynamics of the system (77) on eliminating the reservoir Hamiltonian (79) is governed by the appropriate master equation for the density operator $\hat{\rho}$. The master equation in the interaction picture leads to the following c-number Fokker–Planck equation for the quasidistribution function $\Phi_{(s)}$ [168,169,178]

$$\frac{\partial \Phi_{(s)}}{\partial \tau} = L_{\text{class}} + L_{\text{quant}} \tag{80}$$

where

$$\begin{aligned} L_{\text{class}} = \sum_{i=1}^{2} \Bigg[& \gamma_i \frac{\partial}{\partial \alpha_i} (\alpha_i \Phi_{(s)}) + \gamma_i \frac{\partial}{\partial \alpha_i^*} (\alpha_i^* \Phi_{(s)}) + \frac{\partial}{\partial \alpha_i} (D_i \Phi_{(s)}) \\ & + \frac{\partial}{\partial \alpha_i^*} (D_i^* \Phi_{(s)}) + \gamma_i \langle n_i \rangle \frac{\partial^2 \Phi_{(s)}}{\partial \alpha_i^* \partial \alpha_i} \Bigg] \end{aligned} \tag{81}$$

$$L_{\text{quant}} = \left(\frac{1-s}{2} \right) \sum_{i=1}^{2} \gamma_i \frac{\partial^2 \Phi_{(s)}}{\partial \alpha_i^* \partial \alpha_i} - \frac{s}{2} \frac{\partial^2}{\partial \alpha_1^2} (D_{11} \Phi_{(s)}) - \frac{s}{2} \frac{\partial^2}{\partial \alpha_1^{*2}} (D_{11}^* \Phi_{(s)}) \tag{82}$$

The quasidistribution function $\Phi_{(s)}$ is defined as follows: $\Phi_{(s=1)} = P$ and $\Phi_{(s=-1)} = Q$. The function $\Phi_{(s)}$ is determined in the complex plane $(\alpha_1, \alpha_2, \alpha_1^*, \alpha_2^*)$, where α_i is an eigenvalue of the annihilation operator $\hat{a}_i$, namely, $\hat{a}_i |\alpha_i\rangle = \alpha_i |\alpha_i\rangle$. Here, $|\alpha_i\rangle$ is a coherent state. The initial condition for the Fokker–Planck equation is given by

$$\Phi_{(s)}(\alpha_1(\tau), \alpha_2(\tau); \tau)|_{\tau=0} = \Phi_{(s)}(\alpha_1(0) = \alpha_{10}, \alpha_2(0) = 0; 0) \tag{83}$$

which means that the amplitude of the fundamental mode initially differs from zero whereas the amplitude of the second harmonic equals zero. The coefficients D_i and D_{11} are given by

$$\begin{aligned} D_1 &= -\mathscr{F} - \alpha_1^* \alpha_2 \\ D_1^* &= -\mathscr{F} - \alpha_1 \alpha_2^* \\ D_2 &= 0.5\alpha_1^2 \\ D_2^* &= 0.5\alpha_1^{*2} \\ D_{11} &= \frac{\partial D_1}{\partial \alpha_1^*} = -\alpha_2 \\ D_{11}^* &= \frac{\partial D_1^*}{\partial \alpha_1} = -\alpha_2^* \end{aligned} \tag{84}$$

The general relations among the coefficients D_i and D_{ij} are presented elsewhere [179]. The quantities γ_1 and γ_2 are the damping constants for the fundamental and second- harmonic modes, respectively. In Eq.(82) we shall restrict ourselves to the case of zero-frequency mismatch between the cavity and the external forces ($\omega_1 - \omega_L = 0$). In this way we exclude the rapidly oscillating terms. Moreover, the time τ and the external amplitude $\mathscr{F}$ have been redefined as follows: $\tau = \kappa t$ and $\mathscr{F} = \frac{F}{\kappa}$. The s ordering in Eq.(80) which is responsible for the operator structure of the Hamiltonian allows us to contrast the classical and quantum dynamics of our system. If the Hamiltonian (77)–(79) is classical (i.e., if it is a c number), then the equation for the probability density has the form of Eq.(80) without the s terms:

$$-\frac{s}{2}\frac{\partial^2}{\partial\alpha_1^2}(D_{11}\Phi_{(s)}) - \frac{s}{2}\frac{\partial^2}{\partial\alpha_1^{*2}}(D_{11}^*\Phi_{(s)}), \qquad \gamma_i\left(\frac{1-s}{2}\right)\frac{\partial^2\Phi_{(s)}}{\partial\alpha_i^*\partial\alpha_i}$$

The s terms distinguish the classical and quantum dynamics quite naturally. If they do not appear, the difference between P and Q vanishes.

The Fokker–Planck equation (80) generates an infinite and hierarchic set of equations for the statistical moments (see Section IV.A.1). Below, we restrict ourselves to a Gaussian approximation. The cumulants are defined by the following relations:

$$\xi_i = \langle\hat{a}_i\rangle \tag{85}$$

$$B_i = \langle\hat{a}_i^\dagger\,\hat{a}_i\rangle - \langle\hat{a}_i^\dagger\rangle\langle\hat{a}_i\rangle \tag{86}$$

$$B_{12} = \langle\hat{a}_1^\dagger\,\hat{a}_2\rangle - \langle\hat{a}_1^\dagger\rangle\langle\hat{a}_2\rangle \tag{87}$$

$$C_i = \langle\hat{a}_i^2\rangle - \langle\hat{a}_i\rangle^2 \tag{88}$$

$$C_{12} = \langle\hat{a}_1\,\hat{a}_2\rangle - \langle\hat{a}_1\rangle\langle\hat{a}_2\rangle \tag{89}$$

Integration per partes of the Fokker–Planck equation for the quasidistribution $\Phi_{(s=1)} = P$ (the choice of a particular s is a question of taste only) allows us to write the appropriate equations for the cumulants. In what follows, we assume that damping is included only by way of coupling to the reservoir at zero temperature, that is, $\langle n_i\rangle = 0$. The first truncation (the cumulants higher than first-order vanish) leads to the classical limit. Then, from Eq. (80), we get the classical Bloembergen equations [102] [see Eqs. (1)]:

$$\frac{d\xi_1}{d\tau} = -\gamma_1\,\xi_1 + \mathscr{F} + \xi_1^*\,\xi_2 \tag{90}$$

$$\frac{d\xi_2}{d\tau} = -\gamma_2\,\xi_2 - 0.5\xi_1^2 \tag{91}$$

The initial conditions have the following form:

$$\xi_1(0) = \xi_{10}, \qquad \xi_2(0) = 0 \tag{92}$$

The s terms in Eq. (80) contribute nothing to the preceding equations. The second-order truncation (Gaussian approximation) leads to the following set of equations:

$$\frac{d\xi_1}{d\tau} = -\gamma_1\xi_1 + \mathscr{F} + \xi_1^*\xi_2 + B_{12} \tag{93}$$

$$\frac{d\xi_2}{d\tau} = -\gamma_2\xi_2 - 0.5(\xi_1^2 + C_1) \tag{94}$$

$$\frac{dB_1}{d\tau} = -2\gamma_1 B_1 + B_{12}^*\xi_1 + B_{12}\xi_1^* + C_1^*\xi_2 + C_1\xi_2^* \tag{95}$$

$$\frac{dB_2}{d\tau} = -2\gamma_2 B_2 - B_{12}^*\xi_1 - B_{12}\xi_1^* \tag{96}$$

$$\frac{dC_1}{d\tau} = -2\gamma_1 C_1 + 2(C_{12}\xi_1^* + B_1\xi_2) + \xi_2 \tag{97}$$

$$\frac{dC_2}{d\tau} = -2\gamma_2 C_2 - 2C_{12}\xi_1 \tag{98}$$

$$\frac{dC_{12}}{d\tau} = -(\gamma_1 + \gamma_2)C_{12} + B_{12}\xi_2 - C_1\xi_1 + C_2\xi_1^* \tag{99}$$

$$\frac{dB_{12}}{d\tau} = -(\gamma_1 + \gamma_2)B_{12} + C_{12}\xi_2^* + \xi_1(B_2 - B_1) \tag{100}$$

The set of equations (93)–(100), proposed for the first time by Peřina et al. [169], is a development of the Bloembergen equations (90)–(91). The initial conditions with respect to (83) are given by

$$\begin{gathered} \xi_1(0) = \xi_{10}, \qquad \xi_2(0) = \xi_{20} = 0 \\ B_{1,2}(0) = B_{12}(0) = C_{1,2}(0) = C_{12}(0) = 0 \end{gathered} \tag{101}$$

The s terms in Eq. (80) contribute only the term ξ_2 in Eq. (97). Thus, the term ξ_2 represents the quantum diffusional s-terms in the Fokker–Planck equation. The other terms in Eqs. (93)–(100) originate in the drift terms of the Fokker–Planck equation. The terms B_{12} and C_1 in Eqs. (93)–(94) play the role of feedback terms that pump quantum fluctuations into the classical Bloembergen equations. If the s terms in Eq. (80) do not appear (the classical case), the term ξ_2 in Eq. (97) does not appear, either. In this case the subset (95)–(100) with zero initial conditions has zero solutions and in consequence leads to the first truncation [171].

Let us consider the driving field amplitude in the form $\mathscr{F} = \mathscr{F}_0(1 + \sin\Omega\tau)$, meaning that the external pump amplitude is modulated with a frequency Ω

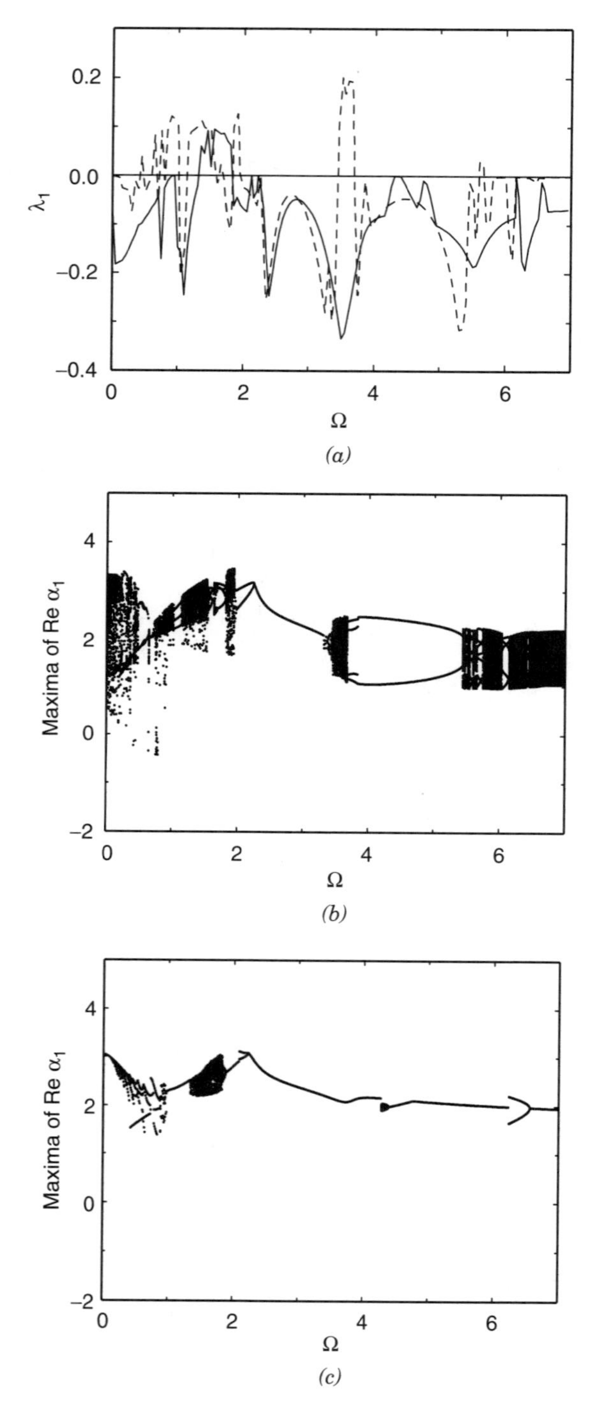

Figure 38. The calssical (dashed) and quantum (solid line) maximal Lyapunov exponents (a) and the appropriate bifurcation maps (b,c) versus the modulated parameter Ω. The parameters are

around $\mathscr{F}_0$. For the time-independent field $\mathscr{F} = \mathscr{F}_0$ ($\Omega = 0$), the system does not manifest chaotic behavior. However, a change of Ω in the range $0 < \Omega < 7$ leads the system from periodic to chaotic motion or vice versa. The dynamical behavior of our system is reflected by the Lyapunov exponents. The maximal Lyapunov exponents as a function of the modulation parameter Ω for the classical case [Eqs. (90)–(91)] (dashed line) and for the quantum case [Eqs. (93)–(100)] (solid line) is plotted in Fig. 38a. For the classical case, one observes several regions where the system behaves chaotically ($\lambda_1 > 0$) whereas elsewhere it behaves orderly ($\lambda_1 < 0$). For the quantum case we observe only one region of chaos $1.3 < \Omega < 1.72$, which does not overlap exactly any classical region of chaos. Generally, as is seen in Fig. 38, the quantum correction reduces chaos in the system but does not eliminate it completely. For example, for $\Omega = 1.4$, both the classical and quantum versions of the system behave chaotically whereas the classical maximal Lyapunov exponent is greater than quantum. This means a reduction of chaos in the classical system due to the quantum correction. The reduction is also reflected by the appropriate bifurcation diagrams (Fig. 38b,c). Another useful way to visualize the reduction of chaos is to analyze the motion in the phase space. However, in our case, the classical phase space is four-dimensional ($\mathrm{Re}\,\xi_1$, $\mathrm{Im}\,\xi_1$, $\mathrm{Re}\,\xi_2$, $\mathrm{Im}\,\xi_2$). This means that we can compare only the motion in the reduced phase space. For physical interpretation it is convenient to consider the motion in two-dimensional intensity space ($I_1 = |\xi_1|^2, I_2 = |\xi_2|^2$). Then, instead of a typical *phase portrait*, we deal with *an intensity portrait*. In the quantum case the intensities are the average numbers of photons determined by $\langle \hat{a}_i^+ \hat{a}_i \rangle = |\xi_i|^2 + B_i$, where B_i is the quantum correction to the classical intensity $I_i = |\xi_i|^2$.

The reduction of chaos for $\Omega = 1.45$ is presented in the intensity portraits of Fig. 39. However, as is seen in Fig. 38a, there is a small region ($1.68 < \Omega < 1.80$) where the system behaves orderly in the classical case but the quantum correction leads to chaos. By way of an example for $\Omega = 1.75$, the classical system, after quantum correction, loses its orderly features and the limit cycle settles into a chaotic trajectory. Generally, Lyapunov analysis shows that the transition from classical chaos to quantum order is very common. For example, this kind of transition appears for $\Omega = 3.5$ where chaos is reduced to periodic motion on a limit cycle. Therefore a global reduction of chaos can be said to take place in the whole region of the parameter $0 < \Omega < 7$.

As we see in Fig. 38, transitions leading from classical order to quantum order are also possible. For example, for $\Omega = 6.7$ the quasiperiodic classical motion is reduced to periodic motion after the quantum correction.

C. Final Remarks

Using a cumulant expansion, we have shown how to obtain *quantum corrections* to purely classical equations of motion. Quantum correction reduces chaos in

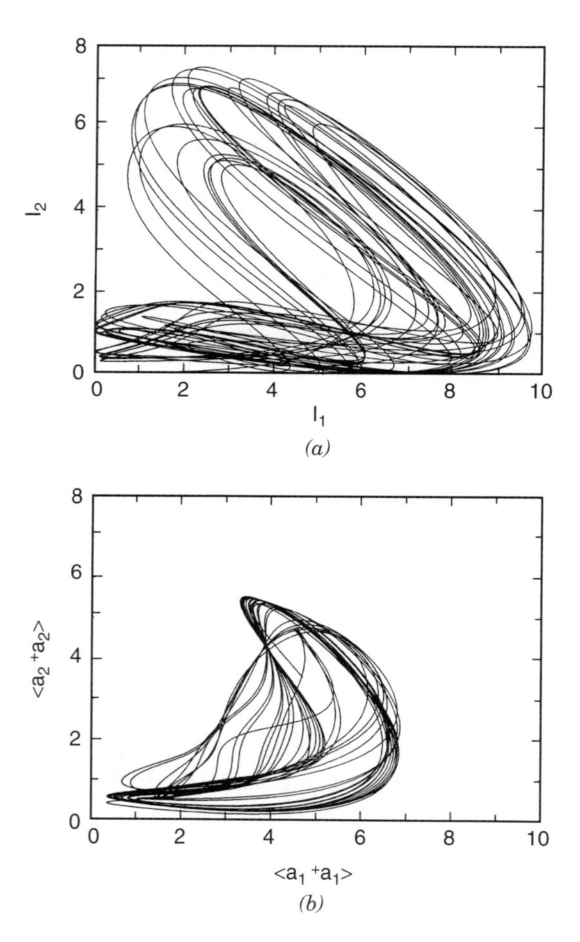

Figure 39. Transition from classical chaos (a) to quantum chaos (b). The parameters are those of Fig. 38 but with $\Omega = 1.4$.

the classical systems. The Lyapunov analysis and bifurcation maps show that after the first quantum correction, the number of chaotic regions is reduced, although not eliminated fully. The question is what happens if third-order or higher-order corrections are taken into account?. Let us note that, for example, the set (72)–(74) consists of 5 equations in real variables. If third-order truncation is performed, the set (72)–(74) is additionally modified and supplemented with four equations in real variables, thus leading to 9 equations. The fourth truncation leads to 15 equations in real variables, and so on. From the formal point of view, the quantum corrections become more and more rigorous with higher and higher order of the approximation. On the other hand, even if the numerical calculations are performed in extended precision, computer errors can accumulate significantly, leading to spurious high-order quantum corrections due to the increasing numbers of equations and iterations. The quantum Lyapunov whose classical counterpart is positive has to be calculated with a finite time, empirically expressed. The time is of the rank $(\lambda)^{-1}$, where λ is the classical Lyapunov exponent [158].

References

1. H. Poincaré, *Les methodes nouvelle de la mécanique céleste*, Gauthier-Villars, 1892.
2. E. N. Lorenz, *J. Atmos. Sci.* **20**, 130 (1963).
3. P. Cvitanović, *Universiality in Chaos*, Adam Hilger Ltd, Bristol, UK, 1984.
4. G. Casati (ed), *Chaotic Behaviour in Quantum Systems*, Plenum Press, New York, 1985.
5. H. G. Schuster, *Deterministic Chaos. An Introduction*, VCH Verlagsgesellschaft, Weinheim, 1988.
6. M. C. Gutzwiller, *Chaos in Classical and Quantum Systems*, Springer-Verlag, New York, 1990.
7. E. Ott, *Chaos in Dynamical Systems*, Cambridge Univ. Press, New York, 1993.
8. R. C. Hilborn, *Chaos and Nonlinear Dynamics*, Oxford Univ. Press, New York, 1994.
9. Special Issue, *Instabilities in Active Optical Media, J. Opt. Soc. Am. B* **2**(1) (1985).
10. Special Issue, *Spatio-temporal Coherence and Chaos in Physical Sytems, Physica D* **23** (1986).
11. R. W. Boyd, M.G. Raymer, and L.M. Narducci (Eds.), *Optical Instabilities*, Cambridge Univ. Press, London, 1986.
12. F. T. Arecchi and R. G. Harrison, *Instabilities and Chaos in Quantum Optics*, in *Springer Series in Synergetics*, Vol. 34, Springer-Verlag, Berlin, 1987.
13. P. W. Milonni, M.L. Shih, and J.R. Ackerhalt, *Chaos in Laser-Matter Interactions*, World Scientific, Singapore, 1987.
14. L. W. Casperson, *Spontaneous pulsations in Lasers*, in J. D. Harvey and D. F. Walls (Eds.), *Laser Physics, Lecture Notes in Physics*, Vol. 182, Springer-Verlag, Berlin, 1983.
15. E.R. Buley and F.W. Cummings, *Phys. Rev.* **134**, A1454 (1964).
16. H. Haken, *Phys. Lett. A* **53**, 77 (1975).
17. R. Graham, *Phys. Lett. A* **58**, 440 (1976).
18. F. Yamada and R. Graham, *Phys. Rev. Lett.* **45**, 1322 (1980).
19. M.-L. Shih and P.W. Milloni, *Opt. Commun.* **49**, 155 (1984).
20. H. Zeghlache and P. Mandel, *J. Opt. Soc. Am. B* **2**,18 (1985).
21. T. Ogawa, *Phys. Rev. A* **37**, 4286 (1988).
22. R. G. Harrison and D. J. Biswas, *Prog. Quantum Electron.* **10**, 147 (1985).
23. P. W. Milonni, J. R. Ackerhalt, and M.-L. Shih, *Opt. Commun.* **49**, 155 (1984).
24. M.-L. Shih, P.W. Milonni, and J. R. Ackerhalt, *J. Opt. Soc. Am. B* **2**, 130 (1985).
25. L. W. Casperson and A. Yariv, *Appl. Phys. Lett.* **17**, 259 (1970).
26. L. W. Casperson, *Phys. Rev. A* **21**, 911 (1980).
27. M. Mayr, H. Risken, and H.D. Vollmer,, *Opt. Commun.*, **36**, 480 (1981).
28. P. Mandel, *Opt. Commun.* **44**, 400 (1983).
29. D. K. Bandy, L. M. Narducci, L.A. Lugiato, and N.B. Abraham, *J. Opt. Soc. Am. B* **2**, 56 (1985).

30. F. T. Arecchi, R. Meucci, G. Puccioni, and J. Tredicce, *Phys. Rev. Lett.* **49**, 1217 (1982).
31. L. W. Casperson, *IEEE J. Quantum Electron.* **QE-14**, 756 (1978).
32. C. O. Weiss and H. King, *Opt. Commun.* **44**, 59 (1982).
33. R. S. Gioggia and N.B. Abraham, *Phys. Rev. Lett.* **51**, 650 (1983).
34. L. M. Hoffer, T.H. Chyba, and N. B. Abraham, *J. Opt. Soc. Am. B* **2**, 102 (1985).
35. N. B. Abraham, T. Chyba, M. Coleman, R. S. Gioggia, N. J. Halas, L. M. Hoffer, S.-N. Liu, M. Maeda, and J.C. Wesson, in J. D. Harvey and D.F. Walls (Eds.), *Laser Physics, Lecture Notes in Physics* Vol. 182, Springer-Verlag, Berlin, 1983.
36. C. O. Weiss and W. Klische, *Opt. Commun.* **51**, 47 (1984).
37. C. O. Weiss, W. Klische, P.S. Ering, and M. Cooper, *Opt. Commun.* **52**, 405 (1985).
38. C. O. Weiss and J. Brock, *Phys. Rev. Lett.* **57**, 2804 (1986).
39. C. O. Weiss, N. B. Abraham, and V. H. Hübner, *Phys. Rev. Lett.* **61**, 1587 (1988).
40. D. J. Biswas and R.G. Harrison, *Phys. Rev. A* **32**, 3835 (1985).
41. T. Midavaine, D. Dangoisse, and P. Glorieux, *Phys. Rev. Lett.* **55**, 1989 (1985).
42. F. Hollinger and C. Jung, *J. Opt. Soc. Am. B* **2**, 218 (1985).
43. W. Klische, H. R. Telle, and C. O. Weiss, *Opt. Lett.* **9**, 561 (1984).
44. Y. C. Chen, H. G. Winful, and J. M. Liu, *Appl. Phys. Lett.* **47**, 208 (1985).
45. H. G. Winful, Y. C. Chen, and J. M. Liu, *Appl. Phys. Lett.* **48**, 616 (1986).
46. T. Baer, *J. Opt. Soc. Am. B* **3**, 1175 (1986).
47. C. Barcikowski and R. Roy, *Phys. Rev. A* **43**, 6455 (1991).
48. C. Barcikowski, R. F. Fox, and R. Roy, *Phys. Rev. A* **45**, 403 (1992).
49. J. Ye, H. Li, and J. G. McInerney, *Phys. Rev. A* **47**, 2249 (1993).
50. E. Ott, C. Grebogi, and J. A. Yorke, *Phys. Rev. Lett.* **64**, 1196 (1990).
51. R. Roy, T.W. Murphy, Jr., T. D. Maier, Z. Gills, and E. R. Hunt, *Phys. Rev. Lett.* **68**, 1259 (1992).
52. K. Pyragas, *Phys. Lett. A* **170**, 421 (1992).
53. S. Bielawski, D. Derozier, and P. Glorieux, *Phys. Rev. E* **49**, R971 (1994).
54. R. Meucci, M. Ciofini, and R. Labate, *Phys. Rev. E* **53**, R5537 (1996).
55. R. Meucci, R. Labate, and M. Ciofini, *Phys. Rev. E* **56**, 2829 (1997).
56. P. Glorieux, *Int. J. Bifure Chaos* **8**, 1749 (1998).
57. R. Roy, and K. S. Thornburg, *Phys. Rev. Lett.* **72**, 2009 (1994).
58. P. Colet and R. Roy, *Opt. Lett.* **19**, 2056 (1994).
59. P. M. Alsing, A. Gavrielidies, V. Kovanis, R. Roy, and K. S. Thornburg, *Phys. Rev. E* **56**, 6302 (1997).
60. G. D. Van Wiggeren and R. Roy, *Phys. Rev. Lett.* **81**, 3547 (1998).
61. S. Bocatelli, C. Grebogi, Y.-C. Lai, H. Mancini, and D. Maza, *Phys. Rep.* **329**, 103 (2000).
62. P. I. Bielobrov, G. M Zaslavski, and G.T. Tartakovski, *JETP* **44**, 945 (1977).
63. P. W. Milonni, J. R. Ackerhalt, and H. W. Galbraith, *Phys. Rev. Lett.* **50**, 966 (1983).
64. M. Munz, *Z. Phys. B* **53**, 311 (1983).
65. M. Kuś, *Phys. Rev. Lett.* **54**, 1343 (1985).
66. R. F. Fox and J. Edison, *Phys. Rev. A* **34**, 482 (1986).
67. J. Edison and R .F. Fox, *Phys. Rev. A* **34**, 3288 (1986).
68. A. Nath and D. S. Ray, *Phys. Rev. A* **36**, 431 (1987).

69. C. Boden, M. Dämmig, and F. Mitschke, *Phys. Rev. A* **45**, 6829 (1992).
70. K. Ikeda, *Opt. Commun.* **30**, 257 (1979).
71. K. Ikeda, H. Daido, and O. Akimoto, *Phys. Rev. Lett.* **45**, 709 (1980).
72. L. A. Lugiato, *Opt. Commun.* **33**, 108 (1980).
73. H. J. Carmichael, R. R. Snapp, and W. C. Schieve, *Phys. Rev. A* **26**, 3408 (1982).
74. R. R. Snapp, H. J. Carmichael, and W. C. Schieve, *Opt. Commun.* **40**, 68 (1981).
75. H. M. Gibbs, F. A. Hopf, D. L. Kaplan, and R. L. Shoemaker, *Phys. Rev. Lett.* **46**, 474 (1981).
76. H. Nakatsuka, S. Asaka, H. Itoh, K. Ikeda, and M. Matsuoka, *Phys. Rev. Lett.* **50**, 109 (1983).
77. F. A. Hopf, D. L. Kaplan, H. M. Gibbs, and R. L. Shoemaker, *Phys. Rev. A* **25**, 2172 (1982).
78. K. Ikeda and O. Akimoto, *Phys. Rev. Lett.* **48**, 617 (1982).
79. L. A. Lugiato, L. M. Narducci, D. K. Brandy, and C. A. Pennise, *Opt. Commun.* **43**, 281 (1982).
80. C. M. Savage, H. J. Carmichael, and D. F. Walls, *Opt. Commun.* **42**, 211 (1982).
81. H. J. Carmichael, C. M. Savage, and D. F. Walls, *Phys. Rev. Lett.* **50**, 163 (1983).
82. C. Parigger, P. Zoller, and D. F. Walls, *Opt. Commun.* **44**, 213 (1983).
83. C. M. Savage, and D. F. Walls, *Optica Acta* **30**, 557 (1983).
84. G. J. Milburn, *Phys. Rev. A* **41**, 6567 (1990).
85. J. R. Ackerhalt, H. W. Galbraith, and P.W. Milonni, *Phys. Rev. Lett.* **51**, 1259 (1983).
86. J. R. Ackerhalt and P. W. Milonni, *Phys. Rev. A* **34**, 1211 (1986).
87. J. R. Ackerhalt and P. W. Milonni, *Phys. Rev. A* **37**, 1552 (1987).
88. H. W. Galbraith, J. R. Ackerhalt, and P.W. Milonni, *J. Chem. Phys.* **79**, 5345 (1983).
89. P. W. Milonni, J. R. Ackerhalt, and H. W. Golbraith, *Phys. Rev. A* **28**, 887 (1983).
90. A. Nath and D. S. Ray, *Phys. Rev. A* **35**, 1959 (1987).
91. C. J. Candall and J. R. Albritton, *Phys. Rev. Lett.* **52**, 1887 (1984).
92. W. Lu and R. G. Harrison, *Europhys. Lett.* **16**, 655 (1991).
93. J. M. Wesinger, J. M. Finn, and E. Ott, *Phys. Rev. Lett.* **44**, 453 (1980).
94. D. S. Ray, *Phys. Rev. A* **29**, 3440 (1984).
95. W. Krolikowski, M. R. Belic, M Cronin-Golomb, and A. Bledowski, *J. Opt. Soc. Am.* **B7**, 1204 (1990).
96/97. S. Wabnitz, *Phys. Rev. Lett.* **58**, 1415 (1987).
98. P. A. Franken, A. E. Hill, C. W. Peters, and G. Weinreich, *Phys. Rev. Lett.* **7**, 118 (1961).
99. J. A. Armstrong, N. Bloembergen, J. Ducuing, and P. S. Pershan, *Phys. Rev.* **127**, 1918 (1962).
100. D. S. Chemla, *Rep. Prog. Phys.* **43**, 1191 (1980).
101. S. Kielich, *Molecular Nonlinear Optics*, Nauka, Moscow, 1986 (in Russian).
102. N. Bloembergen, *Nonlinear Optics*, Benjamin, New York, 1965.
103. R. Tanas, "Quantum noise in nonlinear optical phenomena," first chapter in Part 1 of this three-volume set.
104. P. D. Drummond, K. J. McNeil and D. F. Walls, *Optica Acta* **27**, 321 (1980).
105. P. Mandel and T. Erneux, *Optica Acta* **29**, 7 (1982).
106. L. A. Lugiato, C. Oldano, C. Fabre, E. Giacobino, and R. J. Horowicz, *Nuovo Cim. D* **10**, 959 (1988).
107. K. Grygiel and P. Szlachetka, *Opt. Commun.* **78**, 177 (1990).
108. K. Grygiel and P. Szlachetka, *Opt. Commun.* **91**, 241 (1992).

109. N. V. Alekseeva, K. N. Alekseev, G. P. Berman, A. K. Popov, and V. Z. Yakhnin, *Quantum Opt.* **3**, 323 (1991).
110. K. Grygiel and P. Szlachetka, *Opt. Commun.* **158**, 112 (1998).
111. K. Grygiel, manuscript in preparation.
112. G. Benettin, L. Galgani, and J.-M. Strelcyn, *Phys. Rev. A* **14**, 2338 (1976).
113. J. P. Eckmann and D. Ruelle, *Rev. Mod. Phys.* **57**, 617 (1985).
114. A. Wolf, J. B. Swift, H. L. Swinney, and J. A. Vastano, *Physica* **D16**, 285 (1985).
115. K. Grygiel and P. Szlachetka, *Acta Phys. Polon. B* **26**, 1321 (1995).
116. O. E. Roessler, *Phys. Lett. A* **71**, 155 (1979).
117. T. Kapitaniak and W.-H. Steeb, *Phys. Lett. A* **152**, 33 (1991).
118. S. P. Dawson, *Phys. Rev. Lett.* **76**, 4348 (1996).
119. E. Barreto, B. H. Hunt, C. Grebogi, and J.A. York, *Phys. Rev. Lett.* **78**, 4561 (1997).
120. K. Stefański, *Chaos Solit. Fract.* **9**, 83 (1998).
121. M. A. Harrison and Y.-C. Lai, *Phys. Rev. E* **59**, R3799 (1999).
122. L. Yang, Z. Liu, and J. Mao, *Phys. Rev. Lett.* **84**, 67 (2000).
123. P. Szlachetka and K. Grygiel, in W. Florek, D. Lipiński, and T. Lulek, (Eds.), *Symmetry and Structural Properties of Condensed Matter*, 2nd Int. School of Theoretical Physics (Poznań, 1992), World Scientific, Singapore, 1993, pp. 221–236.
124. S. Hayes, C. Grebogi, E. Ott, and A. Mark, *Phys. Rev. Lett.* **73**, 1781 (1994).
125. L. M. Pecora and T. L. Carroll, *Phys. Rev. Lett.* **64**, 821 (1990).
126. L. M. Pecora and T. L. Carroll, *Phys. Rev. A* **44**, 2374 (1991).
127. T. Kapitaniak, *Phys. Rev. E* **50**, 1642 (1994).
128. S. P. Raj, S. Rajasekar, and K. Murali, *Phys. Lett. A* **264**, 283 (1999).
129. N. Platt, E. A. Spiegel, and C. Tresser, *Phys. Rev. Lett.* **70**, 279 (1993).
130. Y. C. Lai and C. Grebogi, *Phys. Rev. E* **52**, R3313 (1995).
131. H. Sun, S. K. Scott, and K. Showalter, *Phys. Rev. E* **60**, 3876 (1999).
132. F. Heagy, T. L. Carroll, and L. M. Pecora, *Phys. Rev. Lett.***73**, 3528 (1994).
133. J. H. Eberly, N. B. Narozhny, and J. J. Sanchez-Mondrogon, *Phys. Rev. Lett.* **44**, 1323 (1980).
134. B. W. Shore and P.L. Knight, *J. Mod. Opt.* **40**, 1195 (1993).
135. M. Kozierowski and S. M. Chumakov, *Phys. Rev. A* **52**, 4194 (1995).
136. S. M. Jensen, *IEEE J. Quantum Electron.* **QE-18**, 1580 (1982).
137. V. M. Kenkre and D.K. Campbell, *Phys. Rev. B* **34**, 4959 (1986).
138. A. Chefles and S.M. Barnett, *J. Mod. Opt,* **43** ,709 (1996).
139. J. Fiurášek, J. Křepelka, and J. Peřina, *Opt. Commun.* **167**, 115 (1999).
140. V. Peřinová and M. Karská, *Phys. Rev. A* **39**, 4056 (1989).
141. E. T. Whittaker, *A Treatise on the Analytical Dynamics of Particles and Rigid Bodies with an Introduction to the Problem of Three Bodies*, Cambridge Univ. Press, Cambridge, UK, 1952.
142. K. Grygiel and P. Szlachetka, *Opt. Commun.* **177**, 425 (2000).
143. N. Minorsky, *Nonlinear Oscillations*, Van Nostrand, Princeton, NJ, 1962, p. 285.
144. L. Gammaitoni, F. Marchesoni, and S. Santucci, *Phys. Lett. A* **195**, 116 (1994).
145. L. Gammaitoni, P. Hänggi, P. Jung, and F. Marchesoni, *Rev. Mod. Phys.* **70**, 223 (1998).
146. N. Korolkowa and J. Peřina, *Opt. Commun.* **136**, 135 (1996).

147. S. R. Friberg, A. M. Weiner, Y. Silberberg, B. G. Sfez, and P. S. Smith, *Opt. Lett.* **13**, 904 (1998).

148. Lj. Kocarev, K. Halle, K. Eckert, and L. O. Chua, *Int. J. Bifurc. Chaos* **2**, 706 (1992).

149. K. M. Cuomo and A.V. Oppenheim, *Phys. Rev. Lett.* **71**, 65 (1993).

150. P. D. Townsend, G. L. Baker, J. L. Jackel, J. A. Shelburne III, and S. Etemand, SPIE Vol. 1147, *Nonlinear Properties of Organic Materials II*, 1989, p. 256.

151. G. Casati, B. Chirkov, J. Ford, and F.M. Izrailev, in G. Casati and J. Ford, (Eds.), *Stochastic Behavior in Classical and Quantum Hamiltonian Systems of Lecture Notes in Physics*, Vol. 93, Springer, Berlin, 1979.

152. D. L. Shepelyansky, *Physica D* **28**, 103 (1987).

153. G. Casati, B.V. Chirikov, and D. L. Shepelyansky, *Phys. Rev. Lett.* **53**, 2525 (1984).

154. G. Casati, I. Guarneri, and D. L. Shepelyansky, *IEEE Quant. Electron* **QE-24**, 1420 (1988).

155. G. Casati and L. Molnari, *Progr. Theor. Phys.* **98**, 287 (1989).

156. P. W. Milonni, J. R. Ackerhalt and M. E. Goggin, *Phys. Rev. A* **35**, 1714 (1987).

157. G. M. Zaslavsky, *Phys. Rep.* **80**, 175 (1981).

158. G. P. Berman and A. R. Kolovsky, *Physica D* **8**, 117 (1983).

159. F. Haake, *Quantum Signatures of Chaos*, Springer, Berlin, 1991.

160. L. R. Reichl, *The Transition to Chaos: Quantum Manifestations*, Springer, Berlin, 1992.

161. G. Casati and B.V. Chirikov, *Physica D* **86**, 220 (1995).

162. C. C. Gerry and E. R. Vrscay, *Phys. Rev. A* **39**, 5717 (1989).

163. G. J. Milburn and C. A. Holms, *Phys. Rev. A* **44**, 4704 (1991).

164. W. Leoński, *Physica A* **233**, 365 (1996).

165. P. Szlachetka, K. Grygiel, and J. Bajer, *Phys. Rev. E* **48**, 101 (1993).

166. K. Grygiel, and P. Szlachetka, *Phys. Rev. E* **51**, 36 (1995).

167. K.N. Alekseev and J. Peřina, *Phys. Rev. E* **57**, 4023 (1998).

168. J. Peřina, *Quantum Statistic of Linear and Nonlinear Optical Phenomena*, Kluwer, Dordrecht, 1991.

169. J. Peřina, J. Křepelka, R. Horák, Z. Hradil, and J. Bajer, *Czech. J. Phys. B* **37**, 1161 (1987).

170. R. Schack and A. Schenzle, *Phys Rev. A* **41**, 3847 (1990).

171. P. Szlachetka, K. Grygiel, J. Bajer, and J. Peřina, *Phys. Rev. A* **46**, 7311 (1992) .

172. K. Grygiel, W. Leoński, and P. Szlachetka, *Acta Phys. Slovaca* **48**, 379 (1998).

173. B. Sundaram and P. W. Milonni, *Phys. Rev. E* **51**, 1971 (1995).

174. L. E. Ballentine and S. M. McRae, *Phys. Rev. E* **58**, 1799 (1998).

175. M. Toda and K. Ikeda, *Phys. Lett. A* **124**, 165 (1987).

176. G. J. Milburn *Phys. Rev. A* **33**, 674 (1986).

177. K. Vogel and H. Risken, *Phys. Rev. A* **38**, 2409 (1988).

178. P. D. Drummond, K. J. McNeil, and D. E. Walls, *Optica Acta* **28**, 211 (1981).

179. P. Szlachetka, *J. Phys. A* **20**, 1455 (1987).

NON-ABELIAN STOKES THEOREM

BOGUSŁAW BRODA

Department of Theoretical Physics, University of Łódź, Łódź, Poland

CONTENTS

Modern Nonlinear Optics, Part 2, Second Edition, Advances in Chemical Physics, Volume 119,
Edited by Myron W. Evans. Series Editors I. Prigogine and Stuart A. Rice.
ISBN 0-471-38931-5

I. INTRODUCTION

For many years the (standard, i.e., Abelian) *Stokes* theorem has been one of the central points of (multivariable) analysis on manifolds. Lower-dimensional versions of this theorem, known as the (proper) Stokes theorem, in dimensions 1 and 2, and the *Gauss* theorem, in dimensions 2 and 3, respectively, are well known and extremely useful in practice, such as in classical electrodynamics (Maxwell equations). In fact, it is difficult, if not impossible, to imagine lectures on classical electrodynamics without intensive use of the Stokes theorem. The standard Stokes theorem is also called the *Abelian* Stokes theorem, as it applies to (ordinary, i.e., Abelian) differential forms. Classical electrodynamics is an Abelian [i.e., $U(1)$] gauge field theory (gauge fields are Abelian forms), therefore its integral formulas are governed by the Abelian Stokes theorem. But many interesting and physically important phenomena are described by non-Abelian gauge theories. Hence it would be very desirable to have at our disposal a non-Abelian version of the Stokes theorem. Since non-Abelian differential forms need necessitate a somewhat different treatment, one is forced to use a more sophisticated formalism to deal with this new situation.

The aim of this chapter is to present a short review of the non-Abelian Stokes theorem. At first, we will give an account of different formulations of the non-Abelian Stokes theorem and next of various applications of thereof.

A. Abelian Stokes Theorem

Before we engage in the non-Abelian Stokes theorem it seems reasonable to recall its Abelian version. The (Abelian) Stokes theorem says (see, e.g., Ref. 1 for an excellent introduction to the subject) that we can convert an integral around a closed curve C bounding some surface S into an integral defined on this surface. Specifically, in three dimensions

$$\oint_C \vec{A} \cdot d\vec{s} = \int_S \operatorname{curl} \vec{A} \cdot \vec{n}\, d\sigma \tag{1}$$

where the curve C is the boundary of the surface S, that is, $C = \partial S$ (see Fig. 1), $\vec{A}$ is a vector field (e.g., the vector potential of electromagnetic field) and $\vec{n}$ is a unit outward normal at the area element $d\sigma$.

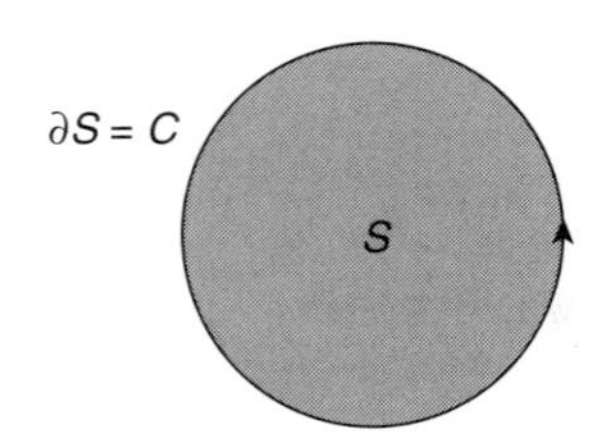

Figure 1. Integration areas for the lowest-dimensional (nontrivial) version of the Abelian Stokes theorem.

More generally, in any dimension

$$\int_{\partial N} \omega = \int_N d\omega \tag{2}$$

where now N is a d-dimensional submanifold of the manifold M, ∂N is its $(d-1)$-dimensional boundary, ω is a $(d-1)$-form, and $d\omega$ is its differential, a d form. We can also rewrite Eq. (1) in the spirit of Eq. (2)

$$\oint_{\partial S = C} A_i \, dx^i = \frac{1}{2} \int_S (\partial_i A_j - \partial_j A_i) dx^i \wedge dx^j \tag{3}$$

where A_i $(i = 1, 2, 3)$ are components of the vector $\vec{A}$, and the Einstein summation convention after repeating indices is assumed.

In electrodynamics, we define the strength tensor of electromagnetic field

$$F_{ij} = \partial_i A_j - \partial_j A_i$$

and the magnetic induction, its dual, as

$$B_k = \frac{1}{2} \varepsilon_{ijk} F_{ij}$$

where ε_{ijk} is the totally antisymmetric (pseudo) tensor. Thus the right-hand side (r.h.s.) of Eq. (3) represents the magnetic flux through S, and we can rewrite Eq. (3) in the form of (1):

$$\begin{aligned} \oint_{\partial S = C} A_i \, dx^i &= \frac{1}{2} \int_S F_{ij} \, dx^i \wedge dx^j \\ &= \int_S B_i n^i \, d\sigma \end{aligned} \tag{4}$$

In turn, in geometry $\vec{A}$ plays the role of connection (it defines the parallel transport around C) and F is the curvature of this connection. A "global version" of the Abelian Stokes theorem

$$\exp\left(i \oint_{\partial S = C} A_i(x) \, dx^i \right) = \exp\left(\frac{i}{2} \int_S F_{ij}(x) \, dx^i \wedge dx^j \right) \tag{5}$$

which is a rather trivial generalization of Eq. (4) is a very good starting point for our discussion of the non-Abelian Stokes theorem. The object on the left-hand side (l.h.s.) of (5) is called the holonomy, and more generally, for open curves C, global connection.

B. Historical Remarks

The birth of the ideas related to the (non-)Abelian Stokes theorem dates back to the ninetenth century, with the emergence of the Abelian Stokes theorem. The Abelian Stokes theorem can be treated as a prototype of the non-Abelian Stokes theorem or a version of thereof when we confine our discussion to an Abelian group.

A work closer to the proper non-Abelian Stokes theorem, by Schlesinger [2], which generally considered noncommuting matrix value functions, appeared in 1927. In fact, no work on the genuine non-Abelian Stokes theorem appeared before the concept of non-Abelian gauge fields emerged in the early 1950s. The first papers on the non-Abelian Stokes theorem appeared in the late 1970s. At first, the non-Abelian Stokes theorem emerged in the operator version [3–5] and later on, in the very end of 1980s, in the path integral one [6,7].

C. Contents

The substantive part of this chapter consists of two sections. Section II is devoted to the non-Abelian Stokes theorem itself. In the beginning, we introduce the necessary notions and conventions. The operator version of the non-Abelian Stokes theorem is formulated in Section II.A. Section II.B concerns the path integral versions of the non-Abelian Stokes theorem: coherent-state approach and holomorphic approach. Section II.C describes generalizations of the non-Abelian Stokes theorem: topologically more general situations (Section II.C.1) and higher-degree forms (Section II.C.2). Section III is devoted to applications of the non-Abelian Stokes theorem in mathematical and theoretical physics. Section III.A presents an approach to the computation of Wilson loops in two-dimensional Yang–Mills theory. Section III.B deals with the analogous problem for three-dimensional (topological) Chern–Simons gauge theory. Other possibilities, including higher-dimensional gauge theories and QCD, are mentioned in Section III.C.

II. NON-ABELIAN STOKES THEOREM

What is the non-Abelian Stokes theorem? To answer this question, we should first recall the form of the well-known Abelian Stokes theorem [see Eq. (2)]

$$\int_{\partial M} \omega = \int_{M} d\omega \tag{6}$$

where the integral of the form ω along the boundary ∂N of the submanifold N is equated to the integral of the differential $d\omega$ of this form over the submanifold N. The differential form ω is usually an ordinary (i.e., Abelian) differential form, but it could also be something more general, such as a connection one-form A. Thus, the non-Abelian Stokes theorem should be a version of (6) for non-Abelian (say, Lie-algebra valued) forms. Since it could be risky to directly integrate Lie-algebra-valued differential forms, the generalization of (6) may be a nontrivial task. We should not be too ambitious perhaps from the very beginning, and not try to formulate the non-Abelian Stokes theorem in full generality at once, as this could be difficult or even impossible. The lowest nontrivial dimensionality of the objects entering (6) is as follows: $\dim N = 2$ $(\dim \partial N = 1)$ and $\deg \omega = 1$ $(\deg d\omega = 2)$. A short reflection leads us to the first candidate for the l.h.s. of the non-Abelian Stokes theorem, the Wilson loop

$$P \exp\left(i \oint_C A \right) \tag{7}$$

called the holonomy, in mathematical context, where P denotes the, so-called path ordering, A is a non-Abelian connection one-form, and C is a closed loop, a boundary of the surface S $(\partial S = C)$. Correspondingly, the r.h.s. of the non-Abelian Stokes theorem should contain a kind of integration over S. Therefore, the actual Abelian prototype of the non-Abelian Stokes theorem is of the form (5) rather than of (4). More often, the trace of Eq. (7) is called the *Wilson loop*

$$W_R(C) = \mathrm{Tr}_R P \exp\left(i \oint_C A \right) \tag{8}$$

or even the "normalized" trace of it

$$\frac{1}{\dim R} \mathrm{Tr}_R P \exp\left(i \oint_C A \right) \tag{9}$$

where the character R means a(n irreducible) representation of the Lie group G corresponding to the given Lie algebra $\mathbf{g}$. Of course, one can easily pass from (7) to (8) and finally to (9). In fact, the operator (7) is a particular case of a more general parallel-transport operator

$$U_L = P \exp\left(i \int_L A \right) \tag{10}$$

where L is a smooth path, which for the L a closed loop $(L = C)$ yields (7). Eq. (10) could be considered as an ancestor of Eq. (7). As the l.h.s. of the non-Abelian Stokes theorem, we can assume any of the formulas given above for the closed-loop C [i.e., Eqs. (7)–(9)], possibly yielding various versions of the non-Abelian Stokes theorem. For some reason, it is sometimes more convenient to use the Wilson loop in the operator version (7) rather than in the version with the trace (8). Sometimes it does not make any greater difference. The r.h.s. should be some expression defined on the surface S, and essentially constituting the non-Abelian Stokes theorem.

A physicist would view the expression (10) as typical in quantum mechanics and as corresponding to the evolution operator. Equations (8) and (9) are, incidentally, very typical in gauge theory, such as in QCD. Thus, guided by our intuition, we can reformulate our chief problem as a quantum-mechanical one. In other words, the approaches to the l.h.s. of the non-Abelian Stokes theorem are analogous to the approaches to the evolution operator in quantum mechanics. There are the two main approaches to quantum mechanics, especially to the construction of the evolution operator: opearator approach and path-integral approach. Both can be applied to the non-Abelian Stokes theorem successfully, and both provide two different formulations of the non-Abelian Stokes theorem.

The *conventions* are as follows. Sometimes, especially in a physical context, a coupling constant, denoted, for instance, e, appears in front of the integral in Eqs. (7)–(10). For simplicity, we will omit the coupling constant in our formulas.

The non-Abelian curvature or the strength field on the manifold M is defined by

$$F_{ij}(A) = \partial_i A_j - \partial_j A_i - i[A_i, A_j]$$

Here, the connection or the gauge potential A assuming values in a(n irreducible) representation R of the compact, semisimple Lie algebra $\mathbf{g}$ of the Lie group G is of the form

$$A_i(x) = A_i^a(x) T^a, \qquad i = 1, \ldots, \dim M$$

where the Hermitian generators, $T^{a\dagger} = T^a$, $T^a = T^a_{kl}$, $k, l = 1, \ldots, \dim R$, fulfill the commutation relations

$$[T^a, T^b] = if^{abc} T^c, \qquad a, b, c = 1, \ldots, \dim G \tag{11}$$

The line integral (10) can be rewritten in more detailed (as it is frequently used in our further analysis) forms, such as

$$U(x'', x') = P \exp\left[i \int_{x'}^{x''} A_i(x)\, dx^i \right]$$

or

$$U_{kl} = P \exp\left[i \int_{x'}^{x''} A_i^a(x) T^a\, dx^i \right]$$

without parametrization, and

$$U(t'',t') = P\exp\left\{ i\int_{x'}^{x''} A_i^a[x(t)]T^a \frac{dx^i(t)}{dt}\, dt \right\}$$

with an explicit parametrization, or some variations thereof. Here, the oriented smooth path L starting at the point x' and endng at the point x'' is parametrized by the function $x^i(t)$, where $t' \leq t \leq t''$ and $x'^i = x^i(t')$, $x''^i = x^i(t'')$.

A. Operator Formalism

Unfortunately, it is not possible to automatically generalize the Abelian Stokes theorem [e.g., Eq. (4)] to the non-Abelian one. In the non-Abelian case one faces a qualitatively different situation because the integrand on the l.h.s. assumes values in a Lie algebra **g** rather than in the field of real or complex numbers. The picture simplifies significantly if one switches from the "local" language to a global one [see Eq. (5)]. Therefore we should consider the holonomy (7) around a closed curve C:

$$P\exp\left(i\oint_C A_i dx^i \right)$$

The holonomy represents a parallel-transport operator around C assuming values in a non-Abelian Lie group G. (Interestingly, in the Abelian case, the holonomy has a physical role; it is an object playing the role of the phase that can be observed in the Aharonov–Bohm experiment, whereas A_i itself does not have such an interpretation.)

The *non-Abelian Stokes theorem* is as follows. The non-Abelian generalization of Eq. (5) should read as

$$P\exp\left(i\oint_{\partial S=C} A_i(x)dx^i \right) = \mathscr{P}\exp\left(\frac{i}{2}\int_S \mathscr{F}_{ij}(x)dx^i \wedge dx^j \right)$$

where the l.h.s. has been already roughly defined. As far as the r.h.s. is concerned, the symbol $\mathscr{P}$ denotes some "surface ordering," whereas $\mathscr{F}_{ij}(x)$ is a "path-dependent curvature" given by the formula

$$\mathscr{F}_{ij}(x) \stackrel{\text{def}}{=} U^{-1}(x,O)F_{ij}(x)U(x,O)$$

where $U(x,O)$ is a parallel-transport operator along the path L in the surface S joining the base point O of ∂S with the point x:

$$U(x,O) = P\exp\left(i\int_L A_i(y)dy^i \right)$$

See Fig. 2, and later sections for more details.

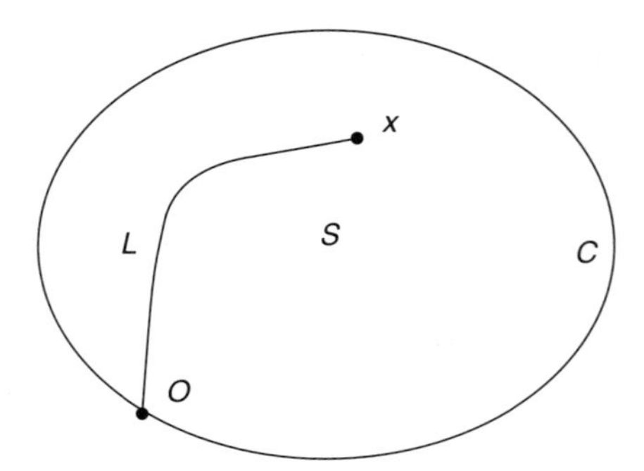

Figure 2. Parallel-transport operator along the path L in the surface S.

1. *Calculus of Paths*

As a kind of a short introduction for properly manipulating parallel-transport operators along oriented curves, we recall a number of standard facts. It is obvious that we can perform some operations on the parallel-transport operators. We can superimpose them, we can introduce an identity element, and finally, we can find an inverse element for each element.

The structure is roughly similar to the structure of the group with the following standard postulates satisfied: (1) associativity, $(U_1U_2)U_3 = U_1\ (U_2U_3)$; (2) existence of an identity element, $IU = UI = U$; (3) existence of an inverse element U^{-1}, $U^{-1}U = UU^{-1} = I$. But let us note that not all elements can be superimposed. Although parallel-transport operators are elements of a Lie group G, their geometric interpretation has been lost in the notation above. We can superimpose two elements only when the endpoint of the first element is the initial point of the second one. Thus U_1U_2 could be meaningful in the form (Fig. 3)

$$U(x_1, x)U(x, x_2) = U(x_1, x_2)$$

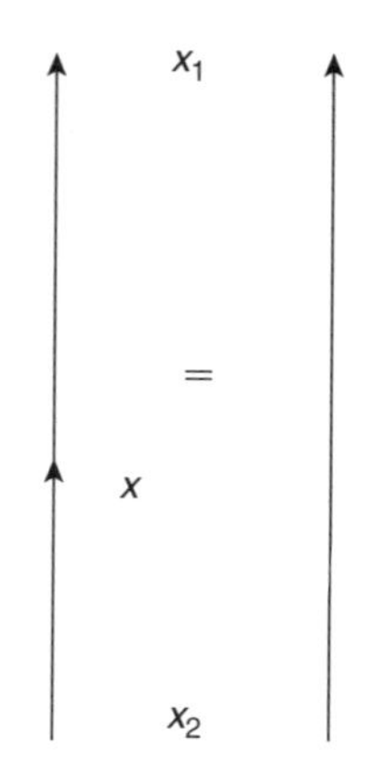

Figure 3. Allowable composition of elements.

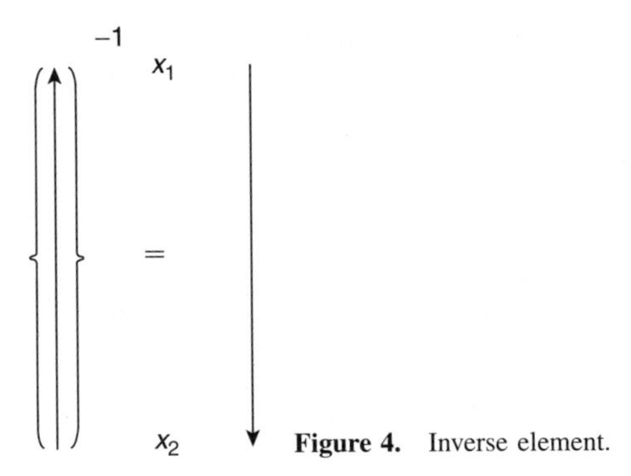

Figure 4. Inverse element.

Obviously

$$I = U(x, x)$$

and (Fig. 4)

$$U^{-1}(x_1, x_2) = U(x_2, x_1)$$

These formulas become particularly convincing in graphical form. Perhaps one of the most useful facts is expressed by Fig. 5.

It appears that this structure fits into the structure of the so-called grouppoid.

2. *Ordering*

There are a lot of different ordering operators in our formulas which have been collected in this section.

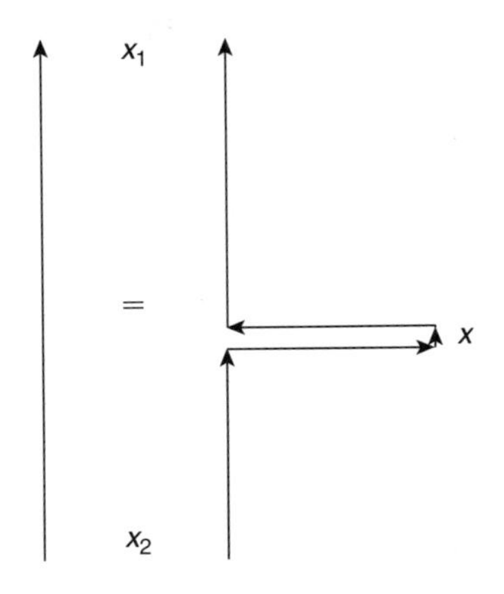

Figure 5. Deformation of a path.

We know from quantum theory that the path-ordered exponent of an operator $\hat{A}$ can be expressed by the power series called the *Dyson series*:

$$
\begin{aligned}
P\exp\left[i\int_{t'}^{t''}\hat{A}(t)\,dt\right]
&= P\left[1+\sum_{n=1}^{\infty}\frac{i^n}{n!}\int_{t'}^{t''}dt_1\int_{t'}^{t''}dt_2\cdots\int_{t'}^{t''}dt_n\,\hat{A}(t_1)\hat{A}(t_2)\cdots\hat{A}(t_n)\right]\\
&= 1+\sum_{n=1}^{\infty}\frac{i^n}{n!}\int_{t'}^{t''}dt_1\int_{t'}^{t''}dt_2\cdots\int_{t'}^{t''}dt_n\,\mathrm{P}\left[\hat{A}(t_1)\hat{A}(t_2)\cdots\hat{A}(t_n)\right]\\
&= 1+\sum_{n=1}^{\infty}i^n\int_{t'}^{t''}dt_1\int_{t'}^{t_1}dt_2\cdots\int_{t'}^{t_{n-1}}dt_n\,\hat{A}(t_1)\hat{A}(t_2)\cdots\hat{A}(t_n)
\end{aligned}
$$

where

$$
P[\hat{A}_1(t_1)\cdots\hat{A}_n(t_n)]\stackrel{\text{def}}{=}\hat{A}_{\sigma(1)}\left(t_{\sigma(1)}\right)\cdots\hat{A}_{\sigma(n)}\left(t_{\sigma(n)}\right)),\qquad t_{\sigma(1)}\geq\cdots\geq t_{\sigma(n)}
$$

For example, for two operators

$$
P[\hat{A}_1(t_1)\hat{A}_2(t_2)]\equiv\theta(t_1-t_2)\hat{A}_1(t_1)\hat{A}_2(t_2)+\theta(t_2-t_1)\hat{A}_2(t_2)\hat{A}_1(t_1)
$$

where θ is the step function.

Since the operators and matrices appearing in our considerations are, in general, noncommutative, we assume the following *conventions*:

$$
\prod_{n=1}^{N}X_n\stackrel{\text{def}}{=}X_NX_{N-1}\cdots X_2X_1 \tag{12}
$$

$$
P_t(X_1X_2,\ldots,X_{N-1}X_N)\stackrel{\text{def}}{=}X_NX_{N-1}\cdots X_2X_1 \tag{13}
$$

whereas for two parameters

$$
\begin{aligned}
P_{s,t}\prod_{m,n=1}^{N}X_{m,n}&\stackrel{\text{def}}{=}\prod_{n=1}^{N}\prod_{m=1}^{N}X_{m,n}\\
&\equiv X_{N,N}X_{N-1,N}\cdots X_{1,N}X_{N,N-1}\cdots X_{1,2}X_{N,1}\cdots X_{2,1}X_{1,1}
\end{aligned}
$$

3. *Theorem*

The non-Abelian Stokes theorem in its original operator form roughly claims that the holonomy around a closed curve $C=\partial S$ equals a surface-ordered exponent of the *twisted* curvature, namely

$$P\exp\left(i\oint_C A\right) = \mathscr{P}\exp\left(i\int_S \mathscr{F}\right) \tag{14}$$

where $\mathscr{F}$ is the twisted curvature:

$$\mathscr{F} \equiv U^{-1}FU$$

A more precise form of the non-Abelian Stokes theorem as well as an exact meaning of the notions appearing in the theorem will be given in the course of the proof.

Proof. Following Aref'eva [3] and Menski [8], we will present a short, direct proof of the non-Abelian Stokes theorem.

In our parametrization, the first step consists of the decomposition of the initial loop (see Fig. 6) into small lassos according to the rules given in the Section II.A.1

$$P\exp\left(i\oint_C A\right) = \lim_{N\to\infty} (\mathrm{P}_{s,t}) \prod_{m,n=1}^{N} U_{m,n}^{-1} W_{m,n} U_{m,n}$$

where the objects involved are defined as follows. Parallel-transport operators from the reference point to the point with coordinates $(\frac{m}{N}, \frac{n}{N})$ consists of two segments (see Fig. 7)

$$U_{m,n} \stackrel{\text{def}}{=\!=} P\exp\left(i\int_{(0,\frac{n}{N})}^{(\frac{m}{N},\frac{n}{N})} A\right) P\exp\left(i\int_{(0,0)}^{(0,\frac{n}{N})} A\right)$$

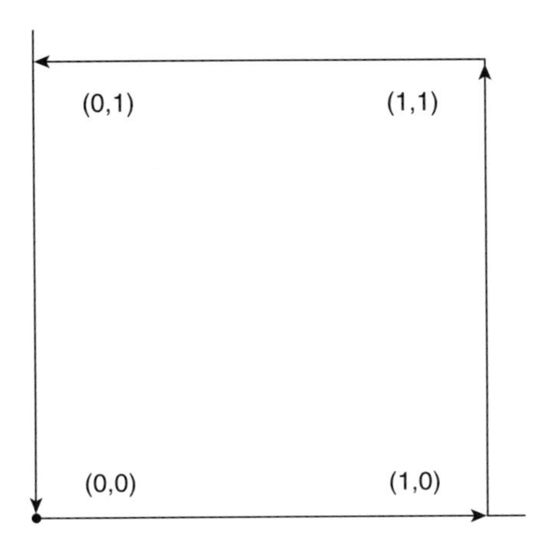

Figure 6. The parametrized loop C as a boundary of the "big" square S with the coordinates $\{(0,0), (1,0), (1,1), (0,1)\}$.

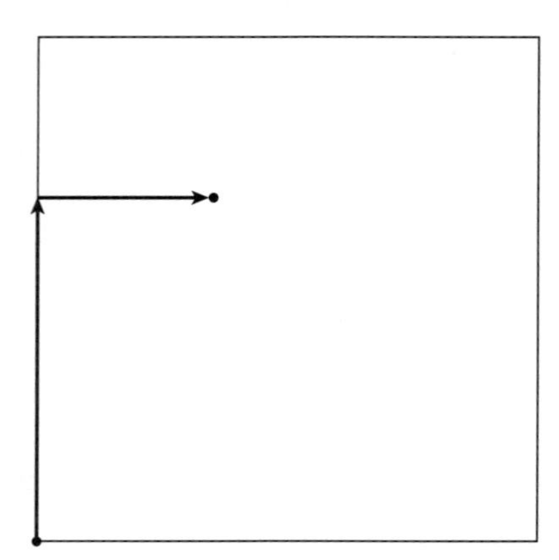

Figure 7. Approaching the "small" plaquette inside the "big" square.

parallel-transport operator round a small plaquette $S_{m,n}$ (see Fig. 8)

$$W_{m,n} \stackrel{\text{def}}{=} \operatorname{Tr} P\exp\left(i\oint_{S_{m,n}} A\right)$$

where $S_{m,n}$ is a boundary of a (small) square with coordinates

$$S_{m,n} = \left\{\left(\frac{m}{N},\frac{n}{N}\right),\left(\frac{m-1}{N},\frac{n}{N}\right),\left(\frac{m-1}{N},\frac{n-1}{N}\right),\left(\frac{m}{N},\frac{n-1}{N}\right)\right\}$$

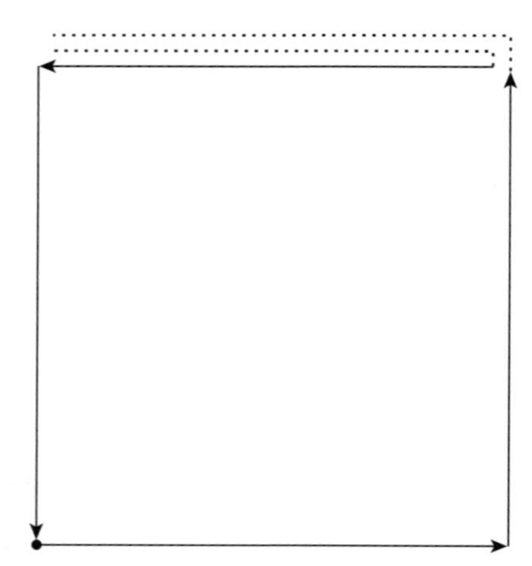

Figure 8. The small plaquette.

Now

$$\begin{aligned}
U_{m,n}^{-1} W_{m,n} U_{m,n} &= U_{m,n}^{-1}\left[1 + \frac{i}{N^2} F_{m,n} + O\left(\frac{1}{N^3}\right)\right] U_{m,n} \\
&= 1 + \frac{i}{N^2} U_{m,n}^{-1} F_{m,n} U_{m,n} + O\left(\frac{1}{N^3}\right) \\
&= 1 + \frac{i}{N^2} \mathscr{F}_{m,n} + O\left(\frac{1}{N^3}\right) \\
&= \mathscr{W}_{m,n} + O\left(\frac{1}{N^3}\right)
\end{aligned}$$

where $W_{m,n}$ has been calculated in the next (and last) paragraph of this section [see Eq. (15)]

$$\mathscr{F}_{m,n} \overset{\text{def}}{=\!=} U_{m,n}^{-1} F_{m,n} U_{m,n}$$

and

$$\mathscr{W}_{m,n} \overset{\text{def}}{=\!=} \exp\left(\frac{i}{N^2} \mathscr{F}_{m,n}\right)$$

Then

$$\begin{aligned}
P\exp\left(i \oint_C A\right) &= \lim_{N\to\infty} (\mathrm{P}_{s,t}) \prod_{m,n=1}^{N} \left[\mathscr{W}_{m,n} + O\left(\frac{1}{N^3}\right)\right] \\
&= \lim_{N\to\infty} (\mathrm{P}_{s,t}) \prod_{m,n=1}^{N} \exp\left[\frac{i}{N^2} \mathscr{F}_{m,n} + O\left(\frac{1}{N^3}\right)\right] \\
&= \lim_{N\to\infty} (\mathrm{P}_{t}) \prod_{m,n=1}^{N} \exp\left[\frac{i}{N^2} \mathscr{F}_{m,n} + O\left(\frac{1}{N^3}\right)\right].
\end{aligned}$$

The last equality follows from the fact that operations corresponding to the change of the order of the operators yield the commutator

$$\left[\frac{1}{N^2} \mathscr{F}_{m',n'}, \frac{1}{N^2} \mathscr{F}_{m'',n''}\right] = O\left(\frac{1}{N^4}\right)$$

and there are maximum $N - 1$ transpositions possible in the framework of s-ordering, so

$$(N-1)\, O\left(\frac{1}{N^4}\right) = O\left(\frac{1}{N^3}\right)$$

Thus, we arive at the final form of the non-Abelian Stokes theorem:

$$P\exp\left(i\oint_C A\right) = \lim_{N\to\infty}(P_t)\prod_{m,n=1}^{N}\exp\left[\frac{i}{N^2}\mathscr{F}_{m,n}\right] = \mathscr{P}\exp\left(i\int_S \mathscr{F}\right)$$

In this paragraph we will perform a fairly standard calculus and derive the contribution coming from a small loop $W_{m,n}$:

$$\begin{aligned}
W_{m,n} &= P\exp\left(i\int_{\left(\frac{m}{N},\frac{n-1}{N}\right)}^{\left(\frac{m}{N},\frac{n}{N}\right)} A\right) P\exp\left(i\int_{\left(\frac{m-1}{N},\frac{n-1}{N}\right)}^{\left(\frac{m}{N},\frac{n-1}{N}\right)} A\right) \\
&\quad P\exp\left(i\int_{\left(\frac{m}{N},\frac{n-1}{N}\right)}^{\left(\frac{m-1}{N},\frac{n-1}{N}\right)} A\right) P\exp\left(i\int_{\left(\frac{m}{N},\frac{n}{N}\right)}^{\left(\frac{m}{N},\frac{n-1}{N}\right)} A\right) \\
&= \left[1 + \frac{i}{N}A^y_{m,n} + \frac{1}{2}\left(\frac{i}{N}A^y_{m,n}\right)^2 + O\left(\frac{1}{N^3}\right)\right] \\
&\quad \cdot \left[1 + \frac{i}{N}A^x_{m,n-1} + \frac{1}{2}\left(\frac{i}{N}A^x_{m,n-1}\right)^2 + O\left(\frac{1}{N^3}\right)\right] \\
&\quad \cdot \left[1 - \frac{i}{N}A^y_{m-1,n} + \frac{1}{2}\left(\frac{i}{N}A^y_{m-1,n}\right)^2 + O\left(\frac{1}{N^3}\right)\right] \\
&\quad \cdot \left[1 - \frac{i}{N}A^x_{m,n} + \frac{1}{2}\left(\frac{i}{N}A^x_{m,n}\right)^2 + O\left(\frac{1}{N^3}\right)\right] \\
&= 1 + \frac{i}{N}\left(A^y_{m,n} - A^y_{m-1,n}\right) - \frac{i}{N}\left(A^x_{m,n} - A^x_{m,n-1}\right) \\
&\quad + \left(\frac{i}{N}\right)^2\left(A^y_{m,n}\right)^2 + \left(\frac{i}{N}\right)^2\left(A^x_{m,n}\right)^2 + \left(\frac{i}{N}\right)^2 A^y_{m,n}A^x_{m,n} \\
&\quad - \left(\frac{i}{N}\right)^2\left(A^y_{m,n}\right)^2 - \left(\frac{i}{N}\right)^2 A^y_{m,n}A^x_{m,n} - \left(\frac{i}{N}\right)^2 A^x_{m,n}A^y_{m,n} \\
&\quad - \left(\frac{i}{N}\right)^2\left(A^x_{m,n}\right)^2 + \left(-\frac{i}{N}\right)^2 A^y_{m,n}A^x_{m,n} + O\left(\frac{1}{N^3}\right) \\
&= 1 + \frac{i}{N^2}\left(\partial_x A^y - \partial_y A^x\right) - \frac{i^2}{N^2}\left(A^x A^y - A^y A^x\right) + O\left(\frac{1}{N^3}\right) \\
&= 1 + \frac{i}{N^2}F_{m,n} + O\left(\frac{1}{N^3}\right)
\end{aligned} \tag{15}$$

where

$$F = \partial_x A_y - \partial_y A_x - i[A_x, A_y]$$

(occasionally, we put x and y in the upper position), and

$$A = A_{m,n} + O\left(\frac{1}{N}\right)$$

Then

$$W_{m,n} = \exp\left[\frac{i}{N^2} F_{m,n} + O\left(\frac{1}{N^3}\right)\right]$$

There are may other approaches to the (operator) non-Abelian Stokes theorem, which are more or less interrelated, including an analytical approach advocated by Bralić [4] and Hirayama and Ueno [9]. An approach using product integration [10], and last, but not least, a (very interesting) coordinate gauge approach [11,12].

B. Path Integral Formalism

There are two main approaches to the non-Abelian Stokes theorem in the framework of the path integral formalism: coherent-state approach and holomorphic approach. In the literature, both approaches occur in a few, and slightly different, incarnations. Also, both have found applications in different areas of mathematical and/or theoretical physics, and therefore both are useful. The first one is formulated more in the spirit of group theory, whereas the second one follows from traditional path integral formulation of quantum mechanics or rather quantum field theory. Similar to the situation in quantum theory, the path integral formalism is easier in some applications and more intuitive than the operator formalism, but traditionally it is mathematically less rigorous. In the same manner as quantum mechanics, initially formulated in the operator language and next reformulated in the path integral one, we can translate the operator form of the non-Abelian Stokes theorem into the path integral language.

In order to formulate the non-Abelian Stokes theorem in the path integral language, we will perform the following three steps:

1 We will determine a coherent-state/holomorphic path integral representation for the parallel-transport operator, deriving an appropriate transition amplitude [a path integral counterpart of the l.h.s. in Eq. (14)].

2 For a closed curve C we will calculate the trace of the path integral form of the parallel-transport operator in quantum theory in an external gauge field A.

3 We will apply the Abelian Stokes theorem to the exponent of the integrand of the path integral yielding the r.h.s. of the non-Abelian Stokes theorem [a counterpart of the r.h.s. in Eq (8)].

Preliminary formulas are presented in Eqs. (16) and (17). Since both path integral derivations of the transition amplitude have a common starting point that is independent of the particular approach, we present it here:

$$P\exp\left[i\int_{t'}^{t''} A(t)dt\right] = \lim_{N\to\infty}\prod_{n=1}^{N}(1+i\epsilon A_n) \tag{16}$$

In this equation

$$A_n = A(t_n), \quad \epsilon = \frac{t''-t'}{N}, \quad t_n = n\epsilon + t', \quad t_N = t'', \quad t_0 = t' \tag{17}$$

From this moment on, both approaches differ.

1. Coherent-State Approach

a. Group-Theoretic Coherent States. According to Zhang et al. [13] (see also Ref. 14) [Per86] the group-theoretic coherent states emerge in the following construction:

1 For **g**, a semisimple Lie algebra of a Lie group G, we introduce the standard Cartan basis $\{H_i, E_\alpha, E_{-\alpha}\}$:

$$\begin{aligned}[H_i, H_j] &= 0\\ [H_i, E_\alpha] &= \alpha_i E_\alpha\\ [E_\alpha, E_{-\alpha}] &= \alpha^i H_i\\ [E_\alpha, E_\beta] &= N_{\alpha;\beta}E_{\alpha+\beta}\end{aligned} \tag{18}$$

2 We chose a unitary irreducible representation R of the group G, as well as a normalized state the, so-called, reference state $|R\rangle$. The choice of the reference state is in principle arbitrary but not unessential. Usually it is an "extremal state" (the highest-weight state), the state anihilated by E_α, namely, $E_\alpha|R\rangle = 0$.

3 A subgroup of G that consists of all the group elements h that will leave the reference state $|R\rangle$ invariant up to a phase factor is the maximum-stability subgroup H. Formally, this is

$$h|R\rangle = |R\rangle e^{i\phi(h)}, \quad h \in H$$

The phase factor is unimportant here because we shall generally take the expectation value of any operator in the coherent state.

4 For every element $g \in G$, there is a unique decomposition of g into a product of two group elements, one in H and the other in the quotient G/H:

$$g = \xi h, \quad g \in G, \quad h \in H, \quad \xi \in \frac{G}{H}$$

In other words, we can obtain a unique coset space for a given $|R\rangle$.

5 One can see that the action of an arbitrary group element $g \in G$ on $|R\rangle$ is given by $g|R\rangle = \xi h|R\rangle = \xi|R\rangle e^{i\phi(h)}$. The combination $|\xi, R\rangle \stackrel{\text{def}}{=} \xi|R\rangle$ is the general group definition of the coherent states. For simplicity, we will denote the coherent states as $|g, R\rangle$.

The coherent states $|g, R\rangle$ are generally nonorthogonal but are normalized to unity:

$$\langle g, R\,|\,g, R\rangle = 1.$$

Furthermore, for an appropriately normalized measure $d\mu(g)$, we have a very important for our furher analysis identity the so-called, resolution of unity:

$$\int |g, R\rangle d\mu(g) \langle g, R| = I \tag{19}$$

b. Path Integral. Our first aim is to calculate the "transition amplitude" between the two coherent states $|g', R\rangle$ and $|g'', R\rangle$

$$\begin{aligned}
&\langle g'', R|P \exp\left[i \int_{t'}^{t''} A(t)\, dt\right] |g', R\rangle \\
&= \lim_{N\to\infty} \int \cdots \int \langle g_N, R|(1 + i\epsilon A_N)|g_{N-1}, R\rangle d\mu(g_{N-1}) \\
&\quad \langle g_{N-1}, R|(1 + i\epsilon A_{N-1})|g_{N-2}, R\rangle d\mu(g_{N-2}) \\
&\quad \cdots d\mu(g_1) \langle g_1, R\,|(1 + i\epsilon A_1)|g_0, R\rangle
\end{aligned} \tag{20}$$

where we have used (16) and (19). To continue, one should evaluate a single amplitude (i.e., the amplitude for an infinitesimal "time" ϵ):

$$\begin{aligned}
&\langle g_n, R|(1 + i\epsilon A_n)|g_{n-1}, R\rangle \\
&= \langle g_n, R|g_{n-1}, R\rangle + i\langle g_n, R|A_n|g_{n-1}, R\rangle\epsilon
\end{aligned}$$

Now

$$
\begin{aligned}
\langle g_n, R | g_{n-1}, R\rangle &= \langle g(t_n), R | g(t_{n-1}), R\rangle \\
&= \langle R | g^\dagger(t_n) g(t_{n-1}) | R\rangle = \langle R | g^\dagger(t_n) g(t_n - \epsilon) | R\rangle \\
&= \langle R | g^\dagger(t_n)[g(t_n) - \dot{g}(t_n)\epsilon + O(\epsilon^2)] | R\rangle \\
&= \langle R | g^\dagger(t_n) g(t_n) | R\rangle - \langle R | g^\dagger(t_n)\dot{g}(t_n) | R\rangle\epsilon + O(\epsilon^2) \\
&= 1 - \langle R | g^\dagger(t_n)\dot{g}(t_n) | R\rangle\epsilon + O(\epsilon^2)
\end{aligned}
$$

whereas

$$
\begin{aligned}
i\langle g_n, R | A_n | g_{n-1}, R\rangle\epsilon &= i\langle g(t_n), R | A(t_n) | g(t_{n-1}), R\rangle\epsilon \\
&= i\langle R | g^\dagger(t_n) A(t_n) g(t_n - \epsilon) | R\rangle\epsilon \\
&= i\langle R | g^\dagger(t_n) A(t_n) g(t_n) | R\rangle\epsilon + O(\epsilon^2)
\end{aligned}
$$

Then

$$
\begin{aligned}
\langle g_n, R | (1 + i\epsilon A_n) | g_{n-1}, R\rangle &= 1 - \langle R | g^\dagger(t_n)\dot{g}(t_n) | R\rangle\epsilon \\
&\quad + i\langle R | g^\dagger(t_n) A(t_n) g(t_n) | R\rangle\epsilon + O(\epsilon^2) \\
&= \exp[\langle R | - g^\dagger(t_n)\dot{g}(t_n) \\
&\quad + i g^\dagger(t_n) A(t_n) g(t_n) | R\rangle\epsilon + O(\epsilon^2)]
\end{aligned}
$$

Returning to (20), we obtain

$$
\langle g'', R | P\exp\left[i\int_{t'}^{t''} A(t)\,dt\right] | g', R\rangle = \int [D\mu(g)] \exp\left(i\int_{t'}^{t''} L\,dt\right)
$$

where the "Lagrangian" appearing in the path integral is defined as

$$
L = \langle R | [i g^\dagger(t)\dot{g}(t) + g^\dagger(t) A(t) g(t)] | R\rangle \stackrel{\text{def}}{=} R | A^g(t) | R\rangle \tag{21}
$$

and

$$
[D\mu(g)] = \prod_{t'<t<t''} d\mu[g(t)]
$$

According to Hirayama and Ueno [15], we can transform L in Eq. (21) to the form originally proposed by Diakonov and Petrov [16]. Specifically, for any Lie algebra element K we have in the Cartan basis (18)

$$\langle R|K|R\rangle = \langle R|\sum_i \lambda_i H_i + \cdots |R\rangle = \sum_i \lambda_i \langle R|H_i|R\rangle$$

where the ellipses represent $E_{\mp\alpha}$ generators vanishing between $|R\rangle$ states. Since

$$H_i|R\rangle = m_i|R\rangle$$

then

$$\begin{aligned}\sum_i \lambda_i \langle R|H_i|R\rangle &= \sum_i \lambda_i \langle R|R_i|R\rangle = \sum_i \lambda_i m_i \\ &= \frac{1}{\kappa}\sum_i m_i \mathrm{Tr}(H_i K) = \frac{1}{\kappa}\,\mathrm{Tr}(\underline{m}\cdot\underline{H}\,K)\end{aligned}$$

where the normalization

$$\mathrm{Tr}(H_i H_j) = \kappa\delta_{ij}$$

has been assumed. Thus

$$\begin{aligned}L &= \langle R|A^g(t)|R\rangle \\ &= \frac{1}{\kappa}\sum_i \mathrm{Tr}[m_i H_i A^g(t)] \\ &= \frac{1}{\kappa}\mathrm{Tr}\{\underline{m}\cdot\underline{H}\,[ig^\dagger(t)\dot{g}(t) + g^\dagger(t)A(t)g(t)]\}\end{aligned} \tag{22}$$

c. Non-Abelian Stokes Theorem. Finally, the l.h.s. of the non-Abelian Stokes theorem reads as

$$\int D\mu(g)\exp\left(i\oint_{t'}^{t''} L\,dt\right)$$

where

$$D\mu(g) = \prod_{t'\le t<t''} d\mu[g(t)]$$

and $g(t') = g(t'')$. Or, in the language of differential forms

$$\int D\mu(g)\exp\left(i\oint_C L\right)$$

where now

$$D\mu(g) = \prod_{x \in C} d\mu[g(x)]$$

and

$$L = \frac{1}{\kappa} \sum_i \mathrm{Tr}[\underline{m} \cdot \underline{H} A_i^g(x)] dx^i \stackrel{\text{def}}{=} B_i \, dx^i$$

with

$$A_i^g(x) = i g^\dagger(x) \partial_i g(x) + g^\dagger(x) A_i(x) g(x)$$
$$A^g(t) = A_i^g[x(t)] \frac{dx^i(t)}{dt}$$

Here B_i is an Abelian differential form, so obviously

$$\int D\mu(g) \exp\left(i \oint_{C = \partial S} B \right) = \int D\mu(g) \exp\left(i \int_S dB \right)$$

2. *Holomorphic Approach*

a. Quantum-Mechanics Background. For further convenience, let us formulate an auxiliary "Schrödinger problem" governing the parallel-transport operator (10) for the Abelian gauge potential A

$$i \frac{dz}{dt} = -\dot{x}^i A_i z \tag{23}$$

which expresses the fact that the "wavefunction" z should be covariantly constant along the line L

$$D_t z \stackrel{\text{def}}{=} \left(\frac{d}{dt} - i \dot{x}^i A_i \right) z = 0$$

where D_t is the absolute covariant derivative.

First, let us derive the path integral expression for the parallel-transport operator U along L. To this end, we should consider the non-Abelian formula (differential equation) analogous to Eq. (23)

$$i \frac{dz_k}{dt} = -\dot{x}^i(t) A_i^a[x(t)] T^a_{kl} z_l \tag{24}$$

or

$$(D_t z)_k \stackrel{\text{def}}{=\!=} \left\{ \delta_{kl} \frac{d}{dt} - i\dot{x}^i A_i^a[x(t)] T^a_{kl} \right\} z_l = 0$$

where z is an auxiliary "wavefunction" in an irreducible representation R of the gauge Lie group G, which is to be parallelly transported along L parametrized by $x^i(t)$, $t' \leq t \leq t''$. Formally, Eq. (24) can be instantaneously integrated out, yielding

$$z_k(x'') = U_{kl}(x'', x') z_l(x')$$

where $x'' = x(t'')$, $x' = x(t')$, and

$$U(x'', x') \equiv U(t'', t') = P \exp\left(i \int_{t'}^{t''} \dot{x}^i(t) A_i[x(t)]\, dt \right)$$

as expected.

Let us now consider the following auxiliary classical mechanics problem with the classical Lagrangian

$$L(\bar{z}, z) = i\bar{z} D_t z \tag{25}$$

The equation of motion for z following from Eq. (25) reproduces Eq. (24) and yields the classical Hamiltonian:

$$H = i\dot{x}^i(t) A_i^a[x(t)] T^a_{kl} \pi_k z_l = -\dot{x}^i(t) A_i^a[x(t)] T^a_{kl} \bar{z}_k z_l \tag{26}$$

The corresponding auxiliary quantum-mechanics problem is given, according to Eq. (26), by the Schrödinger equation

$$i \frac{d}{dt} |\Phi\rangle = \hat{H}(t) |\Phi\rangle \tag{27}$$

with

$$\hat{H}(t) = -\hat{A}(t) = -\dot{x}^i(t) A_i^a[x(t)] \hat{T}^a = -\dot{x}^i(t) A_i^a[x(t)] T^a_{kl} \hat{a}_k^+ \hat{a}_l \equiv H_{kl}(t)\, \hat{a}_k^+ \hat{a}_l$$

where the creation and annihilation operators satisfy the standard commutation $(-)$ or anticommutation $(+)$ relations:

$$[\hat{a}_k, \hat{a}_l^+]_\mp = \delta_{kl}, \quad [\hat{a}_k^+, \hat{a}_l^+]_\mp = [\hat{a}_k, \hat{a}_l]_\mp = 0$$

It can be easily checked by direct computation that we have really obtained a realization of the Lie algebra **g** in a Hilbert (Fock) space, $[\hat{T}^a, \hat{T}^b]_- = if^{abc}\hat{T}^c$, in accordance with (11), where $\hat{T}^a = T^a_{kl}\hat{a}^+_k \hat{a}_l$. For an irreducible representation R, the second-order Casimir operator C_2 is proportional to the identity operator I, which, in turn, is equal to the number operator $\hat{N}$ in our Fock representation, that is, if $T^a \to \hat{T}^a$, then $I \to \hat{N} = \delta_{kl}\hat{a}^+_k \hat{a}_l$. Thus we obtain an important for our further considerations constant of motion $\hat{N}$:

$$[\hat{N}, \hat{H}]_- = 0 \tag{28}$$

It is interesting to note that this approach works equally well for commutation relations as well as for anticommutation relations.

b. *Path Integral.* Let us now derive the holomorphic path integral representation for the kernel of the parallel-transport operator:

$$\begin{aligned}
&\langle \bar{z}'' | P \exp\left[i \int_{t'}^{t''} \hat{A}(t)\, dt \right] | z' \rangle \\
&= \lim_{N\to\infty} \int \cdots \int \langle \bar{z}_N | (1 + i\epsilon \hat{A}_N) | z_{N-1} \rangle e^{-\bar{z}_{N-1} z_{N-1}} \frac{d\bar{z}_{N-1}\, dz_{N-1}}{2\pi i} \\
&\langle \bar{z}_{N-1} | (1 + i\epsilon \hat{A}_{N-1}) | z_{N-2} \rangle e^{-\bar{z}_{N-2} z_{N-2}} \frac{d\bar{z}_{N-2} dz_{N-2}}{2\pi i} \\
&\cdots e^{-\bar{z}_1 z_1} \frac{d\bar{z}_1\, dz_1}{2\pi i} \langle \bar{z}_1 | (1 + i\epsilon \hat{A}_1) | z_0 \rangle
\end{aligned} \tag{29}$$

Now we should calculate the single expectation value:

$$\langle \bar{z}_n | (1 + i\epsilon\, \hat{A}_n) | z_{n-1} \rangle = \langle \bar{z}_n | z_{n-1} \rangle + i \langle \bar{z}_n | \hat{A}_n | z_{n-1} \rangle \epsilon$$

Here

$$\langle \bar{z}_n | z_{n-1} \rangle = e^{\bar{z}_n z_{n-1}}$$

whereas

$$\begin{aligned}
i \langle \bar{z}_n \,|\, \hat{A}_n \,|\, z_{n-1} \rangle \epsilon &= i \langle \bar{z}_n \,|\, \bar{z}_n A_n z_{n-1} \,|\, z_{n-1} \rangle \epsilon \\
&= i\epsilon \bar{z}_n A_n z_{n-1} \langle \bar{z}_n \,|\, z_{n-1} \rangle
\end{aligned}$$

Thus

$$\begin{aligned}
\langle \bar{z}_n \,|\, (1 + i\epsilon \hat{A}_n) \,|\, z_{n-1} \rangle &= e^{\bar{z}_n z_{n-1}} + i\epsilon \bar{z}_n A_n z_{n-1} e^{\bar{z}_n z_{n-1}} \\
&= e^{\bar{z}_n z_{n-1}} [1 + i\epsilon \bar{z}_n A_n z_n + O(\epsilon^2)] \\
&= e^{\bar{z}_n z_{n-1}} e^{+i\epsilon \bar{z}_n A_n z_n + O(\epsilon^2)}
\end{aligned}$$

Combining this expression with the exponent in (29), we obtain

$$e^{-\bar{z}_n z_n} e^{\bar{z}_n z_{n-1} + i\epsilon \bar{z}_n A_n z_n + O(\epsilon^2)} = e^{-\bar{z}_n (z_n - z_{n-1}) + i\epsilon \bar{z}_n A_n z_n + O(\epsilon^2)}$$
$$= \exp[(-\bar{z}_n \dot{z}_n + \epsilon \bar{z}_n A_n z_n)\epsilon + O(\epsilon^2)]$$

Finally

$$U(\bar{z}'', z'; t'', t') = \langle \bar{z}'' | P \exp\left[-i \int_{t'}^{t''} \hat{H}(t) dt\right] | z' \rangle$$
$$\equiv \langle \bar{z}'' | P \exp\left[i \int_{t'}^{t''} \hat{A}(t)\, dt\right] | z' \rangle$$
$$= \int [D^2 z] \exp\left\{ \bar{z}(t'') z(t'') + i \int_{t'}^{t''} [i\bar{z}(t)\dot{z}(t) + \bar{z}(t) A(t) z(t)] dt \right\}$$
$$\equiv \int [D^2 z] \exp\left[\bar{z}(t'') z(t'') + i \int_{t'}^{t''} L\, dt\right]$$

where

$$[D^2 z] = \prod_{t' < t < t''} \frac{d\bar{z}(t)\, dz(t)}{2\pi i}$$

and L is of "classical" form (25).

Let us confine our attention to the one-particle subspace of the Fock space. As the number operator $\hat{N}$ is conserved by virtue of Eq. (28), if we start from the one-particle subspace of the Fock space, we shall remain in this subspace during all the evolution. The transition amplitude $U_{kl}(t'', t')$ between the one-particle states $|1_k\rangle = \hat{a}_k^+ |0\rangle$ and $|1_l\rangle = \hat{a}_l^+ |0\rangle$ is given by the following scalar product in the holomorphic representation

$$U_{kl} = \langle 1_k | P \exp\left[i \int_{t'}^{t''} \hat{A}(t)\, dt\right] | 1_l \rangle$$
$$= \int \langle 1_k | z'' \rangle \langle \bar{z}'' | P \exp\left[i \int_{t'}^{t''} \hat{A}(t)\, dt\right] | z' \rangle \langle \bar{z}' | 1_l \rangle\, e^{-\bar{z}'' z'' - \bar{z}' z'} \frac{d\bar{z}''\, dz''\, d\bar{z}'\, dz'}{(2\pi i)^2}$$
$$= \int D^2 z\, z_k(t'') \bar{z}_l(t') \exp\left[-\bar{z}(t') z(t') + i \int_{t'}^{t''} L\, dt\right] \tag{30}$$

where now

$$D^2 z = \prod_{t' \le t \le t''} \frac{d\bar{z}(t)\, dz(t)}{2\pi i}$$

Depending on the statistics, there are two ($\mp$) possibilities (fermionic and bosonic):

$$z_k \bar{z}_l \mp \bar{z}_l z_k = \bar{z}_k \bar{z}_l \mp \bar{z}_l \bar{z}_k = z_k z_l \mp z_l z_k = 0$$

Both are equivalent in terms of one-particle subspace of the Fock space which we will discuss in further detail later.

One can easily check that Eq. (30) represents the object we are looking for. Namely, from the Schrödinger equation [Eq. (27)] it follows that for the general one-particle state $\alpha_k \hat{a}_k^+ |0\rangle$ (summation after repeating indices) we have

$$i \frac{d}{dt} (\alpha_m \hat{a}_m^+ |0\rangle) = H_{kl}(t) \hat{a}_k^+ \hat{a}_l \alpha_m \hat{a}_m^+ |0\rangle = H_{kl}(t) \alpha_l (\hat{a}_k^+ |0\rangle) \tag{31}$$

Using the property of linear independence of Fock space vectors in Eq. (31), and comparing Eqs. (31) and (24), we can see that Eq. (30) really represents the matrix elements of the parallel-transport operator. For closed paths, $x(t') = x(t'')$ $= x$, Eq. (30) gives the holonomy operator $U_{kl}(x)$ and U_{kk} is the Wilson loop. Interestingly, the Wilson loop, which is supposed to describe a quark–antiquark interaction, is represented by a "true" quark and antiquark field, z and $\bar{z}$, respectively. So, the mathematical trick can be interpreted "physically."

Obviously, the "full" trace of the kernel in Eq. 3.9 is obtained by imposing appropriate boundary conditions, and integrating with respect to all the variables without the boundary term. Analogously, one can also derive the parallel-transport operator (a generalization of the one just considered) for symmetric n tensors (bosonic n-particle states) and for n forms (fermionic n-particle states).

c. The Non-Abelian Stokes Theorem. Let us now define a (bosonic or fermionic) Euclidean two-dimensional "topological" quantum field theory of multicomponent fields $\bar{z}, z$ transforming in an irreducible representation R of the Lie algebra $\mathbf{g}$ on the compact surface S, $\dim S = 2$, $\partial S = C$, $S \subset M$, $\dim M = d$, in an external non-Abelian gauge field A, by the classical action

$$S_{\text{cl}} = \int_S \left(i D_i \bar{z} D_j z + \frac{1}{2} \bar{z} F_{ij} z \right) dx^i \wedge dx^j, \qquad i, j = 1, \ldots, d \tag{32}$$

or in a parametrization $x^i(\sigma^1, \sigma^2)$, by the action

$$S_{\text{cl}} = \int_S \mathcal{L}_{\text{cl}}(\bar{z}, z) d^2\sigma$$
$$= \int_S \varepsilon^{AB} \left(iD_A \bar{z} D_B z + \frac{1}{2} \bar{z} F_{AB} z \right) d^2\sigma, \qquad A, B = 1, 2 \tag{33}$$

where

$$D_A = \partial_A x^i D_i, \qquad D_i \stackrel{\text{def}}{=} \partial_i - iA_i, \qquad F_{AB} = \partial_A x^i \partial_B x^j F_{ij}$$

At present, we are prepared to formulate a holomorphic path-integral version of the non-Abelian Stokes theorem

$$\int D^2 z\, z_k'' \bar{z}_k' \exp\left(-\bar{z}' z' + i \oint_C i \bar{z} D z \right)$$
$$= \int D^2 z\, z_k'' \bar{z}_k' \exp(-\bar{z}' z' + i S_{\text{cl}}) \tag{34}$$

or in the polar parametrization $x^i(\sigma^1, \sigma^2) t' \le \sigma^1 \equiv t \le t''$, $0 \le \sigma^2 \equiv s \le 1$

$$\int D^2 z\, z_k(t'') \bar{z}_k(t') \exp\left\{ -\bar{z}(t') z(t') + i \int_{t'}^{t''} L[\bar{z}(t), z(t)] dt \right\}$$
$$= \int D^2 z\, z_k(t'', 1) \bar{z}_k(t', 1) \exp\left[-\bar{z}(t', 1) z(t', 1) + i \int_0^1 \int_{t'}^{t''} \mathcal{L}_{\text{cl}}\, dt\, ds \right] \tag{35}$$

where $L(z, \bar{z})$ and $\mathcal{L}_{\text{cl}}(\bar{z}, z)$ are defined by Eqs. (25) and (33), respectively. The measure on both sides of Eqs. (34) and (35) is the same; specifically, it is concentrated on the boundary ∂S, and the imposed boundary conditions are free.

It should be noted that the surface integral on the r.h.s. of Eqs. (34) and (35) depends on the curvature F as well as on the connection A entering the covariant derivatives, which is reminiscent of the path dependence of the curvature $\mathcal{F}$ in the operator approach.

A quite different formulation of the holomorphic approach to the non-Abelian Stokes theorem has been proposed in [17].

d. Appendix. For completness of our derivations, we will remind the reader of a few standard facts being used above. First, we assume the following (non-quite standard, but convenient) definition

$$|z\rangle \stackrel{\text{def}}{=} e^{za^+ + \bar{z}a + \frac{1}{2}|z|^2} |0\rangle = e^{za^+} e^{\bar{z}a} |0\rangle$$
$$= e^{za^+} |0\rangle = \sum_{k=0}^{\infty} \frac{(z^n a^+)^n}{n!} |0\rangle = \sum_{k=0}^{\infty} \frac{z^n}{\sqrt{n!}} |n\rangle$$

where $|0\rangle$ is the Fock vacuum, namely, $\hat{a}|0\rangle = 0$, and the Baker–Campbell–Hausdorff (BRS) formula has been applied in the first line. We could also treat $|z\rangle$ as a coherent state for the Heisenberg group. We can easily calculate

$$\langle \bar{z}|z\rangle = \sum_{m,n=0}^{\infty} \frac{\bar{z}^m z^n}{\sqrt{m!\,n!}} \langle m|n\rangle = \sum_{n=0}^{\infty} \frac{(\bar{z}z)^n}{n!} = e^{\bar{z}z}$$

The identity operator is of the form

$$I = \int |z\rangle\langle \bar{z}| e^{-\bar{z}z} \frac{d\bar{z}\,dz}{2\pi i}$$

Actually

$$\begin{aligned} I|z'\rangle &= \int |z\rangle\langle \bar{z}|z'\rangle e^{-\bar{z}z} \frac{d\bar{z}\,dz}{2\pi i} = \int |z\rangle e^{\bar{z}z' - \bar{z}z} \frac{d\bar{z}\,dz}{2\pi i} \\ &= \int |z\rangle \delta(z' - z)\, dz = |z'\rangle \end{aligned}$$

All these formulas for a single pair of creation and annihilation operators obviously apply to a more general situation of $\dim R$ pairs. The matrix elements are

$$\begin{aligned} \langle \bar{z}|\hat{A}|z\rangle &= \langle 0|e^{\sum_k \bar{z}_k \hat{a}_k} \sum_a A^a \sum_{k,l} T^a_{kl} \hat{a}^+_k \hat{a}_l e^{\sum_k z_k \hat{a}^+_k}|0\rangle \\ &= \langle 0|e^{\sum_k \bar{z}_k \hat{a}_k} \sum_a A^a \sum_{k,l} T^a_{kl} \bar{z}_k z_l e^{\sum_k z_k \hat{a}^+_k}|0\rangle \\ &= \langle \bar{z}| \sum_a A^a \sum_{k,l} T^a_{kl} \bar{z}_k z_l |z\rangle \\ &= \langle \bar{z}|\bar{z}Az|z\rangle \end{aligned}$$

where we have used the formula

$$[\hat{a}, e^{z\hat{a}^+}]_{\mp} = z e^{z\hat{a}^+}$$

Now

$$\begin{aligned} \langle 1_k|(1 + i\epsilon\hat{A})|1_l\rangle &= \delta_{kl} + i\epsilon \langle 0|\hat{a}_k \sum_a A^a \sum_{i,j} T^a_{ij} \hat{a}^+_i \hat{a}_j \hat{a}^+_l|0\rangle \\ &= \delta_{kl} + i\epsilon \sum_a A^a T^a_{kl} = (1 + i\epsilon A)_{kl} \end{aligned}$$

Also

$$\langle 1_k | z \rangle = z_k$$

3. Measure

The theory described above possesses the following "topological" gauge symmetry

$$\delta z(x) = \theta(x), \qquad \delta \bar{z}(x) = \bar{\theta}(x) \tag{36}$$

where $\theta(x)$ and $\bar{\theta}(x)$ are arbitrary except at the boundary ∂S, where they vanish. The origin of the symmetry (36) will become clear when we convert the action (32) into a line integral. Integrating by parts in Eq. (32) and using the Abelian Stokes theorem, we obtain

$$S_{\rm cl} = i \oint_{\partial S} \bar{z} D_i z \, dx^i$$

or in a parametrized form

$$S_{\rm cl} = i \oint_{\partial S} \bar{z} D_t z \, dt$$

To covariantly quantize the theory, we shall introduce the BRS operator s. According to the form of the topological gauge symmetry (36), the operator s is easily defined by $sz = \phi, s\bar{z} = \bar{\chi}, s\phi = 0, s\bar{\chi} = 0, s\bar{\phi} = \bar{\beta}, s\chi = \beta, s\bar{\beta} = 0,$ and $s\beta = 0$, where ϕ and $\bar{\chi}$ are ghost fields in the representation R, associated with θ and $\bar{\theta}$, respectively; $\bar{\phi}$ and χ are the corresponding antighosts; and $\bar{\beta}$, β are Lagrange multipliers. All the fields possess a suitable Grassmann parity correlated with the parity of $\bar{z}$ and z. Obviously $s^2 = 0$, and we can gauge-fix the action in Eq. (32) in a BRS-invariant manner by simply adding the following s-exact term:

$$\begin{aligned} S' &= s\left(\int_S (\bar{\phi} \triangle z \pm \bar{z} \triangle \chi) \, d^2\sigma \right) \\ &= \int_S (\bar{\beta} \triangle z \pm \bar{\phi} \triangle \phi \pm \bar{\chi} \triangle \chi + \bar{z} \triangle \beta) \, d^2\sigma \end{aligned}$$

The upper (resp. lower) sign stands for the fields $\bar{z}$,z of bosonic (fermionic) statistics. Integration after the ghost fields yields some numerical factor and the quantum action

$$S = S_{\rm cl} + \int_S (\bar{\beta} \triangle z + \bar{z} \triangle \beta) \, d^2\sigma \tag{37}$$

If necessary, one can insert $\sqrt{g}$ into the second term, which is equivalent to change of variables. Thus the partition function is given by

$$Z = \int e^{iS} \, D\bar{z} \, Dz \, D\bar{\beta} \, D\beta,$$

with the following boundary conditions: $\bar{\beta}|_{\partial S} = \beta|_{\partial S} = 0$.

One can observe that the job that the fields $\bar{\beta}$ and β are supposed to do consists in eliminating a redundant integration inside S. The gauge-fixing condition following from Eq. (37) imposes the following constraints:

$$\triangle z = 0, \qquad \triangle \bar{z} = 0$$

Since values of the fields z and $\bar{z}$ are fixed on the boundary $\partial S = C$, we deal with the two well-defined two-dimensional (2D) Dirichlet problems. The solutions of the Dirichlet problems fix values of z and $\bar{z}$ inside S. Another, more singular, gauge-fixing term has been proposed [18].

The issue of the measure has been also discussed [15].

C. Generalizations

1. *Topology*

Up to now we have investigated the non-Abelian Stokes theorem for a topologically trivial situation. The term *topologically trivial situation* means, in this context, that the loop we are integrating along in the non-Abelian Stokes theorem is "unknotted" in the sense of theory of "knots" [19]. It appears that in contradistinction to the Abelian case, the non-Abelian one is qualitatively different. If the loop C is topologically nontrivial and the bounded surface S ($\partial S = C$) is not simply connected, the parameter space given in the form of a unit square (as in the proof of the non-Abelian Stokes theorem) is not appropriate. The non-Abelian Stokes theorem presented in the original form applies only to a surface S homeomorphic to a disk (square). But still, of course, the standard (topologically trivial) version of the non-Abelian Stokes theorem makes sense locally. The meaning of *locally* in this context will become clear in due course. The non-Abelian Stokes theorem for knots (and also for links—multicomponent loops) was formulated by Hirayama et al. in 1998 [20]. Interestingly, it follows from this new version of the non-Abelian Stokes theorem that the value of the line integral along C can be nontrivial (different from 1) even for the field strenght $F_{\mu\nu}(x)$ vanishing everywhere on the surface S. This is an interesting result that could have some applications in physics. One can speculate that it could give rise to a new version of the Aharonov–Bohm effect.

To approach the non-Abelian Stokes theorem for knots, we should recall a necessary portion of the standard lore of theory of knots. Since the first task is to find an oriented surface S whose boundary is C, we should construct the, so-called Seifert surface, satisfying the abovementioned condition by definition. It appears that the Seifert surface for any knot assumes a standard form homeomorphic to a (flat) disk with $2g$ ("thin") strips attached. The number g is called the genus. The strips may, of course, be horribly twisted and intertwined [19].

Now we should decompose C and next S into pieces that can be put together to form slices that are topologically trivial and thus subject to the standard non-Abelian Stokes theorem. Such decomposition is shown in Fig. 9. Explicitly, this reads as

$$\begin{aligned} C &= (C_{10(g-1)+9} \cdots C_{11})(C_9 C_7 C_4 C_1) C_0 \\ &= \left(\prod_{k=0}^{g-1} C_{10k+9} C_{10k+7} C_{10k+4} C_{10k+1} \right) C_0, \qquad \text{for} \quad g \geq 1 \end{aligned} \tag{38}$$

and $C = C_0$ for $g = 0$. Next

$$\begin{aligned} C = \Big[& C_{10(g-1)+6} \underbrace{\big(C^{-1}_{10(g-1)+6} \overbrace{}^{C^{-1}_{10(g-1)+10}} C_{10(g-1)+9} C_{10(g-1)+8} \big)}_{S_{4g}} C_{10(g-1)+3} \\ & \cdots C_{13} \underbrace{\big(C^{-1}_{13} C^{-1}_{12} C_{11} C_{10} \big)}_{S_5} \Big] \cdot \Big[C_6 \underbrace{\big(C^{-1}_6 C^{-1}_{10} C_9 C_8 \big)}_{S_4} \\ & C^{-1}_3 \underbrace{\big(C_3 C^{-1}_8 C_7 C_5 \big)}_{S_3} C^{-1}_6 \underbrace{\big(C_6 C^{-1}_5 C_4 C_2 \big)}_{S_2} C_3 \underbrace{\big(C^{-1}_3 C^{-1}_2 C_1 C_0 \big)}_{S_1} \Big] \\ = \prod_{k=0}^{g-1} & C_{10k+6} \underbrace{\big(C^{-1}_{10k+6} \underline{C^{-1}_{10k+10}} C_{10k+9} C_{10k+8} \big)}_{S_{4k+4}} \\ & C^{-1}_{10k+3} \underbrace{\big(C_{10k+3} C^{-1}_{10k+8} C_{10k+7} C_{10k+5} \big)}_{S_{4k+3}} \\ & C^{-1}_{10k+6} \underbrace{\big(C_{10k+6} C^{-1}_{10k+5} C_{10k+4} C_{10k+2} \big)}_{S_{4k+2}} \\ & C_{10k+3} \underbrace{\big(C^{-1}_{10k+3} C^{-1}_{10k+2} C_{10k+1} C_{10k} \big)} S_{4k+1} \\ = \prod_{k=0}^{g-1} & C_{10k+6} S_{4k+4} C^{-1}_{10k+3} S_{4k+3} C^{-1}_{10k+6} S_{4k+2} C_{10k+3} S_{4k+1} \end{aligned}$$

where $C_{10k+10} \,|_{k=g-1} \equiv C_{10g} = 1$.

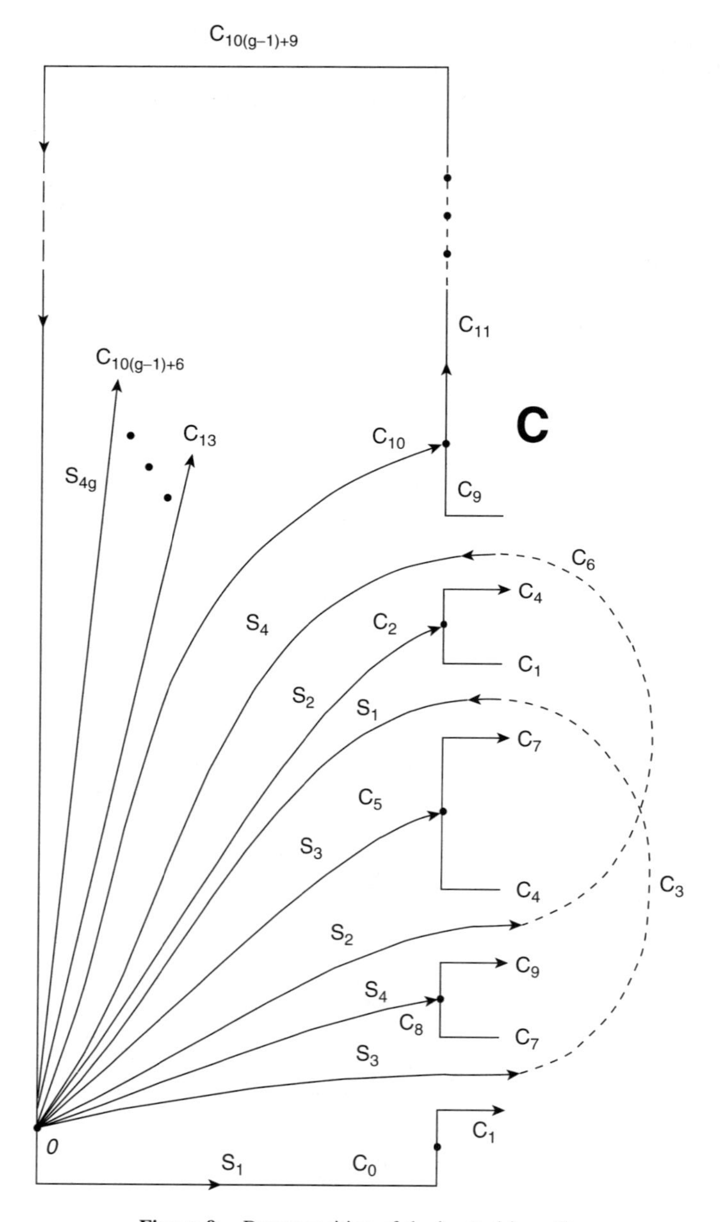

Figure 9. Decomposition of the knotted loop C.

Further generalization, to multicomponent loops (links) is described in the original paper [20].

2. *Higher-Dimensional Forms*

The theorem considered up to now is a very particular, although seemingly the most important, non-Abelian version of the Stokes theorem. It connects a non-Abelian differential 1-form in dimension 1 and a 2-form in dimension 2. The forms are of a very particular shape, namely, the connection 1-form and the curvature 2-form. Now, we would like to discuss possible generalizations to arbitrary, higher-dimensional differential forms in arbitrary dimensions. Since there may be many variants of such generalizations depending on a particular mathematical and/or physical context; we will start giving a general recipe.

Our idea is very simple. First, working in the framework of the path integral formalism, we should construct a topological field theory of auxiliary topological fields on ∂N, the boundary of the d-dimensional submanifold N, in (an) external (gauge) field(s) in which we are interested. Next, we should quantize the theory, namely, build the partition function in the form of a path integral, where auxiliary topological fields are properly integrated out. Thus the LHS of the non-Abelian Stokes theorem has been constructed. Applying the Abelian Stokes theorem to the (effective) action (in the exponent of the path-integral integrand) we obtain the "r.h.s." of the non-Abelian Stokes theorem. If we also wish to extend the functional measure to the whole N, we should additionaly quantize the theory on the r.h.s. to eliminate the redundant functional integration inside N.

The example candidate for the topological field theory defining the l.h.s. of the non-Abelian Stokes theorem could be given by the (classical) action

$$S_{\mathrm{Top}} = \frac{1}{2}\oint_{\partial N} (\bar{z} d_A \zeta + d_A \bar{\zeta} z + \bar{z} B z) \tag{39}$$

where z and $\bar{z}$ are 0-forms, ζ and $\bar{\zeta}$ are $(d-2)$-forms (all the forms are in an irreducible representation $R(G)$), and d_A is the exterior covariant derivative

$$d_A \zeta \equiv d\zeta + A\zeta, \qquad d_A \bar{\zeta} \equiv d\bar{\zeta} - A^T \bar{\zeta}$$

The non-Abelian B field naturally appears in the context of (topological) gauge theory [see Eq. (45)]. Now, the Abelian Stokes theorem should suffice.

Generalization of the non-Abelian Stokes theorem to higher-degree forms in the operator language seems more difficult and practically has not been attempted (see, however, Ref. 8 for an introductory discussion of this issue).

III. APPLICATIONS

The number of applications of the non-Abelian Stokes theorem is not as large as in the case of the Abelian Stokes theorem; nevertheless, it is the main motivation for formulating the non-Abelian Stokes theorem. It is interesting to note that in contradistinction to the Abelian Stokes theorem, whose formulation is "homogenous" (unique), different formulations of the non-Abelian Stokes theorem are useful for particular purposes and applications. From a purely techincal point of view, one can classify applications of the non-Abelian Stokes theorem as exact and approximate. The term *exact applications* means that one can perform successfully an "exact calculus" to obtain an interesting result, whereas the term *approximate application* means that a more or less controllable approximation (typically, perturbative) is involved in the calculus. Since exact applications seem to be more convincing and more illustrative for the subject, we will basically confine our discussion to presentation few of them.

Since the non-Abelian Stokes theorem applies to non-Abelian gauge theories, and non-Abelian gauge theories are nonlinear, it is not surprising that exact applications are scarce. In fact they are limited to low-dimensional cases and/or topological models, which are usually exactly solvable. The first case that we consider is pure, two-dimensional ordinary (almost topological) Yang–Mills gauge theory. But a rich source of applications of the non-Abelian Stokes theorem comes from topological field theory of the Chern–Simons type. The path integral procedure makes it possible to obtain skein relations for knot and link polynomial invariants. In particular, it appears that only the path integral version of the non-Abelian Stokes theorem permits us to nonperturbatively and covariantly generalize the method of obtaining topological invariants [21].

As a byproduct of our approach, we have computed the parallel-transport operator U in the holomorphic path integral representation. In this way, we have solved the problem of saturation of Lie algebra indices in the generators T^a. This issue appears, for example, in the context of equation of motion for Chern–Simons theory in the presence of Wilson lines (an interesting connection with the Borel–Weil–Bott theorem and quantum groups has been also suggested). Our approach enables us to write those equations in terms of $\bar{z}$ and z purely classically. Incidentally, in the presence of Chern–Simons interactions the auxiliary fields $\bar{z}$ and z acquire fractional statistics, which could be detected by braiding. To determine the braiding matrix, one should, in turn, find the so-called monodromy matrix, making use, for example, of non-Abelian Stokes theorem.

A. Two-Dimensional Yang–Mills Theory

There is a vast literature on the subject of the two-dimensional Yang–Mills theory, approaching it from different points of view. One of the latest papers is that by Aroca and Kubyshin [22], who list of references to earlier papers. Two-dimensional Yang–Mills theory is a specific theory. From the dynamical point of view it is almost trivial—there are no local degrees of freedom, as a standard canonical analysis indicates. In fact, it is "semitopological" field theory; that is, it only roughly describes combinatorial–topological phenomena and surface areas.

There are many important and interesting aspects in two-dimensional Yang–Mills theory. One of them is the issue of determination of "physical" observables: Wilson loops [Eq. (8)]. Calculation of the Wilson loops $W_R(C)$ in two-dimensional Yang–Mills theory can be facilitated by the use of the non-Abelian Stokes theorem.

A nice feature of (Euclidean) two-dimensional Yang–Mills gauge theory defined by the action

$$S_{2\mathrm{dYM}}(A) = \frac{1}{4}\int_M F^a_{AB}(A)F^{a\,AB}(A)\sqrt{g}\,d^2x, \qquad A,B = 1,2 \tag{40}$$

is the possibility of recasting the action, and next, and more importantly, the whole partition function in the form

$$Z = \int DF\, e^{-S_{2\mathrm{dYM}}(F)} \tag{41}$$

where now F is an independent field, and the action $S_{2\mathrm{dYM}}$ is of the same form as the original Eq. 40 but this time without A dependence (the subscript denotes two-dimensional Yang–Mills theory).

Let us now consider "physical" observables, namely, Wilson loops. Confronting the partition function (41) with the form of the Wilson loop transformed by the non-Abelian Stokes theorem to a surface expression (14), we can see that a kind of a Gaussian functional integral emerges. For an Abelian theory, we would exactly obtain an easy Gaussian functional integral, but in a non-Abelian case we should be more careful because $\mathscr{F}$ is a path-dependent object. The fact that $\mathscr{F}$ is path-dependent can be ignored in the case of a single loop because of the commutativity of the infinitesimal surface integrals (see below). Since, according to the non-Abelian Stokes theorem

$$W_R(C) = \mathrm{Tr}_R\mathscr{P}\exp\left(i\int_S \mathscr{F}\sqrt{g}\,d^2x\right)$$

where $\mathscr{F} = \mathscr{F}_{12}$. For the expectation value

$$\langle W_R(C)\rangle = Z^{-1}\int DF\exp(-S_{2\mathrm{dYM}})W_R(C)$$

we obtain [7]

$$
\begin{aligned}
\langle W_R(C)\rangle &\propto \int DF \exp\left(-\frac{1}{2}\int_M F^a F^a \sqrt{g}\, d^2x\right) \mathrm{Tr}_R \mathscr{P} \exp\left(i\int_S F\sqrt{g}\, d^2x\right)\\
&= \int \prod_{x\in M\setminus S} dF(x) \exp\left(-\frac{1}{2}\int_{M\setminus S} F^a F^a \sqrt{g}\, d^2x\right)\\
&\quad \cdot \mathrm{Tr}_R \mathscr{P} \int \prod_{x\in S} dF(x)\, \exp\left[i\int_S \left(-\frac{1}{2}F^a F^a + iF^a T^a\right)\sqrt{g}\, d^2x\right]\\
&\propto \mathrm{Tr}_R \exp\left(-\frac{1}{2}T^a T^a \int_S \sqrt{g}\, d^2x\right)
\end{aligned}
$$

Thus, finally

$$
\langle W_R(C)\rangle = \mathrm{Tr}_R \exp\left[-\frac{1}{2}C_2(R)S\right] = \dim R \exp\left[-\frac{1}{2}C_2(R)S\right]
$$

where

$$
S = \int_S \sqrt{g}\, d^2x
$$

and

$$
T^a T^a = C_2(R), \qquad \mathrm{Tr}_R I = \dim R
$$

In the case of n nonoverlapping regions $\{S_i\}$, $i = 1, \ldots, n$, $C_i = \partial S_i$, $S_i \cap S_j = \emptyset$ for $i \neq j$, and n irreducible representations R_i of the group G with the generators T_i, we immediately obtain–literally repeating the last derivation—the formula for the expectation value of the product of the n Wilson loops:

$$
\left\langle \prod_{i=1}^{n} W_{R_i}(C_i) \right\rangle = \prod_{i=1}^{n} \dim R_i \, \exp\left[-\frac{1}{2}C_2(R_i)S_i\right]
$$

The case of the overlapping regions $\{S_i\}$ is a bit more complicated [7]. First, one has to decompose the union of all regions $\{S_i\}$, $\partial S_i = C_i$ into a disjoint union of connected (i.e., not intersected by the loops) regions $\{S_\alpha\}$. Each loop C_i is next deformed into an equivalent loop C_i', which is a product of "big" (not infinitesimal) lassos independently (a lasso per a region) covering each connected region S_{α_i}, $S_{\alpha_i} = S_\alpha \cap S_i$ ($S_{\alpha_i} \in \{S_\alpha\}$). The lassos coming from the different loops C_i' but covering the same connected region S_α should necessarily be arranged in such a way to enter the region S_α at the same basepoint O_α. Consequently, the connected region S_α, $S_\alpha \subset S_{i_1} \cap \cdots \cap S_{i_k}$, $i \leq k \leq n$, can be covered with the k identical copies of the net of "small" (infinitesimal) lassos. Every Gaussian functional integration with respect to the infinitesimal area δS enclosed by an infinitesimal lasso can be easily performed, yielding

$$\exp\left(-\frac{1}{2}\delta S\, T_\alpha^2\right) \tag{42}$$

where

$$T_\alpha = \sum_{i \in \alpha} T_i, \qquad T_i = I \otimes \cdots \otimes T_i \otimes \cdots \otimes I$$

Integration with respect to the consecutive infinitesimal areas gives the terms of the form (42). Since T_α is a generator of $\mathbf{g}$ in a product representation R_α, namely, $R_\alpha = R_{i_1} \otimes \cdots \otimes R_{i_k}$, it follows that T_α^2 is a Casimir operator. Accordingly, (42) commutes with the product of the parallel-transport operators acting in the product representation R_α. Since the products in the pairs connect every infinitesimal area δS with the basepoint O_α, they cancel each other. This fact means that the integral with respect to the whole region S_α is given only by the infinite product of the terms (42) and reads

$$M_\alpha = \exp\left(-\frac{1}{2}S_\alpha T_\alpha^2\right) \tag{43}$$

The full expectation value of the n loops $\{C_i\}$ consists of the trace of a product of M_α blocks (43) joined with the parallel-transport operators, which are remnants of the primary decomposition of the loops. These joining curves enclose zero areas, and can be deformed into points (without destroying M_α blocks) giving some "linking" operators L_α. An operator L_α is of a very simple form; specifically, it is a product of the Kronecker deltas, which contract indices belonging to the same representation but to different M matrices. Thus L causes the matrix multiplication of M matrices to be performed in a prescribed order in each representation sector independently. In other words, M mixes, with some weights, indices of different representations (braiding), whereas L sets the order of the matrix multiplications in a representation sector. M depends on the metric (area of S) and group-theoretic quantities, while the concrete form of L depends on the topology of the overlaps. Thus the expectation value of the product of the n Wilson loops is finally given by

$$\left\langle \prod_{i=1}^{n} W_{R_i}(C_i) \right\rangle = \prod_\alpha L_\alpha M_\alpha$$

This analysis is a bit simplified and shortened but gives the flavor of the power of the non-Abelian Stokes theorem in practical instances.

B. Three-Dimensional Topological Quantum Field Theory

Topological quantum field theory has become a fascinating and fashionable subject in mathematical physics. At present, the main applications of topological field theory are in mathematics (topology of low-dimensional manifolds) rather than in physics. Its application to the issue of classification of knots and links is one of the most interesting. To approach this problem, one usually tries to somehow encode the topology of a knot or link . As was first noted by Witten [23], the problem can be attacked by means of standard theoretical physics techniques of quantum field theory. In particular, using three-dimensional Chern–Simons gauge theory, one can derive not only all the well-known polynomial invariants of knots and links but also many of their generalizations. Most authors working in the topological field theory description of polynomial invariants follow Witten's original approach, which relies heavily on the underlying conformal field theory structure. There is also a genuinely three-dimensional covariant approach advocated in its perturbative version [24,25] using the non-Abelian Stokes theorem in its operator formulation. We shall sketch an application of the non-Abelian Stokes theorem to a genuinely three-dimensional, nonperturbative, covariant path integral approach to *polynomial invariants* of knots and links in the framework of (topological) quantum Chern–Simons gauge field theory.

To begin, we introduce the classical topological Chern–Simons action on the three-dimensional sphere S^3

$$
\begin{aligned}
S_{\mathrm{CS}} &= \frac{k}{4\pi}\int_{S^3} \mathrm{Tr}\left(A \wedge dA + \frac{2}{3} A \wedge A \wedge A\right) \\
&= \frac{k}{4\pi}\int_{S^3} d^3x\, \varepsilon^{ijk}\, \mathrm{Tr}\left(A_i \partial_j A_k + \frac{2}{3} A_i A_j A_k\right)
\end{aligned}
\tag{44}
$$

where $k \in \mathbf{Z}^{\pm}$. The use of this equation is not obligatory. One could as well choose the action of the so-called BF theory

$$
S_{BF} = \frac{k}{4\pi}\int_{S^3} d^3x\, \varepsilon^{ijk} \mathrm{Tr}(B_i F_{jk})
\tag{45}
$$

where $B_i = B_i^a(x) T^a$ is an auxiliary gauge field, and now $k \in \mathbf{R}^{\pm}$.

To encode the topology of a link $\mathscr{L} = \{C_i\}$ into a path integral, we introduce an auxiliary one-dimensional topological field theory (topological quantum mechanics) in an external gauge field A, living on the corresponding loop C_i.

The classical action of this theory is chosen in the form [see Eq. (25)]

$$S_{C_i}(A) = i\oint_{C_i} dt\,\bar{z}_i D_t z_i \tag{46}$$

where the multiplet of scalar fields $\bar{z}_i, z_i$ transforms into an irreducible representation R_i. The partition function corresponding to (46) has the following standard form:

$$Z_i(A) = \int D^2 z_i \exp[iS_{C_i}(A)]$$

It is obvious that the observables $S_{C_i}(A)$ are akin to the Wilson loops.

We define the topological invariant of the link $\mathscr{L}$ as the (normalized) expectation value

$$\left\langle \prod_i Z_i(A) \right\rangle \equiv \left[\int d\mu \; \exp\,(iS)\right]^{-1} \int d\mu \; \exp\,(iS) \prod_i Z_i(A) \tag{47}$$

where S consists of S_{CS} plus quantum terms, and the measure should also contain the auxiliary fields. We can calculate (47) recursively, using the *skein relations*. Thus, our present task reduces to the derivation of the corresponding skein relation. To this end, we consider a pair of loops, say, C_1 and C_2, where a part of C_1, forming a small loop ℓ ($\ell = \partial N$), is wrapped round C_2 (see Fig. 10). In other words, C_2 pierces N at a point P. Such an arrangement can be interpreted as a preliminary step toward finding the corresponding monodromy matrix $\mathbf{M}$. Having given the loop ℓ, we can utilize the non-Abelian Stokes theorem [actually the Abelian Stokes theorem for (46)] in its holomorphic version, obtaining Eq. (32). In a general position, N and C_2 can intersect in a finite number of points,

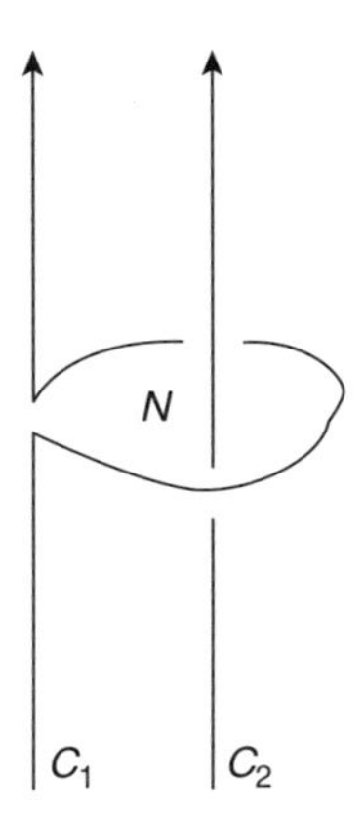

Figure 10. C_1 and C_2, where a part of C_1, forming a small loop $\ell(\ell = \partial N)$, is warpped around C_2.

and the contribution to the path integral coming from these points can be explicitly calculated. We replace the curvature in (32) by the functional derivative operator

$$F_{ij}^{a}(x) \rightarrow \frac{4\pi}{ik}\varepsilon_{ijk}\frac{\delta}{\delta A_{k}^{a}(x)}$$

This substitution yields an equivalent expression, provided the order of terms in (47) is such that the functional derivative can act on S_{CS} to produce F. Using formal translational invariance of the product measure DA, and functionally integrating by parts in (47) with respect to A, we obtain, for each intersection point P, the *monodromy operator*

$$M = \exp\left[\frac{4\pi}{ik}(\bar{z}_1 T_1^a z_1)(\bar{z}_2 T_2^a z_2)(P)\right] \tag{48}$$

To calculate the *matrix elements* of (48), one utilizes the following scalar product:

$$(f, g) = \frac{1}{2\pi i}\int fg \exp(\bar{z}z)\, d\bar{z}dz$$

This kind of the scalar product is implicit in our derivations of the path integral. Expanding (48) in a power series, multiplying with respect to this scalar product, and resumming, we get the *monodromy matrix*:

$$\mathbf{M} = (\bar{z}_1\bar{z}_2, M\, z_2 z_1) = \exp\left(\frac{4\pi}{ik} T_1^a \otimes T_2^a\right)$$

The square root of the monodromy matrix gives rise to the so-called braiding matrix **B** responsible for a proper form of skein relations, yielding knot or link invariants.

C. Other Applications

We could continue to develop the idea of the previous section and try to generalize it to higher-dimensional (topological) theories. To this end we should apply a generalization of the non-Abelian Stokes theorem to non-Abelian forms of higher degree, for example, following the approach proposed in Eq. (39), and yielding the resluts obtained in an earlier study [26].

Quite a different story is the possibility of applying the non-Abelian Stokes theorem (in the coherent-state version) to computations in QCD (QCD string, area low, etc.) [6,27,28] or gravity [29]. Since such calculations are posssible only perturbatively, their results are not rigorously controllable and thus are uncertain.

IV. SUMMARY

In this short review we have addressed the main issues related to the non-Abelian Stokes theorem. The two principal approaches (operator and path integral) to the non-Abelian Stokes theorem have been formulated in their simplest possible forms. A generalization for a knotted loop as well as a suggestion concerning higher-degree forms have also been presented. Only nonperturbative applications of the non-Abelian Stokes theorem (to low-dimensional gauge theories) have been described. The review is not comprehensive; rather, it is directed toward topological aspects reflecting the author's interests.

Acknowledgments

The author is grateful to Professors M. Blau and P. Kosiński for interesting discussions at the very early stages of development of some of these ideas. The work has been supported by the grant of the University of Łódź.

References

1 . M. Spivak, *Calculus on Manifolds. A Modern Approach to Classical Theorems of Advanced Calculus*, Benjamin, New York, 1965.

2 . L. Schlesinger, *Math. Ann.* **99**, 413 (1927).

3 . I. Ya. Aref'eva, *Theor. Math. Phys.* **43**, 353 (1980) [transl. from *Teor. Mat. Fiz.* **43**, 111 (1980)].

4 . N. E. Bralić, *Exact Computation of Loop Averages in Two Dimensional Yang–Mills theory*, Ph.D. thesis, Chicago Univ. 1980 [reprinted in *Phys. Rev. D* **22**, 3090 (1980)].

5 . P. M. Fishbane, S. Gasiorowicz, and P. Kaus, *Phys. Rev. D* **24**, 2324 (1981).

6 . D. I. Diakonov and V. Yu. Petrov, "How to introduce monopole-quark interactions in a gauge-invariant way," *Nonperturbative Approaches to QCD, Proc. Int. Workshop at ECT*, Trento, Italy, 1995.

7 . B. Broda, *Phys. Lett. B* **244**, 444 (1990).

8 . M. B. Menski, *Group of Paths: Measurements, Fields, Particles*, Nauka, Moscow, 1983, Chaps. N-6–7 (in Russian).

9 . M. Hirayama and S. Matsubara, *Progr. Theor. Phys.* **99**, 691 (1998).

10 . R. L. Karp, F. Mansouri, and J. S. Rno, *J. Math. Phys.* **40**, 6033 (1999).

11 . M. B. Halpern, *Phys. Rev. D* **19**, 517 (1979).

12 . V. I. Shevchenko and Yu. A. Simonov, *Phys. Lett. B* **437**, 146 (1998).

13 . W.-M. Zhang, D. H. Feng, and R. Gilmore, *Rev. Mod. Phys.* **62**, 867 (1990).

14 . A. Perelomov, Generalized Coherent States and Their Applications (Texts and Monographs in Physics) Springer Verlag, 1986.

15 . M. Hirayama and M. Ueno, *Progr. Theor. Phys.* **103**, 151 (2000).

16 . D. I. Diakonov and V. Yu. Petrov, *Phys. Lett. B* **224**, 131 (1989).

17 . F. A. Lunev, *Nucl. Phys. B* **494**, 433 (1997).

18 . B. Broda, *J. Math. Phys.* **33**, 1511 (1992).

19 . D. Rolfsen, *Knots and Links*, Publish or Perish, Wilmington, DE, 1976.

20 . M. Hirayama, M. Kanno, M. Ueno, and H. Yamakoshi, *Progr. Theor. Phys.* **100**, 817 (1998).

21 . B. Broda, *Mod. Phys. Lett. A* **5**, 2747 (1990).
22 . J. M. Aroca and Yu. Kubyshin, *Ann Phys.* **283**, 11 (2000).
23 . E. Witten, *Commun. Math. Phys.* **121**, 351 (1989).
24 . L. Smolin, *Mod. Phys. Lett. A* **4**, 1091 (1989).
25 . P. Cotta-Ramusino, E. Guadagnini, M. Martellini, and M. Mintchev, *Nucl. Phys. B* **330**, 557 (1990).
26 . B. Broda, *Mod. Phys. Lett. A* **9**, 609 (1994).
27 . D. V. Antonov and D. Ebert, *Mod. Phys. Lett. A* **12**, 2047 (1997).
28 . Yu. A. Simonov, *Cluster Expansion, Non-Abelian Stokes Theorem and Magnetic Monopoles*, Preprint ITEP 88-110, 1988.
29 . D. Diakonov and V. Petrov, Non-Abelian Stokes theorems in Yang-Mills and gravity theories, preprint NORDITA-2000-65-HE and e-Print Archive:hep-th/0008035.

THE LINK BETWEEN THE SACHS AND O(3) THEORIES OF ELECTRODYNAMICS

M. W. EVANS

CONTENTS

I. INTRODUCTION

In this volume, Sachs [1] has demonstrated, using irreducible representations of the Einstein group, that the electromagnetic field can propagate only in curved spacetime, implying that the electromagnetic field tensor can exist only when there is a nonvanishing curvature tensor $\kappa_{\mu\nu}$. Using this theory, Sachs has shown that the structure of electromagnetic theory is in general non-Abelian. This is the same overall conclusion as reached in O(3) electrodynamics [2], developed in the second chapter of this volume. In this short review, the features common to Sachs and O(3) electrodynamics are developed. The $\boldsymbol{B}^{(3)}$ field of O(3) electrodynamics is extracted from the quaternion-valued $B^{\mu\nu}$ equivalent in the Sachs theory; the most general form of the vector potential is considered in both theories, the covariant derivatives are compared in both theories, and the possibility of extracting energy from the vacuum is considered in both theories.

Modern Nonlinear Optics, Part 2, Second Edition, Advances in Chemical Physics, Volume 119, Edited by Myron W. Evans. Series Editors I. Prigogine and Stuart A. Rice.
ISBN 0-471-38931-5

II. THE NON-ABELIAN STRUCTURE OF THE FIELD TENSOR

The non-Abelian component of the field tensor is defined through a metric q^{μ} that is a set of four quaternion-valued components of a 4-vector, a 4-vector each of whose components can be represented by a 2×2 matrix. In condensed notation:

$$q^{\mu} = (q^{\mu 0}, q^{\mu 1}, q^{\mu 2}, q^{\mu 3}) \tag{1}$$

and the total number of components of q^{μ} is 16. The covariant and second covariant derivatives of q^{μ} vanish [1] and the line element is given by

$$ds = q^{\mu}(x)dx_{\mu} \tag{2}$$

which, in special relativity (flat spacetime), reduces to

$$ds = \sigma^{\mu}dx_{\mu} \tag{3}$$

where σ^{μ} is a 4-vector made up of Pauli matrices:

$$\sigma^{\mu} = \left(\begin{bmatrix} 1 & 0 \\ 0 & 1 \end{bmatrix}, \begin{bmatrix} 0 & 1 \\ 1 & 0 \end{bmatrix}, \begin{bmatrix} 0 & -i \\ i & 0 \end{bmatrix}, \begin{bmatrix} 1 & 0 \\ 0 & -1 \end{bmatrix} \right) \tag{4}$$

In the limit of special relativity

$$q^{\mu}q^{\nu *} - q^{\nu}q^{\mu *} \rightarrow \sigma^{\mu}\sigma^{\nu} - \sigma^{\nu}\sigma^{\mu} \tag{5}$$

where * denotes reversing the time component of the quaternion-valued q^{μ}. The most general form of the non-Abelian part of the electromagnetic field tensor in conformally curved spacetime is [1]

$$F^{\mu\nu} = \frac{1}{8}QR(q^{\mu}q^{\nu *} - q^{\nu}q^{\mu *}) \tag{6}$$

To consider magnetic flux density components of $F^{\mu\nu}$, Q must have the units of weber and R, the scalar curvature, must have units of inverse square meters. In the flat spacetime limit, $R = 0$, so it is clear that the non-Abelian part of the field tensor, Eq. (6), vanishes in special relativity. The complete field tensor $F^{\mu\nu}$ vanishes [1] in flat spacetime because the curvature tensor vanishes. These considerations refute the Maxwell–Heaviside theory, which is developed in flat spacetime, and show that O(3) electrodynamics is a theory of conformally curved spacetime. Most generally, the Sachs theory is a closed field theory that, in principle, unifies all four fields: gravitational, electromagnetic, weak, and strong.

There exist generally covariant four-valued 4-vectors that are components of q^{μ}, and these can be used to construct the basic structure of O(3) electrodynamics in terms of single-valued components of the quaternion-valued metric q^{μ}. Therefore, the Sachs theory can be reduced to O(3) electrodynamics, which is a Yang–Mills theory [3,4]. The empirical evidence available for both the Sachs and O(3) theories is summarized in this review, and discussed more extensively in the individual reviews by Sachs [1] and Evans [2]. In other words, empirical evidence is given of the instances where the Maxwell–Heaviside theory fails and where the Sachs and O(3) electrodynamics succeed in describing empirical data from various sources. The fusion of the O(3) and Sachs theories provides proof that the $\boldsymbol{B}^{(3)}$ field [2] is a physical field of curved spacetime, which vanishes in flat spacetime (Maxwell–Heaviside theory [2]).

In Eq. (5), the product $q^{\mu}q^{\nu*}$ is quaternion-valued and non-commutative, but not antisymmetric in the indices μ and ν. The $\boldsymbol{B}^{(3)}$ field and structure of O(3) electrodynamics must be found from a special case of Eq. (5) showing that O(3) electrodynamics is a Yang–Mills theory and also a theory of general relativity [1]. The important conclusion reached is that Yang–Mills theories can be derived from the irreducible representations of the Einstein group. This result is consistent with the fact that all theories of physics must be theories of general relativity in principle. From Eq. (1), it is possible to write four-valued, generally covariant, components such as

$$q_X = \left(q_X^0, q_X^1, q_X^2, q_X^3\right) \tag{7}$$

which, in the limit of special relativity, reduces to

$$\sigma_x = (0, \sigma_x, 0, 0) \tag{8}$$

Similarly, one can write

$$q_Y = \left(q_Y^0, q_Y^1, q_Y^2, q_Y^3\right) \to (0, 0, \sigma_Y, 0) \tag{9}$$

and use the property

$$q_X q_Y^* - q_Y q_X^* \to \sigma_X \sigma_Y - \sigma_Y \sigma_X \tag{10}$$

in the limit of special relativity. The only possibility from Eqs. (7) and (9) is that

$$\begin{gathered} q_X^1 q_Y^{2*} - q_Y^2 q_X^{1*} = 2iq_Z^3 \\ \downarrow \\ \sigma_X \sigma_Y - \sigma_Y \sigma_X = 2i\sigma_Z \end{gathered} \tag{11}$$

where q_x^1 is single valued. In a 2×2 matrix representation, this is

$$q_X^1 = \begin{bmatrix} 0 & q_X^1 \\ q_X^1 & 0 \end{bmatrix} \rightarrow \sigma_X = \begin{bmatrix} 0 & 1 \\ 1 & 0 \end{bmatrix} \tag{12}$$

Similarly

$$q_Y^{2*} = \begin{bmatrix} 0 & -iq_Y^2 \\ iq_Y^2 & 0 \end{bmatrix} \rightarrow \sigma_Y = \begin{bmatrix} o & -i \\ i & 0 \end{bmatrix} \tag{13}$$

$$q_Y^3 = \begin{bmatrix} q_Z^3 & 0 \\ 0 & -q_Z^3 \end{bmatrix} \rightarrow \sigma_Z = \begin{bmatrix} 1 & 0 \\ 0 & -1 \end{bmatrix} \tag{14}$$

Therefore, there exist cyclic relations with O(3) symmetry

$$\begin{aligned} q_X^1 q_Y^{2*} - q_Y^2 q_X^{1*} &= 2iq_Z^3 \\ q_Y^2 q_Z^{3*} - q_Z^3 q_Y^{2*} &= 2iq_X^1 \\ q_Z^3 q_X^{1*} - q_X^1 q_Z^{3*} &= 2iq_Y^2 \end{aligned} \tag{15}$$

and the structure of O(3) electrodynamics [2] begins to emerge. If the space basis is represented by the complex circular ((1),(2),(3)) then Eqs. (15) become

$$\begin{aligned} q_X^{(1)} q_Y^{(2)*} - q_Y^{(2)} q_X^{(1)*} &= 2iq_Z^{(3)} \\ q_Y^{(2)} q_Z^{(3)*} - q_Z^{(3)} q_Y^{(2)*} &= 2iq_X^{(1)} \\ q_Z^{(3)} q_X^{(1)*} - q_X^{(1)} q_Z^{(3)*} &= 2iq_Y^{(2)} \end{aligned} \tag{16}$$

These are cyclic relations between single-valued metric field components in the non-Abelian part [Eq. (6)] of the quaternion-valued $F^{\mu\nu}$. Equation (16) can be put in vector form

$$\begin{aligned} \boldsymbol{q}^{(1)} \times \boldsymbol{q}^{(2)} &= i\boldsymbol{q}^{(3)*} \\ \boldsymbol{q}^{(2)} \times \boldsymbol{q}^{(3)} &= i\boldsymbol{q}^{(1)*} \\ \boldsymbol{q}^{(3)} \times \boldsymbol{q}^{(1)} &= i\boldsymbol{q}^{(2)*} \end{aligned} \tag{17}$$

where the asterisk denotes ordinary complex conjugation in Eq. (17) and quaternion conjugation in Eq. (16).

Equation (17) contains vector-valued metric fields in the complex basis ((1),(2),(3)) [2]. Specifically, in O(3) electrodynamics, which is based on the existence of two circularly polarized components of electromagnetic radiation [2]

$$\boldsymbol{q}^{(1)} = \frac{1}{\sqrt{2}}(i\boldsymbol{i} + \boldsymbol{j}) \exp(i\phi) \tag{18}$$

$$\boldsymbol{q}^{(2)} = \frac{1}{\sqrt{2}}(-i\boldsymbol{i} + \boldsymbol{j}) \exp(i\phi) \tag{19}$$

giving

$$\boldsymbol{q}^{(3)*} = \boldsymbol{k} \tag{20}$$

and

$$\boldsymbol{B}^{(3)} = \frac{1}{8} QR\boldsymbol{q}^{(3)} \tag{21}$$

Therefore, the $\boldsymbol{B}^{(3)}$ field [2] is proved from a particular choice of metric using the irreducible representations of the Einstein group [1]. It can be seen from Eq. (21) that the $\boldsymbol{B}^{(3)}$ field is the vector-valued metric field $\boldsymbol{q}^{(3)}$ within a factor $\frac{1}{8}\, QR$. This result proves that $\boldsymbol{B}^{(3)}$ vanishes in flat spacetime, because $R = 0$ in flat spacetime. If we write

$$B^{(3)} = \frac{1}{8} QR \tag{22}$$

then Eq. (17) becomes the B cyclic theorem [2] of O(3) electrodynamics:

$$\begin{gathered} \boldsymbol{B}^{(1)} \times \boldsymbol{B}^{(2)} = iB^{(0)}\boldsymbol{B}^{(3)*} \\ \cdots \end{gathered} \tag{23}$$

Since O(3) electrodynamics is a Yang–Mills theory [3,4], we can write

$$\boldsymbol{q} = q^{(1)}\boldsymbol{i} + q^{(2)}\boldsymbol{j} + q^{(3)}\boldsymbol{k} \tag{24}$$

from which it follows [5] that

$$D^{\mu}(D_{\mu}\boldsymbol{q}) = \boldsymbol{0}; \quad D_{\mu}\boldsymbol{q} = \boldsymbol{0} \tag{25}$$

Thus the first and second covariant derivatives vanish [1].

The Sachs theory [1] is able to describe parity violation and spin–spin interactions from first principles [6] on a classical level; it can also explain several problems of neutrino physics, and the Pauli exclusion principle can be derived from it classically. The quaternion form of the theory [1], which is the basis of this review chapter, predicts small but nonzero masses for the neutrino and photon; describes the Planck spectrum of blackbody radiation classically; describes the Lamb shifts in the hydrogen atom with precision equivalent to quantum electrodynamics, but without renormalization of infinities; proposes grounds for charge quantization; predicts the lifetime of the muon state; describes electron–muon mass splitting; predicts physical longitudinal and time-like photons and fields; and has built-in *P*, *C*, and *T* violation.

To this list can now be added the advantages of O(3) over U(1) electrodynamics, advantages that are described in the review by Evans in Part 2 of this three-volume set and by Evans, Jeffers, and Vigier in Part 3. In summary, by interlocking the Sachs and O(3) theories, it becomes apparent that the advantages of O(3) over U(1) are symptomatic of the fact that the electromagnetic field vanishes in flat spacetime (special relativity), if the irreducible representations of the Einstein group are used.

III. THE COVARIANT DERIVATIVE

The covariant derivative in the Sachs theory [1] is defined by the spin–affine connection:

$$D^{\rho} = \partial^{\rho} + \Omega^{\rho} \tag{26}$$

where

$$\Omega_{\mu} = \frac{1}{4}(\partial_{\mu} q^{\rho} + \Gamma^{\rho}_{\tau\mu} q^{\tau}) q^{*}_{\rho} \tag{27}$$

and where $\Gamma^{\rho}_{\tau\mu}$ is the Christoffel symbol. The latter can be defined through the reducible metrics $g_{\mu\nu}$ as follows [1]:

$$\Gamma^{\rho}_{\mu\alpha} = \frac{1}{2} g^{\rho\lambda}(\partial_{\mu} g_{\lambda\alpha} + \partial_{\alpha} g_{\mu\lambda} - \partial_{\lambda} g_{\alpha\mu}) \tag{28}$$

In O(3) electrodynamics, the covariant derivative on the classical level is defined by

$$D_{\mu} = \partial_{\mu} - igA_{\mu} = \partial_{\mu} - ig\mathrm{M}^{a} A^{a}_{\mu} \tag{29}$$

where M^{a} are rotation generators [2] of the O(3) group, and where a is an internal index of Yang–Mills theory. The complete vector potential in O(3) electrodynamics is defined by

$$\boldsymbol{A} = A^{(1)}\boldsymbol{e}^{(2)} + A^{(2)}\boldsymbol{e}^{(1)} + A^{(3)}\boldsymbol{e}^{(3)} \tag{30}$$

where $\boldsymbol{e}^{(1)}, \boldsymbol{e}^{(2)}, \boldsymbol{e}^{(3)}$ are unit vectors of the complex circular basis ((1),(2),(3)) [2]. If we restrict our discussion to plane waves, then the vector potential is

$$\boldsymbol{A}^{(1)} = \frac{A^{(0)}}{\sqrt{2}} (i\boldsymbol{i} + \boldsymbol{j}) \exp(i\phi) \tag{31}$$

where ϕ is the electromagnetic phase. Therefore, there are O(3) electrodynamics components such as

$$A_X^{(1)} = \frac{iA^{(0)}}{\sqrt{2}} e^{(i\phi)}; \quad A_Y^{(1)} = \frac{A^{(0)}}{\sqrt{2}} e^{(i\phi)} \tag{32}$$

In order to reduce the covariant derivative in the Sachs theory to the O(3) covariant derivative, the following classical equation must hold:

$$-igA_\mu = \frac{1}{4}(D_\mu q^\rho) q_\rho^* \tag{33}$$

This equation can be examined component by component, giving relations such as

$$-igA_X^{(1)} = -\frac{1}{4}(D_X q_Y^{(1)}) A_Y^{(1)} \tag{34}$$

where we have used

$$q_Y^{(1)} = -iq_X^{(1)} \tag{35}$$

Using [2]

$$g = \frac{\kappa}{A^{(0)}} \tag{36}$$

we obtain

$$i\kappa q_X^{(1)} = \frac{1}{4}(D_X q_Y^{(1)}) q_Y^{(1)} = -\frac{i}{4}(D_X q_Y^{(1)}) q_X^{(1)} \tag{37}$$

so that the wave number κ is defined by

$$\kappa = -\frac{1}{4} D_X q_Y^{(1)} \tag{38}$$

Therefore, we can write

$$D_X q_Y^{(1)} = D_1 q^{1(1)} = \partial_1 q^{1(1)} + \Gamma_{\lambda 1}^{1} q^{\lambda(1)} \tag{39}$$

and the wave number becomes the following sum:

$$\kappa = -\frac{1}{4}(\Gamma^1_{11}q^{1(1)} + \Gamma^1_{21}q^{2(1)}) \tag{40}$$

Using the identities

$$q^{1(1)} = q^{(1)}_x = \frac{i}{\sqrt{2}}e^{i\phi} \tag{41}$$

$$q^{2(1)} = q^{(1)}_Y = \frac{1}{\sqrt{2}}e^{i\phi} \tag{42}$$

the wave number becomes

$$\kappa = -\frac{1}{4}\left(\frac{i\Gamma^1_{11}}{\sqrt{2}}e^{i\phi} + \frac{\Gamma^1_{21}}{\sqrt{2}}e^{i\phi}\right) \tag{43}$$

Introducing the definition (28) of the Christoffel symbol, it is possible to write

$$\begin{aligned}\Gamma^1_{11} &= \frac{1}{2}g^{1\lambda}(\partial_1 g_{\lambda 1} + \partial_1 g_{1\lambda} - \partial_1 g_{11}) \\ &= \frac{1}{2}g^{13}\partial_Z g_{11} + \cdots\end{aligned} \tag{44}$$

so that

$$\kappa = -\frac{i}{8\sqrt{2}}g^{13}\partial_Z g_{11}e^{i\phi} + \cdots \tag{45}$$

This equation is satisfied by the following choice of metric:

$$g_{11} = \frac{1}{2}; \quad g^{13} = -8\sqrt{2}\, e^{-i\phi} \tag{46}$$

Similarly

$$\begin{aligned}\Gamma^1_{21} &= \frac{1}{2}g^{1\lambda}(\partial_2 g_{\lambda 1} + \partial_1 g_{2\lambda} - \partial_\lambda g_{12}) \\ &= \frac{1}{2}g^{13}\partial_Z g_{12} + \cdots\end{aligned} \tag{47}$$

so that the wave number can be expressed as

$$\kappa = \frac{i\kappa}{8\sqrt{2}}g^{13}g_{12}e^{i\phi} \tag{48}$$

an equation that is satisfied by the following choice of metric:

$$g_{12} = \frac{i}{2}; \quad g^{13} = -8\sqrt{2}\, e^{-i\phi} \tag{49}$$

Therefore, it is always possible to write the covariant derivative of the Sachs theory as an O(3) covariant derivative of O(3) electrodynamics. Both types of covariant derivative are considered on the classical level.

IV. ENERGY FROM THE VACUUM

The energy density in curved spacetime is given in the Sachs theory by the quaternion-valued expression

$$En_d = A^{\mu} j_{\mu}^{*} \tag{50}$$

where A^{μ} is the quaternion-valued vector potential and J_{μ}^{*} is the quaternion-valued 4-current as given by Sachs [1]. Equation (50) is an elegant and deeply meaningful expression of the fact that electromagnetic energy density is available from curved spacetime under all conditions; the distinction between field and matter is lost, and the concepts of "point charge" and "point mass" are not present in the theory, as these two latter concepts represent infinities of the closed-field theory developed by Sachs [1] from the irreducible representations of the Einstein group. The accuracy of expression (50) has been tested [1] to the precision of the Lamb shifts in the hydrogen atom without using renormalization of infinities. The Lamb shifts can therefore be viewed as the results of electromagnetic energy from curved spacetime.

Equation (50) is geometrically a scalar and algebraically quaternion-valued equation [1], and it is convenient to develop it using the identity [1]

$$q_{\gamma} q^{\kappa *} + q^{\kappa} q_{\gamma}^{*} = 2\sigma_0 \delta_{\gamma}^{\kappa} \tag{51}$$

with the indices defined as

$$\gamma = \kappa = \mu \tag{52}$$

to obtain

$$q^{\mu} q_{\mu}^{*} = \sigma_0 \delta^{\mu} \tag{53}$$

Using summation over repeated indices on the right-hand side, we obtain the following result:

$$q^{\mu} q_{\mu}^{*} = 4\sigma_0 \tag{54}$$

In the limit of flat spacetime

$$q^{\mu}q_{\mu}^{*} \rightarrow \sigma^{\mu}\sigma_{\mu} = 4\sigma_{0} \tag{55}$$

where the right-hand side is again a scalar invariant geometrically and a quaternion algebraically.

Therefore, the energy density (50) assumes the simple form

$$A^{\mu}J_{\mu}^{*} = 4A_{0}J_{0}^{*}\sigma_{0} \tag{56}$$

A_0 and J_0^* are magnitudes of A^{μ} and J_{μ}^*. In flat spacetime, this electromagnetic energy density vanishes because the curvature tensor vanishes. Therefore, in the Maxwell–Heaviside theory, there is no electromagnetic energy density from the vacuum and the field does not propagate through flat spacetime (the vacuum of the Maxwell–Heaviside theory) because of the absence of curvature. The $\boldsymbol{B}^{(3)}$ field depends on the scalar curvature R in Eq. (21), and so the $\boldsymbol{B}^{(3)}$ field and O(3) electrodynamics are theories of conformally curved spacetime. To maximize the electromagnetic energy density, the curvature has to be maximized, and the maximization of curvature may be the result of the presence of a gravitating object. In general, wherever there is curvature, there is electromagnetic energy that may be extracted from curved spacetime using a suitable device such as a dipole [7].

Therefore, we conclude that electromagnetic energy density exists in curved spacetime under all conditions, and devices can be constructed [8] to extract this energy density.

The quaternion-valued vector potential A^{μ} and the 4-current J_{μ}^* both depend directly on the curvature tensor. The electromagnetic field tensor in the Sachs theory has the form

$$F_{\mu\nu} = \partial_{\mu}A_{\nu}^{*} - \partial_{\nu}A_{\mu}^{*} + \frac{1}{8}QR(q_{\mu}q_{\nu}^{*} - q_{\nu}q_{\mu}^{*}) \tag{57}$$

where the quaternion-valued vector potential is defined as

$$A_{\gamma} = \frac{Q}{4}q_{\gamma}^{*}\int(\kappa_{\rho\lambda}q^{\lambda} + q^{\lambda}\kappa_{\rho\lambda}^{+})\,dx^{\rho} \tag{58}$$

The most general form of the vector potential is therefore given by Eq. (58), and if there is no curvature, the vector potential vanishes.

Similarly, the 4-current J_{μ}^* depends directly on the curvature tensor $\kappa_{\rho\lambda}$ [1], and there can exist no 4-current in the Heaviside–Maxwell theory, so the 4-current cannot act as the source of the field. In the closed-field theory, represented by the irreducible representations of the Einstein group [1], charge and current are manifestations of curved spacetime, and can be regarded as the results of the field. This is the viewpoint of Faraday and Maxwell rather than that of Lorentz. It follows that there can exist a vacuum 4-current in general relativity, and the implications of such a current are developed by Lehnert [9]. The vacuum 4-current also exists in O(3) electrodynamics, as demonstrated by Evans and others [2,9]. The concept of vacuum 4-current is missing from the flat spacetime of Maxwell–Heaviside theory.

In curved spacetime, both the electromagnetic and curvature 4-tensors may have longitudinal as well as transverse components in general and the electromagnetic field is always accompanied by a source, the 4-current J_μ^*. In the Maxwell–Heaviside theory, the field is assumed incorrectly to propagate through flat spacetime without a source, a violation of both causality and general relativity. As shown in several reviews in this three-volume set, Maxwell–Heaviside theory and its quantized equivalent appear to work well only under certain incorrect assumptions, and quantum electrodynamics is not a physical theory because, as pointed out by Dirac and many others, it contains infinities. Sachs [1] has also considered and removed the infinite self-energy of the electron by a consideration of general relativity.

The O(3) electrodynamics developed by Evans [2], and its homomorph, the SU(2) electrodynamics of Barrett [10], are substructures of the Sachs theory dependent on a particular choice of metric. Both O(3) and SU(2) electrodynamics are Yang–Mills structures with a Wu–Yang phase factor, as discussed by Evans and others [2,9]. Using the choice of metric (17), the electromagnetic energy density present in the O(3) curved spacetime is given by the product

$$En_d = \boldsymbol{A} \cdot \boldsymbol{j} \tag{59}$$

where the vector potential and 4-current are defined in the ((1),(2),(3)) basis in terms of the unit vectors similar to those in Eq. (2), and as described elsewhere in this three-volume set [2]. The extraction of electromagnetic energy density from the vacuum is also possible in the Lehnert electrodynamics as described in his review in the first chapter of this volume (i.e., here, in Part 2 of this three-volume set). The only case where extraction of such energy is not possible is that of the Maxwell–Heaviside theory, where there is no curvature.

The most obvious manifestation of energy from curved spacetime is gravitation, and the unification of gravitation and electromagnetism by Sachs [1] shows that electromagnetic energy emanates under all circumstances from spacetime curvature. This principle has been tested to the precision of the Lamb shifts of H as discussed already. This conclusion means that the electromagnetic field does not emanate from a "point charge," which in general relativity can be present only when the curvature becomes infinite. The concept of "point charge" is therefore unphysical, and this is the basic reason for the infinite electron self-energy in the Maxwell–Heaviside theory and the infinities of quantum electrodynamics, a theory rejected by Einstein, Dirac, and several other leading scientists of the twentieth century. The electromagnetic energy density inherent in curved spacetime depends on curvature as represented by the curvature tensor discussed in the next section. In the Einstein field equation of general relativity, which comes from the reducible representations of the Einstein group [1], the canonical energy momentum tensor of gravitation depends on the Einstein curvature tensor.

Sachs [1] has succeeded in unifying the gravitational and electromagnetic fields so that both share attributes. For example, both fields are non-Abelian under all conditions, and both fields are their own sources. The gravitational field carries energy that is equivalent to mass [11], and so is itself a source of gravitation. Similarly, the electromagnetic field carries energy that is equivalent to a 4-current, and so is itself a source of electromagnetism. These concepts are missing entirely from the Maxwell–Heaviside theory, but are present in O(3) electrodynamics, as discussed elsewhere [2,10]. The Sachs theory cannot be reduced to the Maxwell–Heaviside theory, but can be reduced, as discussed already, to O(3) electrodynamics. The fundamental reason for this is that special relativity is an asymptotic limit of general relativity, but one that is never reached precisely [1]. So the Poincaré group of special relativity is not a subgroup of the Einstein group of general relativity.

In standard Maxwell–Heaviside theory, the electromagnetic field is thought of as propagating in a source-free region in flat spacetime where there is no curvature. If, however, there is no curvature, the electromagnetic field vanishes in the Sachs theory [1], which is a direct result of using irreducible representations of the Einstein group of standard general relativity. The empirical evidence for the Sachs theory has been reviewed in this chapter already, and this empirical evidence refutes the Maxwell–Heaviside theory. In general relativity [1], if there is mass or charge anywhere in the universe, then the whole of spacetime is curved, and all the laws of physics must be written in curved spacetime, including, of course, the laws of electrodynamics. Seen in this light, the O(3) electrodynamics of Evans [2] and the homomorphic SU(2) electrodynamics of Barrett [12] are written correctly in conformally curved spacetime, and are particular cases of Einstein's general relativity as developed by Sachs [1]. Flat spacetime as the description of the vacuum is valid only when the whole universe is empty.

From everyday experience, it is possible to extract gravitational energy from curved spacetime on the surface of the earth. The extraction of electromagnetic energy must be possible if the extraction of gravitational energy is possible, and the electromagnetic field influences the gravitational field and vice versa. The field equations derived by Sachs [1] for electromagnetism are complicated, but can be reduced to the equations of O(3) electrodynamics by a given choice of metric. The literature discusses the various ways of solving the equations of O(3) electrodynamics [2,10], analytically, or using computation. In principle, the Sachs equations are solvable by computation for any given experiment, and such a solution would show the reciprocal influence between the electromagnetic and gravitational fields, leading to significant findings.

The ability of extracting electromagnetic energy density from the vacuum depends on the use of a device such as a dipole, and this dipole can be as simple as battery terminals, as discussed by Bearden [13]. The principle involved in this device is that electromagnetic energy density $A^{\mu} J_{\mu}^{*}$ exists in general relativity under all circumstances, and electromagnetic 4-currents and 4-potentials emanate form spacetime curvature. Therefore, the current in the battery is not driven by the positive and negative terminals, but is a manifestation of energy from curved spacetime, just as the hydrogen Lamb shift is another such manifestation. A battery runs down because the chemical energy needed to form the dipole dissipates.

In principle, therefore, the electromagnetic energy density in Eq. (50) is always available whenever there is spacetime curvature; in other words, it is always available because there is always spacetime curvature.

V. THE CURVATURE TENSOR

The curvature tensor is defined in terms of covariant derivatives of the spin–affine connections Ω_{ρ}, and according to Section (III), has its equivalent in O(3) electrodynamics.

The curvature tensor is

$$\begin{aligned}\kappa_{\rho\lambda} &= -\kappa_{\lambda\rho} = \Omega_{\rho;\lambda} - \Omega_{\lambda;\rho} \\ &= \partial_{\lambda}\Omega_{\rho} - \partial_{\rho}\Omega_{\lambda} + \Omega_{\lambda}\Omega_{\rho} - \Omega_{\rho}\Omega_{\lambda}\end{aligned} \tag{60}$$

and obeys the Jacobi identity

$$D_{\gamma}\kappa_{\rho\lambda} + D_{\rho}\kappa_{\lambda\gamma} + D_{\lambda}\kappa_{\gamma\rho} \equiv 0 \tag{61}$$

which can be written as

$$D_{\mu}\tilde{\kappa}^{\mu\nu} \equiv 0 \tag{62}$$

where

$$\tilde{\kappa}^{\mu\nu} = \frac{1}{2}\varepsilon^{\mu\nu\rho\sigma}\kappa_{\rho\sigma} \tag{63}$$

is the dual of $\kappa_{\rho\sigma}$.

Equation (4) has the form of the homogeneous field equation of O(3) electrodynamics [2,10]. If we now define

$$\begin{aligned}\kappa^{\rho\lambda} &= \Omega^{\rho;\lambda} - \Omega^{\lambda;\rho} \\ &= (\partial^{\lambda} + \Omega^{\lambda})\Omega^{\rho} - (\partial^{\rho} + \Omega^{\rho})\Omega^{\lambda}\end{aligned} \tag{64}$$

then

$$\begin{aligned}D_{\rho}\kappa^{\rho\lambda} &= (\partial_{\rho} + \Omega_{\rho})((\partial^{\lambda} + \Omega^{\lambda})\Omega^{\rho} - (\partial^{\rho} + \Omega^{\rho})\Omega^{\lambda}) \\ &\equiv L^{\lambda} \neq 0\end{aligned} \tag{65}$$

has the form of the inhomogeneous field equation of O(3) electrodynamics with a nonzero source term L^{λ} in curved spacetime.

The curvature tensor can be written as a commutator of covariant derivatives

$$\begin{aligned}\kappa_{\mu\nu} &= -\kappa_{\nu\mu} = -[D_{\mu}, D_{\nu}] = -[\partial_{\mu} + \Omega_{\mu}, \partial_{\nu} + \Omega_{\nu}] \\ &= \Omega_{\mu;\nu} - \Omega_{\nu;\mu}\end{aligned} \tag{66}$$

and is the result of a closed loop, or holonomy, in curved spacetime. This is the way in which a curvature tensor is also derived in general gauge field theory on the classical level [11]. If a field ϕ is introduced such that

$$\phi'(x) = S\phi(x) \tag{67}$$

under a gauge transformation, it follows that

$$\delta\phi = \Omega_{\mu}dx^{\mu}\phi \tag{68}$$

and that

$$\partial_{\mu}\phi' = S(\partial_{\mu}\phi) + (\partial_{\mu}S)\phi \tag{69}$$

The expression equivalent to Eq. (68) in general gauge field theory is [11]

$$\delta\psi = ig\mathrm{M}^{a}A_{\mu}^{a}dx^{\mu}\psi \tag{70}$$

where M^{a} are group rotation generators and A_{μ}^{a} are vector potential components with internal group indices a. Under a gauge transformation

$$(\partial_{\mu} + \Omega'_{\mu})\phi' = S(\partial_{\mu} + \Omega_{\mu})\phi \tag{71}$$

leading to the expression

$$\Omega'_{\mu} = S\Omega_{\mu}S^{-1} - (\partial_{\mu}S)S^{-1} \tag{72}$$

The equivalent equation in general gauge field theory is

$$A'_{\mu} = SA_{\mu}S^{-1} - \frac{i}{g}(\partial_{\mu}S)S^{-1} \tag{73}$$

Equations (72) and (73) show that the spin–affine connection Ω_{μ} and vector potential A_{μ} behave similarly under a gauge transformation. The relation between covariant derivatives has been developed in Section III.

VI. GENERALLY COVARIANT 4-VECTORS

The most fundamental feature of O(3) electrodynamics is the existence of the $\boldsymbol{B}^{(3)}$ field [2], which is longitudinally directed along the axis of propagation, and which is defined in terms of the vector potential plane wave:

$$\boldsymbol{A}^{(1)} = \boldsymbol{A}^{(2)*} \tag{74}$$

From the irreducible representations of the Einstein group, there exist 4-vectors that are generally covariant and take the following form:

$$\begin{aligned} B_1^{\mu} &= (B_X^{(0)}, B_X^{(1)}, B_X^{(2)}, B_X^{(3)}) \\ B_2^{\mu} &= (B_Y^{(0)}, B_Y^{(1)}, B_Y^{(2)}, B_Y^{(3)}) \\ B_3^{\mu} &= (B_Z^{(0)}, B_Z^{(1)}, B_Z^{(2)}, B_Z^{(3)}) \end{aligned} \tag{75}$$

All these components exist in general, and the $\boldsymbol{B}^{(3)}$ field can be identified as the $B_z^{(3)}$ component. In O(3) electrodynamics, these 4-vectors reduce to

$$\begin{aligned} B_1^{\mu} &= (0, B_X^{(1)}, B_X^{(2)}, 0) \\ B_2^{\mu} &= (0, B_Y^{(1)}, B_Y^{(2)}, 0) \\ B_3^{\mu} &= (B_Z^{(0)}, 0, 0, B_Z^{(3)}) \end{aligned} \tag{76}$$

so it can be concluded that O(3) electrodynamics is developed in a curved spacetime that is defined in such a way that

$$\boldsymbol{B}^{(3)} = -ig\boldsymbol{A}^{(1)} \times \boldsymbol{A}^{(2)} \tag{77}$$

In O(3) electrodynamics, there exist the cyclic relations (23), and we have seen that in general relativity, this cyclic relation can be derived using a particular choice of metric. In the special case of O(3) electrodynamics, the vector

$$B_3^{\mu} = (B_Z^{(0)}, B_Z^{(1)}, B_Z^{(2)}, B_Z^{(3)}) \tag{78}$$

reduces to

$$B_3^{\mu} = (B_Z^{(0)}, 0, 0, B_Z^{(3)}) \tag{79}$$

Similarly, there exists, in general, the 4-vector

$$A_3^{\mu} = (A_Z^{(0)}, A_Z^{(1)}, A_Z^{(2)}, A_Z^{(3)}) \tag{80}$$

which reduces in O(3) electrodynamics to

$$A_3^{\mu} = (A_Z^{(0)}, 0, 0, A_Z^{(3)}) \tag{81}$$

and that corresponds to generally covariant energy–momentum.

The curved spacetime 4-current is also generally covariant and has components such as

$$\begin{aligned} j_1^{\mu} &= (j_X^{(0)}, j_X^{(1)}, j_X^{(2)}, j_X^{(3)}) \\ j_2^{\mu} &= (j_Y^{(0)}, j_Y^{(1)}, j_Y^{(2)}, j_Y^{(3)}) \\ j_3^{\mu} &= (j_Z^{(0)}, j_Z^{(1)}, j_Z^{(2)}, j_Z^{(3)}) \end{aligned} \tag{82}$$

which, in O(3) electrodynamics, reduce to

$$\begin{aligned} j_1^{\mu} &= (0, j_X^{(1)}, j_X^{(2)}, 0) \\ j_2^{\mu} &= (0, j_Y^{(1)}, j_Y^{(2)}, 0) \\ j_3^{\mu} &= (j_Z^{(0)}, 0, 0, j_Z^{(3)}) \end{aligned} \tag{83}$$

The existence of a vacuum current such as this is indicated in O(3) electrodynamics by its inhomogeneous field equation

$$D_{\mu}\boldsymbol{G}^{\mu\nu} = \boldsymbol{J}^{\nu} \tag{84}$$

which is a Yang–Mills type of equation [2]. The concept of vacuum current was also introduced by Lehnert and is discussed in his review (first chapter in this volume; i.e., in Part 2).

The components of the antisymmetric field tensor in the Sachs theory [1] are

$$\begin{aligned} B^3 &= F^{21} = -F^{12} = (B_Z^{(0)}, B_Z^{(1)}, B_Z^{(2)}, B_Z^{(3)}) \\ B^1 &= F^{32} = -F^{23} = (B_X^{(0)}, B_X^{(1)}, B_X^{(2)}, B_X^{(3)}) \\ B^2 &= F^{13} = -F^{31} = (B_Y^{(0)}, B_Y^{(1)}, B_Y^{(2)}, B_Y^{(3)}) \\ E^1 &= F^{01} = -F^{10} = (E_X^{(0)}, E_X^{(1)}, E_X^{(2)}, E_X^{(3)}) \\ E^2 &= F^{02} = -F^{20} = (E_Y^{(0)}, E_Y^{(1)}, E_Y^{(2)}, E_Y^{(3)}) \\ E^3 &= F^{03} = -F^{30} = (E_Z^{(0)}, E_Z^{(1)}, E_Z^{(2)}, E_Z^{(3)}) \end{aligned} \tag{85}$$

each of which is a 4-vector that is generally covariant. For example

$$B_Z^{\mu} B_{\mu Z} = \text{invariant} \tag{86}$$

So, in general, in curved spacetime, there exist longitudinal and transverse components under all conditions. In O(3) electrodynamics, the upper indices ((1),(2),(3)) are defined by the unit vectors

$$\begin{aligned} \boldsymbol{e}^{(1)} &= \frac{1}{\sqrt{2}}(\boldsymbol{i} - i\boldsymbol{j}) \\ \boldsymbol{e}^{(2)} &= \frac{1}{\sqrt{2}}(\boldsymbol{i} + i\boldsymbol{j}) \\ \boldsymbol{e}^{(3)} &= \boldsymbol{k} \end{aligned} \tag{87}$$

which form the cyclically symmetric relation [2]

$$\begin{aligned} \boldsymbol{e}^{(1)} \times \boldsymbol{e}^{(2)} &= i\boldsymbol{e}^{(3)*} \\ &\cdots \end{aligned} \tag{88}$$

where the asterisk in this case denotes complex conjugation. In addition, there is the time-like index (0). The field tensor components in O(3) electrodynamics are therefore, in general

$$\begin{aligned} F^{01} &= -F^{10} = (0, E_X^{(1)}, E_X^{(2)}, 0) \\ F^{02} &= -F^{20} = (0, E_Y^{(1)}, E_Y^{(2)}, 0) \\ F^{03} &= -F^{30} = (E_Z^{(0)}, 0, 0, E_Z^{(3)}) \\ F^{21} &= -F^{12} = (B_Z^{(3)}, 0, 0, B_Z^{(3)}) \\ F^{13} &= -F^{31} = (0, B_Y^{(1)}, B_Y^{(2)}, 0) \\ F^{32} &= -F^{23} = (0, B_X^{(1)}, B_X^{(2)}, 0) \end{aligned} \tag{89}$$

and the following invariants occur:

$$\begin{aligned} B_Y^{(1)} B_Y^{(2)} + B_Y^{(2)} B_Y^{(1)} &= B^{(0)2} \\ B_X^{(1)} B_X^{(2)} + B_X^{(2)} B_X^{(1)} &= B^{(0)2} \\ E_Y^{(1)} E_Y^{(2)} + E_Y^{(2)} E_Y^{(1)} &= E^{(0)2} \\ E_X^{(1)} E_X^{(2)} + E_X^{(2)} E_X^{(1)} &= E^{(0)2} \\ B_Z^{(0)2} - B_Z^{(3)2} = E_Z^{(0)2} - E_Z^{(3)2} &= 0 \end{aligned} \tag{90}$$

From general relativity, it can therefore be concluded that the $\boldsymbol{B}^{(3)}$ field must exist and that it is a physical magnetic flux density defined to the precision of the Lamb shift. It propagates through the vacuum with other components of the field tensor.

VII. SACHS THEORY IN THE FORM OF A GAUGE THEORY

The most general form of the vector potential can be obtained by writing the first two terms of Eq. (57) as

$$F_{\rho\gamma,1} = \partial_\rho A_\gamma^* - \partial_\gamma A_\rho^* \tag{91}$$

The vector potential is defined as

$$A_\gamma^* = \frac{Q}{4}\int (\kappa_{\rho\lambda} q^\lambda + q^\lambda \kappa_{\rho\lambda}^+) q_\gamma^* \, dx^\rho \tag{92}$$

and can be written as

$$A_\gamma^* = \frac{Q}{4} q_\gamma^* \int (\kappa_{\rho\lambda} q^\lambda + q^\lambda \kappa_{\rho\lambda}^+) \, dx^\rho \tag{93}$$

In order to prove that

$$\int q_\gamma^* \, dx^\rho = q_\gamma^* \int dx^\rho \tag{94}$$

we can take examples, giving results such as

$$\begin{aligned} q_Z^* &= (-q_Z^{(0)}, q_Z^{(1)}, q_Z^{(2)}, q_Z^{(3)}) \\ &= (-q_Z^{(0)}, 0, 0, q_Z^{(3)}) \\ \int q_Z^* \, dX &= q_Z^* \int dX \end{aligned} \tag{95}$$

because q_z^* has no functional dependence on X. The overall structure of the field tensor, using irreducible representations of the Einstein group, is therefore

$$F_{\rho\gamma} = C(\partial_\rho q_\gamma^* - \partial_\gamma q_\rho^*) + D(q_\rho q_\gamma^* - q_\gamma q_\rho^*) \tag{96}$$

where C and D are coefficients. This equation has the structure of a quaternion valued non-Abelian gauge field theory. The most general form of the field tensor and the vector potential is quaternion-valued. If the following constraint holds

$$\frac{D}{C^2} \equiv -ig \tag{97}$$

the structure of Eq. (96) becomes

$$F_{\rho\gamma} = \partial_\rho A^*_\gamma - \partial_\gamma A^*_\rho - ig[A^*_\rho, A^*_\gamma] \tag{98}$$

which is identical with that of gauge field theory with quaternion-valued potentials. However, the use of the irreducible representations of the Einstein group leads to a structure that is more general than that of Eq. (98). The rules of gauge field theory can be applied to the substructure (98) and to electromagnetism in curved spacetime.

VIII. ANTIGRAVITY EFFECTS IN THE SACHS THEORY

Sachs' equations (4.16) (in Ref. 1)

$$\begin{aligned} \frac{1}{4}(\kappa_{\rho\gamma}q^\lambda + q^\lambda\kappa^+_{\rho\lambda}) + \frac{1}{8}Rq_\rho &= kT_\rho \\ -\frac{1}{4}(\kappa^+_{\rho\gamma}q^{\lambda*} + q^{\lambda*}\kappa_{\rho\gamma}) + \frac{1}{8}Rq^*_\rho &= kT^*_\rho \end{aligned} \tag{99}$$

are 16 equations in 16 unknowns, as these are the 16 components of the quaternion-valued metric. The canonical energy-momentum T_ρ is also quaternion-valued, and the equations are factorizations of the Einstein field equation. If there is no linear momentum and a static electromagnetic field (no Poynting vector), then

$$\mathrm{T}_\rho = (\mathrm{T}^0_\rho, 0, 0, 0) \tag{100}$$

so we have the four components $\mathrm{T}^0_0, \mathrm{T}^0_1, \mathrm{T}^0_2$, and T^0_3. The T^0_0 component is a component of the canonical energy due to the gravitoelectromagnetic field represented by q^0_0. The scalar curvature R is the same with and without electromagnetism, and so is the Einstein constant k.

Considering T^0_0 In Eq. (99), we obtain

$$k\mathrm{T}^0_0 = \frac{1}{8}Rq^0_0 + \frac{1}{4}(\kappa_{0\lambda}q^\lambda + q^\lambda\kappa^+_{0\lambda}) \tag{101}$$

and if we choose a metric such that all components go to zero except q^0_0, then

$$k\mathrm{T}^0_0 \rightarrow \frac{1}{8}Rq^0_0 \tag{102}$$

However, R also vanishes in this limit, so

$$T_0^0 \rightarrow 0 \tag{103}$$

So, in order to produce antigravity effects, the gravito-electromagnetic field must be chosen so that only q_0^0 exists in a static situation. Therefore, antigravity is produced by q_1^0, q_2^0, and q_3^0 all going to zero asymptotically, or by

$$q_0^0 \gg (q_1^0 \approx q_2^0 \approx q_3^0) \tag{104}$$

This result is consistent with the fact that the curvature tensor $\kappa_{0\lambda}$ must be minimized, which is a consistent result. The curvature is

$$\kappa_{\rho\lambda} = -\kappa_{\lambda\rho} = \Omega_{\rho;\lambda} - \Omega_{\lambda;\rho} \tag{105}$$

and is minimized if

$$\Omega_{\rho;\lambda} \approx \Omega_{\lambda;\rho} \tag{106}$$

If $\rho = 0$, then $\Omega_{0;\lambda} \approx \Omega_{\lambda;0}$. This minimization can occur if the spin–affine connection is minimized. We must now investigate the effect of minimizing $\kappa_{0\lambda}$ on the electromagnetic field

$$F_{\rho\gamma} = Q\left[\frac{1}{4}(\kappa_{\rho\lambda}q^{\lambda}q_{\gamma}^{*} + q_{\gamma}q^{\lambda*}\kappa_{\rho\lambda} + q^{\lambda}\kappa_{\rho\lambda}^{+}q_{\gamma}^{*} + q_{\gamma}\kappa_{\rho\lambda}^{+}q^{\lambda*}) + \frac{1}{8}(q_{\rho}q_{\gamma}^{*} - q_{\gamma}q_{\rho}^{*})R\right] \tag{107}$$

We know that $R \rightarrow 0$ and $\rho = 0$, so

$$F_{0\gamma} = Q\left[\frac{1}{4}(\kappa_{0\lambda}q^{\lambda}q_{\gamma}^{*} + \cdots)\right] \tag{108}$$

and the $F_{0\gamma}$ component must be minimized. This is the gravito-electric component. Therefore, the gravito-magnetic component must be very large in comparison with the gravito-electric component.

IX. SOME NOTES ON QUATERNION-VALUED METRICS

In the flat spacetime limit, the following relation holds:

$$q^{\mu}q^{\nu*} - q^{\nu}q^{\mu*} \rightarrow \sigma^{\mu}\sigma^{\nu} - \sigma^{\nu}\sigma^{\mu} \tag{109}$$

where

$$\sigma^{\mu} = \left(\begin{bmatrix} 1 & 0 \\ 0 & 1 \end{bmatrix}, \begin{bmatrix} 0 & 1 \\ 1 & 0 \end{bmatrix}, \begin{bmatrix} 0 & -i \\ i & 0 \end{bmatrix}, \begin{bmatrix} 1 & 0 \\ 0 & -1 \end{bmatrix} \right) \tag{110}$$

Therefore, the quaternion-valued metric can be written as

$$q^{\mu} = \left(\begin{bmatrix} q^{\mu 0} & 0 \\ 0 & q^{\mu 0} \end{bmatrix}, \begin{bmatrix} 0 & q^{\mu 1} \\ q^{\mu 1} & 0 \end{bmatrix}, \begin{bmatrix} 0 & -iq^{\mu 2} \\ iq^{\mu 2} & 0 \end{bmatrix}, \begin{bmatrix} q^{\mu 3} & 0 \\ 0 & -q^{\mu 3} \end{bmatrix} \right) \tag{111}$$

with components

$$\begin{aligned}
q^{0} &= \left(\begin{bmatrix} q_0^0 & 0 \\ 0 & q_0^0 \end{bmatrix}, \begin{bmatrix} 0 & q_0^1 \\ q_0^1 & 0 \end{bmatrix}, \begin{bmatrix} 0 & -iq_0^2 \\ iq_0^2 & 0 \end{bmatrix}, \begin{bmatrix} q_0^3 & 0 \\ 0 & -q_0^3 \end{bmatrix} \right) \\
q_X &= \left(\begin{bmatrix} q_X^0 & 0 \\ 0 & q_X^0 \end{bmatrix}, \begin{bmatrix} 0 & q_X^1 \\ q_X^1 & 0 \end{bmatrix}, \begin{bmatrix} 0 & -iq_X^2 \\ iq_X^2 & 0 \end{bmatrix}, \begin{bmatrix} q_X^3 & 0 \\ 0 & -q_X^3 \end{bmatrix} \right) \\
q_Y &= \left(\begin{bmatrix} q_Y^0 & 0 \\ 0 & q_Y^0 \end{bmatrix}, \begin{bmatrix} 0 & q_Y^1 \\ q_Y^1 & 0 \end{bmatrix}, \begin{bmatrix} 0 & -iq_Y^2 \\ iq_Y^2 & 0 \end{bmatrix}, \begin{bmatrix} q_Y^3 & 0 \\ 0 & -q_Y^3 \end{bmatrix} \right) \\
q_z &= \left(\begin{bmatrix} q_Z^0 & 0 \\ 0 & q_Z^0 \end{bmatrix}, \begin{bmatrix} 0 & q_Z^1 \\ q_Z^1 & 0 \end{bmatrix}, \begin{bmatrix} 0 & -iq_Z^2 \\ iq_Z^2 & 0 \end{bmatrix}, \begin{bmatrix} q_Z^3 & 0 \\ 0 & -q_Z^3 \end{bmatrix} \right)
\end{aligned} \tag{112}$$

In the flat spacetime limit

$$\begin{aligned}
q^0 \rightarrow \sigma^0 &= \left(\begin{bmatrix} 1 & 0 \\ 0 & 1 \end{bmatrix}, 0, 0, 0 \right) \\
q_X \rightarrow \sigma_X &= \left(0, \begin{bmatrix} 0 & 1 \\ 1 & 0 \end{bmatrix}, 0, 0 \right) \\
q_Y \rightarrow \sigma_Y &= \left(0, 0, \begin{bmatrix} 0 & -i \\ i & 0 \end{bmatrix}, 0 \right) \\
q_Z \rightarrow \sigma_Z &= \left(0, 0, 0, \begin{bmatrix} 1 & 0 \\ 0 & -1 \end{bmatrix} \right)
\end{aligned} \tag{113}$$

This means that in the flat spacetime limit

$$\begin{aligned}
&q_0^0 \to 1; \quad q_0^1 \to 0; \quad q_0^2 \to 0; \quad q_0^3 \to 0 \\
&q_X^0 \to 0; \quad q_X^1 \to 1; \quad q_X^2 \to 0; \quad q_X^3 \to 0 \\
&q_Y^0 \to 0; \quad q_Y^1 \to 0; \quad q_Y^2 \to 1; \quad q_Y^3 \to 0 \\
&q_Z^0 \to 1; \quad q_Z^1 \to 0; \quad q_Z^2 \to 0; \quad q_Z^3 \to 1
\end{aligned} \tag{114}$$

Checking with the identity:

$$q_\gamma q^{\kappa *} + q^\kappa q_\gamma^* = 2\sigma_0 \delta_\gamma^\kappa \tag{115}$$

then

$$\begin{aligned}
q_X q^{X*} + q^X q_X^* = 2\sigma_0 \delta_X^X = 2\sigma_0 \\
(q_X^0)^2 + (q_X^1)^2 + (q_X^2)^2 + (q_X^3)^2 = \sigma_0
\end{aligned} \tag{116}$$

which is a property of quaternion indices in curved spacetime. In flat spacetime:

$$(q_X^1)^2 = \sigma_0 \tag{117}$$

that is

$$\begin{pmatrix} 1 & 0 \\ 0 & 1 \end{pmatrix} = \begin{pmatrix} 1 & 0 \\ 0 & 1 \end{pmatrix} \tag{118}$$

The reduction to O(3) electrodynamics takes place using products such as

$$\begin{aligned}
q_X q_Y^* - q_Y q_X^* &= \begin{bmatrix} 0 & q_X^1 \\ q_X^1 & 0 \end{bmatrix} \begin{bmatrix} 0 & -iq_Y^2 \\ iq_Y^2 & 0 \end{bmatrix} - \begin{bmatrix} 0 & -iq_Y^2 \\ iq_Y^2 & 0 \end{bmatrix} \begin{bmatrix} 0 & q_X^1 \\ q_X^1 & 0 \end{bmatrix} \\
&= \begin{bmatrix} iq_X^1 q_Y^2 & 0 \\ 0 & -iq_Y^2 q_X^1 \end{bmatrix} \\
&= i \begin{bmatrix} q_Z^3 & 0 \\ 0 & q_Z^3 \end{bmatrix}
\end{aligned} \tag{119}$$

that is

$$q_Z^3 = q_X^1 q_Y^2 \tag{120}$$

In flat spacetime, this becomes

$$1 = 1 \tag{121}$$

If the phases are defined as

$$q_X^1 = e^{i\phi}; \quad q_Y^{2*} = e^{-i\phi} \tag{122}$$

then the $\boldsymbol{B}^{(3)}$ field is recovered as

$$B^{(3)} = \frac{1}{8} QR \tag{123}$$

Applying Eq. (99), it is seen that T^{μ} has the same structure as q^{μ}:

$$\mathrm{T}^{\mu} = \left(\begin{bmatrix} \mathrm{T}^{\mu 0} & o \\ 0 & \mathrm{T}^{\mu 0} \end{bmatrix}, \begin{bmatrix} 0 & \mathrm{T}^{\mu 1} \\ \mathrm{T}^{\mu 1} & 0 \end{bmatrix}, \begin{bmatrix} 0 & -i\mathrm{T}^{\mu 2} \\ i\mathrm{T}^{\mu 2} & 0 \end{bmatrix}, \begin{bmatrix} \mathrm{T}^{\mu 3} & 0 \\ 0 & -\mathrm{T}^{\mu 3} \end{bmatrix} \right) \tag{124}$$

Therefore, the energy momentum is quaternion-valued. The vacuum current is

$$j_{\gamma} = \frac{Qk'}{4\pi} (\mathrm{T}_{\rho}^{;\rho} q_{\gamma}^{*} - q_{\gamma} \mathrm{T}_{\rho}^{;\rho *}) \tag{125}$$

where Q and $\kappa'/4\pi$ are constants. We may investigate the structure of the 4-current j_{γ} by working out the covariant derivative:

$$\mathrm{T}_{\rho}^{;\rho} = \partial^0 \mathrm{T}_0 + \partial^1 \mathrm{T}_1 + \partial^2 \mathrm{T}_2 + \partial^3 \mathrm{T}_3 + \Gamma_{0\rho}^{\rho} \mathrm{T}^0 + \Gamma_{1\rho}^{\rho} \mathrm{T}^1 + \Gamma_{2\rho}^{\rho} \mathrm{T}^2 + \Gamma_{3\rho}^{\rho} \mathrm{T}^3 \tag{126}$$

The partial derivatives and Christoffel symbols are not quaternion-valued, so we may write

$$\mathrm{T}_{\rho}^{;\rho} = (\partial^0 + \Gamma_{0\rho}^{\rho}) \mathrm{T}_0 - (\partial^1 + \Gamma_{1\rho}^{\rho}) \mathrm{T}_1 - (\partial^2 + \Gamma_{2\rho}^{\rho}) \mathrm{T}_2 - (\partial^3 + \Gamma_{3\rho}^{\rho}) \mathrm{T}_3 \tag{127}$$

Therefore the vacuum current in general relativity is defined by

$$\begin{aligned} j_{\gamma} = \frac{Qk'}{4\pi} (((\partial^0 + \Gamma_{0\rho}^{\rho}) \mathrm{T}_0 - (\partial^1 + \Gamma_{1\rho}^{\rho}) \mathrm{T}_1 - (\partial^2 + \Gamma_{2\rho}^{\rho}) \mathrm{T}_2 - (\partial^3 + \Gamma_{3\rho}^{\rho}) \mathrm{T}_3) q_{\gamma}^{*} \\ + q_{\gamma} ((\partial^0 + \Gamma_{0\rho}^{\rho}) \mathrm{T}_0 + (\partial^1 + \Gamma_{1\rho}^{\rho}) \mathrm{T}_1 + (\partial^2 + \Gamma_{2\rho}^{\rho}) \mathrm{T}_2 + (\partial^3 + \Gamma_{3\rho}^{\rho}) \mathrm{T}_3)) \end{aligned} \tag{128}$$

This current exists under all conditions and is the most general form of the Lehnert vacuum current described elsewhere in this volume, and the vacuum current in O(3) electrodynamics. In the Sachs theory, the existence of the electromagnetic field tensor depends on curvature, so energy is extracted from curved spacetime. The 4-current j_μ contains terms such as

$$\begin{aligned} j_{\gamma,0} &= \frac{Qk'}{4\pi}((\partial^0 + \Gamma^\rho_{0\rho})\mathrm{T}_0 q^*_\gamma + q_\gamma(\partial^0 + \Gamma^\rho_{0\rho})\mathrm{T}_0) \\ &= \frac{Qk'}{4\pi}(\partial^0 + \Gamma^\rho_{0\rho})\left(\begin{bmatrix} \mathrm{T}^0_0 & 0 \\ 0 & \mathrm{T}^0_0 \end{bmatrix} q^*_\gamma + q_\gamma \begin{bmatrix} \mathrm{T}^0_0 & 0 \\ 0 & \mathrm{T}^0_0 \end{bmatrix}\right) \end{aligned} \tag{129}$$

We may now choose $\gamma = 0, 1, 2, 3$ to obtain terms such as

$$\begin{aligned} j_{0,0} &= (\partial^0 + \Gamma^\rho_{0\rho})\left(-\begin{bmatrix} \mathrm{T}^0_0 & 0 \\ 0 & \mathrm{T}^0_0 \end{bmatrix}\begin{bmatrix} q^0_0 & 0 \\ 0 & q^0_0 \end{bmatrix} + \begin{bmatrix} q^0_0 & 0 \\ 0 & q^0_0 \end{bmatrix}\begin{bmatrix} \mathrm{T}^0_0 & 0 \\ 0 & \mathrm{T}^0_0 \end{bmatrix}\right) = 0 \\ j_{1,0} &= -(\partial^0 + \Gamma^\rho_{0\rho})\left(\begin{bmatrix} \mathrm{T}^0_0 & 0 \\ 0 & \mathrm{T}^0_0 \end{bmatrix}\begin{bmatrix} 0 & q^1_1 \\ q^1_1 & 0 \end{bmatrix} + \begin{bmatrix} 0 & q^1_1 \\ q^1_1 & 0 \end{bmatrix}\begin{bmatrix} \mathrm{T}^0_0 & 0 \\ 0 & \mathrm{T}^0_0 \end{bmatrix}\right) \\ &= -(\partial^0 + \Gamma^\rho_{0\rho})(q^1_1\mathrm{T}^0_0(\sigma_X + \sigma_0)) \\ &\neq 0 \end{aligned} \tag{130}$$

There are numerous other components of the 4-current density j_γ that are nonzero under all conditions. These act as sources for the electromagnetic field under all conditions. In flat spacetime, the electromagnetic field vanishes, and so does the 4-current density j_γ.

A check can be made on the interpretation of the quaternion-valued metric if we take the quaternion conjugate:

$$q^{\mu*} = \left(-\begin{bmatrix} q^{\mu 0} & 0 \\ 0 & q^{\mu 0} \end{bmatrix}, \begin{bmatrix} 0 & q^{\mu 1} \\ q^{\mu 1} & 0 \end{bmatrix}, \begin{bmatrix} 0 & iq^{\mu 2} \\ iq^{\mu 2} & 0 \end{bmatrix}, \begin{bmatrix} q^{\mu 3} & 0 \\ 0 & -q^{\mu 3} \end{bmatrix}\right) \tag{131}$$

which must reduce, in the flat space-time limit, to:

$$\sigma^\mu = \left(\begin{bmatrix} 1 & 0 \\ 0 & 1 \end{bmatrix}, \begin{bmatrix} 0 & 1 \\ 1 & 0 \end{bmatrix}, \begin{bmatrix} 0 & -i \\ i & 0 \end{bmatrix}, \begin{bmatrix} 1 & 0 \\ 0 & -1 \end{bmatrix}\right) \tag{132}$$

This means that the flat spacetime metric is

$$\begin{bmatrix} -1 & 0 & 0 & 0 \\ 0 & 1 & 0 & 0 \\ 0 & 0 & 1 & 0 \\ 0 & 0 & 0 & 1 \end{bmatrix} = -g^{\mu\nu} \tag{133}$$

which is the negative of the metric $g^{\mu\nu}$ of flat spacetime, that is, Minkowski spacetime.

If we define

$$q^{\mu *} = \left(\begin{bmatrix} q^{\mu 0} & 0 \\ 0 & q^{\mu 0} \end{bmatrix}, \ -\begin{bmatrix} 0 & q^{\mu 1} \\ q^{\mu 1} & 0 \end{bmatrix}, \ -\begin{bmatrix} 0 & -iq^{\mu 2} \\ iq^{\mu 2} & 0 \end{bmatrix}, \ -\begin{bmatrix} q^{\mu 3} & 0 \\ 0 & -q^{\mu 3} \end{bmatrix} \right) \tag{134}$$

then we obtain

$$g^{\mu\nu} = g_{\mu\nu} = \begin{bmatrix} 1 & 0 & 0 & 0 \\ 0 & -1 & 0 & 0 \\ 0 & 0 & -1 & 0 \\ 0 & 0 & 0 & -1 \end{bmatrix} \tag{135}$$

in the flat spacetime limit. This is the usual Minkowski metric.

To check on the interpretation given in the text of the reduction of Sachs to O(3) electrodynamics, we can consider generally covariant components such as

$$\begin{aligned} q_X &= (q_X^0, q_X^1, q_X^2, q_X^3) \rightarrow (\sigma^0, \sigma^1, \sigma^2, \sigma^3) \\ q_Y &= (q_Y^0, q_Y^1, q_Y^2, q_Y^3) \rightarrow (\sigma^0, \sigma^1, \sigma^2, \sigma^3) \\ q_Y^* &= (-q_Y^0, q_Y^1, q_Y^2, q_Y^3) \rightarrow (-\sigma^0, \sigma^1, \sigma^2, \sigma^3) \end{aligned} \tag{136}$$

It follows that

$$q_X q_Y^* - q_Y q_X^* \rightarrow \sigma_X \sigma_Y - \sigma_Y \sigma_X = 2i\sigma_Z \tag{137}$$

and that:

$$\begin{aligned} \sigma_X &= (0, \sigma_X, 0, 0) \\ \sigma_Y &= (0, 0, \sigma_Y, 0) \end{aligned} \tag{138}$$

Note that products such as $\sigma_X \sigma_Y$ must be interpreted as single-valued, because products such as

$$[0 \ \sigma_x \ 0 \ 0] \begin{bmatrix} 0 \\ 0 \\ \sigma_y \\ 0 \end{bmatrix} = \begin{bmatrix} 0 & 0 & 0 & 0 \\ 0 & 0 & 0 & 0 \\ 0 & 0 & 0 & 0 \\ 0 & 0 & 0 & 0 \end{bmatrix} \tag{139}$$

give a null matrix. Therefore, the quaternion-valued product $q_X q_Y^*$ must also be interpreted as

$$q_X q_Y^* - q_Y q_X^* \rightarrow \sigma_X \sigma_Y - \sigma_Y \sigma_X = 2i\sigma_Z \tag{140}$$

as in the text.

Acknowledgments

The U.S. Department of Energy is acknowledged for its website: http://www.ott.doe.gov/electromagnetic/. This website is reserved for the Advanced Electrodynamics Working Group.

References

1. M. Sachs, 11th chapter in Part 1 of this three-volume set.
2. M. W. Evans, 2nd chapter in this volume (i.e., Part 2 of this compilation).
3. M. W. Evans, *O(3) Electrodynamics*, Kluwer Academic, Dordrecht, 1999.
4. T. W. Barrett and D. M. Grimes, *Advanced Electromagnetism*, World Scientific, Singapore, 1995.
5. A demonstration of this property in O(3) electrodynamics is given by Evans and Jeffers, 1st chapter in Part 3 of this three-volume set; see also Ref(s).—by Vigier in Bibliography at end of that chapter.
6. M. W. Evans, J. P. Vigier, and S. Roy (eds.), *The Enigmatic Photon*, Kluwer Academic, Dordrecht, 1997, Vol. 4.
7. T. E. Bearden, *Energy from the Active Vacuum*, World Scientific, Singapore, in press.
8. T. E. Bearden, 11th and 12th chapters in this volume (i.e., Part 2).
9. M. W. Evans and S. Jeffers, 1st chapter in Part 3 of this three-volume set; see also Ref(s).—by Vigier listed in that chapter.
10. T. W. Barrett, in A. Lakhtakia (Ed.), *Essays on the Formal Aspects of Electromagnetic Theory*, World Scientific, Singapore, 1993.
11. L. H. Ryder, *Quantum Field Theory*, 2nd ed., Cambridge Univ. Press, Cambridge, UK, 1987.
12. T. W. Barrett, *Apeiron* **7**(1), (2000).
13. T. E. Bearden et al., *Phys. Scripta* **61**, 513 (2000).

THE LINK BETWEEN THE TOPOLOGICAL THEORY OF RAÑADA AND TRUEBA, THE SACHS THEORY, AND O(3) ELECTRODYNAMICS

M. W. EVANS

CONTENTS

I. REVIEW OF THE LITERATURE AND GENERAL CONCEPTS

The topological approach of Rañada and Trueba, the general relativistic approach of Sachs, and the O(3) electrodynamics are interlinked and shown to be based on the concept of Faraday's lines of force.

In the review by Rañada and Trueba [1], electric and magnetic lines of force were discussed as real, physical entities, based on the original concepts of Faraday. These authors discussed Kelvin's suggestion of 1868 that atoms are knots of links of vortex lines of the ether, a topological concept, and that Kelvin found the concept of point particle to be extremely unsatisfactory. Point particles are eliminated from consideration in the Sachs [2] theory of electrodynamics, and are replaced by curvature of spacetime. The O(3) electrodynamics of Evans [3] has been demonstrated [4] to be a subtheory of the Sachs theory. Rañada and Trueba discuss the fact that, in contemporary topology, invariant numbers characterize configurations that can deform, distort, or warp. These concepts are similar to the curving of spacetime in general relativity [2], of which O(3) electrodynamics [3] is a subtheory, and also a gauge theory. Topology [1] shows that the variety of chemical elements is due to the way in which curves can be knotted and linked, transmutability of the elements

Modern Nonlinear Optics, Part 2, Second Edition, Advances in Chemical Physics, Volume 119, Edited by Myron W. Evans. Series Editors I. Prigogine and Stuart A. Rice.
ISBN 0-471-38931-5

is due to the breaking and reconnection of lines, and the quantum character of the spectrum is due to the natural topological configurations of the vector field. Rañada and Trueba [1] reinstated lines of force, which they describe as having been relegated in importance in Maxwell's treatise and replaced by the concepts of field and vector potential. Lines of force are integral lines of the magnetic and electric fields, and so exist also in the Sachs [2] and Evans [3] theories.

The Aharonov–Bohm effect requires topological consideration [1], (i.e., a structured vacuum), and there exist conservation laws of topological origin, the simplest one is given by the sine–Gordon equation, which also appears in the discussion of O(3) electrodynamics by Evans and Crowell [5].

All topological theories are nonlinear, a feature of both the Sachs and Evans theories, and the whole of quantum theory can be replaced by topology [1], which reduces in some circumstances to the Yang–Mills theory [1], of which O(3) electrodynamics [3] is an example. O(3) electrodynamics has been developed into an O(3) symmetry quantum field theory by Evans and Crowell [5], and Witten [1] has developed a topological quantum field theory. In the theory of Rañada and Trueba [1], the Maxwell equations are linearized by change of variables of a set of nonlinear equations, and are compatible with topological constants of motion nonlinear in A^{μ} and $F_{\mu\nu}$. One of these is $\boldsymbol{B}^{(3)}$ [3], whose rigorous form in general relativity is the quaternion-valued equivalent [6] of the Sachs theory. One of these constants of motion is the electromagnetic helicity of a knot, which has been discussed elsewhere [7] in terms of $\boldsymbol{B}^{(3)}$. However, helicity is not conserved in the Sachs theory [2] because the latter contains parity violation as an intrinsic feature. The electromagnetic helicity of a knot is defined by Rañada and Trueba [1] as

$$H = \frac{1}{2}(\boldsymbol{A}\cdot\boldsymbol{B} + \boldsymbol{C}\cdot\boldsymbol{E})d^3\boldsymbol{r} \tag{1}$$

where, in the Maxwell–Heaviside theory, the magnetic and electric fields are defined as

$$\boldsymbol{B} = \nabla \times \boldsymbol{A}; \qquad \boldsymbol{E} = \nabla \times \boldsymbol{C} \tag{2}$$

The helicity H of a knot can, however, be defined as the $\boldsymbol{B}^{(3)}$ field as follows:

$$\begin{aligned} H &= \left| -ig \int \nabla\cdot(\boldsymbol{A}^{(1)} \times \boldsymbol{A}^{(2)})\, dZ \right| = \left| -ig \int \frac{\partial}{\partial Z}\boldsymbol{A}^{(1)} \times \boldsymbol{A}^{(2)}\, dZ \right| \\ &= |-ig\boldsymbol{A}^{(1)} \times \boldsymbol{A}^{(2)}| = |\boldsymbol{B}^{(3)}| \end{aligned} \tag{3}$$

This definition can be rewritten in the form (1) using:

$$\nabla\cdot(\boldsymbol{A}^{(1)} \times \boldsymbol{A}^{(2)}) = \boldsymbol{A}^{(2)}\cdot(\nabla \times \boldsymbol{A}^{(1)}) - \boldsymbol{A}^{(1)}\cdot(\nabla \times \boldsymbol{A}^{(2)}) \tag{4}$$

This definition is related to the difference between left- and right-handed photons because $\boldsymbol{B}^{(3)}$ switches sign between left and right circularly polarized electromagnetic radiation. Therefore, H and $\boldsymbol{B}^{(3)}$ constitute electromagnetic helicities of a knot, and there is also a link between $\boldsymbol{B}^{(3)}$ and the Sachs theory [1], as shown in the review [6] by Evans, linking O(3) electrodynamics and the Sachs theory.

As discussed by Rañada and Trueba [1], Faraday thought of lines of force as real and physical, and these authors represent magnetic lines of force by a complex function $\phi(t, X, Y, Z)$ of time and Cartesian coordinates. Any magnetic field [1] is therefore defined by

$$\boldsymbol{B} = g(\phi, \phi^*)\nabla\phi^* \times \nabla\phi \tag{5}$$

and any electric field by

$$\boldsymbol{E} = -g(\phi, \phi^*)(\partial_0\phi^*\nabla\phi - \partial_0\phi\nabla\phi^*) \tag{6}$$

so that the product

$$\boldsymbol{E} \cdot \boldsymbol{B} = 0 \tag{7}$$

vanishes. Electromagnetism by evolution of magnetic lines is discussed by Rañada and Trueba [1] in terms of level curves of a complex function, and leads to the appearance of a rich topological structure.

The $\boldsymbol{B}^{(3)}$ field [3] of O(3) electrodynamics is defined in terms of the cross-product of plane wave potentials $\boldsymbol{A}^{(1)} = \boldsymbol{A}^{(2)}$

$$\boldsymbol{B}^{(3)} = -ig\boldsymbol{A}^{(1)} \times \boldsymbol{A}^{(2)} \tag{8}$$

where the parameter g is

$$g = \frac{\kappa}{|\boldsymbol{A}^{(1)}|} \equiv \frac{\kappa}{A^{(0)}} \tag{9}$$

and where κ is the wave number. The modulus of the $\boldsymbol{B}^{(3)}$ field is therefore defined as

$$|\boldsymbol{B}^{(3)}| \equiv -ig(A_X^{(1)}A_Y^{(2)} - A_X^{(2)}A_Y^{(1)}) \tag{10}$$

which can be written in terms of lines of force as

$$|\boldsymbol{B}^{(3)}| = g_0\left(\frac{\partial\phi_X^{(1)}}{\partial Z}\frac{\partial\phi_Y^{(2)}}{\partial Z} - \frac{\partial\phi_X^{(2)}}{\partial Z}\frac{\partial\phi_Y^{(1)}}{\partial Z}\right) \tag{11}$$

where the phi functions are

$$\begin{aligned} \phi_Y^{(1)} &= i\exp(-i(\omega t - \kappa Z)); \qquad \phi_Y^{(2)} = -i\exp(i(\omega t - \kappa Z)) \\ \phi_X^{(1)} &= \exp(-i(\omega t - \kappa Z)); \qquad \phi_X^{(2)} = \exp(i(\omega t - \kappa Z)) \end{aligned} \tag{12}$$

and where

$$\gamma \equiv \omega t - \kappa Z \tag{13}$$

is the electromagnetic phase. Using the results:

$$\frac{\partial \phi_X^{(2)}}{\partial Z} = -i\kappa\phi_x^{(2)}; \qquad \frac{\partial \phi_Y^{(1)}}{\partial Z} = -\kappa\phi_Y^{(1)} \tag{14}$$

the modulus of the $\boldsymbol{B}^{(3)}$ field becomes

$$|\boldsymbol{B}^{(3)}| = -2ig_0\kappa^2 \tag{15}$$

and the constant g_0

$$g_0 = i\frac{A^{(0)}}{2\kappa^2} \tag{16}$$

is a function of ϕ and ϕ^* as required.

Therefore, the $\boldsymbol{B}^{(3)}$ field can be defined in terms of lines of force, and the topological considerations of Rañada and Trueba [1] can be extended to O(3) electrodynamics, and thence to the Sachs theory.

It is also shown straightforwardly that the electric equivalent of $\boldsymbol{B}^{(3)}$, the putative $\boldsymbol{E}^{(3)}$ field, vanishes, as discussed elsewhere [8]. The demonstration uses the definition:

$$\boldsymbol{E}^{(3)} = -g(\phi^{(1)}, \phi^{(2)})\left(\frac{1}{c}\frac{\partial \phi^{(1)}}{\partial t}\frac{\partial \phi^{(2)}}{\partial Z} - \frac{1}{c}\frac{\partial \phi^{(2)}}{\partial t}\frac{\partial \phi^{(1)}}{\partial Z}\right)\boldsymbol{k} \tag{17}$$

and considers a component such as the X component of the phi field. The result is

$$\boldsymbol{E}^{(3)} = -g(\phi, \phi^*)\left(\frac{1}{c}\frac{\partial \phi_X^{(1)}}{\partial t}\frac{\partial \phi_X^{(2)}}{\partial Z} - \frac{1}{c}\frac{\partial \phi_X^{(2)}}{\partial t}\frac{\partial \phi_X^{(1)}}{\partial Z}\right)\boldsymbol{k} \tag{18}$$

and the same result is obtained by considering the Y component of the phi field.

II. SUMMARY

In this short review, we have extended the topological considerations of Rañada and Trueba [1] to O(3) electrodynamics [3] and therefore also linked these concepts to the Sachs theory reviewed elsewhere in this three-volume compilation [2]. In the same way that topology and knot theory applied to the Maxwell–Heaviside theory produce a rich structure, so does topology applied to the higher-symmetry forms of electrodynamics such as the Sachs theory and O(3) electrodynamics.

Acknowledgments

The U.S. Department of Energy gratefully acknowledged for the website http://www.ott.doe.gov/electromagnetic/, which is reserved for the Advanced Electrodynamics Working Group.

References

1. A. F. Rañada and J. L. Trueba, 2nd chapter in Part 3 of this three-volume set.
2. M. Sachs, 11th chapter in Part 1 of this compilation.
3. M. W. Evans, this volume.
4. M. W. Evans, this volume.
5. M. W. Evans and L. B. Crowell, *Classical and Quantum Electrodynamics and the* $\boldsymbol{B}^{(3)}$ *Field*, World Scientific, Singapore, 2000.
6. M. W. Evans, this volume.
7. M. W. Evans, J. P. Vigier, and S. Roy, *The Enigmatic Photon*, Kluwer, Dordrecht, 1997, Vol. 4.
8. M. W. Evans, J. P. Vigier, S. Roy, and S. Jeffers, *The Enigmatic Photon*, Kluwer, Dordrecht, 1994–1999, in five volumes.

BEYOND NONCAUSAL QUANTUM PHYSICS

J. R. CROCA

Departamento de Fisica, Faculdade de Ciências, Universidade de Lisboa, Lisboa, Portugal

CONTENTS

Modern Nonlinear Optics, Part 2, Second Edition, Advances in Chemical Physics, Volume 119,
Edited by Myron W. Evans. Series Editors I. Prigogine and Stuart A. Rice.
ISBN 0-471-38931-5

I. INTRODUCTION

The purpose of this work is to review the evidence produced thus far demonstrating that there are sound theoretical and experimental reasons to believe that it is now possible for anybody familiar with science to choose a causal epistemology to interpret natural phenomena. It is true that the noncausal interpretation, where space and time play no significative role, which was developed in the 1920s by Niels Bohr and his school, has been accepted by the scientific community as the last word, as the perfect final theory to explain the experimental data at atomic and molecular levels. However, if one asks people who use quantum mechanics in their daily work as the basic tool, just as a structural engineer uses classical mechanics, whether they believe that external reality is, ultimately, a creation of our minds, what could the answer be? Most of them, for sure, would neither believe in nor acknowledge the basic underlying statement. That the (earth's) moon or any other material entities exist only because we, the observers, "created" these entities out of a bunch of probabilities without physical meaning is far from being easy to accept. Since the advent of Greek science, we have been familiar with and accustomed to causality. This means that we believe that every natural phenomenon can, in principle, be explained as a sequence of effects evolving in space and time. This assumption has proved to be very useful to explain and predict empirical evidence for more than two millennia.

II. NONCAUSAL INTERPRETATIONS

It is not easy to follow the deep reasons that led to the noncausal paradigm, also known as the "Copenhagen interpretation of quantum physics." Some tenuous links can be established, but, of course, they are no more than that. On the other hand, it is still a hard unsolved question to know why the scientific community adopted, apparently so enthusiastically, a noncausal way of thinking even without fully understanding it. Plenty of learned works deal with these problems. In a general way, each work mainly reflects the views of its writer and so far, no agreement has been found. Although very interesting, this delicate question it is of no importance for our own purposes; we shall therefore skip it.

In the late nineteenth century, a whole set of experiments progressively lead to the conclusion that classical physics, namely, Newtonian mechanics, thermodynamics, and nascent electromagnetism, were unable to explain empirical evidence gathered by experimentalists. Scientists of that time were unable to conciliate two apparent contradictory aspects exhibited by radiation and matter. Some experiments demonstrated that light behaved like a wave, while others showed a rather corpuscular nature. On the other hand, electrons, protons, and the other massive particles would manifest wave-like properties in certain experimental conditions.

In order to fully grasp the magnitude of the problem then faced by the creators of quantum physics, let us consider the now classic double-slit experiment. It is common knowledge that if a wave coming from a nearly point source passes through a screen with two holes, two waves are generated. These two waves overlap in their course, giving birth to an interference pattern. Interferometric properties, shown in this experiment, constitute the basic proof for the undulatory properties of any entity. Now suppose that the source emits massive particles, such as electrons, neutrons, protons, and atoms. The source emits quantum particles one by one in such a way that we deal with only one single particle at a time in the experimental apparatus.

If two detectors are placed after the screen, one in front of each slit S_1 and S_2, what results from the experiment are to be expected? Since the source emits quantum particles, one by one, at any particular time there is only one particle in the apparatus; therefore one can say that sometimes one detector "sees" the particle while at other times it is the other that is activated. In either case, the two detectors are never activated at the same time; therefore no coincidences are to be expected from this experiment. In such circumstances it seems reasonable to say that sometimes the particle goes across slit 1; other times, throughout slit 2. Everybody agrees with this description, which does not seem to generate any particular trouble. Problems begin to arise when we remove the two detectors and place an array of detectors far away from the slits. After a certain elapsed time, which is necessary to detect a sufficient number of particles on the array of detectors, what shall the distribution of those quantum particles coming from the two slits be? If the former statement were true, that is, if the particles sometimes went through one slit, and another time through the other slit, a continuous distribution should be expected. Nevertheless, what experiments clearly show is an interferometric pattern distribution. This observable interferometric pattern seems to imply that the entity we call a photon, an electron, a neutron, or similar, has somehow gone through the two slits at the same time.

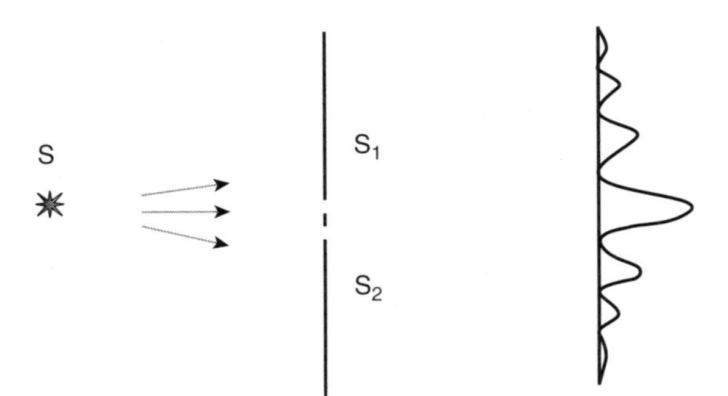

Figure 1. Double-slit experiment.

Therefore, we are facing two contradictory statements. On one hand, the entity called a "quantum particle" has gone through one slit or the other, exhibiting corpuscular properties. On the other hand, the entity called a quantum particle has gone through the two slits at the same time, exhibiting its undulatory nature. This problem was, and is even now, so important for the development of quantum physics that its solution gave birth to two main schools of thought.

Before the Solvay Congress of October 1927, basic quantum research (both theoretical and experimental) had already being done: Heisenberg had already elaborated his matrix mechanics, emphasizing the corpuscular aspect of quantum entities, and had derived the inequalities that were later publicized under his name, the so-called Heisenberg uncertainty relations. Schrödinger, inspired by the work of de Broglie on the duality particle wave, developed his wave mechanics. In his papers, he derived the evolution equation for quantum systems that, as due, got his name: the Schrödinger equation. At the Solvay Conference, it was Niels Bohr who, maybe inspired by the Heisenberg uncertainty relations and by the nonlocal Fourier analysis, was able to present a consistent global picture capable of integrating into a coherent whole the apparent contradictory behavior and properties of quantum entities. It is true that Bohr was able to make a consistent general synthesis to resolve the contradictions, but it is no less true that the vision he proposed was very far from the causal way of thinking that we were used to, from the golden Greek times. In his proposal, he denied the existence of an objective reality and stated that the usual concepts of space and time were mere auxiliary tools, therefore loosing the relevance they have played in the causal paradigm where every phenomenon had a past history and a future. In this sense, the Copenhagen interpretation of quantum physics radically breaks away from causality. We must insist that this attitude was in deep contradiction with the relativity paradigm developed by Einstein, where spacetime played the primordial role.

III. CAUSAL MODELS

Even if the great majority of the scientific community accepted the Copenhagen interpretation of quantum physics, some of the most important physicists of the twentieth century, such as Einstein, de Broglie, Max Planck, Schrödinger, and many others, never adhered to "indeterminism." Throughout their lifetimes, they always fought one way or another the Copenhagen paradigm. In the beginning, they limited themselves to the so-called quantum paradoxes from which the well-known Schrödinger cat is probably the most famous. The construction of a causal model able to give a good account of quantum phenomena was theoretically blocked by von Newman's theorem. Only after the important work of David Bohm [1], who proved in 1952 that this mathematical theorem was, in fact, no so general as usually claimed, was the path for building causal theories finally cleared. The proof was obtained in a very interesting way: David Bohm built a causal model precisely achieving what the theorem of von Newman forbade!

From that time on, many causal theories were developed. In this work, we shall only refer to the causal theory proposed by de Broglie [2] and known as the "double-solution theory." This theory, as well as Bohm's theory, are the most developed of all causal theories; they are both able to explain and predict, practically all quantum phenomena.

A. de Broglie Causal Theory

The starting point of de Broglie's theory is the belief that the reality is observer-independent even if the observer interacts and therefore modifies in greater or lesser degree the external reality. Therefore in this model it is assumed that the matter waves ϕ are real physical waves different from the common statistical wave Ψ, fictive and arbitrarily normalized. This real wave ϕ is composed of an extended, yet finite wave θ, plus a singularity ξ, such that

$$\phi = \theta + \xi \tag{1}$$

is solution of a nonlinear equation. This singularity represents a relatively large concentration of energy, is very well localized in space, and is responsible for the usual detection process. Since the accompanying wave θ has very small amplitude, it turns out to be unable, by itself alone, to trigger a normal quadratic detector. The singularity ξ is coupled to the extended wave θ, by a nonlinear chaotic process such that even if its presence on the wave cannot be predicted precisely, it nevertheless has the tendency to be localized on the points where the intensity of accompanying θ is greater. The particle is graphically represented by the sketch in Fig. 2.

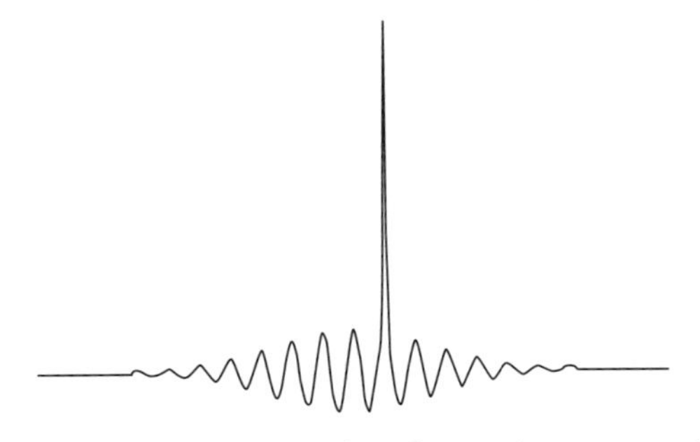

Figure 2. Graphic representation of a causal quantum particle.

In the linear approximation, which was the version presented by de Broglie, equation (1) reads as

$$\phi = v + \xi \tag{2}$$

where ϕ is the solution of a nonlinear equation and v is also a real physical wave, but now, in this linear approximation, is the solution of the usual Schrödinger equation. Since v represents a real wave, it cannot be normalized at will just like the usual statistical wave Ψ that can be, arbitrarily normalized. Nevertheless, the two waves are solutions of the same linear evolution equation; therefore they must be somehow related. This relationship is given by a constant C such that

$$\Psi = Cv \tag{3}$$

The physical extended wave is related to the localized singularity by the guiding principle

$$\vec{p} = -\nabla\varphi \tag{4}$$

where φ represents the phase of the wave. This principle states that the wave, practically devoid of energy, guides preferentially the singularity, through a nonlinear process, to the points of higher wave intensity.

Let us now see how this causal model explains the double-slit experiment. In this model, since any quantum particle is composed by a wave with the singularity, it follows that when the particle encounters the screen with the two slits, the physical real wave, being widespread, is able to cross the two slits simultaneously, while the singularity passes through one slit only. The two real partial waves, coming from the two slits, one with the singularity, and the other without it, superimpose themselves in their natural course, giving birth to an interferometric pattern. The singularity according to the guiding principle goes preferentially to the points where the intensity of the wave is higher. So after a sufficient number of particle hits, an interference pattern progressively appears in the detection region. This situation is shown in Fig. 3.

Therefore, in this way, by devising a more ingenious model of the quantum particle, the double-slit experiments can be explained in a very simple and elegant manner with no need to reject the objective reality of quantum beings. The apparent contradiction resulting from the requirement, for one particle to cross the two slits at the same time, while also being able to cross either one or the other, is easily surpassed by the causal model. The wave crosses the two slits at the same time while the singularity goes through either one or the other.

As discussed previously, Bohr and his followers chose to reject objective reality to explain the same experiment, instead of looking for a subtler model of

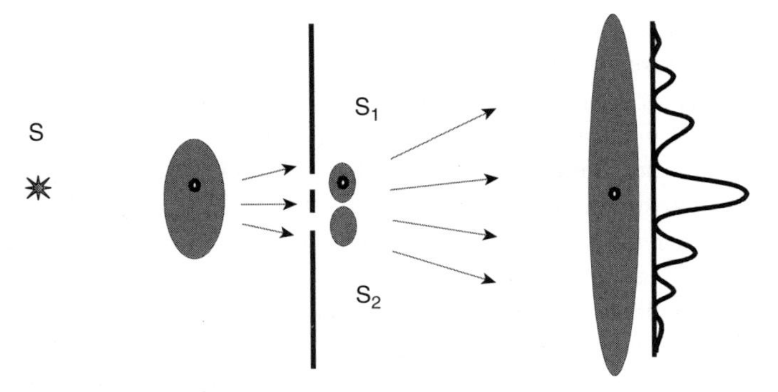

Figure 3. Double-slit experiment in de Broglie causal picture.

the particle. In their picture, the particle having no real existence crosses potentially the two slits and also interferes potentially at the overlapping region. If two detectors are each placed in front of each slit, then one of the potentialities or probabilities materializes itself at one of the detectors.

Since, in this causal model, the extended wave θ represents a real physical finite wave with well-defined energy, it seems natural to represent it by a suitable mathematical form. At the time when de Broglie put forth his causal interpretation of quantum mechanics, it was necessary for him to construct a finite wave using the Fourier analysis, namely, the multiplicity of harmonic plane waves, infinite in space and time, summing up and giving origin to a wavepacket.

$$\psi(\xi) = \int_{-\infty}^{+\infty} g(\eta)\, e^{i\eta\xi}\, d\eta \tag{5}$$

This means that the localized entity, the wavepacket, resulting from the infinite sum of waves also has an infinite number of energies or velocities. Each infinite harmonic plane wave composing the packet has a proper energy and therefore its own velocity. With this mathematical description, which implies basically a nonlocal character in both time and space (since the wavepacket in reality is composed of on infinitude of waves), what shall be the energy of the single particle? To answer this question, the Copenhagen School assumed that the particle had no real energy at all. The particle does not even exist prior to the measurement; it is only an ensemble of potentialities or probabilities that could eventually be put into existence. Everything we can say, using this paradigm, is that when a measurement is performed, one out of this infinite number of possible energies materializes, or collapses, into one single energy. In a causal picture it is possible to say that the wavepacket represents only a statistical distribution of probabilities of real particles. But in such case each particle needs to be represented mathematically by a wave, infinite in time and space. This solution is, of course, not a good one, since we know that the real physical particles are not infinite. Indeed, we know, on the contrary, that they are finite, with a precisely definite energy, and localized in both space and time.

Now, thanks to the development of the local wavelet analysis [3], it is possible to represent mathematically, according to the observations, finite waves with a well-defined energy. Therefore it is possible to represent the θ wave, the extended yet localized part of the particle, with a defined energy by a wavelet. The wavelets, or finite waves, were developed in geophysics by Morlet in the early 1980s, to avoid some shortcomings of the nonlocal Fourier analysis.

It seems quite natural to describe the extended part of a quantum particle not by wavepackets composed of infinite harmonic plane waves but instead by finite waves of a well-defined frequency. To a person used to the Fourier analysis, this assumption—that it is possible to have a finite wave with a well-defined frequency—may seem absurd. We are so familiar with the Fourier analysis that when we think about a finite pulse, we immediately try to decompose, to analyze it into the so-called pure frequencies of the harmonic plane waves. Still, in nature no one has ever seen a device able to produce harmonic plane waves. Indeed, this concept would imply real physical devices existing forever with no beginning or end. In this case it would be necessary to have a perfect circle with an endless constant motion whose projection of a point on the centered axis gives origin to the sine or cosine harmonic function. This would mean that we should return to the Ptolemaic model for the Havens, where the heavenly bodies localized on the perfect crystal balls turning in constant circular motion existed from continuously playing the eternal and ethereal harmonic music of the spheres.

These harmonic plane waves, which supposedly existed over the full spectrum of time, whether past, present, or future, and also in the whole gamut of infinite space, have, of course, no real physical existence. They are mere abstract entities, existing only in our minds. It would be necessary to dispose of the whole infinity of space and time in order to produce them. This logical requirement certainly appears impossible to achieve as a goal since practical real physical devices are always finite in space and time. Now, the time is ripe for improving our description of nature with more appropriate tools.

As common observation tells us, in nature we have devices that produce, in finite time and finite space, waves with a fairly well-defined frequency. For instance, when the hammer strikes the piano chord, a wave of fairly well-defined frequency is produced. Furthermore, this real physical sound wave has, as we very well know, a beginning and also an end. So why not describe this sound, produced by the piano, by a wave, with a beginning and an end, a finite wave, with a well-defined frequency. For what physical reason one has to say that this finite wave, of a well-defined frequency, is composed of a infinity of frequencies, each corresponding to harmonic plane waves having no real physical existence? Even if it is true that, in both the past and the present, these abstract entities have been quite helpful, mainly because of their conceptual and practical simplicity, the time has come to improve our description of nature with more appropriate tools. Before the development of the wavelet local analysis, we had to describe mathematically these finite physical waves, as a sum of infinite waves deprived of physical reality. Now, since we handle a more appropriately tuned mathematical tool enabling us to avoid the nonlocal epistemological problems arising from the Fourier analysis, we are free to use finite waves, with a well-definite frequency, to represent the real physical waves.

From the basic concept of finite localized waves, inspired by the ideas originating from de Broglie, one concludes that a good way to represent the quantum particle would be to use wavelets. Under these conditions, it seems quite natural to describe the single wave not by a sum of infinite harmonic plane waves, without physical reality, as done in nonlocal Fourier analysis, but instead with the help of local finite wavelets. One nonminor advantage of this choice is that it allows one to get rid of certain inconveniences arising in the usual quantum mechanics, such as instantaneous actions at a distance, retroaction in time, and other strange effects, delighting, for sure, all those enamored with wilderness and weirdness.

Under such conditions it seems reasonable to describe the extended part of a quantum particle by a generalized Gaussian Morlet wavelet [4]

$$\theta(x,t) = A\, e^{-\beta^2(\xi_1-\varepsilon_1')^2 + i(\xi_2-\varepsilon_2')} \tag{6}$$

which, as we shall see, is solution to a nonlinear Schrödinger equation. The symbols in the formula have the following meaning:

$$\xi_1 = \frac{1}{\hbar}(px - 2Et), \quad \xi_2 = \frac{1}{\hbar}(px - Et), \quad \varepsilon_1' = \frac{\varepsilon_1}{\hbar}, \quad \varepsilon_2' = \frac{\varepsilon_2}{\hbar} \tag{7}$$

B. Nonlinear Quantum Mechanics

The usual quantum mechanics, just like the most important physical theories of our time, is a linear theory. Still, we have the feeling that these linear theories, even if they have helped us to understand and predict an astonishing quantity of phenomena, must correspond in reality to some sort of statistical approximation of deeper and more general nonlinear theories. It would be logical to assume that these nonlinear theories must generate the usual linear theories as a particular case. Natural phenomena are certainly very complex. Therefore, the deeper one goes in the level of description of nature, the farther one stands from linear behavior. The well-known rule that the whole is equal to the sum of the individual parts is indeed a very good rule; above all, it is relatively easy to apply. Nevertheless, it has some shortcomings; for example, it holds true for only a very limited number of single real situations or when large statistical simplifications and assumptions are made. In everyday life, in the majority of situations we face, this simplistic linear rule fails to apply because we know too much about the problems to accept oversimplistic linear approximations. This necessity was pointed out by Einstein, de Broglie, and many others in their works. They always believed that a more complete theory, for describing the quantum phenomena, must be a nonlinear theory that should, of course, have as a linear approximation the usual linear quantum theory.

Even if de Broglie never proposed a nonlinear master equation for the quantum mechanics, he always thought that quantum mechanics needs, for a more complete description, a nonlinear approach, and his double solution theory is, as we have seen, a good example of it. Here we shall propose a possible way to tackle this problem by means of an example. Let us select

$$-\frac{\hbar^2}{2m}\nabla^2\psi + \frac{\hbar^2}{2m}\frac{\nabla^2|\psi|}{|\psi|}\psi + V\psi = i\hbar\frac{\partial\psi}{\partial t} \tag{8}$$

for the master nonlinear wave equation, for the case of spinless particle and in the nonrelativistic approximation [5]. Versions of this equation have already been presented in the scientific literature by some authors, including Gueret and Vigier [6]. It is easy to see that this equation corresponds to the usual Schrödinger equation plus the so-called quantum potential

$$Q(\psi) = \frac{\hbar^2}{2m}\frac{\nabla^2|\psi|}{|\psi|} \tag{9}$$

Writing the solution of the nonlinear master equation in the general form

$$\psi(\vec{r}, t) = a(\vec{r}, t)e^{(i/\hbar)\varphi(\vec{r}, t)} \tag{10}$$

one gets, after some calculations and separating the real and the imaginary parts

$$\begin{cases} \dfrac{1}{2m}(\nabla\varphi)^2 + V = -\dfrac{\partial\varphi}{\partial t} \\ \dfrac{1}{m}\nabla(a^2\nabla\varphi) + \dfrac{\partial a^2}{\partial t} = 0 \end{cases} \tag{11}$$

where, if we identify the classical action S with the phase φ of the wave, the first equation represents the classical Hamilton–Jacobi equation and the second, the continuity equation. Using the usual known relations

$$\nabla\varphi = \vec{p}; \quad \frac{\partial\varphi}{\partial t}; \quad \rho = a^2; \quad \vec{J} = \rho\vec{v}; \quad \vec{v} = \frac{\nabla\varphi}{m}$$

it is possible to write

$$\begin{cases} T + V = E \\ \nabla\vec{J} + \dfrac{\partial\rho}{\partial t} = 0 \end{cases} \tag{12}$$

which shows that the nonlinear equation corresponds, in a certain sense, to two classical equations: the equation of conservation of energy plus the fluid continuity equation. This choice of (8) for the non-linear master equation, for describing the quantum phenomena, seems quite natural since it has the corresponding classical analog in the two fundamental balance equations of the classical physics: one for the particles and the other for the fluids. By a symmetric procedure we can also say that the proposed nonlinear equation for describing the quantum reality results from the fusion of the two fundamental equations of the classical mechanics: again, one for the particles the other for the fluids. In this perspective, it is worth mentioning here that the usual linear Schrödinger equation corresponds to adding the nonlocal quantum potential to the Hamilton–Jacobi equation.

One basic difficulty with the nonlinear equation arises from the following. Consider a physical situation where a source of particles is composed of many emitters, each emitting a particle at a time. If considered alone, each particle would be described by a localized wave ψ_i solution of the master equation. Now, what happens if, instead of emitting the particles one by one, the source emits many particles at the same time? If the master equation were a linear equation, like the usual Schrödinger equation, the answer would be trivial. The general solution would be simply the sum of all particular solutions.

$$\psi = \psi_1 + \psi_2 + \cdots + \psi_n$$

In a nonlinear framework, the general solution must factor in the relative interactions among the different particles, and, consequently, the composition rule, which is generally unknown:

$$\psi = \psi(\psi_1, \psi_2, \ldots, \psi_n)$$

Attempts have been made to find a general composition rule for this equation, and some interesting results have even be derived by Rica da Silva. It is relatively easy to show that in some particular cases, for some solutions, this composition rule for a large number of solutions transforms into the linear sum rule. For the one-dimensional case and in free space, $V = 0$, a particular solution of the nonlinear equation can be written in the form

$$\psi = A e^{-\beta^2(\xi_1 - \varepsilon_1')^2 + i(\xi_2 - \varepsilon_2')} \tag{13}$$

which, as we have already seen, is the Gaussian Morlet wavelet. Recalling the meaning of the symbols

$$\xi = \frac{1}{\hbar}(px - 2Et), \quad \xi_2 = \frac{1}{\hbar}(px - Et), \quad \varepsilon_1' = \frac{\varepsilon_1}{\hbar}, \quad \varepsilon_2' = \frac{\varepsilon_2}{\hbar} \tag{14}$$

one gets, by substitution, the following generic solution

$$\psi(x,t) = A \exp\left[-\frac{\beta^2}{\hbar^2}(px - 2Et - \varepsilon_1)^2 + \frac{i}{\hbar}(px - Et - \varepsilon_2)\right] \tag{15}$$

For $\varepsilon_2 = 0$, and remembering that for null potential $E = T = p^2/2m$, a continuous sum of these solutions can be written as

$$\psi = \iint_D g(p,\varepsilon) \exp\left[-\frac{\beta^2}{\hbar^2}\left(px - \frac{p^2}{m}t - \varepsilon\right)^2 + \frac{i}{\hbar}\left(px - \frac{p^2}{2m}t\right)\right] dp\, d\varepsilon \tag{16}$$

where D is the domain of the integration variables. In order to proceed with the integration, it is necessary give an explicit form for the coefficient function. Assuming that the beam is practically monochromatic, and that the translation parameter follows a Gaussian variation

$$g(p,\varepsilon) = \delta(p) e^{-\alpha^2\varepsilon^2} \tag{17}$$

where $\delta(p)$ stands for the Dirac delta function, extending the integration domain to plus and minus infinity, one gets

$$\psi(x,t) = \sqrt{\frac{\pi}{\alpha^2 + \beta^2}} \exp\left[-\frac{\alpha^2}{\alpha^2 + \beta^2}\frac{\beta^2}{\hbar^2}(px - 2Et)^2 + \frac{i}{\hbar}(px - Et)\right] \tag{18}$$

This final formula, resulting from the coherent infinite sum of solutions, representing the undulatory part of a finite wave of de Broglie, is also a solution of the master nonlinear equation.

In order to investigate the solutions of the nonlinear wave equation (8) in a more extensive manner, it is useful to look at the nonlinear term. The quantum potential for the solution expressed in terms of its amplitude and phase, Eq. (10), assumes the form

$$Q(\psi) = \frac{\hbar^2}{2m}\frac{\nabla^2 a}{a}$$

from which we gather that if the amplitude of the wave is such that

$$\nabla^2 a = fa \tag{19}$$

Then the nonlinear term becomes proportional to a constant f. In such situations the nonlinear equation transforms into a linear equation:

$$-\frac{\hbar^2}{2m}\nabla^2\psi + \left(\frac{\hbar}{2m}f + V\right)\psi = i\hbar\frac{\partial\psi}{\partial t} \tag{20}$$

Looking at this equation, one sees that in free space, where $V = 0$, the nonlinear master equation transforms into the usual linear Schrödinger equation with a nonnull potential:

$$-\frac{\hbar^2}{2m}\nabla^2\psi + \frac{\hbar^2}{2m}f\psi = i\hbar\frac{\partial\psi}{\partial t} \tag{21}$$

Another possible approach to obtain some information on the solutions of the nonlinear equation, for a free particle, consists in rewriting Eqs. (11) for the one-dimensional case:

$$\begin{cases} \dfrac{1}{2m}\varphi_x^2 = -\varphi_t \\ \dfrac{1}{2m}\left[2\dfrac{a_x}{a}\varphi_x + \varphi_{xx}\right] = -\dfrac{a_t}{a} \end{cases} \tag{22}$$

Assuming that the phase of the wave is of the form

$$\varphi = px - Et - \varepsilon_2 \tag{23}$$

one gets by substitution in (22)

$$\frac{\partial a}{\partial x} + \frac{1}{v_g}\frac{\partial a}{\partial t} = 0 \tag{24}$$

where we have made. $v_g = p/m$. The general solution of this equation is

$$a = a(x - v_g t + \varepsilon_1) \tag{25}$$

Under such conditions, for that particular form of the phase, given by (23), it is possible to write a set of solutions for the nonlinear master equation

$$\psi(x,t) = a\left(x - \frac{p}{m}t - \varepsilon_1\right)e^{(i/\hbar)(px - Et - \varepsilon_2)} \tag{26}$$

IV. EXPERIMENTS TO TEST THE NATURE OF THE QUANTUM WAVES

Since the advent of quantum physics, the very nature of quantum waves has been source and subject of controversy. As we have seen, two major interpretations fuelled this debate. The Copenhagen indeterministic school claims that quantum waves are mere probability waves devoid of any physical correspondence. On the contrary, the causal current of de Broglie sustains that quantum waves are real waves. Since both interpretations are able, to a greater or lesser extent, to explain quantum observations, how should we decide which of these is the "good" interpretation to follow? Situations of this kind are not new in the history of science. It is well known that when, Boltzmann, Gibbs, Maxwell, and others were developing statistical physics, the standard accepted theory was thermodynamics. Even if the statistical physics were able to explain practically all heat experiments in terms of interactions among atoms or molecules, many prominent scientists of that time insisted that such a theory was totally useless and devoid of any meaning since it assumed as a basic premise the real existence of the atoms. Only after the experiments of Perrin, which were related to Brownian motion, and after the subsequent development of a newborn "quantum" physics, could statistical physics apply for a the right of citizenship in the family of other, accepted theories. Therefore, what we need, to make the choice between causality and indeterminism, is crucial feasible experiments. For a long period of time, this necessity could not be satisfied. Nevertheless, since the mid-1980s, thanks to the school of Lisbon, and under the influence of Andrade e Silva, former student of de Broglie, and the school of Bari under the direction of Selleri, the first experimental proposals were made [7]. These early proposals were only conceptual; nevertheless, they soon gave origin to proposals of concrete feasible experiments [8]. These experiments can be grouped in to two main families: those based on the collapse of the wavefunction in a measurement, and the ones using the so-called autoreduction of the wavefunction.

A. Experiments Based on Collapse of the Wavefunction

The fundamental idea behind this series of experiments is related to the collapse of the wavefunction when the measurement is performed. The orthodox interpretation of quantum mechanics assumes that the wavefunction is a probability wave devoid of any physical meaning. As a consequence, the quantum measurement implies the collapse of the multiple probabilities into a single one. In the causal interpretation of quantum mechanics, there cannot be a collapse because we always deal with a function representing a real entity.

In order to fully understand the meaning and physical implications of these different assumptions, let us consider the experimental setup depicted in Fig. 4, where S represents a source of particles emitting quantum particles, one at a time. This requirement is of primary concern; without sources of this kind, real experiments would be meaningless. Still, this requirement poses no problem today. For neutrons, electrons, and atoms, it is relatively easy to design and build such sources. However, for photons, this requirement generates many problems since photons are bosons and thus tend to join together, with bunching effect. Monophotonic sources were designed and built mainly to satisfy the requirements of experiments related to the EPR (Einstein, Podolsky, and Rosen) paradox. A lot of work was devoted to the design and construction of such sources [9]. Probably the simplest and most reliable are those of the type developed by Mandel.

The single particle, emitted by the source, is directed toward a beamsplitter. This quantum system is described, in the orthodox theory, by probability wave ψ containing all the available information on the system.

On reaching the beamsplitter, the wave ψ is divided into two wavepackets: one is reflected (ψ_r) and the other is transmitted (ψ_t), and the two trajectories are independent. If a detector placed in the path of the wavepacket ψ_r is triggered, it switches on a light that is seen by the observer. Then, what can be concluded about the other wavepacket ψ_t that has been transmitted?

The answer to this question depends on the chosen, underlying theory. According to the Copenhagen interpretation of quantum physics, the wavefunc-

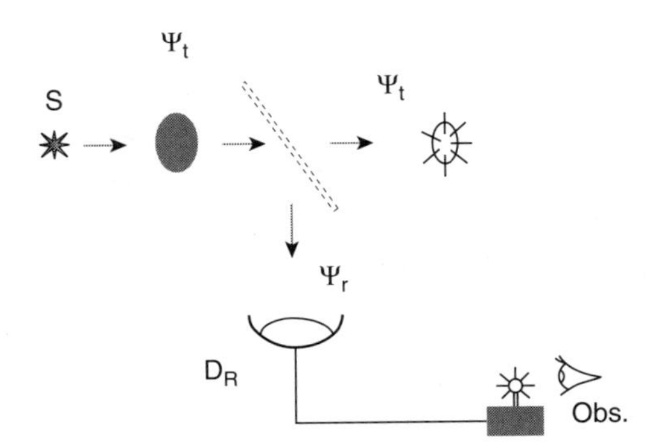

Figure. 4. A single quantum system emitted by the monoparticle source is split into two at the beamsplitter. When the detector is triggered, the observer sees the light.

tion is no more than a mere mathematical probability whose sole function is to help predict the results of the measurements. Therefore, if the particle is seen at the detector, since the light is on, the probability of being at the other path turns instantly to zero, that is, $\psi_t = 0$. This is the well-known reduction postulate of the usual quantum mechanics, also called "collapse" of the initial wavefunction into one of the multiple possible states. Briefly stated, and according to the usual interpretation, after the particle has been detected at the detector, the probability of being in the transmitted path turns to zero and consequently, nothing remains up there.

The answer to this question put forth by proponents of the causal model is fairly different. The source emits a quantum particle one at a time, meaning that a wave plus a singularity $\phi = \psi$, move in the direction of the beamsplitter. Here the physical wave is divided into two, while the singularity is reflected or transmitted. When reflected, the case shown in Fig. 4, in its way the singularity interacts with the detector and triggers it, turning the light on. In the meantime the real wave $\psi_t = \theta$ continus on its path and remains on unaffected because no physical action is taking place on it.

As could be expected, both models give different answers to the question. The orthodox model claims that, after the measurement is performed, nothing remains in the transmission path. The other, the causal approach, predicts that along that way follows a real physical wave carrying a very small amount of energy. In such circumstances, it looks as if we had obtained the situation we were looking for, namely, a practical feasible experiment capable of testing the two competing models for the quantum physics, and chose between them.

This situation can be fully developed by slightly improving the previous experiment. Consider that instead of being connected to the lightbulb, the detector is linked to a fast shutter, as shown in Fig. 5.

When activated by the reflected singularity, the detector sends a pulse opening the fast gate G for a short time necessary enough to let the theta wave pass. If it happens that instead of being reflected, at the beamsplitter, the singularity is transmitted, then at the detector arrives only a wave without singularity unable to trigger the detector. Since this wave, devoid of singularity, has not enough energy to trigger the detector, no pulse is sent to open the gate. So, in this case, the gate remains closed and as a consequence nothing leaves the device. If the preceding causal assumptions hold, this device is no more than a generator of real physical θ waves carrying a very small amount of energy. This energy is so small that it is not sufficient to trigger the usual quadratic sensors and detectors.

Naturally, in terms of the Copenhagen paradigm, nothing at all comes from such a device. In this model the theta waves do not even exist.

The problem now is to find some kind of "special detector" capable of revealing waves having a small amount of energy. Two techniques have been

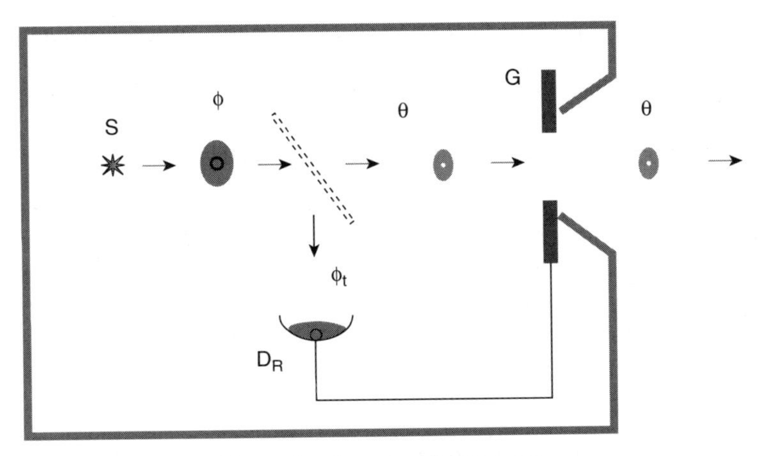

Figure 5. Generator of theta waves.

discussed in the scientific literature to test the very existence of these waves carrying almost no energy at all. The first one is related to de Broglie guiding principle; the other is based on some different possible properties of theta waves.

1. Direct Detection

As already stated, the energy of the waves theta devoid of singularity does not reach the threshold to trigger usual quadratic detectors. Therefore direct detection is impossible with them. In 1983 [10], Selleri proposed a very promising possibility for the case of photonic waves. The method for detection of the theta waves is based on the idea that these waves modify the decay probability of unstable systems. These waves, devoid of singularity and practically without energy, emitted by the generator, are injected into a laser gain tube, as (shown in Fig. 6), where they have the possibility of revealing their real existence by generating a zero energy-transfer stimulated emission. A positive answer from this experiment would prove the existence of some real physical entity emerging from the generator of theta waves. The principal inconvenience of this

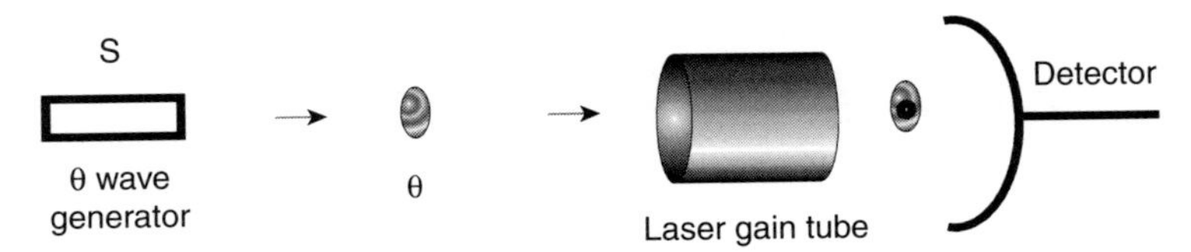

Figure 6. Selleri's experiment.

experiment is that if the answer is no, that is, if no photons are revealed (besides the usual noise), no conclusion can be drawn concerning the existence or nonexistence of waves devoid of singularity.

2. *Experiments Based on de Broglie Guiding Principle*

According to the guiding principle of de Broglie, the single singularity is coupled to the wave theta through a nonlinear process such that its probability distribution is given by the intensity of the accompanying wave. Naturally this wave, in the linear approximation, is the sum of all theta waves that happen to be in the same region of space at that particular time:

$$\phi = \theta_1 + \theta_2 + \cdots + \theta_n \tag{27}$$

It must be pointed out that this formula holds true even when the waves come from independent sources, as shown either by photonic interference with independent sources, or by fourth-order interference. As known, fourth-order interference is observed with incoherent wave overlapping resulting from independent sources. Under these conditions the probability distribution for the detection of singularities, in a long run of similar procedures, is given by the guiding principle

$$P = \alpha|\phi|^2 \tag{28}$$

where the multiplicative constant alpha depends on the detection process.

From expression (28) and recalling the generator of theta waves, a conceptually simple process for detecting such waves (which are not seen directly by a common detector) can be easily devised. Consider a monoparticle source of photons, or any other particle, where this principle is valid for any quantum particle, emitting photons on a single basis, that is, one by one. This particle in its way reaches an array of detectors giving origin to a Gaussian continuous distribution of arrivals. Suppose now that at the same region of space, the wave with the particle is joined by a theta wave, emitted by the special generator, as shown in Fig. 7.

The two waves—one ϕ, from the common source S the other, θ from the generator of theta waves S′—overlap at the detection region. The expected intensity at the array of detectors D_R, after n arrivals of particles, is given by the squared modulus of the superposition of two waves at each instant of time, summed for all n arrivals:

$$I = \sum_{i=1}^{n} |\phi_{t_i} + \theta_{t_i}|^2 \tag{29}$$

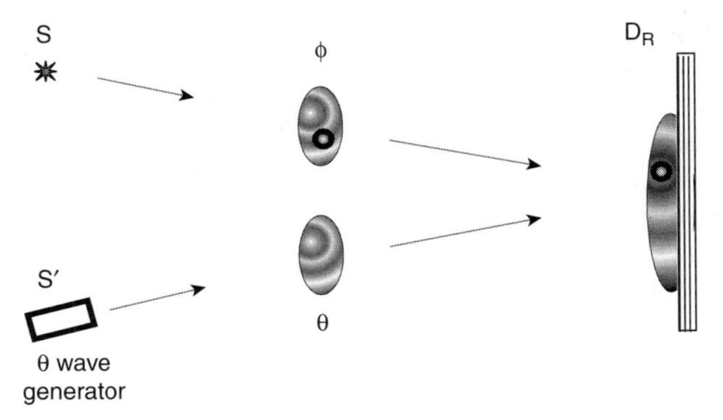

Figure 7. Principle for detection of the waves theta practically without energy.

Developing, one gets

$$I = |\phi_{t_1}|^2 + |\theta_{t_1}|^2 + 2|\phi_{t_1}||\theta_{t_1}|\cos\delta_{t_1} + |\phi_{t_2}|^2 + |\theta_{t_2}|^2 + 2|\phi_{t_2}||\theta_{t_2}|\cos\delta_{t_2} + \cdots + |\phi_{t_n}|^2 + |\theta_{t_n}|^2 + 2|\phi_{t_n}||\theta_{t_n}|\cos\delta_{tn}$$

or, assuming that the intensity of the individual waves are equal

$$I_0 = |\phi_{t_i}|^2 = |\theta_{t_i}|^2 \tag{30}$$

one gets by substitution

$$I = 2I_0(n + \cos\delta_{t_1} + \cos\delta_{t_2} + \cdots + \cos\delta_{t_n}) \tag{31}$$

If the two independent sources S and S′ are coherent, the relative phase difference is constant in time

$$\delta = \delta_{t_i} \tag{32}$$

the expected intensity is given by

$$I = 2nI_0(1 + \cos\delta) \tag{33}$$

This result indicates that the existence theta wave is made evident by the presence of the interference pattern. In the usual Copenhagen interpretation, the waves theta do not exist and therefore the expected result would be simple the one produced by the single source

$$I = nI_0 \tag{34}$$

which shows no interference pattern.

Calculations were performed assuming that the two independent sources are coherent, and emit the pulses at the same time precisely (this synchronicity requirement is, of course, extremely important). These experimental requirements are very difficult to put into practice. Indeed, two independent sources are rarely coherent, even if the emission is at the same time. In such conditions, the relative phase differences are randomly distributed and so their sum cancels out. As a consequence, no interference is to be expected at the detection region. In this case the experiment is not conclusive, either.

Garrucio, et al. [11] devised an experimental proposal to meet the coherence requirements. Their idea was based on the hypothesis that it is the singularity that stimulates the laser emission. This hypothesis is opposite that of Selleri. Nevertheless, because of noise and related problems, the experiment may be not entirely conceptually conclusive, either.

Since the direct and the coherent processes are not truly conclusive, it is necessary to look for another conceptually clear method that could, in principle, avoid the stated experimental difficulties. A conclusive experiment to decide on such important issue as the true nature of the quantum waves, regardless of whether they are real, must be conceptually very simple and, above all considerations, conclusive in principle. This process is known as "incoherent interferometric detection of the theta waves."

In fact, incoherent detection was the first conceptual process developed by J. and M. Andrade e Silva [7] in the early 1980s. The basic idea relates to a variant of the Young double-slit experiment where a θ wave produced by an independent incoherent source is mixed to the usual two coherent waves, producing a blurring of the interference pattern. A sketch of this conceptual three-slit experiment is shown in Fig. 8.

Those early conceptual proposals were developed later [12] into concrete feasible experiments extended to all quantum particles. These experiments are

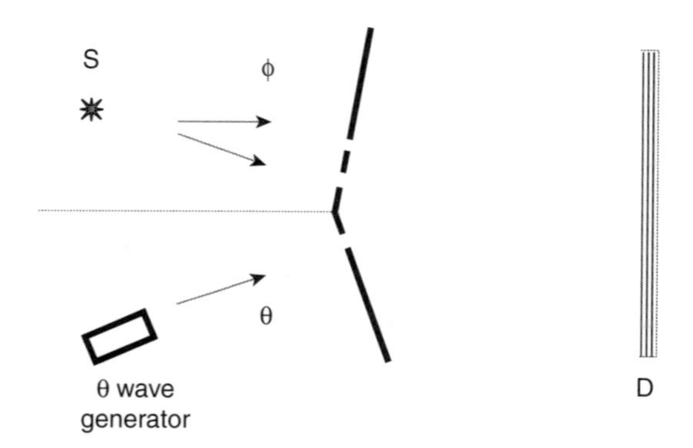

Figure 8. Three-slit experiment for to test the reality of the quantum waves.

conceptually variants of the same idea, namely, mixing the θ waves, produced by the special generator, coherent or not, with the usual waves.

For the case indicated in Fig. 8 the expected intensity predicted at the detector, assuming that the quantum waves are real, is

$$I_c \propto |\phi_1 + \phi_2 + \theta|^2 \tag{35}$$

since the waves ϕ are coherent between themselves and incoherent with the wave θ, the last expression gives

$$I_c \propto |\phi_1 + \phi_2|^2 + |\theta|^2 \tag{36}$$

Assuming, for sake of simplification of the final formula, that the waves have the same amplitude, one has

$$I_c \propto \left(1 + \frac{2}{3}\cos\delta\right) \tag{37}$$

where δ stands for the relative phase shift between the two coherent waves ϕ_1 and ϕ_2. The fringe visibility calculated by the known formula

$$V = \frac{I_M - I_m}{I_M + I_m} \tag{38}$$

gives $V = \frac{2}{3}$.

If it is assumed, as the usual paradigm does, that the quantum waves have no real existence, then the predicted intensity at the detection zone is

$$I_u \propto |\phi_1 + \phi_2|^2 \tag{39}$$

or

$$I_u \propto (1 + \cos\delta) \tag{40}$$

which corresponds to a visibility of one $V = 1$.

These results indicate that the presence of the wave theta is revealed by the value of the expected visibility. If the visibility is one, the θ waves do not exist, meaning that the quantum waves are mere mathematical probability waves devoid of any physical meaning. If the fringe pattern is blurred and the visibility decreased by a factor of $\frac{2}{3}$, then it would imply that quantum waves, just like any ordinary wave, are real.

Another possibility, a variant of the previous experiment, is shown in Fig. 9, where the wave θ from the generator is halved at the beamsplitter and the two

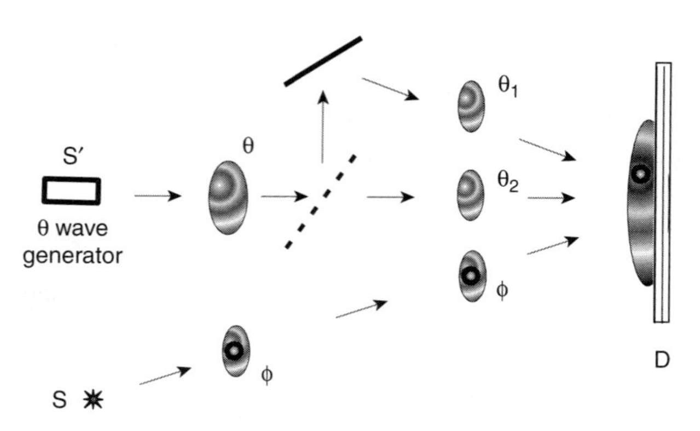

Figure 9. Incoherent detection of θ waves.

coherent waves θ_1 and θ_2 devoid of singularity overlap with the common wave ϕ at the array of detectors. Since the sources are independent, the full wave is incoherent with the waves θ_1 and θ_2 devoid of singularity. In such a scenario in the region of detection, three waves overlap, giving birth to a resulting wave

$$\phi_T = \theta_1 + \theta_2 + \phi \tag{41}$$

guiding the single singularity according to the guiding principle. The expected intensity is given by

$$I_c \propto |\theta_1 + \theta_2 + \phi|^2 \tag{42}$$

or

$$I_c \propto |\theta_1 + \theta_2|^2 + |\phi|^2 \tag{43}$$

because the two theta waves are coherent and incoherent, respectively, to the full wave. Assuming, as before, the same amplitude for the three overlapping waves, one gets the same expression (37) for the expected intensity corresponding to the same visibility, $V = \frac{2}{3}$.

A different result is predicted by the Copenhagen school for this specific experiment. Since the theta waves do not exist, the expected intensity is given simply by

$$I_c \propto |\phi|^2 \tag{44}$$

which means that in this case, since there is no interference term, no interference pattern is to be either expected or observed, $V = 0$.

This experiment is, to a certain extent, the "symmetric" counterpart of the one we discussed earlier. In the first experiment, the theta waves were supposed to show their presence by blurring a clear interference pattern, while in this last experiment, the waves produce an interference pattern where no waves were expected.

Here is the possibility of a concrete, feasible experiment. This experiment acknowledges that in an ideal experiment, all waves must completely overlap in the detection zone. Of course, in a real experiment, things are slightly different [13]. Since the sources are independent, perfect synchronization is difficult to obtain because the relative length, the coherence length, of the waves is small. In general, complete and permanent overlapping is impossible to achieve. The experiment proposed in Fig. 10 is designed to address those problems.

As shown in Fig. 10, two sources, one a common source and the other consisting of θ waves, feed the input ports of a Mach–Zehnder interferometer. The intensity predicted by the usual theory for the two output ports results only from the input source S, and is given by

$$\begin{cases} I_u^1 \propto |\phi_{TR} + \phi_{RT}|^2 \\ I_u^2 \propto |\phi_{RR} + \phi_{TT}|^2 \end{cases} \tag{45}$$

where ϕ_{TR} represents the wave reflected at the first beamsplitter and transmitted at the second. If the relative phase shift of the two coherent waves is δ_ϕ and for the ideal case of no absorption and 50% beamsplitters

$$|\phi_{TR}|^2 = |\phi_{RT}|^2 = |\phi_{RR}|^2 = |\phi_{TT}|^2 = \frac{1}{4}|\phi|^2, \quad I_0 = b|\phi|^2 \tag{46}$$

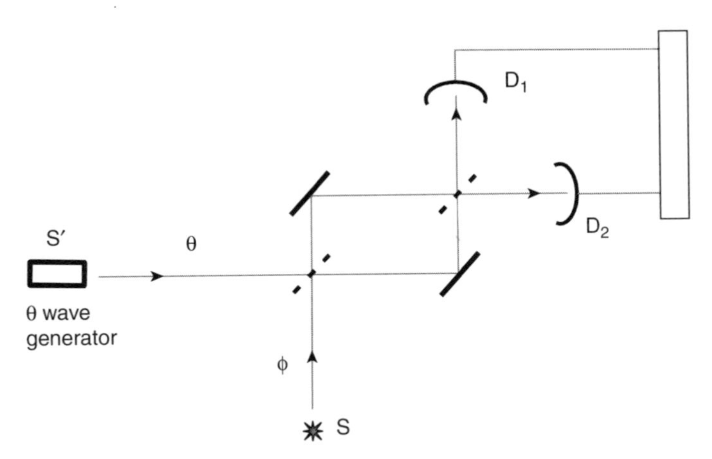

Figure 10. Incoherent detection of θ waves with correction for synchronization.

where b is a proportionality constant, one gets, after the usual calculations

$$\begin{cases} I_u^1 = I_0(1 + \cos\delta_\phi) \\ I_u^2 = I_0(1 - \cos\delta_\phi) \end{cases} \tag{47}$$

In the causal approach of de Broglie one must also consider the θ waves coming from the special generator. If in a first step it is assumed that the two independent sources S and S′ emit finite waves at the same time, the predicted intensity shall be given by the expression

$$\begin{cases} I_u^1 \propto |\phi_{TR} + \phi_{RT} + \theta_{TT} + \theta_{RR}|^2 \\ I_c^2 \propto |\phi_{RR} + \phi_{TT} + \theta_{RT} + \theta_{TR}|^2 \end{cases} \tag{48}$$

and since the relative phase difference of the two independent varies randomly in time, incoherent sources, one is allowed to write

$$\begin{cases} I_c^1 \propto |\phi_{TR} + \phi_{RT}|^2 + |\theta_{TT} + \theta_{RR}|^2 \\ I_c^2 \propto |\phi_{RR} + \phi_{TT}|^2 + |\theta_{RT} + \theta_{TR}|^2 \end{cases} \tag{49}$$

Developing these relations with the usual simplifying assumptions of equal wave intensity

$$|\theta_{TR}|^2 = |\theta_{RT}|^2 = |\theta_{RR}|^2 = |\theta_{TT}|^2 = \frac{1}{4}|\phi|^2 \tag{50}$$

and for a constant phase shift δ_θ between the two coherent θ waves, one gets

$$\begin{cases} I_c^1 = \frac{1}{2}I_0(1 + \cos\delta_\phi - \cos\delta_\theta) \\ I_c^2 = \frac{1}{2}I_0(1 - \cos\delta_\phi + \cos\delta_\theta) \end{cases} \tag{51}$$

Recalling that the experimental conditions were set such that

$$\delta_\phi = \delta_\theta = 0 \tag{52}$$

which correspond to an equal optical path for all the waves, the usual expressions (47) become

$$\begin{cases} I_u^1 = I_0 \\ I_u^2 = 0 \end{cases} \tag{53}$$

and, for the same token, the causal formula (51) gives

$$\begin{cases} I_c^1 = \frac{1}{2} I_0 \\ I_c^2 = \frac{1}{2} I_0 \end{cases} \tag{54}$$

For the particular choice of experimental conditions, given by expression (52), and according to the usual interpretation, the interferometer behaves like a pure transmissible medium. This is a consequence of the fact that at output port 1 the ϕ waves are in phase, while at port 2, the waves are in phase opposition. In the causal interpretation, where the real nature of quantum waves is assumed, predictions are completely different; the Mach–Zehnder interferometer behaves like a 50% beamsplitter.

These predictions, as stated, are valid whenever the two sources emit at the same time. This is en extremely tough experimental requirement to achieve. Usually, the two independent sources emit particles in a random way. This means that sometimes, the two independent waves arrive precisely at the same time, corresponding to a complete overlapping, while they don't even partially mix at other times. If the independent waves do not overlap at the mixing region, no inference about the reality of the quantum waves can be drawn. Between these two extreme cases there are, of course, all the intermediate cases of partial superposition.

A simplified model for factoring in the true emission nature of the two independent sources can be easily developed. Let n_ϕ be the total output coming from common source S registered per second at the output ports, and let n_c be the coincidence rate. The number per second of ϕ waves arriving at the detection zone without the corresponding θ waves is $(n_\phi - n_c)$. The total counting rate, assuming the reality of the quantum waves, is given in two parts: (1) For cases when the ϕ waves arrive alone at the beamsplitter without the corresponding θ waves—here all particles hit detector D_1; (2) For cases when the waves from independent sources arrive at the mixing region at the same time, resulting in the same counting rate at the detectors. Symbolically, this is

$$\begin{cases} I_c^1 = \frac{1}{2} n_c + (n_\phi - n_c) \\ I_c^2 = \frac{1}{2} n_c \end{cases} \tag{55}$$

Let γ be a factor characterizing the mean overlapping coincidence rate of the two independent emitting sources, such that

$$n_c = \gamma n_\phi, \quad 0 \leq \gamma \leq 1 \tag{56}$$

By substitution into (55), one has

$$\begin{cases} I_c^1 = \left(1 - \dfrac{1}{2}\gamma\right) n_\phi \\ I_c^2 = \dfrac{1}{2}\gamma n_\phi \end{cases} \tag{57}$$

Representing Δ as the difference between the counts per second of the two detectors

$$\Delta = I^1 - I^2 \tag{58}$$

one gets

$$\Delta_u = I_u^1 - I_u^2 = n_\phi \tag{59}$$

$$\Delta_c = I_c^1 - I_c^2 = (1 - \gamma) n_\phi \tag{60}$$

These different previsions of the two interpretations are shown in Fig. 11 for the particular case of $\gamma = \frac{1}{2}$.

In the most unfavorable case when there is no overlapping between the waves emitted by the two independent sources $\gamma = 0$, the predictions are the same for the two theories $\Delta_c = \Delta_u$. For different values of the mean overlapping factor $0 \leq \gamma \leq 1$, all intermediary cases are obtained. Considering the best overlapping situation $\gamma = 1$, the difference in the predictions of the two theories is maximal: $\Delta_u = 0$ and $\Delta_c = n_\phi$.

The great advantage of these experiments results from the fact that for one side they are inherently conclusive, while for the other they can be put into practice with today's available technology. In the following, we shall briefly present one concrete practical execution of one of these proposals.

Mandel and his group at the University of Rochester performed a concrete experiment [14] to test the nature of quantum waves. Their experiment was based on a proposal presented by Croca et al. [15], and is sketched in Fig. 12.

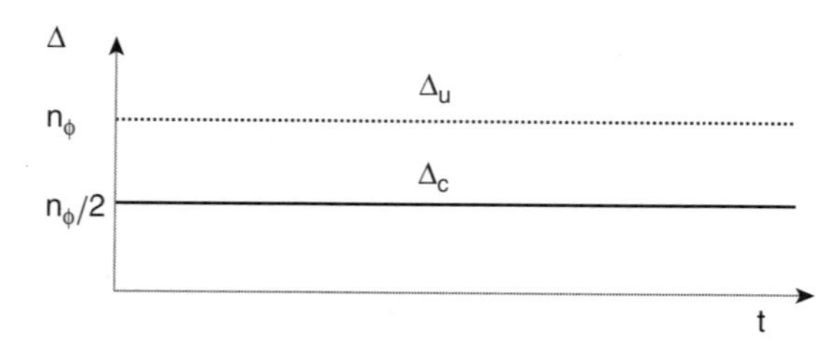

Figure 11. Predictions for the two theories for the particular case of $\gamma = \frac{1}{2}$ (dotted line—usual theory; solid line—causal theory.

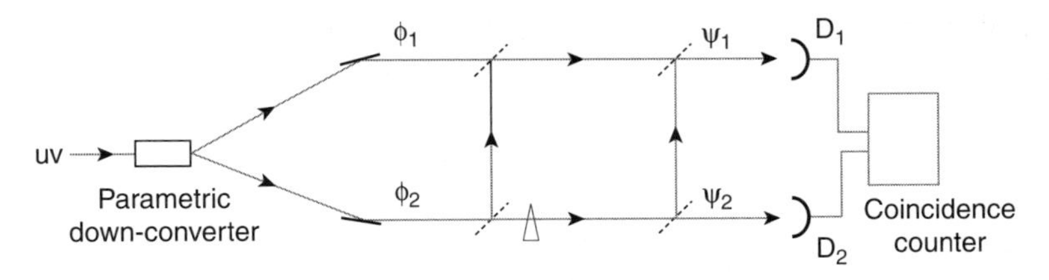

Figure 12. Schematic representation of the experiment done at Rochester University to test the nature of the quantum waves.

As shown in Fig. 12, an UV laser beam of frequency ω_0 impinges on a nonlinear crystal given origin to a parametric downconversion, producing for each arriving photon two photons ω_1 and ω_2 such that $\omega_0 = \omega_1 + \omega_2$. To calculate the joint probability detection at the two detectors for loose less beamsplitters, it is worthwhile to recall that the transmission and reflection coefficients satisfy the usual relation

$$\begin{aligned} |r|^2 + |t|^2 &= 1 \\ rt^* + r^*t &= 0 \end{aligned} \tag{61}$$

Therefore the waves incoming at the detectors can be written in terms of input waves from the nonlinear crystal

$$\begin{aligned} \psi_1 &= t^2\phi_1 + r^2 t\phi_2 + tr^2\phi_2 e^{i\delta} \\ \phi_2 &= t^2\phi_2 e^{i\delta} \end{aligned} \tag{62}$$

where δ represents the phase shift introduced by the phase shifting device. If detector D_2 is triggered, then, at the other detector, D_1, only the other photon can be detected. Recalling the guiding principle, expressed in formula (28), the conditional probability of detection at D_1 is proportional to $|\psi_1|^2$

$$P(D_1|D_2) = \alpha_1|\psi_1|^2 \tag{63}$$

and the coincidence probability becomes

$$P(D_1, D_2) = P(D_1|D_2)P(D_2) = \alpha_1\alpha_2|\psi_1|^2|\psi 2|^2 \tag{64}$$

which gives, by substitution

$$P(D_1, D_2) = \alpha_1\alpha_2|t|^4|\phi_2|^2[|t|^4|\phi_1|^2 + 2|r|^4|t|^2|\phi_2|^2(1 + \cos\delta)] \tag{65}$$

Assuming that the two waves have the same amplitude and for 50% beamsplitters, this gives simply

$$P(\mathrm{D}_1, \mathrm{D}_2) \propto 1 + \frac{1}{2}\cos\delta \tag{66}$$

This expression shows that if the quantum waves are real and the stated assumption is valid, the joint probability detection depends, as expected, on the phase shift.

For the usual interpretation, once the photon is detected at D_2, nothing more from ϕ_2 remains in the interferometer because of the collapse of the wavefunction. Therefore at detector D_1 only the wave ϕ_2 from the usual source arrives. Since the path of this wave does not cross the phase shifting device and, even more, is only one wave, the coincidence count does not depend on the phase.

A version of this experiment was done by Mandel and his group at Rochester University. The experimental apparatus was slightly different from the one shown in Fig. 11, but the principle was essentially the same. The experimentalists concluded that the results do not confirm the existence of the θ waves, because the visibility obtained was not 50%. The statistics of the experiment was very poor; the mean coincidence rate was about six counts per second, but even so it was possible to fit the results with a visibility of 10%. This indicates that the experiment was not conclusive and therefore should have been performed again under better conditions, with more significant statistics, and with a larger coherence length for the overlapping pulses.

B. Experiments Based on Autoreduction of the Wavepacket

Another possibility [16,17] for testing the reality of the quantum waves derives directly from de Broglie causal theory. As we have seen, in this approach, the quantum particle is composed of a wave plus a singularity. These two composing entities have different properties when interacting with matter or with the surrounding subquantum medium.

When the real wave devoid of singularity impinges on a beamsplitter, part of it is reflected and part of it transmitted. The ratio depends on the particularities of the beamsplitter. Consider the incidence of the θ wave on a perfect 50% beamsplitter. Since we are dealing with a real physical wave, half of it is reflected and half is transmitted. Suppose that next we place a succession of equal beamsplitter in the direction of transmission as shown in Fig. 13.

As the wave devoid of singularity crosses the successive beamsplitters it progressively looses its amplitude until it totally vanishes. This natural reduction of amplitude can be represented by

$$\theta = \theta_0 e^{-\mu n} \tag{67}$$

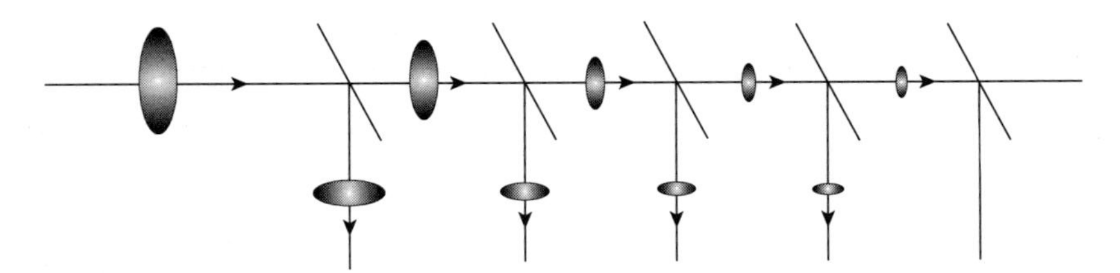

Figure 13. The theta wave, devoid of singularity, looses amplitude as it crosses the successive beamsplitters until nothing of it remains.

where n represents the number of beamsplitters and μ is the attenuation factor related to the transmission coefficient t by the relation $\mu = -\ln t$. The expression (67) essentially remains valid when the wave crosses the void, that is, the subquantum medium, only in this case the attenuation factor changes, and the number n of beamsplitters is changed by the path x traveled by the photon.

Things happen to be rather different with a full wave. Suppose that the full wave hits the beamsplitter and the singularity is transmitted. Next, it impinges on a similar beamsplitter, and the singularity is also transmitted, and this process is repeated endlessly. How can we describe the entire process? Since the singularity is either reflected or transmitted, and the wave is halved in equal parts, after a certain number of beamsplitters the wave completely disappears, mixed with the surrounding subquantum medium. Should this statement be true, we would then have to deal only with the singularity without the accompanying wave, and the wave–particle dualism would be broken. In order to preserve the wave–particle dualism contained in de Broglie's basic affirmation, stating that any quantum particle is composed of a wave and the singularity, it is necessary to make some different assumptions for the behavior of the wave with the singularity. Let us consider the following situation.

When the full wave strikes the first, lossless, 50% beamsplitter, the transmitted amplitude of the wave is halved and one considers only the case when the singularity is transmitted, and the occurs, as a chain reaction, in the remaining beamsplitters. This process keeps continues until a certain beamsplitter number k is reached. After this point the accompanying wave θ, having reached the minimum level of energy compatible with the existence of the quantum particle, starts regenerating itself at the expenses of the energy of the singularity. This situation corresponds to the simplest fundamental state for the free particle having the minimum possible size in which it can manifest interferometric properties. From this point on, the amplitude of the accompanying wave remains essentially constant, no matter the number of beamsplitters it crosses. The process continues as long as the singularity has enough energy to keep feeding the accompanying wave. This situation is illustrated in Fig. 14.

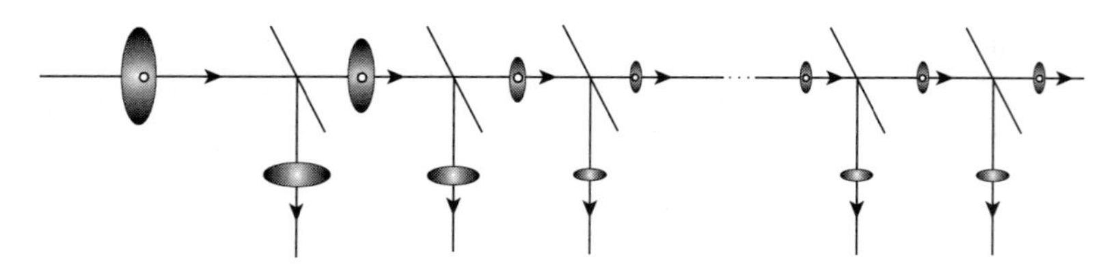

Figure 14. The amplitude of the associated wave decreases till a certain point. From this point on the amplitude of the wave remains constant.

Analytically the amplitude of the associated wave can be represented as follows:

$$\begin{cases} \theta = \theta_0 e^{-\mu n} & n < k \\ \theta = \theta_0 e^{-\mu n} = \theta_0 e^{-\mu k} = \text{constant}, & n \geq k \end{cases} \qquad (68)$$

Here it is possible to see that after a number of $n = k$ beamsplitters is reached, having attained the minimum possible intensity, while still compatible with its existence, the amplitude of the accompanying wave remains constant for all practical purposes.

For the case of particles of light, after crossing the necessary number of beamsplitters to reach the threshold, the energy of the photon singularity begins to decrease.

1. Experiment to Test de Broglie Tired-Light Model for the Photon and Its Implications for the Cosmological Expanding Model of the Big Bang in the Universe

Since the energy is related to the frequency by the usual formula, the very small progressive energy loss of the photon singularity, resulting from the feeding of the associated wave, relates to a correlated frequency shift toward the red. If, instead of beamsplitters, one uses a very long astronomical optical path, the photon as it crosses the space looses energy. This space, as is well known, is not void but filled up with what is usually called the "zero-point field," which de Broglie more appropriately termed the "subquantum medium." In this model, it is possible to interpret, as de Broglie did, the cosmological redshift, not as a Doppler effect, but as a consequence of a decrease of energy resulting from to the interaction of the photon with an all-pervading subquantum medium. This model, sometimes referred to as the "de Broglie tired-light model," is capable of explaining the cosmological redshift phenomenon without any reference to the ordinary Doppler effect. In this situation, there is no logical necessity for invoking or assuming, as if it were unavoidable, the cosmic big bang model of the universe since it is possible to build up a sound causal model where the observable redshift does not imply an expanding universe.

Even if it were possible to explain the cosmological redshift phenomenon by de Broglie's aging photon model, the practical feasible direct test of the model is not easy to perform because the minimum necessary distance for the aging effect to be noted is considerable. Let us make a rough estimate of this distance, assuming that the cosmological redshift is due only to the interaction of the photon with the subquantum medium.

Recalling the previous characteristics of the photon in de Broglie model, it is reasonable to write

$$E = E_0 e^{-\alpha x} \tag{69}$$

for the progressive loss of energy of the singularity due to feeding of the accompanying wave so that it could keep the minimum value compatible with its existence. The mean attenuating factor is given by α, and x is the distance traveled by the photon in cosmic space. Or, equivalently

$$\nu = \nu_0 e^{-\alpha x} \tag{70}$$

using the linear approximation and for $\alpha x \ll 111$, one gets

$$\frac{\Delta\nu}{\nu_0} \approx -\alpha x, \quad \Delta\nu = \nu - \nu_0 \tag{71}$$

or

$$\frac{\Delta\lambda}{\lambda_0} \approx \alpha x \tag{72}$$

where λ represents the wavelength. The fractional increase in the wavelength [18], namely, the redshift, of the photon can be expressed in such way that

$$\alpha \cong \frac{k}{c} \tag{73}$$

where k is the Hubble constant, which has an estimate value of about $k = 1.6 \times 10^{-18}\,\mathrm{s}^{-1}$. By substitution of the values of the constants, one gets $\alpha \approx 10^{-26}$ and by replacing in (70), one gets

$$\nu = \nu_0 e^{-10^{-26} x} \tag{74}$$

From this approximate expression it is possible to estimate the traveling distance through the "empty space" necessary for a relative decrease in the frequency of 10^{-5}, $\Delta\nu/\nu_0 \approx 10^{-5}$. For this small change in frequency, one gets a distance of $x \approx 10^{21}$ m for the path that the photon has to travel. This figure corresponds to about the diameter of our galaxy. This means that even for this very small change in frequency, and assuming that the cosmological redshift is due only to the aging of the photon, the required distance reaches a galactic (i.e., astronomical) scale.

Jeffers et al. [19] performed an experiment to test the reduction in amplitude of the θ wave devoid of singularity, as it travels in the "empty space." The results of this experiment were not conclusive for two basic reasons:

1. The experiment was performed using a laser source, instead of a tested monoparticle source. It is known that even at low laser beam intensity, the probability of having only one single photon at a time is extremely weak because of to the bunching effect.
2. The second reason is related to the minimum necessary distance for the effect to be noticeable, that is, the reduction in amplitude of the wave to be of any experimental significance. The photon in the abovementioned experiment traveled along a path of 258.1 cm, which is a little more 2.5 m. Clearly, this is figure is negligible when compared to the astronomical value we have obtained. Even in the case of an optical path, whose magnitude would be in the kilometer range, the amplitude damping of the wave would be hardly noticeable.

Nevertheless, a laboratory scale experiment is possible if instead of the subquantum medium one uses beamsplitters or other similar absorbers to reduce, in a significant way, the amplitude of the accompanying wave. In this case, it is possible to rely on the equations derived from the assumed properties for de Broglie's causal particles. For a transmission factor of 1% and according to the formula (67), for 10 equal beamsplitters, the final transmitted wave amplitude would be reduced by 20 orders of magnitude, or a factor of about 10^{-20}. It is reasonable to expect that for such a significant overall damping factor, the threshold level k could be achieved. Thus, in such a situation, the size of the experimental apparatus could be small enough to be put in a laboratory of fairly acceptable dimensions.

The concrete experiment consists of using a modified Mach–Zehnder interferometer. In each arm of the interferometer we place the same number of similar beamsplitters, as shown in Fig. 15. The monophotonic source S emits

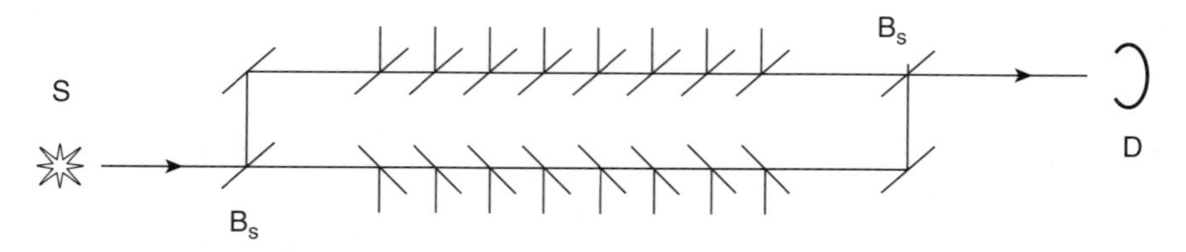

Figure 15. Laboratory-scale experiment for studying the behavior of de Broglie waves.

photons one at a time. This quantum particle enters the input port of the Mach–Zehnder interferometer. Since the two independent arms of the interferometer have the same number of similar beamsplitters, the total amplitude reduction is the same in each arm.

Let us now see the prediction for the results of this experiment.

a. Orthodox Interpretation. According to the usual interpretation of quantum mechanics, the expected intensity at the detection region is given by

$$I_u = |\psi_1 + \psi_2|^2 \tag{75}$$

with

$$\psi_1 = rtA\psi, \quad \psi_1 = trA\psi e^{i\delta} \tag{76}$$

where r and t are respectively the reflection and transmission factors at the Mach–Zehnder interferometer beamsplitters B_s, $A = e^{-\mu n}$ represents the overall absorbing factor due to the presence of the attenuating beamsplitters, and δ is the relative phase difference of the two overlapping waves.

Developing (75), one gets the usual formula

$$I_u \propto 1 + \cos\delta \tag{77}$$

Since the two probability waves ψ undergo the same attenuation along the arms of the interferometer, the predicted visibility for this experiment is one: $V = 1$.

b. Causal Interpretation. Needless to say, predictions assuming the validity of de Broglie's causal model of the photon are completely different. These predictions can be halved in two classes of results. The first one relates to the first expression of (68), which means that the threshold k was not yet attained. In this case the predictions of the two approaches are the same because both waves are attenuated in the same proportions. After this point, the predictions are given, recalling (68), and taking into consideration the cases in which the singularity goes along path 2

$$\begin{cases} \theta_1 = \theta_0 e^{-\mu n}, & n \geq k \\ \theta_2 = \theta_0 e^{-\mu k} = \text{constant}, & n \geq k \end{cases} \tag{78}$$

by

$$I_c = |\theta_1 + \theta_2|^2 = |\theta_0|^2 (e^{-2\mu k} + e^{-2\mu n})(1 - \gamma \cos\delta) \tag{79}$$

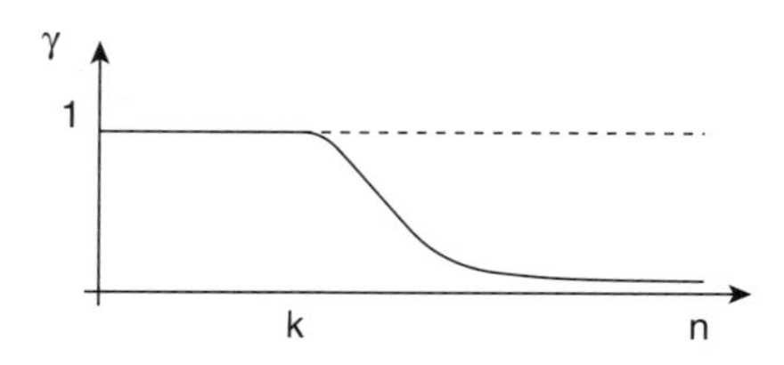

Figure 16. Visibility predicted by the two theories: dashed line, usual model; continuous line, causal model.

where the expected visibility γ is given by

$$\gamma = \operatorname{sech} \mu(n - k), \quad n \geq k \tag{80}$$

The predictions are precisely the same if the singularity goes through the other path. Obviously, in order to carry out the concrete experiment, some practical precautions indicated in previous works [16] must be taken.

The results of the predictions of the two approaches are presented in the next plot, shown in Fig. 16.

V. THE UNCERTAINTY RELATIONS

The uncertainty relations have played a central role since the field of quantum mechanics has been created. Prior to the existence of this theory, experimentalist knew, from their work, that every concrete measurement would necessarily carry an associated error. Yet, it was generally believed that this error was of no fundamental nature, and that one could, in principle, approach the "true" value by filtering out from a huge amount of measurements. Errors were part of the experimental process. With the advent of quantum physics, the error of measurements assumes a new, ontological status, rooted in the very heart of the theory. The theory itself would be built on this unavoidable "error" process.

In order to make any direct measurement, one has to interact with the object; the smaller the interacting particle, the smaller the uncertainty in the final measurement. In any case there is always some degree of error or uncertainty associated with any real concrete measurement, and the best we can do is trying to minimize it.

A. The Usual Heisenberg–Bohr Indeterminacy Relations

The usual incertainty relations were first derived by Heisenberg, in a paper presented in March 1927. Nevertheless, Bohr was the one who fully understood their significance. He presented his views in the fall of the same year in a now-famous lecture delivered at Lake Di Como, Italy. In this lecture, Bohr derived the uncertainty relations as an epistemological consequence of his principle of complementarity, and mathematically as a direct result of the nonlocal Fourier analysis.

It is well known that any reasonable function can be Fourier-represented as a sum of infinite, in space and time, harmonic plane waves (i.e, sinus and cosinus). The more localized the function representing the particle, the more waves the needed to reconstruct it. In the limiting case, when the particle is precisely localized, $\Delta x = 0$, corresponding to a Dirac delta function, the number of waves necessary to build it up reaches infinite values. Since each wave is associated with one velocity, this means that a precisely localized particle has an associated infinitude of velocities, that is, an infinite error for the velocity $\Delta v = \infty$. If, instead of a well-defined position, one wishes to have a particle with a precise velocity $\Delta v = 0$, only one single wave is to be used. Since the harmonic wave with a well-defined velocity is infinite in either space or time, this means that the particle is somehow spread over all space, implying that is its position is completely unknown, $\Delta x = \infty$.

Summarizing these two extreme cases, it is possible to establish an inverse relationship, for these situations between the two conjugated observables: position and velocity:

$$\Delta x = 0 \Leftrightarrow \Delta v = \infty$$
$$\Delta x = \infty \Leftrightarrow \Delta v = 0$$

The following statement is one form of enunciating Bohr's principle of complementarity: *The better the position of a particle is known, the lesser its velocity is known, and vice versa.*

This descriptive reasoning can be made more precise using mathematical formalism and following Niels Bohr, practically step by step. From Fourier analysis, we know that it is possible to represent a well-behaved function as an infinite sum of infinite monochromatic plane waves, that is

$$f(x) = \int_{-\infty}^{+\infty} g(k)\, e^{-ikx}\, dk \tag{81}$$

where $k = 1/\lambda$ represents the wavenumber. Choosing, as Bohr did, for the coefficient function a Gaussian form

$$g(k) = e^{-[k^2/2(\Delta k)^2]} \tag{82}$$

by substitution into (81)

$$f(x) = \int_{-\infty}^{+\infty} e^{-[k^2/2(\Delta k)^2]} e^{ikx} dk \tag{83}$$

and by integration one gets

$$f(x) = \sqrt{2\pi}\Delta k e^{-(x^2/\{2[1/(\Delta k)^2]\})} \tag{84}$$

which implies that

$$\Delta x \Delta k = 1 \tag{85}$$

If, instead of the spatial information on the quantum system, one considers the temporal dependence, and follows a similar line of reasoning, the time frequency relation will be obtained:

$$\Delta t \Delta v = 1 \tag{86}$$

Recalling the fundamental relations of quantum mechanics, Planck–Einstein, and de Broglie

$$E = h\nu; \quad p = \frac{h}{\lambda} = hk$$

and by substitution in (85) and (86), only finally gets

$$\begin{aligned} \Delta x \Delta p_x &= h \\ \Delta t \Delta E &= h \end{aligned} \tag{87}$$

This was, in essence, how Niels Bohr obtained his uncertainty relations. The derivation clearly points out the deep connection between Bohr's principle of complementarity and the Fourier nonlocal analysis. Instead of being simple dual Fourier relations, they got the ontological meaning of representing the duality character manifested by the quantum entities. To do justice to these fundamental contributions, they should be called Heisenberg–Bohr uncertainty relations. While Heisenberg can be remembered as the person who derived them for the very first time, one must also acknowledge the fact that it was Bohr who grasped their full importance, establishing them as a cornerstone of the measurement theory in quantum physics.

The Heisenberg–Bohr uncertainty relations given by (87) relate to the ideal case of Gaussian distributions with minimum uncertainty in the conjugated variables. In most cases and the equality turns into the habitual inequalities

$$\begin{aligned} \Delta x \Delta p_x &\geq h \\ \Delta t \Delta E &\geq h \end{aligned} \tag{88}$$

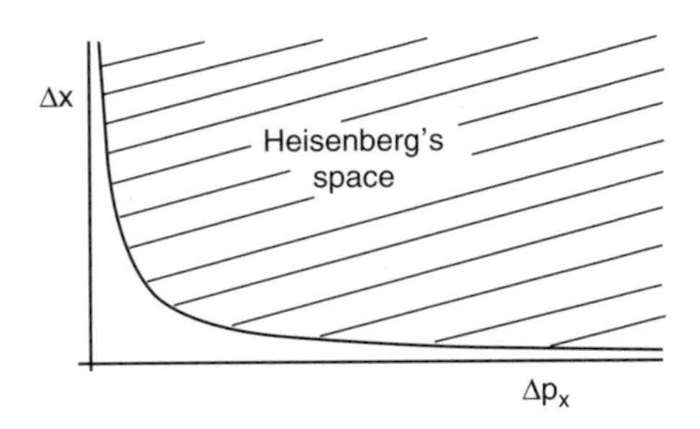

Figure 17. Plot of the Heisenberg's measurement prediction space.

These relations, $\Delta x \Delta p_x \geq h$, define a measurement uncertainty space, the Heisenberg space, shown in Fig. 17.

The Heisenberg space defines the available uncertainty space where, in quantum mechanics, it is possible to perform, direct or indirect, measurements. Outside this space, in the forbidden region, according to the orthodox quantum paradigm, it is impossible to make any measurement prediction. We shall insist that this impossibility does not result from the fact that measuring devices are inherently imperfect and therefore modify, due to the interaction, in an unpredictable way what is supposed to be measured. This results from the fact that, prior to the measurement process, the system does not really possess this property. In this model for describing nature, it is the measurement process itself that, out of a large number of possibilities, creates the physical observable properties of a quantum system.

B. A More General Form of the Uncertainty Relations

The usual uncertainty relations are a direct mathematical consequence of the nonlocal Fourier analysis; therefore, because of this fact, they have necessarily nonlocal physical nature. In this picture, in order to have a particle with a well-defined velocity, it is necessary that the particle somehow occupy equally all space and time, meaning that the particle is potentially everywhere without beginning nor end. If, on the contrary, the particle is perfectly localized, all infinite harmonic plane waves interfere in such way that the interference is constructive in only one single region that is mathematically represented by a Dirac delta function. This implies that it is necessary to use all waves with velocities varying from minus infinity to plus infinity. Therefore it follows that a well-localized particle has all possible velocities.

If, instead of the nonlocal Fourier analysis, one uses the local wavelet analysis to represent a quantum particle, the uncertainty relationships may change in form. On the other hand, this process has the advantage of containing the usual uncertainty relations when the size of the basic gaussian wavelet increases indefinitely.

To make the process of derivation of the new uncertainty relations more comprehensible, it is convenient to do it in parallel and step wise fashion with the usual derivation presented before. This process is shown in Fig. 18.

From Fig. 18 it is seen that the new the uncertainty relations derived with the local wavelet analysis exhibit the form

$$\Delta x^2 = \frac{1}{\Delta k^2 + 1/\Delta x_0^2} \tag{89}$$

Derivation of the uncertainty relations

Non-local de Fourier analysis	Wavelet local analysis
Kernel–Sinus and cosinus	Kernel–Gaussian wavelet
	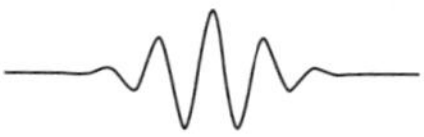
$f_0(x) = e^{ikx}$	$f_0(x) = e^{(x^2/2\Delta x_0^2)+ikx}$

Representation of the particle

$f(x) = \int_{-\infty}^{+\infty} g(k) e^{ikx}\, dk$	$f(x) = \int_{-\infty}^{+\infty} g(k) e^{(x^2/2\Delta x_0^2)+ikx}\, dk$

Coefficient function

$$g(k) = e^{(k^2/2\Delta k^2)}$$

by substitution and integration

$$f(x) \propto e^{-(x^2/2\Delta x^2)}$$

or

$\Delta x = \frac{1}{\Delta k}$	$\Delta x^2 = \frac{1}{\Delta k^2 + 1/\Delta x_0^2}$

by substitution for the quantum value: $p = hk$

$\Delta x\, \Delta p_x = h$	$\Delta x\, \Delta p_x = \sqrt{1 - \frac{\Delta x^2}{\Delta x_0^2}} h$

Figure 18. Derivation of the new form for the uncertainty relations presented with respect to with the usual ones.

which can also be written as

$$\Delta x^2 = \frac{h^2}{\Delta p^2 + h^2/\Delta x_0^2} \tag{90}$$

As already stated, by using local finite wavelets instead of infinite plane waves of the nonlocal Fourier analysis, a new form for the uncertainty relations was obtained. In the derivation process, leading to the new uncertainty relations, we used the reduced Morlet Gaussian wavelet as the basic wavelet. The same analysis could, in principle, be performed with other mother wavelets. The Gaussian wavelet was selected from among the other possibilities, because of its interesting properties. Indeed, from a mathematical point of view, it exhibits a very simple form. As a consequence, the necessary calculations can be fully carried out without approximations. Other very interesting properties of this wavelet result from the fact that, when its size increases indefinitely, it transforms itself into the kernel of the Fourier transform. In this sense, this local analysis contains the nonlocal Fourier analysis as a particular case. On the other hand, and as already pointed out before, they enforce the possibility of representing a reasonable localized particle with a well-defined velocity.

It is easily seen from (90) that when the size of the basic wavelet Δx_0 is large enough, the new relation turns itself into the old, usual Heisenberg relations, which is a very satisfactory result. This situation corresponds to the limiting case when the wavelet analysis transforms into the nonlocal Fourier analysis.

In Fig. 19 the new more general uncertainty relations are represented in a solid line for three finite values of the size of the basic wavelet. The same picture also shows the plot of the usual Heisenberg uncertainty relations

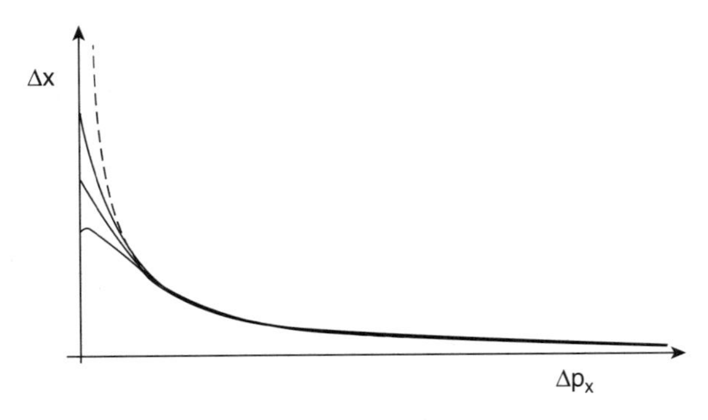

Figure 19. Plot for the new more general uncertainty relations (solid lines) for three different values of size the basic wavelet. The usual Heisenberg uncertainty relations are also represented (dashed line).

represented by a dashed line, which corresponds to an infinite size for the width of the mother wavelet.

From the plot in Fig. 19, it is easily seen that in only a small region, near the origin, the two relations lead to different results. This means that in the great majority of the cases the two relations predict exactly the same results.

The derivation presented above was performed using the basic wavelet with constant size. The same derivation is possible [20] using wavelets where the size depends on the wavelength

$$\sigma = M\lambda \tag{91}$$

where M is a universal constant relating the size of the particle with its wavelength. The size of the particle means, in this context, the longitudinal region of space in which a free single particle exhibits interferometric properties. The derivation, though more complex mathematically, leads to the same final result.

C. Beyond Heisenberg's Uncertainty Relations

The new, more general uncertainty relations (90) were derived in a causal framework assuming that the physical properties of a quantum system are observer-independent, and even more, that they exist before the measurement process occurs. Naturally, because of the unavoidable physical interaction taking place during the measurement process, when the other conjugated observable is to be measured, the quantum system may not remain in the same state. In any case, in the last instance, the precision of a direct concrete measurement for a nonprepared system depends on the relative size between the measurement basic apparatus and the system on which the measurement is being performed.

Until now, the most sensible basic interacting quantum device known to us is the photon. Nevertheless, if the photon possesses an inner structure, as assumed in de Broglie's model, it would imply measurements beyond the photon limit. Since it was assumed that the quantum systems are to be described by local finite wavelets in the derivation of the new uncertainty relations, the measurement space resulting from those general relations must depend on the size of the basic wavelet used. As the width of the analyzing wavelet changes, the measurement scale also changes. This can be seen in the plot in Fig. 20.

From Fig. 20 one sees that as the width of the basic wavelet Δx_0 changes, all the measurement-accessible space is browsed. This space is limited only by Heisenberg's space. The smaller is Δx_0, the greater is the precision of the measurement of the position, that is, the smaller is the uncertainty Δx, for any value of the error in the momentum. Given that the new relation contains the usual as a particular case, it implies that the measurement space available to the

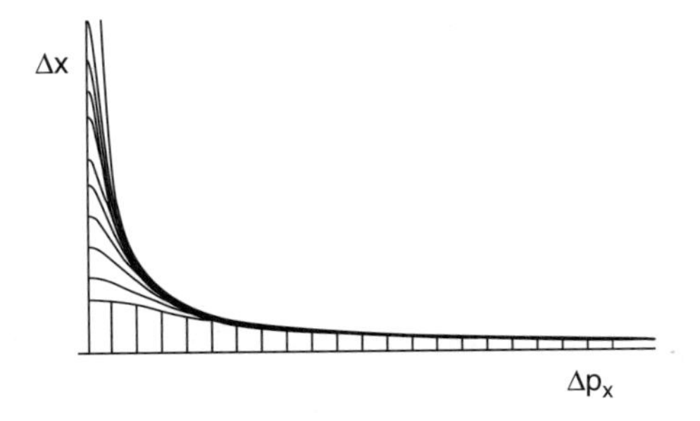

Figure 20. Wavelet measurement space.

general uncertainty relations is the whole space, as showon in Fig. 21. In this causal local interpretation, the precision of any measurement only depends, as expected, on the smallest available interacting basic device with which we make our concrete measurement.

In the Copenhagen interpretation of quantum theory, this standard (see Fig. 21) for the measurement cannot be changed at will since it is composed of sinus waves infinite in length.

The results presented above are rather satisfactory because in this new paradigm the quantum measurement process depends, in the last instance, on the standard used. We are, in principle, free to choose the size, or the scale, of the mother wavelet Δx_0 more suitable for the measurement precision that we want to attain.

In the previous derivation of the new uncertainty relations, we were concerned only with conjugate observables: space and momentum. The same process can be used step by step to derived the relations for the conjugate observables energy and time. It is sufficient to change the variables

$$x \longrightarrow t$$

$$k \longrightarrow \omega$$

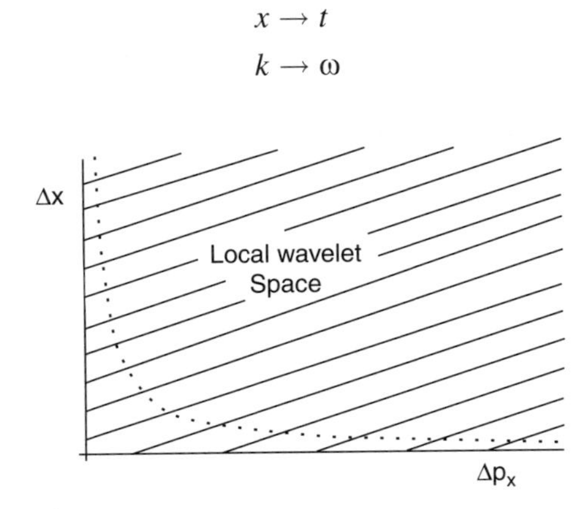

Figure 21. Uncertainty measurement space available to the general uncertainty relation.

to make the usual quantum transformations, and get the general time–energy uncertainty relation:

$$\Delta t^2 = \frac{h^2}{\Delta E^2 + \frac{h^2}{\Delta t_0^2}} \tag{92}$$

D. Experiments to Test the General Validity of the Usual Uncertainty Relations

Starting from the local causal conceptual framework originated by de Broglie and using finite wavelets to represent a quantum particle, rather than infinite harmonic plane waves, it was possible to derive a more general form for the uncertainty relations containing the ordinary Heisenberg relations as a particular case. The problem we now face is to understand the meaning of these new uncertainty relations.

Since the theoretical conclusion drawn from the nonlinear theory of de Broglie [2] allow the anticipation that, in certain very special cases, the usual Heisenberg uncertainty relations are not as general as claimed, the hunting for these experimental conditions began some time ago. Scheer and his group [21] presented some interesting proposals of experiments to test those relations. Here, due to the lack of space, I shall not consider those early proposals; instead I shall discuss two proposals: one the photon ring experiment and the other based on the spreading of matter wavepackets [22].

1. *Photon Ring Experiment*

The idea for this experiment was presented, for the first time, in 1990 at the International Congress on Quantum Measurements in Optics, at Cortina d'Apezzo, Italy. The basic idea is as follows. It is well known that, unlike the electrons, photons have not been passively stored in a kind of photonic condenser. Nevertheless, it is possible to realize the feasibility of building such a device. Consider, for instance, the continuous pumping of a light beam in a transparent nonabsorbing medium with the property of bending the beam so that it follows a closed path converging to a limit circle (see Fig. 22).

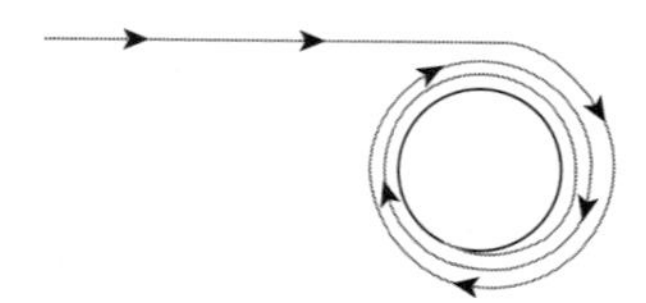

Figure 22. The incident beam converges to a closed orbit.

If such a geometric configuration (Fig. 22) for the medium could be found, the accumulation of light would be infinite for a constant pump beam. Nevertheless, nonlinear effects, which would unavoidably arise at such levels of intensity, would prevent this from happening. On the other hand, even at the linear level, each medium is characterized by some degree of absorption. As a consequence, the number of turns the light pulse can actually achieve in the medium until complete absorption is limited. Therefore, as expected, the amount of light stored in the light condenser is finite.

There seem to be many different ways to manufacture such a "light condenser," such as a ring made of an optical fiber having a large section as well as a variable refractive index, as shown in Fig. 23. The light entering the ring cannot leave it except as a strong short pulse when the gate G touches the ring at the exit region.

Consider the experimental setup sketched in Fig. 23. A beam of light with frequency dispersion $\Delta\nu_0$ feeds the light condenser where the light is trapped untill it reaches equilibrium. Suppose, then, that the shutter is closed and blocks incoming light. After this, the gate contacts the light condenser, thus breaking the condition of total internal reflection. As a direct consequence, and almost instantaneously, a pulse of light is released. This chain of events generates a puzzling question: For these light pulses, what shall be the expected values for $\Delta\nu$ and Δt?

The temporal length Δt of the pulse depends on the specific geometry of the ring and on the working time of the gate. In any case, it appears that Δt depends only on the characteristics of the light condenser.

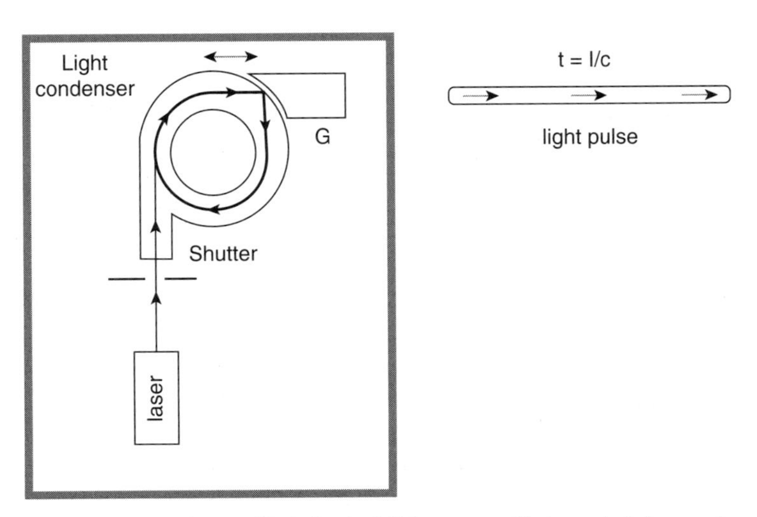

Figure 23. Outline of a new kind of pulsed light source with the optical ring condenser.

The value of the variable $\Delta\nu$, which can be determined on an experimental basis, does not seem to depend on the properties of the light condenser. The light condenser always behaves as a passive medium; that is, color changes of the incident light are neither expected nor observed. Therefore, at least in principle, it looks as if we were allowed to make $\Delta\nu \cong \Delta\nu_0$.

If these statements were correct, we would then be in trouble. When the frequency dispersion $\Delta\nu_0$ is large, no problem is to be expected. But what happens if the pump laser exhibits a high stability with a very short $\Delta\nu_0$? In this situation, and for certain geometries of light condensers, it would also be possible to have Δt short, since they are independent quantities. Therefore, it looks as if it were possible, in such situation, to have $\Delta\nu\Delta t \cong \Delta\nu_0\Delta t \ll 1$. This conclusion contradicts the common uncertainty relations.

2. *Limitless Expansion of Matter Wavepackets*

The problem of the limitless spreading of matter wavepackets leading to a continuous increase in size as the wavepacket travels in space has always been a source of discomfort. The mathematical expression describing this spread–increase as a function of time can be found in any textbook of quantum mechanics [23]. It has the form

$$\Delta x(t) = \Delta x_0 \sqrt{1 + \frac{\hbar t^2}{m_2(\Delta x_0)^4}} \tag{93}$$

or approximately

$$\Delta x(t) \approx \frac{\hbar}{m\Delta x_0} t \tag{94}$$

Working within the framework of orthodox theory, and assuming that this relation holds true in all cases, an electron ejected by the sun, described by a wavepacket of about the size of an atom, $\Delta x_0 \sim 10^{-8}$ cm, would exhibit a dimension greater than the earth's diameter itself when impacting the earth!

On the other hand, it is known that experiments performed with electrons [24] and neutrons [25] demonstrated that the spreading of matter wavepackets has no influence on the interference properties for all practical situations. However surprising, this result had been originally predicted by de Broglie himself. It means that, if two matter wavepackets having an initial size of about $\Delta x_0 \sim 10^{-8}$ cm, are separated by 10^{-8} cm, they do not overlap at the initial time. Later on, when their size reaches a magnitude of about one kilometer, their overlapping is practically complete (i.e., except by a small piece of 10^{-8} cm, in a distance of one lilometer. But even in this case, no interference is observed. This lack of interference is, of course, explained by some authors [26]

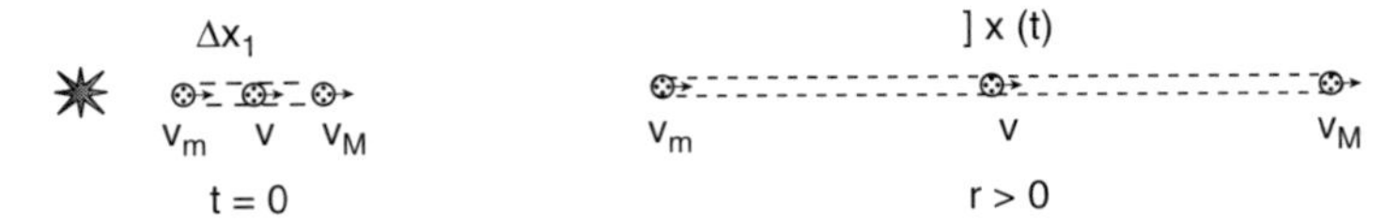

Figure 24. At initial time $t = 0$, particles with different velocities are localized in the burst of small length Δx_0; as time passes, they will be spread continuously as a function of time $\Delta x(t)$.

in the framework of the orthodox theory. Nevertheless, it is my belief that if experiments had shown an interference pattern, the supporters of the orthodox interpretation of quantum mechanics would have demonstrated, with the same theory, that there must be interference. By the habitual use of introducing ad hoc parameters and supplementary hypothesis, it is possible a posteriori to fit almost any experimental result.

The explanation for the absence of interference comes naturally from the causal model of a particle whose undulatory part is described by a finite localized wavelet. In this situation, the limitless spreading of matter wavepackets originates from the fact that in the initial burst, coming from the source, each individual localized quantum particle travels at a different velocity. Therefore, as the time increases the distance among them also increases, as shown in Fig. 24.

In this situation (Fig. 24) two wavepackets of macroscopic size overlapping almost completely except by a small region, of angstrom order, do not interfere because the small wavelets corresponding to each particle do not superimpose on each other. This means that the spreading of matter wavepackets has no true meaning for the individual particles, as it represents only a mathematical description for the time of increasing separation between particles with different velocities. A possible way to test the general validity of the usual uncertainty relations comes directly from this causal interpretation for the infinite spreading of the matter wavepackets.

Consider the following experiment. An electron source emits, at fixed known time t_0, electrons with an uncertainty in position $\Delta x_0 \sim 100\text{Å}$, which, according to Heisenberg uncertainty relations, corresponds to a minimum velocity dispersion $\Delta v_0 \sim 10^{-6}$ cm. A millisecond later, the initial wavepacket will have spread to a size of about 10 m. Let us now suppose that from each electron wavepacket one cuts a piece of size $\delta x_0 \sim 100$ Å (see Fig. 25).

t = 0
$\Delta x_0 = 100$ Å
$\Delta v_0 = 10^{-6}$ cm/s
t
$\Delta x = 100$ Å
$\Delta v = ?$

Figure 25. The uncertainty in position for the electrons is $\Delta x_0 \sim 100$ Å at the time of emission. A certain time t later, a piece of size $\delta x_0 \sim 100$ Å is chopped from the extended packet.

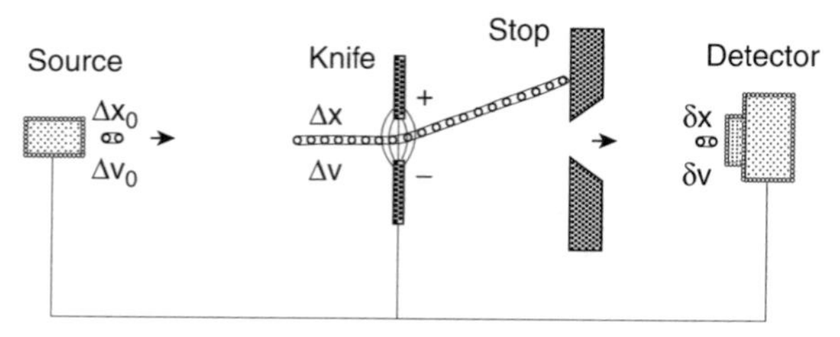

Figure 26. Experiment for testing the general validity of the usual uncertainty relations based in the limitless spreading of matter wavepackets.

This slice, as can be seen in Fig. 26, can be made with an electrostatic field triggered by the emission of the electrons from the source.

The following question now arises: If one measures the dispersion in velocity δv of those slices from the expanded wavepacket of size $\delta x = \Delta x_0$, what shall expect? There are two possible answers, given below.

a. Usual Theory. As Heisenberg uncertainty relations hold true in every circumstance, we must have at least $\delta xv = \hbar/2\text{m}$. Within the framework of this paradigm this conclusion is perfectly natural, because when the cut, is performed in the extended wavepacket, one interacts with all the Fourier components of the packet. The shorter the slice δx, the larger δv, Fourier relations. Afterward, the usual uncertainty relations always hold. Since $\delta x = \Delta x_0$, it means that one must have $\delta v \geq \delta v_0$, because $\Delta x_0 \Delta v_0 = \hbar/2\,\text{m}$.

b. Causal Approach. Assuming that a real quantum particle is to be described by a finite localized wavelet, the spreading of matter "wavepackets" has no intrinsic meaning for the quantum particles. Nonlocal Fourier analysis represents only a statistical description for the average separation among particles having different velocities. In the experiment, one really slices nothing. What really happens is a selection, from the expanded "packet," of a small group of particles having a smaller range of velocities. Only those particles falling into that small range of velocities have the chance to be detected. Therefore the dispersion δv of the selected particles must be smaller than the initial one Δv_0, that is, $\delta v \ll \Delta v_0$ as shown in Fig. 27.

The experimental conditions were set such that

$$\Delta x_0 \Delta v_0 = \frac{\hbar}{2m}$$

and also that

$$\delta x = \Delta x_0; \quad \delta v \ll \Delta v_0$$

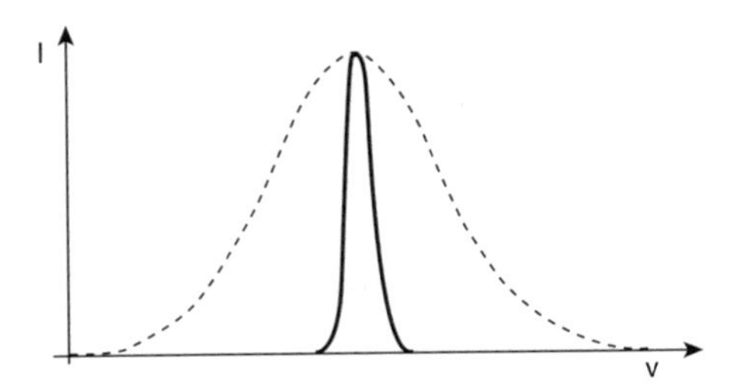

Figure 27. Expected results for the velocity distribution of the electrons. The dashed line refers to the usual theory; the solid line, to causal theory.

and therefore

$$\delta x \delta v \ll \frac{\hbar}{2m}$$

In conclusion, if the measured velocity dispersion δv, of the electron from the slices, is less than the initial one Δv_0, the usual uncertainty relations are not appropriate for describing this experiment, which would precisely be a special case where they do not apply!

E. A New Generation of Microscopes

Till the mid-1990s [27], there were some hints that the usual uncertainty relations were unlikely to have the general validity that they were claimed to exhibit and be endowed with. Experiments designed to test these usual relations were proposed, but regrettably, none of them were performed, as far as we know today. This surprising lack of experimentation did not result from the impossibility to put into practice the principles at work to perform the experiments themselves, since some of them are relatively easy to perform using presently available technology. We suggest that this situation could be explained as the fear of some players of the experimentalist community, to oppose Heisenberg's uncertainty relations, which have become institutionalized, and almost "sacred," with the passing of time.

Causal quantum theories have been developed to handle the empirical quantum evidence, and some of these theories, such as de Broglie's theory in its linear approximation, are almost as good as the usual orthodox quantum theory. A relative large number of experiments were even developed to test de Broglie's causal theory and other alternative theories as well, but only a few limited number of these proposed experiments were carried out effectively. And even so they were not carried out thoroughly. As a consequence, the results obtained were not conclusive, and no solid conclusion in favor or against the completeness of the usual orthodox interpretation of the Copenhagen School could be drawn from them. In other words, until the mid-1990s whose who believed in causality could only hope that the near future would bring some experimental evidence showing that the orthodox quantum theory would indeed demonstrate its limits of applicability.

Now, it is my belief that for the first time ever in nearly century, we have clear, consistent, empirical evidence showing that there is every reason to search for a better and more general quantum theory. After all, usual quantum mechanics, just like any other human construction, was built up from empirical evidence and mathematical tools available at the time, and must have inherent limitations. These limitations came mainly from the explosive technological development of our times, which has greatly enlarged our empirical universe. Experimental available evidence is now much different from the picture physicists could draw of nature when the usual quantum mechanics originated. This empirical evidence comes from the new generation of microscopes developed since the early 1980s. This generation of microscopes has a concrete practical resolution going well beyond Abbe's usual criteria for the theoretical resolution limit of half wavelength.

Until the advent of superresolution microscopes the only way to observe the minute world was based on the common Fourier-type microscopes. The maximum theoretical resolution limit for these apparatuses was established by Abbe, applying the Rayleigh–Fourier diffraction resolution rule [28]. The basic principle underlying the operational working of these ordinary microscopes is, in the eyes of Niels Bohr, a textbook example of Heisenberg uncertainty relations.

The new generation of these superresolution microscopes was initiated by the scanning tunneling electron microscope. This new imaging device was developed by Benning and Roher [29] technicians working at IBM in Zurich. They won the Nobel prize in 1986 precisely for this discovery. The superresolution electronic microscopic was immediately followed by the forcefield scanning apertureless microscope, opening a whole new world for the imaging techniques. Soon, the same principles were applied to the optical domain as well, so that in 1984 Pohl et al. [30], also working for IBM, developed a superresolution optical microscope with a spatial resolution of about $\lambda/20$. Ten years of progressive refinement [31] lead them to resolutions of $\lambda/50$ or even better.

There are different types of scanning optical microscopes; here one is interested only in the most common type. In these microscopes the light emitted by the sample is simply collected by the probe as can be seen in Fig. 28.

The superresolution optical microscope shown in Fig. 28 is basically made up of a sensor or light detector, a scanning system, (not shown in the sketch), designed to control the position of the probe over the sample, and a computer with a display device. Naturally it also has, as does any conventional

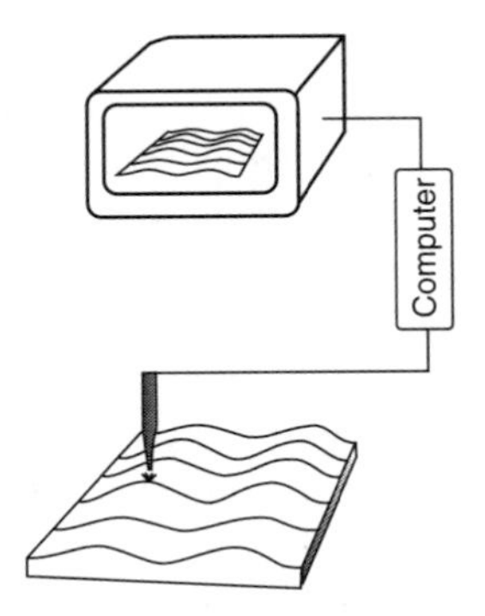

Figure 28. Superresolution optical microscope.

microscope, a source of light to illuminate the sample. In some of these new microscopes the source of light consists practically of a point source and plays another role, more important than simply illuminating the sample. Nevertheless, in this simple type microscope, the role of the source is only to illuminate the sample.

The light detector is usually made of an extremely thin optical fiber, whose tip is much smaller than the diameter of a human hair. The light diffused by the sample is collected by the tip of the needle and next conduced to an electronic detector. The detector converts the incoming light intensity into an electrical pulse feeding the computer. In some cases, instead of the optical fiber, the sensor extremity can be just a simple very small solid-state detector directly converting directly the light into an electric pulse.

The scanning system is a very important part of these microscopes, and is commonly composed of a cantilever whose arms are usually made of piezoelectric quartz crystal. The electric field applied by the computer to the arms of the scanning device controls the position of the tip of the sensor to within a great spatial precision. The right variation of the electric field allows the complete scanning of the sample.

The information received by the computer in the form of an electric signal of varying intensity gives birth, after a suitable processing of the information collected, to a final amplified image of the sample shown on the display device. This final image results from the following process. The sample is illuminated, and its points diffuse light in all directions. The tip of the sensor, positioned over one point of the sample, collects some of the diffused light, and transforms it into an electric pulse proportional to the light intensity. The light intensity captured depends on the distance between the sensor tip and the surface of the sample, and also on the collecting area of the sensor tip. Therefore, during the scanning process, the computer records the variation of light intensity in a scanning line. By scanning successive lines over the whole sample and after an adequate treatment of the information the desired amplified image of the sample is seen on the display device. Suppose, for instance, that only one point object of the sample diffuses light, or equivalently, that the sample consists of only one point; the image shown on the display device, would exhibit a continuous uniform background surface with a single discontinuity. This discontinuity represents, precisely, the image of that single point object.

The experimental resolution of the apparatus depends on the dimensions of the sensor tip, the spatial accuracy of the scanning device (which is obviously better for smaller steps), and the minimum possible distance between the sample to be observed and the sensor's extremity (the smaller, the better).

For this type of superresolution optical microscope, experiments have shown that it is possible to have spatial resolutions of the order of $\lambda/50$. It is believed that the technique can be improved so to allow spatial resolutions of over $\lambda/100$.

F. A Measurement Process that Goes beyond Heisenberg's Uncertainty Relations

Let us now consider the well-known Heisenberg microscope experiment, with both the common Fourier microscope and the new-generation superresolution optical microscope.

If we want to show that there are physical concrete situations not described by Heisenberg's uncertainty relations, it is necessary to predict the uncertainties, for the two conjugate noncommutative observables, for example, position, Δx, and the uncertainty in momentum, Δp, for the microparticle M, after the interaction with the photon, and then make their product and see whether they are contained in Heisenberg uncertainty measurement space.

The uncertainty for the momentum of the particle M, after interaction with the photon, can be predicted in many different ways, as can bee seen in a variety of textbooks on quantum mechanics. Each author tries a slightly different approach, taking into account more or fewer factors, but at the end, of course, all of them unavoidably find the same formula. The main reason why all of these authors find the same final formula, even when they follow different approaches, results from the known fact that the uncertainty for the position is fixed and given by the microscope theoretical resolution. Therefore, since the uncertainty for the position is fixed, there is no liberty for the expression of the uncertainty in momentum if one whishes, as is always the case, to stay in agreement with Heisenberg's uncertainty relations.

In any case, just for the sake of exemplification, we shall derive here the formula for the uncertainty in momentum, following de Broglie's demonstration [32] almost step by step. In order to clarify the process, in Fig. 29 shows the detection region of the two types microscopes facing each other, for the case of the an horizontal incidence of light. (The same derivation could be obtained, of course, for any other incidence angle.) We must keep in mind that the reasoning

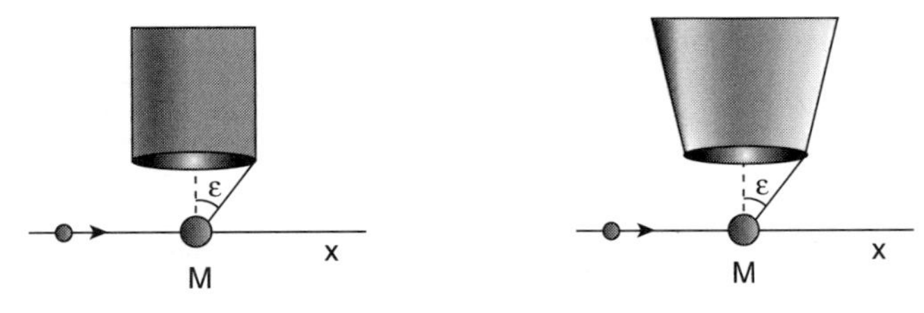

Figure 29. Detection region of the two microscopes: (a) usual Fourier microscope; (b) superresolution microscope.

leading to the prediction of the final uncertainty for the momentum, for the small microparticle particle M after interaction with the impinging photon, is identical for both these microscopes. This result arises because the physical situation, in both cases, is precisely the same—the incident photon strikes the microparticle and is diffused by it.

In the following, it is assumed that before the interaction the particle M is at rest and that during the photon–particle interaction, conservation of momentum is respected. Before the interaction the total momentum is the one due to the traveling photon only

$$\vec{p}' \rightarrow |\vec{p}'| = p' = p'_x = \frac{h}{\lambda} \tag{95}$$

After the interaction we get

$$p'_x = p''_x + p_x \tag{96}$$

The x component of the momentum of the particle M is

$$p_x = p'_x - p''_x = p' - p'' \sin \epsilon \cong \frac{h}{\lambda}(1 - \sin \epsilon) \tag{97}$$

Therefore the value of the momentum of the particle M along the x-axis lies between

$$\frac{h}{\lambda}(1 - \sin \epsilon) \leq p_x \leq \frac{h}{\lambda}(1 + \sin \epsilon) \tag{98}$$

and the uncertainty for the momentum along the x axis is given by

$$\delta p_x = 2\frac{h}{\lambda} \sin \epsilon \tag{99}$$

which maximum corresponds a diffusion angle of $\pi/2$;

$$\delta p_x = 2\frac{h}{\lambda} \tag{100}$$

This value, as expected, relates to the maximum possible momentum transferred from the photon to the microparticle, even if some values of the diffusion angle obviously have a very low or even zero probability. As stated before, this formula for the uncertainty in the momentum of the small particle M after the measurement is precisely the same for both microscopes. In either case, it is necessary to keep in mind that, in this step of the measuring process of the error of the two conjugated observables, the interacting photon behaves like a corpuscle.

The uncertainty for the position, after the interaction, for the common Fourier microscope is given by the maximum theoretical resolution limit derived by Abbe for these imaging systems, which is

$$\delta x = \frac{\lambda}{2} \tag{101}$$

Naturally, as expected, the practical resolution of the real classical Fourier microscopes is always worst.

The product of the two uncertainties (100) and (101) gives

$$\delta x \delta p_x = h \tag{102}$$

which are the current mathematical form for Heisenberg uncertainty relations. These results correspond to the theoretical ideal case lying in the boundary of Heisenberg's uncertainty measurement space. The real measurements achieved with these common Fourier microscopes, in which the experimental realistic resolution is always worst, is related to the values described by

$$\delta x \delta p_x \geq h$$

lying well inside Heisenberg's measurement space.

For the case of the non-Fourier microscope there is no mathematical formula available, derived from first principles, to calculate its resolution limit. Nevertheless, one knows the practical experimental resolution, which will certainly be smaller than the theoretical limit. For the microscope under consideration, experiments have demonstrated that the resolution is of the order of

$$\delta x = \frac{\lambda}{50} \tag{103}$$

In such circumstances the product of the uncertainties in momentum (100) and for the position (103) is

$$\delta x \delta p_x = \frac{1}{25} h \tag{104}$$

which leads to a discrepancy with the usual Heisenberg relations by a significant factor of $1/25$. While result lies outside Heisenberg's space, in the forbidden region, it still belongs to the more general wavelet measurement space, as can be seen in Fig. 21.

These results are summarized in Fig. 30, where the predictions for the two types of microscopes are shown side by side.

Some may argued that these superresolution optical microscopes work only with a large number of photons and, consequently, are no good, that is, appropriate, when only a single photon is diffused. If this claim had any grounds, then it should also be applied to the common Fourier microscope. Nevertheless, it can easily be shown that, in principle, these two types of microscopes can

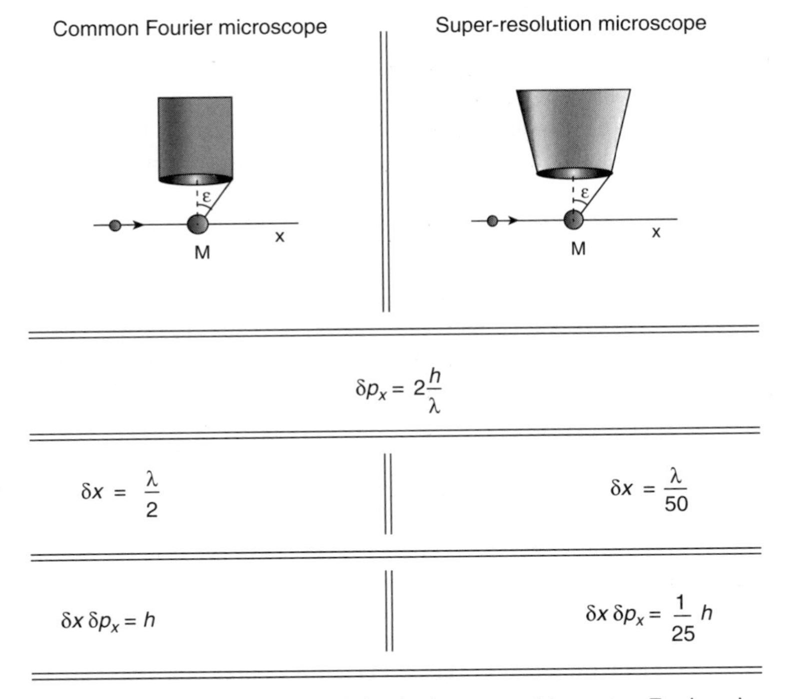

Figure 30. Predictions for the two kinds of microscopes: (a) common Fourier microscope; (b) superresolution microscope.

operate with intensities of light so low that in the limit, one only has a single diffused photon in the apparatus.

The superresolution microscope, in essence, just like the common microscope, is no more than a device for measurement of position, for the mapping of material points. Essentially both types work in the following way. The point forming the object are illuminated, generating diffused light that is eventually captured by the microscope. In this conceptual analysis, the microscope must be treated like a blackbox, since there is no need to go into the particulars of its working.

What is true is that, in any case, whether with the common microscope, or with the superresolution microscope, in order to be observed, the object points must be submitted to some kind of interaction. Since we are dealing with optical microscopes, the interaction occurs with photons. In such circumstances the photon, on interacting with the microparticle, is diffused by it. As a result of this interaction, which is fundamental in all direct concrete quantum measurements, a certain amount of momentum is transferred from the photon to the microparticle, leading to an uncertainty in the momentum of the microparticle.

On the other hand, since the real microscope is not a "perfect" apparatus, the measurement of the position of the small particle inevitably exhibits an associated error.

The product of these uncertainties in momentum and in position, lies in the case of the common Fourier microscopes in the Heisenberg uncertainty measurement space, while for the superresolution optical microscope, the same product lies in the more general wavelet uncertainty measurement space.

CONCLUSION

In this chapter it has been shown, in a manner as close as possible to the method initially presented by Bohr—and since then followed by most authors of textbooks on quantum mechanics—that the uncertainty relations are not as general as claimed. The modus operandi of superresolution optical non-Fourier-type microscopes is not described by Bohr's complementarity principle since, the photon always behaves like a corpuscle all along the measurement process, both when it strikes the microparticle and is diffused by it, and when it is caught by the small probe detector giving origin to an electric pulse. Therefore it is only natural to expect that Heisenberg's uncertainty relations, which are a direct consequence of this photon behavior do not also apply, either. The experimental discrepancy by a significant factor $(1/25)$ is perfectly natural. This means that, in certain very unusual experimental setups, it is possible to predict, before an actual measurement (interaction) takes place, the future uncertainties of conjugate observables position and momentum of a particle in such a way that their product is less than h.

The facts reported indicate that, finally, after such a long struggle, experimental evidence clearly shows that the orthodox interpretation of quantum mechanics is not indeed a complete theory as claimed, and therefore has reached its limits of applicability. In these circumstances it must be replaced by a new, more general quantum theory. Of course, this new theory, aimed at describing quantum phenomena, must be capable of handling on a formal basis—and containing as a particular case—the old theory, nonetheless empowering a new meaning to the mathematical formalism.

References

1. D. Bohm, *Phys. Rev.* **85**, 186 (1952).
2. L. de Broglie, *The Current Interpretation of Wave Mechanics: A Critical Study*, Elsevier, Amsterdam, 1969.
3. A. Grossmann and J. Morlet, SIAM *J. Math. Anal.* **3**, 723 (1989); C. K. Chui, *An Introduction to Wavelets*, Academic New York, 1992.
4. P. Kumar, *Rev. Geophys.* **35**, 385 (1997).
5. A. R. Silva, J. S. Ramos, J. R. Croca, and R. N. Moreira, in *Causality and Locality in Modern Physics and Astronomy*, S. Jeffers et al. (Eds.), Kluwer, Dordrecht, 1998.
6. Ph. Gueret and J. P. Vigier, *Lett. Nuovo Cimento* **38**,125 (1983); J. P. Vigier, *Phys. Lett. A* **135**, 99 (1989).
7. J. and M. Andrade e Silva, *C. R. Acad. Sci. (Paris)* **290**, 501 (1980).
8. J. R. Croca, in A. van der Merwe and A. Garuccio (Eds.), *Waves and Particles in Light and Matter*, Plenum, 1994.
9. M. Dagenais and L. Mandel, *Phys. Rev. A* **18**, 2217 (1978); S. Friber, C. Hong, and L. Mandel, *Phys. Rev. Lett.* **1**, 2011 (1985); P. Grangier, G. Roger, and A. Aspect, *Europhys. Lett. A* **90**, 173 (1986).
10. F. Selleri, *Found. Phys.* **12**, 1087 (1983); **17**, 739 (1987).
11. A. Garuccio, V. Rapizarda, and J. P. Vigier, *Phys. Lett. A* **90**, 17 (1982).
12. J. R. Croca, in A. Van der Merwe, F. Selleri, G. Tarozzi (Eds.), *Microphysical Reality and Quantum Formalism*, Kluwer, Dordrecht, 1988.
13. J. R. Croca, *Ann. Found. L. de Broglie*, **14**, 323 (1989).
14. X. Y. Zou, T. Grayson, and L. Mandel, *Phys. Rev. Lett.* **24**, 3667 (1992).
15. J. R. Croca, A. Garuccio, V. Lepori, and R. N. Moreira, *Found. Phys. Lett.* **3**, 557 (1990).
16. J. R. Croca, M. Ferrero, A. Garuccio, and V. L. Lepore, *Found. Phys. Lett.* **10**, 441 (1997).
17. J. R. Croca, *Apeiron* **4**, 41 (1997).
18. A. Gosh, *Apeiron* **9–10**, 35 (1991).
19. S. Jeffers and J. Sloan, *Found. Phys. Lett.* **333**, 341 (1994).
20. J. R. Croca, *Apeiron* **6**(3–4), 151–165 (1999).
21. J. Scheer and M. Schmidt, *Phys. Lett. A* **115**, 322 (1986); T. Götch, T. Coch, G. Luning, J. Scheer, and M. Schmidt, L. Kostro, A. Posiewnik, J. Pykacz, and M. Zukowski (Eds.), *Problems in Quantum Physics*, World Scientific, Singapore, 1987.
22. J. R. Croca, *Phys. Essays*, **1**, 71 (1990); J. R. Croca, in M. Ferrero and A. van der Merwe (Eds.), *Fundamental Problems in Quantum Physics*, Kluwer, Dordrecht, 1995.

23. D. Bohm, *Quantum Theory*, Prentice-Hall, New York, 1955.
24. G. Möllenstedt and G. Wohland, in P. Bredero and G. Boom (Eds.), *Electron Microscopy*, Leiden, 1980, Vol. I.
25. H. Keiser, S. A. Werner, and E. A. George, *Phys. Rev. Lett.* **50**, 560 (1986); H. Rauch, 2nd ISQM–Meeting, Tokyo, Sept. 1986.
26. G. Comsa, *Phys. Rev. Lett.* **51**, 1105 (1983); A. G. Klein, G. Opat, and W. A. Hamilton, *Phys. Rev. Lett.* **50**, 563 (1983).
27. J. R. Croca, "Experimental violation of Heisenberg uncertainty relations," talk at 5th UK Conf. on Conceptual and Philosophical Problems in Physics, Oxford, Sept. 1996.
28. J. W. Goodman, *Introduction to Fourier Optics*, McGraw-Hill, New York, 1996.
29. G. Benning and H. Roher, *Rev. Mod. Phys.* **59**, 615 (1987).
30. D. W. Pohl, W. Denk, and M. Lanz, *Appl. Phys. Lett.* **44**, 651 (1984).
31. H. Heiselmann and D. W. Pohl, *Appl. Phys. A* **59**, 89 (1994).
32. L. de Broglie, *Les Incertitudes d'Heisenberg et l'Interprétation Probabiliste de la Mécanique Ondulatoire*, Gauthier-Villars, Paris, 1988.

ELECTRODYNAMICS AND TOPOLOGY

PATRICK CORNILLE

Advanced Electromagnetic Systems, S.A., 4 Rue de la Pommeraie, 78470 St. Rémy-Lès-Chevreuse, France

CONTENTS

I. INTRODUCTION

Topology is the discipline within mathematical science dealing with the intuitive concepts of continuity and limits. This discipline is itself composed of several distinct fields of interest. Within the scope of this particular chapter, we shall investigate differential topology, with the understanding that this discipline aims

Modern Nonlinear Optics, Part 2, Second Edition, Advances in Chemical Physics, Volume 119, Edited by Myron W. Evans. Series Editors I. Prigogine and Stuart A. Rice.
ISBN 0-471-38931-5

at studying both the classification of surfaces and the problems usually associated with knots. The theory of knots is now reaching the general public interest, as illustrated by an ever-growing list of cover stories, special issues, and reports published in popular science literature on this particularly "knotty" subject [1–4]. Of course, this chapter does not explore every aspect of knot physics, but parts of it only.

More specifically, we shall first investigate topology in classical physics (i.e., classical electromagnetism, quantum mechanics, and plasma physics) before introducing the concept of helicity and extending it to the study of Beltrami fields. Beltrami fields are more often dealt with in hydrodynamics [5] and in so-called "force-free" magnetic fields [6,7]. Beltrami fields have become a key issue in contemporary physics. Among others, they are being employed to study "pinch" stability [8, p. 137] using the energy method.

Magnetic helicity is also a fundamental concept to investigate how vortices are generated, spring up and grow in magnetohydrodynamic systems. Several series of experiments [9] have been performed in order to demonstrate the existence of these vortices and measure their related helicity, thus enabling us to measure how entangled electromagnetic field lines actually are.

Topology also plays a key role in classical electromagnetism. Take, for example, the voltage measured around a solenoid. The outcome of this seemingly very basic experiment has been demonstrated to depend on the relative position of the measuring device with respect to the solenoid in the surrounding space. This surprising effect is well handled and well explained, provided the topological configuration of the experimental setup is properly taken into account.

This chapter, while reviewing topological effects in classical physics, also introduces the reader to new elements of reasoning. For instance, and to the best of our knowledge, the very fact that rotational fields, as a generic family of fields, was composed of ordinary fields, Beltrami fields, and helicoidal fields, has never been acknowledged before in the available literature. Additionally, we shall insist on the role played by Newton's third principle in a wide range of phenomena, and how this principle can be employed for a better understanding of the Aharonov–Bohm effect.

II. HELMHOLTZ THEOREM

Any vectorial field $\mathbf{B}$ can be split into a sum of two different kinds of vectorial fields known respectively as the *longitudinal field* $\mathbf{B}_{\parallel}$ and the *transverse field* $\mathbf{B}_{\perp}$. These fields satisfy, anywhere in space, the following conditions:

$$\nabla \wedge \mathbf{B}_{\parallel} = 0 \qquad \nabla \cdot \mathbf{B}_{\parallel} \neq 0 \tag{1}$$

$$\nabla \wedge \mathbf{B}_{\perp} \neq 0 \qquad \nabla \cdot \mathbf{B}_{\perp} = 0 \tag{2}$$

This set of conditions implies that the $\mathbf{B}_{\parallel}$ field is intrinsically irrotational (without vortex) and derives from a scalar potential Φ, while the $\mathbf{B}_{\perp}$ field is a solenoidal field (without divergence) defined on the basis of the rotational of a vector potential $\mathbf{A}$:

$$\mathbf{B}_{\parallel} + \mathbf{B}_{\perp} = \nabla\Phi + \nabla \wedge \mathbf{A} \tag{3}$$

Therefore the $\mathbf{B}$ field results from the addition of two distinct fields: a polar field $\mathbf{B}_{\parallel}$ and an axial field $\mathbf{B}_{\perp}$. In the available literature, the expressions "longitudinal" and "transverse," which respectively refer to the $\mathbf{B}_{\parallel}$ and $\mathbf{B}_{\perp}$ vectors, do not necessarily imply that these vectors are projections of $\mathbf{B}$ on orthogonal directions. Indeed (if we leave Beltrami fields apart that fulfill the condition $\mathbf{A} \cdot \nabla \wedge \mathbf{A} \neq 0$), for $\mathbf{A} \cdot \nabla \wedge \mathbf{A} = 0$, the vector $\mathbf{B}_{\perp} = \nabla \wedge \mathbf{A}$ is normal to $\mathbf{B}_{\parallel}$ if $\mathbf{A}$ is parallel to $\mathbf{B}_{\parallel}$. If not, that is, when $\mathbf{A} \perp \mathbf{B}_{\parallel}$, both $\mathbf{B}_{\parallel}$ and $\mathbf{B}_{\perp}$ have the same direction. This analysis demonstrates the $\nabla\wedge\,\nabla\wedge$ operator, as applied to the vector $\mathbf{A}$, is itself a vector having either the same direction as $\mathbf{A}$ or having components in directions parallel to $\mathbf{A}$ and perpendicular to the plan originating from vectors $\mathbf{A}$ and $\nabla \wedge \mathbf{A}$.

For time derivative operators, the time derivative of $\mathbf{A}$ is known to have a direction different from $\mathbf{A}$. However, in some cases, and as observed in spacelike derivations, we may expect that second-order derivations of $\mathbf{A}$ with respect to time lead to a vector having the same direction as $\mathbf{A}$. This, obviously, is the case of $\mathbf{A}$ vectors, which are time-harmonic functions. These remarks can be extended, and therefore applied, to the case of the electromagnetic field, whose calculation is based on the following set of relations:

$$\nabla \wedge \nabla \wedge \mathbf{E} + \frac{1}{c^2}\frac{\partial^2 \mathbf{E}}{\partial t^2} = -\frac{4\pi}{c^2}\frac{\partial \mathbf{J}}{\partial t} \tag{4}$$

$$\nabla \wedge \nabla \wedge \mathbf{B} + \frac{1}{c^2}\frac{\partial^2 \mathbf{B}}{\partial t^2} = \frac{4\pi}{c}\nabla \wedge \mathbf{J} \tag{5}$$

On study of the equation related to the magnetic field, and taking into account all the previous remarks, both current density lines and magnetic field lines appear to be orthogonal with one another.

Here, Φ and $\mathbf{A}$ potentials satisfy the following equations:

$$\Delta\Phi = \nabla \cdot \mathbf{B} \qquad \nabla \wedge \nabla \wedge \mathbf{A} = \nabla \wedge \mathbf{B} \tag{6}$$

The vector potential $\mathbf{A}$ is defined only up to the gradient of a scalar function; this additional degree of freedom can be used to turn $\mathbf{A}$ into a solenoidal potential

$\nabla \cdot \mathbf{A} = 0$ based on the application of relation (3). We therefore obtain the following relations:

$$\Delta\Phi = \nabla \cdot \mathbf{B} \qquad \Delta\mathbf{A} = -\nabla \wedge \mathbf{B} \tag{7}$$

The decomposition expressed here is not unique. Indeed, it is always possible to define new sets of functions such as

$$\Phi_1 = \Phi + \Phi_0 \qquad \mathbf{A}_1 = \mathbf{A} + \mathbf{A}_0 \tag{8}$$

in order to replace the **B** field by $\mathbf{B}_1 = \mathbf{B} + \nabla\Phi_0 + \nabla \wedge \mathbf{A}_0$. The $\mathbf{B}_1$ vector would bear the same physical significance as **B** if functions Φ_0 and $\mathbf{A}_0$ fulfill an additional set of requirements:

$$\Delta\Phi_0 = 0 \qquad \nabla \wedge \nabla \wedge \mathbf{A}_0 = 0 \tag{9}$$

Transforms (8) are shown to respect the partition of **B** and $\mathbf{B}_1$ fields into their respective longitudinal and transverse components.

A. Integral Spatial Solution

In order to solve the preceding set of equations, Poisson's method is applied to the integral formulation of the **F** vector

$$\mathbf{F}(\mathbf{r}) = \frac{1}{4\pi}\int_{V_s} \frac{\mathbf{B}_s}{R}\,\delta\mathbf{r}_s^3 \tag{10}$$

where $\mathbf{B}_s$ and **R** are respectively defined as $\mathbf{B}_s = \mathbf{B}(\mathbf{r}_s)$ and $\mathbf{R} = \mathbf{r} - \mathbf{r}_s$.

The divergence and rotational of the **F** vector are respectively

$$\nabla \cdot \mathbf{F} = -\frac{1}{4\pi}\int_{V_s} \mathbf{B}_s \cdot \nabla_s\left(\frac{1}{R}\right)\delta\mathbf{r}_s^3 \tag{11}$$

$$\nabla \wedge \mathbf{F} = \frac{1}{4\pi}\int_{V_s} \mathbf{B}_s \wedge \nabla_s\left(\frac{1}{R}\right)\delta\mathbf{r}_s^3 \tag{12}$$

Using the following set of identifies

$$\nabla_s \cdot \left(\frac{\mathbf{B}_s}{R}\right) = \frac{1}{R}\nabla_s \cdot \mathbf{B}_s + \mathbf{B}_s \cdot \nabla_s\left(\frac{1}{R}\right) \tag{13}$$

$$\nabla_s \wedge \left(\frac{\mathbf{B}_s}{R}\right) = \frac{1}{R}\nabla_s \wedge \mathbf{B}_s + \nabla_s\left(\frac{1}{R}\right) \wedge \mathbf{B}_s \tag{14}$$

integral relations (11) and (12) can be rewritten as

$$4\pi\nabla \cdot \mathbf{F} = \int_{V_s} \frac{1}{R} \nabla_s \cdot \mathbf{B}_s \, \delta\mathbf{r}_s^3 - \int_{S_s} \frac{1}{R} \mathbf{B}_s \cdot \delta\mathbf{r}_s^2 \tag{15}$$

$$4\pi\nabla \wedge \mathbf{F} = \int_{V_s} \frac{1}{R} \nabla_s \wedge \mathbf{B}_s \, \delta\mathbf{r}_s^3 + \int_{S_s} \frac{1}{R} \mathbf{B}_s \wedge \delta\mathbf{r}_s^2 \tag{16}$$

The uniqueness of the solution is guaranteed if the value of the $\mathbf{B}_s$ field is zero over the surface S_s surrounding the volume V_s. If Φ and $\mathbf{A}$ are respectively defined as $\Phi = -\nabla \cdot \mathbf{F}$ and $\mathbf{A} = \nabla \wedge \mathbf{F}$, integral relations spring up readily:

$$\Phi(\mathbf{r}) = -\frac{1}{4\pi} \int_{V_s} \frac{1}{R} \nabla_s \cdot \mathbf{B}_s \, \delta\mathbf{r}_s^3 \tag{17}$$

$$\mathbf{A}(\mathbf{r}) = \frac{1}{4\pi} \int_{V_s} \frac{1}{R} \nabla_s \wedge \mathbf{B}_s \, \delta\mathbf{r}_s^3 \tag{18}$$

This results in a new identity, $\mathbf{B} = \mathbf{B}_{\parallel} + \mathbf{B}_{\perp} = \nabla\Phi + \nabla \wedge \mathbf{A} = -\Delta\mathbf{F}$, enabling us to determine the integral relation:

$$4\pi\mathbf{B}(\mathbf{r}) = -\nabla \int_{V_s} \frac{1}{R} \nabla_s \cdot \mathbf{B}_s \, \delta\mathbf{r}_s^3 + \nabla \wedge \int_{V_s} \frac{1}{R} \nabla_s \wedge \mathbf{B}_s \, \delta\mathbf{r}_s^3 \tag{19}$$

The equation given above can also be rewritten as

$$-4\pi\mathbf{B}(\mathbf{r}) = \int_{V_s} \left[(\nabla_s \cdot \mathbf{B}_s) \nabla \left(\frac{1}{R} \right) + (\nabla_s \wedge \mathbf{B}_s) \wedge \nabla \left(\frac{1}{R} \right) \right] \delta\mathbf{r}_s^3 \tag{20}$$

Using the same process, Eq. (19) can also be rewritten as follows:

$$-4\pi\mathbf{B}(\mathbf{r}) = \int_{V_s} \frac{1}{R} \Delta_s \mathbf{B}_s \, \delta\mathbf{r}_s^3 \tag{21}$$

B. Fourier Analysis

Helmholtz' theorem is a direct consequence of Fourier's formalism. In order to demonstrate this claim, we shall first consider the direct and inverse transforms of the $\mathbf{B}(\mathbf{r})$ function, respectively known as

$$\mathbf{B}(\mathbf{r}) = \frac{1}{(2\pi)^3} \int_{-\infty}^{+\infty} \int_{-\infty}^{+\infty} \int_{-\infty}^{+\infty} e^{-j\mathbf{k}\cdot\mathbf{r}} \, \mathbf{B}(\mathbf{k}) \, d\mathbf{k}^3 \tag{22}$$

$$\mathbf{B}(\mathbf{k}) = \int_{-\infty}^{+\infty} \int_{-\infty}^{+\infty} \int_{-\infty}^{+\infty} e^{j\mathbf{k}\cdot\mathbf{r}} \, \mathbf{B}(\mathbf{r}) \, \delta\mathbf{r}^3 \tag{23}$$

The following identity is given by definition:

$$\mathbf{B}(\mathbf{k}) = \mathbf{B}_{\parallel}(\mathbf{k}) + \mathbf{B}_{\perp}(\mathbf{k}) = \frac{(\mathbf{k}\cdot\mathbf{B})\mathbf{k} - \mathbf{k}\wedge(\mathbf{k}\wedge\mathbf{B})}{\mathbf{k}^2} \tag{24}$$

This identity results in a new set of relations:

$$\mathbf{B}_{\parallel}(\mathbf{r}) = \frac{j}{(2\pi)^3}\nabla\int_{-\infty}^{+\infty}\int_{-\infty}^{+\infty}\int_{-\infty}^{+\infty} e^{-j\mathbf{k}\cdot\mathbf{r}}\,(\mathbf{k}\cdot\mathbf{B})\,\frac{d\mathbf{k}^3}{\mathbf{k}^2} \tag{25}$$

$$\mathbf{B}_{\perp}(\mathbf{r}) = -\frac{j}{(2\pi)^3}\nabla\wedge\int_{-\infty}^{+\infty}\int_{-\infty}^{+\infty}\int_{-\infty}^{+\infty} e^{-j\mathbf{k}\cdot\mathbf{r}}\,(\mathbf{k}\wedge\mathbf{B})\,\frac{d\mathbf{k}^3}{\mathbf{k}^2} \tag{26}$$

By definition, we also have the following relations:

$$-j\mathbf{k}\cdot\mathbf{B}(\mathbf{k}) = \int_{-\infty}^{+\infty}\int_{-\infty}^{+\infty}\int_{-\infty}^{+\infty} e^{j\mathbf{k}\cdot\mathbf{r}_s}\,\nabla_s\cdot\mathbf{B}_s\,\delta\mathbf{r}_s^3 \tag{27}$$

$$-j\mathbf{k}\wedge\mathbf{B}(\mathbf{k}) = \int_{-\infty}^{+\infty}\int_{-\infty}^{+\infty}\int_{-\infty}^{+\infty} e^{j\mathbf{k}\cdot\mathbf{r}_s}\,\nabla_s\wedge\mathbf{B}_s\,\delta\mathbf{r}_s^3 \tag{28}$$

Thus enabling us to recover relation (19)

$$4\pi\mathbf{B}(\mathbf{r}) = -\nabla\int_{V_s}\frac{1}{R}\,\nabla_s\cdot\mathbf{B}_s\,\delta\mathbf{r}_s^3 + \nabla\wedge\int_{V_s}\frac{1}{R}\,\nabla_s\wedge\mathbf{B}_s\,\delta\mathbf{r}_s^3 \tag{29}$$

provided the following spectral decomposition is used:

$$\frac{1}{4\pi R} = \frac{1}{(2\pi)^3}\int_{-\infty}^{+\infty}\int_{-\infty}^{+\infty}\int_{-\infty}^{+\infty} e^{-j\mathbf{k}\cdot\mathbf{R}}\,\frac{d\mathbf{k}^3}{\mathbf{k}^2} \tag{30}$$

Helmholtz' theorem can also be applied to Dirac's dyadic distribution, written as

$$\delta(\mathbf{R})\overleftrightarrow{\mathbf{I}} = \delta\delta_{\parallel}(\mathbf{R}) + \delta\delta_{\perp}(\mathbf{R}) \tag{31}$$

For a function $\Phi = 1/4\pi R$ verifying

$$\Delta\left(\frac{1}{4\pi R}\right) = -\delta(\mathbf{R}) \tag{32}$$

the identity $\Delta\Phi\overset{\leftrightarrow}{\mathbf{I}} = \nabla\nabla\Phi - \nabla\wedge\nabla\wedge(\Phi\overset{\leftrightarrow}{\mathbf{I}})$, allows us to obtain the following couple of relations:

$$\nabla\nabla\left(\frac{1}{4\pi R}\right) = -\delta\delta_{\parallel}(\mathbf{R}) \qquad \nabla\wedge\nabla\wedge\left(\frac{1}{4\pi R}\overset{\leftrightarrow}{\mathbf{I}}\right) = \delta\delta_{\perp}(\mathbf{R}) \tag{33}$$

The spectral decomposition given above for the function Φ, enables us to obtain the following equations:

$$\nabla\nabla\left(\frac{1}{4\pi R}\right) = -\frac{1}{(2\pi)^3}\int_{-\infty}^{+\infty}\int_{-\infty}^{+\infty}\int_{-\infty}^{+\infty} e^{-j\mathbf{k}\cdot\mathbf{R}}\,\frac{\mathbf{kk}}{\mathbf{k}^2}\,d\mathbf{k}^3 \tag{34}$$

$$\nabla\wedge\nabla\wedge\left(\frac{1}{4\pi R}\overset{\leftrightarrow}{\mathbf{I}}\right) = \frac{1}{(2\pi)^3}\int_{-\infty}^{+\infty}\int_{-\infty}^{+\infty}\int_{-\infty}^{+\infty} e^{-j\mathbf{k}\cdot\mathbf{R}}\left(\overset{\leftrightarrow}{\mathbf{I}} - \frac{\mathbf{kk}}{\mathbf{k}^2}\right)d\mathbf{k}^3 \tag{35}$$

After calculation, Dirac's dyadic distributions are expressed as

$$\delta\delta_{\parallel}(\mathbf{R}) = \frac{1}{3}\,\delta(\mathbf{R})\overset{\leftrightarrow}{\mathbf{I}} + \frac{1}{4\pi R^3}\left(\overset{\leftrightarrow}{\mathbf{I}} - \frac{3}{R^2}\,\mathbf{RR}\right) \tag{36}$$

$$\delta\delta_{\perp}(\mathbf{R}) = \frac{2}{3}\,\delta(\mathbf{R})\overset{\leftrightarrow}{\mathbf{I}} - \frac{1}{4\pi R^3}\left(\overset{\leftrightarrow}{\mathbf{I}} - \frac{3}{R^2}\,\mathbf{RR}\right) \tag{37}$$

These dyadic distributions do not exhibit a null value for $\mathbf{R} \neq 0$. However, these distributions satisfy the following properties:

$$\mathbf{B}_{\parallel}(\mathbf{r}) = \int_{-\infty}^{+\infty}\int_{-\infty}^{+\infty}\int_{-\infty}^{+\infty}\mathbf{B}(\mathbf{r}_s)\cdot\delta\delta_{\parallel}(\mathbf{r}-\mathbf{r}_s)\,\delta\mathbf{r}_s^3 \tag{38}$$

$$\mathbf{B}_{\perp}(\mathbf{r}) = \int_{-\infty}^{+\infty}\int_{-\infty}^{+\infty}\int_{-\infty}^{+\infty}\mathbf{B}(\mathbf{r}_s)\cdot\delta\delta_{\perp}(\mathbf{r}-\mathbf{r}_s)\,\delta\mathbf{r}_s^3 \tag{39}$$

C. Integral Solution in Spacetime

Helmholtz theorem for a function $\mathbf{B}_s(\mathbf{r}_s, t_s)$ that depends on space and on retarded time $t_s = t - R/c$ can be generalized as follows:

$$4\pi\mathbf{B}(\mathbf{r},t) = -\nabla\int_{V_s}\frac{1}{R}\,\nabla_s\cdot\mathbf{B}_s\,\delta\mathbf{r}_s^3 + \nabla\wedge\int_{V_s}\frac{1}{R}\,\nabla_s\wedge\mathbf{B}_s\,\delta\mathbf{r}_s^3 + \frac{1}{c^2}\frac{\partial}{\partial t}\int_{V_s}\frac{1}{R}\frac{\partial\mathbf{B}_s}{\partial t_s}\,\delta\mathbf{r}_s^3 \tag{40}$$

Heras [10] has shown how to use the relation written above to recover Jefimenko's formulas [11,12]. Helmholtz' relation generalized to spacetime can be demonstrated by the following relation:

$$-4\pi\mathbf{B}(\mathbf{r},t) = \left(\Delta - \frac{1}{c^2}\frac{\partial^2}{\partial t^2}\right)\int_{V_s}\frac{1}{R}\,\mathbf{B}_s\,\delta\mathbf{r}_s^3 \tag{41}$$

Equation (41) is verified after a series of lengthy calculations, provided the following identities are used:

$$\nabla\cdot\left(\frac{\mathbf{R}}{R^3}\right) = 4\pi\delta(\mathbf{R}) \qquad \nabla\nabla\left(\frac{1}{R}\right) = -4\pi\delta\delta_{\parallel}(\mathbf{R}) \tag{42}$$

This results in the following equalities:

$$\Delta\mathbf{F} = -\mathbf{B}(\mathbf{r},t) + \frac{1}{4\pi c^2}\int_{V_s}\frac{1}{R}\frac{\partial^2\mathbf{B}_s}{\partial t_s^2}\,\delta\mathbf{r}_s^3 \tag{43}$$

$$\frac{1}{c^2}\frac{\partial^2\mathbf{F}}{\partial t^2} = \frac{1}{4\pi c^2}\int_{V_s}\frac{1}{R}\frac{\partial^2\mathbf{B}_s}{\partial t_s^2}\,\delta\mathbf{r}_s^3 \tag{44}$$

Another identity, which is the generalization to spacetime of Eq. (21), is also demonstrated:

$$-4\pi\mathbf{B}(\mathbf{r},t) = \int_{V_s}\frac{1}{R}\left(\Delta_s - \frac{1}{c^2}\frac{\partial^2}{\partial t_s^2}\right)\mathbf{B}_s\,\delta\mathbf{r}_s^3 \tag{45}$$

We must underline the fact that the condition $\mathbf{B}(\mathbf{r},t) = 0$ does not necessarily imply the condition

$$\left(\Delta_s - \frac{1}{c^2}\frac{\partial^2}{\partial t_s^2}\right)\mathbf{B}_s(\mathbf{r}_s,t_s) = 0 \tag{46}$$

in the integral formulation written above. Indeed, the preceding equality is known to be generally verified by the electromagnetic field localized outside the sources. On the other hand, if the equality (46) is verified everywhere in space, then the integral formulation written above necessarily implies the condition $\mathbf{B}(\mathbf{r},t) = 0$. However, the condition written above is fulfilled in the whole space for plane waves of a nonnull amplitude. This results in a contradiction with the integral formulation asserting that the **B** field must be null. The contradiction can be solved if a relevant point is underlined; the integral formulation written above was obtained with the condition

$$-\int_{S_s}\frac{1}{R}\,\nabla_s\cdot\mathbf{B}_s\,\delta\mathbf{S}_s + \int_{S_s}\frac{1}{R}\,\delta\mathbf{S}_s\cdot(\nabla_s\wedge\mathbf{B}_s) \to 0 \quad \text{pour} \quad S\to\infty \tag{47}$$

satisfied for a field $\mathbf{B}$ decreasing faster than $1/R$ toward infinity. Such a condition is obviously not verified for a plane wave. This problem is far from being trivial, particularly if we assume that the vacuum is a wave-like continuum represented by plane wave whose respective amplitudes are non null at an infinite distance.

Once again, Heras [13] gave another formulation of Helmholtz' theorem now based on Green's retarded function $G = \delta(t - t_s - R/c)/R$ for a value $\mathbf{B}_s(\mathbf{r}_s, t_s)$ function of space and retarded time $t_s = t - R/c$. Heras obtained the following equation:

$$4\pi\mathbf{B}(\mathbf{r}, t) = \int_V \int_T \left[(\nabla_s \cdot \mathbf{B}_s)\nabla_s G + (\nabla_s \wedge \mathbf{B}_s) \wedge \nabla_s G + \frac{G}{c^2} \frac{\partial^2 \mathbf{B}_s}{\partial t_s^2} \right] \delta \mathbf{r}_s^3 \delta t_s \qquad (48)$$

D. Application to Maxwell–Ferrier Equations

Helmholtz' decomposition, defined by Eq. (40), can be readily applied to Maxwell's set of equations with magnetic monopoles, provided these equations are written as

$$\frac{4\pi}{c} \mathbf{J}_e = \nabla P_e + \nabla \wedge \mathbf{B} - \frac{1}{c} \frac{\partial \mathbf{E}}{\partial t} \qquad (49)$$

$$\frac{4\pi}{c} \mathbf{J}_m = \nabla P_m - \nabla \wedge \mathbf{E} - \frac{1}{c} \frac{\partial \mathbf{B}}{\partial t} \qquad (50)$$

where P_e and P_m are, respectively, the electric and magnetic polarizations proposed by an ever-growing list of authors [14–23]. However, the presence of these polarization terms appears to be a natural consequence of the Helmholtz theorem. Indeed, if the electromagnetic field, defined on the basis of the potentials

$$\mathbf{E} = -\nabla\Phi - \frac{1}{c} \frac{\partial \mathbf{A}}{\partial t} \qquad \mathbf{B} = \nabla \wedge \mathbf{A} \qquad (51)$$

is introduced in the preceding set of equations, the wave equation for the vector potential with source term becomes

$$\Delta\mathbf{A} - \frac{1}{c^2} \frac{\partial^2 \mathbf{A}}{\partial t^2} = -\frac{4\pi}{c} \mathbf{J}_e \qquad \nabla P_m = \frac{4\pi}{c} \mathbf{J}_m \qquad (52)$$

provided Ferrier's gauge is employed:

$$\frac{1}{c} \frac{\partial \Phi}{\partial t} + \nabla \cdot \mathbf{A} = -P_e \qquad (53)$$

On the other hand, an additional set of equations, namely

$$\nabla \cdot \mathbf{E} - \frac{1}{c}\frac{\partial P_e}{\partial t} = 4\pi\rho_e \qquad \nabla \cdot \mathbf{B} - \frac{1}{c}\frac{\partial P_m}{\partial t} = 4\pi\rho_m \tag{54}$$

leads us to the equations:

$$\Delta\Phi - \frac{1}{c^2}\frac{\partial^2\Phi}{\partial t^2} = -4\pi\rho_e \qquad \frac{1}{c}\frac{\partial P_m}{\partial t} = -4\pi\rho_m \tag{55}$$

Jackson demonstrated [24, p. 252] that an appropriate set of transforms could always be found so as to get rid of magnetic terms ρ_m and P_m.

III. STUDY OF ROTATIONAL FIELDS

The rotational of an **A** field expresses the vortical feature of the **A** field in the neighborhood of the region placed under close scrutiny. More generally speaking, the **A** field for $\nabla \wedge \mathbf{A} \neq 0$ verifies the following identity:

$$\mathbf{A}^2(\nabla \wedge \mathbf{A})^2 = (\mathbf{A} \cdot \nabla \wedge \mathbf{A})^2 + (\mathbf{A} \wedge \nabla \wedge \mathbf{A})^2 \tag{56}$$

This identity enables us to distinguish between three different kinds of fields behind the usual, generic **A** field. These newly fields are:

1. *Ordinary Fields.* Ordinary fields obey the relations

$$\mathbf{A} \cdot \nabla \wedge \mathbf{A} = 0 \Rightarrow \nabla \wedge \mathbf{A} = \mathbf{K} \wedge \mathbf{A} \Rightarrow \mathbf{A} \wedge \nabla \wedge \mathbf{A} \neq 0 \tag{57}$$

where the vector **K** is a function of both space and time in the general case. Hence, for a velocity field $\mathbf{U}(\mathbf{r}, t) = \omega(t) r \mathbf{n}_\theta$, the rotational of **U**, namely, $\nabla \wedge \mathbf{U} = 2\omega(t)\mathbf{n}_z$, defines the rotation of the field. The equalities written above imply the condition $\mathbf{U} \cdot \nabla \wedge \mathbf{U} = 0$, which correctly describes an ordinary field **U**. In fluid mechanics, this family of fields is known as a "complex laminar flow," where the velocity function is laid down under the form $\mathbf{U} = \Phi\nabla\Psi$, which implies the condition $\mathbf{U} \cdot \nabla \wedge \mathbf{U} = 0$. We remind our readers here that laminar flows are known to verify the condition $\nabla \wedge \mathbf{U} = 0$. Another example might be helpful. Consider now the case of the vector potential $\mathbf{A}(z, t)$ which has a single component along the x axis:

$$\mathbf{A} = \mathbf{n}_x \cos(\omega t - k_z) \Rightarrow \nabla \wedge \mathbf{A} = \mathbf{n}_y k \sin(\omega t - k_z) \tag{58}$$

For this particular situation, the **A** field is best described as rectilinearly polarized. But the **A** field is also an ordinary-type field because it fulfills the set of relations given above.

2. *Beltrami Fields*. Beltrami fields are defined on the basis of the following conditions:

$$\mathbf{A} \wedge \nabla \wedge \mathbf{A} = 0 \Rightarrow \nabla \wedge \mathbf{A} = k\mathbf{A} \Rightarrow \mathbf{A} \cdot \nabla \wedge \mathbf{A} \neq 0 \tag{59}$$

where the scalar function $k(\mathbf{r}, t)$ depends on space and time in the general case. Now, let us consider a vector potential $\mathbf{A}(z, t)$ admitting two components

$$\mathbf{A} = \mathbf{n}_x \cos(\omega t - k_z) + \epsilon \mathbf{n}_y \sin(\omega t - k_z) \tag{60}$$

$$\nabla \wedge \mathbf{A} = \mathbf{n}_x \epsilon k \cos(\omega t - k_z) + \mathbf{n}_y k \sin(\omega t - k_z) \tag{61}$$

The $\mathbf{A}$ potential is a Beltrami field $\nabla \wedge \mathbf{A} = \epsilon k\mathbf{A}$ describing a left-handed circular polarization for $\epsilon = +1$ and a right-handed circular polarization for $\epsilon = -1$.

3. *Helicoidal Fields*. Helicoidal fields verify the following set of inequalities:

$$\mathbf{A} \cdot \nabla \wedge \mathbf{A} \neq 0 \qquad \mathbf{A} \wedge \nabla \wedge \mathbf{A} \neq 0 \tag{62}$$

For instance, a velocity field $\mathbf{U}$ will be of an helicoidal field if the motion is, itself, helicoidal, too. Indeed, point coordinates of the trajectory have the value

$$x = r\cos\theta \qquad y = r\sin\theta \qquad z = b\theta \tag{63}$$

where the helicoidal motion as a function of time is determined by the function $\theta(t)$. Knowing that r and b are constants, we can write Cartesian components of velocity as

$$U_x = -\omega r \sin\theta \qquad U_y = \omega r \cos\theta \qquad U_z = \omega b \tag{64}$$

where $\omega[\theta(t), t] = d\theta/dt$. Cylindrical components of velocity have the following values:

$$U_r = U_x \cos\theta + U_y \sin\theta = 0 \qquad U_\theta = -U_x \sin\theta + U_y \cos\theta = \omega r \qquad U_z = \omega b \tag{65}$$

As a consequence, rotational components are defined by the following set of relations:

$$(\nabla \wedge \mathbf{U})_r = \frac{b}{r}\frac{d\omega}{d\theta} \qquad (\nabla \wedge \mathbf{U})_\theta = 0 \qquad (\nabla \wedge \mathbf{U})_z = 2\omega \tag{66}$$

from which the following set of inequalities can be determined:

$$\mathbf{U} \cdot \nabla \wedge \mathbf{U} = 2\omega^2 b \neq 0 \qquad \mathbf{U} \wedge \nabla \wedge \mathbf{U} \neq 0 \tag{67}$$

Helicoidal motions mix ordinary fields with Beltrami fields. Indeed, the **U** field can always be split into a longitudinal field $\mathbf{U}_{\parallel}$ along a given direction (in the present case, direction z), and a transverse field $\mathbf{U}_{\perp}$ along a direction normal to the previous one (in the present case, a direction normal to z), thus ensuring the following set of identities:

$$\mathbf{U} \cdot \nabla \wedge \mathbf{U} = \mathbf{U}_{\parallel} \cdot (\nabla \wedge \mathbf{U})_{\parallel} + \mathbf{U}_{\perp} \cdot (\nabla \wedge \mathbf{U})_{\perp} \tag{68}$$

$$\mathbf{U} \wedge \nabla \wedge \mathbf{U} = \mathbf{U}_{\parallel} \wedge (\nabla \wedge \mathbf{U})_{\perp} + \mathbf{U}_{\perp} \wedge (\nabla \wedge \mathbf{U})_{\parallel} \tag{69}$$

As a consequence of the definitions given above the **U** field, both the longitudinal and transverse fields satisfy the following set of conditions

$$(\nabla \wedge \mathbf{U})_{\parallel} = k\mathbf{U}_{\parallel} \qquad (\nabla \wedge \mathbf{U})_{\perp} = \mathbf{K} \wedge \mathbf{U}_{\perp} \tag{70}$$

which implies another series of relations:

$$\mathbf{U}_{\parallel} \cdot (\nabla \wedge \mathbf{U})_{\parallel} = 2\omega^2 b \neq 0 \qquad \mathbf{U}_{\perp} \cdot (\nabla \wedge \mathbf{U})_{\perp} = 0 \tag{71}$$

$$\mathbf{U} \wedge \nabla \wedge \mathbf{U} = k(\mathbf{U}_{\perp} \wedge \mathbf{U}_{\parallel}) - (\mathbf{U}_{\parallel} \cdot \mathbf{K})\mathbf{U}_{\perp} \neq 0 \tag{72}$$

The preceding inequality will be verified if the condition $K \neq k$ is satisfied.

We remind our readers that vortical lines are curves whose points are tangential to the $\nabla \wedge \mathbf{A}$ field vectors at a any given instant. These curves will change in time. In the same spirit, the expressions "current lines" or "field lines" refer to the class of curves that are tangential in any of their points to the **A** field vectors at a given instant. Vortical lines usually differ from current lines. However, these lines sometimes match with each other and merge. This is the case when $\nabla \wedge \mathbf{A} = k\mathbf{A}$, where the **A** field defines the so-called Beltrami field, most often dealt with in hydrodynamics and in plasma physics.

The classification of rotational fields identified here is not universally acknowledged in the available literature, even when the subject of topology, as applied to electromagnetism, is completely investigated. Among others, Evans, who pioneered the $\mathbf{B}^{(3)}$ field theory, defines the magnetic field as the rotational of the **A** vector potential (see Ref. 25, formula 11, p. 6), here intended as a Beltrami field, as opposed to the conventional definition of the magnetic field encountered in most classical electromagnetism textbooks.

The classification of fields that we have identified can also be related to the concepts of curvature and torsion of a curve [6,26]. Indeed, time derivatives of unitary vectors $\mathbf{s}, \mathbf{t}, \mathbf{b}$ of a Frénet trihedron whose origins are set onto a point $\mathbf{r}$ in a curve, are expressed by the following relations:

$$\frac{d\mathbf{s}}{ds} = \frac{\mathbf{t}}{R} \qquad \frac{d\mathbf{t}}{ds} = -\frac{\mathbf{s}}{R} - \frac{\mathbf{b}}{T} \qquad \frac{d\mathbf{b}}{ds} = \frac{\mathbf{t}}{T} \tag{73}$$

where R and T are respectively the curvature radius and the torsion radius of the curve defined at the point **r**. These values can be calculated from the following set of equations

$$\frac{1}{R^2} = \frac{a^2}{A^6} \qquad \frac{1}{T^2} = \frac{1}{a^4}\left[\mathbf{A}\cdot\left(\frac{d\mathbf{A}}{dt}\wedge\frac{d^2\mathbf{A}}{dt^2}\right)\right]^2 \tag{74}$$

where

$$\mathbf{A} = A\mathbf{s} \qquad a^2 = \left(\mathbf{A}\wedge\frac{d\mathbf{A}}{dt}\right)^2 \tag{75}$$

The following set of equations

$$\frac{d\mathbf{s}}{ds} = (\mathbf{s}\cdot\nabla)\mathbf{s} \qquad \frac{1}{2}\,\nabla(\mathbf{s}^2) = \mathbf{s}\wedge\nabla\wedge\mathbf{s} + (\mathbf{s}\cdot\nabla)\mathbf{s} = 0 \tag{76}$$

lead us to a first identity:

$$\frac{\mathbf{t}}{R} = -\mathbf{s}\wedge\nabla\wedge\mathbf{s} \Rightarrow (\mathbf{s}\wedge\nabla\wedge\mathbf{s})^2 = \frac{1}{R^2} \tag{77}$$

The second identity in Eq. (74) is related to the torsion of the curve. Take, for example, the case of the helicoidal motion, as defined above. After proper calculation, The following expression is obtained for $\omega = \text{constant}$:

$$(\mathbf{s}\cdot\nabla\wedge\mathbf{s})^2 = \frac{4}{T^2} \tag{78}$$

As a consequence, the identity $(\nabla\wedge\mathbf{s})^2 = (\mathbf{s}\cdot\nabla\wedge\mathbf{s})^2 + (\mathbf{s}\wedge\nabla\wedge\mathbf{s})^2$ associates the module of the rotational $\nabla\wedge\mathbf{s}$ both to the curvature and to the torsion of the curve associated with this field. Of course, another line of reasoning can be explored if we now chose to investigate directly the **A** field, starting from the following identity:

$$\nabla\wedge\mathbf{A} = (\mathbf{s}\cdot\nabla\wedge\mathbf{A})\mathbf{s} + (\mathbf{s}\wedge\nabla\wedge\mathbf{A})\wedge\mathbf{s} \tag{79}$$

Knowing that $\mathbf{s}\cdot\nabla\wedge\mathbf{A} = A\mathbf{s}\cdot\nabla\wedge\mathbf{s}$, we find that the preceding equation becomes:

$$\nabla\wedge\mathbf{A} = A(\mathbf{s}\cdot\nabla\wedge\mathbf{s})\mathbf{s} + (\mathbf{s}\wedge\nabla\wedge\mathbf{A})\wedge\mathbf{s} \tag{80}$$

In order to calculate the last term of this equation, the following couple of equalities will be employed:

$$\nabla\wedge\mathbf{A} = A\nabla\wedge\mathbf{s} + \nabla A\wedge\mathbf{s} \tag{81}$$

$$\mathbf{s}\wedge\nabla\wedge\mathbf{A} = A(\mathbf{s}\wedge\nabla\wedge\mathbf{s}) + \nabla A - (\mathbf{s}\cdot\nabla A)\mathbf{s} \tag{82}$$

By definition, we have

$$\nabla A = (\mathbf{s} \cdot \nabla A)\mathbf{s} + (\mathbf{t} \cdot \nabla A)\mathbf{t} + (\mathbf{b} \cdot \nabla A)\mathbf{b} \tag{83}$$

thus resulting in the relation

$$\mathbf{s} \wedge \nabla \wedge \mathbf{A} = -\frac{A}{R}\,\mathbf{t} + (\mathbf{t} \cdot \nabla A)\mathbf{t} + (\mathbf{b} \cdot \nabla A)\mathbf{b} \tag{84}$$

which gives us

$$(\mathbf{s} \wedge \nabla \wedge \mathbf{A}) \wedge \mathbf{s} = \frac{A}{R}\,\mathbf{b} - (\mathbf{t} \cdot \nabla A)\mathbf{b} + (\mathbf{b} \cdot \nabla A)\mathbf{t} \tag{85}$$

Finally, we obtain

$$\nabla \wedge \mathbf{A} = A(\mathbf{s} \cdot \nabla \wedge \mathbf{s})\mathbf{s} + (\mathbf{b} \cdot \nabla A)\mathbf{t} + \left(\frac{A}{R} - \mathbf{t} \cdot \nabla A\right)\mathbf{b} \tag{86}$$

This results in the relations

$$\mathbf{A} \cdot \nabla \wedge \mathbf{A} = \mathbf{A}^2\,(\mathbf{s} \cdot \nabla \wedge \mathbf{s})\mathbf{s} \tag{87}$$

$$\mathbf{A} \wedge \nabla \wedge \mathbf{A} = -\left[\frac{\mathbf{A}^2}{R} - \mathbf{t} \cdot \nabla\left(\frac{\mathbf{A}^2}{2}\right)\right]\mathbf{t} - \mathbf{b} \cdot \nabla\left(\frac{\mathbf{A}^2}{2}\right)\mathbf{b} \tag{88}$$

The preceding set of relations demonstrates that the classification of rotational fields can result from the "coupling" of unitary vectors $\mathbf{s}, \mathbf{t}, \mathbf{b}$ in Frénet's trihedron, to the concepts of curvature and torsion of a curve.

A. Study of Beltrami and Trkal Fields

The relation $\nabla \wedge \mathbf{A} = k\mathbf{A}$ implies that a Beltrami field [27,28] satisfies the conditions

$$\mathbf{A} \cdot \nabla \wedge \mathbf{A} = k\mathbf{A}^2 \neq 0 \qquad \nabla(\mathbf{A}^2) = 2(\mathbf{A} \cdot \nabla)\mathbf{A} \tag{89}$$

where the identity $\nabla(\mathbf{A}^2)/2 = (\mathbf{A} \cdot \nabla)\mathbf{A} + \mathbf{A} \wedge \nabla \wedge \mathbf{A}$ has been used to define the preceding equality.

Starting with the condition $\nabla \wedge \mathbf{A} = k\mathbf{A}$, the rotational of a Beltrami field $\mathbf{B} = \nabla \wedge \mathbf{A}$ is shown to be also a Beltrami field satisfying the following relations:

$$\mathbf{B} \wedge \nabla \wedge \mathbf{B} = 0 \Rightarrow \mathbf{B} \cdot \nabla \wedge \mathbf{B} \neq 0 \tag{90}$$

The inequality in this equation results from the following identity:

$$(\nabla \wedge \mathbf{A}) \cdot \nabla \wedge \nabla \wedge \mathbf{A} = k(\nabla \wedge \mathbf{A})^2 \tag{91}$$

Indeed, the condition $\nabla \wedge \mathbf{A} = k\mathbf{A}$ implies three different relations:

$$\nabla \wedge \nabla \wedge \mathbf{A} = k^2\mathbf{A} + \nabla k \wedge \mathbf{A} \tag{92}$$

$$\nabla \cdot (\nabla \wedge \mathbf{A}) = k\nabla \cdot \mathbf{A} + \mathbf{A} \cdot \nabla k = 0 \tag{93}$$

$$\mathbf{A} \cdot (\nabla \wedge \nabla \wedge \mathbf{A}) = k^2\mathbf{A}^2 = (\nabla \wedge \mathbf{A})^2 \neq 0 \tag{94}$$

The last equation here implies that the directions of a Beltrami field and the direction of its double rotational counterpart form an acute angle. If the **A** field is solenoidal, $\nabla \cdot \mathbf{A} = 0$, then Eq. (93) implies the condition $\mathbf{A} \cdot \nabla k = 0$ for any kind of $k(\mathbf{r}, t)$ function. This results in the fact that current lines and vortex lines are located on surfaces $k(\mathbf{r}, t) = Ct$ for any given t.

Equation (93) illustrates the fact that Coulomb gauge $\nabla \cdot \mathbf{A} = 0$ is necessarily verified when the value k does not depend on space. In this specific case, Beltrami fields turn into fields of the Trkalian type, which are solutions of Helmholtz equations in Coulomb's gauge:

$$\Delta\mathbf{A} + k_0^2\mathbf{A} = 0 \tag{95}$$

A decomposition of a Trkalian field into Fourier modes

$$\mathbf{A}(\mathbf{r}) = \frac{1}{(2\pi)^3} \int_{-\infty}^{+\infty} \int_{-\infty}^{+\infty} \int_{-\infty}^{+\infty} e^{-j\mathbf{k}\cdot\mathbf{r}}\, \mathbf{A}(\mathbf{k}) d\mathbf{k}^3 \tag{96}$$

is a solution of Helmholtz equation if $|\mathbf{k}| = k_0 = \omega/c$, hence the loss of a degree of freedom. Moreover, conditions $\nabla \wedge \mathbf{A} = k_0\mathbf{A}$ and $\nabla \cdot \mathbf{A} = 0$ imply that spectral component of the $\mathbf{A}(\mathbf{r})$ field verify the following equalities:

$$-j\mathbf{k} \wedge \mathbf{A}(\mathbf{k}) = k_0\mathbf{A}(\mathbf{k}) \qquad \mathbf{k} \cdot \mathbf{A}(\mathbf{k}) = 0 \tag{97}$$

Both real and imaginary parts of the spectral component $\mathbf{A}(\mathbf{k}) = \mathbf{A}_r(\mathbf{k}) + j\mathbf{A}_i(\mathbf{k})$ for a real $\mathbf{A}(\mathbf{r})$ field fulfill the couple of conditions $\mathbf{A}_r(\mathbf{k}) = \mathbf{A}_r(-\mathbf{k})$ and $\mathbf{A}_i(\mathbf{k}) = -\mathbf{A}_i(-\mathbf{k})$, thus giving birth to the following solutions:

$$\mathbf{k} \cdot \mathbf{A}_r(\mathbf{k}) = 0 \qquad \mathbf{k} \cdot \mathbf{A}_i(\mathbf{k}) = 0 \tag{98}$$

$$\mathbf{k} \wedge \mathbf{A}_r(\mathbf{k}) = -k_0\mathbf{A}_i(\mathbf{k}) \qquad \mathbf{k} \wedge \mathbf{A}_i(\mathbf{k}) = k_0\mathbf{A}_r(\mathbf{k}) \tag{99}$$

Vectors $\mathbf{A}_r, \mathbf{A}_i, \mathbf{k}$ form an orthogonal trihedron. Henceforth, for a real vector $\mathbf{A}_1(\mathbf{k})$ noncollinear to $\mathbf{k}$, the solution of the set of equations written above is

$$\mathbf{A}_r(\mathbf{k}) = \mathbf{A}_1(\mathbf{k}) \wedge \mathbf{k} \qquad -k\mathbf{A}_i(\mathbf{k}) = \mathbf{k} \wedge [\mathbf{A}_1(\mathbf{k}) \wedge \mathbf{k}] \tag{100}$$

The equality $\mathbf{A}_r^2(\mathbf{k}) = \mathbf{A}_i^2(\mathbf{k})$ can be easily demonstrated. As a consequence of this demonstration, Fourier's integral written above represents the addition of scalar waves associated with the vector $\mathbf{A}(\mathbf{k})e^{j(\omega t - \mathbf{k}\cdot\mathbf{r})}$, whose polarization is circular.

Let us start with the Trkalian field $\mathbf{A}$ as defined above, with its time-like component:

$$\mathbf{A}(\mathbf{r}, t) = \mathbf{A}_0(\mathbf{r}) \cos(\omega t) \tag{101}$$

where the $\mathbf{A}$ field is a vector potential representing a stationary transverse wave with respect to a given direction. The electromagnetic field is also transverse, since we have

$$\mathbf{E} = -\frac{1}{c}\frac{\partial \mathbf{A}}{\partial t} = k_0 \mathbf{A}_0 \sin(\omega t) \qquad \mathbf{B} = \nabla \wedge \mathbf{A} = k_0 \mathbf{A} = k_0 \mathbf{A}_0 \cos(\omega t) \tag{102}$$

These relations imply that the electromagnetic field is itself Trkalian in origin. Indeed, the scalar product $\mathbf{E} \cdot \mathbf{B}$ verifies the condition $\mathbf{E} \cdot \mathbf{B} \neq 0$ [29–31], which is different from the classical case of a stationary wave associated with an ordinary electromagnetic field:

$$\mathbf{E} = -\frac{1}{c}\frac{\partial \mathbf{A}}{\partial t} = k_0 \mathbf{A}_0 \sin(\omega t) \qquad \mathbf{B} = \nabla \wedge \mathbf{A} = \mathbf{K}_0 \wedge \mathbf{A}_0 \cos(\omega t) \tag{103}$$

where $\mathbf{E} \cdot \mathbf{B} = 0$.

Both ordinary and Trkalian electromagnetic fields are solutions to Maxwell's equations outside the sources:

$$\nabla \wedge \mathbf{E} = -\frac{1}{c}\frac{\partial \mathbf{B}}{\partial t} \qquad \nabla \cdot \mathbf{B} = 0 \tag{104}$$

$$\nabla \wedge \mathbf{B} = \frac{1}{c}\frac{\partial \mathbf{E}}{\partial t} \qquad \nabla \cdot \mathbf{E} = 0 \tag{105}$$

The production and study of stationary waves for $\mathbf{E} \cdot \mathbf{B} = 0$ and $\mathbf{E} \cdot \mathbf{B} \neq 0$ were subject to a series of experiments [32] where a gas of atoms confined in a cavity was crossed by different types of stationary waves originating from a laser.

B. Force-Free Field and Virial Theorem

When $\mathbf{A}(\mathbf{r})$, as defined above, represents the $\mathbf{B}(\mathbf{r})$ magnetostatic field and when the relation $\nabla \wedge \mathbf{B} = 4\pi\mathbf{J}/c = k\mathbf{B}$ is verified, the magnetic force density is null since $\mathbf{f}_b = \mathbf{J} \wedge \mathbf{B}/c = 0$. Hence the origin of the expression "force-free field" sometimes attributed to Beltrami fields in the literature [7,33-36]. The conservation of density of electrical current $4\pi\nabla \cdot \mathbf{J} = c\mathbf{B} \cdot \nabla k = 0$ is naturally verified as a consequence of the relations given above. The concept of force-free field is employed to a greater extent in the virial theorem. Indeed, the virial on a volume V is defined by the following integral relation:

$$\int_V \mathbf{f} \cdot \mathbf{r}\, \delta V = \int_V \mathbf{r} \cdot \nabla \cdot \overleftrightarrow{\mathbf{T}}\, \delta V \tag{106}$$

where $\mathbf{f} = \nabla \cdot \overleftrightarrow{\mathbf{T}}$ is a magnetic force density. If we now substitute the identities:

$$\int_V \nabla \cdot (\overleftrightarrow{\mathbf{T}} \cdot \mathbf{r})\, \delta V = \int_S \mathbf{r} \cdot (\delta\mathbf{S} \cdot \overleftrightarrow{\mathbf{T}}) \tag{107}$$

$$\int_V \mathbf{f} \cdot \nabla \cdot \overleftrightarrow{\mathbf{T}}\, \delta V = \int_V [\nabla \cdot (\overleftrightarrow{\mathbf{T}} \cdot \mathbf{r}) - \overleftrightarrow{\mathbf{T}} \cdot \cdot \overleftrightarrow{\mathbf{I}}]\delta V \tag{108}$$

in relation (106), this integral relation can be rewritten as

$$\int_V \mathbf{f} \cdot \mathbf{r}\, \delta V = \int_S \mathbf{r} \cdot (\delta\mathbf{S} \cdot \overleftrightarrow{\mathbf{T}}) - \int_V \overleftrightarrow{\mathbf{T}} \cdot \cdot \overleftrightarrow{\mathbf{I}}\, \delta V \tag{109}$$

For a force density magnetic in origin, the magnetic stress tensor is written as

$$\overleftrightarrow{\mathbf{T}} = \frac{1}{8\pi} \mathbf{B}^2 \overleftrightarrow{\mathbf{I}} - \frac{1}{4\pi} \mathbf{B}\mathbf{B} \tag{110}$$

thus the relations

$$\overleftrightarrow{\mathbf{T}} \cdot \cdot \overleftrightarrow{\mathbf{I}} = \frac{1}{8\pi} \mathbf{B}^2 \qquad \nabla \cdot \overleftrightarrow{\mathbf{T}} = \frac{1}{4\pi} \mathbf{B} \wedge \nabla \wedge \mathbf{B} \tag{111}$$

from which the virial relation follows:

$$8\pi \int_V \mathbf{f} \cdot \mathbf{r}\, \delta V = \int_S \mathbf{B}^2(\mathbf{r} \cdot \delta\mathbf{S}) - 2\int_S (\mathbf{r} \cdot \mathbf{B})(\mathbf{B} \cdot \delta\mathbf{S}) - \int_V \mathbf{B}^2\, \delta V \tag{112}$$

If the magnetic field $\mathbf{B}$ is a Beltrami field, the left-handed part of the equation above is equal to zero, thus resulting into the following equality:

$$\int_V \mathbf{B}^2\, \delta V = \int_S \mathbf{B}^2(\mathbf{r} \cdot \delta\mathbf{S}) - 2\int_S (\mathbf{r} \cdot \mathbf{B})(\mathbf{B} \cdot \delta\mathbf{S}) \tag{113}$$

The left-handed side of this equation represents a magnetic energy included in the volume V, which is a positively defined quantity. As a consequence, the magnetic field cannot have a null value on the surface S.

C. Ordinary Fields and the Superposition Principle

Ordinary fields fulfill the relation $\nabla \wedge \mathbf{A} = \mathbf{K} \wedge \mathbf{A}$, where vector $\mathbf{K}$ is a function of space and time. This results in the following identity:

$$\mathbf{P} = \mathbf{A} \wedge \nabla \wedge \mathbf{A} = \mathbf{A} \wedge (\mathbf{K} \wedge \mathbf{A}) = \mathbf{A}^2 \mathbf{K} - (\mathbf{A} \cdot \mathbf{K})\mathbf{A} \tag{114}$$

The vector $\mathbf{P}$ satisfies the condition $\mathbf{P} \neq 0$. In order to demonstrate this claim, we must only prove that this vector cannot be null-valued. Indeed, if it were null, the equality $\mathbf{A}^2\mathbf{K} = (\mathbf{A} \cdot \mathbf{K})\mathbf{A}$ would be verified, or $A^2K^2 = (\mathbf{A} \cdot \mathbf{K})^2$, specifically, $\cos^2\theta = (\mathbf{A} \cdot \mathbf{K})^2/(AK)^2 = 1$, which contradicts the very fact that $\mathbf{K} \wedge \mathbf{A} \neq 0$.

As already illustrated with Beltrami fields, the rotational of an ordinary field $\mathbf{B} = \nabla \wedge \mathbf{A}$ is also demonstrated to be an ordinary field satisfying the following relations:

$$\mathbf{B} \cdot \nabla \wedge \mathbf{B} = 0 \Rightarrow \mathbf{B} \wedge \nabla \wedge \mathbf{B} \neq 0 \tag{115}$$

To demonstrate the condition $\mathbf{B} \cdot \nabla \wedge \mathbf{B} = 0$, knowing that $\mathbf{B} = \mathbf{K} \wedge \mathbf{A}$, the following set of identities is employed:

$$\mathbf{B} \cdot \nabla \wedge \mathbf{B} = \mathbf{B} \cdot \nabla \cdot (\mathbf{AK} - \mathbf{KA}) \tag{116}$$

$$\mathbf{D} = (\mathbf{AK} - \mathbf{KA}) \cdot \mathbf{B} = \mathbf{B} \wedge \mathbf{B} = 0 \tag{117}$$

$$\nabla \cdot \mathbf{D} = \mathbf{B} \cdot \nabla \cdot (\mathbf{AK} - \mathbf{KA}) + (\mathbf{AK} - \mathbf{KA}) \cdot \cdot \nabla \mathbf{B} \tag{118}$$

But the equality $\mathbf{AK} \cdot \cdot \cdot \nabla \mathbf{B} = \mathbf{KA} \cdot \cdot \nabla \mathbf{B}$ implies the condition $\mathbf{B} \cdot \nabla \wedge \mathbf{B} = 0$.

Beltrami fields can also result from the superposition of ordinary vectorial fields. Indeed, a field $\mathbf{A} = \mathbf{A}_1 + \mathbf{A}_2$, which results from the addition of two ordinary scalar fields, is not necessarily an ordinary field:

$$\mathbf{A} \cdot \nabla \wedge \mathbf{A} = \mathbf{A}_1 \cdot \nabla \wedge \mathbf{A}_2 + \mathbf{A}_2 \cdot \nabla \wedge \mathbf{A}_1 \neq 0 \tag{119}$$

Figure 1 illustrates the different possibilities resulting from the superposition of $\mathbf{A}_1$ and $\mathbf{A}_2$ vectors. We remark that Beltrami fields do not follow the principle of field superposition. Indeed, if $\mathbf{A}_1$ and $\mathbf{A}_2$ vectors were Beltrami fields, the following relations would immediately follow:

$$\nabla \wedge (\mathbf{A}_1 + \mathbf{A}_2) = k_1\mathbf{A}_1 + k_2\mathbf{A}_2 \neq (k_1 + k_2)(\mathbf{A}_1 + \mathbf{A}_2) \tag{120}$$

However, the principle of field superposition is verified if the condition $k_1 = k_2$ is imposed. On the other hand, the superposition of two Trkalian fields generates a

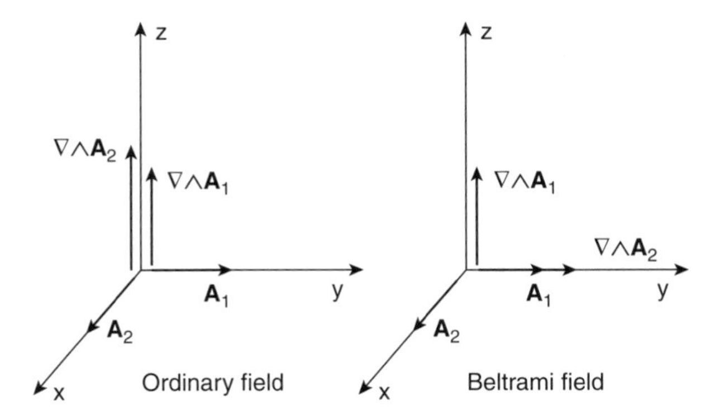

Figure 1. Beltrami field as the sum of ordinary fields.

Trkalian field. For example, the field $\mathbf{B} = \mathbf{A} + k^{-1}\nabla \wedge \mathbf{A}$ is Trkalian if the $\mathbf{A}$ field also belongs to this category.

Let us now consider the Trkalian field $\mathbf{B} = \nabla \wedge \mathbf{A} + k^{-1}\nabla \wedge \nabla \wedge \mathbf{A}$ and try to determine whether the $\mathbf{A}$ field is also a Trkalian field. The rotational of $\mathbf{B}$ is written as

$$\nabla \wedge \mathbf{B} = \nabla \wedge \nabla \wedge \mathbf{A} + k^{-1}\nabla \wedge \nabla \wedge \nabla \wedge \mathbf{A} \tag{121}$$

With the condition $\nabla \wedge \mathbf{B} = k\mathbf{B} = k\nabla \wedge \mathbf{A} + \nabla \wedge \nabla \wedge \mathbf{A}$, the following identity is found:

$$\nabla \wedge \nabla \wedge \nabla \wedge \mathbf{A} - k^2\nabla \wedge \mathbf{A} = 0 \tag{122}$$

which can be rewritten as

$$\nabla \wedge [\nabla(\nabla \cdot \mathbf{A}) - \Delta\mathbf{A} - k^2\mathbf{A}] = 0 \tag{123}$$

The solution of the preceding equation has the following expression:

$$\Delta\mathbf{A} + k^2\mathbf{A} = \nabla\Psi \tag{124}$$

The $\mathbf{A}$ field will be a Trkalian field for $\Psi = 0$ if the condition $\nabla \cdot \mathbf{A} = 0$ is verified. If this condition is not verified, when $\Psi \neq 0$, we can carry out a gauge transform $\mathbf{A} = \mathbf{A}_0 + \nabla\Phi$ with the condition $\nabla \cdot \mathbf{A}_0 = 0$. The preceding equation is now verified, provided each of the following equations is verified as well:

$$\Delta\Phi + k^2\Phi = \Psi \qquad \Delta\mathbf{A}_0 + k^2\mathbf{A}_0 = 0 \tag{125}$$

As a consequence, the gauge transform written above have the property to turn an ordinary field $\mathbf{A}$ into a Trkalian field $\mathbf{A}_0$.

D. Hansen Decomposition and Beltrami Field

In Helmholtz theorem, both Φ and $\mathbf{A}$ potentials form a group composed of four scalar components in order to describe a vectorial field having three scalar components. As a consequence, the four scalar components are not independent. Their dependence can be expressed in several ways:

The first one aims at defining a potential $\mathbf{A} = \Psi\nabla\Pi$ relying on two scalar functions Π and Ψ known as "Euler potentials" or "Clebsch parameters." In this case, the $\mathbf{B}$ field depends on a set of three potentials known as Monge potentials:

$$\mathbf{B} = \nabla\Phi + \nabla\wedge\mathbf{A} = \nabla\Phi + \nabla\Psi\wedge\nabla\Pi \tag{126}$$

The $\mathbf{A}$ field is an ordinary field verifying the condition $\mathbf{A}\cdot\nabla\wedge\mathbf{A} = 0$, which is also a necessary and sufficient condition for the differential form $d\Pi = \nabla\Pi\cdot d\mathbf{r} = (\mathbf{A}/\Psi)\cdot d\mathbf{r}$ to be a total differential relation. Interestingly, we must underline the fact that $\mathbf{B}_\perp$-field lines are the intersections of $\Psi = Ct$ and $\Pi = Ct$ surfaces since $\nabla\Psi$ and $\nabla\Pi$ gradients are normal both to $\mathbf{B}_\perp$ and to these surfaces. Indeed, the definition of the $\mathbf{B}_\perp$ field implies the equalities $\nabla\Pi\cdot\mathbf{B}_\perp = 0$ and $\nabla\Psi\cdot\mathbf{B}_\perp = 0$.

When the $\mathbf{B}_\perp$ field is an ordinary field, the $\mathbf{B}$ field admits the so-called Clebsch decomposition, written as

$$\mathbf{B} = \nabla\Phi + \nabla\wedge\mathbf{A} = \nabla\Phi + \Psi\nabla\Pi \tag{127}$$

Clebsch's decomposition allows us to identify the $\mathbf{B}$ field either as an ordinary field or as a Beltrami field if its calculated density of helicity $\mathbf{B}\cdot\nabla\wedge\mathbf{B} = \nabla\Phi\cdot(\nabla\Psi\wedge\nabla\Pi)$ is respectively equal to or different from zero. Interestingly, two authors [37,38] recently proposed a dual Clebsch decomposition along the following lines of reasoning

$$\mathbf{B}_r = \nabla\Phi + \Psi\nabla\Pi \qquad \mathbf{B}_i = \nabla\Psi - \Phi\nabla\Pi \tag{128}$$

where the $\mathbf{B}_r$ and $\mathbf{B}_i$ fields were Trkalian fields fulfilling the conditions:

$$\nabla\wedge\mathbf{B}_r = k\mathbf{B}_r \qquad \nabla\wedge\mathbf{B}_i = k\mathbf{B}_i \tag{129}$$

From the preceding relations, one obtains the following set of identities:

$$\mathbf{B}_r\cdot\nabla\wedge\mathbf{B}_r = k\mathbf{B}_r^2 = \mathbf{B}_i\cdot\nabla\wedge\mathbf{B}_i = k\mathbf{B}_i^2 = \nabla\Phi\cdot(\nabla\Psi\wedge\nabla\Pi) \neq 0 \tag{130}$$

Starting from the dual decomposition given above, a complex field $\mathbf{B}_c = \mathbf{B}_r + j\mathbf{B}_i$ can be found to verify the conditions:

$$\nabla \wedge \mathbf{B}_c = k\mathbf{B}_c \Rightarrow \mathbf{B}_c \cdot \nabla \wedge \mathbf{B}_c = k\mathbf{B}_c^2 \tag{131}$$

However, the preceding conditions, as opposed to what usually happens in the real case, are not sufficient conditions to define a complex Trkalian field. Indeed, the couple of authors we referred to, demonstrated that for $\mathbf{B}_r^2 \neq 0$ and $\mathbf{B}_i^2 \neq 0$, the complex field $\mathbf{B}_c$ is an ordinary field if we chose $\mathbf{B}_c = e^{j\Pi}\nabla F$, where F is a complex quantity having the expression $F = (\Phi + j\Psi)e^{-j\Pi}$. This results from the equality $\nabla \wedge \mathbf{B}_c = j\nabla\Pi \wedge \mathbf{B}_c$ which implies $\mathbf{B}_c \cdot \nabla \wedge \mathbf{B}_c = k\mathbf{B}_c^2 = 0$ where k is a constant different from zero. In this case, the definition of the $\mathbf{B}_c$ field implies the following relations:

$$\nabla\Pi \wedge \nabla F = -jk\nabla F \qquad (\nabla F)^2 = 0 \tag{132}$$

The second way to proceed is to decompose the $\mathbf{A}$ vector potential as the addition of two terms having the form:

$$\mathbf{A} = \Psi\mathbf{r} + \nabla \wedge (\Pi\mathbf{r}) \tag{133}$$

As a consequence, the vectorial field $\mathbf{B}$ is defined on the basis of three independent scalar fields:

$$\mathbf{B} = \nabla\Phi + \nabla \wedge (\Psi\mathbf{r}) + \nabla \wedge \nabla \wedge (\Pi\mathbf{r}) \tag{134}$$

Vectorial fields $\mathbf{M} = \nabla \wedge (\Psi\mathbf{r})$ and $\mathbf{N} = \nabla \wedge \nabla \wedge (\Pi\mathbf{r})$ are respectively known as toroidal and poloidal fields. The scalar potentials Ψ and Π are called *Debye potentials*, while the vector potentials $\mathbf{F} = \Psi\mathbf{r}$ and $\mathbf{G} = \Pi\mathbf{r}$ are known as *Whittaker potentials.*

Another decomposition of the vectorial field $\mathbf{B}$ can be found in the literature. This decomposition is formally identical to the previous one, with a notable difference, though, the vector $\mathbf{r}$ is now replaced by a constant unitary vector $\mathbf{n}$. Still another decomposition of the $\mathbf{B}$ field was proposed by Rowe [39]. This decomposition is written as

$$\mathbf{B} = \mathbf{B}_{\parallel} + \mathbf{B}_{\perp} = \nabla\Phi + \Pi\mathbf{r} + \nabla \wedge (\Psi\mathbf{r}) \tag{135}$$

The choice of scalar potentials Φ, Ψ, Π is unique if we demand that the mean values of functions $\Phi(\mathbf{r}), \Psi(\mathbf{r})$ over a sphere centered on the origin fulfill the conditions:

$$\int_0^{4\pi} \Phi(r\mathbf{n})\, d\Omega = 0 \qquad \int_0^{4\pi} \Psi(r\mathbf{n})\, d\Omega = 0 \tag{136}$$

If we force the condition $\nabla \wedge \mathbf{B}_{\parallel} = 0$, the condition $\nabla\Pi \wedge \mathbf{r} = 0$ must result. This latter condition is fulfilled for a function having the form $\Pi(r)$. In this specific case, Eq. (135) can be rewritten as

$$\mathbf{B} = \mathbf{B}_{\parallel} + \mathbf{B}_{\perp} = \nabla\Phi_0 + \nabla \wedge (\Psi\mathbf{r}) \tag{137}$$

where the potential Φ_0 is defined as

$$\Phi_0(\mathbf{r}) = \Phi(\mathbf{r}) + \int_0^r \Pi(r')r'dr' \tag{138}$$

This potential is decomposed into a potential Φ, the mean value of which, on a spherical surface of radius r, is reduced to zero by means of condition (136) and an integral term that represents the contribution of Π inside the sphere.

In the case when $\Phi = \Psi$ and $k\Pi = \Psi$, knowing k is a parameter that does not depend on space, we obtain the decomposition of the $\mathbf{B}$ field introduced by Hansen [40]:

$$\mathbf{B} = \mathbf{L} + \mathbf{M} + \mathbf{N} = \nabla\Psi + \nabla \wedge (\Psi\mathbf{r}) + k^{-1}\,\nabla \wedge \nabla \wedge (\Psi\mathbf{r}) \tag{139}$$

If the scalar field Ψ verifies Helmholtz equation

$$\Delta\Psi + k^2\Psi = 0 \tag{140}$$

then in this specific case, the vectors $\mathbf{L}, \mathbf{M}, \mathbf{N}$ are solutions of Helmholtz vectorial equation:

$$\Delta\mathbf{Q} + k^2\mathbf{Q} = 0 \tag{141}$$

As a consequence of the preceding definitions, the following set of equalities can be found:

$$\mathbf{M} = \mathbf{L} \wedge \mathbf{r} \qquad \nabla \wedge \mathbf{M} = k\mathbf{N} \qquad \nabla \wedge \mathbf{N} = k\mathbf{M} \tag{142}$$

Fields $\Psi\mathbf{r}$ and $\nabla \wedge (\Psi\mathbf{r})$ are ordinary fields but one cannot conclude anything concerning the field $\mathbf{A} = \Psi\mathbf{r} + \nabla \wedge (\Psi\mathbf{r})/k$. This field is not Trkalian, even though it fulfills the condition $\mathbf{A} \cdot \nabla \wedge \mathbf{A} \neq 0$ since $\nabla \cdot \mathbf{A} \neq 0$. This confirms the claim above concerning the superposition principle of ordinary or Beltrami fields. On the other hand, the solenoidal field $\mathbf{B}_{\perp} = \nabla \wedge \mathbf{A} = \mathbf{M} + \mathbf{N}$ is a Trkalian field because it can be shown that $\nabla \wedge \mathbf{B}_{\perp} = k\mathbf{B}_{\perp}$ by using the equalities given above. We shall also notice that a toroidal field $\mathbf{M} = \nabla \wedge (\Psi\mathbf{r})$ can be formulated as a function of two Eulerian potentials:

$$\mathbf{M} = \nabla \wedge (\Psi\mathbf{r}) = \nabla\Psi \wedge \mathbf{r} = \nabla\Psi \wedge \nabla\frac{r^2}{2} \tag{143}$$

E. Hansen Decomposition in Different Coordinate Systems

1. Case of Cartesian Coordinates

Hansen's decomposition for a constant unitary vector **n** is written as

$$\mathbf{L} + \mathbf{M} + \mathbf{N} = \nabla\Psi + \nabla\wedge(\Psi\mathbf{n}) + k^{-1}\,\nabla\wedge\nabla\wedge(\Psi\mathbf{n}) \tag{144}$$

For a scalar potential having the form $\Psi = e^{\epsilon j\mathbf{k}\cdot\mathbf{r}}$, vectorial functions $\mathbf{L}, \mathbf{M}, \mathbf{N}$ have the expression

$$\mathbf{L}(\mathbf{k},\mathbf{r}) = \epsilon j\mathbf{k}\Psi \qquad \mathbf{M}(\mathbf{k},\mathbf{r}) = \epsilon j\mathbf{k}\wedge\mathbf{n}\Psi \qquad k\mathbf{N}(\mathbf{k},\mathbf{r}) = -\mathbf{k}\wedge(\mathbf{k}\wedge\mathbf{n})\Psi \tag{145}$$

This results in the fact that the $\mathbf{L}, \mathbf{M}, \mathbf{N}$ vectors are orthogonal to one another:

$$\mathbf{L}\cdot\mathbf{M} = \mathbf{M}\cdot\mathbf{N} = \mathbf{N}\cdot\mathbf{L} = 0 \tag{146}$$

An arbitrary field $\mathbf{E}(\mathbf{r})$ can be represented by a linear combination of vectorial functions given above, namely

$$\mathbf{E}(\mathbf{r}) = \frac{1}{(2\pi)^3}\int_{-\infty}^{+\infty}\int_{-\infty}^{+\infty}\int_{-\infty}^{+\infty}(a\mathbf{L} + b\mathbf{M} + c\mathbf{N})\,d\mathbf{k}^3 \tag{147}$$

where the quantities a, b, c are unknown functions of $\mathbf{k}$.

In the same spirit, a Green dyadic function can be rewritten in the following form:

$$\mathbf{GG}(\mathbf{r},\mathbf{r}') = \frac{1}{(2\pi)^3}\int_{-\infty}^{+\infty}\int_{-\infty}^{+\infty}\int_{-\infty}^{+\infty}(\mathbf{AL} + \mathbf{BM} + \mathbf{CN})\,d\mathbf{k}^3 \tag{148}$$

We shall stress the fact that the scalar function Ψ fulfills the condition of orthogonality:

$$\int_{-\infty}^{+\infty}\int_{-\infty}^{+\infty}\int_{-\infty}^{+\infty}\Psi(\mathbf{k},\mathbf{r})\Psi(-\mathbf{k}',\mathbf{r})\delta\mathbf{r}^3 = \delta(\mathbf{k}-\mathbf{k}') \tag{149}$$

In the same spirit, vectorial functions $\mathbf{L}, \mathbf{M}, \mathbf{N}$ respect the relations of normality:

$$\int_{-\infty}^{+\infty}\int_{-\infty}^{+\infty}\int_{-\infty}^{+\infty}\mathbf{L}(\mathbf{k},\mathbf{r})\cdot\mathbf{L}(-\mathbf{k}',\mathbf{r})\delta\mathbf{r}^3 = k^2\,\delta(\mathbf{k}-\mathbf{k}') \tag{150}$$

$$\int_{-\infty}^{+\infty}\int_{-\infty}^{+\infty}\int_{-\infty}^{+\infty}\mathbf{M}(\mathbf{k},\mathbf{r})\cdot\mathbf{M}(-\mathbf{k}',\mathbf{r})\delta\mathbf{r}^3 = k_t^2\,\delta(\mathbf{k}-\mathbf{k}') \tag{151}$$

$$\int_{-\infty}^{+\infty}\int_{-\infty}^{+\infty}\int_{-\infty}^{+\infty}\mathbf{N}(\mathbf{k},\mathbf{r})\cdot\mathbf{N}(-\mathbf{k}',\mathbf{r})\delta\mathbf{r}^3 = k_t^2\,\delta(\mathbf{k}-\mathbf{k}') \tag{152}$$

with the definition $k_t^2 = k_x^2 + k_y^2$ for a vector **n** directed along the z direction.

Moreover, the vectorial functions **L**,**M**,**N** are orthogonal to each other:

$$\int_{-\infty}^{+\infty}\int_{-\infty}^{+\infty}\int_{-\infty}^{+\infty} \mathbf{L}(\mathbf{k},\mathbf{r})\cdot\mathbf{M}(-\mathbf{k}',\mathbf{r})\delta\mathbf{r}^3 = 0 \tag{153}$$

with two other relations obtained while performing a circular permutation of the $\mathbf{L},\mathbf{M},\mathbf{N}$ vectors. The orthogonal relations written above are obtained by applying the Gauss theorem to calculate these integrals, a procedure that is demonstrated in detail by Chew [41]. These relations of orthogonality enable us to determine the unknown functions $\mathbf{A}_0,\mathbf{B}_0,\mathbf{C}_0$ associated with Dirac's dyadic distribution over space written as

$$\delta(\mathbf{r}-\mathbf{r}')\overleftrightarrow{\mathbf{I}} = \frac{1}{(2\pi)^3}\int_{-\infty}^{+\infty}\int_{-\infty}^{+\infty}\int_{-\infty}^{+\infty}(\mathbf{A}_0\mathbf{L}+\mathbf{B}_0\mathbf{M}+\mathbf{C}_0\mathbf{N})\,d\mathbf{k}^3 \tag{154}$$

The calculation of the $\mathbf{A}_0(\mathbf{k},\mathbf{r}')$ vector results from the following sequence:

$$\begin{aligned}&\int_{-\infty}^{+\infty}\int_{-\infty}^{+\infty}\int_{-\infty}^{+\infty}\delta(\mathbf{r}-\mathbf{r}')\overleftrightarrow{\mathbf{I}}\cdot\mathbf{L}(-\mathbf{k}',\mathbf{r})\delta\mathbf{r}^3\\ &=\frac{1}{(2\pi)^3}\int_{-\infty}^{+\infty}\int_{-\infty}^{+\infty}\int_{-\infty}^{+\infty}\mathbf{A}_0(\mathbf{k},\mathbf{r}')d\mathbf{k}^3\int_{-\infty}^{+\infty}\int_{-\infty}^{+\infty}\int_{-\infty}^{+\infty}\mathbf{L}(\mathbf{k},\mathbf{r})\cdot\mathbf{L}(-\mathbf{k}',\mathbf{r})\delta\mathbf{r}^3\end{aligned} \tag{155}$$

The preceding equation can also be rewritten as

$$\mathbf{L}(-\mathbf{k}',\mathbf{r}') = \frac{1}{(2\pi)^3}\int_{-\infty}^{+\infty}\int_{-\infty}^{+\infty}\int_{-\infty}^{+\infty}k^2\delta(\mathbf{k}-\mathbf{k}')\mathbf{A}_0(\mathbf{k},\mathbf{r}')\,d\mathbf{k}^3 \tag{156}$$

Thus resulting in the equality $\mathbf{L}(-\mathbf{k}',\mathbf{r}') = k^2\mathbf{A}_0(\mathbf{k}',\mathbf{r}')$. After a similar calculation for $\mathbf{B}_0$ and $\mathbf{C}_0$ functions, the spectral component of Dirac's dyadic distribution is finally found to have the following expression:

$$\begin{aligned}\mathbf{A}_0\mathbf{L}+\mathbf{B}_0\mathbf{M}+\mathbf{C}_0\mathbf{N} = \frac{1}{k^2}\,\mathbf{L}(\mathbf{k},\mathbf{r})\mathbf{L}(-\mathbf{k},\mathbf{r}') + \frac{1}{k_t^2}\,[\mathbf{M}(\mathbf{k},\mathbf{r})\mathbf{M}(-\mathbf{k},\mathbf{r}')\\ +\mathbf{N}(\mathbf{k},\mathbf{r})\mathbf{N}(-\mathbf{k},\mathbf{r}')]\end{aligned} \tag{157}$$

The Green dyadic function is a solution of the following equation:

$$\nabla\wedge\nabla\wedge\mathbf{GG} - k_0^2\,\mathbf{GG} = \delta(\mathbf{r}-\mathbf{r}')\overleftrightarrow{\mathbf{I}} \tag{158}$$

In order to calculate the unknown functions $\mathbf{A}, \mathbf{B}, \mathbf{C}$, which are part of the integral expression of Green's dyadic function, this function is substituted in the left-hand side of the preceding equation, and supplemented with the definitions written above to obtain, after proper calculations:

$$\mathbf{AL} + \mathbf{BM} + \mathbf{CN} = -\frac{1}{k_0^2 k^2}\,\mathbf{L}(\mathbf{k},\mathbf{r})\mathbf{L}(-\mathbf{k},\mathbf{r}') + \frac{1}{k_t^2(k^2 - k_0^2)}\,[\mathbf{M}(\mathbf{k},\mathbf{r})\mathbf{M}(-\mathbf{k},\mathbf{r}') + \mathbf{N}(\mathbf{k},\mathbf{r})\mathbf{N}(-\mathbf{k},\mathbf{r}')] \tag{159}$$

We notice that the integration of the preceding term with an $\mathbf{L}$, over any component of $\mathbf{k}$ cannot be performed using the residual method since the term in $\mathbf{L}$ does not satisfy Jordan's lemma. As a consequence, the term in $\mathbf{L}$ must be modified by successively subtracting from it and adding to it, a term that has the form $k^2\overleftrightarrow{\boldsymbol{\Omega}}(\mathbf{r}')e^{\epsilon j\mathbf{k}\cdot(\mathbf{r}-\mathbf{r}')}$. This operation enables us to withdraw the singularity from the integral, now calculated as the principal value, and to use it again in order to calculate the integral:

$$-\frac{1}{k_0^2}\,\overleftrightarrow{\boldsymbol{\Omega}}(\mathbf{r}')\int_{-\infty}^{+\infty}\int_{-\infty}^{+\infty}\int_{-\infty}^{+\infty} e^{\epsilon j\mathbf{k}\cdot(\mathbf{r}-\mathbf{r}')}\,d\mathbf{k}^3 = -\frac{1}{k_0^2}\,\delta(\mathbf{r}-\mathbf{r}')\overleftrightarrow{\boldsymbol{\Omega}}(\mathbf{r}') \tag{160}$$

2. *Case of Cylindrical Coordinates*

In cylindrical coordinates, the scalar field Ψ admits a Fourier–Bessel mode decomposition under the form

$$\Psi_n(\mathbf{k},\mathbf{r}) = e^{\epsilon jk_z z + jn\theta}\,J_n(k_r r) \tag{161}$$

with the definition $k^2 = k_r^2 + k_z^2$.

An arbitrary field $\mathbf{E}(\mathbf{r})$ can be represented as a linear combination of vectorial functions $\mathbf{L}_n, \mathbf{M}_n, \mathbf{N}_n$ written as

$$\mathbf{E}(\mathbf{r}) = \frac{1}{(2\pi)^3}\sum_{n=-\infty}^{+\infty}\int_{-\infty}^{+\infty}\int_{0}^{+\infty}(a_n\mathbf{L}_n + b_n\mathbf{M}_n + c_n\mathbf{N}_n)k_r\,dk_z\,dk_r \tag{162}$$

where the vectorial functions $\mathbf{L}_n, \mathbf{M}_n, \mathbf{N}_n$ depend on variables $k_r, k_z, \mathbf{r}$ while factors a_n, b_n, c_n are unknown functions of k_r and k_z.

In the same spirit, a dyadic Green function in a cylindrical coordinate system is written as

$$\mathbf{GG}(\mathbf{r},\mathbf{r}') = \frac{1}{(2\pi)^3}\sum_{n=-\infty}^{+\infty}\int_{-\infty}^{+\infty}\int_{0}^{+\infty}(\mathbf{A}_n\mathbf{L}_n + \mathbf{B}_n\mathbf{M}_n + \mathbf{C}_n\mathbf{N}_n)k_r\,dk_z\,dk_r \tag{163}$$

where unknown vectors $\mathbf{A}_n, \mathbf{B}_n, \mathbf{C}_n$ are now functions of variables $k_r, k_z, \mathbf{r}'$.

The relations of orthogonality in a cylindrical coordinate system are now given by the following relations:

$$\int_{-\infty}^{+\infty}\int_{0}^{+\infty}\int_{0}^{2\pi}\Psi_n(k_r,k_z,\mathbf{r})\Psi_{-n'}(-k_r',-k_z',\mathbf{r})r\,dz\,dr\,d\theta=\frac{1}{k_r}\,\delta_{nn'}\delta(k_r-k_r')\delta(k_z-k_z') \tag{164}$$

$$\int_{-\infty}^{+\infty}\int_{0}^{+\infty}\int_{0}^{2\pi}\mathbf{L}_n(k_r,k_z,\mathbf{r})\mathbf{L}_{-n'}(-k_r',-k_z',\mathbf{r})r\,dz\,dr\,d\theta=\frac{k^2}{k_r}\,\delta_{nn'}\delta(k_r-k_r')\delta(k_z-k_z') \tag{165}$$

$$\int_{-\infty}^{+\infty}\int_{0}^{+\infty}\int_{0}^{2\pi}\mathbf{M}_n(k_r,k_z,\mathbf{r})\mathbf{M}_{-n'}(-k_r',-k_z',\mathbf{r})r\,dz\,dr\,d\theta=k_r\,\delta_{nn'}\delta(k_r-k_r')\delta(k_z-k_z') \tag{166}$$

$$\int_{-\infty}^{+\infty}\int_{0}^{+\infty}\int_{0}^{2\pi}\mathbf{N}_n(k_r,k_z,\mathbf{r})\mathbf{N}_{-n'}(-k_r',-k_z',\mathbf{r})r\,dz\,dr\,d\theta=k_r\,\delta_{nn'}\delta(k_r-k_r')\delta(k_z-k_z') \tag{167}$$

Dirac's dyadic distribution represented in a cylindrical coordinate system is written as

$$\delta(\mathbf{r}-\mathbf{r}')\overleftrightarrow{\mathbf{I}}=\frac{1}{(2\pi)^3}\sum_{n=-\infty}^{+\infty}\int_{-\infty}^{+\infty}\int_{0}^{+\infty}(\mathbf{A}_n\mathbf{L}_n+\mathbf{B}_n\mathbf{M}_n+\mathbf{C}_n\mathbf{N}_n)k_r\,dk_z\,dk_r \tag{168}$$

After performing a series of calculations quite similar to the one already employed in the Cartesian case, the unknown function $\mathbf{A}_n$ is expressed as

$$\mathbf{A}_{n'}(k_r',k_z',\mathbf{r}')=\frac{2\pi}{k^2}\,\mathbf{L}_{n'}(k_r',-k_z',\mathbf{r}') \tag{169}$$

where the identity $\mathbf{L}_{n'}(k_r',-k_z',\mathbf{r}')=\mathbf{L}_{-n'}(-k_r',-k_z',\mathbf{r}')$ has been used.

Once completed, this calculation results in the following equation:

$$\begin{aligned}\mathbf{A}_n\mathbf{L}_n+\mathbf{B}_n\mathbf{M}_n+\mathbf{C}_n\mathbf{N}_n=&\frac{2\pi}{k^2}\,\mathbf{L}_n(k_r,k_z,\mathbf{r})\mathbf{L}_n(k_r,-k_z,\mathbf{r}')\\&+\frac{2\pi}{k_r^2}\,[\mathbf{M}_n(k_r,k_z,\mathbf{r})\mathbf{M}_n(k_r,-k_z,\mathbf{r}')\\&+\mathbf{N}_n(k_r,k_z,\mathbf{r})\mathbf{N}_n(k_r,-k_z,\mathbf{r}')]\end{aligned} \tag{170}$$

The Green's dyadic function is a solution of

$$\nabla\wedge\nabla\wedge\mathbf{GG}-k_0^2\,\mathbf{GG}=\delta(\mathbf{r}-\mathbf{r}')\overleftrightarrow{\mathbf{I}} \tag{171}$$

In accordance with the methodology employed while working with Cartesian coordinates, the spectral decomposition of Green's function and Dirac's distributions are both exchanged with their cylindrical coordinates in Eq. (171) written above in order to obtain the spectral component of Green's function:

$$\begin{aligned}\mathbf{A}_n\mathbf{L}_n + \mathbf{B}_n\mathbf{M}_n + \mathbf{C}_n\mathbf{N}_n = &-\frac{2\pi}{k_0^2k^2}\,\mathbf{L}_n(k_r, k_z, \mathbf{r})\mathbf{L}_n(k_r, -k_z, \mathbf{r}') \\ &+ \frac{2\pi}{k_r^2(k^2 - k_0^2)}\,[\mathbf{M}_n(k_r, k_z, \mathbf{r})\mathbf{M}_n(k_r, -k_z, \mathbf{r}') \\ &+ \mathbf{N}_n(k_r, k_z, \mathbf{r})\mathbf{N}_n(k_r, -k_z, \mathbf{r}')]\end{aligned} \tag{172}$$

The term in $\mathbf{L}$ exhibits a singularity identical to the one we already dealt with in the Cartesian case. Therefore, we can employ the same process to extract the singularity. We shall find in Chew's treatise [41] a detailed study concerning the extraction of singularity in the cylindrical case.

IV. INVESTIGATION OF TOPOLOGICAL EFFECTS IN PHYSICS

Topology is the discipline within mathematical science in charge of studying the seemingly intuitive concepts of continuity and limit. This discipline is itself composed of several fields of investigation. Within the scope of this particular section, we shall now focus on differential topology, with the understanding that this discipline is tasked with classifying surfaces and dealing with problems usually connected with knots. We shall first remind our readers of some basic definitions that might be helpful, if not necessary, for following the course of this discussion:

Connected Region. A region of space enclosed within the boundaries of a closed surface is described as a "connected region" if it is possible to skip from one point to an other using an infinite number of paths, all of them located within this specific region of space. For example, each region located inside and outside a closed surface is, individually, a connected region.

Reconcilable Circuits. Whether closed or open, two circuits are mutually reconcilable if they can be brought to match one another by continuous distortion, without leaving the connected region of space. If not, these circuits are described as irreconcilable.

Loops or Reducible Surfaces. Any loop or closed surface located within a connected region of space is described as "reducible" if it can be continuously contracted into a point without leaving the aforementioned region of space, as illustrated by loop C_1 in Fig. 2. In the opposite case,

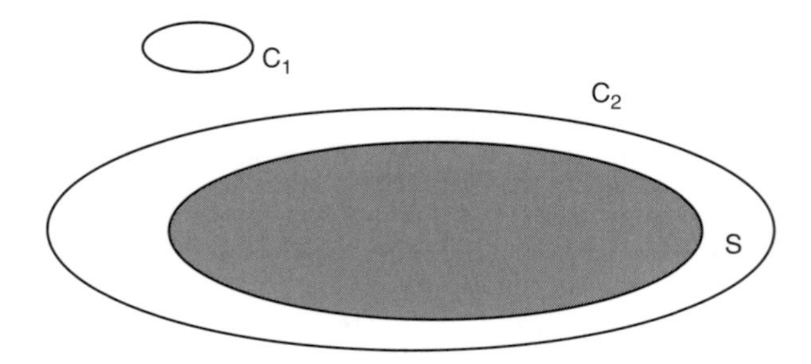

Figure 2. Topological definitions.

the loop or surface is termed "irreducible," as illustrated by loop C_2 in Fig. 2. As a consequence, reconcilable paths form a reducible loop when they are combined. Conversely, two irreconcilable paths form an irreducible loop.

Simply Connected Region. A region of space is described as "simply connected" when all circuits joining any two points are reconcilable or any loop drawn within the region is reducible. For instance, regions located inside or outside a finite surface are individually, simply connected spaces.

Doubly Connected Region. A region of space is called "doubly connected" when there exists only two irreconcilable paths in this region. In this case, a single irreducible loop can be identified in this region.

Multiply Connected Region. A region of space is described as "multiply connected" if there are n irreconcilable paths or $n - 1$ irreducible loops in this aforementioned region of space. A multiply connected region can be forced into a simply connected region by introducing an appropriate cut capable of preventing any loop to close its path along a contour crossing this cut.

A. Study of Helicity

Helicity [26,42-45] is a pseudoscalar K whose definition is based on the following integral relation:

$$K = \int_V \mathbf{A} \cdot (\nabla \wedge \mathbf{A})\, \delta V \tag{173}$$

Helicity is defined as a qualitative measurement of how a topological configuration is linked, knotted, or twisted. If $\mathbf{A}$ is the vector potential of electromagnetism, the quantity K then defines "magnetic helicity." In order to obtain a nonnull value for K, the condition $\mathbf{A} \cdot (\nabla \wedge \mathbf{A}) \neq 0$ must be verified in the volume V. As a consequence, the $\mathbf{A}$ field is necessarily a Beltrami field.

For a simply connected medium, K is invariant in the gauge transform $\mathbf{A}_0 = \mathbf{A} + \nabla\phi$ if the integral term given below is equal to zero:

$$\int_V \nabla\phi \cdot (\nabla \wedge \mathbf{A}_0)\, \delta V = \int_S \phi(\nabla \wedge \mathbf{A}_0) \cdot \delta\mathbf{S} = 0 \tag{174}$$

For a finite surface S, the equation given above is verified for the condition $\mathbf{B} \cdot \delta\mathbf{S} = 0$, that is, if the normal component of the field $\mathbf{B} = \mathbf{B}_0 = \nabla \wedge \mathbf{A}_0$ vanishes on the surface S. On the other hand, for a spherical surface S whose radius tends toward infinity, the equation will be satisfied provided the quantity $\phi|\mathbf{B}_0| \to 0$ faster than $1/R^2$.

For a multiply connected medium, such as the volume V of a torus bounded by the surface S is a magnetic surface, the preceding integral becomes

$$\int_V \nabla \cdot (\phi\mathbf{B})\, \delta V = \int_S \phi\mathbf{B} \cdot \delta\mathbf{S} + \int_{S_1} [\phi]\, \mathbf{B} \cdot \delta\mathbf{S}_1 \tag{175}$$

where the surface S_1, a cross section of the torus bounded by a loop C_1 represented in Fig. 3, is used as a cut to make the scalar potential ϕ single-valued. The quantity $[\phi]$ defines the "jump" of the potential function through the surface S_1.

Since the bounding surface S of the torus is a magnetic surface, the first term on the right-hand side of Eq. (175) vanishes.

For a loop C_2 located on the torus crossing the surface S_1, we have

$$[\phi] = \int_{C_2} \nabla\phi \cdot \delta\mathbf{r} = \int_{S_2} \mathbf{B} \cdot \delta\mathbf{S}_2 = F_2 \tag{176}$$

The value of $[\phi]$ does not depend on the choice of the loop C_2 provided this loop is situated on the torus. In this case, the value of $[\phi]$ is shown to be constant over

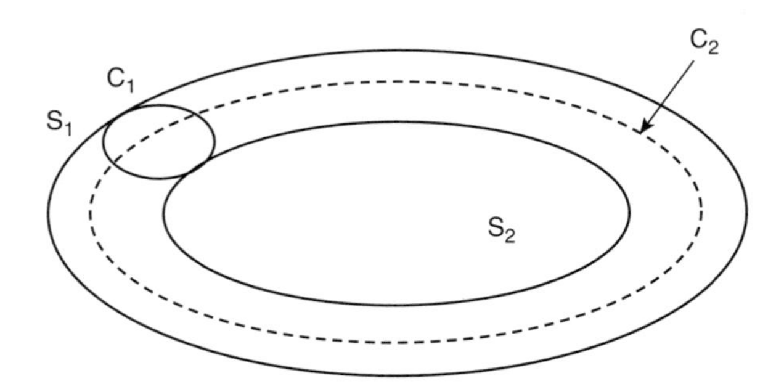

Figure 3. The cross-sectional surface S_1 of the torus is the cut needed to make the function ϕ single-valued.

S_1. We can therefore chose the loop C_2 bounding the surface S_2 inside of the torus. In this case, the value F_2 can now be interpreted as the flux of magnetic field passing through the hole of the torus.

In other words, the gauge transform in a multiply connected medium can be represented by a modification of the helicity that denotes how loops C_1 and C_2 are linked, since we have

$$\int_V \nabla \cdot (\phi \mathbf{B})\, \delta V = [\phi] \int_{S_1} \mathbf{B} \cdot \delta \mathbf{S}_1 = F_1 F_2 \tag{177}$$

We will now demonstrate the relation between helicity and the Gauss linking number.

Let us frist consider the vector potential $\mathbf{A}$ defined by the equation:

$$\mathbf{A}_j(\mathbf{r}_i) = \frac{1}{4\pi} \int_{C_j} \frac{\delta \mathbf{r}_j}{R} \tag{178}$$

where $\mathbf{R} = \mathbf{r}_i - \mathbf{r}_j$. The rotational of the potential $\mathbf{A}$ defines a $\mathbf{B}$ field:

$$\mathbf{B}_j(\mathbf{r}_i) = \frac{1}{4\pi} \int_{C_j} \nabla_i \left(\frac{1}{R}\right) \wedge \delta \mathbf{r}_j \tag{179}$$

The field $\mathbf{B}$ fulfills the condition $\nabla \cdot \mathbf{B} = 0$. This result is consistent with the fact that the curve C_j is a closed loop. The integral written above is identical to Biot–Savart law used to calculate the magnetic field generated by a current loop, provided this integral is multiplied by $4\pi I/c$.

The Gauss integral is an integer number N_{ij} that can be calculated by considering the field $\mathbf{B}_j$ along the curve C_i:

$$N_{ij} = \int_{C_i} \mathbf{B}_j \cdot \delta \mathbf{r}_i = \frac{1}{4\pi} \int_{C_i} \int_{C_j} \left[\nabla_i \left(\frac{1}{R}\right) \wedge \delta \mathbf{r}_j\right] \cdot \delta \mathbf{r}_i \tag{180}$$

where curves C_i and C_j are closed loops endowed with the following features:

1. They are oriented in the counterclockwise direction.
2. Open surfaces S_i and S_j respectively lean on them.

This relation can also be rewritten under another form in order to explain why N_{ij} is an integer number:

$$N_{ij} = \frac{1}{4\pi} \int_{C_i} \int_{C_j} \nabla_j \left(\frac{1}{R}\right) \cdot (\delta \mathbf{r}_i \wedge \delta \mathbf{r}_j) \tag{181}$$

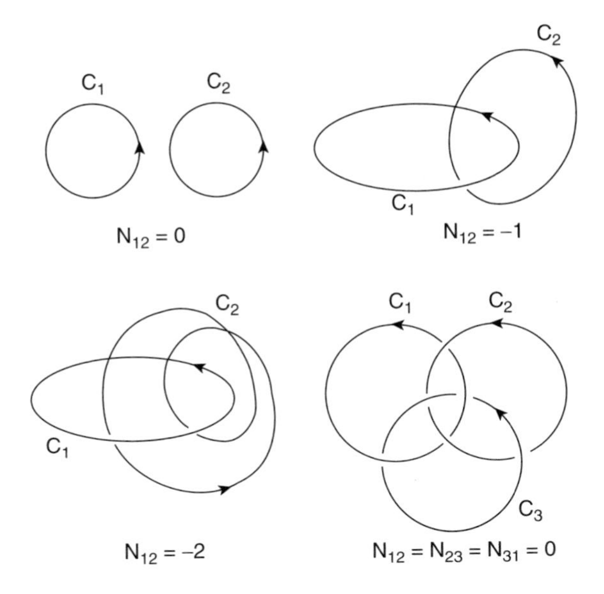

Figure 4. Examples of linking numbers N_{ij} as defined by Eq. (181).

Indeed, the integrand in the previous integral has the dimension of a solid angle $d\Omega = (\delta\mathbf{r}_i \wedge \delta\mathbf{r}_j) \cdot \mathbf{R}/R^3$. It follows that the double integral must give a multiple of 4π. The number $N_{ij} = N_{ji}$ represents the way in which loop C_j is wrapped around the C_i loop and is defined as the total number of piercing of surface S_i by loop C_j. Each piercing contributes to the total number for a quantity $+1$ if intersection of loop C_j is performed in the direction parallel to $\delta\mathbf{S}_i$ or -1 in the opposite case. The number N_{ij}, associated to wrapping, will be null if no piercing is practiced, or when the algebraic sum of piercing gives a null result too. Figure 4 gives a few examples of wrappings for different types of topological configurations.

If the Gauss integral is multiplied by $4\pi I/c$, Ampère's theorem is recovered. Therefore, the Gauss integral allows us to understand why a conducting wire wound as a solenoid, in which a current I flows, originates a magnetic field whose intensity is proportional to the number $n = N/L$ of turns by units of length. Therefore, the topological effect of winding has the following consequence: a significant rise of amplitude of the magnetic field with respect to a rectilinear wire in which the same current flows. Indeed, the amplitude of the magnetic field at a distance R of a rectilinear wire is $B = 2I/cR$ while the magnetic field at the extremity of a solenoid is written as $B = 2\pi nI/c$.

In order to introduce the notion of mutual helicity matching Moffatt's approach [42], the vector potential $\mathbf{A}$ is now defined on the basis of a volumic distribution:

$$\mathbf{A}_j(\mathbf{r}_i) = \frac{1}{4\pi} \int_{V_j} \frac{\mathbf{J}(\mathbf{r}_j)}{R} \, \delta V_j \tag{182}$$

If the definition $\mathbf{J} = \nabla \wedge \mathbf{B}$ is included in the integral written above, it becomes, after transformation:

$$\mathbf{A}_j(\mathbf{r}_i) = \frac{1}{4\pi} \int_{V_j} \mathbf{B}_j(\mathbf{r}_j) \wedge \nabla_j \left(\frac{1}{R} \right) \delta V_j \tag{183}$$

Mutual helicity K_{ij} is defined on the basis of the following relationship:

$$K_{ij} = \frac{1}{4\pi} \int_{V_i} \mathbf{B}_i(\mathbf{r}_i) \cdot \mathbf{A}_j(\mathbf{r}_i) \, \delta V_i \tag{184}$$

After substituting the $\mathbf{A}_j$ potential in the preceding relation, mutual helicity K_{ij} between volumes V_i and V_j is written as follows:

$$K_{ij} = \frac{1}{4\pi} \int_{V_i} \int_{V_j} [\mathbf{B}_i(\mathbf{r}_i) \wedge \mathbf{B}_j(\mathbf{r}_j)] \cdot \nabla_j \left(\frac{1}{R} \right) \delta V_i \delta V_j \tag{185}$$

Each curve C_i is associated to a tube, whose volume is V_i, coaxial with the curve C_i. In addition, we suppose that $\mathbf{B}_i$ field lines are parallel to C_i. As a consequence, the general integral term

$$F_i = \int_{S_i} \mathbf{B}_i(\mathbf{r}_i) \cdot \delta \mathbf{S}_i \tag{186}$$

represents the axial flux of $\mathbf{B}_i$ field throughout a cross section of the tube of surface S_i. Knowing that $\delta V_i = \delta \mathbf{S}_i \cdot \delta \mathbf{r}_i$, for each field $\mathbf{B}_i$, we find the equality:

$$\mathbf{B}_i(\mathbf{r}_i) \delta V_i = [\mathbf{B}_i(\mathbf{r}_i) \cdot \delta \mathbf{S}_i] \delta \mathbf{r}_i \tag{187}$$

Starting with the equalities given above and the definitions of K_{ij} and N_{ij}, we obtain a relation $K_{ij} = N_{ij} F_i F_j$ relating mutual helicity to the linking of $\mathbf{B}_i$ and $\mathbf{B}_j$ field lines and to their respective fluxes.

B. Temporal Derivation of Helicity

Let us now consider the following couple of equations:

$$\frac{1}{c} \frac{\partial \mathbf{A}}{\partial t} + \nabla \Phi = -\mathbf{E} \qquad \frac{1}{c} \frac{\partial \mathbf{B}}{\partial t} + \nabla \wedge \mathbf{E} = 0 \tag{188}$$

and multiply the first equation by $\mathbf{B}$ and the second equation by $\mathbf{A}$. We then merge these two equations by adding them together. Each term of the resulting equation is supplemented with the quantity $\Phi\nabla \cdot \mathbf{B} - \mathbf{E} \cdot \mathbf{B}$. We get

$$\frac{1}{c}\frac{\partial}{\partial t}(\mathbf{A} \cdot \mathbf{B}) + \nabla \cdot \mathbf{T} = -2\mathbf{E} \cdot \mathbf{B} \tag{189}$$

knowing that $\mathbf{T} = \Phi\mathbf{B} + \mathbf{E} \wedge \mathbf{A}$ is the torsion density and $\mathbf{A} \cdot \mathbf{B}$ is the helicity density of the magnetic field.

In the same spirit, using the following couple of relations

$$\frac{1}{c}\frac{\partial \mathbf{A}}{\partial t} + \nabla\Phi = -\mathbf{E} \qquad \frac{1}{c}\frac{\partial \mathbf{E}}{\partial t} - \nabla \wedge \mathbf{B} = -\frac{4\pi}{c}\mathbf{J} \tag{190}$$

let us multiply the first equation by $\mathbf{E}$ and the second equation by $\mathbf{A}$. Let us then add them up and add the quantity $\Phi\nabla \cdot \mathbf{E} + \mathbf{B}^2$ to each part of this equation. We then obtain the relation

$$\frac{1}{c}\frac{\partial}{\partial t}(\mathbf{A} \cdot \mathbf{E}) + \nabla \cdot \mathbf{S} = \mathbf{B}^2 - \mathbf{E}^2 + 4\pi\left(\rho\Phi - \frac{1}{c}\mathbf{J} \cdot \mathbf{A}\right) \tag{191}$$

knowing that $\mathbf{S} = \Phi\mathbf{E} + \mathbf{A} \wedge \mathbf{B}$ is the spin density and $\mathbf{A} \cdot \mathbf{E}$ the helicity density of the electric field.

Since Stratton's work [46, p. 28], it has been known that Maxwell's equations be can be supplemented with any particular solution of homogeneous equations:

$$\mathbf{E}_0 = \nabla \wedge \mathbf{C} \qquad \mathbf{B}_0 = \frac{1}{c}\frac{\partial \mathbf{C}}{\partial t} + \nabla\Psi \tag{192}$$

These new potentials are solutions of wave equations including inside the sources. To obtain the general solution, one must add a particular solution of the inhomogenous potential equations. Usually, the electromagnetic fields $\mathbf{E}_0, \mathbf{E}_0$ and the potentials $\Psi, \mathbf{C}$ are discarded for the following reasons. Either (1), they represent transient solutions of Maxwell's equations that decay rapidly to zero or (2) the new potentials and the fields vanish at infinity and therefore their values must be zero.

We may contest such claims, especially if vacuum is a fluctuating wave medium. In that case, ordinary plane waves that are solutions of wave equations do not vanish at infinity and therefore can be associated with the so-called zero-point energy. We can also assume that helicoidal fields are associated with zero-point energy. This question is not trivial since many authors consider that the inertia of bodies might be a consequence of the existence of the zero-point energy. However, one can use the Helmholtz theorem Eq. (40), in order to obtain a general solution for the definition of the electromagnetic field:

$$\mathbf{E} = -\nabla\phi + \nabla \wedge C - \frac{1}{c}\frac{\partial \mathbf{A}}{\partial t} \tag{193}$$

$$\mathbf{B} = \nabla\Psi + \nabla \wedge \mathbf{A} + \frac{1}{c}\frac{\partial \mathbf{C}}{\partial t} \tag{194}$$

The electromagnetic field defined above is a solution of Maxwell's equations, provided the new potentials $\Psi, \mathbf{C}$ are solutions of wave equations and satisfy the Lorenz gauge.

The electromagnetic field $\mathbf{E}_0, \mathbf{E}_0$, which was studied by Afanasiev and Stepanovsky [45], Rañada [43,44] and Kiehn [47] admits definitions similar to the ones given above

$$\mathbf{T}_0 = \Psi\mathbf{E}_0 + \mathbf{C} \wedge \mathbf{B}_0 \qquad \mathbf{S}_0 = \Psi\mathbf{B}_0 + \mathbf{E} \wedge \mathbf{C} \tag{195}$$

together with the two continuity relations:

$$\frac{1}{c}\frac{\partial}{\partial t}(\mathbf{C} \cdot \mathbf{E}_0) + \nabla \cdot \mathbf{T}_0 = 2\mathbf{E}_0 \cdot \mathbf{B}_0 \tag{196}$$

$$\frac{1}{c}\frac{\partial}{\partial t}(\mathbf{C} \cdot \mathbf{B}_0) + \nabla \cdot \mathbf{S}_0 = \mathbf{B}_0^2 - \mathbf{E}_0^2 \tag{197}$$

In classical electromagnetism, the condition $\mathbf{E} \cdot \mathbf{B} = 0$ and the condition $\mathbf{E}^2 = \mathbf{B}^2$ in the radiation zone are satisfied; therefore the electric and magnetic helicity densities are conserved. However, the preceding equations are used when the electromagnetic field is endowed with a topological structure far more sophisticated than what is usually acknowledged within the framework of classical electromagnetic theory. The interested reader is referred to a series of interesting theoretical works by Barrett [48], Kiehn [47], and Evans and Vigier [25].

We now recall the identity $(\mathbf{E}^2 - \mathbf{B}^2)^2 + 4(\mathbf{E} \cdot \mathbf{B})^2 = (\mathbf{E}^2 + \mathbf{B}^2)^2 - 4(\mathbf{E} \wedge \mathbf{B})^2$. This identity is important for reading through the following discussion:

1. When we are far from sources in the "radiation zone," the amplitudes of electric and magnetic fields verify the equality $\mathbf{E}^2 = \mathbf{B}^2$.
2. The identity $\mathbf{E} \cdot \mathbf{B} = 0$ is observed within the framework of classical electromagnetism.

In this case, for a volume $V_0(t)$ limitated by the surface $S_0(t)$ moving at a speed $\mathbf{U}(\mathbf{r}, t)$ in a given frame of reference, equation 189 can be rewritten in a form often dealt with in plasma physics [9,49]:

$$\frac{1}{c}\frac{d}{dt}\int_{V_0}\mathbf{A}\cdot\mathbf{B}\,\delta V_0+\int_{S_0}\left[\mathbf{T}-\frac{1}{c}\,(\mathbf{A}\cdot\mathbf{B})\mathbf{U}\right]\cdot\delta\mathbf{S}_0=0 \tag{198}$$

The identity $\mathbf{A}\wedge(\mathbf{U}\wedge\mathbf{B})=(\mathbf{A}\cdot\mathbf{B})\mathbf{U}-(\mathbf{A}\cdot\mathbf{U})\mathbf{B}$ can be utilized to rewrite the preceding equation under the form

$$\frac{1}{c}\frac{dK}{dt}+\int_{S_0}\left[\left(\mathbf{E}+\frac{1}{c}\,\mathbf{U}\wedge\mathbf{B}\right)\wedge\mathbf{A}+\left(\Phi-\frac{1}{c}\,\mathbf{U}\cdot\mathbf{A}\right)\right]\cdot\delta\mathbf{S}_0=0 \tag{199}$$

This relation can be simplified in the case of an infinitely conducting plasma where the condition $\mathbf{E}+\mathbf{U}\wedge\mathbf{B}/c=0$ is observed. This simplification, which differs from the demonstration given below, can either apply to the case when $\mathbf{U}=\mathbf{U}_e$ or $\mathbf{U}=\mathbf{U}_i$.

The evolution in time of helicity can be directly calculated if we start with the following definition of K

$$K=\int_{V(t)}F\,\delta V \tag{200}$$

where $F=\mathbf{A}\cdot\mathbf{B}$. This results in the following relation:

$$\frac{dK}{dt}=\int_{V(t)}\left(\frac{dF}{dt}+F\,\nabla\cdot\mathbf{U}\right)\delta V \tag{201}$$

Using the relation of continuity associated with mass density

$$\frac{1}{\rho_m}\frac{d\rho_m}{dt}=-\nabla\cdot\mathbf{U} \tag{202}$$

the term $\nabla\cdot\mathbf{U}$ is suppressed from the integral given in Eq. [201]. Therefore the evolution in time of helicity is written as

$$\frac{dK}{dt}=\int_{V(t)}\rho_m\,\frac{d}{dt}\left(\frac{F}{\rho_m}\right)\delta V \tag{203}$$

The formulation given above is particularly useful in the case of an infinitely conducting plasma when Faraday's law verifies the following condition:

$$\frac{1}{c}\frac{d}{dt}\int_{S(t)}\mathbf{B}\cdot\delta\mathbf{S}=0 \tag{204}$$

In this specific case, the following series of conditions is found:

$$\mathbf{E}_0 = \mathbf{E} + \frac{1}{c}\,\mathbf{U}\wedge\mathbf{B} = 0 \qquad \frac{\partial\mathbf{B}}{\partial t} + \nabla\wedge(\mathbf{B}\wedge\mathbf{U}) = 0 \qquad \nabla\wedge\mathbf{B} = \frac{4\pi}{c}\,\mathbf{J} \quad (205)$$

For an infinitely conducting plasma, the current density has the value of $\mathbf{J} = \rho_e\mathbf{U}_e + \rho_i\mathbf{U}_i = \rho_e\mathbf{V} + \rho\mathbf{U}_i \approx \rho_e\mathbf{V}$. As a result, one must take $\mathbf{U} \approx \mathbf{U}_i \neq \mathbf{U}_e$ in the following demonstration.

The particle derivative of the magnetic field has the following expression:

$$\frac{d\mathbf{B}}{dt} = \frac{\partial\mathbf{B}}{\partial t} + (\mathbf{U}\cdot\nabla)\mathbf{B} = -\nabla\wedge(\mathbf{B}\wedge\mathbf{U}) + (\mathbf{U}\cdot\nabla)\mathbf{B} = -\mathbf{B}(\nabla\cdot\mathbf{U}) + (\mathbf{B}\cdot\nabla)\mathbf{U} \tag{206}$$

The equation of continuity given in Eq. (202) allows us to suppress $\nabla\cdot\mathbf{U}$ in the preceding equation. As a result, we obtain the relation:

$$\frac{d}{dt}\left(\frac{\mathbf{B}}{\rho_m}\right) = \left(\frac{\mathbf{B}}{\rho_m}\cdot\nabla\right)\mathbf{U} \tag{207}$$

If the preceding equation is multiplied by vector potential $\mathbf{A}$, we obtain

$$\mathbf{A}\cdot\frac{d}{dt}\left(\frac{\mathbf{B}}{\rho_m}\right) = \frac{\mathbf{B}}{\rho_m}\cdot[(\mathbf{A}\cdot\nabla)\mathbf{U} + \mathbf{B}\wedge\nabla\wedge\mathbf{U}] \tag{208}$$

The particle derivative of the vector potential is:

$$\frac{d\mathbf{A}}{dt} = \frac{\partial\mathbf{A}}{\partial t} + (\mathbf{U}\cdot\nabla)\mathbf{A} = (\mathbf{U}\cdot\nabla)\mathbf{A} + \mathbf{U}\wedge\nabla\wedge\mathbf{A} - c\nabla\Phi \tag{209}$$

where the condition $\mathbf{E} = -\mathbf{U}\wedge\mathbf{B}/c$ has been used in order to suppress the partial time derivative of the vector potential in the second member of the equation given above. If we now multiply the preceding equation by $\mathbf{B}/\rho_m$, we get

$$\frac{\mathbf{B}}{\rho_m}\cdot\frac{d\mathbf{A}}{dt} = \frac{\mathbf{B}}{\rho_m}\cdot[(\mathbf{U}\cdot\nabla)\mathbf{A} + \mathbf{U}\wedge\nabla\wedge\mathbf{A} - c\nabla\Phi] \tag{210}$$

Adding Eqs. (207) and (210) together and merging them into a single equation gives us the following relation:

$$\frac{d}{dt}\left(\frac{F}{\rho_m}\right) = \frac{\mathbf{B}}{\rho_m}\cdot\nabla(\mathbf{U}\cdot\mathbf{A} - c\Phi) \tag{211}$$

The equation of continuity for helicity then becomes

$$\frac{dK}{dt} = \int_{V(t)} \rho_m \frac{d}{dt}\left(\frac{F}{\rho_m}\right) \delta V = \int_{V(t)} \mathbf{B} \cdot \nabla(\mathbf{U} \cdot \mathbf{A} - c\Phi)\, \delta V \tag{212}$$

Finally, we recover the time derivative of helicity previously demonstrated in the case of an infinitely conducting plasma:

$$\frac{dK}{dt} = \int_{S(t)} (\mathbf{U} \cdot \mathbf{A} - c\Phi)\mathbf{B} \cdot \delta\mathbf{S} \tag{213}$$

Helicity K is an invariant if the condition $\mathbf{B} \cdot \delta\mathbf{S} = 0$ is verified over the surface S surrounding the volume V. Helicity K is different from zero if the $\mathbf{A}$ field is a Beltrami field. As a consequence, the $\mathbf{B}$ field is also a Beltrami field. Therefore, we must conclude that the $\mathbf{B}$ field is also a force-free field since

$$(\nabla \wedge \mathbf{B}) \wedge \mathbf{B} = \frac{4\pi}{c} \mathbf{J} \wedge \mathbf{B} = 0 \tag{214}$$

We have previously demonstrated, using the virial theorem, that the $\mathbf{B}$ field cannot be equal to zero over the surface S if this field is a force-free field. As a consequence, the condition $\mathbf{B} \cdot \delta\mathbf{S} = 0$ is verified only if the magnetic field is normal to the surface S. Less restrictive and more physical conditions can be chosen if the $\mathbf{E}_0$ field exhibits a nonnull irrotational component verifying the condition $\nabla \wedge \mathbf{E}_{0\parallel} = 0$. In this case, the magnetic field is not force-free anymore:

$$(\nabla \wedge \mathbf{B}) \wedge \mathbf{B} = \frac{4\pi}{c} \mathbf{J} \wedge \mathbf{B} + \frac{1}{c} \frac{\partial \mathbf{E}_{0\parallel}}{\partial t} \wedge \mathbf{B} = 0 \tag{215}$$

C. Topological Effect Associated with Voltage Measurement

A lengthy solenoid in which circulates a variable current $I_s(t)$ as a function of time, generates a variable magnetic field and induces an electric field satisfying Faraday's law:

$$\int_C \mathbf{E} \cdot \delta\mathbf{r} = -\frac{1}{c} \frac{d}{dt} \int_S \mathbf{B} \cdot \delta\mathbf{S} \tag{216}$$

In the case of an infinitely long solenoid, space is halved and dissociated into two distinct regions: (1) a simply connected region inside the solenoid and (2) a doubly connected region outside the solenoid.

Faraday's law, written under its differential form

$$\nabla \wedge \mathbf{E} = -\frac{1}{c} \frac{\partial \mathbf{B}}{\partial t} \tag{217}$$

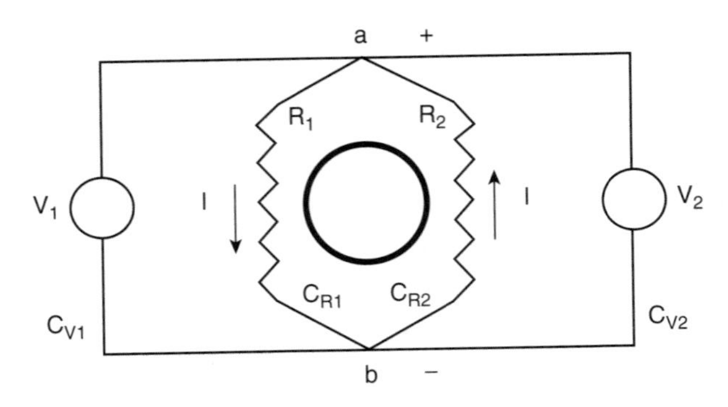

Figure 5. Voltage measurement around a solenoid.

is applied to each of these regions of space, with the following conditions:

$$\text{Inside region: } \mathbf{B} \neq 0 \Rightarrow \mathbf{E} \neq 0$$
$$\text{Outside region: } \mathbf{B} = 0 \Rightarrow \nabla \wedge \mathbf{E} = 0$$

The fact that the solenoid halves space into two regions has observable and measurable consequences. Indeed, let us now consider two resistances R_1 and R_2 connected in a ring and forming a conducting loop C around the solenoid, as illustrated in Fig. 5.

Let us now connect the leads of two voltmeters to two points, say, point a and point b, placed on either side of the loop C as indicated in Fig. 5. We shall now suppose that the voltmeters are characterized by a high internal resistance that enables us to ignore the current taken during the process. We can also expect that the two voltmeters connected to the same points measure the same voltage $V_1 = V_2$. This seemingly logical statement is not verified, i.e. observed, when the experiment is performed [50–52]. Why?

At first glance, the difference may be expected to be a consequence of the fact that $R_1 \neq R_2$ and that, if $R_1 = R_2$, we should measure $V_1 = V_2$. But this is not the case. The differential formulation of Faraday's law cannot be helpful in any manner to get rid of the paradoxes given above while the integral formulation, as we shall now see, can be used to understand the difference of measurements performed by each of our two voltmeters.

Since $\nabla \wedge \mathbf{E} = 0$ in the region outside the solenoid, the electric field may be derived from a scalar potential $\mathbf{E} = -\nabla\Phi$. As a consequence, the measurement of a voltage V given by the formula

$$V = \Phi_a - \Phi_b = -\int_b^a \mathbf{E} \cdot \delta\mathbf{r} \tag{218}$$

would be independent from the path taken by the wires of the voltmeters. This, of course, is not representative of reality since the region located outside the solenoid is doubly connected. It follows that the potential function Φ is a multiform function and the integral given above depends on the adopted path, since we have

$$\int_C \mathbf{E} \cdot \delta\mathbf{r} = -\frac{m}{c}\frac{dF}{dt} \tag{219}$$

where the magnetic flux F flowing throughout the open surface S leaning on the curve C is written as:

$$F = \int_S \mathbf{B} \cdot \delta\mathbf{S} = \int_C \mathbf{A} \cdot \delta\mathbf{r} \tag{220}$$

In the course of this discussion, m is an integer number whose value is $m = 0$ for a reducible loop C not encircling the solenoid, or $\pm m$ if the loop C encircles m times the solenoid. The sign $+$ is defined as such when winding is performed in a counterclockwise direction.

By definition, the voltage measured by a voltmeter between point a and point b can be represented by the following expression

$$V_1 = \int_{C_{v1}} \mathbf{E} \cdot \delta\mathbf{r} \qquad V_2 = \int_{C_{v2}} \mathbf{E} \cdot \delta\mathbf{r} \tag{221}$$

where voltages have positive values if paths $C_{v1}(a, b)$ and $C_{v2}(a, b)$, passing through voltmeters and their associated wires, start from point a to reach point b in the counterclockwise direction. It results from these definitions two different possibilities only:

1. For two reconcilable circuits C_{v1} and C_{v2} forming a reducible loop C, voltages that are measured not only have the same sign but are equal $V_1 = V_2$. This is the classic case when voltmeters measure the same voltage between point a and point b. In this case, we must chose the value $m = 0$ in the integral relation 219.
2. Conversely, for two irreconcilable circuits encircling the solenoid, namely, when $m = 1$, voltages that are measured, are opposite in signs and can be unequal in absolute values.

A further explanation is needed. Using Ohm's law, we can calculate the current $I_r(t)$ flowing in counterclockwise direction, along circuits C_{r1} and C_{r2} passing through resistances R_1 and R_2:

$$I_r = \frac{1}{R_1}\int_{C_{r1}} \mathbf{E} \cdot \delta\mathbf{r} = -\frac{1}{R_2}\int_{C_{r2}} \mathbf{E} \cdot \delta\mathbf{r} \tag{222}$$

Circuits C_{v1} and C_{r1}, as well as circuits C_{v2} and C_{r2}, form reducible loops. These loops do not encircle the solenoid. This situation is illustrated in the following set of equalities:

$$\int_{C_{v1}} \mathbf{E} \cdot \delta\mathbf{r} = \int_{C_{r1}} \mathbf{E} \cdot \delta\mathbf{r} \Rightarrow V_1 = R_1 I_r \tag{223}$$

$$\int_{C_{v2}} \mathbf{E} \cdot \delta\mathbf{r} = \int_{C_{r2}} \mathbf{E} \cdot \delta\mathbf{r} \Rightarrow V_2 = -R_2 I_r \tag{224}$$

Voltage at the source becomes

$$V_s = V_1 - V_2 = (R_1 + R_2) I_r = \int_C \mathbf{E} \cdot \delta\mathbf{r} = 2\pi E_\theta \neq 0 \tag{225}$$

If d is the diameter of the solenoid infinite in length, the $\mathbf{A}(\mathbf{r}, t)$ vector potential is characterized by a single component only

$$A_\theta(r, t) = \frac{d^2}{8r} B_s(t) = \frac{\pi d^2}{4cr} n I_s(t) \tag{226}$$

knowing that the magnetic field is given by the formula $B_s(t) = 2\pi n I_s(t)/c$, where n is the number of turns by unit of length. The flux of the magnetic field therefore has the value $F(t) = \pi^2 d^2 n I_s(t)/2c$. These remarks enable us to determine the value of source-induced voltage:

$$V_s = (R_1 + R_2) I_r = -n\pi^2 \frac{d^2}{2c^2} \frac{dI_s}{dt} \tag{227}$$

Romer [52] performed an experiment where the current I_s was a linear function in time. This experiment indicates that, after the calculation given above, voltages V_1, V_2, V_s are constants. Furthermore, the experiment (52) confirms perfectly the explanations given above.

D. The Aharonov–Bohm Effect

The famous experiment proposed by Aharonov and Bohm [53,54] is schematically represented in Fig. 6. In such an experiment, a source emits an electron beam directed toward a wall in which two slits, located on each side of the beam axis, are located. A photographic plate (film) placed behind the slits "records" impacting electrons. After the emission of a large number of electrons by the source, the aforementioned film exhibits neat, clear, and dark fringes that are parallel to the slits. This result is interpreted as a manifestation of the wave nature of electrons.

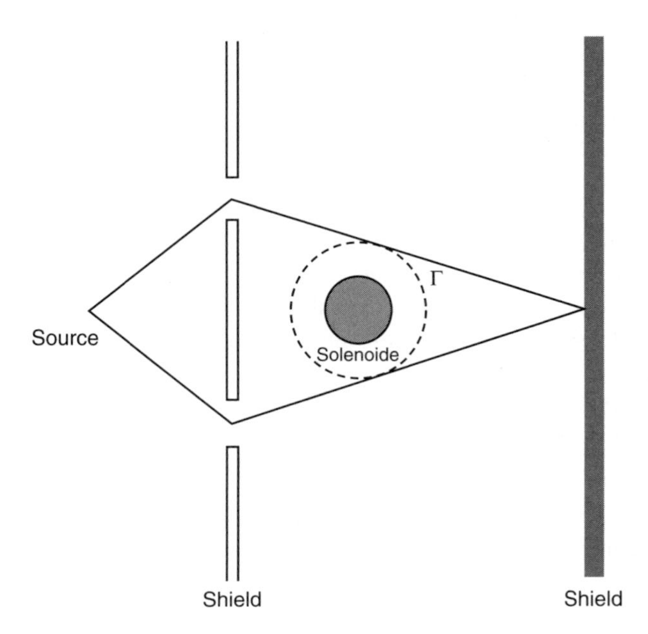

Figure 6. Experimental setup to observe the quantum interference of electrons passing ouside a solenoid.

If electrons were particles only, they should cross the slits and impact the film in such a way that a recorded "peak" of impacts lines up with the source and the slits, with very few impacts on either side of these peaks. There is another possible interpretation where we can imagine that particles behave as waves and are guided by them. Indeed, de Broglie waves associated with these electrons are subdivided in partial waves when they cross the slits before superimposing on each other, impacting on film, and producing an obvious interference pattern (where shining fringes relate to electron impacts).

The so-called Aharonov–Bohm effect is observed with another experimental setup. A solenoid is placed immediately after the plate, between the slits, and its axis is parallel to the slits, and therefore normal to the beam trajectory beam. If the solenoid is long enough, the magnetic field remains confined in it; as a consequence, the magnetic field is shown to have a null value in the region crossed by electrons beamed on either sides of the solenoid. The Lorentz force exerted on the electrons is expected to be null in the absence of any external electrical field.

But observation does not match with this conclusion. Indeed, we observe that the interference pattern is slightly shifted when compared to the interference pattern recorded without a solenoid. If we suppress the magnetic field in the solenoid, the interference pattern recovers its original position. The decision to generate a magnetic field inside the solenoid or to cancel it should not generate a shift of the fringes, since electrons are not acted on by any force from the solenoid, at least any force capable of justifying a phase shift of matter waves associated to electrons.

In search for an explanation, Aharonov and Bohm worked out quantum mechanics equations based on the measurable physical effect of the vector potential, which is nonnull in a region outside the solenoid. Like many other paradoxes in physics, including the twin paradox, the interpretation of this experiment proposed in 1959 was the subject of an intense controversy among researchers. This controversy is well summarized in a review article [55] and in other references of interest [56–67].

The interpretation of the original Aharonov–Bohm experiment implies mathematical considerations to be taken into account concerning:

1. Initial and boundary conditions associated with wavefunctions
2. The existence of topological effects affecting the validity of Stokes' theorem

Needless to say, these mathematical considerations are particularly delicate to handle. However, an experiment performed by Möllenstedt and Bayh [58] with ordinary solenoids, led to an following observation in which the interference pattern moves at a continual pace in perfect synchronization with the evolution in time of the magnetic field in the solenoid. This experiment indicated that the magnetic flux $F = \Delta\varphi c\hbar/q$ was not quantified, while it was in the case of magnetic flux observed with supraconducting magnets.

Many physicists opposed both the interpretation of this effect attributed to the vector potential, and the experimental conditions of Chambers [57] and Möllenstedt [58] experiments. Criticisms insisted the solenoid was not infinite in length. As a consequence, they claimed, the magnetic field leaks out in a region too close to the area crossed by the electrons, to have no effect. This leak was even employed to quantify the F flux. In addition, electron beams can interfere, as we shall soon demonstrate below, with the magnetic field created inside the solenoid.

These are the reasons why Tonomura [60] reproduced a cleansed experiment in 1982 using, this time, a toroidal magnet so as to get rid—or, more accurately almost completely rid—of the leaking magnetic flux. Tonomura demonstrated the effect would persist in this case. Subsequently, Tonomura et al. [61] reproduced the experiment to determine the flux imprisoned in the magnet was quantified. The answer was negative. But the experiment [62] exhibited a flux quantification when the magnet was a superconductor one.

Physicists have proposed several mechanisms to interpret the Aharonov–Bohm effect. We shall focus here on the most interesting of them. We shall first examine the case of the classical solenoid so as to define the problem on a sound basis. Inside a solenoid, infinite in length, in which a constant current circulates, the magnetic field $\mathbf{B}_0$ is also constant and is directed toward the symmetry axis. If r_0 is the radius of the solenoid section, we observe that for $r < r_0$, the $\mathbf{A}$ vector potential can be written as:

$$\mathbf{A}(\mathbf{r}) = \frac{1}{2}\,\mathbf{B}_0 \wedge \mathbf{r} = -\nabla\wedge\left(\frac{r^2}{4}\,\mathbf{B}_0\right) \tag{228}$$

Thus originating the following relations:

$$\nabla \cdot \mathbf{A} = 0 \qquad \mathbf{B} = \nabla\wedge\mathbf{A} = \mathbf{B}_0 \tag{229}$$

Outside the solenoid, the magnetic field has a null value $\mathbf{B} = 0$. Therefore, it would be very tempting to chose $\mathbf{A} = 0$. Unfortunately, this choice is not possible. The reason is simple. Stokes theorem, when applied to an open surface S leaning on a closed contour C encircling the solenoid, gives us

$$F = \pi r_0^2 B_0 = \int_S \mathbf{B}\cdot\delta\mathbf{S} = \int_C \mathbf{A}\cdot\delta\mathbf{r} \neq 0 \Rightarrow \mathbf{A} \neq 0 \tag{230}$$

As a consequence, the vector potential $\mathbf{A}$ inside the solenoid must be written as

$$\mathbf{A}(\mathbf{r}) = \frac{1}{2}\frac{r_0^2}{r^2}\,\mathbf{B}_0 \wedge \mathbf{r} = -\nabla\wedge\left[\frac{r_0^2}{2}\,\ln(r)\mathbf{B}_0\right] \tag{231}$$

This gives us the following equation:

$$\nabla \cdot \mathbf{A} = 0 \qquad \mathbf{B} = \nabla\wedge\mathbf{A} = 0 \tag{232}$$

We must insist that internal and external vector potentials reconnect with each other in a continuous manner on the solenoid boundary for $r = r_0$.

The gauge transform $\mathbf{A}_0 = \mathbf{A} + \nabla\phi$ does not modify the value of the magnetic flux F calculated on the basis of Stokes theorem, as defined above. However, this gauge transform does not allow us to cancel the external vector potential. This statement is in sharp contrast with, and even contradicts, what is usually claimed in the available literature. Indeed, Helmholtz decomposition of the vector potential $\mathbf{A} = \mathbf{A}_\parallel + \mathbf{A}_\perp$ shows that only the longitudinal component $\mathbf{A}_\parallel$ is modified in this process:

$$\mathbf{A}_{0\parallel} = \nabla\phi \qquad \mathbf{A}_\parallel = 0 \qquad \mathbf{A}_{0\perp} = \mathbf{A}_\perp = -\nabla\wedge\left[\frac{r_0^2}{2}\,\ln(r)\mathbf{B}_0\right] \tag{233}$$

As a result, the condition $\mathbf{B}_\perp = \nabla \wedge \mathbf{A}_\perp = 0$ does not necessarily imply that the vector potential $\mathbf{A}_\perp$ derives from a scalar potential, as proved by the definitions given above. Indeed, one can always define a longitudinal field $\mathbf{A}_\parallel = \nabla\phi$ where $\mathbf{A}_\parallel = \mathbf{A}_\perp$. This potential is written as $\phi = r_0^2 B_0 \theta/2$, knowing that $-\pi\theta = \arctan(y/x)\pi$ when $\mathbf{B}_0$ is directed toward the z axis. However, the scalar potential ϕ is not a true potential since the circulation of $\mathbf{A}_\parallel$ on a close curve C encircling the origin is not null. This results in the fact that the ϕ function possess a singularity on the negative part of the x axis. As a consequence, a gauge transform using this potential is not possible since it would imply a change in the value of the magnetic field.

The absence of magnetic monopole implies the conditions $\nabla \cdot \mathbf{B} = 0$ and $\mathbf{B}_\parallel = -\nabla\chi = 0$, which are consistent with the relations given above, since $\mathbf{B}_\parallel = \nabla \wedge \mathbf{A}_\parallel = 0$. However, as with the $\mathbf{A}$ vector potential, the equality $\mathbf{B}_\perp = \mathbf{B}_\parallel = -\nabla\chi \neq 0$ enables us to define a scalar potential χ that can be calculated on the basis of Biot–Savart law for a filiform (filament-shaped) circuit

$$\mathbf{B}_\perp(r) = \frac{I}{c} \nabla \wedge \int_C \frac{\delta\mathbf{r}'}{R} = \frac{I}{c} \int_C \frac{1}{R^3} \delta\mathbf{r}' \wedge \mathbf{R} \tag{234}$$

knowing that $d\chi = \nabla\chi \cdot \delta\mathbf{r} = -\mathbf{B}_\parallel \cdot \delta\mathbf{r}$, we obtain

$$d\chi = \frac{I}{c} \int_C \frac{\mathbf{R}}{R^3} \cdot (\delta\mathbf{r}' \wedge \delta\mathbf{r}) = \frac{I}{c} d\Omega \tag{235}$$

where the variation of solid angle $d\Omega = \nabla\Omega \cdot \delta\mathbf{r}$ observed by an observer at the point $\mathbf{r}$ moving along the distance $\delta\mathbf{r}$ is:

$$d\Omega = \int_{dS'} \frac{\mathbf{R}}{R^3} \cdot \delta\mathbf{S} \tag{236}$$

If the observer, located at point $\mathbf{r}$, observes the curve C as a solid angle Ω, the variation of solid angle $d\Omega$, as illustrated in Fig. 7, can be calculated by assuming this variation to be sustained by the lateral surface dS' resulting from a translation $d\mathbf{r}'' = -\delta\mathbf{r}$ of the curve C where the surface element dS' has the value $\delta\mathbf{S} = d\mathbf{r}'' \wedge \delta\mathbf{r}'$.

The problem of gauge transform also surfaces when Schrödinger's equation is employed to study the Aharonov–Bohm effect. If the transform $\Psi = \Psi_0 e^{j\varphi}$ is performed within Schrödinger's equation:

$$\epsilon j\hbar \frac{\partial\Psi}{\partial t} = \frac{1}{2m_0} \left(-\epsilon j\hbar\nabla - \frac{q}{c} \mathbf{A}\right)^2 \Psi + q\Phi\Psi \tag{237}$$

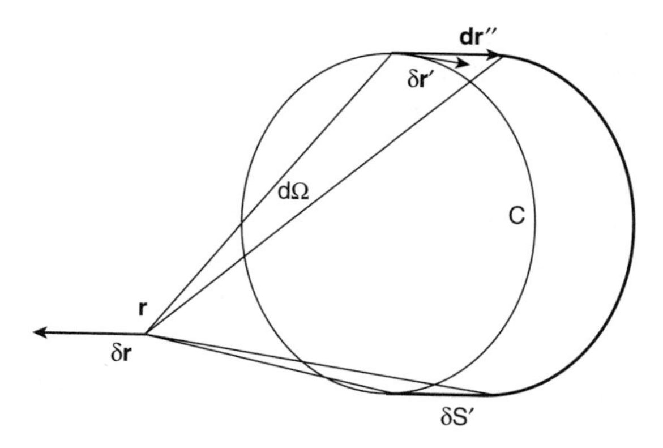

Figure 7. Variation of solid angle viewed at point $\mathbf{r}$ for a translation $\mathbf{dr}'' = -\delta\mathbf{r}$ of the curve C.

knowing that $F = \varphi c\hbar/q$ is an arbitrary function and $\epsilon = \pm 1$, we obtain an identical equation

$$\epsilon j\hbar \frac{\partial \Psi_0}{\partial t} = \frac{1}{2m_0}\left(-\epsilon j\hbar\nabla - \frac{q}{c}\mathbf{A}_0\right)^2 \Psi_0 + q\Phi_0\Psi_0 \tag{238}$$

provided we define two new potentials satisfying the aforementioned gauge transform:

$$\Phi_0 = \Phi - \frac{\epsilon}{c}\frac{\partial F}{\partial t} \qquad \mathbf{A}_0 = \mathbf{A} + \epsilon\nabla F \tag{239}$$

However, the F function is no longer arbitrary anymore if the gauge transform defined above is employed outside the solenoid to associate potentials before $\Phi_0(\mathbf{r}, 0) = 0, \mathbf{A}_0(\mathbf{r}, 0) = 0$ and after having connected the magnetic field source $\Phi(\mathbf{r}, t) \neq 0, \mathbf{A}(\mathbf{r}, t) \neq 0$. Indeed, the conditions

$$\Phi = \frac{\epsilon}{c}\frac{\partial F}{\partial t} \qquad \mathbf{A} = -\epsilon\nabla F \tag{240}$$

and the gauge transform written above, leaving invariant the electromagnetic field

$$\mathbf{E} = -\nabla\Phi - \frac{1}{c}\frac{\partial \mathbf{A}}{\partial t} = 0 \qquad \mathbf{B} = \nabla\wedge\mathbf{A} = 0 \tag{241}$$

imply the following integral relation:

$$\epsilon F(\mathbf{r}, t) = \int_{t_0}^{t} c\Phi(\mathbf{r}, t')\,dt' - \int_{\mathbf{r}_0}^{\mathbf{r}} \mathbf{A}(\mathbf{r}', t_0)\cdot\delta\mathbf{r}' \tag{242}$$

We observe that the gauge transform is unique and cannot allow us to eliminate the vector potential outside the solenoid. In addition, the vector potential $\mathbf{A}$ derives from a multiform scalar potential F. This result contradicts the solenoidal characteristic of the $\mathbf{A}_\perp = -\nabla\wedge[r_0^2 \ln(r)\mathbf{B}_0/2]$ vector potential. Henceforth, the gauge transform given above represents nonobservable stationary waves in vacuum since the Lorentz gauge:

$$\frac{1}{c}\frac{\partial\Phi}{\partial t} + \nabla\cdot\mathbf{A} = 0 \tag{243}$$

together with definitions (238), imply the following wave equation

$$\Delta F - \frac{1}{c^2}\frac{\partial^2 F}{\partial t^2} = 0 \tag{244}$$

which indeed admits standing wave solutions in vacuum.

The first interpretation of the Aharonov–Bohm effect is based on the system of equations given by Olariu and Popescu [55, p. 423] where charge density and current density are suppressed:

$$\nabla\cdot\mathbf{E} = \frac{1}{c^2}\frac{\partial^2\Phi}{\partial t^2} - \Delta\Phi \tag{245}$$

$$\frac{1}{c}\frac{\partial\mathbf{E}}{\partial t} - \nabla\wedge\mathbf{B} = \Delta\mathbf{A} - \frac{1}{c^2}\frac{\partial^2\mathbf{A}}{\partial t^2} \tag{246}$$

In order to obtain these equations, Olariu relies on Noerdlinger's works [68], which explained the Aharonov–Bohm effect as a nonlocal effect in time, rather than a nonlocal effect in space, of the electrical field generated by electrons belonging to the solenoid. In the presence of an electric field, the φ phase defined in Eq. (240) over a C loop encircling the solenoid, is not a total differential anymore. Indeed, using Faraday's law, we can calculate the work performed by the electric field on beamed electrons circulating along the fixed loop C at a given instant:

$$E(t) = q\int_C \mathbf{E}\cdot\delta\mathbf{r} = -\frac{q}{c}\frac{d}{dt}\int_C \mathbf{A}\cdot\delta\mathbf{r} \tag{247}$$

The work performed by the electric field between two different instants would produce a phase shift having the following value:

$$\Delta\varphi = \frac{1}{\hbar}\int_{t_0}^{t_1} E(t)\,dt \tag{248}$$

which can be rewritten as

$$\Delta\varphi = -\frac{q}{c\hbar}\int_C [\mathbf{A}(\mathbf{r}, t_1) - \mathbf{A}(\mathbf{r}, t_0)] \cdot \delta\mathbf{r} \tag{249}$$

However, we must cast some doubts on the pertinence of an Aharonov–Bohm effect explained as a nonlocal action in time of the electric field applied to beamed electrons. Indeed, in an experiment performed by Bayh [58], electrons propagate from source to film, where they impact along a distance of the order of 1 m in about 10 ns. About 1000 electrons arrived at the detector each second; therefore, at any given moment during the experiment, there is likely to be only electron traversing the apparatus. This implies that each electron is influenced by the action of the electric field for a very short period of time only with no possible coherent relationship with another electron. Moreover, the shift of the interference pattern in time generated by the entire population of electrons in Bayh's experiment, happens in a much longer timeframe than the time taken by a single electron to travel from the source to the film. Therefore, the preceding calculation, which applies for at least two electrons, each one passing one side of the loop C, must be calculated on a time lapse of $T = t_1 - t_0$, which is far from greater than the time taken by an electron to traverse the zone where the solenoid is located.

In an interferometry experiment, electron beams incident through the medium's stationary scalar waves, interact with electrons located in the photographic plate itself. This "original" interference process is observed by the interference pattern recorded on the plate. The same situation occurs for electrons contained in the solenoid (whether supraconducting or not). The φ phase of a wave on a loop C is defined at a given instant for a given path direction in space. As a consequence, the phase of the waves associated with wavepackets of electrons circulating through the solenoid (or on its surface for a supraconducting solenoid), is modified when a current appears in the solenoid. By virtue of the superposition principle, this must result in a shift of the interference pattern, as illustrated by Bayh's experiment, and in the appearance of an electric field that can be null locally in space at a given instant, but cannot be permanently null because the potentials are time-dependent. The electric field we alluded to above can also be interpreted as resulting from a Sagnac-like effect [69, p. 482] for waves circulating in opposite directions over the loop C. In addition, we can underline that there also exists an analogy between the phase and Feynman's propagator as far as the interpretation of the Aharonov–Bohm effect is concerned.

In a wave theory of matter and fields, completely isolated systems or "free" particles shall never be considered. Indeed, two kinds of forces are known to exist in a medium: external forces and internal forces [70]. Therefore, if we modify locally in spacetime the equilibrium of stationary waves in the solenoid, the resulting imbalance will be redistributed through the entire spacetime continuum because of the propagation of waves, and as such, will influence the behavior of waves associated to the two electron beams.

Therefore, questioning the physical significance of potential is not relevant here. The new formulation of Maxwell's equations [20–23], where potentials are directly coupled to fields clearly indicates that potentials, play a key role in particle behavior. To make a long story short, the difference in nature between potentials and fields stems from the fact that potentials relate to a state of equilibrium of stationary waves in the medium usually nonaccessible to an observer (except when potentials are used in a measurement process of the interferometric kind, at a given instant in time). Conversely, fields illustrate a nonequilibrium state of the medium as an observable progressive electromagnetic wave, since this wave induces the motion of material particles.

A second interpretation of the Aharonov–Bohm effect was devised by Boyer [65,66], who used matter waves associated to moving electrons. Waves coming from each slit interfere with a phase shift $\Delta\varphi = 2\pi d \sin\theta/\lambda$, where d is the distance between two slits. If P is the impulse of an electron in the beam, the de Broglie relation gives us $P = 2\pi\hbar/\lambda$. This results in the fact that the phase difference can be written in the form $\Delta\varphi = Pd\sin\theta/\hbar$ and must be equal to the phase difference associated with the magnetic flow F:

$$\Delta\varphi = \frac{P}{\hbar} d \sin\theta = \frac{q}{c\hbar} F \tag{250}$$

We notice that the deflexion angle θ associated with the motion of the interference pattern does not depend on Planck's constant, and that this angle depends on classical values, particularly on the magnetic field of the solenoid through the expression of the flow.

Finally, some authors [71–75] proposed an interpretation of the Aharonov–Bohm effect based on the interaction between the electrons of the beam and electrons of the solenoid. The Aharonov–Bohm effect looses its mystery if we acknowledge that Newton's third principle is transgressed here. Indeed, if electrons within the solenoid cannot act on electron of the beams, electron beams can act on the electrons of the solenoid, by means of the momentum equations [70]

$$\frac{d}{dt}\int_{V_f(t)} (\rho_{mf}\mathbf{U}_f + \mathbf{P}_{fs})\delta V = -\int_{V_f(t)} \nabla \cdot (\overleftrightarrow{\mathbf{T}}_{fs} - \mathbf{U}_f\mathbf{P}_{fs})\delta V = 0 \tag{251}$$

$$\frac{d}{dt}\int_{V_s(t)} (\rho_{ms}\mathbf{U}_s + \mathbf{P}_{sf})\delta V = -\int_{V_s(t)} \nabla \cdot (\overleftrightarrow{\mathbf{T}}_{sf} - \mathbf{U}_s\mathbf{P}_{sf})\delta V = 0 \tag{252}$$

where indices f and s, respectively, refer to electrons from the beam and electrons from the solenoid. In addition, we use the following definitions:

$$\mathbf{P}_{ij} = \frac{1}{4\pi c}(\mathbf{E}_i \wedge \mathbf{B}_j) \qquad \mathbf{f}_{ij} = \rho_i \mathbf{E}_j + \frac{1}{c}\mathbf{J}_i \wedge \mathbf{B}_j \tag{253}$$

$$\overleftrightarrow{\mathbf{T}}_{ij} = \frac{1}{4\pi}\left[\frac{1}{2}(\mathbf{E}_i \cdot \mathbf{E}_j + \mathbf{B}_i \cdot \mathbf{B}_j)\overleftrightarrow{\mathbf{I}} - (\mathbf{E}_i\mathbf{E}_j + \mathbf{B}_i\mathbf{B}_j)\right] \tag{254}$$

Since the electromagnetic field generated by electrons within the solenoid is null outside the solenoid, a new set of conditions follows:

$$\mathbf{f}_{fs} = 0 \qquad \mathbf{P}_{fs} = 0 \qquad \overleftrightarrow{\mathbf{T}}_{fs} = 0 \tag{255}$$

while the corresponding values inside the solenoid all differ from zero. This results in the violation of Newton's third principle, since we have $\mathbf{f}_{fs} = 0 \neq -\mathbf{f}_{sf} \neq 0$. As a consequence, the magnetic field generated by electrons located outside the solenoid must modify the velocity $\mathbf{U}_{es}$ of electrons inside the solenoid. However, for a good conductor, current density $\mathbf{J}_s$ inside the solenoid is given by the relation $\mathbf{J}_s = \rho_{es}\mathbf{U}_{es} + \rho_{is}\mathbf{U}_{is} = \rho_{es}\mathbf{V} + \rho\mathbf{U}_{is} \approx \rho_{es}\mathbf{V}$. Since the current density $\mathbf{J}_s$ must remain constant if electrons outside the solenoid are acted on by a null Lorentz force, we must infer that the relative velocity $\mathbf{V} = \mathbf{U}_{is} - \mathbf{U}_{es}$ must remain constant. This implies that the motion of the solenoid in the opposite direction, as underlined by Herman [73]. This results in a phase modification of matter waves in the medium if the wavefunction associated to electrons inside the solenoid can penetrate the region where the electromagnetic field is null. Even in the case when this penetration is excluded, initial and boundary conditions on the boundary of the solenoid must be taken into account.

However, an objection against this line of reasoning, quite similar to the one already mentioned concerning the nonlocal theory of electric field, can be formulated. More specifically, electrons from the beam act on electrons from the solenoid for a very short period of time only. In addition, the shift of the interference pattern is not modified during the period of time during which measurement is performed but rather is fixed by initial constant current circulating in the solenoid. The fact that the interference pattern is "locked" in the case of a constant magnetic field and recovers its initial position when the current feeding the solenoid is turned off, confirms the preceding claim. As a consequence, it may be assumed that the displacement of the interference pattern is induced by the mutual interaction of electrons from the solenoid and electrons from the photographic plate, since the presence of a magnetic field inside the solenoid modifies the topology of space. Therefore, we observe a modification of initial conditions in the medium that can be calculated by means of an hydrodynamical model related to Schrödinger's equation.

We recall here that, in quantum mechanics, each dynamic variable used in classical mechanics is associated with a linear hermitian operator. As a consequence, energy and momentum of a particle placed in an external potential $\mathbf{A}, \Phi = E_P/q$, will respectively be associated to the following operators:

$$E \Rightarrow j\epsilon\hbar \frac{\partial}{\partial t} \qquad \mathbf{P} \Rightarrow -j\epsilon\hbar\nabla - \frac{q}{c}\,\mathbf{A} \tag{256}$$

This association allows us to recover Schrödinger's equation on the basis of the following dispersion law

$$-\epsilon\hbar\omega = \frac{1}{2m_0}\,(\hbar\mathbf{k})^2 + E_P \tag{257}$$

by replacing the values by corresponding operators applied to the Ψ function:

$$j\epsilon\hbar\,\frac{\partial\Psi}{\partial t} = \frac{1}{2m_0}\left(-j\epsilon\hbar\nabla - \frac{q}{c}\,\mathbf{A}\right)^2\Psi + q\Phi\Psi \tag{258}$$

We emphasize that no physical nor mathematical justifications are given concerning the meaning of the complex operators depending on Planck's constant. The equation given above becomes after development of the operator:

$$j\epsilon\hbar\,\frac{\partial\Psi}{\partial t} = -\frac{\hbar^2}{2m_0}\,\Delta\Psi + \frac{j\epsilon\hbar q}{m_0 c}\,\mathbf{A}\cdot\nabla\Psi + \left(\frac{j\epsilon\hbar q}{2m_0 c}\,\nabla\cdot\mathbf{A} + \frac{q^2}{2m_0c^2}\,\mathbf{A}^2 + q\Phi\right)\Psi \tag{259}$$

A hydrodynamical representation of the preceding equation can be found along the same lines as those defined by Madelung, stating $\Psi = \Pi e^{j\varphi}$ where the quantities $\Pi(\mathbf{r}, t)$ and $\varphi(\mathbf{r}, t)$ are real functions. This results in the following relations:

$$\frac{\partial\Psi}{\partial t} = \left(\frac{\partial\Pi}{\partial t} + j\Pi\,\frac{\partial\varphi}{\partial t}\right)e^{j\varphi} \tag{260}$$

$$\Delta\Psi = \left[\Delta\Pi - (\nabla\varphi)^2\Pi + j(2\nabla\Pi\cdot\nabla\varphi + \Pi\Delta\varphi)\right]e^{j\varphi} \tag{261}$$

After having substituted the preceding relations into the Schrödinger's equation (259), we split this equation into its real and imaginary parts, so as to obtain a nonlinear coupled system of equations:

$$\frac{\partial \Pi^2}{\partial t} + \nabla \cdot \left[\frac{\Pi^2}{m_0} \left(\epsilon\hbar \nabla \varphi - \frac{q}{c}\, \mathbf{A} \right) \right] = 0 \tag{262}$$

$$\epsilon\hbar\, \frac{\partial \varphi}{\partial t} + \frac{1}{2m_0} \left(\epsilon\hbar \nabla \varphi - \frac{q}{c}\, \mathbf{A} \right)^2 - \frac{1}{2} \frac{\hbar^2}{m_0} \frac{\Delta \Pi}{\Pi} + q\Phi = 0 \tag{263}$$

We also know that the equations given above can be interpreted as being those of a fluid whose density is $\rho = \Pi^2$, whose local momentum is written as $\mathbf{P} = m_0 \mathbf{U} = \epsilon\hbar \nabla \varphi - q\mathbf{A}/c$. Knowing that $\mathbf{J} = \rho \mathbf{U}$, the first equation turns into an equation of electrical continuity:

$$\frac{\partial \rho}{\partial t} + \nabla \cdot \mathbf{J} = 0 \tag{264}$$

Using the definition $\mathbf{J} = \Psi\Psi^* \mathbf{U}$, the current density $\mathbf{J}$ can also be written as

$$\mathbf{J} = \frac{\epsilon\hbar}{2m_0}\, j(\Psi \nabla \Psi^* - \Psi^* \nabla \Psi) - \frac{q}{m_0 c}\, \Psi\Psi^* \mathbf{A} \tag{265}$$

If we take the gradient of the phase equation (263), we obtain

$$\frac{\partial \mathbf{P}}{\partial t} = q\mathbf{E} + \nabla \left(\frac{\hbar^2}{2m_0} \frac{\Delta \Pi}{\Pi} - \frac{1}{2m_0}\, \mathbf{P}^2 \right) \tag{266}$$

This results in the equation of motion

$$\frac{d\mathbf{P}}{dt} = q\mathbf{E} + \frac{q}{c}\, \mathbf{U} \wedge \mathbf{B} - \nabla E_Q \tag{267}$$

where the classic potential $E_P = q\Phi$ is completed by another quantum potential whose expression is $E_Q = -\hbar^2 \Delta \Pi / 2m_0 \Pi$.

For a wavefunction of the form $\Psi = \Pi e^{j\varphi}$, the approach chosen above leads us to define the following values:

$$\epsilon\hbar \nabla \varphi = m_0 \mathbf{U} + \frac{q}{c}\, \mathbf{A} = \mathbf{b} \tag{268}$$

$$\epsilon\hbar\, \frac{\partial \varphi}{\partial t} = -\frac{1}{2m_0}\, \mathbf{U}^2 - q\Phi + \frac{1}{2} \frac{\hbar^2}{m_0} \frac{\Delta \Pi}{\Pi} = a \tag{269}$$

where the potential functions Φ and $\mathbf{A}$ in the preceding relations are associated with electrons from the solenoid. For the wavefunction Ψ to recover the same value after a complete rotation around the solenoid $\mathbf{r} = \mathbf{r}_0$, we must have

$$\epsilon\hbar\varphi(\mathbf{r}, t) = \int_{t_0}^{t} a(\mathbf{r}, t')dt' + \int_{\mathbf{r}_0}^{\mathbf{r}} \mathbf{b}(\mathbf{r}', t_0) \cdot \delta\mathbf{r}' = n2\pi\hbar \tag{270}$$

where n is an integer number.

By substituting the $\mathbf{b}$ definition (268) into Eq. (270), we get the following integral relation:

$$\int_{t_0}^{t} a\, dt' + \int_{\mathbf{r}_0}^{\mathbf{r}} m_0\mathbf{U} \cdot \delta\mathbf{r}' + \frac{q}{c}\int_S \mathbf{B} \cdot \delta\mathbf{S} = n2\pi\hbar \tag{271}$$

If the velocity $\mathbf{U}$ of an electron within the beam is constant outside the solenoid, the variation of the vector potential $\mathbf{A}$ as a function of time in the medium, and thus also in the solenoid, will induce a modification of the phase, as indicated by the equations written above. This will produce a modification of the boundary conditions on the boundary of the solenoid for the quantities a and $\mathbf{b}$. We must also stress that the modification of the vector potential outside the solenoid is generated by either an external or an internal source feeding the solenoid. This can explain the existence of the Aharonov–Bohm effect for toroidal, permanent magnets. The interpretation of the Aharonov–Bohm effect is therefore classic, but the observation of this effect requires the principle of interference of quantum mechanics, which enables a phase effect to be measured.

Acknowledgments

I would like to thank Dr. Myron Evans for his invitation to contribute a chapter to this book on contemporary optics and electrodynamics and Alexandre David Szames, who helped us translate this chapter from French to English.

References

1. V. Jones, *Pour la Science* **159**, 88 (Jan. 1991).
2. M. Atiyah, *Rev. Mod. Phys.* **67**(4), 977 (1995).
3. *Pour la Science* (Dossier Hors series) (April 1997).
4. R. Gimore, *Rev. Mod. Phys.* **70**(4), 1455 (1998).
5. R. L. Ricca and M. A. Berger, *Phys. Today* 28 (Dec. 1996).
6. D. Reed, in T. W. Barrett and D. M. Grimes (Eds.), *Advanced Electromagnetism: Foundations, Theory and Applications*, World Scientific, Singapore, 1995, p. 217.
7. G. E. Marsh, *Force-Free Magnetic Fields: Solutions, Topology and Applications*, World Scientific, Singapore, 1996.
8. G. Schmidt, *Physics of High Temperature Plasmas*, Academic, New York, 1979.
9. R. L. Stenzel, J. M. Urrutia, and C. L. Rousculp, *Phys. Rev. Lett.* **74**(5), 702 (1995).
10. J. A. Heras, *Am. J. Phys.* **62**(6), 525 (1994).

11. O. D. Jefimenko, *Causality Electromagnetic Induction and Gravitation*, Electret Scientific Company, Star City, WV, 1992.
12. O. D. Jefimenko, *Electricity and Magnetism,* Electret Scientific Company, Star City, WV, 1989.
13. J. A. Heras, *Am. J. Phys.* **63**(10), 928 (1995).
14. R. Ferrier, *C. R. Acad. Sci.* **184**, 585 (1927).
15. R. Ferrier, *C. R. Acad. Sci.* **185**, 533 (1927).
16. R. Ferrier, *C. R. Acad. Sci.* **185**, 104 (1927).
17. R. Ferrier, *C. R. Acad. Sci.* **186**, 1710 (1928).
18. S. R. Milner, *Phil. Trans. Roy. Soc. Lond. A* **253**, 185 (1960).
19. S. R. Milner, *Phil. Trans. Roy. Soc. Lond. A* **253**, 205 (1960).
20. P. Cornille, *J. Phy. D: Appl. Phys.* **23**, 129 (1990).
21. P. Cornille, in A. Lakhtakia (Ed.), *Essays on the Formal Aspects of Electromagnetic Theory*, World Scientific, Singapore, 1993, p. 138.
22. P. Cornille, *J. Electromagn. Waves Appl.* **8**(11), 1425 (1994).
23. P. Cornille, in T. W. Barett and D. M. Grimes (Eds.), *Advanced Electromagnetism: Foundations, Theory and Applications*, World Scientific, Singapore, 1995, p. 148.
24. J. D. Jackson, *Classical Electrodynamics*, Wiley, New York, 1975.
25. M. Evans and J. P. Vigier, *The Enigmatic Photon*, Vol. 1: *The Field* $B^{(3)}$, Kluwer, Dordrecht, 1994.
26. G. E. Marsh, in T. W. Barret and D. M. Grimes (Eds.), *Advanced Electromagnetism: Foundations, Theory and Applications*, World Scientific, Singapore, 1995, p. 52.
27. A. Barnes, *Am. J. Phys.* **45**(4), 371 (1977).
28. J. P. McKelvey, *Am. J. Phys.* **58**(4), 306 (1990).
29. H. Zaghloul and H. A. Buckmaster, *Am. J. Phys.* **56**(9), 801 1988.
30. H. Zaghloul and H. A. Buckmaster, *Ann. Phys.* **15**(1), 21 (1990).
31. H. Zaghloul and H. A. Buckmaster, in A. Lakhtakia (Ed.), *Essays on the Formal Aspects of Electromagnetic Theory*, World Scientific, Singapore, 1993, p. 183.
32. F. Bretenaker and A. Le Floch, *Phys. Rev. A* **43**(7), 3704 (1991).
33. G. F. Freire, *Am. J. Phys.* **34**, 567 (1966).
34. S. K. Antiochos, *AstroPhys. J.* **312**, 886 (1987).
35. H. Zaghloul, *Phys. Lett. A*, **140**(3), 95 (1989).
36. H. Zaghloul and O. Barajas, *Am. J. Phys.* **58**(8), 783 (1990).
37. P. R. Baldwin and R. M. Kiehn, *Phys. Lett. A* **189**, 161 (1994).
38. P. R. Baldwin and G. M. Townsend, *Phys. Rev. E* **51**(3), 2059 (1995).
39. E. G. P. Rowe, *J. Phys. A: Math. Genet.* **12**(1), 145 (1979).
40. W. W. Hansen, *Phys. Rev.* **47**(1), 139 (1935).
41. W. C. Chew, *Waves and Fields in Inhomogeneous Media*, Van Nostrand Reinhold, New York, 1990.
42. H. K. Moffatt, *J. Fluids Mech.* **35**(Part 1), 117 (1969).
43. A. F. Rañada, *J. Phys. A: Math. Genet.* **23**, L815 (1990).
44. A. F. Rañada, *J. Phys. A: Math. Genet.* **25**, 1621 (1992).
45. G. N. Afanasiev and Y. P. Stepanovsky, *Il Nuov. Cimento A* **109**(3), 271 (1996).
46. J. A. Stratton, *Théorie de l'électromagnétisme*, Dunod, Paris, 1961.
47. R. M. Kiehn, *Torsion-Helicity and Spin as Topological Coherent Structures in Plasmas*, available at www.un.edu/rkiehn/car/carhomep.htm.

48. T. W. Barrett, in A. Lakhtakia (Ed.), *Essays on the Formal Aspects of Electromagnetic Theory*, World Scientific, Singapore, 1993, p. 6.
49. P. M. Bellan, *Phys. Rev. Lett.* **57**(19), 2383 (1986).
50. D. R. Moorcroft, *Am. J. Phys.* **37**, 221 (1969).
51. W. Klein, *Am. J. Phys.* **49**(6), 603 (1981).
52. R. H. Romer, *Am. J. Phys.* **50**(12), 1089 (1982).
53. Y. Aharonov and D. Bohm, *Phys. Rev.* **115**(3), 485 (1959).
54. Y. Aharonov and D. Bohm, *Phys. Rev.* **123**(4), 1511 (1961).
55. S. Olariu and I. I. Popescu, *Rev. Mod. Phys.* **17**(2), 339 (1985).
56. S. Olariu and I. I. Popescu, *Phys. Rev. D.* **33**(6), 1701 (1986).
57. R. G. Chambers, *Phys. Rev. Lett.* **5**(1), 3 (1960).
58. J. Woodilla and H. Schwarz, *Am. J. Phys.* **39**(1), 111 (1971).
59. G. Matteucci and G. Pozzi, *Am. J. Phys.* **46**(6), 619 (1978).
60. A. Tonomura, T. Matsuda, R. Suzuki, A. Fukuhara, and N. Osakabe, *Phys. Rev. Lett.* **48**(21), 1443 (1982).
61. A. Tonomura, H. Umezaki, T. Matsuda, N. Osakabe, J. Endo, and Y. Sugita, *Phys. Rev. Lett.* **51**(5), 331 (1983).
62. A. Tonomura, N. Osakabe, T. Matsuda, T. Kawasaki, and J. Endo, *Phys. Rev. Lett.* **56**(8), 792 (1986).
63. A. Tonomura, *Phys. Today* **43**(4), 22 (1990).
64. Y. Imry and R. Webb, *Pour la Science.* **140**(6), 32 (1989).
65. T. H. Boyer, *Am. J. Phys.* **40**(1), 56 (1972).
66. T. H. Boyer, *Phys. Rev. D.* **8**(6), 1679 (1973).
67. P. Bocchieri and A. Loinger, *Il Nuov. Cimento A* **66**(2), 164 (1981).
68. P. D. Noerdlinger, *Il Nuov. Cimento* **23**(1), 158 (1962).
69. E. J. Post, *Rev. Mod. Phys.* **39**(2), 475 (1967).
70. P. Cornille, *Progr. Energy Combust. Sci.* **25**(2), 161 (1999).
71. X. Zhu and W. C. Henneberger, *J. Phys. A: Math. Genet.* **23**, 3983 (1990).
72. G. Spavieri and G. Cavalleri, *Europhys. Lett.* **18**, 301 (1992).
73. R. M. Herman, *Found. Phys.* **22**, 713 (1992).
74. L. O'raifeartaigh, N. Straumann, and A. Wipf, *Comments Nucl. Particle Phys.* **20**, 15 (1991).
75. L. O'raifeartaigh, N. Straumann, and A. Wipf, *Found. Phys.* **23**, 703 (1993).

QUANTUM ELECTRODYNAMICS: POTENTIALS, GAUGE INVARIANCE, AND ANALOGY TO CLASSICAL ELECTRODYNAMICS

CARL E. BAUM

Air Force Research Laboratory, Directed Energy Directorate, Kirtland AFB, New Mexico

CONTENTS

I. INTRODUCTION

As is well known in classical electromagnetics, the fields described by the Maxwell equations can be derived from a vector potential and a scalar potential. However, there are various forms that are possible, all giving the same fields. This is referred to as *gauge invariance*. In making measurements at some point

Modern Nonlinear Optics, Part 2, Second Edition, Advances in Chemical Physics, Volume 119, Edited by Myron W. Evans. Series Editors I. Prigogine and Stuart A. Rice.
ISBN 0-471-38931-5

it is the fields and perhaps current and charge densities that one considers. Potentials are quantities inferred (within the ambiguity of gauge invariance) by integration of the fields along appropriate paths.

In quantum electrodynamics (QED) the potentials asume a more important role in the formulation, as they are related to a phase shift in the wavefunction. This is still an integral effect over the path of interest. This manifests itself in the phase shift of an electron around a closed path enclosing a magnetic field, even though there are no fields (approximately) on the path itself (static conditions). As can be shown the result of such an experiment is gauge-invariant, allowing the use of various choices of the vector potential (all giving the same result).

Generalizing this question somewhat, one can explore the degree to which one can measure vector and scalar potentials, including the implications of gauge invariance [6]. Assuming that there are sources (current and/or charge) in some region of space away from the observer, in what sense can the potentials be distinguished? In particular, one can compare static and dynamic conditions. Under static conditions, it is possible to have zero fields in the vicinity of the observer (away from the source region), while having nonzero potentials. Two different antennas, one emphasizing the vector potential and the other the scalar potential (Lorenz gauge), are discussed in which the same fields are away from the source region.

Then we go on to consider an alternate way of viewing the role of the vector potential in quantum electrodynamics (QED) [7]. The closed-path line integral of the vector potential can be related to the same integral of the time integral of the electric field. Alternately, by looking at a perfectly conducting loop on this path, we can have a static current proportional to the vector potential or associated magnetic flux (i.e., no additional time integral).

II. ELECTROMAGNETIC FIELDS AND POTENTIALS

In standard form we have the Maxwell equations

$$\nabla \times \vec{E} = -\frac{\partial \vec{B}}{\partial t}, \quad \nabla \times \vec{H} = \vec{J} + \frac{\partial \vec{D}}{\partial t} \tag{1}$$

and constitutive relations (for linear media)

$$\vec{D} = \overleftrightarrow{\varepsilon} \cdot \vec{E}, \quad \vec{B} = \overleftrightarrow{\mu} \cdot \vec{H} \tag{2}$$

where $\overleftrightarrow{\varepsilon}$ = permittivity = $\varepsilon_0 \overleftrightarrow{1}$ for free space
$\overleftrightarrow{\mu}$ = permeability = $\mu_0 \overleftrightarrow{1}$ for free space
$Z_0 = [\mu_0/\varepsilon_0]$ = wave impedance of free space
$c = [\mu_0 \varepsilon_0]^{1/2} \equiv$ speed of light
$\overleftrightarrow{1} = \vec{1}_x\vec{1}_x + \vec{1}_y\vec{1}_y + \vec{1}_z\vec{1}_z \equiv$ dyadic identity

The divergence equations are not independent but can be derived from (2.1) under zero initial conditions

$$\nabla \cdot (\nabla \times) \equiv 0, \quad \nabla \cdot \vec{B} = 0 \quad \text{(no magnetic charge)}$$
$$\nabla = \vec{J} = -\frac{\partial}{\partial t}\nabla \cdot D \equiv -\frac{\partial \rho}{\partial t} \quad \text{(equation of continuity)} \tag{3}$$

The fields can be derived from the well-known scalar and vector potentials in free space [8,14,16,17,23,24,26] as

$$\vec{E} = -\frac{\partial \vec{A}}{\partial t} - \nabla \Phi, \quad \vec{H} = \frac{1}{\mu_0}\nabla \times \vec{A} \tag{4}$$

As is well known, these potentials are not unique since one can add a potential χ as

$$\vec{A}' \equiv \vec{A} - \nabla \chi, \quad \Phi' \equiv \Phi + \frac{\partial \chi}{\partial t} \tag{5}$$

giving the same result in (4). Different choices of χ correspond to different gauge conditions.

The most common form taken uses the Lorenz potentials that satisfy the Lorenz gauge:

$$\nabla \cdot \vec{A} + c^{-2}\frac{\partial \Phi}{\partial t} = 0 \tag{6}$$

These are taken as retarded potentials (outgoing waves for zero initial conditions) with explicit forms

$$\vec{A}(\vec{r}, t) = \mu_0 \int_V \frac{\vec{J}\left(\vec{r}\,', t - \dfrac{|\vec{r} - \vec{r}\,'|}{c}\right)}{4\pi|\vec{r} - \vec{r}\,'|} dV', \quad \Phi(\vec{r}, t) = \frac{1}{\varepsilon_0}\int_V \frac{\rho\left(\vec{r}\,', t - \dfrac{|\vec{r} - \vec{r}\,'|}{c}\right)}{4\pi|\vec{r} - \vec{r}\,'|} dV' \tag{7}$$

This form has both potentials propagating away from the source at speed c and is relativistically invariant, and for this reason is often preferred.

A related potential is the Hertz potential

$$\vec{\Pi}(\vec{r}, t) \equiv c^2 \int_{-\infty}^{t} \vec{A}(\vec{r}, t')dt' \tag{8}$$

for which we have

$$\vec{E} = -c^{-2}\frac{\partial^2}{\partial t^2}\vec{\Pi} + \nabla(\nabla \cdot \vec{\Pi}), \quad \vec{H} = \varepsilon_0 \nabla \times \left[\frac{\partial}{\partial t}\vec{\Pi}\right] \tag{9}$$

However, this is basically the same as $\vec{A}$, since in complex frequency domain

$$\tilde{\vec{\Pi}}(\vec{r}, s) = \frac{c^2}{s}\tilde{\vec{A}}(\vec{r}, s), \quad \sim \equiv \text{Laplace transform (two-sided)} \tag{10}$$

$$s \equiv \Omega + j\omega \equiv \text{Laplace transform variable or complex frequency}$$

Note that the retarded potentials can be expressed as a single vector potential via (6) as

$$\Phi(\vec{r}, t) = -c^2 \int_{-\infty}^{t} \nabla \cdot \tilde{\vec{A}}(\vec{r}, t')\, dt' \tag{11}$$

with zero initial conditions. Thus we can define what we might call the electric gauge condition for which

$$\Phi_e(\vec{r}, t) = 0$$

$$\vec{A}_e(\vec{r}, s) = \vec{A}(\vec{r}, s) + \int_{-\infty}^{t} \nabla\Phi(\vec{r}, t')\, dt' = \vec{A}(\vec{r}, t) - c^2 \int_{-\infty}^{t}\int_{-\infty}^{t'} \nabla(\nabla \cdot \vec{A}(\vec{r}, t''))\, dt''dt'$$

$$\vec{E} = -\frac{\partial \vec{A}_e}{\partial t}, \quad \vec{H} = \frac{1}{\mu_0}\nabla \times \vec{A}_e \tag{12}$$

Note that this is also a retarded potential propagating outward at speed c.

Another convenient choice is the Coulomb gauge for which we have [14,17,23]

$$\nabla \cdot \vec{A}_c = 0, \quad \Phi_c(\vec{r}, t) = \frac{1}{\varepsilon_0}\int_V \frac{\rho(\vec{r}, t)}{4\pi|\vec{r} - \vec{r}'|}\, dV'$$

$$\vec{A}_c(\vec{r}, t) = \vec{A}(\vec{r}, t) + \int_{-\infty}^{t} [\nabla\Phi(\vec{r}, t') - \nabla\Phi_c(\vec{r}, t')]\, dt' \tag{13}$$

$$\vec{E} = \frac{\partial}{\partial t}\vec{A}_c - \nabla\Phi_c, \quad \vec{H} = \frac{1}{\mu_0}\nabla \times \vec{A}_c$$

However, note that Φ_c (and hence $\vec{A}_c$) now "propagates" with infinite speed, but that the fields still propagate with the speed of light c.

III. POTENTIALS AND QUANTUM MECHANICS

In formulating quantum electrodynamics (QED), it has been found convenient to introduce the electromagnetic interaction with charged particles via the potentials instead of the fields. Consider a particle of charge q traveling on some path P from $\vec{r}_1$ to $\vec{r}_2$. Then the magnetic change in phase of the wavefunction is just [12,15]

$$\phi_h = \frac{q}{\hbar}\int_P \vec{A}(\vec{r},t)\cdot d\vec{l} \tag{14}$$

where $\hbar = \dfrac{h}{2\pi} = 1.05443 \times 10^{-34}$ joule-seconds

$q_e \equiv$ electron charge $\simeq -1.60206 \times 10^{-19}$ (Coulombs)

$h \equiv$ Planck's constant

and the electric change in phase is only

$$\phi_e = \frac{-q}{\hbar}\int_{-\infty}^{t}\int_P \nabla\Phi(\vec{r},t')\cdot d\vec{l}\,dt' = \frac{-q}{\hbar}\int_{-\infty}^{t}[\Phi(\vec{r}2,t') - \Phi(\vec{r}1,t')]\,dt' \tag{15}$$

where the total change in phase is only

$$\phi = \phi_h + \phi_e \tag{16}$$

Combining these expressions, we have

$$\phi = \frac{q}{\hbar}\int_P [\vec{A}(\vec{r},t) - \int_{-\infty}^{t} \nabla\Phi(\vec{r},t)\,dt']\cdot d\vec{l} \tag{17}$$

This is interpreted in the sense of changing a quantum wavefunction ψ in the form

$$\psi \rightarrow e^{j\phi}\psi \tag{18}$$

Note that if the path is closed, we have

$\vec{r}2 = \vec{r}1$

$$\phi = \frac{q}{\hbar}\oint_P \vec{A}(\vec{r},t)\cdot d\vec{l} = \frac{q}{\hbar}\int_S [\nabla \times \vec{A}(\vec{r},t)]\cdot d\vec{S} = \frac{q}{\hbar}\int_S \vec{B}(\vec{r},t)\cdot d\vec{s} \tag{19}$$

with P the boundary of S and the unit normal $\vec{1}_S$ taken in the usual right-handed sense. Note that this is independent of the scalar potential and of the gauge chosen since $\vec{A}$ enters via the curl. This phase shift is the basis of the shifting of the diffraction pattern of electrons from a common source going through two slits with a confined magnetic field between the slits (Aharonov–Bohm experiment).

For later use, as indicated in Fig. 1, if we have two paths P_1 and P_2 from $\vec{r}_a$ to $\vec{r}_b$, there are in general two different phase changes given by (17). For the same wavefunction ψ at $\vec{r}_a$ for particles traversing both paths we have at $\vec{r}_b$ wavefunctions

$$\psi_n = e^{j\phi_n}\psi \quad \text{for path } P_n \tag{20}$$

The difference in phase is the closed-path integral

$$\begin{aligned}\phi_1 - \phi_2 &= \frac{q}{\hbar}\int_C \vec{A}(\vec{r},t)\cdot d\vec{l} = \frac{q}{\hbar}\int_S [\nabla \times \vec{A}(\vec{r},t)]\cdot d\vec{l}_S\, dS \\ &= \frac{q}{\hbar}\int_S \vec{B}(\vec{r},t)\cdot \vec{l}_S\, dS = \frac{q}{\hbar}\Phi_m(t) \\ \Phi_m(t) &\equiv \text{magnetic flux through S (with boundary } C) \\ \vec{1}_S &\equiv \text{unit normal to } S\end{aligned} \tag{21}$$

Note that quantum phase is not a physical observable (at least in current formulations). The observable is $\psi\psi^*$ or $|\psi|^2$. If, however, we have two (or more)

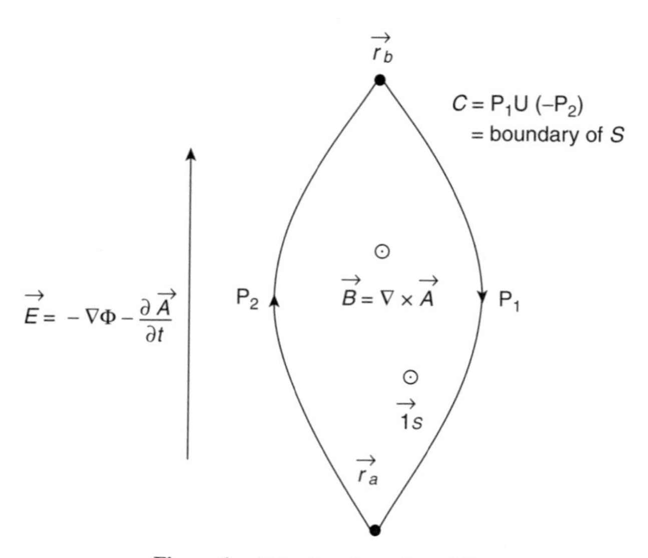

Figure 1. Paths for charged particle.

quantum wavefunctions, such as ψ_1 and ψ_2 at $\vec{r}_b$, we obtain interference at $\vec{r}_b$ from the relative phase with appropriate normalization via

$$|\psi_1 + \psi_2|^2 = |e^{j\phi_1} + e^{j\phi_2}|^2|\psi|^2 = 4\cos^2(\phi_1 - \phi_2)|\psi|^2 \tag{22}$$

where the phase difference, as in (21), is the operative parameter. Note that if P_1 and P_2 become the same, then $\Phi_m = 0$, and there is no phase difference. The absolute phase at $\vec{r}_b$ can be made whatever one wishes by a gauge transformation involving an arbitrary scalar potential, but this has no effect on the phase difference.

It is also possible to have this phase difference with negligible electric and magnetic fields on C. It is essential that there be a magnetic field and associated flux Φ_m passing through S. A solenoid (with current) or magnetic materials (permanent magnet) can be used to confine the magnetic fields away from C. However, as we shall later see (Section V), in setting up such a field, we have

$$\begin{aligned} \Phi_m(t) &= -\oint_C \left[\int_{-\infty}^{t} \vec{E}(\vec{r}, t')\, dt' \right] d\vec{l} = \oint_C \vec{A}(\vec{r}, t)\, d\vec{l}, \\ \Phi_m(-\infty) &= 0 \text{ (initial condition)} \end{aligned} \tag{23}$$

Thus, the phase difference is related to the time integral of the electric field (the electric impulse) on the contour. At some time t the electric field (or its contour integral) can be zero, while the corresponding impulse (time integral) is non zero. What some might term action at a distance (from the magnetic field away from C) is mathematically equivalent to action from a previous time (when the electric field was present on C), assuming zero initial conditions.

Electron motion is more generally formulated in a form of the Schrödinger equation, including the spin in the presence of external fields known as the *Pauli equation*. This equation is gauge invariant in the sense that a transformation as in (5) also changes the quantum wavefunction ψ as

$$\psi' = e^{-(q/\hbar)\chi}\,\psi \tag{24}$$

leaving the Pauli equation unchanged [11]. Noting that $\psi\psi^*$ or $|\psi|^2$ is the physical observable, this phase change is not important. In particular, one can choose

$$\chi = -\int_{-\infty}^{t} \Phi(\vec{r}, t')\, dt' \tag{25}$$

and we have the electric gauge in (12) in which only the vector potential $\vec{A}_e$ appears. This can be readily computed from the usual Lorenz vector and scalar potentials from (7) and (12). Other choices, such as the Coulomb gauge, can also be used where convenient.

Another derivation of gauge invariance concerns the time-independent Schrödinger equation [19]. Here it is shown that a zero-curl vector potential can be absorbed into the scalar potential with no change in the observables, that is only a phase change ψ in of the form (24). This is consistent with the previous discussion in that it is only the divergence-free part (nonzero curl) of the vector potential that is associated with the measurable effect of an electron path enclosing a magnetic flux. Note that $\nabla \times \vec{A}$ is gauge-invariant, and this is the part producing the measurable effect.

In formulating QED a least-action principle involving a Lagrangian is often used [9,18,20]. This involves the potentials in various forms. Not only is relativistic invariance (Lorenz potentials) desired, but also gauge invariance. At least in the current state of QED, gauge invariance is included as a fundamental part [21,22].

IV. IMPLICATIONS OF ZERO ELECTRIC AND MAGNETIC FIELDS AWAY FROM SOURCES

Consider a volume (simply connected) V_s with surface S_s of finite dimensions containing the sources as indicated in Fig. 2. The observer at $\vec{r}$ is assumed away from the sources:

$$\vec{r} \notin V_s \bigcup S_s \tag{26}$$

Suppose that $\vec{E}$ and $\vec{H}$ are zero for all times at the observer. Then, from (12), we have

$$\vec{A}_e(\vec{r}, t) = \vec{A}_e(\vec{r}, -\infty) = \text{constant vector}, \ \nabla \times \vec{A}_e(\vec{r}, t) = \nabla \times \vec{A}_e(\vec{r}, -\infty) = \vec{0} \tag{27}$$

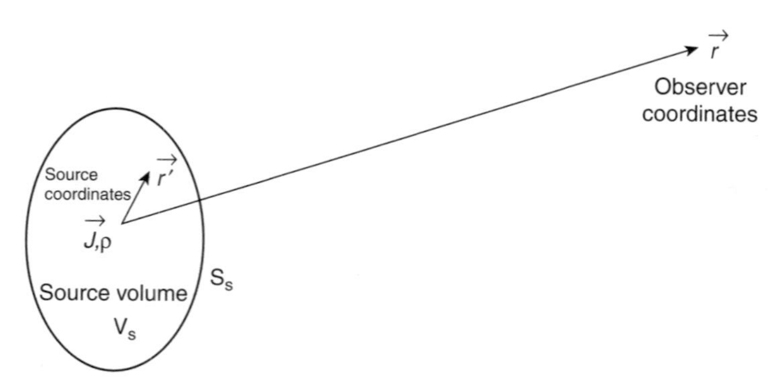

Figure 2. Sources confined to a simply connected volume of finite linear dimensions.

as the only solution, a zero-curl constant vector for all time. Setting $\vec{J}$ and ρ initial conditions to zero, we have

$$\vec{A}_e(\vec{r}, t) = \vec{0} \text{ for all } t \tag{28}$$

as the only solution. In terms of the Lorenz form of the potentials, this gives

$$\vec{A}(\vec{r}, t) + \int_{-\infty}^{t} \nabla\Phi(\vec{r}, t)dt' = \vec{0} \tag{29}$$

Since the initial conditions on $\vec{J}$ and ρ are zero, then

$$\vec{A}(\vec{r}, -\infty) = 0, \quad \Phi(\vec{r}, -\infty) = 0 \tag{30}$$

and constant potentials in this form are also excluded.

This can also be considered in complex-frequency form for which we find

$$\tilde{\vec{A}}_e(\vec{r}, s) = \vec{0} \quad \text{for all } s, \quad \tilde{\vec{A}}(\vec{r}, s) + \frac{1}{s}\nabla\Phi(\vec{r}, s) = 0 \quad \text{for all } s \tag{31}$$

and an appropriate limit for $s \to 0$.

V. CHANGE FROM ONE STATIC POTENTIAL TO A SECOND STATIC POTENTIAL

Let us now consider an initial set of static potentials (subscript 1) followed by a second set (subscript 2) with sufficient time between to allow static conditions to be achieved (at least approximately). Then, without loss of generality, consider zero initial conditions

$$\vec{A}_{e_1}(\vec{r}) = \vec{0}, \quad \vec{A}_1(\vec{r}) = \vec{0}, \quad \Phi_1(\vec{r}) = 0 \tag{32}$$

Actually, for these potentials, this can be taken in a retarded time sense, that is, zero for

$$t < t_1 + \frac{r}{c} \tag{33}$$

since they are formulated in retarded in retarded time via (7). Note that the finite linear dimensions of V_s allow t_1 to be adjusted for propagation through V_s.

Next, turn on the sources $\vec{J}$ and ρ, have them reach constant values $\vec{J}_2$ and ρ_2 after some time t_2 less transit times across V_s. Then, in a retarded time sense for

$$t > t_2 + \frac{r}{c} \tag{34}$$

we have $\vec{A}_2$ and Φ_2 constant with

$$\vec{A}_{e_2}(\vec{r}) = \vec{A}_2(\vec{r}) + \int_{t_1}^{t_2} \nabla\Phi(\vec{r},t')\,dt' \tag{35}$$

From (7) we have constant $\vec{A}_2$ and Φ_2. However, if we constrain that the electric field be zero at this time, then we have

$$\vec{E}_2(\vec{r}) = -\nabla\Phi(\vec{r}) = 0 \tag{36}$$

allowing only a uniform Φ_2 (independent of $\vec{r}$). This is inconsistent with (7) unless

$$\Phi_2 = 0 \tag{37}$$

which, in turn, implies that $\rho_2(\vec{r})$ is constrained to a distribution with no exterior potential. A set of charges inside a constant potential surface (zero potential) such as a conducting cavity with requisite resulting surface charge density on the interior of the surface is an example of such a charge distribution. Furthermore, we have, after t_2

$$\nabla \times \vec{A}_{e_2}(\vec{r}) = \vec{0} = \nabla \times \vec{A}_2(\vec{r}) \tag{38}$$

since we also assume that

$$\vec{H}_2(\vec{r}) = \vec{0} \tag{39}$$

Then, for both Lorenz and electric gauges (and others such as Coulomb as well), the change from initial (1) to final (2) conditions is characterized by (35). Referring to (4), we can see that this change is characterized by the electric impulse

$$\int_{t_1}^{t_2} \vec{E}(\vec{r},t')\,dt' = -\vec{A}_{e_2}(\vec{r}) = -\vec{A}_2(\vec{r}) - \int_{t_1}^{t_2} \nabla\Phi(\vec{r},t')\,dt' \tag{40}$$

subject to (36) and (38) as conditions on the potentials. So, while the fields are zero before t_1 and after t_2, they are not in general zero in between these two times. The electric impulse is characterized by $\vec{A}_{e_2}$ (i.e., the change in $\vec{A}_e$ from initial to final conditions). As indicated by (35), this can be expressed in various gauges, as the electric impulse is gauge-invariant.

VI. POTENTIALS FROM AN ELECTRIC DIPOLE

Now, consider how one might realize the conditions set forth in Section V. Consider first an elementary z-directed electric dipole at the coordinate origin in the source region as

$$\vec{p}(t) = p(t)\vec{1}_z \tag{41}$$

This can be regarded as elementary charges $\pm Q(t)$ placed at $z = \pm d/2$ with the limit taken as $d \to 0$ with the product

$$p(t) = Q(t)d \tag{42}$$

constant.

The usual cylindrical and spherical coordinates are related to the cartesian coordinates via

$$x = \Psi\cos(\phi), \quad y = \Psi\sin(\phi), \quad z = r\cos(\theta), \quad \Psi = r\sin(\theta) \tag{43}$$

The relevant dyadics are

$$\begin{aligned} \overleftrightarrow{1} &\equiv \vec{1}_x\vec{1}_x + \vec{1}_y\vec{1}_y + \vec{1}_z\vec{1}_z \equiv \text{identity} \\ \overleftrightarrow{1}_r &\equiv \overleftrightarrow{1} - \vec{1}_r\vec{1}_r = \vec{1}_\theta\vec{1}_\theta + \vec{1}_\phi\vec{1}_\phi \equiv \text{transverse identity} \end{aligned} \tag{44}$$

The Lorenz potentials for this case are from Ref. 3

$$\begin{aligned} \tilde{\vec{A}}(\vec{r}, s) &= e^{-\gamma r}\frac{\mu_0}{4\pi r}s\tilde{\vec{p}}(s), \quad \gamma \equiv \frac{s}{c} \\ \tilde{\Phi}(\vec{r}, s) &= e^{-\gamma r}\left\{\frac{s}{4\pi c r} + \frac{1}{4\pi r^2}\right\}\vec{1}_r \cdot \tilde{\vec{p}}(s) \\ \gamma &= \frac{s}{c} \equiv \text{propagation constant} \end{aligned} \tag{45}$$

and the fields are

$$\begin{aligned} \tilde{\vec{E}}(\vec{r}, s) &= e^{-\gamma r}\left\{\frac{-\mu_0}{4\pi r}s^2\overleftrightarrow{1}r + \frac{Z_0}{4\pi r^2}s\left[3\vec{1}_r\vec{1}_r - \overleftrightarrow{1}\right] + \frac{1}{4\pi\varepsilon_0 r^3}\left[3\vec{1}_r\vec{1}_r - \overleftrightarrow{1}\right]\right\} \cdot \tilde{\vec{p}}(s) \\ \tilde{\vec{H}}(\vec{r}, s) &= e^{-\gamma r}\left\{\frac{-1}{4\pi c r}s^2 - \frac{1}{4\pi c r^2}s\right\}\vec{1}_r \times \tilde{\vec{p}}(s) \end{aligned} \tag{46}$$

as these are related by (4). Including the electric gauge as in (12), we have

$$\begin{aligned}\tilde{\vec{E}}(\vec{r},s) &= -s\tilde{\vec{A}}(\vec{r},s) - \nabla\Phi(\vec{r},s) = -s\tilde{\vec{A}}_e(\vec{r},s)\\ \tilde{\vec{H}}(\vec{r},s) &= \frac{1}{\mu_0}\nabla\times\tilde{\vec{A}}(\vec{r},s) = \frac{1}{\mu_0}\nabla\times\tilde{\vec{A}}_e(\vec{r},s)\end{aligned} \tag{47}$$

So the electric gauge vector potential is readily expressed in terms of the electric field as in (46).

In changing from one static potential to a second, Section V gives this in terms of the electric impulse

$$\int_{t_1}^{t_2} \vec{E}(\vec{r},t')\,dt' = -\vec{A}_2(r) - \int_{t_1}^{t_2} \nabla\Phi(\vec{r},t')\,dt' = -\vec{A}_{e_2}(\vec{r}) \tag{48}$$

where zero initial conditions are assumed, and hence zero initial electric dipole moment. From (46) and (47) we have

$$-\vec{A}_{e_2}(\vec{r}) = \int_{t_1}^{t_2} \vec{E}(\vec{r},t)\,dt' = \frac{1}{4\pi\varepsilon_0 r^3}\left[3\vec{1}_r\vec{1}_r - \overleftrightarrow{1}\right]\cdot\int_{t_1}^{t_2}\vec{p}(t')\,dt' \tag{49}$$

now with

$$\lim_{t\to\infty}\vec{p}(t) = \vec{0} \tag{50}$$

or zero final condition for the electric dipole moment, with nonzero complete time integral.

In terms of the Lorenz potentials, since there is zero current and charge at late time, we have

$$\begin{gathered}\vec{A}_2(\vec{r}) = \vec{0}, \quad \Phi_2(\vec{r}) = 0\\ \int_{t_1}^{t_2}\nabla\Phi(\vec{r},t')\,dt' = \tilde{\vec{A}}_{e_2}(\vec{r}) = \frac{1}{4\pi\varepsilon_0 r^3}[3\vec{1}_r\vec{1}_r - \overleftrightarrow{1}]\cdot\int_{t_1}^{t_2}\vec{p}(t')\,dt'\end{gathered} \tag{51}$$

So both Lorenz potentials are zero for late time (as well as initially), but are, of course, nonzero for intermediate times. The electric gauge vector potential is, however, nonzero at late times and is the negative of the electric impulse.

VII. LORENZ VECTOR POTENTIAL WITHOUT SCALAR POTENTIAL: EQUIVALENT ELECTRIC DIPOLE

Now consider the case of no scalar potential (Lorenz gauge) for all time. This implies that outside V_s

$$\begin{gathered}\Phi(\vec{r},t) = 0, \quad \vec{A}(\vec{r},t) = \vec{A}_e(\vec{r},t) \\ \nabla \cdot \vec{A}(\vec{r},t) = 0 = \nabla \cdot \vec{A}_e(\vec{r},t) \\ \int_{t_1}^{t_2} \vec{E}(\vec{r},t)\,dt' = \nabla \cdot \vec{A}_{e_2}(\vec{r}) = \vec{A}_2(\vec{r})\end{gathered} \tag{52}$$

For simplicity, one can take

$$\rho(\vec{r},t) = 0 \tag{53}$$

consistent with the preceding equations, although as discussed before, there are special cases of nonzero ρ which produce no fields outside V_s.

There are various forms that $\vec{J}$ can take with

$$\nabla \cdot \vec{J}(\vec{r},t) = 0 \tag{54}$$

One way to synthesize such divergenceless current distributions is to think of the various ways to make closed conducting loops. In a DC (direct-current, i.e., low-frequency) sense there is a static current distribution with no (macroscopic) charge density. Then take this same current distribution to nonzero frequencies, either in an approximate sense if the loop is electrically small, or in a more exact sense by distributing current sources around the loop, all producing the same current (both magnitude and phase), thereby suppressing more and more exactly (in the limit of a large number of current sources) any buildup of charge. As the integrals in (7) make quite clear, a divergenceless current distribution can produce a vector potential with no scalar potential and hence fields (including radiated or $e^{-\gamma r}/r$ fields) via (4). One can also say this directly with the dyadic Green function of free space [25].

An important class of such loops can be referred to as "field containing inductors" [10]. In this type of structure, the loop is constructed as a solenoid, which is closed on itself to form a toroid-like structure. Such a doubly connected surface and higher-order connectedness is also possible. At DC, the structure is designed to produce no external magnetic field. One can synthesize such geometries by considering a closed, perfectly conducting, multiply connected surface with magnetic field inside, but not through the surface. Solving for the resulting surface current density via the inside tangential magnetic field, one

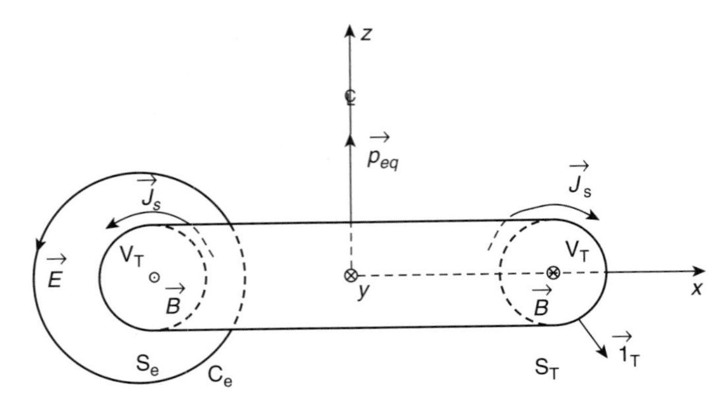

Figure 3. Ideal toroidal divergenceless current distribution.

then constructs this current density (at least approximately) via appropriate spacing of wires on the selected surface with one or more sources to drive the resulting loop(s). Note that there are no exterior fields *at zero frequency.*

As indicaated in Fig. 3, consider the simplest case of such a field-containing inductor, a body of revolution as a toroid. The cross section of the toroid need not be circular. It lies on S_T and contains the volume V_T. With surface current density $\vec{J}_s(\vec{r}_s, t)$ on S_T as indicated, we have the following for zero frequency:

$$\tilde{\vec{B}}(\vec{r},0) = \mu_0 \tilde{\vec{H}}(\vec{r},0) = \begin{cases} \mu_0 \dfrac{I}{\Psi} & \text{for} \quad \vec{r} \in V_T \\ 0 & \text{for} \quad \vec{r} \notin V_T \bigcup S_T \end{cases}$$
$$\tilde{\vec{J}}(\vec{r}_s,0) = \vec{1}_T(\vec{r}_s) \times \tilde{\vec{H}}(\vec{r}_s,0), \qquad (r_s \text{ on inside of } S_T)$$
$$\vec{1}_T(\vec{r}_s) \equiv \text{outward-pointing normal to } S_T \tag{55}$$

Now we constrain

$$\tilde{\vec{J}}_s(\vec{r}_s, s) = \tilde{f}(s)\tilde{\vec{J}}_s(\vec{r}_s, 0) \tag{56}$$

where $\tilde{f}(s)$ is some frequency function to be chosen for convenience. This assures a divergenceless current distribution

$$\nabla_s \cdot \tilde{\vec{J}}_s(\vec{r}_s, 0) = 0 = s\tilde{\rho}_s(\vec{r}_s, s) \tag{57}$$

Appendix A considers the response of such a toroidal antenna in reception and transmission. In reception, we have the open-circuit (OC) voltage from (A.14) for the electrically small case as

$$\tilde{V}_{OC}(s) = \tilde{\vec{h}}_V(s) \cdot \tilde{\vec{E}}^{(\text{inc})}(s), \quad \tilde{\vec{h}}_V(s) = \Xi s^2 \vec{1}_z \Xi = \frac{\Psi_2^2 - \Psi_1^2}{4} \frac{wN}{c^2} \tag{58}$$

where the last parameter represents the specific geometry of the antenna as derived in (A.7) [with N turns and dimensions Ψ_2 (outer radius), Ψ_1 (inner radius), and w (height)] for a rectangular cross section. Other cross sections can be similarly calculated.

In transmission, this type of antenna is described in the electrically small regime by an equivalent electric dipole moment

$$\tilde{\vec{p}}_{\text{eq}}(s) = \frac{1}{2}\vec{h}_V(s)\tilde{I}(s) = \Xi s \vec{1}_z \tilde{I}(s) \tag{59}$$

(where I being the current driving the antenna port) that produces fields

$$\begin{aligned} \tilde{\vec{E}}(\vec{r}, s) &= e^{-\gamma r}\left\{-\frac{\mu_0}{4\pi r}s^2\vec{1}_r + \frac{Z_0}{4\pi r^2}s[3\vec{1}_r\vec{1}_r - \overleftrightarrow{1}] + \frac{1}{4\pi\varepsilon_0 r^3}s[3\vec{1}_r\vec{1}_r - \overleftrightarrow{1}]\right\} \cdot \tilde{\vec{p}}_{\text{eq}}(s) \\ \tilde{\vec{H}}(\vec{r}, s) &= e^{-\gamma r}\left\{-\frac{1}{4\pi c r}s^2 - \frac{1}{4\pi r^2}\right\}\vec{1}_r \times \tilde{\vec{p}}_{\text{eq}}(s) \end{aligned} \tag{60}$$

provided r is large compared to antenna dimensions.

An interpretation of this equivalent electric dipole is indicated in Fig. 3. Consider a contour, C_e, say, on a plane of constant ϕ, enclosing a surface S_e on this plane. Define

$$S'_T \equiv S_e \bigcap V_T \tag{61}$$

Note that C_e cannot be shrunk to zero without passing through V_T since S_T is a multiply connected surface. From the integral form of the first of (1), we have

$$\oint_{C_e} \vec{E} \cdot d\vec{l} = -\frac{\partial}{\partial t}\int_{S_e} \vec{B} \cdot \vec{1}_{S_e}\, ds \tag{62}$$

where on a plane of constant ϕ

$$\vec{1}_{S_e} = \vec{1}\phi \tag{63}$$

By design, $\vec{B}$ is relatively large in V_T, but note that $\vec{E}$ can be zero everywhere outside V_T only if there is no time variation, that is, only for a static situation. This is consistent with (48) due to the factor of s included.

Having the fields from this equivalent dipole, we are in a position to calculate the potentials from the electric impulse for static initial and final conditions as defined in Section V. Again we have

$$-\vec{A}_{e_2}(\vec{r}) = \int_{t_1}^{t_2} \vec{E}(\vec{r}, t')\,dt' = \frac{1}{4\pi\varepsilon_0 r^3}[3\vec{1}_r\vec{1}_r - \vec{1}] \cdot \int_{t_1}^{t_2} \tilde{\vec{p}}_{\text{eq}}(t')\,dt' \tag{64}$$

So we require zero initial and final conditions for $\vec{p}_{\rm eq}$ just as for $\vec{p}$ in Section VI. From (59) this means that we have the following for the current driving the toroidal antenna:

$$I_1 = 0 \quad \text{(initial condition)}, I_2 = \frac{1}{\Xi}\vec{1}_2 \cdot \int_{t_1}^{t_2} \vec{p}_{\rm eq}(t')\,dt' \quad \text{(final condition)} \tag{65}$$

In another form, we have

$$\begin{aligned} \int_{t_1}^{t_2} \vec{p}_{\rm eq}(t')\,dt' &= \Xi \vec{1}_z I_2 \\ -\vec{A}_{e_2}(\vec{r}) &= \frac{1}{4\pi\varepsilon_0 r^3}[3\vec{1}_r\vec{1}_r - \overleftrightarrow{1}] \cdot \vec{1}_z \Xi I_2 \end{aligned} \tag{66}$$

Now, from (52), we have the following for the Lorenz potentials:

$$\begin{aligned} \Phi_2(\vec{r}) = \Phi_1(\vec{r}) &= 0, \quad \vec{A}_1(\vec{r}) = \vec{0} \\ \vec{A}_2(\vec{r}) = \vec{A}_{e_2}(\vec{r}) &= -\frac{1}{4\pi\varepsilon_0 r^3}\left[3\vec{1}_r\vec{1}_r - \overleftrightarrow{1}\right] \cdot \vec{1}_z \Xi I_2 \\ &= -\frac{1}{4\pi\varepsilon_0 r^3}\left[3\vec{1}_r\vec{1}_r - \overleftrightarrow{1}\right] \cdot \int_{t_1}^{t_2} \vec{p}_{\rm eq}(t')dt' \end{aligned} \tag{67}$$

VIII. COMPARISON OF POTENTIALS FOR ELECTRIC DIPOLE AND TOROIDAL ANTENNA EQUIVALENT ELECTRIC DIPOLE

Comparing the two cases in Sections VI and VII, let the two cases be the same in initial and final senses; that is set

$$\int_{t_1}^{t_2} \vec{p}(t')\,dt' = \int_{t_1}^{t_2} \vec{p}_{\rm eq}(t')\,dt' \tag{68}$$

Then we have the same electric impulse in both cases. This gives the same electric gauge vector potential $\vec{A}_e 2$. However, the Lorenz gauge potentials are quite different. For the electric dipole in Section VI, both $\vec{A}_2$ and Φ_2 are zero. For the toroidal antenna equivalent electric dipole in Section VII, while Φ_2 is zero, $\vec{A}_2$ is non zero. How then are these two cases different? Within the gauge condition (6) they are the same.

As long as we stay in the electrically small region for both antennas and equate $\vec{p}(t)$ and $\vec{p}_{\rm eq}(t)$ for all time, we also have the same fields. How, now, can we, away from the source regions, tell the two cases apart? Is there anything inherent in the Lorenz potentials, as distinguished from other potentials, such as derived from the electric gauge that can be measured to make this distinction?

IX. MAGNETIC FIELD MEASUREMENT BY INTEGRATION OF ELECTRIC FIELD AROUND A CLOSED PATH

Now we look at an analogy to classical electrodynamics [7]. The basic way to measure a magnetic field is a loop as indicated in Fig. 4. For simplicity, let this be a thin conductor on the contour C as indicated with a port (at the loop gap) where we can define voltage and current with some load taken as a resistance R. The basic performance for wavelengths large compared to the loop (electrically small loop) is [10]

$$\begin{aligned}
V_{\mathrm{OC}}(t) &= \vec{A}_{h_{\mathrm{eq}}} \cdot \frac{\partial \vec{B}^{(\mathrm{inc})}(t)}{\partial t} \qquad \text{(open-circuit voltage)} \\
I_{\mathrm{SC}}(t) &= \vec{\ell}_{h_{\mathrm{eq}}} \cdot \vec{H}^{(\mathrm{inc})}(t) \qquad \text{(short-circuit current)} \\
\vec{B}^{(\mathrm{inc})}(t) &= \mu_0 \vec{H}^{(\mathrm{inc})}(t) \qquad \text{(incident magnetic field)} \\
\mu_0 \vec{A}_{h_{\mathrm{eq}}} &= L \vec{\ell}_{h_{\mathrm{eq}}}, \quad L \equiv \text{loop inductance} \\
\vec{A}_{h_{\mathrm{eq}}} &\equiv \text{equivalent area}, \quad \vec{\ell}_{h_{\mathrm{eq}}} \equiv \text{equivalent length}
\end{aligned} \tag{69}$$

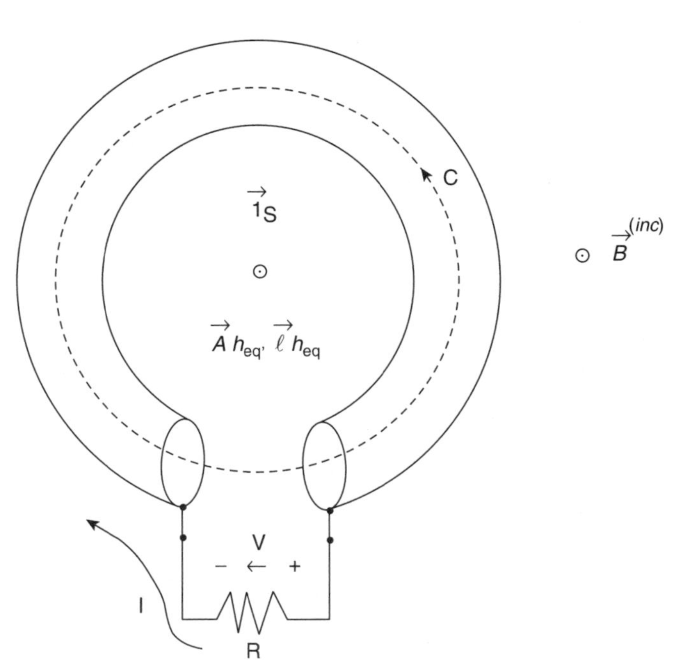

Figure 4. Basic magnetic-field sensor.

If the loop is approximately planar, encompassing an area A_h with $\vec{1}_S$ perpendicular to this plane, then we have

$$\vec{A}_{h_{eq}} = \vec{A}_h \vec{\ell}_s, \quad \vec{\ell}_h = \ell_h \vec{1}_s \tag{70}$$

For general complex frequencies (still electrically small), we have

$$\tilde{V}(s) = R\tilde{I}(s) = \frac{sR}{R + sL} \vec{A}_{h_{eq}} \cdot \vec{B}^{(\text{inc})}(s) \tag{71}$$

Looking at the open-circuit voltage, we have

$$V_{OC}(t) = \frac{d}{dt} \Phi_m^{(\text{inc})}(t) = -\oint_C \vec{E}^{(\text{inc})}(t) \cdot d\vec{\ell}, \Phi_m^{(\text{inc})}(t) = A_h \vec{1}_S \cdot \vec{B}^{(\text{inc})}(t) \tag{72}$$

So the open-circuit voltage here resembles the phase difference in (21) except for a time derivative. The short-circuit current takes the form

$$I_{SC}(t) = \frac{1}{L} \vec{A}_{h_{eq}} \cdot \vec{B}^{(\text{inc})}(t) = \frac{\Phi_m^{(\text{inc})}(t)}{L} \tag{73}$$

Now we have a loop parameter, the short-circuit current, which is proportional to a magnetic flux without a time derivative, like the phase difference in (21). Of course, this is now an incident flux that has been excluded by the closed (perfectly conducting) loop.

The discussion above is in terms of a shorted loop excluding a magnetic flux. Remaining in a classical context, it is also possible to have such a flux in a loop (passing through S) in the absence of an incident field by impressing the current from some source and then short-circuiting the loop gap. For a perfectly conducting loop, the resulting fields on the contour C in Fig. 4 can be zero in such a steady-state condition; the loop current flows on the surface of the conductor.

Suppose we take a long solenoid (with current flowing) or a permanent magnet and place it inside the loop in Fig. 4. The resulting magnetic field on C (after placement) due to the solenoid can be made quite small if the solenoid is long compared to the loop diameter. This is what one does in a QED sense to establish the flux between the two paths in Fig. 1. In so doing, one establishes a current in the perfectly conducting loop. So, phase shift around C in a quantum sense is like establishing a current in a perfectly conducting loop on C.

X. QUANTIZATION OF MAGNETIC FLUX ENCLOSED BY A SUPERCONDUCTING PATH

Returning to a quantum view, the short-circuit current in a loop needs to be quantized in the normal form for a superconducting loop [13]. Basically, one just

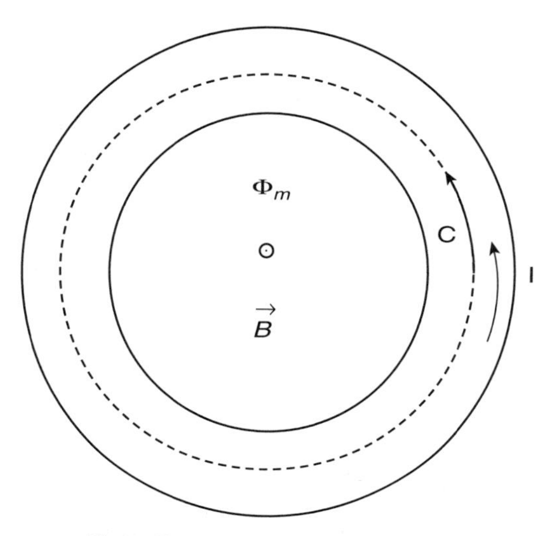

Figure 5. Flux in superconducting loop.

makes the phase difference in (1.5) around a superconducting loop as in Fig. 5 take the form

$$\phi_1 - \phi_2 = \frac{q}{\hbar}\Phi_m = 2\pi n = \text{integer} \tag{74}$$

so that the wavefunction is continuous around the loop. It has also been observed that the appropriate charge is

$$q = 2q_e \tag{75}$$

since the superconducting electrons (which are ordinarily Fermi particles) apparently form bound pairs that act as Bose particles. Noting the negative electron charge, we have [15].

$$\nabla\Phi_m = -\frac{\pi\hbar}{q_e} = -\frac{h}{2q_e} \simeq 2.0677 \times 10^{-15} \text{ webers} \tag{76}$$

as the separation of the quantized flux levels. For typical areas, this can represent a rather small magnetic field, for instance

$$\begin{aligned} A_h &= 10^{-4}m^2 \\ \Delta B &\simeq 2 \times 10^{-11} \text{ (teslas) } (\simeq 2 \times 10^{-7} \text{of earth's magnetic field}) \end{aligned} \tag{77}$$

Larger areas correspond to even smaller magnetic-field increments.

XI. CONCLUDING REMARKS

So now we have the question poased in an interesting form. There are two quite different kinds of antennas, both of which produce electric dipole fields, but different Lorenz potentials, one emphasizing the vector potential and the other, the scalar potential. In a classical electromagnetic sense, one cannot distinguish these two cases by measurements of the fields (the measurable quantities) at distances away from the source region. The gauge invariance of QED implies the same in quantum sense.

Note that as discussed in previous sections, under static conditions, these two antennas give no fields. In going between two static conditions, one can have the same fields at intermediate times, but a change in the electric impulse, this being related to a change in the Lorenz vector potential or to a nonzero time integral of the gradient of the Lorenz scalar potential. However, with no fields, the vector potential has zero curl, which in a QED sense is not measurable.

One can modify the two-antenna experiment in various ways, if one wishes, to give other kinds of antennas. For example, one could enclose each antenna in a conducting shield, perhaps with high permeability as well. One merely redefines the antenna to include the shield as part of it. Under initial and final static conditions, the ideal toroidal coil has no external fields and its shield has no currents, and any residual magnetization is assumed negligible (assumed linear materials). The static Lorenz vector potential is then the same, although it may take more time to achieve static conditions due to the required time for the shield currents and magnetization to decay to zero. The electric dipole inside a shield has its excitation modified, so that the electric dipole-moment time history, including the charges induced on the shield have the previously specified form giving the desired electric impulse. Note that the return of the exterior charges on the conducting shield, even with charges allowed to remain on the interior antenna (and shield interior surface as well).

So our choices of the two antennas is not unique for separately emphasizing the Lorenz vector and scalar potentials. All that is required is for the two to have the same exterior fields (say, electric dipole fields, or more general multipole fields) with different potentials (related by the gauge condition). In a classical electromagnetic sense, these antennas cannot be distinguished by exterior measurements. This is a classical nonuniqueness of sources. In a QED sense, the same is the case due to gauge invariance in its currently accepted form.

There is also another way to think about how the vector potential, specifically its curl part, operates in QED. One can envision, on one hand, a static condition where the phase change of ψ around a closed path, with no electromagnetic fields on the path, can be related to $\vec{A}$. However, to establish this static condition, there is required a net time integral of $\vec{E}$ along the path (from previous times), that is, the electric impulse, to establish these new static conditions. An alternative view can have the electric field always zero on the path if a current is allowed to flow on the path, such as by a perfectly conducting loop. The scattered magnetic field and scattered flux serves to cancel the incident magnetic flux, and a constant current flows under the new static conditions. This constant induced current can be likened to the constant quantum phase shift induced and the resulting shift in the diffraction patterns for, say, electrons, traversing the two paths in Fig. 1. Current in a perfectly conducting loop is then a classical analog of this quantum effect.

Going a step further, a perfectly conducting loop can be replaced by a superconducting loop, a quantum device. This refines the result by introducing a quantization of the flux enclosed by the loop. This, of course, is not the same as an incident flux interacting with a loop that has been superconducting from $t = -\infty$, but assumes an initial magnetic field present before the loop is made superconducting.

APPENDIX A: THE TOROIDAL ANTENNA

To analyze the properties of the toroidal antenna, consider it first as a receiver. As in Fig. 6, let the antenna be a body of revolution with respect to the z axis with the usual cordinates. With the incident electric field $\vec{E}^{(\mathrm{inc})}$ taken initially parallel to the z axis, let the antenna be electrically small. Neglect the field distortion due to the antenna conductors, or equivalently consider the antenna (as in Section VII) as a set of distributed sources in space specified by a surface current density $\vec{J}_s$ with

$$\vec{J}_s(\vec{r}_s, t) \cdot \vec{1}_\phi = 0 \tag{A.1}$$

From

$$\oint_C \vec{H} \cdot d\vec{\ell} = \frac{\partial}{\partial t} \int_S \vec{D} \cdot d\vec{S} \tag{A.2}$$

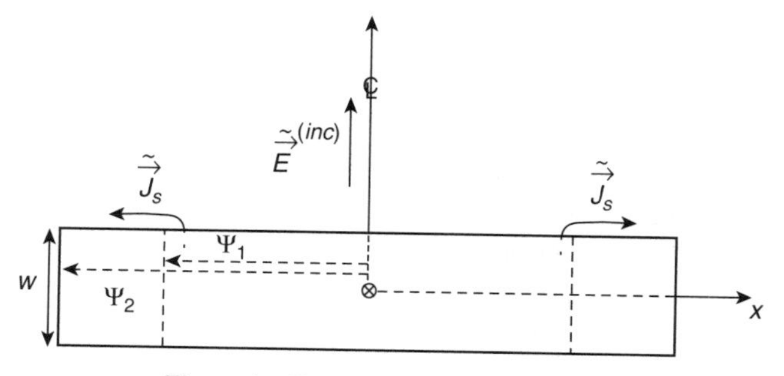

Figure 6. Toroidal antenna as a receiver.

take the contour on constant Ψ, z (a circle), giving

$$\Psi \int_0^{2\pi} H_\phi^{(\text{inc})}(\vec{r}, t) d\phi = \pi \Psi^2 \varepsilon_0 \frac{\partial}{\partial t} E_z^{(\text{inc})} \tag{A.3}$$

where $\vec{E}^{(\text{inc})}$ is taken as uniform over the antenna. The average (over ϕ) magnetic flux density is

$$B_\phi^{(\text{avg})} = \mu_0 H_\phi^{(\text{avg})} = \frac{\Psi}{2} \mu_0 \varepsilon_0 \frac{\partial}{\partial t} E_z^{(\text{inc})} \tag{A.4}$$

The magnetic flux in the toroid is just the integral of $B_\phi^{(\text{avg})}$ over the toroidal cross section at constant ϕ, giving a flux (per turn) as

$$\Upsilon = w \int_{\Psi_1}^{\Psi_2} B_\phi^{(\text{avg})} d\Psi = \frac{\Psi_2^2 - \Psi_1^2}{4} w \mu_0 \varepsilon_0 \frac{\partial}{\partial t} E_z^{(\text{inc})} \tag{A.5}$$

giving a total flux for an N-turn toroidal antennas as $N\Upsilon$.

The actual design of such an antenna has many possibilities ranging from the typical Rogowski coil (large N) to other forms (such as the CPM (circular parallel mutual-inductance) type) involving various parallel arrangements for high-frequency performance [2,4]. This antenna has some similarity to the FMM (flush moebius mutual-inductance) type of sensor for measuring vertical current density (displacement and, if present, conduction) [1,4]. In any event, the geometry of the various turns is made to assure an accurate averaging over the incident magnetic field so that (A.5) applies.

Such toroidal antennas are also used to measure current, say, I positive along the z axis. In this case the magnetic field falls off as Ψ^{-1} and the open-circuit voltage is [2,4]

$$V_{\text{OC}} = M \frac{dI}{dt}, \quad M = \frac{\mu_0 N w}{2\pi} \ln\left(\frac{\Psi_2}{\Psi_1}\right) = \text{mutual inductance} \tag{A.6}$$

Note there is a question of sign convention at the antenna port. By comparison, (A.5) gives

$$V_{\text{OC}} = \frac{d}{dt}(N\Upsilon) = \frac{\Psi_2^2 - \Psi_1^2}{4} w N \mu_0 \varepsilon_0 \frac{\partial^2 E_z^{(\text{inc})}}{\partial t^2} = \frac{\Psi_2^2 - \Psi_1^2}{4} w N \mu_0 \frac{\partial^2 E_z^{(\text{inc})}}{\partial t^2} \tag{A.7}$$

The difference is associated with the fact that displacement current density $\partial D_z^{(\text{inc})}/\partial t$ is allowed in the region $\Psi_1 < \Psi < \Psi_2$. For Ψ_1 near Ψ_2, one can assign an area $\pi\Psi_1^2$ or $\pi\Psi_2^2$ to multiply by the displacement current density to give a current so that we can write

$$V_{\text{OC}} = M \frac{\partial}{\partial t}\left[\vec{A}_e \cdot \frac{\partial}{\partial t} D^{(\text{inc})}\right] \tag{A.8}$$

where now

$$M\vec{A}_e = \frac{\Psi_2^2 - \Psi_1^2}{4} w\mu_0 N \vec{1}_z \tag{A.9}$$

If desired, we can separate these as

$$\begin{aligned}
\vec{A}_e &= \pi\Psi_e^2 \vec{1}_z \equiv \text{equivalent area} \\
\Psi_e &\equiv \text{equivalent radius} = \text{average of } \Psi_1 \text{ and } \Psi_2 \text{ in some} \\
&\quad \text{chosen sense} \\
M &= \frac{\Psi_2^2 - \Psi_1^2}{4\pi\Psi_e^2} w\mu_0 N
\end{aligned} \tag{A.10}$$

In another form, we can write

$$V_{\mathrm{OC}} = \frac{\Psi_2^2 - \Psi_1^2}{4} wN \frac{1}{c^2}\frac{\partial^2}{\partial t^2}\vec{E}_2^{(\mathrm{inc})} \tag{A.11}$$

so that this is a second time derivative electric field sensor. However, in the usual electric field sensor the open-circuit voltage is proportional to the incident electric field (no time derivatives). So, except for the second time derivative (or factor of s^2), this can be regarded as an electric dipole, except that it does not have any electric dipole moment [14]

$$\vec{p} = \int_V \rho(\vec{r})\,\vec{r}\,dV = Q\vec{h}_e \tag{A.12}$$

where

$$\begin{aligned}
\vec{h}_e &\equiv \text{equivalent height} \\
Q &= \text{charge delivered to antenna port (in transmission)}
\end{aligned}$$

since (57) gives zero charge density. One way to look at this is to consider the free-space wave equation for the incident electric field as

$$\nabla \times [\nabla \times \vec{E}^{(\mathrm{inc})}] = -\nabla^2\vec{E}^{(\mathrm{inc})} = \frac{1}{c^2}\frac{\partial^2\vec{E}^{(\mathrm{inc})}}{\partial t^2} \tag{A.13}$$

So we have an equivalent electric dipole to which we can assign the equivalent area $\vec{A}_e$ as discussed above. We can also think of this as an electric dipole of equivalent area $\vec{A}_e$ coupled to the output via a transformer of mutual inductance M. In terms of the usual antenna reception properties [5], we have

$$V_{OC}(s) = \tilde{\vec{h}}_V(s) \cdot \tilde{\vec{E}}^{(\text{inc})}(s), \quad \vec{h}_V(s) = \Xi d^2 \vec{1}_z, \quad \Xi = \frac{\Psi_2^2 - \Psi_2^1}{4} \frac{wN}{c^2} \tag{A.14}$$

This last term is an effective sensitivity parameter for the toroidal antenna in reception.

Now consider such an antenna in transmission by reciprocity. An electric dipole produces fields [3]

$$\tilde{\vec{E}}(\vec{r}, s) = e^{-\gamma r} \left\{ -\frac{\mu_0}{4\pi r} s^2 + \frac{Z_0}{4\pi r^2} s[3\vec{1}_r \vec{1}_r - \overleftrightarrow{1}] + \frac{1}{4\pi\varepsilon_0 r^3} [3\vec{1}_r \vec{1}_r = \overleftrightarrow{1}] \right\} \cdot \tilde{\vec{p}}(s) \tag{A.15}$$

For our case, let us restrict

$$\lambda = \frac{c}{\omega} \gg \Psi_1, \ell, \quad r \gg \Psi_1, \ell \tag{A.16}$$

and assign an equivalent electric dipole moment $\vec{p}_{\text{eq}}$ to go in (A.16) and describe the radiation properties of the antenna. Clearly by symmetry

$$\vec{p}_{\text{eq}}(t) = p_{\text{eq}}(t) \vec{1}_z \tag{A.17}$$

that is, parallel to $\vec{A}_e$.

The far field from such an equivalent electric dipole is

$$\tilde{\vec{E}}_f(\vec{r}, s) = -e^{-\gamma r} \frac{\mu_o}{4\pi r} s^2 \overleftrightarrow{1}_r \cdot \vec{1}_z \tilde{p}_{\text{eq}}(s) \equiv \frac{e^{-\gamma r}}{r} \tilde{\vec{F}}_V(\vec{1}_r, s) \tilde{V}(s) \equiv \frac{e^{-\gamma r}}{r} \tilde{\vec{F}}_I(\vec{1}_r, s) \tilde{I}(s) \tag{A.18}$$

where

$$V(t) = \text{voltage at antenna port}$$

$$I(t) = \text{current out of antenna port}$$

and

$$\tilde{Z}_{\text{in}}(s) = \frac{1}{\tilde{Y}_{\text{in}}(s)} = \frac{\tilde{V}(s)}{\tilde{I}(s)} \quad \text{(in transmission)} \equiv \text{antenna input impedance}$$

This follows the conventions discussed in [5]. Then we have the transmission functions for the far field as

$$\vec{\tilde{F}}_V(\vec{1}_r, s) = -\frac{\mu_0}{4\pi}\frac{s^2}{\tilde{V}(s)}\vec{1}_r \cdot \vec{1}_z \tilde{p}_{\text{eq}}(s), \quad \vec{\tilde{F}}_I(\vec{1}_r, s) = -\frac{\mu_0}{4\pi}\frac{s^2}{\tilde{I}(s)}\vec{1}_r \cdot \vec{1}_z \tilde{p}_{\text{eq}}(s)$$
$$\vec{\tilde{F}}_I(\vec{1}_r, s) = \tilde{Z}_{\text{in}}(s)\vec{\tilde{F}}_V(\vec{1}_r, s) \tag{A.19}$$

In reception antennas can be characterized by effective height for voltage and a related parameter for current [5] with wave incident in direction $\vec{1}_i$ as

$$\tilde{V}_{\text{OC}}(s) \equiv \vec{\tilde{h}}_V(\vec{1}_i, s) \cdot \vec{\tilde{E}}^{(\text{inc})}(s) = \text{open-circuit voltage}$$
$$\tilde{I}_{\text{SC}}(s) \equiv \vec{\tilde{h}}_I(\vec{1}_i, s) \cdot \vec{\tilde{E}}^{(\text{inc})}(s) \equiv \text{short-circuit current}$$
$$\vec{\tilde{h}}_I(\vec{1}_i, s) = -\tilde{Y}_{\text{in}}(s)\vec{\tilde{h}}_V(\vec{1}_i, s) \tag{A.20}$$

where as before the incident field is evaluated at $\vec{r} = \vec{0}$. Since the antenna is electrically small, then only the z component is relevant and

$$\tilde{V}_{\text{OC}}(s) \equiv \tilde{h}_V(s)\vec{\tilde{E}}_z^{(\text{inc})}(s), \quad \tilde{I}_{\text{SC}}(s) \equiv \tilde{h}_I(s)\vec{\tilde{E}}_z^{(\text{inc})}(s) \tag{A.21}$$
$$\vec{\tilde{h}}_V(\vec{1}_i, s) = \tilde{h}_V(s)\vec{1}_z, \quad \vec{\tilde{h}}_I(\vec{1}_i, s) = \tilde{h}_I(s)\vec{1}_z$$

The reciprocity between transmission and reception now establishes [5]

$$\vec{\tilde{F}}_I(\vec{1}_r, s) = -s\frac{\mu_0}{4\pi}\overleftrightarrow{1}_r \cdot \vec{\tilde{h}}_V(-\vec{1}_r, s) = -s\frac{\mu_0}{4\pi}\overleftrightarrow{1}_r \cdot \vec{1}_z \tilde{h}_V(s)$$
$$\vec{\tilde{F}}_V(\vec{1}_r, s) = s\frac{\mu_0}{4\pi}\overleftrightarrow{1}_r \cdot \vec{\tilde{h}}_I(-\vec{1}_s, s) = s\frac{\mu_0}{4\pi}\overleftrightarrow{1}_r \cdot \vec{1}_z \tilde{h}_I(s) \tag{A.22}$$
$$\overleftrightarrow{1}_r \equiv \overleftrightarrow{1} - \vec{1}_r\vec{1}_r \equiv \text{transverse (to } r\text{) dyadic}$$

Combining with (A.18) this gives

$$\tilde{p}_e(s) = \frac{1}{s}\tilde{h}_V(s)\tilde{I}(s) = -\frac{1}{s}\tilde{h}_I(s)\tilde{V}(s) \tag{A.23}$$

and various other combinations.

The equivalent dipole moment in transmission is then as follows, using (A.14):

$$\tilde{p}_{\text{eq}}(s) = \Xi s\tilde{I}(s) = \Xi\frac{s}{\tilde{Z}_{\text{in}}(s)}\tilde{V}(s) \tag{A.24}$$

Such an antenna is clearly inductive, and for low frequencies has

$$\tilde{Z}_{\text{in}}(s) = sL_{\text{in}}, \quad L_{\text{in}} \simeq NM \quad [\text{from}(\text{A.6})] = \frac{\mu_0 N^2 w}{2\pi} \ln\left(\frac{\Psi_2}{\Psi_1}\right) \tag{A.25}$$

Actually the inductance is slightly larger than this since the N discrete windings only approximate a continuous ϕ-independent surface current density.

In terms of a low-frequency voltage drive, we then have

$$\tilde{p}_{\text{eq}}(s) = \frac{\Xi}{L_{\text{in}}} \tilde{V}(s) \tag{A.26}$$

This is like the usual electric dipole except that the current does not go to zero as $s \to 0$, but rather diverges as $1/s$. On the other hand consider a current drive. Then at low frequencies (A.24) indicates that $\tilde{p}_{\text{eq}} \to 0$ (as well as $\tilde{V}(s) \to 0$).

In resolving the apparent paradox of how an antenna with no charge (in free space) can have an electric-dipole moment, one can go back to definitions. In Ref. 14 the fields from a current distribution are evaluated by expanding

$$\tilde{\vec{A}}(\vec{r}, s) = \mu_0 \int_V \frac{e^{-\gamma|\vec{r} - \vec{r}|}}{4\pi|\vec{r} - \vec{r}|} \tilde{\vec{J}}(\vec{r}\,', s)\, dV' \simeq \mu_0 \frac{e^{-\gamma r}}{4\pi r} \int_V \tilde{\vec{J}}(\vec{r}\,', s)\, dV' \tag{A.27}$$

Then, evaluating the volume integral of the current density (with no current crossing S, the boundary of V), we have [3,14]

$$\int_V \tilde{\vec{J}}(\vec{r}\,', s)\, dV' = s \int_V \vec{r}\,' \tilde{\rho}(\vec{r}\,', s)\, dV' = s\tilde{\vec{p}}(s) \tag{A.28}$$

This shows that a divergenceless current distribution gives a zero electric dipole moment in this sense.

However, the expansion in (A.27) as a leading term at low frequencies is only an approximation. Since, with an assumed divergenceless current distribution, this is zero, we need to include higher-order terms. This involves higher-order terms in the expansion of $e^{\gamma|\vec{r}-\vec{r}\,'|+\gamma r}$ around $s=0$ (for an electrically small antenna). By the previous derivation such higher-order terms can give nonzero electric-dipole-like fields. See the factor of s^2 that enters in the results, clearly a higher-order term.

In the usual texts a multipole expansion involving spherical Bessel functions and spherical vector harmonics is also introduced [16,23,23,26]. The fields from electric and magnetic dipoles correspond to the lowest-order terms ($n=1$) in the expansion. If we define dipole by this expansion then our toroidal antenna is an electric dipole. In any event, the fields away from the source are the same. This is perhaps a matter of consistency in definitions.

Acknowledgment

This work was sponsored in part by the Air Force Office of Scientific Research, and in part by the Air Force Research Laboratory, Directed Energy Directorate.

References

1. C. E. Baum, *Two Types of Vertical Current Density Sensors*, Sensor and Simulation Note 33 (1967).[1]
2. C. E. Baum, *Some Considerations for Inductive Current Sensors*, Sensor and Simulation Note 59 (1968).
3. C. E. Baum, *Some Characteristics of Electric and Magnetic Dipole Antennas for Radiating Transient Pulses*, Sensor and Simulation Note 125 (1971).
4. C. E. Baum, in J. E. Thompson and L. H. Luessen (Eds.), *Fast Electrical and Optical Measurements*, Martinus Nijhoff, Dordrecht, 1986, Vol. I, pp. 73–144.
5. C. E. Baum, *General Properties of Antennas*, Sensor and Simulation Note 330 (1991).
6. C. E. Baum, *Vector and Scalar Potentials away from Sources, and Gauge Invariance in Quantum Electrodynamics*, Physics Note 3 (1991).
7. C. E. Baum, *Concerning an Analogy Between Quantum and Classical Electrodynamics*, Physics Note 4 (1992).
8. C. E. Baum and H. N. Kritikos, *Symmetry in Electromagnetics*, Physics Note 2 (1990); in C. E. Baum and H. N. Kritikos (Eds.), *Electromagnetic Symmetry*, Taylor & Francis, New York, 1995, Chap. 1, pp. 1–90.
9. V. B. Berestetskii, E. M. Lifshitz, and L. P. Pitaevskii, *Relativistic Quantum Theory*, Addison-Wesley, Reading, MA, 1971.
10. Y. G. Chen, R. Crumley, C. E. Baum, and D. V. Giri, *Field-Containing Inductors,* Sensor and Simulation Note 287 (1985); *IEEE Trans. EMC* 345–350 (1988).
11. R. P. Feynman, *Quantum Electrodynamics*, Benjamin, New York, 1962.
12. R. P. Feynman, R. B. Leighton, and M. Sands, *The Feynman Lectures on Physics*, Vol. II, Addison-Wesley, Reading, MA, 1964.
13. R. P. Feynman, R. B. Leighton, and M. Sands, *The Feynman Lectures on Physics*, Vol. III, Addison-Wesley, Reading, MA, 1965.
14. J. D. Jackson, *Classical Electrodynamics*, 2nd ed., Wiley, New York, 1975.
15. R. B. Leighton, *Principles of Modern Physics*, McGraw-Hill, New York, 1959.
16. P. M. Morse and H. Feshbach, *Methods of Theoretical Physics*, McGraw-Hill, New York, 1953.
17. C. H. Papas, *Theory of Electromagnetic Wave Propagation*, Dover, New York, 1988.
18. P. Roman, *Introduction to Quantum Field Theory*, Wiley, New York, 1969.
19. J. J. Sakurai, *Advanced Quantum Mechanics*, Addison-Wesley, Reading, MA, 1967.
20. S. S. Schweber, *An Introduction to Relativistic Quantum Field Theory*, Row, Peterson, Evanston, IL, 1964.
21. J. Schwinger, in J. Schwinger (Ed.), *Selected Papers on Quantum Electrodynamics*, Dover, New York, 1958, paper 20.

[1]The various "Notes" are publications of the Air Force Research laboratory, Directed Energy Directorate, Kirtland AFB, NM, USA.

22. J. Schwinger, private communication, Sept. 27, 1991. (unfortunately R. P. Feynman is not currently available for consultation).
23. W. R. Smythe, *Static and Dynamic Electricity*, 3rd ed., Taylor & Francis, New York, 1989.
24. J. A. Stratton, *Electromagnetic Theory*, McGraw-Hill, New York, 1941.
25. C.-T. Tai, *Dyadic Green's Functions in Electromagnetic Theory*, Intext Educational Publishers, Scranton, PA, 1971.
26. J. Van Bladel, *Electromagnetic Fields*, Taylor & Francis, New York, 1985.

EXTRACTING AND USING ELECTROMAGNETIC ENERGY FROM THE ACTIVE VACUUM

THOMAS E. BEARDEN

CEO, CTEC, Inc., Huntsville, Alabama

CONTENTS

Modern Nonlinear Optics, Part 2, Second Edition, Advances in Chemical Physics, Volume 119,
Edited by Myron W. Evans. Series Editors I. Prigogine and Stuart A. Rice.
ISBN 0-471-38931-5

The first principles of things will never be adequately known. Science is an open-ended endeavor, it can never be closed. We do science without knowing the first principles. It does in fact not start from first principles, nor from the end principles, but from the middle. We not only change theories, but also the concepts and entities themselves, and what questions to ask. The foundations of science must be continuously examined and modified; it will always be full of mysteries and surprises [1].

I. INTRODUCTION

In this chapter we present new concepts that provide usable extraction of electromagnetic energy from the vacuum, and appear to be present but previously unrecognized in all electrical power systems.

We argue that generators and batteries do not use their available internal energy to power their external circuits, but dissipate it to separate their internal charges to produce source dipoles as negative resistors. Once formed, the source dipole receives and transduces electromagnetic (EM) energy flow from the vacuum (in particle physics the *dipole* is a particle–antiparticle broken symmetry in its fierce energy exchange with the active vacuum; see Lee's definition [2]), and pours it out along the attached circuit, filling all surrounding space [3]. EM circuits and loads have always been powered by vacuum-provided EM energy. They are not powered by the energy the operator inputs to the generator shaft, nor by the chemical energy available in the battery.

A tiny fraction—the Poynting [4] component (Poynting considered only that part of the energy flow that enters the circuit)—of the huge energy flow transduced from the vacuum is intercepted and diverged into the circuit to power it. The huge remainder of the transduced vacuum energy flow—which we call the *Heaviside* [5,6] *component*—misses the circuit entirely. It is not intercepted and is not diverged into the circuit. It is just wasted.

No one in the 1880s could explain the source of the startlingly large Heaviside energy flow component. (There was no known special or general relativity, quantum mechanics, electron, atom, atomic nucleus, etc. available at the time. There was no theory of the active vacuum, and broken symmetry of the dipole with such an entity was unknown.) Faced with such a giant quandary, Lorentz arbitrarily discarded the Heaviside component as being "physically insignificant"[1] because it does not contribute to powering the circuit of interest. To the contrary, it is highly significant. In this chapter we give its source and mechanism, its significance, and its practical use.

The common dipole breaks 3-symmetry in EM energy flow [2], resulting in a novel and more fundamental negentropic Whittaker [8] *4-symmetry* EM energy flow completely freed from 3-symmetry and 3-space EM energy conservation [9a]. We explain that special 4-symmetry and its giant negentropy in this chapter.

[1]That erroneous notion, expressed in pharaseology such as "can have no physical consequences," is often used by electrodynamicists to this day. [7]

Without the additional 3-symmetry condition, the resulting Whittaker 4-symmetry EM energy flow mechanism resolves the nagging problem of the "source charge" concept in classical electrodynamics theory. Quoting Sen [10]: "The connection between the field and its source has always been and still is the most difficult problem in classical and quantum mechanics." We give the solution to the problem of the source charge in classical electrodynamics.

The giant negentropy mechanism also explains the heretofore unknown source of the enormous "dark[2] (unaccounted) EM energy flow" component that was (1) discovered by Heaviside [5,6], (2) never considered by Poynting [4], and (3) arbitrarily discarded by Lorentz [11].[3] Lorentz' discard of the Heaviside dark energy flow component has continued to be applied by electrodynamicists [12].[4]

This huge nondiverged EM energy flow is a dynamic ordering of the vacuum surrounding every EM circuit and EM field reaction. We present several mechanisms where the dark energy flow can be intercepted to provide excess energy from the locally reordered vacuum. The Bohren [13][5] experiment—along lines pioneered by Letokhov [14]—is cited as proof that excess energy has been experimentally intercepted from the Heaviside energy flow component. We present more than a dozen candidate processes and mechanisms for extracting this available Heaviside EM energy from the vacuum, where asymmetric self-regauging (ASR) [15,16] is the *gauge freedom* key to breaking the Lorentz symmetrical regauging condition first accomplished by L .V. Lorenz [17].[6]

[2]By "dark" we mean nonobserved and nonintercepted, but physically present in space as a real EM energy flow. Obviously we have used the phraseology of the dark-matter problem. We have, in fact, nominated the Heaviside dark energy as the solution to the dark-matter problem, and as responsible for producing the excess gravity in the spiral arms of spiral galaxies that must be present to hold them together. [See T. E. Bearden, *J. New Energy* **4**(4), 4–11. (2000).]

[3]Figure 25 on p. 185 of Ref. 11 shows the Lorentz concept of integrating the energy flow vector around a closed cylindrical surface surrounding any volumetric element of interest. This discarded the Heaviside *nondiverged* component, leaving only the Poynting *diverged* component. I have not yet discovered the original earlier paper where Lorentz first did this procedure circa 1886, and would welcome a citation to it.

[4]Panofsky and Phillips [12] show this Lorentz exercise, as do W. Gough and J. P. G. Richards, *Eur. J. Phys.* **7**, 195 (1986).

[5]H. Paul and R. Fischer, commenting on how a particle can absorb more than the light incident on it, in Ref. 13(b), p. 327, verify the Bohren experiment and results of Ref. 13(a).

[6]Ludwig Valentin Lorenz first effectively regauged the Heaviside–Maxwell equations symmetrically in his paper on the identity of the vibrations of light with electrical currents [17a] not long after Maxwell's seminal 1864 oral presentation of his paper on a dynamical theory of the electromagnetic field [17b]. When the prestigious H. A. Lorentz later adopted the symmetric regauging because it provided simpler equations that were easier to solve, electrodynamicists adopted it quickly. No one seemed to notice that physically this constituted the arbitrary and total discard of all Heaviside–Maxwell systems not in thermodynamic equilibrium with their active vacuum. Even Jackson [7] erroneously states that the Lorentz-regauged equations are identical in every respect.

ASR provides an open EM system far from thermodynamic equilibrium in its violent energy exchange with the active vacuum. As is well known, an open dissipative system in disequilibrium with an active environment is permitted to

1. Self-order
2. Self-rotate or self-oscillate
3. Output more energy than is input by the operator (the excess is taken from the active environment, in this case the active vacuum)
4. Power itself and its load simultaneously (all the energy is taken from the active environment, in this case the active vacuum)
5. Exhibit negentropy

The common dipole exhibits all five functions, as we discuss later.

Finally, we explain why ordinary electrodynamic systems do not exhibit such negentropic functions, and we state the principles necessary to correct the shortcomings in those systems that prevent it.

It appears that a permanent solution to the world energy problem, dramatic reduction of biospheric hydrocarbon combustion pollution, and eliminating the need for nuclear power plants (whose nuclear component is used only as a heater) could be readily accomplished by the scientific community [18]. However, to solve the energy problem, we must (1) update the century-old false notions in electrodynamic theory of how an electrical circuit is powered and (2) correct the classical electrodynamics model for numerous foundations flaws.

We reconsider both items 1 and 2 on the basis of more modern developments in particle physics and gauge field theory well after the foundations of electrodynamics were set by Maxwell. Self-powering systems readily extracting electrical energy from the vacuum to power themselves and their loads can be quickly developed whenever the scientific community will permit their research and development to be funded.

II. LORENTZ SYMMETRIC REGAUGING

A. Two Equal and Opposite Regaugings

For energy flow through space around the circuit, we must use Maxwell's equations as we would for radiating energy, rather than employ only the $\mathbf{j}\phi$ circuit analysis conventionally utilized.

In Gaussian units, Jackson [Ref. 7, pp. 220–223] shows that Maxwell's four equations (vacuum form) can first be reduced to a set of two coupled equations in the $(\mathbf{A},\Phi)$ representation as follows:

$$\nabla^2\Phi + \frac{1}{c}\frac{\partial}{\partial t}(\nabla \cdot \mathbf{A}) = -4\pi\rho \tag{1}$$

$$\nabla^2 \mathrm{A} - \frac{1}{c^2}\frac{\partial^2 \mathbf{A}}{\partial t^2} - \nabla\left(\nabla \cdot \mathbf{A} + \frac{1}{c}\frac{\partial \Phi}{\partial t}\right) = -\frac{4p}{c}\mathbf{J} \tag{2}$$

The result is two coupled Maxwell–Heaviside equations. Jackson shows that potentials $\mathbf{A}$ and Φ in these two equations are "arbitrary" (i.e., yield the same force fields) [19,20][7] in a specific sense, since the $\mathbf{A}$ vector can be replaced with $\mathbf{A}' = \mathbf{A} + \nabla\Lambda$, where Λ is a scalar function and $\nabla\Lambda$ is its gradient. The $\mathbf{B}$ field is given by $\mathbf{B} = \nabla \times \mathbf{A}$, so that the new $\mathbf{B}'$ field becomes

$$\mathbf{B}' = \nabla \times (\mathbf{A} + \nabla\Lambda) = \nabla \times \mathbf{A} + \mathbf{0} = \nabla \times \mathbf{A} = \mathbf{B} \tag{3}$$

The $\mathbf{B}$ field due to the vector potential has remained entirely unchanged, even though the magnetic vector potential has been asymmetrically changed. However, if no other change were made, then the electric field $\mathbf{E}$ would have still been changed because of the gradient $\nabla\Lambda$. In that case the net change would be asymmetric, because one obtained a "free" $\mathbf{E}$-field and excess regauging energy, which could then do work on the system—either beneficially or detrimentally, depending on the specific conditions, geometry, and timing. In order to prevent this excess "free" $\mathbf{E}$ field from appearing, electrodynamicists simultaneously and asymmetrically regauge (transform) the scalar potential Φ so as to precisely offset the $\mathbf{E}$-field change due to the regauging of Eqs. (3) and (4). In short, they also change Φ to Φ', where

$$\Phi' = \Phi - \frac{1}{c}\frac{\partial \Lambda}{\partial t} \tag{4}$$

[7] Nahin [19] states: "In an 1893 letter to Oliver Lodge, Heaviside said of his own work that it represented the 'real and true Maxwell' as Maxwell would have done it if he had not been humbugged by his vector and scalar potentials." The false notion that changing the potentials has no physical significance if the force fields are not changed is due largely to Heaviside. Heaviside considered potentials "mystical" and "not real," and stated they should be "murdered from the theory." Along with Hertz and others, he reduced Maxwell's 20 quaternion equations in 20 unknowns to four equations, getting rid of many potentials in so doing. We point out that an EM potential is a change to the energy density of the vacuum, and hence produces spacetime curvature, which is a gravitational effect. Heaviside's position predated relativity by decades, and is one of the reasons that electrodynamics and general relativity have not been successfully combined in an engineering theory. Also according to Josephs [20], ironically, well after Heaviside's death, his handwritten notes containing a theory of electrogravitation, based on his theory of energy flow, were found beneath the floor boards in his little garret apartment. His trapped EM energy flow loops were gravitational. Possible effects of such loops were found by Sweet and Bearden [21].

With that additional change, now both the *net* **E** and **B** fields remain unchanged[8] even though—in terms of a physical system rather than abstract mathematical models—the fundamental stored (potential) energy of the system has changed twice, and two free excess forces have appeared in the system. "Unchanged net force fields" just mean that all excess new forces vectorially sum to a zero resultant; it does not address the effects of the additive energies of the "zero-summing" field components. This is what has been assumed by electrodynamicists when they state that the net summation of the two *asymmetric* regaugings has been entirely *symmetric*. It is symmetrical with respect to *translation*, not spacetime curvature, gravitation, and system stress.

After Lorentz, electrodynamicists assumed that the system designer can arbitrarily create and change forcefields and potentials. The designer can freely change the potential energy of the system at will, theoretically without any input energy *cost* whatsoever. However, electrodynamicists also assumed that the system designer will be a "gentleman," and cooperate with their resulting *intention*. In short, the received view assumes that the designer will freely change the excess free energy and the excess free forces in the system only in a highly restrictive manner where all excess forces fight themselves to a draw and the excess energy cannot be used to perform useful work. In other words, they assume that the designer will play the "Lorentz condition" game—for a *game* is all that it is. The designer is, of course, free to violate that "gentleman's agreement" at will, and the results have significant implication [16,22].

Here's how the electrodynamicists do it, as Jackson points out. Conventionally, a set of potentials (**A**, Φ) is habitually and *arbitrarily* chosen such that

$$\nabla \cdot \mathbf{A} + \frac{1}{c}\frac{\partial \Phi}{\partial t} = 0 \tag{5}$$

This *net symmetric* regauging operation successfully separates the variables, so that two inhomogeneous wave equations result to yield the new Maxwell

[8]However, so-called "canceling" appositive EM fields are actually produced, which sum to a vector zero, which the electrodynamicists then discard by assumption. We point out, but do not further pursue, the fact that the locally produced *field energies* of the appositive fields remain and add, even though the fields offset each other translationally, since the energy of the field is proportional to its square, and that is always positive regardless of field orientation. Thus "trapped" EM energy has been localized in spacetime in the symmetric regauging of CEM, and this is a local curvature of spacetime a priori. Thus any substantial EM gauge symmetry transformations are accompanied by local gravitational changes in the regauged system. This, in fact, may prove to be the road to practical antigravity processes and devices, achieved by EM means. At least one system—Sweet's vacuum triode device—seems to have accomplished antigravity effects [21].

equations:

$$\nabla^2\Phi - \frac{1}{c^2}\frac{\partial^2\Phi}{\partial t^2} = -4\pi\rho \tag{6}$$

$$\nabla^2\mathbf{A} - \frac{1}{c^2}\frac{\partial^2\mathbf{A}}{\partial t^2} = -\frac{4\pi}{c}\mathbf{J} \tag{7}$$

Thus the two previously coupled Maxwell equations (1) and (2) (potential form) have been changed to the form given by Eqs. (6) and (7), to leave two *much simpler* inhomogeneous wave equations, one for Φ and one for $\mathbf{A}$.

Of course, the equations are simpler, since they arbitrarily discard that very large class of Maxwellian systems in local thermodynamic *dis*equilibrium with the active vacuum! They arbitrarily discard all those Maxwellian systems that are capable of producing electrical circuits and power systems with a coefficient of performance (COP) > 1.0.

B. Breaking the Lorentz Symmetric Regauging Condition

For COP > 1.0, three functions are required in the system:

1. We must first asymmetrically regauge the system, or have it asymmetrically self-regauge itself, in order to freely change its collected energy and obtain a net force to utilize. By gauge freedom, in theory this is cost-free to the system operator.
2. Then we must adroitly utilize the excess force created by the asymmetrical regauging, along with its associated excess potential energy, to perform useful work.
3. Finally, we must dissipate this collected potential energy to do work, but *without* using half of it to more rapidly kill the source dipole of the system itself.

In short, we must violate the Lorentz symmetric regauging condition during the excitation discharge represented by operation 2 of an open system far from thermodynamic equilibrium.

The condition for violating Lorentz symmetric regauging is

$$\nabla \cdot \mathbf{A} + \frac{1}{c}\frac{\partial\Phi}{\partial t} \neq 0 \tag{8}$$

Any regauging of the potentials that complies with Eq. [8] will a priori produce one or more excess forces in the system and freely change the energy of the system as well. By controlling the regauging to keep it asymmetric, the engine designer may then control where, how, and when these excess forces appear, and how much excess energy appears in the system with them. Every system already performs this asymmetric regauging function in its *excitation phase*.

During the *excitation discharge* phase, however, use of the closed current loop circuit results in a "back emf" (electromotive force) across the source dipole interior, precisely equal to the "forward emf" around the external circuit across the external loads and losses. Since "emf" is just a forcefield, this gives symmetric discharge of the free excitation energy of the circuit—half in powering the external loads and losses, and half in destroying the dipole itself. Hence it destroys the source dipole faster than it powers the load alone.

Since the closed current loop circuit is used ubiquitously, present circuits and electrical power systems *a priori* cannot exhibit COP > 1.0, since they violate Eq. (8) during their excitation discharge. Hence they violate condition 3 required for COP > 1.0.

III. REEXAMINING THE COMMON DIPOLE

A. Whittaker Decomposition of the Potential between the End Charges

Any dipole has a scalar potential between its ends, as is well known. Extending earlier work by Stoney [23], in 1903 Whittaker [8] showed that the scalar potential decomposes into—*and identically is*—a harmonic set of bidirectional longitudinal EM wavepairs. Each wavepair is comprised of a longitudinal EM wave (LEMW) and its phase conjugate LEMW replica. Hence the formation of the dipole actually initiates the ongoing production of a harmonic set of such biwaves in 4-space (see Section III.A.1).

We separate the Whittaker waves into two sets: (1) the convergent phase conjugate set, in the imaginary plane; and (2) the divergent real wave set, in 3-space. In 4-space, the fourth dimension may be taken as *-ict*. The only variable in *-ict* is *t*. Hence the phase conjugate waveset in the scalar potential's decomposition is a set of harmonic EM waves converging on the dipole in the time dimension, *as a time-reversed EM energy flow structure inside the structure of time* [24].[9] Or, one can just think of the waveset as converging upon the dipole in the imaginary plane [25][10]—a concept similar to the notion of "reactive power" in electrical engineering.

[9] Time-like currents and flows do appear in the vacuum energy, if extended electrodynamic theory is utilized. For instance, in the received view, the Gupta–Bleuler method removes time-like photons and longitudinal photons. For disproof of the Gupta–Bleuler method, proof of the independent existence of such photons, and a short description of their characteristics, see Evans' AIAS group papers on Whittaker's *F* and *G* fluxes and analysis of the EM entity in Ref. 24a; to see how such entities produce ordinary EM fields and energy in vacuo, see Ref. 24b.

[10] Jones [25] gives a short treatise on the complex Poynting vector. In a sense our present use is similar to the complex Poynting energy flow vector, but in our usage the absolute value of the imaginary energy flow is equal to the absolute value of the real energy flow, and there is a transformation process in between. So we are working in only four dimensions (Minkowski space). This usage is possible because the imaginary flow is into a *transducer*, which takes care of transforming the received *imaginary* EM energy into the output *real* EM energy. We stress that the word "imaginary" is not at all synonymous with *fictitious*, but merely refers to the "dimension" or state in which the EM energy flow actually exists. We also point out that ultimately the concept of "dimension" refers simply to a fundamental mathematical degree of freedom.

The divergent *real* EM waveset in the scalar potential's decomposition is thus a harmonic set of EM longitudinal waves radiating out from the dipole in all directions in 3-space at the speed of light (see Section III.A.2). As can be seen, there is perfect 4-symmetry in the resulting EM energy flow, but there is broken 3-symmetry since there is no observable 3-flow EM energy input to the dipole but there is observable 3-flow of EM energy output.

Further, there is perfect 1:1 correlation between the convergent waveset in the imaginary plane and the divergent waveset in 3-space. This perfect correlation between the two sets of waves and their dynamics represents a *deterministic reordering* of a fraction of the 4-vacuum energy. This reordering is initiated by the formation of the dipole, and spreads radially outward at the speed of light so long as the dipole remains intact.

1. Existence of EM Waves in 4-Space

We emphasize the fact that the waves exist in 4-space. The incoming EM longitudinal waves are in the imaginary plane, and hence are incoming from the time domain *-ict*. Evans has pointed out that Whittaker's method depends on assuming the Lorentz gauge. If the latter is not used, the Whittaker method is inadequate, because the scalar potential becomes even more richly structured, as captured by Sachs' generalization, which removes the necessity for the Lorentz gauge. For the negentropic vacuum-reordering mechanism involving only the dipole or the charge as a *composite dipole*, it appears that the Whittaker method can be applied without problem, at least to generate the minimum negentropic process itself. However, this still leaves the capability for *additional structuring*, so that the actual negentropic reordering of the vacuum energy (and the structure of the outpouring of the EM energy 3-flow from the charge or dipole) may be made much richer than given by the simple Whittaker structure alone. Specifically, 4-flow symmetry may be broken while n-flow symmetry is maintained, where the integer $n > 4$. The Whittaker structure used in this chapter should be regarded as the *simplest* structuring of the negentropic process that can be produced, and hence a lower boundary condition on the process.

2. Existence of EM Waves in 3-Space

We mention in passing that this dramatically alters the conception of how the EM field is thought to exist in spacetime. Take the **E** field as an example. In 3-space the **E**-field **E** exists as an outgoing longitudinal EM wave. At any 3-space point, **E** exists as an ongoing energy flow process in 4-space, where a convergent inflow of longitudinal EM wave energy from the time domain (imaginary plane) enters the point, and the outgoing EM longitudinal wave in 3-space comes from the point.

Maxwell in fact simply assumed the transverse EM wave in space, based on the notion of Faraday's "taut string" field lines (lines of force). When Maxwell wrote his theory, the electron, nucleus, atom, and molecular structure were unknown. He wrote a purely material fluid theory, since the material ether was ubiquitously assumed and nowhere in all the universe was there thought to be any "total absence of matter." Consequently, in applications to currents in wires and to circuits, Maxwell ignored the simultaneous recoil of the positive nuclei, which were not even known. He used only a "unitary electrical material fluid" flowing (from positive to negative) down the wire like water through a pipe. Further, electron drift velocity was unknown since the electron was unknown, and the fluid was thought to move at signal velocity.

Actually, the Drude electrons, during their 360° spin in 3-space, act as gyros and are constrained longitudinally by repulsion of other charges beyond. Hence the gyroelectrons precess laterally to their longitudinal disturbing force, yielding the well-known *measured* transverse wave. Almost all our instruments still are electron-precession detectors.

In any field applied to the wire, the simultaneous appositive recoil of the nuclei—with equal energy to the electron gyroprecession but highly damped in amplitude because of the much greater m/q ratio of the positive nuclei—was ignored (and still is, even though it is known to occur).

In a transmitting wire antenna, the surrounding spacetime is perturbed by two perturbations: (1) the Drude electrons (which are confined to the wire) and (2) the recoiling nuclei. There is a very tiny phase deviation from 180° between the two perturbations. So the total "disturbance" of the surrounding spacetime medium by the two equal energy and nearly opposite direction charge perturbations is very nearly a 3-spatial longitudinal EM wave—in agreement with the Whittaker decomposition of the source dipole potentials between any two separated perturbed charges in the wire.

The ready reaction of the resulting "quasilongitudinal" EM wave with a distant wire antenna results in the leading half (from the electrons) reacting with the distant Drude electrons, stripping off that leading half of the incoming quasilongitudinal wave. The remaining half enters the nuclei, interacting with them and producing the well-known Newtonian third-law recoil. Electrodynamicists accept the third-law recoil but consider it as an effect without an EM cause—which is erroneous.

If that phase conjugate half of the wave does not interact with the nucleus—such as in multiwave interactions before the phase conjugate wave reaches the nucleus, as occurs in a pumped phase conjugate mirror—then the Newtonian third-law recoil does not occur in that pumped mirror material because its electromagnetic causative interaction did not occur with the nuclei. The *results* of that phenomenon are experimentally verified and well known. *As shown by Sweet and Bearden* [21], the results can include an tigravity. However, see reference [98]. Antigravity is associated with extreme COP—e.g., 1.5×10^6—and curved spacetime together with Dirac Sea hole current is involved.

Much of electrodynamic 136-year-old foundations fundamentals are either erroneous or flawed, and a serious reconstruction of the model from the ground up is warranted.

B. Interpreting the 4-Symmetry in Electrical Engineering Terms

The EM energy flow in the imaginary plane is a kind of incoming "reactive power" in the language of electrical engineering, but in the time domain prior to interaction with the observing charge. The outgoing EM energy flow in the real plane (3-space) is "real power." So the dipole is continuously receiving a steady stream of *unobserved* reactive power, transducing it into real power, and outputting it as a continuous outflow of real EM power.

To initiate the hypothesized giant negentropy process, all one has to do is first expend a little energy to form the dipole. Once the dipole is formed, the process is automatically initiated and sustained by the broken 3-symmetry of the dipole [2]. The process continues indefinitely and freely, so long as the dipole remains intact [9a].[11]

Actually, this giant negentropy is to be expected from the further broken symmetry findings in particle physics. The various major symmetries can be individually broken, or multiples can be broken, so long as CPT itself is not broken. Quoting Lee [Ref. 2, pp. 187–188]:

> At present, it appears that physical laws are not symmetrical with respect to *C*, *P*, *T*, *CP*, *PT* and *C*. Nevertheless, all indications are that the joint action of *CPT* (i.e., particle $\leftrightarrow$ antiparticle, right $\leftrightarrow$ left and past $\leftrightarrow$ future) remains a good symmetry.

Indeed, that quotation reveals that one is free to violate *PT* (parity and time) symmetries simultaneously, so long as *C* symmetry is not violated. A little reflection reveals that time can be converted to spatial energy, under such condition. I.e., with *PT* broken but *CPT* conserved, a fraction of "time" flow is being converted to "spatial EM energy" flow or vice versa. It follows that, since the source charge (or source dipole) does not violate *C* symmetry, it is free to violate parity symmetry and time symmetry—which yields the giant negentropy of the common dipole (and, as will be seen, of the common charge as well).

C. How the Reactive Power Is Transduced into Real Power

We suggest a mechanism that accomplishes the transduction or at least models it. The charges constituting the ends of the dipole have a very special

[11] Thus the significance of the closed current loop circuit, ubiquitously utilized in all electrical power systems. Such a circuit utilizes half its collected Poynting energy to destroy the dipole, while using less than the other half to power the load. In short, it shuts off the giant negentropy and free 3-flow of energy, faster than it can freely collect and discharge energy to power the load. Such a circuit exhibits COP < 1.0 a priori. More than a dozen processes and mechanisms for approaching COP > 1.0 systems are given in Ref. 16.

characteristic: Simply modeled, a charge may be said to spin 720° in making one complete rotation, not just 360°. We may say that it spins 360° in the imaginary plane, and spins 360° in the real plane (3-space). Let us examine a dipole charge spinning 720° per rotation in that manner. During its 360° spin in the imaginary plane, it *absorbs* the converging reactive power. During its 360° spin in the real plane (in 3-space), it *reradiates* the EM energy it has absorbed from the imaginary plane, as real power in a steady, divergent, radial 3-flow of EM energy at the speed of light in all directions.

If one does not press it too far, this simple analogy is useful for visualization of the transduction process.

1. Interpreting What This Means

If the hypothesis holds, we have arrived at some interesting findings:

1. As is well-known in particle physics, a dipole is a broken 3-space symmetry in the violent flux exchange between the active vacuum and the dipole.
2. This dipole's broken 3-space symmetry in EM energy flow, provides a *relaxation* to a more fundamental EM energy flow symmetry in 4-space where *P* and *T* symmetries are broken but *CPT* symmetry is maintained.
3. This means that time (or "time-like EM energy flow") is transduced to 3-spatial EM energy flow. Hence we must understand "time as energy" and time-like energy flow as well as 3-spatial EM energy flow.
4. There is no law of nature or physics that requires 3-symmetry of EM energy flow as an additional condition applied to 4-symmetry of EM energy flow.
5. The dipole is a practical and very simple means of "breaking" the additional 3-flow symmetry condition in EM energy flow, and of relaxing to the fundamental 4-flow symmetry *without* 3-flow symmetry.
6. So long as the dipole statically exists (e.g., imagine an electret suddenly formed, or a charged capacitor with no leakage), real usable EM energy will continuously pour from the dipole at light speed in all directions. At the same time, reactive EM power (actually, energy) will continuously flow into the dipole from the time domain (the complex plane), and be transduced into real EM power output in 3-space by the dipole.
7. A dipole and its scalar potential thus represent a *true negative resistor* system of the most fundamental kind. The dipole continually receives EM energy in unusable form (reactive power, which cannot perform real work), converts it to usable form (real power, which can perform real work), and outputs it as usable, real EM energy flow (real power) in 3-space.
8. Simultaneously, at its formation the dipole initiates a continuing *giant negentropy*—a progressive reordering of a substantial and usable portion

of the vacuum energy [9a].[12] Further, this reordering of vacuum energy continuously spreads in all directions from the initiation point, at the speed of light. In original atoms formed shortly after the beginning of the universe, *dipoles*—and, as we shall also see, *charges*—have been pouring out real EM energy in 3-space for some 15 billion years or so. Each has reordered a fraction of the entire vacuum's energy, where the magnitude of the reordering varies inversely as the radial distance from the dipole.

9. If the dipole is destroyed, the ordering of the vacuum energy ceases, leaving a "separated chunk" of reordered vacuum energy that continues to expand and propagate radially at the speed of light in all directions, steadily reducing in local intensity as it expands.
10. At any very small volume in spacetime, from the dipole dynamics of the universe it follows that a great conglomerate of reordered vacuum flows and fluxes—some continuous, some chopped—is continually passing through that 4-volume. Further, the situation is totally nonlinear, so that direct wave-to-wave interactions occur continuously amongst these energy flows and waves. We hypothesize that this is the actual mechanism constituting Puthoff's cosmological feedback mechanism [26,27].[13]
11. Further, in 1904 Whittaker [28] showed that any EM field or wave pattern can be decomposed into two scalar potential functions. Each of these two potential functions, of course, decomposes into the same kind of harmonic longitudinal EM wavepairs as shown in Whittaker [8], plus superposed dynamics. In other words, the interference of scalar[14]

[12] Unfortunately, entropy is one of those concepts in physics for which there are several differing major views. For our work in energy from the vacuum, we take the very simple view that a negentropic process is like a negative resistor—it receives energy in a form unusable to us, transforms it, and outputs it into a form that is usable. We completely avoid the various notions of "information" and attempts to equate information and energy. We do point out, however, that a time reversal process in one form or another is involved. In that sense, for instance, Newton's third law is a negentropic process and involves time reversal. Its cause in electrodynamic interactions is the interaction with mass of that missing half of the EM wave in vacuum, unwittingly omitted by Maxwell.

[13] In Sachs' great generalization of a combined general relativity and electrodynamics, we are also speaking of spacetime curvature functions, and a unified field theory. See also Sachs' chapter on symmetry in electrodynamics: from special to general relativity, macro to quantum domains in this series of volumes on modern nonlinear optics (Part 1, 11th chapter).

[14] As Whittaker showed in 1903 [8], the scalar potential is actually a harmonic set of bidirectional EM longitudinal EM wavepairs, where each pair is composed of a longitudinal EM wave and its phase conjugate replica. Only because classical electrodynamicists have erroneously defined the field and potential as their own reaction cross sections with a unit point static charge, has the "static" potential been misidentified as a *scalar* entity, which it is not. The energy diverged from a uniform potential, around a fixed static point unit charge, is actually the set of divergences around the intercepting charge of the energy flows of all those EM waves constituting the potential. The sum total of all these individual wave divergences indeed has a scalar magnitude, but the magnitude of the total energy divergence from the potential is not the potential itself nor its magnitude.

potentials—each of which is actually a set of longitudinal EM waves, and not a scalar entity[15] at all, but a *multivectorial* entity—produces EM fields and waves and their dynamics. Hence we hypothesize that the Whittaker interference of the propagating reordered EM energy entities, continuously occurring at any point in space, generates the zero-point EM field energy fluctuations of the vacuum itself. Indeed, an AIAS group paper by Evans et al. [29] has already shown that just such "scalar interferometry" produces transverse EM fields and waves in the vacuum.

IV. MECHANISM GENERATING THE FLOW OF TIME

A. Background: Observation as an Operator

Since the nature of time is itself an unresolved question, we take a simple approach in order to arrive at a mechanism generating the "flow of an object through time."

First, it is well known in physics that the choice of fundamental units one chooses for one's model is arbitrary. While most often *mass, length, time*, and *charge* are used, a perfectly valid model can be generated using only a single fundamental unit.

Suppose we use the joule as the single fundamental unit. Then each of the entities we conventionally call "mass," "length," "time," and "charge" will become totally a function of energy. So we can legitimately state that "mass is energy," and we are already comfortable with that statement since the dawn of relativity and the nuclear age. But we can also legitimately state that "time is energy," and be rigorously accurate. We have previously postulated that one second is equal to spatial EM energy compressed by the factor c^2. So time is just extremely compressed EM spatial energy [30]. In that case, time has the same energy density, so to speak, as does mass.

We also point out that any observable is an instantaneous 3-spatial snapshot of a four-dimensional event. "Observation" itself may be taken as a process where a $\partial/\partial t$ operator is invoked on 4-space (spacetime), leaving a purely 3-spatial output. However, no observable "exists in time," since rigorously it is

[15] We point out the obvious—a "scalar" mass in 3-space actually has a time vector since it moves through time continually, just to continue to exist. Further, it is a special form of energy (energy compressed by c^2) moving through time. Since we may choose any form of energy we wish by simple transduction, we may take it as compressed EM energy. So the mere continued existence of any mass proves conclusively that EM energy can and does ubiquitously flow through the time dimension. The combined continued existences of numerous masses proves conclusively that the flow of time can have a myriad of internal electromagnetic energy flows. An equilibrium between (1) an inflow of EM energy to a transducer from the time dimension, and (2) an outflow of EM energy in 3-space from the transducer, will be seen as a discrete excitation (potential energy) associated with the transducer. Hence the notion of the charge.

only a single, frozen "3-slice" at one single instant, forever fixed. So *mass does not exist in time, but masstime does.*

B. The Relationship of Observation to Cause and Effect

We define the 4-entity prior to the invocation of the $\partial/\partial t$ operator (observation) as the *causal side* of the observation process. The most significant concept embedded in the concept of causality would appear to be that an entity exists and acts during time. The most significant content of the concept of the effect is that it was "caused" by an entity existing in and acting in time, but it (the effect) itself does not exist in time. We define the 3-entity that is the output of observation (after the application of the $\partial/\partial t$ operator) as the *effect side* of the observation process. As is well known, all observation is 3-spatial, so an observed entity does not exist or act in time. We also speak of the *intermediary* as the entity—usually a *mass* that is itself an effect, or *masstime*, which is a causal entity—with which the causal entity interacts and to which the $\partial/\partial t$ operator is applied.

We completely accept the Sachs [27] unification approach to a combined general relativity and electrodynamics, generalized from a topological standpoint.

C. Polarizations of Photons and EM Waves

As is known in quantum field theory, there are four polarizations of a photon [31]. These are the x, y, z, and t polarizations, where x, y, z, and t refer to the four dimensions in a 4-space. Thus—at least in theory—there must also be four polarizations of electromagnetic waves, even though not all these waves are yet experimentally known.

The x and y polarizations (or any combination) are the familiar transverse photon and the transverse wave. The z polarization along the line of propagation gives the longitudinal photon and the longitudinal EM wave.

Without further elaboration, we speak of a mass in which a small portion exists as *masstime* rather than mass, as having been "time-charged" or "time-excited."

D. Imperfect Longitudinal EM Waves

In attempting to produce longitudinal EM waves (LWs) from transverse EM waves (TWs) that are input to a polarization transduction process, it is reasonable that only imperfect LWs are produced, and that a residue of TW content remains. The resulting imperfect LW might be referred to as an *undistorted progressive wave* (UPW). Some work has been done on UPWs [32]. Such waves are theorized to have remarkable characteristics including wave velocities either slower or faster than standard light velocity [33].

The t-polarization wave in the time dimension is quite unique: The *spatial* energy of the wave is in equilibrium and not vibrating at all; instead, the photons comprising the wave are vibrating in their *time* components. That is called a "time-polarized" photon or a "*scalar* photon." Its wave version does not yet seem to be known in the literature.

E. EM Waves and Photons Carry Both Spatial Energy and Time Energy

On the other hand, the entire question of "EM waves in spacetime" may be in need of a thorough overhaul. A photon is a "piece of angular momentum" in the form of $(\Delta E)(\Delta t)$. Hence the photon carries an increment of spatial energy ΔE and also an increment of time–energy Δt. The time–energy component (Δt) may be regarded as ordinary spatial energy that has been compressed by the factor c^2.

So the photon transports two types of energy: (1) a "weak spring" spatial energy ΔE, so to speak and (2) a "very stout spring" time-energy Δt, so to speak. [*Note*: Since waves consist of photons composed of both spatial energy and time-energy, the wave propagation must transport not only spatial energy but time energy. The "*spatial energy only*" EM transverse wave, composed of oscillating $\boldsymbol{E}$ and H fields, cannot exist in space*time* prior to interaction with charged mass. Instead, what exists in spacetime prior to interaction must be $\boldsymbol{E}t$ and $\boldsymbol{H}t$ dimensionally—in short, *impulse field* waves. After interaction (observation), the $\partial/\partial t$ operator has been applied, converting the impulse fields (the causes) into force fields (the effects). Present electrodynamics seems substantially confused between the *effect entity* side of the observation process and the *causal entity* side. The causative side (input) entities must occupy LLLT, while the effects side entities are restricted to LLL. No LLL entity can be a cause a priori.]

When a mass m absorbs a photon $(\Delta E)(\Delta t)$, the (ΔE) component is compressed spatially by the factor c^2, turning it into an extra amount of mass Δm, so that the mass becomes $(m + \Delta m)$. At the same time, the (Δt) component is joined, so that what results is $(m + \Delta m)\Delta t$. In short, mass m is changed to masstime mt by photon absorption. So in the absorption of a photon γ by a mass m, we have

$$\gamma + m \rightarrow (\Delta E)(\Delta t) + m \rightarrow (m + \Delta m)\Delta t \tag{9}$$

In short, the mass turns into masstime, and masstime mt is as different from mass m as impulse Ft is from force F. We point out that "mass" m alone does not even exist in time; masstime mt does exist in time. This is proposed as a simple but fundamental correction to much of present physics.

In the simplest case, in the next instant a photon is reemitted, and so we have

$$(m + \Delta m)\Delta t \rightarrow (\Delta E)(\Delta t) + m \rightarrow \gamma + m \tag{10}$$

So emission of a photon changes masstime back to mass, in the simplest case.

In passing, what we call "observable" change must involve

$$(\Delta E)(\Delta t) \geq \frac{h}{4\pi} \tag{11}$$

F. Photon Interaction as the Mechanism Generating Time Flow and Duality

From the foregoing, mass cannot continuously exist as mass; mass rigorously exists only at a single point in time, and never again. It only appears to exist in time because of the reactions in Eqs. (9) and (10) above. Rigorously, a mass does not really "travel through time" continuously per se, but proceeds with an overall serial change mechanism, driven by its total virtual and observable photon interactions, as

$$m \rightarrow mt \rightarrow m \rightarrow mt \rightarrow m \rightarrow \cdots \tag{12}$$

We propose that this may account for the duality of particle and wave. When a mass is observed, time has been stripped away, leaving a frozen 3-spatial snapshot, which we will see as (having been) a particle (simplest case). That occurs just after major (observable) photon emission from the masstime state. Immediately another observable photon is absorbed, and so state *mt* occurs. The particle of mass actually oscillates at a very high rate between the *m* and *mt* states—so high a rate that by arranging the interaction conditions one may interact with it either as a wave (react predominantly in the *mt* state) or as a corpuscle (react predominately in the *m* state). "Mass" as it exists is actually an oscillation or wave between *m* and *mt* states. Every differential piece of the mass is also in oscillation between (dm) and (dm)(dt) states.

G. The Overall Flow of Time Is Internally and Dynamically Structured

During the transition in any mass to masstime state by reaction of the mass with an "observable" photon, a myriad of tiny virtual photon interactions involving very tiny $(\Delta E)(\Delta t)$ components of size

$$(\Delta E)(\Delta t) \ll \frac{h}{4\pi} \tag{13}$$

occurs with the mass *m*. These tinier increments of time and their increments of energy, constitute internal structures in the overall time flow process. Therefore they are "energy currents" or "time-like energy currents" and *dynamic* structures inside the flow of time.

So the Δt component of masstime has a myriad of energy–time dynamics infolded within it. Hence the *mt* state is very dynamic in time, particularly for fundamental particles. The *mt* state is in fact a "collection of time–energy dynamics" and therefore "wave-like."

The major point is that mass does not emit photons; masstime does. Mass "travels through time" by an extremely high oscillation between corpuscle-like state *m* and wave-like state *mt*.

The concept can be extended, but this suffices for our concept of energy currents in time, and the interaction of such energy currents with mass in a mass system.

V. CONNECTION BETWEEN FIELD AND SOURCE

A. The Problem of the Charge as a Source

We use the foregoing hypothesis to propose a solution to a previously unsolved major foundations problem in electrodynamics. Quoting Sen [10]:

> The connection between the field and its source has always been and still is the most difficult problem in classical and quantum electrodynamics.

The problem really lies in how we approach the notion of the "source charge," since the usual classical electrodynamics does not model the interaction of the vacuum and the charge.[16] With no active vacuum input to the charge,

[16] For example, the notion of charge is much more complicated in gauge field theory than is usually assumed in more classical EM theory. In gauge-theoretic electrodynamics, the field is a curvature in spacetime and so is charge, so that the field intrinsically possesses charge. Further, the charge is a curvature in *spacetime*, it is inextricably connected to both the time coordinate and the 3-space coordinates. A priori, field changes thus may involve changes in the very nature of charge as we observe it, and correspondingly charge changes may involve changes in the very nature of the field effects we observe. As a crude example, changes in the "time" portion of the charge-as-spacetime-curvature can readily affect changes in the "spatial energy" aspect. It is not too difficult, then, to visualize that an inflow of EM energy into the time portion of the charge-as-spacetime-curvature alters the time aspects—which, in turn, causes a corresponding canonical alteration of the 3-space aspects of the charge, producing an outflow of 3-space EM energy from the charge. Indeed, conservation of 4-energy would require such. A joint paper by Evans and Bearden is being prepared in this symmetry area, and Sachs' magnificent paper on symmetry in electrodynamics (cited in footnote 13, above) is included in these volumes.

the received crippled and fragmentary model of electrodynamics implies that the charge not only creates the fields and potentials that surround it but also creates *out of nothing* all that EM energy comprising those associated fields and potentials. Since energy can be neither created nor destroyed, but only changed in form, the conventional notion that the source charge produces its associated fields and potentials and EM energy, in the absence of any interaction with the vacuum, is a nonsequitur. This is the vexing and hitherto unresolved problem referred to by Sen.

The real difficulty is that the conventional model eliminates the vacuum interaction. Conservation of energy flow is an expression of symmetry, and there can be no symmetry without the vacuum interaction. Hence by not modeling the vacuum interaction, the conventional EM model must grossly violate the conservation of energy law. Its view of the charge as the source of fields and potentials and their energy, with no input of energy, essentially reduces the notion of source charge to a *perpetuum mobile*. In short, it simply posits an output of EM energy without any energy input at all and thus the *creation* of energy from nothing.

1. What Experiment Shows

Experimentally, of course, it is easily shown that 3-spatial EM energy does continuously diverge *out of* that charge, creating all its associated fields and potentials that do appear around it. Just create a charge (e.g., as in pair production), and measure the resulting outflow of the fields and potentials and EM energy from it, at the speed of light in all directions.

Experiment also shows that there is no detectable (observable) 3-space EM energy flow that converges *into* the charge. Hence we are left with a quandary—experiment shows that there is a broken symmetry in the conservation of EM energy 3-flow, directly associated with the source charge.

So the dilemma is: Where *does* the energy come from that is pouring out of the charge continuously?

The charge alone cannot be a true *source*, since rigorously there can be no such thing. As Semiz [34] puts it:

> The very expression ‘energy source’ is actually a misnomer. As is known since the early days of thermodynamics, and formulated as the first law, energy is conserved in any physical process. Since energy cannot be created or destroyed, nothing can be an energy source, or sink. Devices we call energy sources do not create energy, they convert it from a form not suitable for our needs to a form that is suitable, a form we can do work with.

We really do not have *energy sources* as such in nature, even though we sloppily use that term. Instead, we actually have *energy transducers*. Else we must discard the conservation of energy law itself: *Energy can neither be created nor destroyed, but only changed in form.*

A priori, since we can measure no real 3-space input of EM energy to the unchanging charge but can measure real 3-space EM energy pouring from it, energy must be input to it from the active vacuum in a *nonobservable* (other than 3-space) form, and converted by it into an *observable* (3-space) form that is reemitted, is usable, and produces what we call the "fields and potentials" and their energy, associated with that "source charge."

Since it is common usage, we will continue to use the sloppy term "source charge" or "source dipole," but with the understanding that we refer to a special kind of energy transducer. Our problem is in showing where the energy comes from. So far, we know that (1) it does not come from 3-space, and hence must come from the imaginary plane or the time dimension; and (2) it must converge on the source charge (transducer) in a flow whose magnitude is precisely equal to that of the emitted divergent 3-space EM energy flow.

2. *The Charge as a Composite Dipole*

To solve the source charge problem, we first point out that there exists no such thing as an *isolated charge*. As is well known in quantum electrodynamics, clustered around any "isolated charge" in the vacuum are virtual charges of opposite sign. We take one of the separated virtual charges, and a correspondingly small piece of the observable charge of opposite sign, and call the pair a *composite dipole*. So the so-called "isolated charge" is actually a set of composite dipoles. Any of the clustering virtual charges and any of the pieces of the observable charge thus comprise such a composite dipole. The "isolated" observable charge is thus seen as a great entanglement of composite dipoles.

Further, each composite dipole has its own scalar potential between its end charges. With the previously stated reservation (see Section III.A.1), this scalar potential decomposes per Whittaker [8] and thus initiates a giant negentropic reordering of the vacuum energy as previously discussed. So any charge is really an entire set of composite dipoles, composite negative resistors, and broken 3-symmetries in the vacuum flux exchange. Energy flow 4-symmetry must rigorously apply.

The charge is a dipolar system (actually it is a *great set* of dipoles). It pours out a continuous flowset of real EM power in 3-space, radially at the speed of light in all directions. The composite dipoles making up the charge system are being fed by a continuous converging flowset of reactive power from the imaginary plane, as we mentioned previously.

The real EM wave energy flow pouring out radially in all directions in 3-space from the charge system forms the well-known fields and potentials associated with that "source charge." The actual source of the EM energy flow from the charge is the hypothesized negentropic reordering of the 4-vacuum energy into a giant 4-circulation of EM energy flow.

The 4-symmetry in EM energy flow is conserved at all times. *Energy is not created by the charge*—which creation has been implied in classical EM theory without the vacuum interaction, without the charge as a composite dipole, and without the Whittaker [8] decomposition of the scalar potential between the poles of every dipole. Instead of the present "creation of energy" nonsequitur in the conventional model, the charge's received EM energy flow in unusable form is *transduced* by the charge's spin into usable form and output continuously. We note but do not further pursue the fact that "charge" transduces the normal 3-symmetry energy flow into a 4-symmetry energy flow without concomitant 3-symmetry energy flow. This may eventually form the basis for a more vigorous definition of *charge*.

In short, as a dipolar entity, the charge is an *open system far from thermodynamic equilibrium in 3-space EM energy flow*. Indeed, it has no input energy flow in 3-space, but instead has an input energy flow from the imaginary plane (from the time dimension). Hence classical 3-equilibrium thermodynamics does not apply.

The charge is simultaneously in perfect energy flow equilibrium in 4-flow. It continuously receives EM energy from the time dimension (imaginary plane), transduces the energy into real 3-space, and radiates it radially outward in 3-space as a real EM energy flow, producing the fields and potentials associated with that "source charge."

As a dipolar system, the charge's broken 3-symmetry in EM energy flow has allowed the system to relax to a more fundamental 4-symmetry energy flow without the arbitrary additional condition of 3-symmetry energy flow. The charge (composite dipole set) and the dipole are thus the ultimate and universal *negative 4-resistors*. [*Note*: The conventional notion of the negative 3-resistor is that 3-spatial energy is received in unusable form, converted to usable form, and output as usable 3-spatial energy flow. The negative 4-resistor receives energy from the time dimension (the imaginary plane), and not from 3-space. The output of the negative 4-resistor is real 3-spatial EM energy flow.]

As we shall see, the source dipole furnishes the energy to power every electrical system and circuit, since all EM systems and circuits must involve charge that is merely a set of composite dipoles receiving reactive power and pouring out real power (real EM 3-energy flow).

B. Entropic Engineering

When we "make entropy," we must do work. We wrestle nature fiercely to the mat, so to speak, by brute force. All the while, nature protests our entropic brutality by providing the Newtonian third-law reaction force (see Section V.B.1) back on our causative wrestler performing the "forcing." To do entropic engineering, we have to continually input energy to the wrestling mechanism or engine, losing a bit of the input energy in the inefficiencies, and fighting the "back emf," "back mmf" (magnetomotive force), or Newtonian third-law reaction that is nature's protest all the while. Those are nature's penalties for imposing 3-space EM energy flow symmetry on her as—to nature—an *additional and highly undesired condition.*

In short, with 3-symmetry we have to provide the continual input energy to our entropic processes by burning fuel, damming rivers, erecting windmills, building waterwheels, erecting solar cell arrays, building nuclear power plants, building and charging chemical batteries, and so on. In the process, we destroy and pollute the biosphere on a giant scale as we rip down forests, strip-mine and drill the earth, and spill pollutants into the atmosphere, the rivers, the oceans, and so on. We do all that biospheric destruction because we inexplicably insist on 3-space energy flow symmetry, and thus adamantly require nature's adherence to classical equilibrium thermodynamics.[17]

We have to pay and pay continuously, for insisting on doing such atrocious entropic work. In so doing, we "tie nature's feet down" with that added arbitrary requirement for 3-symmetry in energy flow. We ourselves prohibit nature from performing the giant negentropy she so dearly loves and very much prefers. We also arbitrarily and *meanly* discard the bountiful electromagnetic energy 3-flow that nature loves to furnish us so freely when we indulge her vast preference for negentropy.

1. Newton's Third Law

Newton's third-law reaction in mechanics is usually "demonstrated" in elementary fashion by colliding balls. Note that time must be continuously interacting with the balls, in the mass/masstime/mass.... manner, if the balls are even to exist. Taking one ball to observe, the incoming ball's momentum is the "cause" and the resulting momentum acquired by the struck ball is the "effect," so to speak. Note that by "cause" and "effect" we are speaking of the input and output of the observation process itself. So there exists a backreaction from the change in the "struck" ball (the change in the effect side of observation) on the "causal" side, altering the momentum of the incoming ball. The point is that, with respect to the observation process, cause interacts on an intermediary to produce a change on the effect (output) side, and the effect acts back through the intermediary to produce a change in the cause (input) side.

[17] Equilibrium thermodynamics is usually interpreted in terms of 3-spatial energy symmetry anyway, to begin with, and then one "loses some control" steadily and hence loses some ordering. Actually, the thermodynamics of systems far from equilibrium in 3-spatial energy flow, must always be in symmetry in energy 4-flow; "3-space disequilibrium" thermodynamics and 4-space equilibrium thermodynamics are postulated as different views of the same thing.

Newton's third law is a description of what happens, but does not contain the mechanism producing what it describes. In mechanics and electrodynamics the Newtonian third law is assumed, while in general relativity its causal mechanism exists in the theory. Simply put, any change in the curvature of spacetime (the change in the causal input to the observation process) causes a change in the mass energy (to the effects side or to the output of the observation process). Also, any change in the effects side of observation (any change in mass-energy) produces a simultaneous change in the causal (spacetime curvature) side.

C. Negentropic Engineering

A far better way is to cooperate with nature and "let nature make copious negentropy." To do that, we now can see the startlingly simple mechanism. We simply make a little dipole, entropically. So we have to pay for *making* the dipole, *once*, and we have to do a little gentle violence to nature, *once*. Then we need do no more violence, if we just leave the dipole intact and do not destroy it.

When we make the dipole, we make a little bit of "broken 3-symmetry" in the universe's energy flow. Voilá! Nature sings for joy at finally having her feet freed a little from the shackles of 3-symmetry energy flow. In great glee, she instantly sets to reordering a substantial and usable portion of the vacuum energy, in all directions at the speed of light. As long as we do not destroy the dipole (the broken 3-symmetry) that breaks the shackles, nature's feet remain freed from the 3-space symmetry, and she delightedly continues to reorganize a portion of the vacuum energy, with the reordering spreading radially outward at the speed of light.

Simultaneously, in great gratitude, nature pours out an immense real EM energy 3-flow from that little dipole. She will continue to pour it out forever, if we do not destroy the dipole.

D. Entropic versus Negentropic Engineering

To summarize: If we make *entropy*, we tie nature's feet and she forces us to pay for it, and pay continually.

If we make *negentropy*, we only pay a very tiny "one-time initiation fee." From then on a delighted nature pays us for our thoughtfulness, and pays us copiously, continuously, and freely.

The smart thing to do is make just a little bit of entropy wisely, using it to break 3-space energy flow symmetry (basically, to make a dipole). Then we should—adapting a phrase—*leave that mother of all negative resistors and free energy generators alone*! We should concentrate on intercepting, extracting, and using the free energy copiously flowing forth from the ongoing giant negentropy, without destroying the dipole that is freely providing it.

VI. HOW THE EXTERNAL CIRCUIT OF A BATTERY OR GENERATOR IS POWERED

A. EM Energy from the Vacuum Powers Every Circuit

Neither the shaft energy introduced into a generator nor the chemical energy present in a battery is used to power the external circuit. The internal energy in a generator or battery is dissipated only to perform work on the internal charges, to separate them and form a source dipole between the terminals, with some of the energy dissipated in other internal losses. A battery also uses its internal chemical energy to separate its charges, forming (or reforming) the source dipole.

Once formed, the source dipole is a broken symmetry [2] in the vacuum's energy flux along the lines experimentally shown by Wu et al. in 1957 [35]. As Lee points out, the asymmetry between opposite signs of electric charge is called C violation, or *charge conjugation violation*, or sometimes *particle–antiparticle asymmetry*. As Nobelist Lee [2] further states, "Since non-observables imply symmetry, these discoveries of asymmetry must imply observables."

The broken symmetry of a dipole in its vacuum flux exchange has been known in particle physics since the late 1950s. In classical electrodynamics (CEM) the active vacuum and its exchange are omitted altogether, even though experimentally established for many years. As Lee also pointed out, there can be no symmetry of any *observable* system anyway, unless the vacuum interaction is included.

Further, by the definition of broken symmetry, the proven asymmetry of the source dipole in the vacuum flux *must* receive virtual energy and output observable energy. Since we see only the 3-spatial output, to us *it appears* that the source dipole somehow "extracts" from the vacuum some unobservable energy, transduces it, and then pours it out as the observable EM energy that we do observe.

Since the term "source charge" and "source dipole" are widely used, we will continue to use them, but with the clear understanding that in each case we are really speaking of 4-space energy transduction of received virtual energy into output observable energy.

B. Negative Resistor Function of the Source Dipole

To summarize the total energy flow in space surrounding the conductors has two components as follows:

1. A tiny *Poynting component* [3,4] of the energy flow directly along the surface of the conductors strikes the surface charges [36][18] and is diverged (deviated) into the conductors to power the circuit.

   The interacting surface charges are forced to move axially (mostly), since they can move longitudinally down the conductor only with a small drift velocity, which nominally may be a few inches per hour. As the

[18] In Ref. 36, Jackson points out the decisive role played by the surface charges in the circuit. His earlier book [7], did not cover circuits except for some minor capacitive and inductive effects.

struck surface charges are forced inward, the withdrawal of their interacting field energy (Poynting energy) follows them from surrounding space into the wire, as shown by Kraus [3], in terms of the numbers on his energy flow contours in a plane perpendicular to the conductors.

2. The huge nondiverted *Heaviside component* [5,6] filling all space around the circuit, misses the circuit entirely and is wasted in all those circuits using only a single pass of the energy flow. After having passed the circuit, the Heaviside energy flow can still furnish additional energy to the circuit if retroreflected to again pass back over the surface charges. However, conventional power systems use only a single pass of the energy flow, and completely ignore this enormous "dark" (unaccounted for) energy accompanying every circuit. Other methods of extracting energy from the neglected Heaviside component are discussed later.

The perspicacious reader will note that we have also completely replaced "statics" in electrodynamics with "steady-state dynamics."

VII. WHY LORENTZ ELIMINATED THE HEAVISIDE FLOW COMPONENT

A. The Nondiverged Heaviside Component

The Heaviside component is enormous, and often some 10^{13} times as great in magnitude as the Poynting component [15]. The Heaviside nondiverged energy flow component was arbitrarily discarded by H. A. Lorentz [11], who integrated the energy flow vector itself around a closed surface enclosing any volumetric element of interest. This discards any *nondiverted* (nondiverged) energy flow components, regardless of how large, and retains only the *diverted* (diverged) component, regardless of how small.

Effectively Lorentz arbitrarily changed the *energy flow* vector into its *diverted flow component* vector—a fundamental nonsequitur. In one stroke he discarded the bothersome Heaviside component, reasoning that it was "physically insignificant" because—in single-pass circuits—it does not enter the circuit and power it.

This is rather like arguing that all the wind on the ocean that does not strike the sails of a single sailboat is "physically insignificant." A moment's reflection shows that the "insignificant" remaining wind can power a large number of additional sailing vessels. A very large amount of energy can be extracted and used to do work, if that "physically insignificant" wind is intercepted by additional sails.[19]

[19] The Heaviside component represents a huge region of dynamic organization of the vacuum energy. There is no limit to such vacuum organization, as shown by the giant negentropy operation initiated by the broken 3-symmetry of the dipole. We stress again that, prior to its interaction with charge, the Heaviside energy flow is in the complex plane.

B. No Apparent Source of the Enormous Heaviside Energy Flow Component

Suppose that Lorentz had not arbitrarily discarded the huge Heaviside energy flow component surrounding the circuit and not contributing to its power. In that case, electrodynamicists in the 1880s would have been confronted with the dilemma of explaining where such an enormous flow of energy—pouring forth out of the terminals of every generator and battery—could possibly have come from. There was then no conceivable source for such a startling profusion of energy flow. Obviously the operator does not input such enormous energy, since the Heaviside flow is often some 10^{13} times as large in magnitude [37,38][20–21]. (see also Section VII.B.1) as the retained Poynting flow. Neither does a battery contain such enormous chemical energy to provide such a flow for even one second by chemical-to-electrical energy conversion.

To avoid strong attack and suppression from the scientific community on grounds of advocating perpetual motion and violation of energy conservation, in the 1880s there was no other choice but to discard the Heaviside component on some pretext. So Lorentz simply discarded the vexing component. He could not *solve* the problem, so he *got rid* of it.

1. Reordering of Vacuum Energy

In simple language, to see that reordering of vacuum energy (negentropy) does not require work; the organization of the vacuum represents a change to the "primal cause" or "primal energy." Organization of virtual state energy without the involvement of mass effects does not require observable work, because force is not involved and observable work ultimately involves the forcible translation of a resisting mass. So one can organize the "potential for doing work" without having to perform work in doing so. This is in fact what "regauging" or "gauge freedom" actually involves.

[20] But see Hibert [37]. It may surprise or even shock the reader that in general relativity there are really no *conservation of energy laws* as we know them, as was pointed out by Hilbert shortly after Einstein published his general theory. Hilbert wrote: "I assert...that for the general theory of relativity, i.e., in the case of general invariance of the Hamiltonian function, energy equations...corresponding to the energy equations in orthogonally invariant theories do not exist at all. I could even take this circumstance as the characteristic feature of the general theory of relativity."

[21] Commenting on Hilbert's remarkable assessment, Logunov and Loskutov [38] made the following statement: "Unfortunately, this remark of Hilbert was evidently not understood by his contemporaries, since neither Einstein himself nor other physicists recognized the fact that in general relativity conservation laws for energy, momentum, and angular momentum are in principle impossible." It remains largely unrecognized to date. We hypothesize that we may recover the conservation laws in 4-space if we extend them to include time-energy, time momentum, time potential, time force, and the giant negentropy induced by broken 3-symmetry.

Any local region of the vacuum is, after all, an open system *microscopically* fluctuating far from equilibrium with the surrounding rest of the vacuum, as shown by quantum fluctuations. So that local region can exhibit (1) self-ordering; (2) self-oscillation, self-spinning, and so on; and (3) negentropy.

To use this principle in practice, the trick is to "tickle" the local vacuum into performing the exact type of reordering and self-structuring that one wishes. One does this by adroitly changing the *effect* side of the observation process, thereby altering the interacting *causative* side as well—and changing the effect side precisely so that the desired set of changes occurs on the causal side. A discussion of this process is well beyond the extent of this chapter. In mechanics and electrodynamics, the interaction of the effect back on the cause has been erroneously omitted, but is included as Newton's third-law reaction—an effect conventionally posited without a cause. But its cause is present in general relativity since curvature of spacetime (cause) acts on mass energy (effect) to change it, and a change in mass energy (effect) interacts back on spacetime curvature (cause) to change it accordingly. That backreaction—missing from classical electrodynamics—is the cause of the Newton third-law reaction.

A logical mess exists in electrodynamics, where the effect has been rather universally confused with the cause. All illustrations continue to show the *E–H* planar (*X–Y*) wave in 3-space, which is an effect existing after the interaction with charge. What exists in spacetime before interaction must be *Et–Ht*, since observation itself is a $\partial/\partial t$ operator imposed upon the LLLT entity and producing an LLL entity—in this case, *E–H*. The integration of *E–H* along z does not add the missing time dimension, but merely represents a "3-spatial composite" of many frozen *X–Y* slices. It is thus the *spatial spread of the serial effects*. In *Am. J. Phys.* 69(2), 107–109 (2001), in endnote 24, editor Romer finally points out that the standrd plane wave diagram is horrible and wrong. Consideration of the *Et–Ht* "impulse" or causal wave in spacetime prior to the interaction with matter, particularly in phase conjugation pairs, leads to many very interesting new EM phenomena not covered in this paper.

VIII. SOME CHARACTERISTICS OF POWER SYSTEMS

A. The Deadly Closed Current Loop Circuit

In conventional systems, a closed current[22] loop contains the generator or battery source dipole as well as the external circuit's loads and losses. This

[22] In the generator this current is unitary, consisting of charges having the same *m*/*q* ratio. In the battery this is not true, since the internal lead ion current between the plates has an *m*/*q* ratio far greater than that of the electrons in the current between the external surfaces of the plates and the external circuit. Accordingly, in the storage battery it is possible to adroitly dephase the massive ion currents in charge mode, from the much less massive external electron currents that can be in load-powering mode. The result is the Bedini overunity battery switching process, with a negative resistor created right on the surface of the plates between the two dephased currents. (See Ref. 9b.)

arrangement requires that half the collected energy in the circuit forcibly pump spent electrons in the ground return line back through the back emf of the source dipole. Specifically, for every electron passing through the voltage drop across the loads and losses in the external circuit, an electron—so to speak—must be forcibly rammed back up through the source dipole against the same voltage.

Forcing the spent electrons through the source dipole's back emf performs work on the end charges of the dipole to forcibly scatter them. This destroys the dipole and cuts off its free extraction of energy from the vacuum. In a charged battery, this "back emf work" chemically scatters the dipole charges, whose restoration as a source dipole again is performed by dissipation of some of the chemical energy. This dissipation of part of the chemical energy causes a partial reversal of the normal chemistry [39] of the electrolyte–plate system, which reduces the chemical energy available by the battery to reestablish the source dipole. The battery's remaining chemical energy is expended to continually restore the source dipole as it is continually destroyed, until the chemical energy is exhausted. Then one must introduce additional energy into the battery to "recharge" it by forcing the chemistry back to its initial fully charged condition.

Electrical loads are—and always have been—powered by energy extracted and converted from the vacuum by the source dipole, not by shaft energy furnished to the generator or by the chemical energy in the battery. For *unitary current* (i.e., a current whose basic flowing charges all possess the same m/q ratio) closed-loop circuits, half the Poynting energy collected in the external circuit is expended in the circuit loads and losses (forward emf direction), and half is expended against the back emf of the source dipole (in the back emf direction), destroying the dipole.

Another way of seeing this is to simply examine the scalar potential existing between the two charges of a dipole, as we mentioned previously. A "scalar" potential is not really a scalar *entity*, although it has a scalar reaction cross section for reaction with a static charge (see footnote 14, above). Instead, it is a harmonic set of bidirectional phase conjugate longitudinal EM wavepairs, as shown by Whittaker [8] in 1903. Thus any dipole (negative 4-resistor) or charge (composite dipole set) has an enormous set of longitudinal EM wave energy that flows into it from the time domain as we discussed above, and a corresponding enormous set of longitudinal EM wave energy flows out from it in 3-space, in all directions. Once the source dipole is formed in the generator or battery, this energy flow exchange between source dipole and the universal active vacuum is established and ongoing, as is the broken symmetry of the dipole in that energy flux exchange with the active vacuum. At any point in the universe where the negentropic reordering has reached at its light speed, a charge will interact with the flow and extract energy from it. We emphasize that *any scalar potential itself is such a negative 4-resistor, and the negative 4-resistor process is induced upon any charge or dipole placed in that potential.*

B. Present Power System Design Forcibly Applies Lorentz Regauging

A conventional (unitary closed current loop) circuit's energy dissipation potential is separated into two equal halves, each in opposite direction. One that provides the force for discharge through the external circuit of the source dipole uses what is called the "forward emf." The other, which represents a resistance force back through the internal dipole, provides what is called the "back emf." The purely forward emf discharge represents energy dissipation in the load and losses. The forward emf discharge proceeds, however, only by simultaneously—so to speak—ramming the spent charge carriers in the ground return line back through the dipole, scattering the charges and destroying the dipole. Precisely as much energy is used in *destroying the source dipole* as is dissipated in the external loads and losses combined. Hence the circuit destroys its source dipole—and the negentropic reordering of the vacuum energy from which the circuit's excitation energy is extracted. And it does this faster than it powers the load itself.

The closed current loop circuit thus enforces a special sort of dynamic *Lorentz symmetric self-regauging* during discharge of the circuit's excitation energy. The energy rate being destructively returned to the vacuum in destroying the source dipole, is equal to the energy rate being constructively returned to the vacuum from the external loads and losses. The excited system forcibly kills its free input of energy from the vacuum as fast as it powers the combined loads and losses—and thus faster than it powers its loads.

To restore the destroyed dipole, the operator must input as much energy as was required to destroy it. But with the closed current loop circuit, this operator input a priori is greater than the useful output of work in the load. Hence the coefficient of performance (COP) of this closed current loop system (with unitary m/q of the charge carriers) is self-limited to COP < 1.0.

Classical thermodynamics, with its infamous second law, rigorously applies during the excitation discharge phase of the closed current loop system, since the system itself is diabolically designed to continuously and forcibly restore itself into equilibrium with its active environment by killing its own source dipole gusher of vacuum energy flow as fast as it powers its loads and losses.

In a generator-powered system, continual input of energy to the generator shaft is required to continually add energy to perform work on the scattered charges, in order to restore the source dipole that the closed current loop continually destroys. Thus our present *self-crippling* vacuum-powered generator circuits and systems exhibit COP < 1.0 a priori, as do our self-crippling battery-powered circuits and systems.

We must pay for the initial energy input to the generator to establish the source dipole. *Once formed*, the dipole continuously extracts and pours out enormous observable EM energy flow from the vacuum—if we do not foolishly destroy the dipole. However, all our conventional circuits are deliberately designed to do just that: destroy their source dipoles faster than they power their loads. They are deliberately—although unwittingly—designed specifically as systems that self-enforce COP < 1.0.

The typical closed current loop circuit receives only a single pass of the energy flow, and therefore intercepts, collects, and utilizes only the very small Poynting component, simply wasting the enormous Heaviside component that misses the circuit altogether. Our present *single-pass* power systems nominally waste some 10^{13} times as much energy as they catch and utilize. Scientists can easily do better than this if they (1) remove Lorentz' arbitrary and erroneous discarding of the Heaviside energy flow, (2) develop circuits and circuit functions to catch and use much of that available but *presently neglected* huge energy flow, and (3) develop and utilize circuits that destroy their source dipole slower than they power their loads.

Those are, in fact, the requirements for electrical power systems exhibiting COP > 1.0. Such open systems in disequilibrium with their active vacuum are permitted; indeed, every dipolar circuit already is in such disequilibrium. Such a system can also be "close-looped" to power itself and its load. For instance, an open dissipative system with COP $= 2.0$, can use 1.0 of its COP to power itself, and the other 1.0 to power the loads and losses [98]. This is no different from the operation of a windmill, except that the electrical system operates in an EM energy wind initiated from the vacuum by the source dipole. We point out that "powering a system" actually need only be "powering its internal losses" if the source dipole is maintained.

C. What We Pay the Power Company to Do

Essentially we pay the power company to engage in a giant Sumo wrestling match inside its generators and to *lose* by killing the free extraction of energy from the vacuum faster than the wrestling process powers the loads.

We pay the power company to use only a "single pass" of the energy flow along its transmission lines and the consumer power circuits, and thereby to just "waste" some 10^{13} times as much available EM energy as the company allows us to "use."

Present electrical power systems simply repeat this travesty over and over, so that we are continually inputting external energy to the generator to restore the source dipole, and having to input more than we get back out as work in the load. That is why all conventional EM power systems exhibit COP < 1.0 a priori. The system is specifically designed to force itself to do precisely that, by killing itself faster than it powers its load.

Such an inane power system continually forms a marvelous extractor of vacuum energy, then attacks itself suicidally. In an oil derrick analogy, the system continually destroys its own energy flow "wellhead" (source dipole) and does not capitalize on it. That is rather like drilling an oil well, bringing in a great gusher, catching a little oil in barrels, burning half of the barreled oil to deliberately cap the well, then drilling another well beside the first one, forcibly recapping the second one, and so on.

This is what keeps those coal trains running, the fleets of oil tankers steaming, the natural-gas lines flowing with gas and the oil pipelines flowing with oil, and gasoline and diesel engines powering our transport. It keeps enormously expensive nuclear power plants being built so that their nuclear reactors can produce heat to boil water to make steam to run turbines to input shaft power to the electrical generators for the generators to restore their continually killed source dipoles [18].

This insanity keeps our energy costs high, economically burdens every citizen and every nation, impoverishes many undeveloped and developing nations along with their peoples, and pollutes the planet to the limit of its tolerance and beyond. On our present course, we are embarked on destroying our biosphere and ourselves along with it.

Eerily, our scientific community ignores the terrible >135-year-old foundation errors in classical electromagnetics and assures us that this is the best that electrodynamics can do. In fact, the scientific community has not yet even recognized the problem, much less the solution. As Bunge [40] so poignantly stated in 1967: "it is not usually acknowledged that electrodynamics, both classical and quantal, are in a sad state."

Since that statement, not very much has been done to alleviate the problem, particularly in the electrodynamics model utilized to design and build electrical power systems. Heartbreakingly, the community itself seems predominately bent on defending nonsequiturs and the status quo, rather than correcting a remarkable but aged electrodynamics discipline that is seriously flawed and in great need of revision from the foundations up.

IX. REQUIREMENTS FOR MAXWELLIAN EM POWER SYSTEMS EXHIBITING COP > 1.0

Along with some suggestions, the characteristics for *permissible* electrical power systems that exhibit COP > 1.0 are listed and discussed in the following paragraphs.

A. Open Thermodynamic System

Particularly during its excitation discharge, the system must be an open thermodynamic system far from equilibrium in its energetic exchange with the active vacuum. In that case classical equilibrium thermodynamics does not apply, and such a system is permitted to

1. Self-order
2. Self-oscillate or self-rotate

3. Output more energy than the operator inputs (the excess energy is received from the vacuum
4. Power itself and its loads simultaneously (all the input energy is received from the vacuum)
5. Exhibit negentropy

B. Circuit Energy Loads and Losses

The external circuit's loads and losses must not be *completely* coupled into the same closed unitary current loop with the source dipole in the generator. One suggestion is to develop and use proven *energy shuttling* in circuits. This discovery by Tesla [41][23] can only be seen (and designed) by electrodynamics theory embedded in an algebra of higher topology than tensors. (Barrett later improved Tesla's mechanism for use in communication systems and obtained patents [42].)

C. Additional Energy Collection

The system must iteratively collect additional energy from the available but normally wasted enormous Heaviside energy flow component.

1. A primary way to do this is to iteratively *retroreflect* the nondiverted Heaviside energy flow component after each pass, reflecting it back and forth across the surface charges in the circuit's conductors, collecting additional EM energy in the circuit on each repass.
2. A second avenue is to intensively reinvestigate and develop Kron's [43] discovery of the "open path" for EM networks as a dual of the conventional closed path.
3. A third suggestion is to further investigate and develop (in higher-topology algebra) Tesla's energy shuttling in EM circuits as shown and improved by Barrett [41,42].
4. A fourth suggestion is to utilize intensely scattering optically active media (ISOAM) and develop self-excitation processes in the medium. With output in the infrared region, such a process could use the excess heat to provide the heater portion of conventional power plants, allowing relatively straightforward phasein of clean vacuum energy powering of most present major power systems. Previous experiments with such ISOAM have utilized *external* excitation of the medium and thus have COP < 1.0. However, *self*-excitation looms in the mechanisms being

[23] Several of Tesla's patented circuits exhibit this effect, as analyzed and rigorously shown by Barrett [41]. However, this can only be seen when the circuits are examined in a higher-topology electrodynamics. Barrett's analysis is in quaternionic electrodynamics. Tensor analysis will not show it.

uncovered in the most recent experiments [44], which have shown positive-feedback loops, trapping of light flow energy in large random walks of over 1000 individual interactions, weak Anderson-type localization, and constructive interference of forward-time and reversed-time light paths. These experiments point toward a potential *"vacuum-energy-powered heater."* With additional research, such a heater can become self-powering by the presence of sufficient positive feedback (which will allow excess collection from the Heaviside energy flow component). We have pointed out [16] that this ISOAM process—with the self-excitation occurring spontaneously as a "kick-in" process in an exploding gas—probably accounts for the phenomena observed in the gamma-ray burster. Reignition, afterglow, and similar effects are observed in both the gamma-ray burster and also in the latest ISOAM experiments. Similar phenomena occur in X-ray bursters as well, and perhaps even in the confirmed gamma ray emissions from intense storm clouds.

5. A fifth suggestion is to reopen the intensive investigation of true negative resistors such as those by Kron [43] and Chung [45]—and the potential and the dipole as negative 4-resistors as given in this chapter—adding the consideration of vacuum energy interaction into the electrodynamics utilized for the investigation. (Chung and colleagues found that the carbon fiber composite can be produced as either a negative or a positive resistance by controlling the production process [45]. Chung's experiment has been replicated by Naudin (htt://jnaudin.free.fr/cnr/enrevp1.html/. Naudin has also derived a much simpler version easily replicated.) Indeed, the original point-contact transistor often behaved in true negative resistor fashion, but was never understood. As Burford and Verner [46] state:

> The theory underlying their function is imperfectly understood even after almost a century... although the very nature of these units limits them to small power capabilities, the concept of small-signal behavior, in the sense of the term when applied to junction devices, is meaningless, since there is no region of operation wherein equilibrium or theoretical performance is observed. Point-contact devices may therefore be described as sharply nonlinear under all operating conditions.

(This quote is from p. 281 of Ref. 46. We comment that point-contact transistors can be easily developed into true negative resistors enabling COP $>$ 1.0 circuits.) The point-contact transistor was simply bypassed by advancing to other transistor types more easily manufactured and with less manufacturing variances.

6. As a sixth suggestion, we point out that all semiconductor materials are also optically active materials, and that a point discharge into such

materials represents a very sharp regauging discharge due to the increase in potential at the tip. This means that the junction includes asymmetric self-regauging, iterative time-reversal retroreflection, increased Poynting and Heaviside energy flow components, and optical scattering processes inside the junction materials. The Fogal semiconductor [47] is expected to be in production shortly. This semiconductor exhibits the desired characteristics for proper negative-resistor work. Fogal has rigorously demonstrated to several large communications companies—now funding him—and is filing patents upon the use of his unique semiconductor to "infold" and "outfold" EM signals of extremely wide bandwidths into and out of a DC potential via a proprietary process. This is believed to be a direct application of the Sachs [27] unified general relativity and electrodynamics approach, along lines indicated by Evans [48], whereby the internal structure of a DC potential may be very much richer than is given by Whittaker [8] decomposition. Ziolkowski [49] has previously added the product waveset in addition to Whittaker's sum set. Since wave products are modulations, this has direct signal infolding implications. Mathematically, specialized semiconductors and their circuits should be able to perform Whittaker–Ziolkowski infolding of EM signals inside a DC potential, as the Fogal chip and circuits have now experimentally demonstrated.

7. As a seventh suggestion, intense sudden discharges in ionized gases are especially of interest because of the presence of optical frequency components and the involvement of iterative optical retroreflection and other processes. These processes seem to be involved in several investigations and inventions [50–52].[24]
8. As an eighth possibility, the present author [53,54][25] has advanced an engineerable mechanism—still largely proprietary—for altering the rate of flow of a mass particle (or a set of them, constituting a mass) through time, including time-reversing the particle back to a previous state. The mechanism provides for exciting and discharging a charge with a *time–charge* excitation that is pumping in the time domain—imaginary plane—where the absolute value of the time–charge (time–energy) is ordinary spatial energy compressed by the factor c^2. Hence absolute

[24] For instance, the anomalous quenching of the Hall effect generates a negative resistance effect. The Hall voltage across a narrow current-carrying channel in the presence of a perpendicular magnetic field **B** behaves anomalously around $\mathbf{B} = 0$. The Hall resistance fluctuates about zero and is "quenched," and then rises to a plateau at higher fields and eventually recovers and exhibits normal behavior beyond that region. For further details, see patents by Correa and Correa [50], Mills [51], and Shoulders [52].

[25] Although Ref. 53 is proprietary, some details have been given; see Ref. 54.

value of time–charge (time–energy excitation) has equal energy density to mass. Because it is in the time domain, however, this highly compressed EM energy exists in the complex plane. In a small time-reversal zone (TRZ) created by the time pumping process, like electrical charges attract and unlike electrical charges repel. This phenomenon thus allows like charges to attract in even numbers, without violation of the Pauli exclusion principle. We believe this process or a similar one[26] may be involved in the intense clusters of like charges demonstrated by Shoulders [52,55] (see also Ref. 56)[27] and in cold fusion reactions. The law of attraction and repulsion of charges is reversed in a TRZ, so that even numbers of fermions there may act as quasibosons and thus be time-reversed. The TRZ then decays away, providing new and different *excitation decay* reactions of the quasibosons by quark flipping; these decay reactions do not exist in normal forward-time particle physics. An entirely new class of "inside–outside" nuclear interactions is available at low spatial energy (but high time–energy) that are not achievable by present "outside–inside" collision physics. As the TRZ decays, energetic decay changes are initiated which *start* from every point in spacetime inside the TRZ—including inside nucleons located in the zone—and move outward, interacting first with the nearly time-reversed quarks and gluons so that quark flipping—and change of proton to neutron and vice versa—become favored reactions and not formidable.[28] In the highly localized TRZ the quarks are nearly unglued by the time reversal anyway, so that alteration of quarks is not formidable. We have proposed novel new *time–energy* reactions [53,54] that are consistent with most of the observed low *spatial energy* transmutations of the electrolyte experiments. The mechanisms involved in these reactions are also consistent with the anomalous phenomena experienced in the instruments occurring for several years in electrolyte experiments at China Lake [57]. In addition to a vast new set of highly localized nuclear reactions of extremely high *time*–energy but extremely low

[26] For example, one might extend our notion of the isolated fermion (such as an electron) as a composite dipole. Each dipole would thus be acted on in a TRZ, such that like virtual charges would now cluster around the observable fermion while the former unlike virtual charges would be repelled. Since the energy density of time is so great, sufficient energy is readily available for lifting whole fermions of like sign out of the Dirac Sea. If valid, this approach would indeed yield intense clusters of like observable charges, perhaps explainable in no other way. One thing is certain—Shoulders has experimentally shown that the clusters do indeed exist, in discharge situations strongly suggestive of intense phase conjugate actions and thus TRZ formations. The actual mechanism responsible for these demonstrated clusters is as yet purely speculative.

[27] See patents and paper by Shoulders. A theory has been proposed by Jin and Fox [56].

[28] We note that Sachs' epochal unification of general relativity and electrodynamics [27] does cover the quarks and gluons causally, as well as fermions and bosons. We point out that curvature of spacetime involves both positive and negative curvatures—with time involved as well as space. Certainly the theory is compatible with the consideration of time as a special form of EM energy.

spatial energy, the TRZ mechanism would seem to allow the production of true negative resistors—for instance, to be used as an external circuit bypass shunt around the source dipole in the generator, transformer, or battery. If so, once the process is developed and shown to be valid, EM circuits exhibiting COP > 1.0 will hopefully become a standard development, as will direct engineering of the atomic nucleus and nucleons in that nucleus.

9. As a ninth mechanism, application by Kawai [58] of adroit self-switching of the magnetic path in magnetic motors results in approximately doubling the COP. Modification of an ordinary magnetic engine of COP < 0.5 will not produce COP > 1.0. However, modification of available high efficiency COP $= 0.6$–0.8 magnetic engines to use the Kawai process does result in engines exhibiting COP $= 1.2$–1.6. Two Kawai-modified Hitachi engines were rigorously tested by Hitachi engineers and produced COP $= 1.4$ and COP $= 1.6$, respectively. The Kawai process and several other Japanese overunity systems have been blocked by the Yakuza from further development and marketing.

10. As a 10th suggestion, the magnetic Wankel engine (for details, see an article by this author on the master principle of EM overunity and the Japanese overunity engines [59]) should also be capable of COP > 1.0 and closed-loop self-powering, but apparently it has also been suppressed, as have all present Japanese COP > 1.0 EM systems. The Wankel engine simply wraps a linear magnetic motor around most of a circular path, with only a few degrees open between the ends. The back mmf upon a rotating rotor magnet is thus confined to that few degrees of the rotation. A small external coil with a continuous small trickle current has its current sharply interrupted just as the rotor enters the back mmf region. The resulting sharp Lenz law effect temporarily overrides the back mmf magnetic field, reversing the net magnetic field in the region and converting it to a forward mmf region. This boosts the rotor through the critical region, so that it continues to drive forward, and requires only the minute expenditures in the coil and the switching costs. This produces a magnetic rotary motor with no net back mmf, with less energy input by the operator in comparison to the work produced by the engine. The dipolarity's extraction of energy from the vacuum is what actually powers the engine anyway, as we have discussed.

11. As an 11th suggestion, multivalued magnetic potentials arise naturally in magnetics theory [60,61] as well as in other potential theory. This is particularly true during phase transitions, where multivalued potentials seem to be the rule rather than the exception [62]. Theoreticians do all in their power to minimize or eliminate their consideration [61]. However, if deliberately used and optimized, the multivalued magnetic potential

can provide a nonconservative field, where $\int \mathbf{F} \cdot \mathbf{ds} \neq 0$ around a rotary permanent magnet loop. In theory, this can enable a "self-powering" permanent magnet rotary engine [63]. Nonlinear effects (e.g., the magnetic Wankel external use of Lenz law and Johnson's internal use of deliberately initiated and controlled exchange forces) may be evoked to provide the multivalued potentials and net nonconservative fields.

12. As a 12th suggestion, certain passive nonlinear circuit components such as ferroelectric capacitors [64,65][29] have multiple nonlinear current processes ongoing inside. In principle it is possible to utilize such components only during the time they pass the current against the applied voltage. By adroit switching, in theory one can intermittently connect and utilize such passive components as true negative resistors.
13. As a 13th suggestion, feedback systems with a multipower open loop chain can produce COP > 1.0 performance [66]. Indeed, a frequency converter using 64 transistor stages and similar sophisticated feedforward and feedback mechanisms was placed in the original Minuteman missile,[30] then deliberately modified to stop its demonstrated COP > 1.0 performance. Very quietly, Westinghouse engineers then obtained several patents [67] surrounding the technology, but no further mention of it appears in the literature. The particular germanium transistor involved, was later removed from production.
14. As a 14th approach, Johnson [68] has built many novel linear and rotary motors and at least one self-powering magnetic rotary device—later stolen in a mysterious break-in at his laboratory—personally tested by the present author. Johnson uses a bidirectional "two particle" theory of magnetic flux lines that can be justified by Whittaker's earlier work showing the internal bidirectional energy flows in all potentials and fields. He also utilizes controlled spin waves and self-initiated precise exchange forces, which are known to momentarily produce bursts of very strong forcefields [69]. His approach is to use highly nonlinear assemblies of magnets that initiate the foregoing phenomena at very precise points in the rotation cycle. In short, he seeks to produce precisely located and directed sudden additional magnetic forces, using self-initiated nonlinear magnetic phenomena. This is analogous to what the Wankel engine did using the Lenz law effect by sharply interrupting a weak current in a external coil. We point out that the Lenz law effect and other very abrupt field changes momentarily produce not only an

[29] See Diestelhorst et al. [64]. In particular, multivalued conjugate reflectivities may become involved in some ferroelectric capacitors; see also Itoh et al. [65].

[30] This information was obtained in private conversations with engineers directly involved with the project and involved with the frequency converter both before and after its modification. The converter exhibited COP $= 1.05$–1.15, prior to modification to prevent it.

amplified *Poynting* energy flow component, but also an amplified *Heaviside* energy flow component as well.

15. As a 15th approach, we previously proposed a patent-pending mechanism whereby a degenerate semiconductor alloy (say, of a bit of iron in aluminum wire) is utilized for the conductors of the external circuit. By obtaining an electron relaxation time of, say, a millisecond, one can excite the circuit with potential alone, then switch away the excitation source prior to decay of the excitation potential (i.e., prior to the flow of any appreciable current). In this way, almost pure asymmetric regauging is used to excite the circuit, without requiring work (except for switching, which can be made very efficient). The excited circuit then discharges in Lorentz symmetrical fashion, but all the work in the load is "free." If LE is load energy and SE is switch energy utilized, this approach yields $\mathrm{COP} = \mathrm{LE} \div \mathrm{SE}$ and $\mathrm{COP} > 1.0$ is possible.
16. As a 16th approach, at Magnetics Energy Ltd. we are presently working on a patent-pending process whereby a permanent magnet is given a "memory" at will. By adroitly manipulating the memory, most of the magnetic flux from the magnet can be made to prefer and take a desired magnetic path among several available. Then the memory (and preference) is adroitly switched. Once one controls what path the flux "prefers" and when it prefers it, obviously $\mathrm{COP} > 1.0$ is possible. This in fact is a special kind of "Maxwell Demon" [99]. A Maxwell Demon is indeed possible, if switching (actually, directed asymmetrical regauging) can be accomplished at a level deeper than the energy process utilized. The reader will recall that gauge freedom guarantees the ability to change the potential energy of the system at will. If one does that deterministically, then also regauges so as to asymmetrically discharge the excess potential energy deterministically in the load, a priori one has built a permissible Maxwell Demon [70].[31] Our work has culminated in the motionless electromagnetic generator (MEG) [99] which extracts EM energy from the vacuum [see Chap. 12].

[31] For typical objections see Ref. 70a; For a list of numerous references on the 120-year debate, see Ref. 70b. The most extant arguments against Maxwell's Demon invoke ad hoc assumptions requiring either classical thermodynamics, information "costs," or both. There is nothing in gauge freedom that requires expenditure of energy as work, and there is nothing prohibiting it from being invoked deterministically at will and without cost. Simply moving energy from one place to another is not work. If we can do this sufficiently accurately and cheaply in the real world, then we can permissibly build a *Maxwell's Demon* in the real world. The regauging process (change of potential) is actually a change in the local vacuum potential, so that in the regauged systems a priori one creates an open system not in equilibrium with its environment—the active vacuum. Such open disequilibrium systems [71,72] are permitted to exhibit $\mathrm{COP} > 1.0$. Perpetual-motion critics already recognize that one, but call it "fictitious perpetual motion." This "fictitious perpetual motion" energy production from a simple dipole, has been going for some 15 billion years in those extant dipoles formed in the original creation of the universe. That is quite a persistent Maxwell's Demon indeed!

17. As a 17th approach, Bedini [9b] has perfected a remarkable process for dephasing—and freely overpotentializing—the ion currents inside a storage battery between its plates, from the electron currents circulating between the outside of the plates and the external circuit including the load. He takes advantage of the fact that the loop current is nonunitary, and that the internally confined ion currents have an m/q ratio several hundred times greater than that of the external electron currents. Consequently a significant hysteresis (relatively speaking) between ion current response and electron current response can be obtained by adroit sharp switching. He also creates an *overpotential* (a true negative resistor) on the interface of the plates between the two currents. This excess dipolarity thus extracts reactive power from the active vacuum (as we explained previously for a dipolarity), and blasts it out in both directions, onto the ion currents in charge mode internal to the battery and also onto the electrons in the external circuit in load powering mode. Hence the battery is simultaneously blast-charged with excess energy (the ions collect greater than normal energy) while the load is powered with excess energy (the Drude electrons are also overpotentialized and thus excited with excess energy). Bedini has produced working models (several personally tested, e.g., by this author) and is moving toward the market with his patent-pending process.

D. Energy Dissipation

The system must dissipate its excess collected energy (its asymmetrically regauged excitation energy) in the load (and in the external circuit losses) *without* dissipating the source dipole—or at least so that the discharge is asymmetric and dissipates the source dipole much slower than it powers the load. For a two-wire circuit, one method might be to utilize a *true* negative-resistor shunt[32] in parallel with the primary source dipole but in its external circuit. This splits the return current in the secondary into two parts: (1) one part passes back through the secondary, causing back-field coupling into the primary for that component; and (2) one part which does not pass back through the secondary, so that this component does not cause back-field coupling into the primary. In that way, the net back-field coupling into the primary is reduced, and the primary does not have to dissipate as much energy as does the secondary. It of course has to pass the energy flow to the secondary, but that need not be dissipated in the primary. With that arrangement, the primary will furnish all energy dissipated in the secondary circuit, but will not dissipate in itself as much

[32] Negative-resistor candidates for such a shunt may arise from point-contact transistors, from the Fogal transistor, and from the work of Wang and Chung [45].

energy as is dissipated in the secondary circuit. Hence a transformer-coupled system using such a nonlinear transformer permissibly exhibits COP > 1.0. We point out that the negative resistor represents excess energy input into the secondary circuit from the surrounding organized Heaviside dark energy flow component.

E. Self-Powering

For self-powering of Maxwellian COP > 1.0 systems once developed, clamped positive energy flow feedback [98] from output side to input side and excess collection from the Heaviside component can be used to power a motor turning the generator shaft, with the remainder of the output dissipated in a load. We stress that no laws of physics, electrodynamics, or thermodynamics are violated. Nor are the Maxwell–Heaviside equations violated, *before their arbitrary Lorentz regauging*. The conservation of energy law is obeyed at all times. Such an open dissipative Maxwellian system—which is what is being described—rigorously is permitted to self-power itself in that fashion, as shown by Prigogine [71][33] and others [72] in the study of nonlinear systems far from thermodynamic equilibrium. But following Lorentz, electrodynamicists have arbitrarily discarded all such permissible Maxwellian systems *merely because it greatly simplifies the mathematics*!

X. PROOF OF THE AVAILABLE, NEGLECTED HEAVISIDE ENERGY FLOW COMPONENT

A. Bohren's Experiment

To prove the ubiquitous existence of the Heaviside energy flow component, and to demonstrate that it can easily be tapped, one can refer to Bohren's [13a] demonstration that a resonant particle collects and emits up to 18 times as much energy as is input to it by conventional accounting (i.e., in the Poynting component of the true energy input). Resonant particle absorption and emission is a COP > 1.0 process already proven and standard in the literature for decades; For example see the pioneering work by Letokhov [14]. The effect reported by Bohren was confirmed and verified, for instance, by Paul and Fischer (see Ref. 13(b) and footnote 5, above). Bohren, Paul, Fischer, and other electrodynamicists are unaware that their energy input actually included the huge *unaccounted for* Heaviside energy flow component as well as the *accounted* Poynting flow defined by reaction with a *static* unit point charge.

[33] In 1977, Russian-born Belgian chemist Ilya Prigogine received the Nobel Prize for chemistry for contributions to nonequilibrium thermodynamics, especially the theory of dissipative structures.

B. Explanation for Bohren's COP = 18

The reason for the COP > 1.0 in this process is that the *resonant* particle sweeps out a greater *geometrical reaction cross section* in the total energy flow than is included in Poynting's theory for a standard *static* particle's interception. In short, it proves that the neglected Heaviside component is present and can be readily intercepted to obtain real expendable energy. We did a back-of-the-envelope calculation for the relative magnitude in a simple DC circuit of the Heaviside component compared to the Poynting component. The neglected Heaviside component for a nominal simple circuit was on the order of 10^{13} times as great in magnitude as the feeble Poynting component. A more exact calculation and a functional theoretical model would be welcomed, but we could not locate such in the literature (see Section III.A.1).

C. The Heaviside Energy Flow Component Was Arbitrarily Discarded

Practical EM power systems exhibiting COP > 1.0 are included in the Maxwell–Heaviside equations prior to Lorentz symmetric regauging (see footnote 6, above), which changed the equations to a small subset of the Maxwell–Heaviside theory. Specifically, the Lorentz procedure arbitrarily discards that entire class of Maxwellian systems that are not in equilibrium with their active vacuum environment. It is precisely that discarded class of Maxwellian systems that contains all Maxwellian EM power systems exhibiting COP > 1.0, by functioning as open dissipative systems freely receiving and using excess energy from the active vacuum.

XI. PROPOSED SOLUTION FOR THE "DARK MATTER" GRAVITATIONAL ENERGY

A. Background of the Problem

As is well known, the observable or accountable matter in distant spiral galaxies is insufficient for generation of sufficient gravity in the spiral arms to keep the matter in the arms from flying away [73,74]. It has been conjectured for some time that various types of exotic new "matter" never observed (and therefore "dark"), must be responsible for the excess gravity.

In general relativity, it is not mass per se, but mass–energy that is assumed to be responsible for gravitation, by the energy acting upon spacetime to curve it. In short, if appreciable "dark energy" can be discovered that has been previously unaccounted for, that could well explain the extra gravity.

So, one is rigorously seeking excess, unaccounted-for "dark energy" in some form—which may or may not be in the form of "exotic new matter."

B. A Proposed "Dark Energy" Candidate

We have found what may be the perfect candidate.

Lorentz arbitrarily discarded the vast Heaviside energy flow component accompanying every EM *field or potential* reacting with charge. (Here we are specifically considering the "charge" as a set of composite dipoles, since there really is no such thing as an "isolated charge," anyway.) The previous calculations of the fields, potentials, and energy radiations for all such reactions in the universe—including in those distant spiral galaxies—have grossly underestimated the actual EM energy involved, using only the reaction cross section of the field or potential to a unit point static charge rather than the field or potential itself.

It follows that throughout the observed universe a myriad of negative 4-resistor interactions are pouring forth very large amounts of unaccounted Heaviside EM field energy flow, across the universe in all directions. Consequently, at any location in space, there exists a vast flux of these Heaviside "dark radiation" energy flow components. Indeed, in our view the nonlinear wave and field interactions of these unaccounted-for dark energy flows may be taken as what is "driving" the EM vacuum fluctuation of "zero-point" energy, essentially what is included in Puthoff's cosmological feedback principle [75].

C. Some Observations of Interest

Three facts [73] are of interest: (1) the local gravitational potential from the distribution of stars perpendicular to the galactic plane seems greater than can be provided by the masses of known types of stars; (2) because of the decrease in luminosity to mass (or energy) in the outward direction from the center of galaxies, there must be some form of missing "dark" (non-Poynting radiant) matter (or alternatively, unaccounted and therefore "dark" energy flow) in the outer galactic regions that contributes to the gravity; and (3) in clusters of galaxies it is known that there must be more mass (or dark energy) present than is contained in the visible (by Poynting detection) parts of galaxies.

D. The Dark Energy Is Present at Every EM Field Interaction with Matter

We point out that the Heaviside component of radiation does, in fact, represent a "dark" and massive form of radiated EM energy that is physically always present, is missed by standard detectors, is arbitrarily excluded from the EM theory, and has been completely unaccounted for in astrophysics as well as elsewhere. Certainly the EM dark-energy radiation is gravitational, so one may hypothesize it as a candidate or major contributor to resolving the dark-matter problem. In short, the dark-matter problem may arise not because of missing *matter*, but because of *unaccounted for, undetected, and theoretically discarded dark EM radiation of Heaviside form.*

As with any other hypothesis, of course, this one requires falsification or validation by future experimental and theoretical investigations. We hope to see such definitive experiments in the future.

XII. THE "SCALAR" POTENTIAL IS A MULTIVECTORIAL, MULTIWAVE ENTITY

There is, of course, a scalar potential established between the two end charges of a source dipole. Let us examine what kind of energy flows actually comprise a "scalar" potential, and whether it is a scalar entity or actually a set of multiwave multivector EM energy flows.

When a "scalar" potential is set upon a transmission line, it speeds down the line at nearly light speed, revealing its vector nature. When it is set onto the middle of the transmission line, it speeds off in both directions simultaneously, revealing its *bidirectional* vector nature. In addition to this observation, there is rigorous mathematical proof as well.

In 1903 Whittaker [8] (see also Ref. 76)[34] showed that the scalar potential *identically is* a harmonic set of longitudinal EM bidirectional wavepairs, where each wavepair is comprised of a coupled longitudinal EM wave and its phase conjugate replica. Hence the potential is a bidirectional, multiwave, multivectorial entity and an equilibrium condition in a myriad bidirectional flows of longitudinal EM wave energy. There is thus a vast, bidirectional, longitudinal electromagnetic wave "infolded electrodynamics" inside every potential and comprising it.

In 1904 Whittaker [28] showed that any EM field or wave consists of two scalar potential functions, initiating what is known as *superpotential* theory [77]. By Whittaker's [8] 1903 paper, each of the scalar potential functions is derived from internally structured scalar potentials. Hence all EM fields, potentials, and waves may be expressed in terms of sets of more primary "interior" or "infolded" longitudinal EM waves and their impressed dynamics.[35] This is indeed a far more fundamental electrodynamics than is presently utilized, and one that provides for a vast set of new phenomenology presently unknown to conventional theorists.

[34] In addition to Whittaker's sum set of waves comprising the "scalar" potential, Ziolkowski [49] added the product set. See also Refs. 76a and 76b.

[35] One might appropriate the Russian term "information content of the field" for this more fundamental interior EM, from which all other EM is made. The "infolded" electrodynamics is largely ignored in the Western scientific community, which heretofore has erroneously equated "information content of the field" as mere spectral analysis. In so doing, it has dismissed an engineerable unified field theory of great power. The Sachs combined GR/EM theory allows this information content of the field to be rigorously dealt with. Evans' melding of O(3) electrodynamics into a special subset of the Sachs unified field theory allows direct engineering to be developed.

XIII. DEEPER NEGENTROPY OF THE "ISOLATED CHARGE" IN SPACE

A. A Charge Is a Set of Composite Dipoles

From quantum electrodynamics and particle physics, it is known that "empty space" is filled with intense virtual particle activity. An "isolated charge in space" must interact with the fleeting virtual charges that appear and disappear in accordance with the uncertainty principle of quantum mechanics. Consequently, virtual charges of opposite sign will be drawn toward the observable charge, before they disappear. The result is a formation of denser virtual charges of opposite sign, surrounding the observable charge, and a polarization of the local vacuum.

We may take a tiny "piece" of the observable charge, coupled with a nearby virtual charge of opposite sign during its existence, and consider the pair to be a *dipole* in a special "composite" (coherent virtual and observable) sense. So the "unit point charge" often used in electrodynamics to interact with the fields and potentials—and erroneously "define" them as their own reaction cross sections—is not really a point charge at all but is a set of *composite dipoles*. Further, it occupies the "neighborhood of a point" rather than a point.

B. Decomposing the Dipole's Potential

Each little composite dipole also has a "scalar potential" between its ends. We may decompose that potential into a harmonic set of bidirectional EM longitudinal wave (LW) pairs [8], where each pair consists of an outgoing LW and an incoming LW. Now, however, the incoming (convergent) LWs are virtual; i.e., comprised of organization and dynamics in the virtual flux of the vacuum [9(a)].

We may repeat this analysis for each of the composite dipoles constituting the so-called "isolated observable charge."

So any "isolated charge" in fact organizes and dynamicizes a fraction of the entire vacuum potential of the universe. The simple charge imposes a fraction of negentropy and organization on the vacuum, spreading at light speed across the universe. A vast set of "energy circulations" in the form of LWs and virtual LWs is established by charge–vacuum interaction, where a set of convergent virtual LWs feeds virtual energy continuously into the "charge," and the charge organizes some of its received energy into observable LW energy radiated out to the ends of the universe.

The charge, scalar potential, and dipole are all true negative 4-resistors of extraordinary magnitude. They order the virtual state energy flux of the vacuum, and bridge the gap between virtual and observable state, extending into the entire macroscopic universe level.

C. Deepening the Structure and Dynamics

Each of the virtual particles (virtual charges) contained in the composite end of the dipole, for instance, will also be accompanied by an organization of much finer, localized virtual particles of opposite sign. Hence another set of even finer composite dipoles is formed, each of which can again be decomposed into finer harmonic composite bidirectional LW wavesets. Thus there is "structuring within structuring" to as deep a level as we care to examine. The *organization of the vacuum potential* continues at ever finer levels without limit.

So even a single electron organizes a fraction of the vacuum energy of the universe, to a very surprising depth and degree. The vast, ever-changing interactions of the vacuum organization and dynamics, with particle dynamics, simply stretches one's imagination. But it is real, and the total energy content affected by each "reorganization" is enormous. This is an indication of the vast extent and dynamics of the "self-ordering" that the entire energetic vacuum performs, in response to the slightest stimulation by a charge. It also illustrates that the vacuum is a special kind of scalar potential, with internal Whittaker [8] structuring and dynamics [28]. [An even more primary vacuum (spacetime curvature) electrodynamics, not limited by the Lorentz regauging, is given by Mendel Sachs in the chapter on symmetry dynamics in this three-volume series (see 11th chapter in Part 1).] In essence this favors an already chaotic statistics, resolving the quantum mechanics problem of the missing chaos. It also means that spacetime itself—and each of its curvatures—possesses remarkable internal structuring.

D. Vacuum Engines as Deterministic Spacetime Curvature Sets

We argue that now we have uncovered in what manner the concepts of vacuum, spacetime, and potential are just different names for the same entity.

Once the internal structuring of a potential is formed, that internal structuring consists of patterned sets of spacetime curvatures with impressed deterministic dynamics. This is the spacetime curvature engine concept—or "engine" or "vacuum engine," for short. The engine and its dynamics will then work freely, both externally and internally at any and all levels, on any matter placed in it and exposed to it. It will work indefinitely, without any additional input of energy by the operator once the engine is made.

Virtual energy that appears and disappears in the dynamics of the engine's actions, need exhibit no inertia in this reordering, since the reordering occurs "between" the extinction of one virtual particle and the appearance of another. There is no "change of an ordering" in the classical sense, but only the "emergence of a new ordering." In short, in the *causal domain* (such as the active vacuum) prior to the invocation of the $\partial/\partial t$ observation operator, negentropy is readily and freely obtained on a massive scale.

We point out that such engines also may be considered to consist of sets of bidirectional longitudinal EM waves with impressed dynamics.

E. Causal System Robots (CSRs)

We provide an example of some of the startling implications of the Sachs–Evans approach.

In theory it is possible to form complete functioning *causal systems* of such spacetime curvature engines, to include almost any set of functions that will deterministically act on matter, energy, fields, and other entities. In other work we have referred to such a system as a *causal system robot* (CSR). In theory, such a functioning robot system in the 4-space causal domain can be designed and produced to perform most conceivable functions on matter, regardless of level or complexity. The reason is simple: any physical system already identically is a dynamic mass assembly in mutual interaction with a resident set of spacetime curvature engines and their dynamics. Any system proves that a causal system robot performing the engines functions already exists and is therefore possible.

Further, ordinary EM fields, potentials, and waves appear to be "superhighways" for such systems to travel in. The conventional EM fields, potentials, and waves consist of nothing but bundles of such longitudinal EM waves and their dynamics, anyway. So the "propagation" of a CSR "inside" ordinary electrodynamics, is simply the propagation of a set of LW wave dynamics of special kind, in the "inner LW medium" of the electrodynamic fields, waves, and potentials.

According to general relativity, for any observed function of an observable system, there must correspond such a precise functioning "engine set" of spacetime curvatures (see Section XIII.E.1). It follows that a CSR can be structured to perform the spacetime curvature analog of any physical function. This includes communications and signal processing. Eerily, if one has developed longitudinal EM wave technology, one could even communicate with such systems via LW communication, if the necessary communication and signal processing functions are built into the CSR system.

It would be very difficult to make and "debug" such a functioning CSR system the first time. However, once one is made and debugged until it is sufficiently accurate, any number of copies could be replicated with ease and very cheaply. Merely insert the CSR inside an ordinary EM signal and record the resulting "internally structured" signal on a diskette or CD-ROM. Then replicate the diskette or CD-ROM, and the replicated signal will also have an internal replica of the CSR. So clones could be made for pennies per copy, without limit.[36]

[36]At least two nations appear to have weapons programs in this area, but that is beyond the scope of this chapter. Nevertheless, a good theoretical basis for such systems can be taken from the Sachs unified field theory approach presented by Sachs in this series and other places.

1. A corresponding "Engine Set" of Spacetime Curvatures Is Needed for Any Observed Function of an Observable System

The cellular regeneration system, for example, uses this fact plus an extension of phase conjugate mirror theory to heal damaged cells and restore them back to normal. Weak longitudinal EM waves are used to "pump" the damaged cell and all its internal components at every level. The cellular nonlinearities add a coupled phase conjugate and extend the pumping to include pumping in the time domain, since the coupled phase conjugate, per Whittaker [8], is incoming from the time dimension (the complex plane). The resident "engine" for that damaged cell consists of the engine for a normal cell and a "delta" engine representing the exact damage. The resident engine serves as the input or "signal wave" analog (in standard NLO pumping). Every part of the pumped cell is highly nonlinear and acts as a pumped phase conjugate mirror to any and all time-domain frequencies. An amplified antiengine precisely specific for that cellular disease or damage or genetic change (as in AIDS) is formed in the pumped cell and every part of it. The action of the antiengine produces a time-reversed propagation of the cell and all its parts in the time domain rather than the spatial domain. The cell and its parts "dedifferentiate" (biology term) or "time-reverse" (physics term) back to a previous physical state, healing the cell. The body does this within its capabilities, and that is the long-sought mechanism for healing.

The Prioré effort in France demonstrated an amplification of this exact action in cells, in thousands of successful lab animal tests in the 1960s and early 1970s, but no one could understand the mechanism. At the time, phase conjugate optics as we know it today had not been developed—much less its extension into NLO pumping in the time dimension. Nonetheless, revolutionary cures of terminal tumors, infectious trypanosomiasis, and atherosclerosis were rigorously demonstrated by the scientists working with the Prioré method.

Suppressed immune systems were also restored if the treated animal was sufficiently mature to have possessed a developed immune system at the time of its suppression. The immune system of a very young animal with an immature immune system at the time of suppression could only be restored back to the immature state that it had at the time prior to immune suppression. This strongly exhibited the direct time-reversal effect that was occurring.

The work was suppressed when the French government changed in the early 1970s. The revolutionary results of the Priore experiments were presented to the assembled French Academy by Robert Courrier, head of the biology section of the Academy and also its *Secretaire Perpetuel* [78]. Many of the results of the lab animal tests are contained in Prioré's doctoral thesis [79]. The thesis was then rejected by the university during the intense suppression of the project after the government changed. The original thesis is in the files of the present author.

Eleven years later, after Prioré was dead, the university did finally approve a doctoral thesis on the subject [80]. Historical popular coverage of the entire affair is given by Graille [81].

U.S. scientists who examined the work and the scientific reports in the French literature also could not comprehend the mechanism. For instance, Bateman [82] reports on the Prioré device and its use in treating and curing cancer and leukemia, including terminal cases in numerous laboratory animals. Bateman is not particularly sympathetic, but realizes that somehow, something extraordinary has been uncovered. He comes very close when he states that "The possibility that some hitherto unrecognized feature of the radiation from a rotating plasma may be responsible for the Prioré effects should not be dismissed out of hand." That "unrecognized feature" is in fact the emission of longitudinal EM waves, which were impressed inside ordinary but very strong pulsed magnetic fields. The magnetic fields guaranteed the interaction of all the cells of the animal's body, and all the parts of the cells down to and including the atomic nuclei in the atoms. In this way the transported LWs pumped every part of every cell in the treated animal's body. Normal cells just got a little younger. The damaged and diseased cells were time-reversed back to an earlier, healthy state. The immune system's ability to recognize the pathogen was also restored, so that the revitalized immune system destroyed the pathogen. A cancerous cell was just time-reversed back into a normal cell. We strongly point out the implications for such a methodology for the treatment and prompt cure of AIDS even in its early stages, metastasized cancer, leukemia, and similar disorders. In principle, any detrimental condition of the body is marked by the presence in that body's resident engine of a specific engine delta for that condition. And an amplified antiengine, acting upon that body at all levels, is in theory capable of directly reversing the condition. We also suggest the possibility of rejuvenation of the aged.

F. Remarks on Vacuum Engineering

It is therefore not surprising that the "self-organization action" of a small source dipole in a generator or battery should produce such an enormous reorganization of vacuum energy and such great negentropy as is demonstrated in the Heaviside component. It should also not be surprising that, with no available theory dealing with or even touching such matters, Lorentz simply chose to resolve the "Heaviside energy flow component" problem by eliminating it altogether.

One result of the Lorentz integration of the energy flow vector around a closed surface [11] was to eliminate all that intense negentropic self-reorganization of the local vacuum that *did not* interact immediately with the circuit. In today's terminology, he effectively eliminated *vacuum energy engineering* from electrodynamics.

Decades later, the vision of vacuum engineering was glimpsed by modern physicists such as Lee.[37] But vacuum engineering by electrodynamic means, although theoretically straightforward in extended electrodynamics based on the Sachs approach, is still missing from conventional electrodynamics by arbitrary exclusion.

XIV. AIAS CONTRIBUTIONS TO A NEW ELECTRODYNAMICS

A. The Institute and Its Noted Director

The Alpha Foundation's Institute for Advanced Study (AIAS) is a novel scientific organization directed by Dr. Myron W. Evans, a noted scientist who has published nearly 600 papers in the refereed literature. Other noted scientists such as Dr. Lehnert of the Alfven Laboratory in Sweden and Dr. Vigier in the Laboratoire de Gravitation et Cosmologie Relativistes, Université Pierre et Marie Curie, Paris, France, and Dr. Mendel Sachs constitute the Fellows of the AIAS.

A major effort has been under way by AIAS theorists (and a few other scientists as well) to extend electrodynamics into a non-Abelian electrodynamics in O(3) symmetry using gauge field theory (e.g., see the AIAS group paper Evans et al. on a general non-Abelian electrodynamics theory [83]). Numerous failings of the present U(1) electrodynamics have been pointed out by the AIAS in a series of papers published in the literature and others presently in the referee process. Some 100 AIAS extended electrodynamics papers are presently carried on a controlled Department of Energy (DOE) website for reference by DOE scientists. The papers are being published in leading journals as rapidly as possible.

In a 1999 AIAS group paper [84] on inconsistencies of the U(1) theory of electrodynamics, specifically, the stress energy momentum tensor, it is shown that the Poynting vector in the received view is identically zero: reductio ad absurdum (M. W. Evans, AIAS correspondence). In the new method, based on equating ϕ with $\mathbf{A}$, the Poynting flow in vacuo is unlimited, simply because the $\mathbf{A}_\mu$ drawn from the vacuum defines the Lehnert charge current density in the vacuum. A new paper in this area of vacuum energy, treating the subject in greater depth, has been completed [85] at this writing. The results appear directly from local gauge invariance. In the new method, it is only assumed that there is an $\mathbf{A}$ present in the internal gauge space, and that $\mathbf{A}$ can be subjected in vacuo to a local gauge transform. Several other papers dealing with extraction of electrical energy from the vacuum are in preparation or published.

[37]See Lee [2], p. 380–381. Also on p. 383 Lee points out that the microstructure of the scalar vacuum field (i.e., of vacuum charge and polarization structuring) is not utilized. Lee indicates the possibility of using vacuum engineering in 'Chap. 25, "Outlook: Possibility of vacuum engineering" [2], pp. 824–828.

B. O(3) EM Is a Subset of Sachs' Generalized Unified Field Theory

The O(3) electrodynamics is now being further extended as a very important subset of Sachs' [27] generalization of unified general relativity and electrodynamics.

Thus the vacuum is indeed a very active and engineerable medium, filled with many kinds of real EM energy currents, and these energy currents may and do interact with EM circuits in such a manner that the circuits extract usable EM energy from the vacuum. As we have argued, conventional circuits receive *all* their EM energy from the vacuum interaction with the source dipole and not from the generator or battery. As is slowly being developed and published, there is a rigorous theoretical basis for extracting and using electrical energy directly from the vacuum. It is a concept whose time has come.

C. Other Scientists and Inventors Are Recognized Also

We also recognize the enormous contributions made by other advanced theorists outside the AIAS such as Barrett [42,86], Cornille [87],[38] Ziolkowski [49,89], Letokhov [14], Cole [90], and Puthoff [26,90] as well as many others. Happily, Sachs [27] is now a Fellow Emeritus of AIAS. We also specifically recognize inventors, including Mills [51,91],[39] Shoulders [52,55], Johnson [68], Kawai [58], Patterson [92], Lawandy [44b,93],[40] Mead and Nachamkin [94], Sweet [21,95][41] (now deceased), Bedini [9b], Fogal [47],[42] Chung [45], Paula

[38] Quoting Cornille [87, p. 168]: "The calculation concerning the electromagnetic conservation laws given in most textbooks, for example, in Jackson [7, p. 239] is not correct, as noted by Selak et al. [88], because it is not permissible to substitute a convective time derivative for an Eulerian time derivative even when we have a constant volume of integration."

[39] Because of outside pressure, the U.S. Patent and Trademark Office abruptly canceled a patent being awarded to Mills, stating that it was reviewing his granted patents again. Mills' corporation subsequently sued the U.S. PTO for such unparalleled discriminatory action.

[40] Lawandy's epochal experiment is described in his 1994 paper [44b].

[41] Sweet's solid-state vacuum triode used specially conditioned barium ferrite magnetics whose H field was in self-oscillation. The device produced COP $= 1.2 \times 10^6$, outputing some 500 W for an input of only 33 mW. Sweet never revealed his complete ELF self-oscillation conditioning procedure for the magnets. However, in ferromagnets, self-oscillations of (1) magnetization, (2) spin waves above spin-wave instability threshold, and (3) magnons are known at frequencies from ~ 1 kHz to ~ 1 MHz. For an entry into this technical area with detailed reference citations, see Ref. 95.

[42] Fogal has also invented and demonstrated to several major communications corporations a process and mechanism for "infolding" EM signals as longitudinal EM waves inside DC potentials, so that essentially unlimited bandwidth can be transmitted through the "interior" of a DC potential. He has also invented the process of "outfolding" the signals again, into an ordinary video signal. Presently a patent has been filed on the process by Fogal, and this revolutionary new communication process will be heading for the commercial market. By the time this chapter is published, the new Fogal communication process should be going into production. As of this writing, it also appears that, being longitudinal EM waves and thus able to propagate "inside" the intense scalar potential, the signals can be transmitted at superluminal speed, often nearly instantly.

and Alexandra Correa [50], Mandel'shtam et al. [96],[43] and many others [41,97].[44]

XV. CONCLUSION

A. Nonsequiturs in Conventional Electrodynamics

There are many foundations non sequiturs in classical electrodynamics that are sorely in need of correction; we have pointed out only a few.[45] The present energy crisis has occurred largely as a result of continuing to perpetuate these major flaws in electrodynamics theory, and continuing to build our electrical power systems in accord with the flawed theory.

B. Extracting Copious EM Energy from the Vacuum Is Easy

Most electrodynamicists hold the opinion that extracting usable electrical energy from the vacuum is extraordinarily difficult. To the contrary, it is a very simple thing to do and has always been done by our conventional power systems anyway. Just collect some charge (a composite dipole) or form a dipole, and the "scalar" potential between its end charges represents an organized, enormous, bidirectional 4-flow of EM energy, being established over the entire vacuum at the speed of light. Real EM 3-space energy flows outward in all directions from the dipole, and reactive power (i.e., EM energy in the imaginary plane) flows into the dipole continuously from the imaginary plane, as shown by reinterpreting Whittaker [8]. Since the beginning, every electrical load has been powered by energy extracted directly from the vacuum, and not by the heat energy

[43] In the 1930s Russian scientists at the University of Moscow and supporting agencies developed and tested parametric oscillator generators exhibiting COP > 1.0. The theory, results, pictures, and other material are presented in both the Russian and French literature, with many references cited in the particular translation in Ref. 96a. Apparently the work was never resurrected after World War II. Other pertinent references are listed in Ref. 96b.

[44] These include Nikola Tesla, whose patented circuits exhibit energy shuttling (regauging at will) in a fashion similar to a negative resistor, if the circuits are analyzed in an algebra of higher topology than tensor algebra (see Ref. 41 and footnote 23, above), and T. H. Moray, whose work with special multicontact transistors preceded that of the classical discoverers of the transistor. Moray exhibited a self-powering 50-kW device weighing only 55 lb, and extracting its energy directly from the vacuum, prior to World War II [97a]. He also sintered his semiconductors while under pressure in giant presses, reminiscent of the production methods found by Chung to allow production of a true negative resistor. A related medical equipment patent is described in Ref. 97b. Some details of the "Moray valve" may be seen in that patent. There does not appear to be an adequate technical discussion of the Moray device in the literature, since all such discussions have utilized ordinary U(1) electrodynamics at essentially sophomore or junior level.

[45] Indeed, classical U(1) electrodynamics is modeled as a field theory on a flat spacetime. In the more rigorous and general Sachs–Evans unified field approach, this is falsified. In that more fundamental model, EM waves and fields can propagate only through curved spacetime.

produced from all the hydrocarbons burned and nuclear fuel rods consumed, or by the energy from the hydroturbines and waterwheels turned by dams across streams, by windmill-powered generators, by solar cells, or by the chemical energy in batteries, and so on.

C. Collecting and Utilizing the Dark Energy Is the Long-Ignored Problem

The problem is in *collecting and using* the enormous energy easily extracted from the vacuum, not in simply producing the direct Heaviside EM dark-energy flows. In short, the problem is how to obtain much more Poynting (*intercepted*) energy from the easily available and enormous Heaviside (*nonintercepted*) energy. And then the problem is to not use half of the collected energy to destroy the dipole negative 4-resistor furnishing the energy from the vacuum.

One can build a "vacuum energy extractor" for less than a dollar. Simply place a charged capacitor (or electret) upon a permanent magnet, so that the **E** field of the capacitor is at right angles to the **H** field of the magnet, and the energy flow from the magnet (a function of $\mathbf{E} \times \mathbf{H}$) is maximized.[46] The system will extract energy from the vacuum and steadily output it indefinitely as a Heaviside energy flow. It does, however, sharply focus attention on the real problem of *how to collect and use some of the energy from the balanced vacuum energy 4-circulations set up by the system between the local vacuum and the distant, nonlocal vacuum.* Again, the problem is how to convert Heaviside dark-energy-flow to Poynting (intercepted) energy flow.

Once the vacuum energy transducer (generator's source dipole) is in place, it is another matter to intercept, collect, and use the "modified local vacuum circulation energy" pouring from the transducer to power loads, and to do so without destroying the source dipole created in the collecting generator. Unfortunately, our power scientists and engineers have been focusing on the wrong end of the problem for more than a century. They have focused enormous efforts on getting additional energy into the shaft of the generator, and reducing its internal losses. They have not focused on what happens once the source dipole is formed inside the generator, and the giant negentropy 4-resistor and vacuum energy extraction emerge.

D. Burning Fuel Does Not Add a Single Watt to the Power Line

This has lead to one of the greatest ironies in the history of science: All the hydrocarbons ever burned, all the steam turbines that ever turned the shaft of a

[46] For that matter, the charged capacitor, the electret, and the permanent magnet are dipoles. Individually, each extracts and outputs enormous energy flow from the local vacuum, continuously pouring out the extracted energy toward the ends of the universe and thus establishing its fields and potentials by altering the entire ambient vacuum potential of the universe.

generator, all the rivers ever dammed, all the nuclear fuel rods ever consumed, all the windmills and waterwheels, all the solar cells, and all the chemistry in all the batteries ever produced, have not directly delivered a single watt into the external circuit's load. All that incredible fuel consumption and energy extracted from the environment has only been used to continually restore the source dipole that our own closed current loop circuits are deliberately designed to destroy faster than the load is powered.

We strongly urge the rapid, high priority development of permissible COP > 1.0 EM power systems that violate the Lorentz symmetric regauging condition in their discharge of free excitation energy received from the vacuum via the source dipole. We will gladly contribute our own findings to the effort, including citing COP > 1.0 power systems (see Ref. 96 and footnote 43, above) and negative resistors[47] (see Ref. 45 and footnote 32, above) produced by known scientists and documented in the literature, but usually suppressed by scientific resistance to any dramatic change in U(1) electrodynamics and the Lorentz condition.

E. Classical Electrodynamics for Energy Systems Is in Woeful Shape

It is known in particle physics that there can be no symmetry of a mass system without the incorporation of the active vacuum interaction. Yet the vacuum interaction is still missing from the classical electrodynamics model. Symmetry implies nonobservables, and asymmetry implies observables. So every observable mass system that is asymmetric a priori, must be accompanied by nonobservables interacting with it, or else it can have no symmetry (or equilibrium). Yet, classical electrodynamics continues to assume equilibrium and symmetry in observable systems without incorporating the active vacuum.

Wherever we examine classical U(1) electrodynamics, we find nonsequiturs of first magnitude. This alone should be a compelling reason for the scientific community to assign the highest priority, ample funding, and the best theoreticians to the sorely-needed revision of electrodynamics from the foundations level up.

[47]Apparently a true negative resistor was developed by the renowned Gabriel Kron [43a], who was never permitted to reveal its construction or specifically reveal its development. For an oblique statement of Kron's negative-resistor success, we quote: "When only positive and negative real numbers exist, it is customary to replace a positive resistance by an inductance and a negative resistance by a capacitor (since none or only a few negative resistances exist on practical network analyzers)." Apparently Kron was required to insert the words "none or" in that statement. See also Kron [43b, p. 39]: "Although negative resistances are available for use with a network analyzer" Here the introductory clause states in rather certain terms that negative resistors were available for use on the network analyzer, and Kron slipped this one through the censors. It may be of interest that Kron was a mentor of Sweet, who was his protégé. Sweet worked for the same company, but not on the Network Analyzer project. However, he almost certainly knew the secret of Kron's "open path" discovery and the secret of Kron's negative resistor.

F. Vigorous Corrective Action Is Warranted and Imperative

With vigorous and refocused attention by the scientific community to development of the electrodynamics of COP > 1.0 energy systems and circuits, self-powering electrical power systems fueled by vacuum energy can be developed and deployed in a rather straightforward manner. The problem is nowhere near as complex as hot fusion or developing a large new accelerator.[48] The cost of one large hydrocarbon-burning power plant will allow the development to be done. The energy crisis can be solved forever. The present enormous pollution of the earth's environment by hydrocarbon combustion and nuclear wastes can be dramatically lowered. Global warming can be slowed and eventually even reversed.

Our children, the biosphere, and the slowly strangling species on earth will benefit enormously from that sorely needed scientific effort. We desperately need to *do it*, and we need to *do it now*.

References

1. A. O. Barut, *Found. Phys.* **24**(11), 1571 (Nov. 1994).
2. T. D. Lee, *Particle Physics and Introduction to Field Theory*, Harwood, New York, 1981, p. 184.
3. J. D. Kraus, *Electromagnetics*, 4th ed., McGraw-Hill, New York, 1992 (an illustration of the Poynting component that is being withdrawn from that spacefilling energy flow is shown in Fig. 12-59, p. 576. Until the energy interacts with a charge, it is in the complex domain. Note that the force fields **E** and **H** in Poynting's $\mathbf{E} \times \mathbf{H}$ are both defined after such interaction and not as they exist in spacetime, where all non-interacting fields are force-free).
4. J. H. Poynting, *Phil. Trans. Roy. Soc. Lond.* **175**(Part II), 343–361 (1885).
5. O. Heaviside, *The Electrician* (1885, 1886, 1887, and later); *Electrical Papers* **2**, 94, 405, 514 (1887) (groundbreaking articles on electromagnetic induction and its propagation).
6. O. Heaviside, *Phil. Trans. Roy. Soc. Lond.* **183A**, 423–480 (1893) (a classic article on energy forces, stresses, and fluxes in the EM field).
7. J. D. Jackson, *Classical Electrodynamics*, 2nd ed. Wiley, New York, 1975, p. 237.
8. E. T. Whittaker, *Math. Ann.* **57**, 333–355 (1903) (an excellent paper which decomposes the scalar potential into a harmonic set of bi-directional phase conjugate longitudinal wavepairs).
9. T. E. Bearden (a)"Giant negentropy in the common dipole," (b) "Bedini's method for forming negative resistors in batteries," *Proc. IC-2000* (St. Petersburg, Russia, 2000).

[48] Progress in this area is and has been, however, strongly suppressed by both the scientific community and powerful world financial entities. This latter group's suppression actions have included assassination of several researchers, entrapment, harassment, threats to one's family, kidnapping, "sic the crazies on him," mysterious disappearance of an inventor and his family, giving an inventor the "offer he cannot refuse," and blacklisting scientists or engineers so that they become unemployable. The present author has had his full share of experiences from these nefarious operations. The hope, however, is that the scientific community will at long last revise its view of how circuits and loads are powered, realize that all electrical power systems have always utilized electrical energy from the vacuum via the dipole's broken symmetry, change the EM circuitry models and theoretical models to include the vacuum interaction and its broken symmetry, and then get on with rapidly solving the present electrical energy crisis.

10. D. K. Sen, *Fields and/or Particles*, Academic Press, New York, 1968, p. viii.
11. H. A. Lorentz, *Vorlesungen über Theoretische Physik an der Universität Leiden*, Vol. V, *Die Maxwellsche Theorie* (1900–1902), Akademische Verlagsgesellschaft M.B.H., Leipzig, 1931, "Die Energie im elektromagnetischen Feld," pp. 179–186.
12. W. K. H. Panofsky and M. Phillips, *Classical Electricity and Magnetism*, 2nd ed., Addison-Wesley, Reading, MA, 1962, p. 181.
13. (a) C. F. Bohren, *Am. J. Phys.* **51**(4), 323–327 (1983); (b) H. Paul and R. Fischer, ibid., p. 327.
14. V. S. Letokhov, *Zh. Eksp. Teor. Fiz.* **53**, 1442 (1967); *Sov. Phys. JETP* **26**, 835–839 (1968); *Contemp. Phys.* **36**(4), 235–243 (1995).
15. T. E. Bearden, *J. New Energy* **1**(2), 60–78 (1996). [The value 10^{-13} refers to the magnitude of the entire flow integrated over all space. It does not refer to the local flow intensity, which is what is usually calculated for energy flow, and erroneously referred to as its "magnitude."]
16. M. W. Evans, P. K. Anastasovski, T. E. Bearden, et al., *Physica Scripta* **61**(5), 513–517 (2000).
17. (a) L. V. Lorenz, *Phil. Mag.* **34**, 287–301 (1867); (b) J. C. Maxwell, *Phil. Trans. Roy. Soc. Lond.* **155** (1865).
18. T. E. Bearden, "The unnecessary energy crisis: How to solve it quickly," paper presented to Association of Distinguished American Scientists (ADAS), June 24, 2000 (an ADAS position paper on the subject). Also on http://www.cheniere.org.
19. P. Nahin, *Oliver Heaviside: Sage in Solitude*, IEEE Press, New York, 1988. p. 134, n. 37.
20. H. J. Josephs, *The Heaviside Papers Found at Paignton in 1957*, Monograph 319, IEE, Jan. 1959, pp. 70–76.
21. F. Sweet and T. E. Bearden, "Utilizing scalar electromagnetics to tap vacuum energy," *Proc. 26th Intersoc. Energy Conversion Eng. Conf. (IECEC '91)*, Boston, MA, 1991, pp. 370–375.
22. M. W. Evans, P. K. Anastasovski, T. E. Bearden, et al., *Found. Phys. Lett.* **13**(2), 179–184 (2000); "Runaway solutions of the Lehnert equations: The possibility of extracting energy from the vacuum," *Optik* **111**(9), 407–409 (2000); "Vacuum energy flow and Poynting theorem from topology and gauge theory," (in press); *Found. Phys. Lett.* **13**(3), 289–296 (2000); "Electromagnetic energy from curved space-time," (in press); M. W. Evans and T. E. Bearden, "The most general form of the vector potential in electrodynamics," (in press).
23. G. J. Stoney, *Phil. Mag.* **42**, 332 (1896); **43**, 139–142, 273–280, 368–373 (1897) (articles on new theorems in wave Propagation, wave motion, etc.).
24. M. W. Evans et al. (a) *J. New Energy* **4**(3), 68–71; 72–75 (1999); (b) *J. New Energy* **4**(3), 76–78, 82–88 (1999).
25. D. S. Jones, *The Theory of Electromagnetism*, Pergamon Press, Oxford, 1964, pp. 57–58.
26. H. E. Puthoff, *Phys. Rev. A* **40**(9), 4857–4862 (1989).
27. M. Sachs, *General Relativity and Matter*, Reidel, 1982.
28. E. T. Whittaker, *Proc. Lond. Math. Soc.*, Series 2, Vol. 1, 1904, pp. 367–372 (a paper on expression of EM field due to electrons by means of two scalar potential functions; was published in 1904 and orally delivered in 1903).
29. M. W. Evans et al., *J. New Energy* **4**(3), 76–78 (1999).
30. T. E. Bearden, "EM corrections enabling a practical unified field theory with emphasis on time-charging interactions of longitudinal EM waves," paper presented at INE Symp. Univ. Utah, Aug. 14–15, 1998, published in *J. New Energy*, **3**(2/3), 12–28 (1998).
31. L. H. Ryder, *Quantum Field Theory*, 2nd ed., Cambridge Univ. Press, 1996, p. 147ff.
32. W. A. Rodrigues, Jr. and J.-Y. Lu, *Found. Phys.* **27**(3), 435–508 (1997) [paper on the existence of undistorted progressive waves (UPWs) of arbitrary speeds $0 \leq v < \infty$ in nature; a slightly

corrected version is downloadable as hep-th/9606171 on the Los Alamos National Laboratory Website (it includes corrections to the published version)].

33. W. A. Rodrigues, Jr. and J. Vaz Jr., *Adv. Appl. Clifford Algebras* **7**(S), 457–466 (1997).
34. I. Semiz, *Am. J. Phys.* **63**(2), 151 (1995). (on the black hole as the ultimate energy source).
35. C. S. Wu, E. Ambler, R. W. Hayward, D. D. Hoppes, and R. P. Hudson, *Phys. Rev.* **105**, 1413 (1957).
36. J. D. Jackson, *Am. J. Phys.* **64**(7), 855–870 (1996).
37. D. Hilbert, *Gottingen Nachrichten*, Vol. 4, 1917, p. 21.
38. A. A. Logunov and Yu. M. Loskutov, *Sov. J. Part. Nucl.* **18**(3), 179 (1987).
39. D. Linden (Editor in Chief), *Handbook of Batteries*, 2nd ed., McGraw-Hill, New York, 1995 (gives the pertinent battery chemistry); see also C. A. Vincent and B. Scrosati, *Modern Batteries: An Introduction to Electrochemical Power Sources*, 2nd ed., Wiley, New York, 1997.
40. M. Bunge, *Foundations of Physics*, Springer-Verlag, New York, 1967, p. 176.
41. T. W. Barrett, *Annales de la Fondation Louis de Broglie* **16**(1), pp. 23–41 (1991).
42. T. W. Barrett, "Active signalling [sic] systems," U.S. Patent 5,486,833 (Jan. 23, 1996); U.S. Patent 5,493,691 (Feb. 20, 1996).
43. G. Kron (a) *J. Appl. Phys.* **16**, 173 (1945); (b) *Phys. Rev.* **67**(1–2), 39 (1945).
44. (a) V. S. Letokhov, *Zh. Eksp. Teor. Fiz.* **53**, 1442 (1967); *Sov. Phys. JETP* **26**, 835–839 (1968); *Contemp. Phys.* **36**(4), 235–243 (1995) (describe early discoveries); (b) N. M. Lawandy et al., *Nature* **368**(6470), 436–438 (1994); D. S. Wiersma, M. P. van Albada, and A. Lagendijk, *Nature* **373**, 103 (1995) (describe initiation of experiments with external excitation of the medium); (c) D. S. Wiersma and Ad Lagendijk, *Phys. Rev. E* **54**(4), 4256–4265 (1996); D. S. Wiersma, M. P. van Albada, B. A. van Tiggelen, and A. Lagendijk, *Phys. Rev. Lett.* **74**(21), 4193–4196 (1995); D. S. Wiersma, M. P. Van Albada, and A. Lagendijk, *Phys. Rev. Lett.* **75**, 1739 (1995); D. S. Wiersma et al., *Nature* **390**, 671–673 (1997); F. Sheffold et al., *Nature* **398**, 206 (1999); J. Gomez Rivas et al., *Europhys. Lett.* **48**(1), 22–28 (1999); G. van Soest, M. Tomita, and A. Lagendijk, *Opt. Lett.* **24**(5), 306–308 (1999); A. Kirchner, K. Busch, and C. M. Soukoulis, *Phys. Rev. B* **57**, 277 (1998) (describe new effects); (d) D. W. and A. Lagendijk, *Phys. World*, Jan. 1997, pp. 33–37. (presents an excellent overview).
45. S. Wang and D. D. L. Chung, *Composites, Part B* **30**, 579–590 (1999).
46. W. B. Burford III and H. Grey Verner, *Semiconductor Junctions and Devices*, McGraw-Hill, New York, 1965, pp. 281–291.
47. W. J. Fogal, U.S. Patent 5,196,809 (March 23, 1993); U.S. Patent 5,430,413 (July 4, 1995); "Charged barrier semiconductor technology and wave function bipolar designs," *Proc. Int. Symp. New Energy* (Denver, CO, May 12–15, 1994), pp. 109–120.
48. M. W. Evans, "O(3) electrodynamics," a review in M. W. Evans (Ed.), Modern Nonlinear Optics, Part 2 (this volume series).
49. R. W. Ziolkowski, *J. Math. Phys.* **26**(4), 861–863 (1985).
50. P. N. Correa and A. N. Correa, U.S. Patent 5,416,391 (May 16, 1995); U.S. Patent 5,449,989 (Sept. 12, 1995); U.S. Patent 5,502,354 (March 26, 1996).
51. R. L. Mills et al., U.S. Patent 6,024,935 (Feb. 15, 2000) (with 499 claims recognized); R. L. Mills, Austral. Patent 668678 (Nov. 20, 1991).
52. K. R. Shoulders, U.S. Patent 5,018,180 (May 21, 1991); U.S. Patents 5,054,046 (1991); 5,054,047 (1991); 5,123,039 (1992); 5,148,461 (1992).
53. T. E. Bearden, *Formation and Use of Time-Reversal Zones, EM Wave Transduction, Time-Density (Scalar) EM Excitation and Decay, and Spacetime Curvature Engines to Alter Matter and Convert Time Into Energy*, Invention Disclosure Document 446522 (Oct. 26, 1998).

54. T. E. Bearden, *Explore* **8**(6), 7–16 (1998); "Toward a practical unified field theory and a deep experimental example," paper presented at INE Symp., Univ. Utah, Aug. 14–15, 1998.
55. K. Shoulders and S. Shoulders, *J. New Energy* **1**(3), 111–121 (1996).
56. S.-X. Jin and H. Fox, *J. New Energy* **1**(4), 5–20 (1996).
57. M. H. Miles and B. F. Bush, "Radiation measurements at China Lake: Real or artifacts?" *Proc. ICCF-7* (*7th Int. Conf. Cold Fusion*), Vancouver, BC, Canada, April 1998, p. 101.
58. T. Kawai, U.S. Patent 5,436,518 (July 25, 1995).
59. T. E. Bearden, *Infinite Energy* **1**(5–6), 38–55 (1995–1996).
60. J. C. Verite, *IEEE Trans. Magn.* **MAG-23**(3), 1881–1887 (1987).
61. P. R. Kotiuga, *J. Appl. Phys.* **61**(8), 3916–3918 (1987).
62. Y. Huo, *Meccanica* (Netherlands) **30**(5), 475–494 (1995).
63. T. E. Bearden, "Use of regauging and multivalued potentials to achieve overunity EM engines: Concepts and specific engine examples," *Proc. Int. Sci. Conf. "New Ideas in Natural Sciences"* (St. Petersburg, Russia, June 17–22, 1996); Part I: *Problems of Modern Physics*, 1996, pp. 277–297.
64. M. Diestelhorst, H. Beige, and R.-P. Kapsch, *Ferroelectrics* **172**, 419–423 (1995).
65. S. Itoh et al., *Ferroelectrics* **170**, 209–217 (1995).
66. R. M. DeSantis et al., *On the Analysis of Feedback Systems with a Multipower Open Loop Chain*, AD 773188, Oct. 1973, available through the U.S. National Technical Information System.
67. J. H. Andreatta, U.S. Patent 3,239,771 (March 8, 1966); T. L. Dennis, Jr., U.S. Patent 3,239,772 (March 8, 1966); H. J. Morrison, U.S. Patent 3,815,030 (June 4, 1974) (patents on high-efficiency switching amplifiers and square-wave-driven power amplifiers).
68. H. R. Johnson, U.S. Patent 4,151,431, (April 24, 1979); see also Johnson's U.S. Patents 4,877,983 (Oct. 31, 1989) and 5,402,021 (March 28, 1995).
69. B. D. Cullity, *Introduction to Magnetic Materials*, Addison-Wesley, Reading, MA, 1972; A. G. Gurevich and G. A. Melkov, *Magnetization Oscillations and Waves*, CRC Press, Boca Raton, FL, 1996; V. S. L'vov, *Wave Turbulence Under Parametric Excitation: Applications to Magnets*, Springer-Verlag, Berlin, 1994 (expositions on exchange forces and exchange energy); see also V. S. L'vov and L. A. Prozorova, in A. S. Borovik-Romanov and S. K. Sinha (Eds.), *Spin Waves and Magnetic Excitations*, North-Holland, Amsterdam, 1988.
70. (a) S. W. Angrist, *Sci. Am.* **218**, 114–122 (1968); L. Brillouin, *Am. Sci.* **37**, 554–568 (1949); M. Jammer, "Entropy," in P. Wiener (Ed.), *Dictionary of the History of Ideas*, Vol. 2, Scribner's, New York, 1973 (section on Maxwell's Demon); H. S. Leff, *Am. J. Phys.* **55**(8), 701–705 (1987); (b) H. S. Leff, *Am. J. Phys.* **58**(3), 201–209 (1990).
71. I. Prigogine, *From Being to Becoming: Time and Complexity in the Physical Sciences*, Freeman, San Francisco, 1980.
72. L. Brillouin, *Am. Sci.* **37**, 554–568 (1949); G. Nicolis and I. Prigogine, *Exploring Complexity*, Piper, Munich, 1987.
73. M. Longair, in Paul Davies (Ed.), *The New Physics*, Cambridge Univ. Press, New York, 1989 (see specifically "Dark matter in galaxies and clusters of galaxies," p. 163).
74. L. M. Krauss, *Quintessence: The Mystery of Missing Matter in the Universe*, rev. ed. Basic Books, Perseus, New York, 2000 (gives a thoroughly updated account of the entire dark-matter problem).
75. H. E. Puthoff, *Phys. Rev. A* **40**(9), 4857–4862 (1989).
76. (a) I. M. Besieris, A. M. Shaarawi, and R. W. Ziolkowski, *J. Math. Phys.* **30**(6), 1254–1269 (1989); (b) R. Donnelly and R. Ziolkowski, *Proc. Roy. Soc. Lond. A* **437**, 673–692 (1992).

77. M. Phillips, in S. Flugge (Ed.), *Principles of Electrodynamics and Relativity*, Vol. IV of *Encyclopedia of Physics*, Springer-Verlag, 1962. (an overview of much of superpotential theory).
78. R. Courrier, "Exposé par M. le Professeur R. Courrier, Secretaire Perpetuel de L'Academie des Sciences fait au cours d'une reunion a L'Institut sur les effets de la Machine de M. A. Prioré le 26 Avril 1977" (presentation by Professor R. Courrier, Perpetual secretary of the Academy of Sciences, made at the meeting of the Academy on the effects of the machine of M. A. Prioré, April 26, 1977).
79. A. Prioré, *Guérison de la Trypanosomiase Expérimentale Aiguë et Chronique par L'action Combinée de Champs Magnétiques et D'Ondes Electromagnétiques Modulés* (Healing of Intense and Chronic Experimental Trypanosomiasis by the Combined Action of Magnetic Fields and Modulated Electromagnetic Waves), thesis submitted in candidacy for the doctoral degree, Univ. Bordeaux, 1973.
80. E. Perisse, *Effets des Ondes Electromagnètiques et des Champs Magnètiques sur le Cancer et la Trypanosomiase Experimentale* (Effects of Electromagnetic Waves and Magnetic Fields on Cancer and Experimental Trypanosomias), doctoral thesis No. 83, Univ. Bordeaux, March 16, 1984.
81. J.-M. Graille, *Dossier Prioré: A New Pasteur Affair*, De Noel, Paris, 1984 (in French).
82. J. B. Bateman, *A Biologically Active Combination of Modulated Magnetic and Microwave Fields: The Prioré Machine*, Office of Naval Research, London, Report R-5-78, Aug. 16, 1978.
83. M. W. Evans et al., AIAS group paper, *Found. Phys. Lett.* **12**(3), 251–265 (1999); see also particularly M. W. Evans, "O(3) electrodynamics," in this present volume (Part 2).
84. M. W. Evans et al., AIAS group paper, *Found. Phys. Lett.* **12**(2), 187–192 (1999).
85. M. W. Evans et al., "Vacuum energy flow and Poynting theorem from topology and gauge theory (in press).
86. T. W. Barrett and D. M Grimes (Eds.), *Advanced Electromagnetism: Foundations, Theory, & Applications*, World Scientific, Singapore, 1995; see also particularly T. W. Barrett, in A. Lakhtakia (Ed.), *Essays on the Formal Aspects of Electromagnetic Theory*, World Scientific, River Edge, NJ, 1993, pp. 6–86 (on EM phenomena not explained by Maxwell's equations).
87. P. Cornille, in A. Lakhtakia, (Ed.), *Essays on the Formal Aspects of Electromagnetic Theory*, World Scientific, 1993, Chap. 4. pp. 138–182.
88. Selak, *Astrophys. Space Sci.* **158**, 159 (1989).
89. R. W. Ziolkowski, *Phys. Rev. A* **39**, 2005 (1989) see also M. K. Tippett, *Phys. Rev. A* **43**(6), 3066–3072 (1991).
90. D. C. Cole and H. E. Puthoff, *Phys. Rev. E* **48**(2), 1562–1565 (1993).
91. A. Rosenblum, *Infinite Energy* **3**(17), 21–34 (Dec. 1997–Jan. 1998); E. Mallove, *Infinite Energy* **2**(12), 21,35,41 (1997) (articles deseribing Mills' work).
92. J. Patterson, U.S. Patent 5,372,688 (Dec. 13, 1994); see also U.S. Patents 5,318,675; 5,607,563; 5,036,031; and 4,943,355.
93. N. M. Lawandy, U.S. Patent 5,434,878 (July 18, 1995).
94. F. B. Mead and J. Nachamkin, U.S. Patent 5,590,031 (Dec. 31, 1996).
95. A. G. Gurevich and G. A. Melkov, *Magnetization Oscillations and Waves*, CRC Press, Boca Raton, FL, 1996, p. 279; see also particularly V. S. L'vov, *Wave Turbulence Under Parametric Excitation: Applications to Magnets*, Springer-Verlag, Berlin, 1994, pp. 214–218, 226–234, 281–289.
96. (a) L. Mandelstam [L. I. Mendel'shtam (alternate surname transliteration)], N. Papalexi, A. Andronov, S. Chaikin, and A. Witt, "Report on recent research on nonlinear oscillations,"

Translation of "Expose des recherches recentes Sur les oscillations non lineaires," *Tech. Phys. USSR*, Leningrad, Vol. 2, 1935, pp. 81–134; NASA Translation Document TTF-12,678, Nov. 1969; (b) L. I. Mandelstam and N. D. Papaleksi., *Zh. Teknicheskoy Fiziki* **4**(1), 5–29 (1934); *Z. Phys.* **72**, p. 223 (1931); *Zh. Tekhnicheskoi Fiziki* **4**(1), TBD (1934); see also A. Andronov, *Compt. Rend.* **189**, 559 (1929); A. Andronov and A. Witt, *Compt. Rend.* **190**, 256 (1930); *Bull. De l'Acad. Ed Sc. De l"URSS* **189**, (1930); see also S. Chaikin, *Zh. Prikladnoi Fiziki* **7**, 6 (1930); A. Witt, *Zh. Teknicheskoi Fiziki* **1**(5), 428 (1931); N. Kaidanowski, *Zh. Teknicheskoi Fiziki* **3**, (1933).

97. T. H. Moray (a) *The Sea of Energy*, 5th ed., Salt Lake City, 1978 (Foreword by T. E. Bearden); (b) U.S. Patent 2,460,707 (Feb. 1, 1949) (on an electrotherapeutic apparatus).
98. Close-looping an overunity EM system has very special Dirac sea hole current phenomena involved and special techniques are required. Bedini and the present author have filed a patent application on the major process required, and the details will be released a year from that filing.
99. M. W. Evans, P. K. Anastasovski, T. E. Bearden et al., "Explanation of the Motionless Electromagnetic Generator with the Sachs Theory of Electrodynamics," *Found Phys Lett.* **14**(4), 387–393 (2001).

ENERGY FROM THE ACTIVE VACUUM: THE MOTIONLESS ELECTROMAGNETIC GENERATOR

THOMAS E. BEARDEN

CEO, CTEC, Inc., Huntsville, Alabama

CONTENTS

Modern Nonlinear Optics, Part 2, Second Edition, Advances in Chemical Physics, Volume 119,
Edited by Myron W. Evans. Series Editors I. Prigogine and Stuart A. Rice.
ISBN 0-471-38931-5

I. INTRODUCTION

This chapter explains the operational principles of a motionless electromagnetic generator (MEG)[1] experiment and invention where the generator is a system far from equilibrium in its energetic exchange with its active vacuum environment. The broken symmetry of a permanent magnet dipole is used as a transducer of vacuum energy, receiving longitudinal EM wave energy from the time domain (complex plane) and emitting it as real 3-space EM energy, by a process first shown by Whittaker [1][2] in 1903. As is well known, a system in energy exchange disequilibrium with its active environment is permitted to (1) self-order (2) self-oscillate, (3) output more energy than the operator inputs (the excess energy is received from the environment), (4) power itself and its load simultaneously (all the energy is received from the environment), and (5) exhibit negentropy. My colleagues and I achieved coefficient of performance (COP) of 5.0 in one experimental buildup and 10.0 in another. The MEG is in

[1] Covered by formal patent application by coinventors Thomas E. Bearden, James C. Hayes, James L. Kenny, Kenneth D. Moore, and Stephen L. Patrick. Intellectual property rights to the invention are assigned to Magnetic Energy Ltd., Huntsville, Alabama (USA), with Dr. James L. Kenny as Managing Partner. Several additional patent applications are in preparation.

[2] In this paper on the partial differential equations of mathematical physics, Whittaker decomposes any scalar potential into a harmonic set of bidirectional EM longitudinal EM wavepairs, where each wavepair is composed of a longitudinal EM wave and its phase conjugate replica wave. Dividing the overall waveset into two half-sets, we have one half-set consisting of incoming longitudinal EM waves in the complex plane (the time domain) and a second half-set composed of outgoing longitudinal EM waves in real 3-space. Hence the scalar potential represents a giant circulation of EM energy automatically established and maintained from the time domain (complex plane) into the source dipole establishing the potential, with the absorbed complex energy being transduced and reemitted by the dipole in all directions in 3-space as real longitudinal EM wave energy establishing the EM fields and potentials (and their energy) associated with the dipole.

patent-pending status. As of this writing, a variant of the MEG experiment has been independently replicated by Jean-Louis Naudin in France, and other independent replications are planned. Naudin's version produced a COP of 1.76. His results are posted on his Website: http://jnaudin.free.fr/html/megv2.htm. Arrangements were attempted for a team of skilled university scientists to independently test the MEG under U.S. Department of Defense auspices. Once COP > 1.0 was validated by the university team, a full proprietary disclosure was to be made to them under nondisclosure and noncompete agreement, and they would independently replicate the MEG themselves and test the buildup, again under DoD auspices the attempt failed.[3]

We especially wish to thank Jean-Louis Naudin for his prompt replication of the MEG experiment and for his kind permission to include his replication results and two of his illustrations.

II. BACKGROUND

A. Developmental History

For about 10 years the five researchers who invented the motionless magnetic generator (MEG) have been working together as a team, exploring many avenues whereby electromagnetic energy might be extracted from various sources of potential and eventually from the active vacuum itself. This has been arduous and difficult work, since there were no guidelines for such a process whereby the electrical power system becomes an open dissipative system in the manner of Prigogine's theoretical models [2–4] (see also the following paragraph) but using determinism and classical electromagnetics instead of chaos and statistics. There was also no apparent precedent in the patent database or in the scientific database.

Quoting, Prigogine, from p. 70 of his article on irreversibility as a symmetry-breaking process: "Entropy ... cannot in general be expressed in terms of observables such as temperature and density. This is only possible in the neighborhood of equilibrium.... It is only then that both entropy and entropy production acquire a macroscopic meaning." Prigogine received a Nobel Prize in 1977 for his contributions to the thermodynamics of open systems, particularly with respect to open dissipative systems. What he is pointing out here is that, where equilibrium (and hence symmetry) is broken, the usual presumption of entropy and entropic production have no macroscopic meaning. For such systems, the often encountered challenge on classical equilibrium thermodynamics grounds is a nonsequitur, and merely reveals the scientific ignorance of

[3]The team of university scientists observed the MEG in operation and were quite interested in performing full, independent replication and testing. However, the university's administration refused to sign a noncircumvention agreement. Consequently we stopped the process.

the challenger. In short, such a challenger would decry the windmill in the wind, denying that it can turn without the operator cranking it, because classical equilibrium thermodynamics forbids it. However, the windmill turns happily in the wind, without operator input at all, and in total violation of equilibrium thermodynamics because the windmill is not in equilibrium with its active environment, the active atmosphere. At the same time the windmill completely complies with the thermodynamics of open systems far from equilibrium, and energy conservation is rigorously obeyed. The windmill can "power itself and its load" since all the energy needed to power the windmill and power the load comes from the energy freely input by the wind.

It is generally realized that Maxwell's equations are purely hydrodynamic equations and fluid mechanics rigorously applies [5]. Anything a fluid system can do, a Maxwellian system is permitted to do, a priori. So "electrical energy winds" and "electrical windmills" are indeed permitted in the Maxwell–Heaviside model, prior to Lorentz' regauging of the equations to select only that subset of systems that can have no net "electrical energy wind" from the vacuum. Specifically, this arbitrary Lorentz symmetric regauging—while indeed simplifying the resulting equations and making them much easier to solve—also arbitrarily discards all Maxwellian systems not in equilibrium with their active environment (the active vacuum). In short, it chooses only those Maxwellian systems that never use any net "electrical energy wind from the vacuum." Putting it simply, it discards that entire set of Maxwellian systems that interact with energy winds in their surrounding active vacuum environment.

Since the present "standard" U(1) electrodynamics model forbids electrical power systems with COP > 1.0, my colleagues and I also studied the derivation of that model, which is recognized to contain flaws due to its >136-year-old basis. We particularly examined how it developed, how it was changed, and how we came to have the Lorentz-regauged Maxwell–Heaviside equations model ubiquitously used today, particularly with respect to the design, manufacture, and use of electrical power systems.

The Maxwell theory is well known to be a material fluid flow theory [6],[4] since the equations are hydrodynamic equations. In principle, anything that can be done with fluid theory can be done with electrodynamics, since the fundamental equations are the same mathematics and must describe consistent analogous functional behavior and phenomena [5]. This means that EM systems with "electromagnetic energy winds" from their active external "atmosphere"

[4] Quoting Sir Horace Lamb [6], p. 210: "There is an exact correspondence between the analytical relations above developed and certain formulae in Electro-magnetism .... Hence, the vortex-filaments correspond to electric circuits, the strengths of the vortices to the strengths of the currents in these circuits, sources and sinks to positive and negative poles, and finally, fluid velocity to magnetic force."

(the active vacuum) are in theory quite possible, analogous to a windmill in a wind. In such a case, symmetry is broken in the exchange of energy between the environment (the atmosphere) and the windmill. However, the present EM models used to design and build an electrical power system do not even include the vacuum interaction with the system, much less a broken symmetry in that interaction.

So the major problem was that the present classical EM model excluded such EM systems. We gradually worked out the exact reason for their *arbitrary* exclusion that resulted in the present restricted EM model, where and when it was done, and how it was done. It turned out that Ludvig Valentin Lorenz [7] symmetrically regauged Maxwell's equations in 1867, only two years after Maxwell's [8] seminal publication in 1865. So Lorenz first made arbitrary changes that limited the model to only those Maxwellian systems in equilibrium in their energy exchange with their external environment (specifically, in their exchange with the active vacuum). This arbitrary curtailment is not a law of nature and it is not the case for the Maxwell–Heaviside theory prior to Lorenz' (and later H. A. Lorentz') alteration of it. Thus, for electrical power systems capable of COP > 1.0, *removing* this symmetric regauging condition [9–14] is required—particularly during the discharge of the system's excess potential energy (i.e., during discharge of the excitation) in the load.

Later H. A. Lorentz [15],[5] apparently unaware of Lorenz' 1867 work, independently regauged the Maxwell–Heaviside equations so that they represented a system that was in equilibrium with its active environment. This indeed simplified the mathematics, thus minimizing numerical methods. However, it also discarded all "electrical windmills in a free wind"—so to speak—and left only those electrical windmills "in a large sealed room" where there was never any net free wind.

B. Implications of the Arbitrarily Curtailed Electrodynamics Model

Initially an electrical power system is *asymmetrically* regauged by simply applying potential (voltage), so that the system's potential energy is nearly instantly changed. The well-known gauge freedom principle in gauge field theory assures us that any system's potential—and hence potential energy—can be freely changed in such fashion. In principle, this excess potential energy can then be freely discharged in loads to power them, without any further input from the operator. In short, there is absolutely no theoretical law or law of nature that prohibits COP > 1.0 electrical power systems—or else we have to abandon the highly successful modern gauge field theory and deny the gauge freedom principle.

[5]Such was H. A. Lorentz's prestige that, once he advanced symmetrical regauging of the Maxwell–Heaviside equations, it was rather universally adopted by electrodynamicists, who still use it today; see, for example, Jackson [15].

Although the present electrical power systems do not exhibit COP > 1.0, all of them do accomplish the initial *asymmetric* regauging by applying potential. So all of them do freely regauge their potential energy, and the only thing that the additional energy input to the shaft of a generator (or the chemical energy available to a battery) accomplishes is the physical creation of the potentializing entity: the source dipole [16]. [*Note*: The key is to apply Whittaker's 1903 decomposition of the scalar potential existing between the poles of a dipole. Whittaker showed that any EM scalar potential dipolarity continually receives longitudinal EM wave energy from the time domain (complex plane) and outputs real EM energy in 3-space. Thus the potential's dipolarity (voltage) placed on a circuit by a generator or battery actually represents free regauging energy coming from the time domain of the 4-vacuum, and therefore having nothing to do with the 3-space energy input to the shaft of the generator or with the 3-space chemical energy available in a battery.]

It follows that something the present systems or circuits perform in their discharge of their nearly free[6] regauging energy must prevent the subsequent simple discharge of the energy to power the loads unless further work is done on the input section. In short, some ubiquitous feature in present systems must self-enforce the Lorentz symmetry condition (or a version of it) whenever the system discharges its free or nearly free excitation energy.

Lorentz' [15] (see also footnote 5, above) curtailment of the Maxwell–Heaviside equations greatly simplified the mathematics and eased the solution of the resulting equations, of course. But applied to the design of circuits, particularly during their excitation discharge, it also discarded the most interesting and useful class of Maxwellian systems, those exhibiting COP > 1.0.

Consequently, Lorentz [15] (see also footnote 5, above) unwittingly discarded all Maxwellian systems with "net usable EM energy winds" during their discharge into their loads to power them. Thus present electrical power systems—which have all been designed in accord with the Lorentz condition—cannot freely use the EM energy winds that arise in them by simple regauging, as a result of some universal feature in the design of every power system that prevents such action.

We eventually identified the ubiquitous closed current loop circuit [17][7] as the culprit that enforces a special kind of Lorentz symmetry during discharge of

[6] In real systems, for regauging we have to pay for a little switching costs, for example, but this may be minimal compared to the potential energy actually directed or gated upon the system to potentialize it.

[7] More rigorously, this is any closed current loop circuit where the charge carriers in all portions of the loop have the same m/q ratio. For example, battery-powered circuits do not meet that condition, since the internal ionic currents between the battery plates may have m/q ratios several hundred times the m/q ratio of the electrons that pass between the outsides of the two plates and through the external circuit containing the load. With Bedini's process, a battery-powered system can be made to charge its batteries at the same time that it powers its load; see Bearden [17].

the system's excitation energy. With this circuit, the excitation-discharging system must destroy the source of its EM energy winds as fast as it powers its loads and losses, and thus faster than it actually powers its loads.

Also, as we stated earlier, and contrary to conventional notions, batteries and generators do not dissipate their available internal energy (shaft energy furnished to the generator, or chemical energy in the battery) to power their external circuits and loads, but only to restore the separation of their internal charges, thereby forming the source dipole connected to their terminals. Once formed, the *source dipole's giant negentropy* [16] then powers the circuit via its broken symmetry [18,19] [see the following subsection (Section II.B.1) also].

1. Particle Physics, Including Dipole Symmetry

A discussion by Nobelist Lee of particle physics and its findings, includes broken symmetry, which includes the broken symmetry of a dipole. Quoting from Lee [18], p. 184: "... the discoveries made in 1957 established not only right-left asymmetry, but also the asymmetry of the positive and negative signs of electric charge. In the standard nomenclature, right-left asymmetry is referred to as P violation, or parity nonconservation. The asymmetry between opposite signs of electric charge is called C violation, or charge conjugation violation, or sometimes particle-antiparticle asymmetry." "Since non-observables imply symmetry, these discoveries of asymmetry must imply observables."

Simply put, Lee has pointed out the rigorous basis for asserting that the arbitrarily assumed Lorentz 3-symmetry of the Maxwellian system is broken by the source dipole—and in fact by any dipole. Such broken 3-symmetry in the dipole's energetic exchange with the active vacuum is well known in particle physics, but still is not included at all in classical electrodynamics, particularly in the models used to design and build EM power systems. The proven dipole broken 3-symmetry rigorously means that part of the dipole's received virtual energy—continuously absorbed by the dipole charges from the active vacuum—is transduced into observable 3-space energy and reemitted in real (observable) energy form. That this has been well known in particle physics for nearly a half-century, but is still missing from the classical EM model, is scientifically inexplicable and a foundations error of monumental magnitude. Once made, it is the *source dipole* that powers the circuit.

C. Some Overlooked Principles in Electrodynamics

We recovered a major fundamental principle from Whittaker's [1] profound but much-ignored work in 1903. Any scalar potential is a priori a set of EM energy flows, hence a set of "electromagnetic energy winds," so to speak. As shown by Whittaker, these EM energy winds pour in from the complex plane (the time domain) to any x,y,z point in the potential, and pour out of that point in all directions in real 3-space [1,16,20].

Further, in conventional EM theory, electrodynamicists do not actually calculate or even use the *potential itself* as the unending set of EM energy winds or flows that it actually is. Instead, they calculate and use its *reaction cross section with a unit point static charge* only *at a specific point*. How much energy is diverged around a single standard unit point static coulomb is then said to be the "magnitude of the potential" at that point. This is a nonsequitur of first magnitude.[8]

For example, the small "swirl" of water flow diverged to stream around an intercepting rock in a river bottom is not the river's own flow magnitude. It certainly is not the "magnitude of the river." Neither is the standard reaction cross section of the potential a measure of the potential's actual "magnitude," but is representative only of the local intensity of the potential's composite energy flows. Indeed, the potential's "magnitude" with respect to any local interception and extraction of energy from it is limited only by one's ability to (1) intercept the flow and (2) diverge it into a circuit to power the circuit. The energy flows identically constituting the potential [1,16,20] replenish the withdrawn energy as fast as it can be diverged in practical processes, since the vacuum energy flows themselves move at the speed of light.

D. Work–Energy Theorem in a Replenishing Potential Environment

We also came to better understand the conservation of energy law itself. Particularly, the present work–energy theorem assumes only a "single conversion" of energy into a different form, where such "conversion in form" due to a converting agent is what is considered "work" on that agent. No "replenishing of the dissipated or converted energy by a freely flowing energy river or process" is considered. Instead of the ongoing divergence from a flowing stream of energy that it is, a collection of energy is erroneously treated as if it were a "pile of bricks" called *joules*.

On the other hand, in a system operating in a replenishing potential environment, a conversion in the form of the energy may increase the system energy (e.g., the kinetic energy of an electron gas) of the converting agent, since all the field energy and potential energy input to that converting agent may be replenished (from the time domain). If so, a *free regauging* occurs (Indeed, we would hesitantly nominate this process as the fundamental process underlying the gauge freedom principle itself, but will leave to more qualified theoreticians the affirmation or refutation of that speculation.). In that case,

[8] For example, simply replace the assumed unit point static charge assumed at each point with n unit point static charges, and the collected energy around the new point charge will be n times the former collection. If the former calculation had yielded the actual magnitude of the potential at that point, its magnitude could not be increased by increasing the interception. But since the potential is actually a flow process, increasing the reaction cross section of the interception increases the energy collection accordingly.

the original energy can change (e.g., into field energy form, which is not the kinetic energy of the electron gas), and yet a joule of work can have been done on the electron gas to alter its potential energy by as much work as was done on it. Thus the work performed by this change in energy form with simultaneous replenishment of the original form, may increase the energy of the medium while retaining as much field energy and potential energy as was input, just in different form.

This is a profound change to the implicit assumptions used in applying the work–energy theorem. In short, the present work–energy theorem (without replenishment) was found to be a special case of a much more general and extended "replenished energy conversion of form with intermediate work freely performed upon the converter" process. *Conversion of the form of energy* is rigorously what we call *work*. The energy is not consumed in the work process, and the replenished "energy collection" in its original form is also maintained so that it is not "lost" in doing work on the converter. So to speak, the well-established principle of gauge freedom as an energy replenishing mechanism has been arbitrarily excluded in the conventional view of the work–energy theorem. *The conventional form of the work–energy theorem applies only when there is no simultaneous regauging/replenishment from the vacuum (time domain) involved.* With replenishment, the more general work–energy theorem yields system energy amplification by free system regauging.

This extension of the work–energy theorem to a more general case, including the invocation of gauge freedom, has profound implications in physics. With the energy replenishing environment involved, the work–energy theorem becomes an *energy amplifying* process. Electrical energy can be freely amplified at will—anywhere, anytime—by invoking the extended energy–work process, if regauging accompanies the process simultaneously. Indeed, one joule of field energy or potential energy can do joule after joule of work on intermediary converters, increasing the kinetic energy and other forces. upon the converter medium, while the system retains replenished joule for joule of the input energy in differing field or potential forms. In this extended process, always after each joule of work on the converting agent there still remains a joule of field energy or potential energy of altered form, and the original joule of energy was freely replenished as well.

E. The Extended Principles Permit COP $>$ 1.0 Electrical Power Systems

Gradually we realized that (1) electrodynamics without the arbitrary Lorentz regauging did permit asymmetrically self-regauging electrical power systems, freely receiving and converting electrical energy from their vacuum environment; (2) present systems are designed unwittingly to guarantee their reimposition of symmetry during their excitation discharge; (3) this excitation discharge symmetry is what can and must be broken by proper system design; and (4) a magnetic system "powered" by a permanent magnet dipole's ongoing active negentropy processes [16,20] could readily be adapted since the source dipole of the permanent magnet was not destroyed by the circulating magnetic flux. We experimented on various buildups and prototypes, in this vein, for some years.

F. Patenting and Discovery Activity

After several years of experimental work, in 1997 we filed a provisional patent application on the first MEG prototype of real interest. We filed another formal patent application in 2000 after several years of experimentation in which we used multiple extraction of energy from a magnetic dipole. Fundamentally, we sought to extract electromagnetic energy from the magnetic vector potential that pours from a magnetic dipole due to the giant negentropy mechanism [1,16,20]. Originally we suspected that we would deplete the magnetic dipole, and our experiments sometimes seemed to indicate that a very slow depletion might indeed be happening. These indications were eventually found to be due to normal small measurement errors within the measurement tolerance of our instruments.

Since our last formal patent application filing in 2000, additional buildups and experiments have led us to the firm conclusion that one does not deplete the magnetic dipole. Instead, if one draws the energy *directly from the potentials* (primarily from the magnetic vector potential) furnished by the magnetic dipole of the permanent magnets, one can essentially draw as much energy as desired, without affecting the dipole itself.[9] We found that any amount of energy can be *freely withdrawn* from the vector potential without diminishing the vector potential itself, as long as the withdrawn energy is changed in form in the withdrawal. A giant negentropy mechanism—reinterpreted by Bearden [16] from Whittaker's work [1] and further investigated by Evans and Bearden [20]—is associated with the magnet dipole, and in fact with any dipole, since a scalar potential exists between its poles and the scalar potential decomposes via the Whittaker decomposition [1]. This negentropy mechanism [1,16,20] will replenish the magnetic vector potential energy as fast as energy is withdrawn from it and conducted aside in the circuit.

We hit on the stratagem of using a highly specialized magnetic core material, nanocrystalline in nature and in special tape-wound layered structure, to try to

[9] Indeed, from any nonzero potential, any amount of energy can be diverged and withdrawn. This is easily seen for the electrostatic scalar potential by the simple equation $W = \phi q$, where W is the energy diverged or collected in joules, ϕ is the reaction cross section of the potential in joules per interacting coulomb, and q is the number of interacting static point coulombs. As can be seen from the equation, from a potential of given nonzero reaction cross section, the energy that can be collected is limited only by the number of coulombs of interacting charges—or considering repetition, from how many additional times a given amount of collecting charge is "dipped" into the potential to diverge and collect additional energy flow.

extract the energy from the magnetic vector potential (**A** potential) as a magnetic **B** field (curl of the **A** potential) that is locally restricted to the special nanocrystalline material that forms a closed magnetic flux path closed on both poles of the permanent magnet dipole. Because of its nature and also its tape-wound layered construction, the nanocrystalline material (obtained off the shelf as a commerical product from Honeywell) will perform this separation of **B**-field and **A**-potential energy that is heretofore unheard of in such a simple magnetic core mechanism and flux path material.

We point out that a tightly wound, very long coil does a similar thing, as does a good toroid, and the separation of **A** potential from **B** field is known. This forms the basis for the Aharonov–Bohm [21] effect, for example.[10] In such cases, it is not so well known that the curl-free **A** potential remains fully replenished, even though "all" the magnetic field (curled **A** potential) has been diverted. To our knowledge, such effects have not previously been utilized in magnetic core materials themselves, where the **B**-field energy is extracted and moved into a separate flux path through space. Our experimental measurements showed the magnetic field was indeed missing in the space surrounding the closed flux path material, but the **A** potential was still present outside the core and its changes interacted with coils in a dA/dt manner. The **B**-field and associated magnetic flux were rigorously confined internally to the nanocrystalline material flux path.

Now we had two streams of EM energy, each in different form, and *each equal in energy* (determined by the Poynting-type calculation approach, which only accounts for diversion *from* and not *for* the river) *to the original stream!* In short, we had exercised gauge freedom and asymmetric self-regauging to freely achieve energy amplification in the system.

[10]Quoting Aharonov and Bohm in their paper on the significance of electromagnetic potentials in the quantum theory, on p. 485: "contrary to the conclusions of classical mechanics, there exist effects of potentials on charged particles, even in the region where all the fields (and therefore the forces on the particles) vanish." Our comment is this: Indeed, since the field is usually defined as the force per unit charge in classical electrodynamics, then the field as defined does not exist until after the causative "field as a separate entity" interacts with a charged mass. Hence the field as defined is an *effect* of the interaction, not the *cause* of it. Further, being an effect and an observable as defined, it does not exist in spacetime as such, since no observable does. A priori, any observable is the output (effect) of a $\partial/\partial t$ operation upon LLLT, yielding an LLL "frozen snapshot" at an instant in time, which snapshot itself does not exist in time but was only a 3-space fragment of what was existing in the ongoing 4-space interaction at that point in time. The field-free 4-potential, together with its structure and its dynamics, provides the *causes* existing in spacetime before they interact with intermediaries to produce effects. One major problem with the present classical electrodynamics is that much of it still hopelessly confuses cause and effect. This is a residue of the original assumption (including that by Maxwell in his equations) of the material ether. When the material ether was destroyed by the Michelson–Morley experiments in the 1880s, not a single Maxwell–Heaviside equation was changed. Hence the field concept as defined still assumes a material ether, and still is an effect confused as a cause.

This then led to very novel ramifications and phenomenology, which we have been intensely exploring since filing the previous patent when—to be conservative—we assumed possible slow depletion of the magnetic dipole of the permanent magnet. Now we clearly have no depletion of the magnetic dipole, and also we can now explain where the continuous "magnetic energy wind" comes from, what triggers and establishes it, and how to apply the resulting principles.

Consequently, rapid progress in nondepleting versions of our previous invention, as a full extension to both the previous invention and also to the previous process utilized (possible depletion of stored potential energy), has been accomplished. A second patent application has been filed.

Because of the importance of this furiously progressing work and the urgency of the escalating electrical energy crisis worldwide, a paper [22] describing our results was placed on a Department of Energy Website, (see Ref. 22). The paper was later moved to a restricted DOE site reserved for scientists, until such time as independent testing verification and independent replication is obtained, to assure a fully valid scientific experiment.

Happily, independent replication by Naudin has just been accomplished at the time of this writing (Nov. 2000).

G. Results of the Research

It is now clear—by fluid flow analogy and actual experiment—that we have found the perfect magnetic mechanism for (1) producing "magnetic energy winds" at will, furnished freely by nature in natural dipole processes only recently recognized [16] and clarified [20] in the literature; (2) producing a magnetic "windmill" that freely extracts energy from these free winds provided by nature from these newly understood processes; and (3) creating positive energy feedforward and feedback iterative interactions in a coil around a core, resulting from dual energy inputs to the coil from actions (1) in the core inside the coil and (2) in the surrounding altered vacuum containing a continuously replenished field-free magnetic potential **A**, and hence representing a separate source of energy that will react with a coil.

Iterative mutual interactions, occurring between the two interactions in the coil itself, add a third small increase of energy from the resulting convergent energy gain (asymmetric self-regauging) series. The additional amplification of the energy is given by the limit of the resulting convergent series for energy collection in the coil. In this novel new usage, the net result is that the coil is an *energy amplifying* coil, or negative resistor, freely and continuously fed by excess input energy from an external active source. As can be seen, this is a startling extension to conventional generator and transformer theory.

A multiplicity of such positive energy feedforward and feedback loops occur and exist between all components of the new process. The system becomes a true open system receiving excess energy from the free flow of energy established in its vacuum environment by the subprocesses of this invention's system process.

Consequently, we have experimentally established this totally new process and field of technology, and also have experimentally established that it is not necessary to deplete the permanent-magnet dipole after all. With the new techniques, direct replenishment energy from the active vacuum is readily furnished to the permanent magnet and utilized by the system.

Extraction of usable EM energy from the vacuum is not some tremendously difficult technical feat that can only come a century or so from now. Instead, it is a readily obtainable capability right now, once the fundamental principles are understood and applied.

III. THREE IMPORTANT PRINCIPLES AND MECHANISMS

We explain three very important principles and mechanisms necessary to comprehend the new energy amplifying (regauging) process in a replenishing potential environment:

A. Conservation of Energy

The conservation of energy law states that *energy cannot be created or destroyed.* What is seldom realized is that energy can be and is reused (changed in form) to do work, over and over, while being replenished (regauged) each time. If one has one joule of energy collected in one form, then in a replenishing potential environment, one can change all that joule into a different form of energy, thereby performing one joule of work. However, one still has a replenished joule of energy remaining, by the conservation of energy law in such an environment, even though the first joule was removed in different form. Note that present engineering practice has not considered that the input joule in its original form is replenished. If one extracts and holds that converted joule in its new form, and then changes the form of it yet again in the replenishing environment, one does yet another joule of work—and still has a joule of energy left, just in a yet different form. The process is infinitely repeatable, limited only by the ability to hold the changed form of the energy each time it is changed. In a replenishing potential environment, not only will we do joule after joule of work, but we will still have a joule of "input" energy to use, as well as accomplishing the joule of work, as well as having a new joule of energy leaving the work site in different form. This follows without violation of energy conservation, as a result of the continual free regauging and energy amplification.

Further, only two energy forms are needed for endless iterative shifting of form—say, form **A** and form **B**, since **A** changed totally into **B** performs work on the transforming medium equal to **A** energy dissipation, but yields **A**-equivalent amount of energy still remaining because of replenishment. The **B**-form energy can then be changed back from **B** to **A** yet again, wherein the same amount of work is done on the transforming medium for the second time, and one *still* has a joule of **B** energy remaining because of replenishment. This process can be iterated. We call this the *pingpong principle* and use the iterative work done by each replenished change of energy form to continually increase the excitation energy of a receiving entity, the Drude electron gas in the coil and its attached circuit. We emphasize that this is also a novel way to directly generate and utilize free regauging energy.

To do work, one does it at a certain rate (power level), and so the rate of change of the energy form becomes the important factor that must be manipulated quite precisely.

Cyclic transform of the energy by "pingpong" between two different forms of energy or energy states with replenishment—and where meticulous care is observed with rates of change of the energy form in each case—is all that is required to produce as much work as one wishes in the intermediary, from a single joule of operator-input energy. This regauging energy amplifying process is limited only by one's ability to hold the new form of energy after each transformation and not lose it (or not lose all of it). By letting this iterative pingpong work be done on the Drude electron gas, the energy of that gas is excited much more than by the energy we originally input, if the input energy had been *used only once* [meaning that it was dissipated (escaped from system control) in the conversion (work) process] to perform work (if its form were only changed once) and there was no replenishment.

We stress that present electrical power systems deliberately use their collected energy only once, and do not take advantage of energy regauging by free replenishment from the potential environment. So engineers are totally unfamiliar with the pingpong mechanism and do not apply it, and they are totally unfamiliar with the energy amplifying process and do not apply it. In short, they simply do not use the extended (replenishing) energy–work theorem and the pingpong effect to dramatically increase the energy in the Drude electron gas in the external circuits connected to the generator or battery.

B. "Pingpong" Iterative Change of Energy Form

The **A** potential and the **B** field are extraordinarily useful for just such "pingpong" iterative change of energy form, from **B**-field energy to **A**-potential energy and vice versa, back and forth, repeatedly. That is precisely what happens in each coil of the process of the invention, and this results in dual inputs of energy—one in curl-free **A** form and one in **B** form—simultaneously to the coil. Quite simply, one can extract the energy from a volume of **A** potential, in **B**-field form where the **B**-field energy is removed from the volume, and the **A** potential beyond will instantly (at least at the speed of light) simply refill the volume, but with curl-free **A** potential.

The time rate of change $d\mathbf{A}/dt$ of the resulting curl-free **A** potential is an **E** field that will react—in electric field fashion—directly with the charge of the electrons in the Drude electron gas in the output coil, entering from the surrounding space outside the core. The entering $d\mathbf{A}/dt$ electric field energy interacts in the ideal case with the "full" energy of the original vector potential **A**. The magnetic field **B** in the core in the center of the coil simultaneously interacts with the spin of the same Drude electrons in the same coil, and with the "full" energy of the original vector potential **A**. Consequently, the electrons receive an interaction energy gain of 2.0, and the pingpong between the two processes adds about another 0.5 gain, for an overall gain per output coil of 2.5. With two output coils, the energy amplification is about 2.5, and hence the generator exhibits COP = 5.0.

A similar effect is also demonstrated by the well-known Aharonov–Bohm effect [21,23],[11] but usually in only very small effects, without the pingpong effect, and not used in power systems. By analogy, we may compare this iterative process to "dipping several buckets of water in succession" from a mighty rushing river; the river refills the "hole" immediately after each dipping. We can continue to extract bucket after bucket of water from the same spatial volume in the river, because of the continual replenishment of the extracted water by the river's flow.

Any change of **B**-field energy in the center of the coil, interacts with the coil magnetically since the coil's magnetic field is at its greatest strength in its precise center, and the center of the coil is in the center of the core flux path material. This magnetic interaction between core and coil produces voltage and current in the coil (and therefore in a closed loop containing the coil and the external load). The interaction also simultaneously produces an additional equal energy outside the coil in the form of field-free **A** potential. This latter interaction is absolutely permitted since the magnetic energy in **B** form was "dissipated" (transformed) into **A**-potential energy, thereby causing the electrons in the wires to flow by doing work that built voltage and current, which then was a change of form of the **B**-field energy. Simultaneously, the electron current produces the **A**-potential energy around the coil and outside it, which is absolutely permissible since a change in form of the energy is again involved.

[11] Full discussion of the Aharonov–Bohm effect and hundreds of references can be found in Ref. 23. According to Nobelist Feynman, it required 25 years for quantum physicists to clearly face the Aharonov–Bohm issue of the primacy and separate action of the force-field-free potential. Another long period passed before physicists finally accepted it, even though it was experimentally demonstrated as early as 1960.

Since these changes in energy form occur nearly at the speed of light, in a local coil they appear "instantaneous," although in reality they are not quite instant, just very rapid. However, the work produced by each change of form of the energy in that rapid "pingpong" between the several energy states, continually produces *work on the Drude electrons*, producing momentum and motion in the Drude gas, thus resulting in voltage and current. In this way, the increased momentum and motion—involved in the currents flowing in the voltage drop of the coil and external loop—result in increased stored kinetic energy in the moving Drude gas, which is electromagnetic energy of a different form.

As can be seen, the speed of the pingpong energy-state transformations results in each transformation doing cumulating work in the Drude electron gas of the output coil, to increase that gas's excitation and energy. A continuous "collection" of excess energy—caused purely by the change of form of the energy and not by "loss" or "disordering" of the energy—occurs in the Drude electron gas, resulting in increased voltage and current in the circuit containing the coil. This is simply a mechanism for a "regauging" or increase of potential energy of the Drude electron gas system. The Drude electron gas system's increased excitation energy can then be dissipated "all at once" in conventional fashion in the external load, providing more energy dissipated as work in the load than was input to the coil originally by the operator.

Hence an *energy amplifying* action of the coil and its multiplicity of processes is generated. There is no violation of the energy conservation laws, of the laws of physics, or of the laws of thermodynamics since this is an open system far from equilibrium with its source of potential energy (the magnetic dipole of the permanent magnet), which, in turn, continuously receives replenishment energy from the vacuum by a giant negentropy process reinterpreted by Bearden [16] from Whittaker's work [1] and clarified by Evans and Bearden [20].

C. Dissipation of Energy

The dissipation of the final collected regauging energy within the load can permissibly be greater than what we ourselves initially input,[12] because of the iterative change of form of the energy with replenishment. Therefore the iterative interactive work done on the Drude electron gas provides more than one joule of work done on the gas—thereby increasing its potential energy by more than one joule—for each joule of energy input by the operator to the

[12]See Refs. 9–14,16,17,73–75. We comment that the standard calculation of the Poynting energy flow is not the calculation of the total EM energy flow at all, but only a calculation of how much of the total energy flow is intercepted by the surface charges of the circuit and thereby diverged into the conductors to power the Drude electrons. The energy input to the system by the operator has nothing to do with the energy dissipated by the system in the load—only with the energy expended in continually re-forming the source dipole that the conventional system is designed to destroy faster than it powers the load.

system process. The cumulated potential energy in the Drude electron gas is then discharged in loads in normal fashion. Note that, even here, the energy is not lost when dissipated from the load and outside the system, but just flows out of the load in a different form (e.g., as heat radiated from a resistor load). Again, this process involves an open system not in equilibrium with its active vacuum environment; it is comparable to a windmill in a wind, where we input a little wind and find a way to cause the environment to freely add some additional wind to our own input.

The various asymmetric regaugings violate Lorentz' arbitrary symmetric condition, specifically in the discharge or change of form of the energy. Hence, this process restores to electrodynamics one group of those missing Maxwellian systems arbitrarily discarded by first Lorenz and later Lorentz, more than a century ago. A rigorous rebuttal of objections to COP > 1.0 EM systems has been given [75].

IV. THE PROCESS IS THEORETICALLY SUPPORTED

Several rigorous scientific papers [24–37] by the Alpha Foundation's Institute for Advanced Study (AIAS) have been published or are in the publication process, fully justifying the fact that energy currents (energy winds) can readily be established in the vacuum, and that such energy winds do allow the extraction of EM energy from the vacuum. Two rigorous theoretical papers explaining the motionless electromagnetic generator have been published [73,74].

Also, Cole and Puthoff [38] have shown that there is no prohibition in thermodynamics that prevents EM energy being extracted from the vacuum and utilized to power practical systems.

In electrochemistry it has long been known [39][13] that there can be no current or movement in electrodes without the appearance of excess potential (regauging) called the *overpotential*.

Further, in the most advanced model in physics—gauge field theory—the freedom to change gauge (in electrodynamics, to change the potential) at will, is an axiom of the theory. If we freely change the potential of a physical electrical power system, we freely change its potential energy (in a real system, we may have to pay for a little switching energy losses).

[13] Essentially, *the overpotential is a shift in the Fermi level necessary to allow the electron in the electrode metal to have energies overlapping with vacant acceptor levels in molecules adjacent to the electrode in the solution.* It enables the transfer of electrons via quantum transfer (tunneling). Quoting from Ref. 39, p. 356: "*Unless a system exhibits an overpotential, there can be no net reaction* [emphasis in original]." We point out that an overpotential is an advantageous regauging (free change) of the potential energy of the local region where the overpotential appears.

It follows that we can also freely change (regauge) the excess potential of that system yet again, by any means we choose, including discharging that excess potential energy in a load to do work. Thus gauge field theory has for decades already axiomized the rigorous basis for COP > 1.0 electrical power systems—but such systems have remained neglected because of their arbitrary discard by the ubiquitous use of Lorentz symmetrical regauging.

That such COP > 1.0 electrical power systems have not been previously designed or built is therefore not due to a prohibition of nature or a prohibition of the laws physics at all, but to a characteristic used to design and build the systems themselves. Because of their ubiquitous closed current loop circuits, conventional power systems use half their collected energy to destroy their own source dipoles, which destroys any further use of energy from "the potential between the ends of the dipole" since both dipole and potential vanish, as do the negentropy process and the broken 3-symmetry. The potential of the source dipole, after all, is what potentializes the external circuit with additional excitation energy, to be utilized to power the system. In present systems, half that excitation energy is dissipated to destroy the dipole along with the source potential, and less than the remaining half is used to power the load. This rigorously limits such systems to COP < 1.0.

In the MEG, we do not destroy the potentializing source dipole, which is the magnetic dipole of the permanent magnet. We include the vacuum interaction with the system, and we also include the broken symmetry of the source dipole in that vacuum exchange—a broken symmetry proved and used in particle physics for nearly a half century, but still inexplicably neglected in the conventional Lorentz-regauged subset of the Maxwell–Heaviside model. We also use the extended work–energy theorem, as discussed.

Consequently, our work and this novel process are rigorously justified in both theory and experiment, but the principles and phenomenology are still not incorporated in the classical electrodynamics theory utilized to design and produce electrical power systems. These principles are indeed included in the new O(3) electrodynamics being developed by AIAS (Alpha Institute for Advanced Study)[14] that extends the present U(1) electrodynamics model, as shown by some 100 scientific papers carried by the U.S. Department of Energy on one of its private scientific Websites in Advanced Electrodynamics, and by an increasing number of publications in leading journals such as *Foundations of Physics*, *Physica Scripta*, and *Optik*.

[14] A private communication from Dr. Myron Evans, Sept. 30, 2000, rigorously confirms that the magnitudes of the vector potential and the 4-current do in fact provide EM energy from the vacuum, and determine its magnitude as well.

We thus have invented a process that indeed is well-founded and justified, but the basis for it is not yet explained in the texts and university courses. It is our belief that this absence will be rapidly rectified in the universities, in both the physics and electrical engineering departments, on the advent of practical self-powering electrical power systems freely regauging themselves and extracting energy from the magnetic dipole of a permanent magnet, and that the energy will be continuously replenished to the dipole from the active vacuum via the new giant negentropy process [1,16,20].

V. CONSIDERING THE PROCESS

A. A Potential and Field-Free A Potential

This invention relates generally to the field of electromagnetic power generation. Specifically it relates to a totally new field of extracting additional electromagnetic energy in usable form from a permanent-magnet dipole's magnetic vector potential energy, in addition to the electromagnetic energy extracted from its magnetic field energy, wherein the excess magnetic vector potential energy taken from the magnet dipole is continuously replenished to the permanent-magnet dipole from the active vacuum that is a curved spacetime with an ongoing giant negentropy flow process [1,16,20].

With respect to the electrostatic scalar potential, electrodynamicists are familiar with the fact that unlimited energy can be extracted from a potential. The very simple equation

$$W = \phi q \tag{1}$$

gives the amount of energy W in joules, which is collected at any given point x,y,z from the electrostatic scalar potential—whose reaction cross section is given by ϕ, in joules collected per point coulomb—by charges q in coulombs and located at point x,y,z. Note that as much intercepting charge q as desired can be used at any point to increase the energy collection at the point, and collection can be accomplished at as many points x,y,z as is desired.

So any amount of energy can be collected from any nonzero scalar potential, no matter how small the potential's reaction cross section, if sufficient intercepting charge q and collecting points x,y,z are utilized. In short, one can intercept and collect energy from a potential indefinitely and *in any amount*, and *in any form taken by the collecting interaction*, because the potential is actually a set of EM energy flows in the form of longitudinal EM waves, as shown by Whittaker [1] in 1903 and further expounded by Bearden [16,40]. Subsequently, Evans and Bearden [20] have more rigorously interpreted Whittaker's [1] work and extended the principle into power systems. Thus any energy diverged and "withdrawn" from the potential in a given local region of it is immediately replenished to the potential and to that region from the complex plane (the time domain) by the potential's flowing EM energy streams [1,16,20], as rapidly as the energy is deviated and withdrawn.

For the magnetic vector potential, some preliminary comments are necessary. First, for over a hundred years it has been erroneously advanced that the magnetic vector potential $\mathbf{A}$ is "defined" by the equation

$$\mathbf{B} = \nabla \times \mathbf{A} \tag{2}$$

This is easily seen not to be a definition at all, since an equation says nothing about the nature of anything on its right or on its left, but merely states that the entire right side has the same magnitude as does the entire left side. For an expression to be a *definition*, it must contain an identity ($\equiv$) sign rather than an equal ($=$) sign. Hence in seeing what is attempted to be defined, we rewrite equation (2) as

$$\mathbf{B} \equiv \nabla \times \mathbf{A} \tag{3}$$

Now it is seen that it is the magnetic field $\mathbf{B}$ that is being defined as the curl of a swirling $\mathbf{A}$-potential, which swirling component we will call $\mathbf{A}_C$ for the "$\mathbf{A}$ circulating" component of $\mathbf{A}$. Rigorously, this is correct because all fields *as defined in classical electrodynamics* are effects and the observable results of interactions. The potentials and their structural dynamics are the primary causes [21]. The curl of the circulating $\mathbf{A}$ is a magnetic field $\mathbf{B}$, by identity (3). There may, of course, be present additional $\mathbf{A}$ potential that has zero curl, but that additional longitudinal $\mathbf{A}_L$ potential or $\mathbf{A}_L$ current does not produce a magnetic field $\mathbf{B}$ per se. However, its change dA/dt does interact with charges as an $\mathbf{E}$-field.

In 3-space, the field-free $\mathbf{A}_L$ potential may be moving longitudinally, in which case it is identically an electrical potential ϕ that is moving longitudinally and hence no longer really a *scalar* potential ϕ but a vector potential $\boldsymbol{\phi}$. If $\boldsymbol{\phi}$ translates without changing magnitude, there is no $\mathbf{E}$-field and hence $\boldsymbol{\phi}$ is a field-free vector potential. Identity (3) still does not define $\mathbf{A}$, but defines $\mathbf{B}$ in terms of $\mathbf{A}$ and the curl operator. Note particularly that, in identity (3), we may have additional $\boldsymbol{\phi}$ present as a curl-free, longitudinal magnetic vector potential $\mathbf{A}_L$, and we shall refer to this additional curl-free magnetic vector potential as $\mathbf{A}_L$ (for longitudinally translating $\mathbf{A}$ component without swirl).

B. General Relativistic Considerations

We are rigorously using the Sachs [41][15] unified field theory view that energy of whatever form represents a curvature in spacetime (ST). We argue that how we then observe or "see" the energy effects and label them, depends on the factors of physical interaction with that ST curvature. Thus interaction with magnetic charge produces magnetic energy aspects, while interaction with electrical charge produces electrical energy aspects, and so on. Motion of either of the interactions lets us also "see" some of the magnetic energy as electrical energy, and some of the electrical energy as magnetic energy, and so on. Quoting from Evans [42]:

> With respect to O(3): In 1992 it was shown that there exists a longitudinal component of free space electromagnetism, a component which is phaseless and propagates with the transverse components. Later this was developed into a Yang-Mills theory of electromagnetism with O(3) Lagrangian symmetry. This theory is homomorphic with Barrett's SU(2) electrodynamics and has far reaching implications in field theory in general. Recently it has been recognized to be a sub theory of the Sachs theory of electromagnetism, based on the irreducible representations of the Einstein group of general relativity. The Sachs theory produces a non-Abelian structure for the electromagnetic field tensor. The O(3) electromagnetism also has implications for the potential ability of extracting energy from the vacuum, and its topological implications are currently being investigated by Ranada. The O(3) electromagnetism has been tested extensively against empirical data, and succeeds in describing interferometric effects and physical optical effects where the conventional Maxwell Heaviside theory fails. Implicit in both the O(3) and Sachs theories of electromagnetism is the ability to extract electromagnetic energy from curved space-time.

However, when $\mathbf{A}_L$ interacts with electrical charge, the charge may swirl, in which case the swirling component of the ϕ moving with the charge is an $\mathbf{A}_C$ component, and this $\mathbf{A}_C$ swirling component will produce a magnetic field $\mathbf{B}$ by identity (3).

In the unified field theory approach being used [41–45] in spacetime all energy is simply a special curvature of that spacetime, regardless of the form of the energy. (This is believed to resolve the foundations issue pointed out by Feynman [46]: "It is important to realize that in physics today, we have no

[15] Sach provides a great generalization of general relativity and electrodynamics reaching from the quarks and gluons to the entire universe. O(3) electrodynamics forms a very important subset of Sachs' theory, which means that general relativistic effects such as curved spacetime and EM energy from the curved spacetime vacuum can be engineered electromagnetically. The present invention does engineer curved spacetime to obtain excess energy from the active vacuum.

knowledge of what energy *is*." In Feynman's view, energy is a curvature of spacetime, and the "form" of the energy observed is determined by the type of material interaction that occurs with ST curvature.) Hence one can readily visualize the energy being changed from a vector potential to a scalar potential and vice versa, depending simply on whether the potential is moving or stationary with respect to the frame of the observer (the laboratory frame).

It is also a well-known facet of general relativity that any change of energy density in spacetime a priori is associated with a curvature of spacetime (i.e., in Einstein's theory with the single exception or gravitational field energy, and even that exclusion has been challenged [47]; more recent experiments tend to support Yilmaz' predictions over those of the unaltered Einstein Theory). What has been neglected in general relativity (and arbitrarily discarded in electromagnetic theory long before general relativity was born) is the enormous *unaccounted* nondiverged EM energy flow filling the space around every EM circuit [40] — and in fact around every field and charge interaction[16]—with almost all of it missing the circuit entirely, and not being intercepted and diverged into the circuit to power it.

This nonintercepted huge energy flow was recognized by Heaviside [48–50],[17] not even considered by Poynting [51],[18] and arbitrarily discarded by Lorentz [52][19] as having "no physical significance" because it did not strike the

[16] Consider carefully the implication that the so-called "magnitude of the field" at a point has been defined as the magnitude of the result of the interaction of the field with a unit static point charge. In other words, the *effect* of the field has been erroneously defined as the causative *field itself*, a nonsequitur. The effect cannot be the cause.

[17] Heaviside's paper on the forces, stresses, and fluxes of energy in the EM field [48] followed previous publications several years earlier by Heaviside, including papers in *The Electrician*, beginning in 1885. Here Heaviside also credits Poynting with first discovering EM energy flow in space. On the other hand, Poynting credited Heaviside with being first; for instance, see editorial on energy transfer [49] in *The Electrician*.

[18] Poynting, in his paper on energy transfer in the EM field, got the direction of the flow wrong, which was later corrected by Heaviside. Further, Poynting considered only that very minor component of energy flow surrounding the circuit that actually strikes the circuit and enters it to power it. The enormous additional energy flow which is present but misses the circuit entirely and is usually wasted, was not considered by Poynting at all.

[19] Figure 25 on p. 185 of Ref. 52 (on EM field energy) shows the Lorentz concept of integrating the energy flow vector around a closed cylindrical surface surrounding a volumetric element. This is the procedure that arbitrarily selects only a small component of the energy flow associated with a circuit—specifically, the small Poynting component striking the surface charges and being diverged into the circuit to power it—and then treats that tiny component as the "entire" energy flow. Thereby Lorentz arbitrarily discarded all the vast Heaviside energy transport component, which does not strike the circuit at all, and is merely wasted.

circuit and power any part of it.[20] It is still arbitrarily discarded today, using Lorentz's discard method. We quote from Heaviside [50, p. 94]:

> It [the energy transfer flow] takes place, in the vicinity of the wire, very nearly parallel to it, with a slight slope towards the wire .... Prof. Poynting, on the other hand, holds a different view, representing the transfer as nearly perpendicular to a wire, i.e., with a slight departure from the vertical. This difference of a quadrant can, I think, only arise from what seems to be a misconception on his part as to the nature of the electric field in the vicinity of a wire supporting electric current. The lines of electric force are nearly perpendicular to the wire. The departure from perpendicularity is usually so small that I have sometimes spoken of them as being perpendicular to it, as they practically are, before I recognized the great physical importance of the slight departure. It causes the convergence of energy into the wire.

Also, it is still largely unrecognized in Western science that pure general relativity contains no energy conservation equations [53,54][21] of the kind encountered in electrodynamics and mechanics. This is easily seen by considering the impact of gauge freedom, which allows the potential energy of any region of spacetime to be *freely* changed at will. But this is also a form of freedom of spacetime curvature, hence the notion of *fixed* accountability of energy replenishment and dissipation is completely voided by gauge freedom.

Hilbert [54] first pointed out this remarkable absence of energy conservation laws from general relativity, not long after Einstein published his theory.

It also appears that the ultimate energy interaction is the transduction of energy form between the time domain (complex plane) and 3-space. In fact, *all 3-spatial EM energy actually comes from time-like EM energy currents after 3-symmetry breaking* [1,16,20].

C. Indefiniteness Is Associated with the A Potential

A magnetic vector potential **A** produced by a current-carrying coil not tightly wound or closed (or very long), must possess both a swirl component $\mathbf{A}_C$ (from

[20] This is rather like discarding all the wind on the ocean except for that tiny component of it that strikes the sails of one's own sailboat. It is true that the wind missing one's own boat has no further significance for that one boat, but it may, of course, be captured in the sails of an entire fleet of additional sailing vessels to power them quite nicely. Hence the statement of "no physical significance" is a nonsequitur; "no physical significance to that one specific circuit" is better—but even then is incorrect if additional "sails" (interceptors) are added to catch more of the available energy wind and diverge more of it into the circuit.

[21] Quoting from Hilbert [54]: "I assert ... that for the general theory of relativity, i.e., in the case of general invariance of the Hamiltonian function, energy equations ... corresponding to the energy equations in orthogonally invariant theories do not exist at all. I could even take this circumstance as the characteristic feature of the general theory of relativity." As Logunov and Loskutov [53] pointed out, unfortunately this remark of Hilbert was evidently not understood by his contemporaries, since neither Einstein himself nor other physicists recognized the fact that in general relativity conservation laws for energy, momentum, and angular momentum are, in principle, impossible.

the circling of the coil in each turn of the coil) and a longitudinal component $\mathbf{A}_L$ from the longitudinal advance of the current between coils, since a coil is actually a helix and not a set of circles. It will also possess a magnetic field, both inside the coil and outside it.

Hence, considering both curled and curl-free types, the actual magnitude of **A** is always indefinite—and, in fact, the indefinite nature of the potential together with the freedom to change it at will is universally recognized by electrodynamicists [55] (see also Section V.C.1).[22] However, the prevailing argument that change of potential does not affect the system is a nonsequitur.

Further, in 1904 Whittaker [56] (see also Section V.C.2) showed that any electromagnetic field, wave, etc. can be replaced by two scalar potential functions, thus initiating that branch of electrodynamics called *superpotential theory* [58]. Whittaker's two scalar potentials were then extended by electrodynamicists such as Bromwich [59], Debye [60], Nisbet [61], and McCrea [62] and shown to be part of vector superpotentials [58], and hence connected with **A**.

1. Jackson's Studies on EM Potential

In symmetrically regauging the Heaviside–Maxwell equations, electrodynamicists and gauge field theorists assume that the potential energy of any EM system can be freely changed at will (i.e., that the system can first be asymmetrically regauged, due to the principle of gauge freedom). The symmetric regauging actually consists in two asymmetric regaugings carefully chosen so that the net forcefield [electromotive force (EMF)]—available for excitation discharge of the excited system—is zero. In circuits, this means that the back EMF (across the source dipole) is precisely equal and antiphased to the forward emf (across the external circuit with its loads and losses). Jackson's book does not even address circuits.

For operating EM systems, their initial potentialization (application of potential to the system to increase its potential energy available for further discharge) is asymmetric a priori and universally used. Gauge field theory and its assumption of gauge freedom assures us of the validity of this theoretically work-free process of increasing the energy of the system. In real systems, a little switching cost or other expenditure may be required, but is minuscule in relation to the amount of extra potential energy that can be generated in the system at will.

[22] On p. 67 of their paper on Lorentz-invariant potentials and the nonrelativistic limit, Bloch and Crater [57] state: "[It is usually] ... assumed that the magnitude of potential energy is irrelevant, being arbitrary to the extent of an additive constant." We comment: by noting that this "standard" assumption in classical electrodynamics is totally wrong, particularly when one considers (1) conservation of energy and (2) gravitational effects. We have previously nominated this arbitrarily discarded extra potential energy as a solution to the "dark matter" problem in astrophysics, and as being responsible for the extra gravity holding together the arms of the distant spiral galaxies; see Ref. 40.

As shown by Jackson [55], for the conventional EM model electrodynamicists actually select only a subset of the Maxwellian systems and deliberately discard the remaining Maxwellian subset. Following Lorentz, the electrodynamicists arbitrarily select *two* asymmetric regaugings but precisely such that none of the initial excess regauging energy—freely received in the system by its potentialization—can subsequently be dissipated to power loads without equally destroying the system potentialization represented by the source dipole. This inanity occurs because the *net force* is deliberately brought to zero, thus consisting of equal forward and backward EMFs—or MMFs in a magnetic circuit. This custom produces much simpler equations for that remaining *simpler* subset of Maxwellian systems that are in equilibrium in their exchange with the active vacuum during their dissipation of the free regauging energy.

Hence, for more than a century it has been "customary" to *arbitrarily* discard all Maxwellian systems and subsystems that would *asymmetrically* regauge themselves during the discharge of their initial free excitation energy. This arbitrary, self-imposed condition is neither a law of nature nor a law of electrodynamics or thermodynamics. It is purely arbitrary and imposed by system design. It assumes that half the gauge freedom's excess potential energy be dissipated internally (against the source dipole's back EMF) to destroy any further energetic activity of the system by destroying the source dipolarity (any excess potential on the system, and hence any excess potential energy).

The remaining half of the initial free gauge excitation energy is dissipated *usefully* in the system's external loads and losses. This means that this remaining half of the excitation energy is dissipated *detrimentally* by the system to destroy its own energetic operation. Since any real system has losses, the net result is that *half* the gauge freedom potential energy of the excited system is used to destroy the source dipole itself and all potentialization of the system, and *less than half* is used to power the loads. Since it requires as much additional energy to restore the source dipole as it required to destroy it, the operator then must furnish more energy to provide for continually restoring the dipole than the system *permits* to be dissipated in the external loads.

The set of Maxwellian systems arbitrarily discarded by the ubiquitous Lorentz regauging are precisely those open dissipative Maxwellian systems not in thermodynamic equilibrium in their vacuum exchange. Those are precisely the Maxwellian systems that *do not* forcibly and symmetrically regauge themselves in accord with the Lorentz condition during their excitation discharge. Those *arbitrarily discarded* Maxwellian systems are thereby free to dissipate their gauge freedom initial "free excitation" energy primarily in the external loads and losses, with much less being dissipated in the source dipole to destroy it.

The performance of the *arbitrarily discarded* asymmetrically regauging Maxwellian systems is described by the thermodynamics of an open dissipative system not in equilibrium with its active environment, rather than by classical equilibrium thermodynamics. As is well known in the thermodynamics of such systems (for which Prigogine received a Nobel Prize in 1977), such an open dissipative system is permitted to (1) self-order, (2) self-oscillate or self-rotate, (3) output more energy (e.g., to do useful work) than the operator must input (the excess energy is freely received from the external environment, in this case the active vacuum), (4) power itself and its load(s) simultaneously (all the energy is freely received from the external environment, in this case the active vacuum), and (5) exhibit negentropy.

That our normal EM power systems do not exhibit COP > 1.0 is purely a matter of the arbitrary design of the systems. They are all designed with closed current loop circuits, which can readily be shown to apply the Lorentz symmetric regauging condition during their excitation discharge in the load. Hence all such systems — so long as the current in the loop is unitary (its charge carriers have the same m/q ratio) — can exhibit only COP < 1.0 for a system with internal losses, or COP $= 1.0$ for a superconductive system with no internal losses.

2. *Whittaker's Studies on EM Potential*

Whittaker's groundbreaking paper [56] was published in 1904 and orally delivered in 1903. Whittaker shows that all EM fields, potentials, and waves consist of two scalar EM potential functions. Whittaker's method is well known in the treatment of transverse electric and transverse magnetic modes of a cylindrical cavity or a waveguide. The Debye potentials and the Bromwich potentials are essentially radial components of the vector potentials of which Whittaker potentials are the real parts. Our further comment is that, since each of the scalar potentials used for the Whittaker functions has an internal Whittaker 1903 giant negentropic substructure and dynamics, then all present EM waves, fields, and potentials have—and are composed of—vast internal longitudinal EM wave structures and dynamics, and these have been almost entirely neglected in Western electrodynamics. The "internal" or "infolded" structures and dynamics inside normal EM fields, waves, and potentials can be engineered, and this area has startling implications to all of science, particularly to medical science. Discussion of this "infolded" electrodynamics is beyond the scope of this chapter.

D. Applying the Giant Negentropy Mechanism

So let us consider the **A**-potential most simply as being replaced with such a Whittaker [1,56] decomposition. Then each of these scalar potentials—from which the **A** potential function is made—is decomposable into a set of harmonic phase conjugate wavepairs (of longitudinal EM waves). If one takes all the phase conjugate half-set, those phase conjugate waves are converging on each point in the magnetic vector potential **A** from the imaginary plane (from the time domain). At that same point in **A**, the other waveset—composed of the harmonic set of longitudinal EM waves in 3-space—is outgoing. The 4-conservation of EM energy requires that the incoming energy to the point from the complex plane is being transformed at the point (by the assumed unit point charge at that point) into real EM 3-space energy, and radiating outward from that point as real EM energy, in this case in the form of the magnetic vector potential **A** without curl since the curl operator is absent.

We have previously pointed out [16,20] that this energy flow input from the complex plane to every point in the potential, with its output in real 3-space, is a more fundamental symmetry than is the usually assumed 3-symmetry in EM energy flow in 3-space. Further, it is a giant negentropy and a continuous, sustained reordering of a fraction of the vacuum energy, and the reordering continues to expand in space at light speed so long as the source dipole for the potential exists.

So the **A** potential—in either of its components $\mathbf{A}_L$ or $\mathbf{A}_C$—is not to be thought of as having "fixed energy" since it consists of and identically is a myriad energy flow processes ongoing between the time–energy domain (the complex plane) and the real energy domain (real 3-space).

As is any potential including the electrostatic scalar potential ϕ between the poles of an electric dipole and the magnetostatic scalar potential Φ between the poles of a permanent magnet, the **A** potential is an ongoing set of longitudinal EM energy flows between the time domain (imaginary plane) and real 3-space [1,16,20].

We stress that the EM energy flows constituting the so-called scalar potential and all vector potentials violate 3-flow symmetry in energy conservation, but rigorously obey 4-flow symmetry. There is no law of nature that requires that energy be conserved in 3-space! If we work in 4-space as is normal, then the laws of nature require that energy be conserved in 4-space, as is done by the potential. Imposing the arbitrary additional requirement of 3-flow energy conservation imposes a 3-symmetry restoring operation that destroys or nullifies the giant negentropy 4-process[23] of the dipole [16] and results in system 3-equilibrium with the active vacuum. It results in design and production of electrical power systems exhibiting only $\mathrm{COP} < 1.0$. The ubiquitous closed current loop circuit design produces a circuit that deliberately (albeit unwittingly) reimposes the 3-flow symmetry, kills the dipole and the giant

[23] Which, in turn, destroys the ability of any observable to exist (in time). An observable is a priori a 3-space fragment of an ongoing 4-space interaction, torn out at one frozen moment of time. The fact that observables do not persist in time has a profound impact on the foundations of physics, but its implications remain to be explored. A major impact is that physicists have missed the mechanism that generates the "flow of a mass through time." Discussion of that mechanism is beyond the scope of this chapter.

negentropy process, requires at least as much continuous input energy by the operator as was utilized to kill the dipole, and has generated the gigantic burning of hydrocarbons and the pollution of the biosphere.

E. A Negative-Resistance Process

Because of its giant negentropy process [1,16,20], any potential—and even any vanishingly small but finite region of it—is an open EM energy flow system, freely receiving energy from the complex plane in its active vacuum environment, transducing that received reactive power (in electrical engineering terms) into real power, and outputting real EM energy flow in space in all directions at the speed of light [16,20]. An ordering of the local vacuum results from that action.

The vacuum–dipole energy exchange process is negentropic, since there exists total 1 : 1 correlation between the inflowing longitudinal EM waves in the complex plane and the outflowing EM waves in real 3-space [1,16,20]. The potential then may rigorously be regarded as a novel kind of *negative resistor*,[24] constituting an automatic ongoing negative-resistance process. By *negative resistance process* we mean that each spatial point (and its mathematical neighborhood of immediately surrounding points) occupied by the potential continuously

1. Receives EM energy in unusable form (in the form of longitudinal EM waves input from the complex plane, which is the continuous receipt of reactive power)
2. Transduces the absorbed or received energy into usable form (real energy in 3-space)
3. Outputs the received and transduced EM energy as usable EM energy flow in 3-space

Thus, associated with and contained in any potential and any dipolarity—including the dipolarity of a permanent magnet—we have a novel, free source of EM energy from the vacuum's complex plane (reactive power input, in electrical engineering terms, with real power output). That is true whenever we have a potential of any kind, either $\mathbf{A}$ or ϕ, or a dipole of any kind, either electrical or magnetic, or a polarization. Further, any energy that we divert (collect)

[24] We define a *negative resistor* as any component or function or process that receives energy in unusable or disordered form and outputs that energy in usable, ordered form, where that is the *net* function performed. We specifically do not include "differential" negative resistors such as the tunnel diode, thyristor, and magnetron, which dissipate and disorder more energy overall than they reorder in their "negative resistance" regimes. Also, we extend the definition to 4-space, to include input of energy from the time domain to the negative resistor entity, and output of the energy in 3-space.

from this potential by and on intercepting charges, and hold in the localized vicinity of the charge, is an energetic excitation of the perturbing charges.

F. Modeling the Transduction Mechanism

Charges can be thought of as rotating 720° in one "full rotation," that is, 360° rotation in the complex plane followed by 360° rotation in real 3-space. The charges in the source dipole thus absorb the incoming reactive power while rotating in complex space and are excited therein, then reradiate this absorbed EM excitation energy in real 3-space during their subsequent 360° rotation in that 3-space. Further, all the energy diverted from the energy flows representing the potential, is immediately replenished by the vacuum to the source dipole, by the stated giant negentropy mechanism [16,20].

G. Replenishment via Giant Negentropy

It follows that we may collect energy from an **A** potential of a permanent magnet by applying the curl operator to **A**, then withdrawing and holding the resulting $\mathbf{B} = \nabla \times \mathbf{A}$ magnetic field energy in a localized material flux path. That is the withdrawal of $\mathbf{A}_C$ energy from the overall **A** potential in space, which is the withdrawal of $\mathbf{A}_C$ energy from the magnetostatic potential outflow dynamics between the poles of the magnetic dipole of the permanent magnet. This withdrawal and sharp path localization of the $\mathbf{A}_C$ energy from the permanent-magnet dipole's outpouring **A**-potential energy will be continuously replaced at light speed by the giant negentropy process [1,16,20] engendered in 4-space by the magnetic dipole of the permanent magnet. Hence an unlimited amount of energy may be withdrawn from the **A** potential in space around the magnet in this fashion, and the withdrawn energy will be continuously replaced at light speed from the active vacuum via the giant negentropy process. In real systems, the materials and components will impose physical limits so that only a finite amount of excess energy flow can be accomplished, but in real materials these limits still permit system COP $\gg$ 1.0 [40].

The foregoing discussion shows that, in a magnetic apparatus or process functioning as part of an overall electromagnetic power system, we may have one subprocess that continuously withdraws energy from the curled portion of **A** (i.e., holds and localizes the magnetic field **B** and confines it to a given path), and in that case the source (in this case the permanent magnet) of the **A** potential will simply replenish—at light speed—all the **A** energy that was withdrawn and localized. The replenished **A** energy will not be localized, since under a given set of conditions only so much energy is withdrawn and held in the localized condition.

The principle is that, as energy is drawn from the vector potential and then contained and circulated in field form in a localized material region or path, the withdrawn **A**-potential energy in space outside that localized path is continually replenished from the permanent magnet dipolarity to the space surrounding the localized **B**-field energy path as the real EM energy flow output of the giant negentropy process [16,20] engendered by the magnet dipole. Further, the energy drawn from the permanent-magnet dipolarity is continually replenished from the surrounding vacuum by the input EM energy flow to the magnet dipolarity from the vacuum's complex plane in the ongoing giant negentropy process [1,16,20].

H. Regauging Can Be Negentropic or Entropic

Any increase or decrease of energy in the apparatus and process in the local spacetime constitutes (1) self-regauging by the process, whereby the process freely increases the potential energy of the system utilizing the process; and (2) concomitant curvature of spacetime and increase in that spacetime curvature because of the increase of local energy in the system process.

From the standpoint of gauge field theory, free asymmetric regauging is permitted by gauge freedom and is rigorously allowed, in effect allowing the violation of classical equilibrium thermodynamics because the regauged system freely receives EM energy from an external active source, the active vacuum's complex plane in the evoked giant negentropy process.[25] From the standpoint of general relativity, the excess energy from spacetime is freely allowed, since all EM energy moves in curved spacetime [33,36,37,41–43,45,63] a priori, and simple conservation of EM energy as usually stated in classical equilibrium electrodynamics need not apply in a general relativistic situation [53,54].

I. Use of a Nanocrystalline "Energy-Converting" Material

A nanocrystalline material recently available on the commercial market was found and utilized in this process. When utilized as a closed flux path external to and closed on the two poles of a permanent magnet, the special nanocrystalline

[25] It may be that we are *defining* the causative mechanism for gauge freedom itself as being pure entropy (energy dissipation by disordering) or pure negentropy (energy increase by reordering), but we defer to the advanced theoreticians to determine the truth or falsity of such a question. If one considers Whittaker's process [1] in either direction (i.e., energy freely entering 3-space by exiting from the time domain, and energy freely entering the time domain by exiting from 3-space), the conjecture may have merit. At any rate, it appears that all 3-spatial EM energy comes from the time domain (from *ict*) in the first place. It would appear that a more rigorous reexamination of the fundamental concept of energy propagation "in 3-space" should be accomplished. To first order, it appears that what propagates from the source charge (or source dipole) is the process whereby time energy converted into EM energy in 3-space. Apparently both the time energy cause and the 3-space EM energy effect are propagating in iterative quantum form. If so, this perfectly corresponds to F. Mandl and G. Shaw, *Quantum Field Thory*, Wiley, 1984, "Convariant Quantization of the Photon Propagator" in Chapter 5. Mandl and Shaw argue that the longitudinal and scalar polarizations of the photon are not directly observable, but only in combination, where they manifest as the "instantaneous" Coulomb (i.e., electrostatic) potential. Their argument, translated from particle terminology to wave terminology, directly fits my re-interpretation of Whittaker's 1903 decomposition of the scalar potential.

material will contain all the $\mathbf{B} = \nabla \times \mathbf{A}$ field energy (curled potential energy) in the closed flux path containing the magnet itself, while the magnetic dipole of the permanent magnet continuously replenishes and maintains the external circulation of field-free **A**-potential energy filling the space around the nanocrystalline closed flux path containing the withdrawn magnetic field energy.

This performance can, in fact, be measured, since magnetic field detectors detect little or no magnetic field surrounding the flux path (or even around the magnet in the flux path at an inch or two away from it), and yet coils placed in the spatial flux path outside the core interact with the field-free **A** potential that is still there. A coil placed around the flux path so that the flux path constitutes its core, interacts with both the field-free **A** potential outside the material flux path core, and simultaneously—via the magnetic field inside the coil—with the magnetic field flux energy inside the core.

J. Dual Interactions with Pingponging between Them

Further, the two simultaneous interactions also *iteratively* interact with each other, in a kind of iterative retroreflection and interception of additional energy, so that a net amplification of the electrical energy output by the dually interacting coil results. The fact that iterative retroreflection processes can increase the energy collection from a given potential and enable $\text{COP} > 1.0$ has been previously pointed out [28]. In addition, multiple coils placed around the closed material flux path, forming a common core of each and all of them, all exhibit such gains and also mutual interaction with each other, leading to further gain in the energy output by the coils and their interaction processes.

In short, the novel process of this invention takes advantage of the previously unrecognized giant negentropy process [1,16,20] ongoing to and from the permanent magnet's dipole and between the complex plane of the vacuum energy and real 3-space energy flows constituting the magnetic vector potential and the magnetostatic scalar potential, to provide a gain in the total amount of electromagnetic energy being diverted from (drawn from) the permanent magnet by the attached circuit, components, and their processes.

The total collectable energy now drawn from the magnet is the sum of (1) the magnetic field energy (curled **A**-potential energy) flowing in the flux path, (2) the magnetic energy in the uncurled **A**-potential energy flowing in the surrounding space, (3) a further iterative "pingpong" gain component of energy caused by mutual and iterative interactions [28] of the multiple coils and their multiply interacting processes, and (4) additional energy that can be intercepted and diverged (collected) from the flowing uncurled **A**-potential energy flowing in the surrounding space, and converted into output electrical energy as the outputs of coils, by simply adding additional interceptors (separate receiving circuits with loads.).

We have thus discovered a process for amplifying the circuit's *available* output energy extracted from a permanent magnet dipole's energy outflow, where the dipolarity is an open system and a negative resistor, freely receiving excess energy from the surrounding active vacuum, transducing the received energy into usable form, and outputting the energy as a continuous flow of usable excess electromagnetic energy. Thereby, additional energy may be intercepted in a system employing this process, and the process can be used in practical EM power systems and EM power system processes having $\mathrm{COP} > 1.0$ when used in open-loop mode, and self-powering when used in closed-loop mode.

Further, we may utilize a collector or interceptor (such as a common coil wound around the flux path through it so that said flux path constitutes a core) that interacts with both available components of energy flow and with iterative interactions mutually between the two basic interactions. Each turn of the coil constitutes a $\nabla\times$ operator, bathed by the flowing uncurled $\mathbf{A}$ potential outside the line material. Hence the charges in the coil intercept the uncurled $\mathbf{A}$ flow, and curl the energy intercepted to produce a curled $\mathbf{A}$ flow, thus producing additional magnetic field $\mathbf{B} = \nabla \times \mathbf{A}$. This magnetic field is at its maximum in the exact center of the coil, which is in the exact center of the nanocrystalline core material with its retained $\mathbf{B} = \nabla \times \mathbf{A}$ field energy. Hence the coil interacts with two components of energy flow, because (1) the internal $\mathbf{B} = \nabla \times \mathbf{A}$ field energy is retained in the nanocrystalline material in the coil's core, (2) the external uncurled $\mathbf{A}$-potential energy flow striking its outside surface charges and changed into additional magnetic field energy and into additional electrical current flowing in the coil and out of it, and (3) in addition, iterative mutual interaction between the two basic interactions also occurs, increasing the energy gain and the coefficient of performance.

Any additional EM energy input into the core material and flux path increases the $\mathbf{B} = \nabla \times \mathbf{A}$ field energy flowing in the flux path, hence withdrawn from the vector potential $\mathbf{A}$ around the flux path, hence replenished from the permanent magnet dipole, and hence replenished to the magnet dipole from the complex plane, via the giant negentropy process [16,20]. This increased energy collection in the magnetic flux in the core material passes back through the permanent magnet (which is in the path loop and completes it), momentarily altering the effective pole strength of the magnet and thereby increasing the magnitude of the giant negentropy process associated with said dipole of the permanent magnet. In turn, this increases the outflow of $\mathbf{A}$ energy from the magnetic dipole, increasing both its output $\mathbf{B} = \nabla \times \mathbf{A}$ field energy in the flux path and its output uncurled $\mathbf{A}$-flow energy in space outside the flux path. This further increases the spacetime curvature of the local space surrounding the flux path material, since the energy density of said local spacetime has increased.

K. Varying the Pole Strength of a Permanent Magnet

Hence the process is the first known process that deliberately and interactively alters the pole strengths of the poles of a permanent magnet, utilizing the momentary alteration to vary and increase the pole strength and hence the magnitude of the energy density flowing in the giant negentropy mechanism [16,20]. From the general relativity view, it is the first known process that deliberately increases and structures the local curvature of spacetime, by electromagnetic means, so as to momentarily alter and increase the pole strength of a permanent magnet, using the pole strength alteration to increase the flow of energy into and out of the local spacetime, thereby increasing the curvature of the local spacetime and the resulting EM energy extracted therefrom.

Any extra uncurled **A**-flow energy increase outside the nanocrystalline flux path material increases the interaction with this field-free **A**-flow energy of any coil around the flux path, thereby increasing the magnetic **B**-field flux inside the flux path, and so on.

L. Regenerative Energy Gain

In short, the mutual iterative interaction of each coil wound on the flux path of the special nanocrystalline material, with and between the two energy flows, results in special kinds of regenerative energy feedback and energy feedforward, and regauging of the energy of the system and the energy of the system process. This excess energy in the system and in the system process is thus a form of free and asymmetric self-regauging, permitted by the well-known gauge freedom of quantum field theory. Further, the excess energy drawn from the permanent-magnet dipole is continually replenished from the active vacuum by the stated giant negentropy process [1,16,20] associated with the permanent magnet's magnetic dipole due to its broken 3-symmetry [18] in its energetic exchange with the vacuum.

As a result, each coil utilized is an amplifying coil containing an amplifying regenerative process, compared to a normal coil in a normal flux path that does not hold localized the $\mathbf{B} = \nabla \times \mathbf{A}$ field energy within its core material, and does not simultaneously interact with both internal **B**-field flux energy and external excess field-free **A**-potential.

M. Open System far from Equilibrium, Multiple Subprocesses, and Curved Spacetime

The entire system process is thus a self-regauging regenerative system process and an energy-amplifying system process, where the excess energy is freely furnished from the local curved spacetime as energy flows from the magnetic dipole of the permanent magnet and, in turn, is freely replenished to the permanent-magnet dipole by the giant negentropy process established in the active vacuum environment by the broken 3-symmetry of said magnetic dipole [18] and the concomitant locally curved spacetime.

The system process is thus an open electromagnetic process far from thermodynamic equilibrium [2–4] in its active environment (the active vacuum), freely receiving excess energy from said active environment via the broken 3-equilibrium of the permanent-magnet dipole. Each coil is an open system freely receiving excess energy from its active environment (the active field-free **A** potential flowing through the space occupied by the coil and surrounding it), and creating a local curved spacetime by its extra energy density, while also receiving energy from its internal environment, the **B**-field magnetic flux in the material flux path through the center of the coil and making up its core, and also curving the local spacetime by means of the extra energy density of the local spacetime.

The system process is also a general relativistic process [33,36,37,41,45,63] whereby electromagnetic energy is utilized to curve local spacetime, and then the locally curved spacetime continuously acts back on the system and process by furnishing excess energy to the system and process directly from the curved spacetime; the excess energy is continually input to the system from the imaginary plane (time domain) [1,16,20].

VI. SUMMARY OF THE PROCESS FROM VARIOUS ASPECTS

We summarize the many aspects of the overall process as follows, taking advantage of the following facts:

1. The magnetic flux and magnetic vector potential **A** are freely and continuously furnished by a permanent magnet to a material flux path, where the material flux path holds all curled vector potential **A** and thus all magnetic field inside the flux path, and where the permanent magnet freely furnishes additional field-free magnetic vector potential **A** to replenish the **B**-field (curled magnetic vector potential **A**) energy that was confined to the interior of the material flux path, and where multiple intercepting coils and processes are utilized with mutual iterative positive feedforward and positive feedback between the collectors and subprocesses to increase the energy collected and hence increase the COP of the system and system process.

2. A previously unrecognized giant negentropy mechanism is used as shown unwittingly by Whittaker [1] in 1903, recognized by Bearden [16] and further clarified by Evans and Bearden [20], and the active vacuum continuously replenishes all magnetic vector potential **A** (both curled and field-free) that is continuously output by the permanent magnet into the material flux path and into space surrounding the material flux path. Further, the replenishment energy flow from the active vacuum is from the time domain [1,16,20] and thus from the complex plane, constituting the continuous input of reactive power by the active vacuum environment via time-like energy flows. These time-like potentials and energy flows are known in extended electrodynamics [26,27,29,32,33, 36,37,63–65] but were not previously deliberately utilized in electromagnetic

systems, particularly in EM power systems, even though shown by Whittaker [1] as early as 1903.

3. The field-free magnetic vector potential **A** is continually replenished and remains (with replenishment by the vacuum to the permanent-magnet dipole and thence replenishment from the magnet dipole to the space surrounding the material flux path) when a material flux path is utilized wherein the magnetic field associated with a permanent magnet's flux, through the flux path, is held internally and entirely in the material flux path, with the field-free magnetic vector potential **A** remaining in space surrounding the flux path.

4. A coil will interact with either a magnetic field (i.e., the curl of the **A** potential) or a changing **A** potential where no magnetic field (no curl) is present, or simultaneously with a combination of both a curled **A** potential (with magnetic field **B**) and a field-free **A** potential (without magnetic field **B**) if the two are separated. Indeed, there is a "pingpong" reiterative interaction between the two processes in the coil, constituting positive feedforward from each to the others, and positive feedback from each to the others.

5. A simultaneous interaction of a coil with both a magnetic field (curl of **A**) and a field-free **A** potential produces electromagnetic energy in the form of voltage and current in an external circuit connected to the coil, and the net voltage and amperage (power) produced by the coil is a result of the summation of both simultaneous but separated interactions with the coil and its Drude electrons and of the iterative "pingpong" interactions between the two simultaneous interactions, and therefore the summation provides a greater coil output energy than is produced by the coil from either the magnetic field (curled **A**) separately, or the field-free **A** potential separately, or from both when unseparated. Further, the "pingpong" iterative interaction adds additional energy collection and gain to the electrical power output of the coil.

6. Multiple coils are wound on the material flux path, where magnetic flux is input to the material flux path from a permanent magnet, and where the material flux path holds internally all curl of **A** (magnetic field) from the permanent magnet's flux, so that (a) magnetic field and magnetic flux from the permanent magnet are inside the closed material flux path, (b) no magnetic field is outside the closed material flux path, and (c) a field-free magnetic vector potential **A** replenishes the curled **A** potential held in the material flux path, and where the replenished field-free magnetic vector potential **A** occupies the space outside the material flux path and flows through the surrounding space.

7. A broken 3-space symmetry exists of a magnetic dipole [18] of a permanent magnet, well known in particle physics since 1957 but inexplicably not yet added into classical electrodynamics theory, wherein the broken symmetry of the magnetic dipole rigorously requires that the dipole continually absorb magnetic energy from the active vacuum in unusable form, and that the

broken symmetry output (reemit) the magnetic energy in usable form as real magnetic field energy in 3-space and real magnetic vector potential in 3-space. The receipt of unusable EM energy, transduction into usable form, and output of the usable EM energy, constitutes a true negative resistance process [16,20] resulting from the ongoing giant negentropy process engendered by the broken 3-symmetry of the magnetic dipole of the permanent magnet.

8. Whittaker's 1903 mathematical decomposition [1] of any scalar potential applies Whittaker decomposition to the magnetostatic scalar potential existing between the poles of the permanent magnet, revealing that the magnetostatic scalar potential of the permanent magnet is composed of a set of harmonic longitudinal EM wavepairs, where each wavepair consists of a longitudinal EM wave and its phase conjugate replica wave.

9. The incoming half-set of Whittaker decomposition waves consists of the phase conjugate waves, which are all in the imaginary plane [16,20] prior to interaction and continuously converging upon the magnetic charges of the permanent-magnet dipole at the speed of light. The incoming, converging longitudinal EM waves are continuously absorbed from the imaginary plane by the magnetic charges (magnetic poles), so that the permanent magnet dipole is continuously replenished with time-like energy flow from the active vacuum environment, while continuously transducing the received time-like energy into 3-spatial energy, and outpouring real EM energy flow in the form of the longitudinal EM Whittaker waves [1] emitted in 3-space in all directions.

10. The other half-set of the Whittaker decomposition waves, consisting of outgoing real EM Whittaker longitudinal waves [1] in 3-space, is continuously and freely emitted from the permanent-magnet dipole charges (poles) and continuously diverges outward in space in all directions from the permanent-magnet dipole at the speed of light. Thus there is revealed and used a process for a natural, continuous source of magnetic energy from the vacuum: a continuous EM wave energy flow convergence of electromagnetic energy from the vacuum to the magnetic dipole, but in the imaginary plane and hence constituting a continuous energy input in the form of imaginary power [16,20], In this process the absorbed magnetic energy is transduced into real power and reemitted in real 3-space in all directions, whereby the absorption of energy from the vacuum from the imaginary plane (time domain) is in 4-flow equilibrium with the re-emission of the absorbed energy in 3-space, but not in 3-flow equilibrium, and where the outgoing real magnetic energy provides the surrounding magnetic field and the surrounding magnetic vector potential of the permanent magnetic dipole.

11. In this manner the broken 3-symmetry of the magnetic dipole (permanent magnet) allows the dipole to continuously receive reactive power from the vacuum's time domain, transduce the reactive power into real EM power in 3-space, and reemit the absorbed energy as real magnetic energy pouring into

space and consisting of both a magnetic field and a magnetic vector potential. Thus the permanent magnet, together with its Whittaker-decomposed [1] magnetostatic scalar potential between its poles, represents a dynamo and an energy transducer, continuously and freely receiving energy from an external source (the active vacuum) in the complex plane and transducing the received complex plane EM energy into real EM energy [16,20], and radiating the real EM energy into real space as real EM power. EM energy flow conservation in 3-space is permissibly violated because of the broken 3-symmetry of the magnetic dipole, but EM energy flow in 4-space is not violated and is rigorously conserved. There is no law of nature requiring energy conservation in three dimensions and 3-space; instead, energy conservation is required by the laws of nature and physics in 4-space. The additional condition usually assumed—that energy conservation is also always conserved in 3-space—is not required by nature, physics, or thermodynamics, and the additional 3-conservation requirement is removed by this process in any dipole, by the broken 3-symmetry of the dipole. It is this newly recognized giant negentropy process advanced by Bearden [16] and extended by Evans and Bearden [20] that is directly utilized by this new power system process, in conjunction with directing and interacting material flux paths, intercepting coils, separation of curl of the **A** potential (i.e., the **B** field), and the field-free **A**-potential (replenished from the vacuum), and interaction of a coil with a magnetic field and magnetic flux running through a material core through the coil, and with an external field-free magnetic potential reacting with the coil. The foregoing actions provide a magnetic system that receives—via the permanent-magnet dipole—replenishment EM energy from the active vacuum to the dipole, and from the dipole to the circuit and the space surrounding it, to enable the permanent magnet to continuously furnish magnetic field and flux to a flux path in the process, and continuously furnish both the curl energy of the **A** potential and the field-free energy of the **A**—potential replenished from the vacuum. This system should also have multiple coils interacting simultaneously with both curled **A** potential and magnetic flux inside the coils, while also interacting simultaneously with field-free magnetic **A**-potential from the space in which the coil is embedded, such that excess energy is added to the interacting coils by dA/dt from the changing field-free **A**-potential in space, and where the field-free **A**-potential in space is continuously furnished by the permanent magnet dipole and the extra energy for the furnished field-free **A**-potential is continuously received by the permanent magnet dipole from the active vacuum exchange, via the process shown by Whittaker's decomposition [1] and elaborated by Bearden [16].

12. The difficulty heretofore experienced by designers, engineers, and scientists with using the magnetic energy continuously emitted to form the static field and magnetic scalar potential of a permanent magnet dipole is that all schemes for using the magnetic energy have relied on physical motion, energy input to

overcome the field of the permanent magnet, or other brute-force methods. This invention provides a new process for a coil to extract excess EM energy from the magnetic vector potential energy in space from the permanent magnet, while simultaneously interacting with the magnetic field energy of the permanent magnet flowing through a flux path through the center of the coil but not in space surrounding the flux path. The Whittaker decomposition shows that when the system extracts EM energy from the magnetic vector potential **A** and magnetostatic scalar potential Φ, the energy to continuously form and maintain the magnetic vector potential's vector current is continuously replenished from the vacuum by the convergent reactive EM power being input from the imaginary plane (time domain). Evans and Bearden [20] have also shown that, in the most general form of the vector potential deduced from the Sachs unified field theory [41], EM energy from the vacuum is given by the quaternion-valued canonical energy–momentum. Further, the most general form of the vector potential (i.e., flowing EM energy in vector potential form) has been shown by Evans and Bearden [20] to contain *longitudinal and time-like components* (*energy currents*), in agreement with the simpler Whittaker decomposition [1] as a special case, but much richer in available structure than Whittaker's decomposition. Evans and Bearden [20] have also shown that the scalar potential is in general a part of the quaternion-valued vector potential, and can be defined only through suitable choice of metric for a given experimental setup. They have shown the energy current in vacuum in this more advanced treatment in O(3) electrodynamics, and it is this demonstrated vacuum EM energy current that continuously replenishes any excess energy drawn from the permanent magnet dipole's magnetostatic scalar potential to replenish the curl of the **A** potential (the magnetic **B**-field energy) held inside the material flux path powered by the magnet and also to replenish the field-free **A**-potential filling space around the material flux path.

13. A special nanocrystalline material is contained in the closed flux path powered by the permanent magnet, where the nanocrystalline material performs the highly special function of separating and retaining the curl energy of the **A** potential (i.e., retaining the magnetic field energy) inside the material flux path along with the magnetic flux. The nanocrystalline material consists of coiled flat "tape" layers of material, with the layers acting in the fashion of a perfect toroid to retain all magnetic field (curled **A** potential) inside the material, while having the curl-free **A**-potential filling all space outside the nanocrystalline material.

14. This special nanocrystalline material may be further considered in the manner of a "layered" magnetic flux path material, wherein (a) the "layers" are a molecule in thickness; (b) essentially all eddy currents are eliminated or reduced to completely negligible magnitude; (c) as a result, the nanocrystalline material does not dissipate magnetic energy from the flux path; (d) as a result,

the nanocrystalline material does not produce eddy currents; (e) as a result, the nanocrystalline material does not exhibit heating since heat consists of scattered and dissipated energy; and (f) no such scattering or dissipating of the magnetic flux energy occurs in the nanocrystalline material. Thus the system process is able to process significant power and energy without heating of the core flux path material at all, and without requiring cooling of the core material, as these characteristics have a remarkable advantage over other core materials subject to eddy currents, substantial heating, and the need for cooling.

15. The magnetic flux from a permanent magnet provides a source of magnetic flux energy to and within the nanocrystalline material in a flux path, such that the nanocrystalline material holds the magnetic field component (curl energy of the vector potential **A**) in the flux path while the flux path material itself is not further interacting with the field-free magnetic vector potential and its energy that fill the space around the closed material flux path, and where the field-free magnetic vector potential—in space external to the material flux path—geometrically follows the directions and turns of the material flux path but outside it.

16. Any coil immersed in the nanocrystalline material's magnetic vector potential in space, but not wound around a portion of said nanocrystalline flux path, will react to the magnetic vector potential and its energy. Each turn of a coil acts as a curl operator, producing a magnetic field due to the received energy current from the magnetic vector potential. If an electrical current is passed through the coil, and if the magnetic flux produced by the electrical current in the coil is aligned with the magnetic flux that is produced by the coil's interaction with the magnetic vector potential from the nanocrystalline flux path material, then the two magnetic vector potentials will vectorially add, so that the magnetic field produced by the current through the coil will be augmented by the curl operation of the coil now acting on an increased magnetic vector potential summation consisting of the vector sum of the curled magnetic vector potential and the field-free magnetic vector potential.

17. Any time rate of change of a magnetic vector potential, either curled or curl-free, constitutes an electric field, which, in an interacting pulsed coil, produces pulsed voltage across the coil and pulsed current through the coil and produces current and power in a closed circuit loop consisting of the interacting coil and a connected external circuit. The excess magnetic energy is received by the circuit from the magnetic dipole of the permanent magnet (and replenished to the dipole from the time-like Whittaker energy currents from the active vacuum). The process transduces the excess magnetic energy into excess electrical energy, and outputs the excess electrical energy into electrical loads to power them, whereby the excess energy received and replenished from the active vacuum environment via the giant negentropy process of the permanent magnet dipole [16,20] allows system COP > 1.0.

18. Further, with COP > 1.0, a fraction of the output electrical energy from the process and system can be extracted and positively fed back to the operator input of the system and process (such as the electrical energy fed to a driver coil), with governing and clamping control of the positive-feedback energy magnitude, so that the system process becomes self-powering, freely powering itself and its loads, receiving all the energy from an external energy source due to its broken symmetry in its vacuum exchange and the resulting giant negentropy process [16,20] thereof, and thus constituting an open system far from thermodynamic equilibrium with its active environment, easily self-regauging and powering itself and its load simultaneously by dissipation of energy freely received from its active environment.

19. The dual-action effect is increased in an interacting coil if the coil is wound around a portion of the nanocrystalline flux path, due to the permeability of the flux path as a magnetic core and the input of flux from the permanent magnet. In this case, the increased magnetic field produced inside the coil also interacts with the magnetic flux path core in its center, producing an increased change in the magnetic flux in the nanocrystalline material itself. The coil interacting in such dual fashion thus has *iterative* "energy feedforward and feedback" between the two simultaneous processes, one process proceeding outward from inside, and the other proceeding inward from outside. A convergent series of summing energy additions (regaugings) thereby occurs in the coil, thus producing *energy amplification* in the coil.

20. As a result of the combined actions listed above, the coil's energy output increases on receiving and transducing extra energy from the magnetic flux of the permanent magnet, the magnetic field of the permanent magnet, and an extra field-free magnetic potential **A** surrounding the flux path and continuously furnished by the permanent magnet. The production of one or more potentials is also the production of one or more regaugings. As is well known from electrodynamics and gauge field theory, gauge freedom is permitted freely and at will. In electrodynamics, regauging to change the potential of a system simultaneously changes (freely) the potential energy of the system. In the present process, this regauging is physically applied by (a) holding the curl of the **A** potential inside the special nanocrystalline flux path material, (b) furnishing additional field-free magnetic vector potential **A** from the permanent magnet dipole, and (c) continuously replenishing the potential and energy from the vacuum.

21. This process provides for a permissible gain in the magnetic energy output of the interacting coil(s) for a given amount of energy input by the (a) operator to the active coil (used similar to a "primary" coil in a transmitter) or (b) a clamped, governed positive energy feedback of a fraction of the energy output from one output coil back to the input coil. This additional magnetic field

energy output is retained in the nanocrystalline flux path, and the additional magnetic vector potential energy output moves through space surrounding the coil. The system process of the invention thus is a new process for energy amplification, with the excess output energy freely received from the vacuum and thence to the permanent-magnet dipole, and thence from the permanent magnet dipole to the other parts of the system, via the giant negentropy process associated with the magnet dipole as shown by Evans and Bearden [20] and Whittaker [1].

22. Every multiple coil wound around the special nanocrystalline material flux path will exhibit regauging energy gain by the processes described above, with the energy continuously replenished from the vacuum to the source dipole and from the source dipole to process, via the process illustrated by Whittaker decomposition.

23. Any scalar potential such as the magnetostatic scalar potential between the poles of a permanent magnet, and the magnetic vector potential considered as two scalar potential functions in the manner shown by Whittaker [56], are continually replenished energy flow processes [1,16], so that any system utilizing the output flow from the permanent-magnet dipole and containing such dipole, is an open system far from equilibrium in the replenishing vacuum flux, as shown by the Whittaker decomposition [1] and more precisely expanded by Evans and Bearden [20].

24. The entropy of any open system in disequilibrium with its vacuum environment is a priori less than the entropy of the same system in equilibrium, and, in fact, the entropy of such an open system cannot even be computed, as pointed out by Lindsay and Margenau [66][26] and as well known in physics.

25. This process, producing an open system in disequilibrium with a recognized continuous source of energy [2–4,16,20], is permitted to perform any of five functions: (a) self-order, (b) self-oscillate, (c) output more energy than the operator inputs (the excess energy is freely received from the active environment), (d) power itself and its loads and losses (all the energy is freely received from the active environment), and (e) exhibit negentropy. The process specified by this invention permits all five functions. For example, by extracting some of the output energy from an output section (coil) used in a system employing the process, and feeding the extracted energy back to the input section (coil), the energy gain of the system permits it to become self-oscillating

[26] When a system departs from equilibrium conditions, its entropy must *decrease*. Thus the energy of an open system not in equilibrium must always be greater than the energy of the same system when it is closed or in equilibrium, since the equilibrium state is the state of maximum entropy. Thus, broken 3-equilibrium is a broken 3-symmetry between the active vacuum and material systems, and it is a *negentropic* operation.

and hence self-powering, while obeying energy conservation, the laws of physics, and the laws of thermodynamics.

26. Multiple feedforward and feedback subloops exist between the various parts of a complete flux path loop and **A**-potential flow loop, so that regenerative energy collection gains are developed in the various subprocesses of the overall process. The result is an overall feedback summation and overall feedforward summation, whereby the system process regauges itself with **A**-potential flow energy from the magnetic dipole, and the regauged energy is continuously replenished and received from the active vacuum via the stated giant negentropy process. Increasing the number of interacting coils and/or increasing the magnetic flux results in an increase in the COP, limited only by the saturation limit of the core material.

27. The system process consists of a magnetic negative-resistor process, where energy is received freely in unusable form (pure reactive power from the time domain of the spacetime vacuum), transduced into usable form, and output in usable form as real EM energy flow in 3-space [1,16,20].

28. The system process is a permissible local energy gain process and a self-regauging process, freely increasing the process's and system's potential energy and receiving the regauging energy from the active vacuum via the giant negentropy process [16,20], and freely collecting and dissipating the excess regauging potential energy in loads.

29. The system process may be open-loop where the operator inputs some electrical energy and the system outputs more electrical energy than input by the operator; the excess energy is freely received from the permanent-magnet dipole and from the vacuum to it via the giant negentropy process [16,20], with process transductions of the various energy forms between magnetic form and electrical form.

30. The system process may be closed-loop and "self-powering," where a portion of the amplified energy output is extracted, rigidly clamped in magnitude, and positively fed back to the input. This replaces the operator input entirely, and all energy input to the system process is received from the vacuum through the permanent magnet dipolarity's Whittaker decomposition and constituting direct system application of the stated giant negentropy process [16,20].

31. The system process is a magnetic regenerative gain process, outputting more energy than the operator personally inputs, with the excess energy received from the active vacuum via the broken 3-symmetry of the dipole which initiates and sustains the giant negentropy process [16,20], whereby EM energy continuously flows into the system from the complex plane (time domain), is transduced into usable magnetic energy in real 3-space, and is then transduced into ordinary electrical energy by the system process, thereby powering both the system and the loads.

32. In an embodiment using the process, all coils exhibit energy gain and increased performance, as does the overall system. All coils are energy amplifying coils, each with gain in energy output greater than 1.0 compared to the same coil without simultaneous but separate exposure to and interaction with separate inputs of field-free magnetic potential and magnetic field, and without iterative positive feedback between the two simultaneous interactions.

33. For power system processes, the combined process requires using at least one primary (active) coil in dual interaction with iterative feedback between the duals, and one secondary (passive) coil in dual interaction with iterative feedback between the duals, both on a common nanocrystalline flux path. The resulting "minimum configuration" embodiment produces a power system that is an open thermodynamic system, not in equilibrium with its external environmental energy source: to wit, the continuous inflow of EM energy from the complex plane of the active vacuum into the permanent magnet dipole, and the continuous outflow of real magnetic energy from the permanent magnet dipole, with the holding of the magnetic field energy and magnetic flux energy inside the nanocrystalline material in the closed magnetic flux path, and with excess field-free magnetic vector potential filling the surrounding space, as continuously furnished by the permanent-magnet dipole and continuously replenished to the dipole by the active vacuum as a result of the dipole's broken 3-symmetry in its vacuum energy exchange and the giant negentropy 4-space energy flow operation [16,20] initiated thereby.

34. This process permissibly violates 3-symmetry energy conservation, but rigorously obeys 4-symmetry energy conservation and thus it has not been applied in electrical power systems. The basic excess energy input is received from an unusual source: the complex plane (time domain) of the locally curved and active spacetime (vacuum), as shown by the Whittaker decomposition [1] of the permanent magnet's magnetostatic scalar potential between its poles, and as further demonstrated in several AIAS papers [20,36,37,64,65], and as recognized by Bearden [16] and investigated more deeply by Evans and Bearden [20]. The energy into the nanocrystalline flux path material is input directly from the permanent magnet as magnet flux energy. From its dual interaction with two magnetic energy components, the active (driving) coil produces increased magnetic field flux in its center and thus in its interaction with the magnetic flux path and the magnetic flux in the flux path, and also produces increased magnetic flux back through the permanent magnet dipole, thereby momentarily altering the pole strength of the permanent magnet, and also produces increased field-free **A** potential in space surrounding both the permanent magnet and the nanocrystalline flux path, and flowing geometrically in the direction taken by the nanocrystalline flux path. From its dual interaction with two magnetic vector potentials as well as two magnetic field components

superimposed, the passive (driven) coil produces increased EM field energy in the form of current and voltage out of the coil and into any conveniently attached external load for dissipation in the load by conventional means.

35. The process in this invention uses and applies an open system process for receiving excess energy from an external source (the permanent-magnet flux and **A** potentials), and since the permanent-magnet flux and **A** potentials are continuously replenished from the vacuum via the broken 3-symmetry of the magnet dipolarity via its Whittaker decomposition [1], the process is allowed to (a) be adapted in systems to produce $COP > 1.0$ and (b) be close-looped with clamped positive feedback from load output to input, so that the system powers itself and its load simultaneously.

36. The open system far from equilibrium process of this invention thus allows electromagnetic power systems to be developed that permissibly exhibit a coefficient of performance (COP) of $COP > 1.0$. It allows electromagnetic power systems to be developed that permissibly (a) power themselves and their loads and losses, (b) self-oscillate, and (c) exhibit negentropy.

37. No laws of physics or thermodynamics are violated in such open dissipative systems exhibiting increased COP and energy conservation laws are rigorously obeyed. Classical equilibrium thermodynamics does not apply and is permissibly violated. Instead, the thermodynamics of open systems far from thermodynamic equilibrium with their active environment—in this case the active environment-rigorously applies [2–4].

38. This appears to be the first magnetic process deliberately utilizing and separating special energy flow processes—associated in a curved spacetime with the permanent-magnet's dipolarity—to provide true magnetic energy amplification, receiving the excess energy freely from the permanent-magnet dipole, with said energy continually replenished to the dipole from the imaginary plane in spacetime by the giant negentropy process [16,20].

39. This appears to be the first power system process that in open-loop mode receives electrical energy input by the operator or outside normal source, wherein (a) the electrical energy input is transduced into magnetic energy flows, (b) curled **A**-potential flow (magnetic field energy flow) is separated from field-free **A**-potential flow, (c) dual and iterative "pingpong" interactions of energy feedforward and feedback occur in each active component, (d) the feedforward and feedback pingpong interactions create local energy gain in each active component, (e) the magnetic potential energy of the system is self-regauged and increased (receiving the excess by giant negentropy replenishment from the active vacuum), (f) the increased magnetic energy flow is then re-transduced into electrical energy and output to power loads, (g) the output energy powering the loads is greater than the input energy provided by the operator.

40. The foregoing system functions as described, where the system process positively feeds back a clamped fraction of its electrical output to its electrical input, result in a regenerative, energy amplifying, self-regauging open system process that powers itself and its loads, where the powering energy is freely received from the active vacuum curved spacetime via the giant negentropy process [1,16,20].

41. The process therefore appears to be the first process for an electrical power system that permissibly violates EM energy conservation in 3-space, due to the use of the recognized and proven broken 3-symmetry of the dipole [18], but while rigorously conserving electromagnetic energy in 4-space [16,20]. It therefore appears to be the first electrical power system process that enables the use of a clamped positive feedback from output to input in an electrical power system having COP > 1.0 so that the system continuously receives all the energy—to power its loads and losses—from the magnetic energy flow of a permanent magnet, where the energy flow is continually replenished to the permanent magnet by the energy circulation from the imaginary plane (as absorbed reactive power) and transduced into real power output by the magnet's dipole, and where the freely received magnetic flux energy from the dipole is separated into field energy and magnetic vector potential energy.

42. This appears to be the first COP > 1.0 electrical power system process that deliberately takes useful advantage of the fact that any amount of energy can be intercepted and collected from a potential, regardless of its magnitude, if sufficient intercepting charges (in this case, magnetic charges, or pole strengths) are utilized. In this case, coils utilized around the special nanocrystalline core material interact with the field-free magnetic vector potential filling the space occupied by the electron spins in the Drude electron gas in the coils, while simultaneously the produced magnetic field in the coil due to its curl operation interacts with the localized magnetic field flux in the nanocrystalline flux path and core, and vice versa. In this way, the energy interception and collection is effectively multiplied beyond what is obtained by a coil with a core operating in normal magnetic field coupled to its magnetic vector potential in space.

43. This appears to be the first magnetic process that is a proven true negative resistance process, where "negative resistance process" is defined as a process whereby electromagnetic energy is continuously and freely received in unusable form, converted into usable form, and continuously output in usable form.

44. This appears to be the first power system process that deliberately uses energy to perform more than one joule of work per joule of original input energy, by transforming a given amount of replenished energy into a different form, thereby performing work in the same amount on a receiving medium while retaining the energy in its new form, then transforming that energy back

into the first form again, thereby again performing work in the same amount on the receiving medium again while retaining the energy back in its original form, and so on in "pingpong" iterative fashion. The fraction of the energy that is retained from one transformation to the other determines the increase in energy of the medium receiving the work and thus being excited with kinetic energy, and thereby determines the energy gain of the power system in the multiplicity of such regenerative processes used in the system.

45. This appears to be the first magnetic process for EM power systems that deliberately creates and uses curved local spacetime to provide continuous energy and action on the process's active components and subprocesses. Sachs' unified field model [41–45] as implemented by one of its important subsets—O(3) electrodynamics per Evans [63] and Vigier—is implemented in the system to provide several specific local curvatures of spacetime, and excess energy is thereby regauged into the system and used to power loads, including a self-powering system that powers itself and its loads simultaneously, and also including an open-loop system wherein the operator inputs a little EM energy and obtains more EM energy being dissipated as work in the load.

VII. RELATED ART

There is believed to be no prior art in such true magnetic negative resistor processes for

1. Utilizing curvatures of local spacetime to provide excess energy from spacetime input into the various active components of the system process
2. Receiving EM energy from the spacetime vacuum in unusable reactive power form
3. Having the permanent-magnet dipole convert the received unusable EM energy into usable magnetic energy form
4. Splitting the magnetic energy output of the permanent magnet into separate magnetic field flux and both curled and uncurled magnetic vector potential current, each traveling in a different spatial pathway
5. Producing energy amplification by dual interaction of multiple simultaneous processes in a coil, with iterative feedforward and feedback between the simultaneous interactions
6. Producing driving and driven coils both in curved spacetime and with their magnetic flux inside a nanocrystalline core inside said coils and with a field-free magnetic vector potential in the space in which the coil is embedded
7. Transducing the excess magnetic energy available for output, into electrical energy

8. Outputting the excess energy as ordinary electrical energy—consisting of voltage and current—to power circuits and loads
9. Permissibly exhibiting $\mathrm{COP} > 1.0$ while rigorously obeying energy conservation, the laws of physics, and the laws of thermodynamics
10. Being operated in either open-loop or closed-loop fashion. In open loop the operator inputs a lesser EM energy than is dissipated in the load; in closed loop a fraction of the output energy is positively fed back into the input to power the system and system process, while the remainder of the energy is dissipated in the load to power it
11. Using and applying the extended work-energy theorem for a replenishing potential environment

This appears to be the first process to take advantage of the above listings of operations, functions, and processes, in which no heating or eddy-current dissipation is produced in the cores of coils utilized in embodiments of the process, and where said process and embodiments output electrical power in loads without the need to cool the process components.

The closest *somewhat* related work would appear to be several patents of Raymond C. Gelinas [67][27] in that these patents use the curl-free magnetic vector potential. All the Gelinas patents deal with communications and receivers and transmitters, have no application to electrical power systems, do not use additional EM energy extracted from a permanent magnet and replenished by the vacuum, do not use curved local spacetime, do not use the giant negentropy process, do not function as open systems far from equilibrium in their vacuum exchange, do not use iterative pingpong feedforward and feedback in their various components to achieve gain, do symmetrically regauge themselves so that their excitation discharge is symmetric and not asymmetric, do not function as negative resistors, are not self-powering, cannot produce $\mathrm{COP} > 1.0$, cannot self-operate in closed-loop form, and can and do produce only $\mathrm{COP} < 1.0$. They therefore have no application to the field of the present invention.

A. Description of the Figures and System Operation

The process in this invention is described below and in the process gain block diagrams cited below, which are intended to be read in conjunction with the following set of drawings, which include (1) the background Lenz reactions, Poynting and Heaviside energy flow operations, Heaviside energy flow

[27] All these Gelinas patents are assigned to Honeywell. All deal with communications, have no application to electrical power systems, do not use additional EM energy extracted from a permanent magnet and replenished by the vacuum, do not use curved local spacetime, do not use the giant negentropy process, do not function as open systems far from equilibrium in their vacuum exchange, symmetrically regauge themselves so that their excitation discharge is symmetric and not asymmetric, and produce only $\mathrm{COP} < 1.0$.

component, giant negentropy operation, Whittaker's decomposition of the scalar potential, and creation and use of curved local spacetime utilized in the invention; (2) the principles, the functional block diagram, a physical laboratory test and phenomenology device; and the process operation of the invention as well as typical measurements of a laboratory proof-of-principle device; and (3) the replication of the MEG by Jean-Louis Naudin (see footnote 10, above).

Figure 1 graphically shows Whittaker's decomposition [1] of the scalar potential into a harmonic set of phase conjugate longitudinal EM wavepairs. The 3-symmetry of EM energy flow is broken [16,20] by the dipolarity of the potential, and 4-symmetry in energy flow without 3-flow symmetry is implemented [1,16].

Figure 2 expresses this previously unexpected functioning of the scalar potential—or any dipolarity, including the magnetic dipole of a permanent magnet—as a true negative resistor [see footnote 24, above), receiving energy in unusable form, transducing it into usable form, and outputting it in usable form. Any EM potential is itself a true negative resistor process.

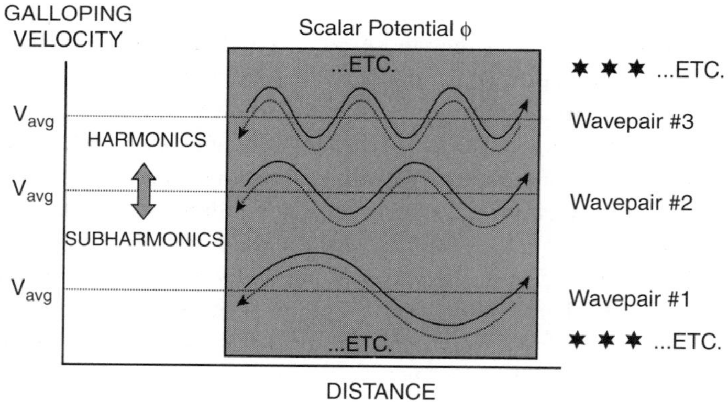

The Structure Is:

- A harmonic set of longitudinal wavepairs.
- In each wavepair the two waves superpose spatially, but travel in opposite directions. The two are phase conjugates and time-reversed replicas of each other.
- The convergent wave set in the imaginary plane, and hence is not observable.
- The charge's spin is 720 degrees, 320 in the real plane and 320 in the imaginary plane.
- Hence the charge receives the complex convergent EM energy, transduces it into real EM energy, and emits enormous energy at the speed of light in all directions.
- This produces the fields and potentials from the "source charge."

Figure 1. The scalar potential is a harmonic set of phase conjugate longitudinal EM waves.

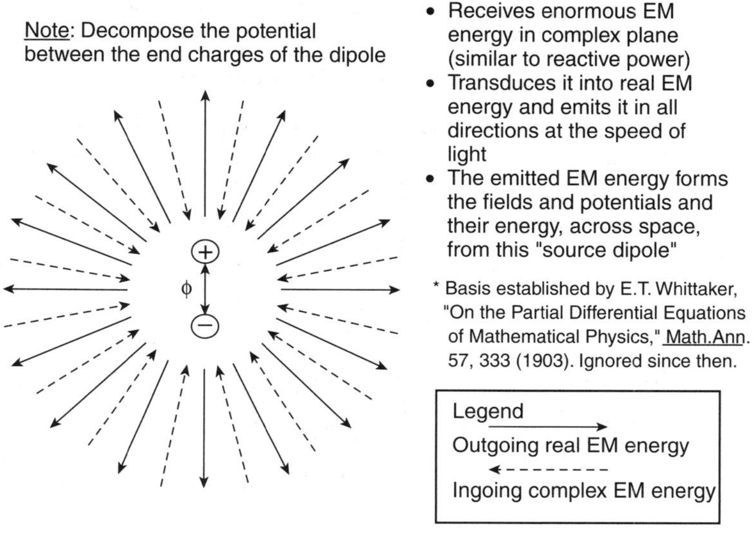

Figure 2. The dipole is a true negative resistor.

Figure 3 shows the startling ramifications of this previously unsuspected process. An ongoing, free, negentropic reordering of a fraction of the local vacuum energy [16] is initiated, spreads at the speed of light in all directions from the moment of formation of the dipole, and continues as long as the dipole and its broken 3-symmetry exists. We have previously stated [19,68,69] that the energy input to the shaft of a generator, and the chemical energy of the battery, have nothing to do with powering the external circuit connected to the battery or the generator. The available internal energy dissipated by the generator or battery does not add a single joule per second of energy flow to the external circuit. Instead, the available internal energy is dissipated internally and only to power its own losses and to force the internal charges apart, forming the internal source dipole connected to the terminals. The energy input to a generator and expended by it, and the chemical energy available by a battery and expended by it, thus are expended only to *continuously re-form the source dipole that the closed current loop circuit continuously destroys*.

Once established, the source dipole applies the giant negentropy process [16,20] shown in Figs. 1–3. Energy is continuously received by the dipole charges from the surrounding active and negentropically reordered vacuum (curved spacetime), transduced into usable form, and output as real EM energy flow in 3-space. The dipole's receipt of this energy as reactive power freely absorbed from the vacuum, does not yet appear in present classical electrodynamics texts. The texts do not include the vacuum interaction, much less the

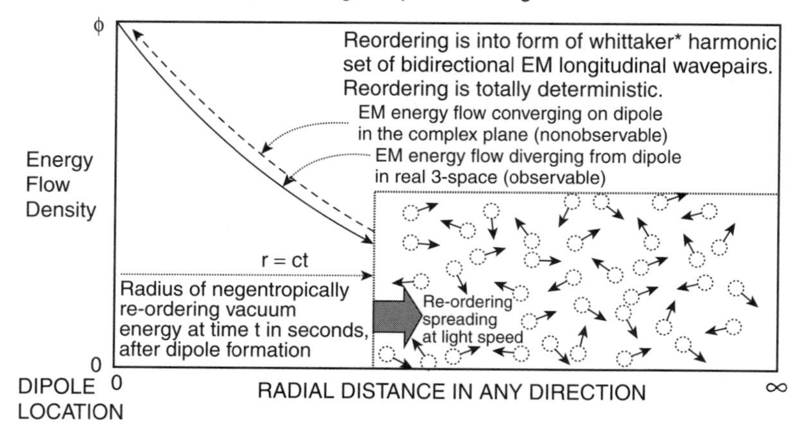

Figure 3. The dipole's broken 3-symmetry initiates a spreading giant negentropic reordering of a fraction of the vacuum's energy.

broken symmetry of the source dipole in that vacuum exchange, even though such has been proved in particle physics since the 1950s. The present invention is believed to be the first applied process using this previously omitted procedure of easily extracting energy from the vacuum and outputting it in usable transduced form as real EM energy flow, via the giant negentropy process [16,20].

The transduced EM energy received from the vacuum by the source dipole, pours out of the terminals of the battery or generator and out through space surrounding the transmission lines and circuits connected to the terminals (Fig. 4) as shown by Kraus [70][28] As is well known, the energy flow (Fig. 4) fills all space surrounding the external circuit conductors out to an infinite lateral radius away [70]. This is an *enormous* EM energy flow—when one includes the space-filling nondiverged component discovered by Heaviside [48–50]. This neglected vast nonintercepted, nondiverged energy flow component was never even considered by Poynting [51], and was arbitrarily discarded by Lorentz [52] as "of no physical significance."

Figure 5 shows that almost all that great EM energy flow—pouring out of the terminals of the generator or battery and out through the surrounding space

[28] Figure 12-59 on p. 576 shows a good drawing of the Poynting flow component that is withdrawn from that huge energy flow filling all space around the conductors, with almost all of it not intercepted, not diverged into the circuit, but just wasted. The amount of Poynting energy being withdrawn into the conductors is given by the numbers Kraus assigns to his contours. As can be seen, the Poynting effect is relatively localized.

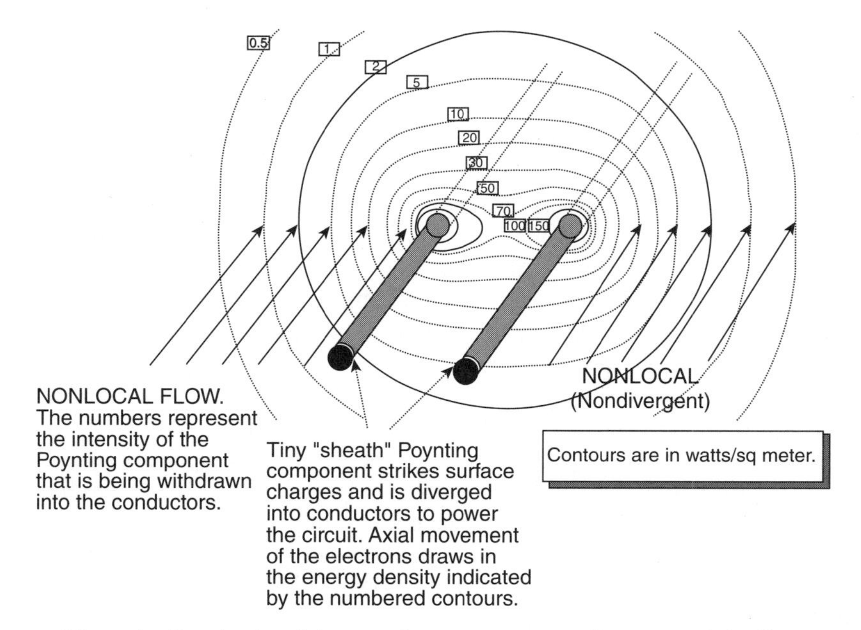

Figure 4. Poynting (caught) energy flow contours surrounding a transmission line.

surrounding the transmission line conductors—misses the circuit entirely and is simply wasted in conventional circuits having no iterative feedback and feed-forward additional collection components and processes. In a simple circuit, for example, the arbitrarily discarded Heaviside nondiverged energy flow

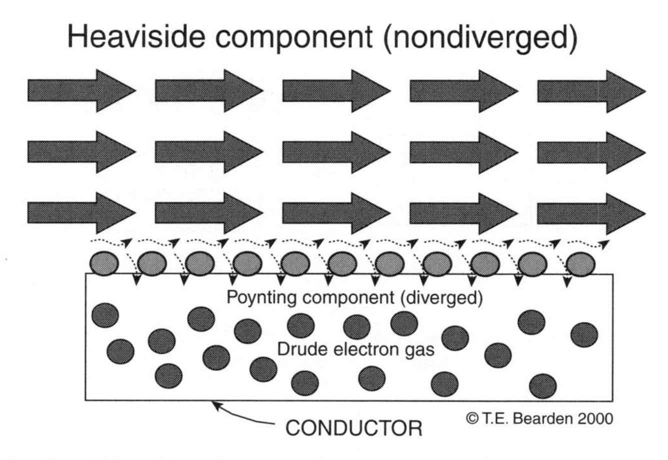

Figure 5. Heaviside and Poynting energy flow components. The Heaviside component is often 10 trillion times the Poynting component, but is simply wasted in ordinary single-pass energy flow circuits.

component may be some 10 trillion times [71][29] in total rate of energy flow as the feeble Poynting component that is intercepted by the surface charges in the circuit conductors and components, and diverged into the wires to power the Drude electrons and the loads and losses.

Figure 6 illustrates the negative-resistor process diagrammatically. The source dipole and the associated scalar potential between its poles act as a true negative resistor, receiving enormous EM energy from the surrounding vacuum in unusable form (via the giant negentropy process shown in Fig. 3). The charges of the dipole absorb this unusable energy and transduce it into usable EM energy form, then reradiate it as usable EM energy. This, of course, is precisely a negative resistor process.

Figure 7 shows the integration trick that Lorentz [52] originated to discard the perplexing and enormous Heaviside nondiverged energy flow component, while retaining the diverged (Poynting) energy flow component. In short, Lorentz' procedure—still utilized by electrodynamicists [72] to discard the embarrassing richness of EM energy poured out of every dipole and not intercepted and used by the attached external circuit—for over a century has specifically and ubiquitously diverted electrodynamicists' attention away from the process described in this invention.

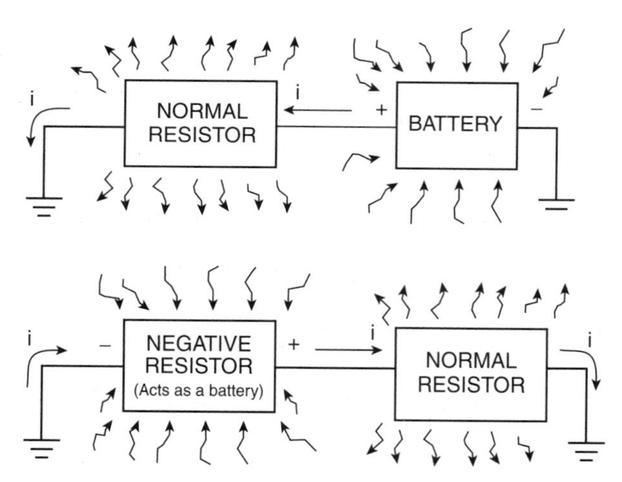

Figure 6. Negative resistance process versus positive resistance process. A negative resistor receives energy in unusable form, transduces it, and outputs it in usable form. A positive resistor receives energy in usable form and scatters it into unusable form.

[29] In Fig. 5 on p. 16 of Ref. 71, the fraction of the energy flow that is intercepted and collected by a nominal circuit (i.e., the Poynting component) is roughly shown to be on the order of 10^{-13} of the entire energy flow available. Thus the Heaviside component that misses the circuit and is nondiverged and wasted is about 10^{13} times as great in magnitude as is the Poynting component that is intercepted and diverged into the circuit to power it.

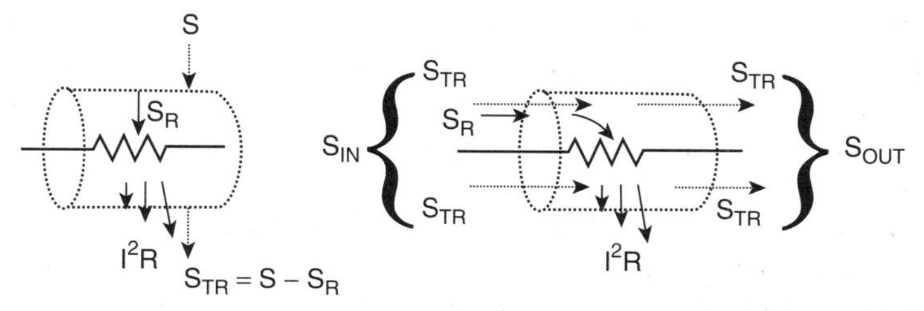

1a. Lorentz surface integration.

1b. Actual S in and S out.

See Panofsky & Phillips, Classical Electricity and Magnetism, 2nd. edn., Addison Wesley, 1962, p. 178–181.

Note: If the S vector is integrated over the closed surface, then all nondiverged energy flow is zeroed, leaving only the very small component of the input S-flow that is powering the joule heating of the resistor. In short, only the small component of the S-flow that is equal in magnitude to the Poynting vector remains. This measures only the tiny portion of the S-flow that is intercepted and diverged into the conductors by their surface charges, powering the electrons and then dissipated out of the resistor as joule heating.

The Lorentz procedure arbitrarily discards the enormous Heaviside component that misses the circuit entirely and is wasted. This results in a non sequitur of first magnitude in energy flow theory.

Figure 7. Lorentz' integration trick to discard the enormous Heaviside nondiverged energy flow component: (a) Lorentz surface integration; (b) actual S_{in} and S_{out}.

We strongly iterate the following point. We have designed the process of this invention and its embodiments by and in accord with Sachs' unified field theory [41–43,45] and the Evans–Vigier O(3) electrodynamics subset thereof [63]. Consequently, all energy in mass-free spacetime is general relativistic in nature, modeling, and interpretation. The general relativity interpretation applies at all times, including that for the electrodynamics. Hence any local change of energy in spacetime is precisely of one and only one nature: a curvature of that local spacetime. A traveling EM wave thus becomes identically a traveling *oscillating curving* of spacetime. Further, wherever the wave exists, its energy a priori curves that part of the spacetime. So EM waves, fields, potentials, and energy flows always involve and identically are spacetime curvatures, structures, and dynamics. We also accent that time is always part of it, since what exists prior to observation is spacetime, not space.[30] Hence "energy currents in time" [16,20] and "electromagnetic longitudinal waves in the time domain" [29,33,36] are

[30] The present notion in EM theory that EM energy travels through a flat spacetime is in fact an oxymoron, since to have had an energy density change in spacetime at all is a priori to curve spacetime.

perfectly rational expressions and facts, albeit strange to the >136-year old classical electrodynamics stripped of its integration with general relativity.

Figure 8 shows the relationship between a linearly moving magnetic vector potential $\mathbf{A}_L$, a swirling or circulating $\mathbf{A}_C$, the implementation of the $\nabla\times$ operator by the interacting coil and its moving charges, and the resulting magnetic field **B**. $\mathbf{A}_L$ can also be defined as the vector potential ϕ_L if desired, where ϕ_L is a vector potential and no longer the familiar scalar potential ϕ since ϕ is in motion. If the coil is wound very tight and is very long (or closed such as in a very tight toroid), then the magnetic field **B** will be retained entirely inside the coil, while the field-free (curl-free) $\mathbf{A}_C$ will remain outside the coil. This illustrates one of the major unrecognized principles of the potential (such as **A**) as a flow process; specifically, what is usually considered to be the energy in the potential in a given volume of space is actually the "reaction cross section" of the potential in that volume. Conventional electrodynamicists and electrical engineers do not calculate magnitudes of either fields or potentials per se, but only their *reaction cross sections*, usually for a unit point static charge assumed fixed at each point. We point out that this procedure calculates the divergence of energy from the potential, and hence the reaction cross section of the potential, but not the potential itself.

The energy so calculated—in this case, the curl of the **A** flow, which is the magnetic field **B**—can in fact be diverted from the **A** potential flow through a volume of space into another different volume of space. The magnitude of the **A** potential flow will continue undiminished through the original volume of space, as long as the source dipole performing the giant negentropy process and thus providing the continuous EM energy flow represented by **A** remain unchanged.

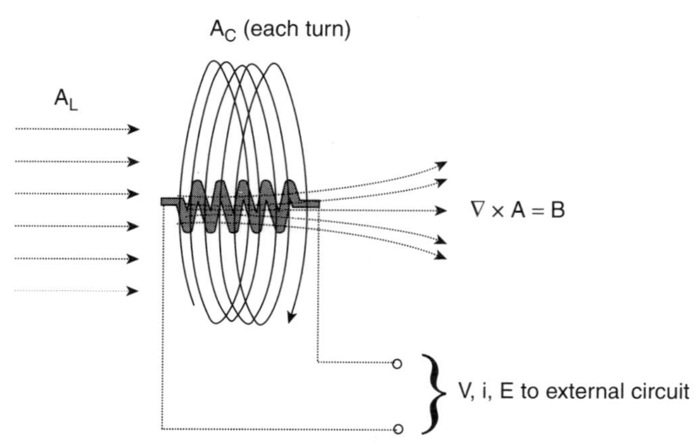

Figure 8. The $\mathbf{A}_L$ and $\mathbf{A}_C$ vector potentials, **B** field, and $\nabla\times$ operator. The $\nabla\times$ operator operates on the $\mathbf{A}_C$ potential energy current, to produce normal **B** field.

In the case used in the process of this invention, we diverge the magnetic field energy from the **A** flow, while simultaneously retaining all the **A**-potential energy flowing through the space outside the tightly wound coil. This is in fact an "energy collecting amplification" subprocess, and is no more mysterious than diverting a tiny flow of water from a nearly infinite river of flowing water, and having the river flow on apparently undiminished. In short, we may deliberately use the *energy flow nature* of the potential **A** in order to simultaneously separate it into two flows of different energy form: curled and uncurled.

If we place a square pulse in the current of the coil in Fig. 8, we also invoke the Lenz law reaction (Fig. 9) to momentarily increase the current and hence the $\mathbf{A}_C$ and the action of $\nabla \times \mathbf{A} = \mathbf{B}$, so that additional $\mathbf{A}_C$ energy and additional **B** energy are obtained. In this way, the energy gain is increased by the Lenz law effect—which is a regauging effect deliberately induced in the invention process by utilizing square pulse inputs. Then when the trailing edge of the pulse appears and sharply cuts off the pulse, a second Lenz law gain effect (Fig. 9) is also produced, further increasing the energy gain in both $\mathbf{A}_C$ and in **B**. We use these two serial Lenz law effects to increase the potential energy of the system twice and also the collected field energy, thus allowing COP > 1.0 since during the regauging process the potential and the potential energy of the system are both increased freely, and so is the diversion of the increased potential energy into **B**-field energy inside the coil. Both the changes increase

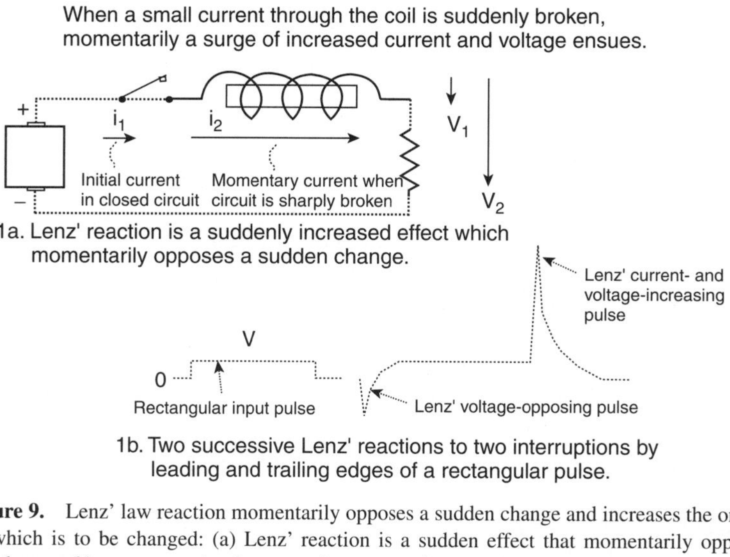

Figure 9. Lenz' law reaction momentarily opposes a sudden change and increases the ongoing action which is to be changed: (a) Lenz' reaction is a sudden effect that momentarily opposes a sudden change; (b) two successive Lenz reactions to two interruptions by leading and trailing edges of a rectangular pulse.

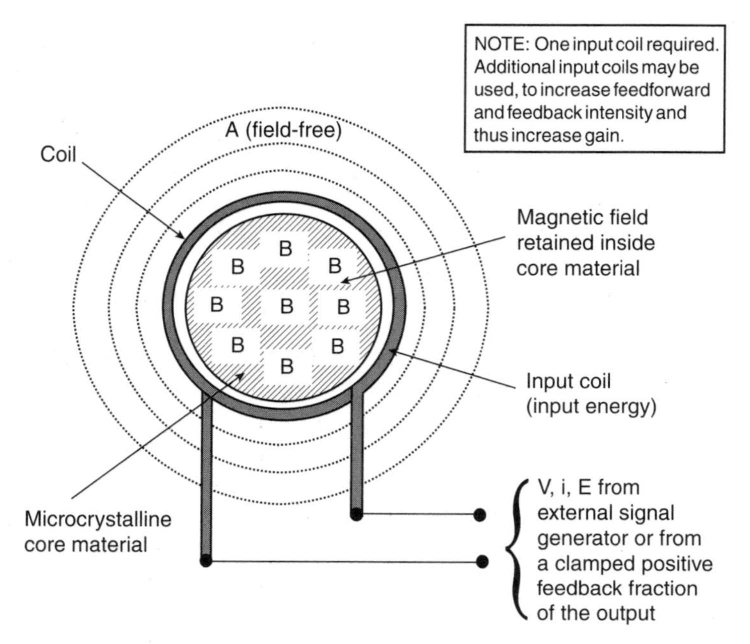

Figure 10. Input coil for either open-loop or closed-loop operation.

the voltage drop across the coil and the current through it, translating the increased magnetic energy into usable electrical energy to power loads and losses.

Figure 10 shows the cross section of an input coil, one form of input device. The input can be from a separate signal generator, in which case the system runs "open loop" and requires continuous input power, but still provides COP > 1.0. Or, a portion of the output power can be extracted, clamped in magnitude, and positively fed back to the input, in which case the system runs "closed loop" and the operator need furnish no external power input once the unit is in operation. In either case, the system is an open system far from thermodynamic equilibrium with its active vacuum environment, freely receiving energy from the active environment to the dipole in the permanent magnet, and from the dipole out into the nanocrystalline material core in the form of magnetic field energy **B**, and in the space outside the core in the form of field-free **A** potential. As can be seen, the **B**-field energy is confined to the core material inside the coil, and the **A** potential outside the core is field-free **A**. Any change in the **B**-field inside the *core* is also a change in the **B** field inside the *coil* and the coil interacts with it to produce current and voltage. Any change in the **A** potential outside the core, also interacts with the coil as $d\mathbf{A}/dt$, which is an **E**-field interaction. The movement of the Drude electrons also applies the $\nabla\times$ operator, thereby producing voltage and current in the coil and also producing additional **B**-field in the core material. In turn, this changes the **B** field in the core, which produces more voltage and current in the coil and additional **A** potential outside the coil, and so on. Hence there are dual iterative retroreflective interactions between the two simultaneous interactions with the coil. The dual interaction increases the performance of the coil, rendering it an *energy amplifying* coil, and also increases the COP of the system process. The output of the input coil is thus due to the alteration and increase of the **B**-field flux and energy in the core material, and an increase and alteration in the field-free **A** potential surrounding the coil and moving around the circuit in the space surrounding the nanocrystalline core material flux path, and interacting simultaneously with the coil in $d\mathbf{A}/dt$ fashion (**E**-field fashion).

Figure 11 shows a cross section of a typical output coil for either open-loop or closed-loop operation. The operation is identical to the operation of the input coil, except this coil outputs energy in the form of voltage and current to an external circuit, external load, and so on, and also outputs energy from its reaction with the **A** potential (i.e., with $d\mathbf{A}/dt$) coming in from outside the flux path. It outputs both an **E**-field energy reaction with the Drude electrons, and also a change in **B**-field energy to the nanocrystalline flux path material in its

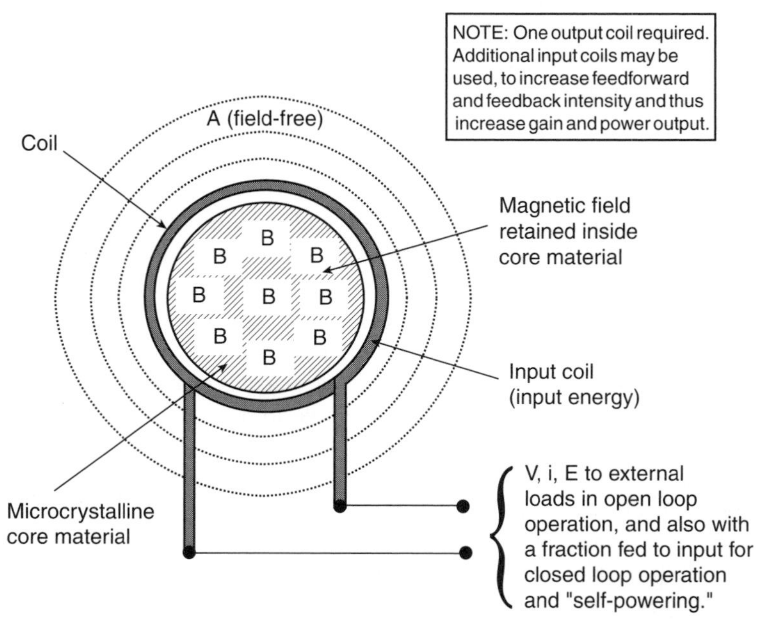

Figure 11. Output coil for either open-loop or closed-loop operation. Multiple output coils may be used in a variety of configurations.

core due to the movement of the electron currents in the coil. The output coil receives its energy input from the field-free **A**-potential outside the nanocrystalline material flux path as well as from the **B**-field energy and magnetic flux inside the nanocrystalline flux path through its core. In short, we "dip" the full Poynting energy flow component from the **A** potential as **B**-field energy, separate this **B**-field energy, and pipe it to the center of the output coil to interact magnetically with the Drude electrons in the coil. The **A** potential is instantly and fully replenished as a curl-free **A**-potential. The time rate of change of this **A** potential thus can be adapted so that a full Poynting energy flow component also interacts with the Drude electrons from outside as a large **E**-field energy reaction. We thus *multiply* the EM energy flow available for interacting with the output coil, compared to conventional "single dipping" and "single Poynting component interacting" systems in a transformer–generator. This "double dipping" provides an energy gain, since it constitutes a free regauging of the potential energy of the system, and an application of the gauge freedom principle of gauge field theory. The lesser additional *mutual interaction* between those primary interactions adds the extra 0.5 gain, so that each output coil now gives a gain of 2.5. With two output coils, the COP of the system is COP = 5.0.

Further, all coils on the core material serve somewhat as both output and input coils, and also have mutual iterative interactions with each other around the loop, coupled by the field-free external **A** potential and the **B**-field and magnetic flux in the nanocrystalline material flux path acting as the cores of the coils. These interactions also provide gain in the kinetic energy produced in the Drude electron gas, due to the iterative summation work performed on the electrons to increase their energy. When more coils are utilized, the gain is affected correspondingly.

Further, these mutual iterative feedback and feedforward energy gains also change the flux back through the permanent magnet, alternating it, so that the pole strength of the magnet alternates and increases. This, in turn, increases the dipolarity of the permanent magnet, which, in turn, increases the magnitude of the associated giant negentropy process [16,20]. This results in more energy received from the active vacuum by the permanent magnet, and also more energy output by said permanent magnet dipole to the core material and to the coils.

Thus we have described a system and process having a multiplicity of iterative feedbacks and feedforwards from each component and subprocess, to every other component and subprocess, all increasing the energy collected in the system and furnished to the load. In open-loop operation, this results in COP > 1.0 permissibly, since the excess energy is freely received from an external source. In closed-loop operation, the COP concept does not apply except with respect to operational efficiency. In that case, the operational efficiency is increased because more energy is obtained from the broken symmetry of the permanent dipole, and therefore additional energy is provided to the loads, compared to what the same permanent magnet can deliver when such iterative feedback and feedforward actions in such multiplicity are not utilized. In closed-loop operation, the system powers itself and its loads and losses simultaneously, and all the energy is freely supplied from the active vacuum by the giant negentropy process of the permanent magnet dipole and the iterative asymmetric self-regauging processes performed in the system processes.

Figure 12 is another view showing the major energy flows in an output coil section and subprocess, and the iterative dual inputs and interactions, of the basic scheme of operation of the process and its active component subprocesses.

Figure 13 is another view showing the dual energy flows in an input coil section and subprocess.

Figure 14 is a block diagram illustration of the components and processes in the system and system process, with the dual feedforward and feedbacks shown. It accents the overall system process gain due to the multiplicity of interactions and iterative interactions between the various system components and subprocesses, and further interactions with the dual local interactions and iterative feedforwards and feedbacks, thus providing a multiplicity of individual energy gain processes and an overall energy gain process.

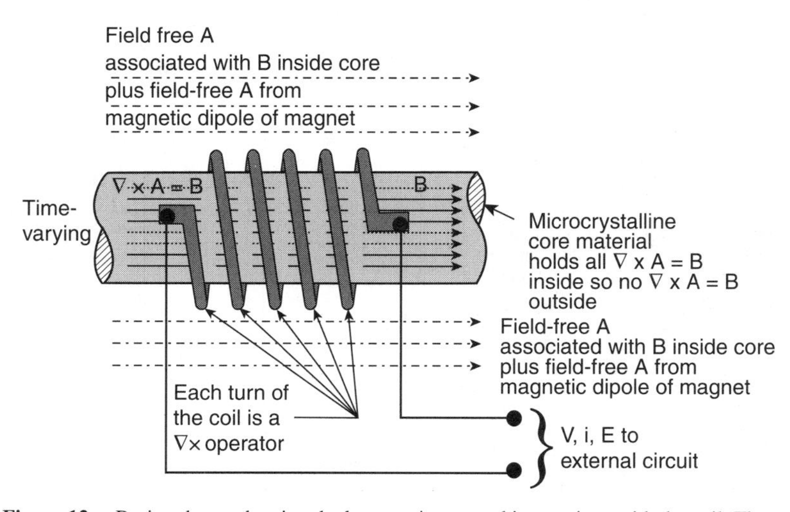

Figure 12. Basic scheme showing dual energy inputs and interactions with the coil. The output of each of these two interactions also "feeds forward" to the other interaction as an additional input to it, resulting in interative "pingpong" of additional energy collection in the circuit, providing energy gain.

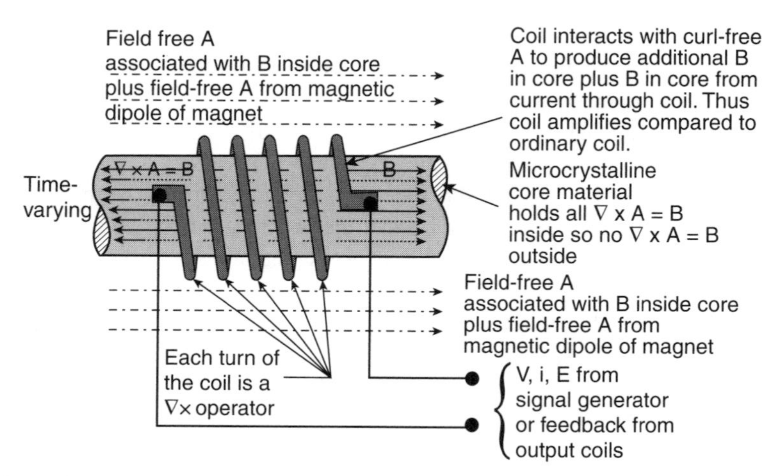

Figure 13. Dual energy inputs to the coil result in amplifying coil–core interaction. Not shown are the other feed loops providing extra curl-free A input from the surrounding space and extra B input in the core.

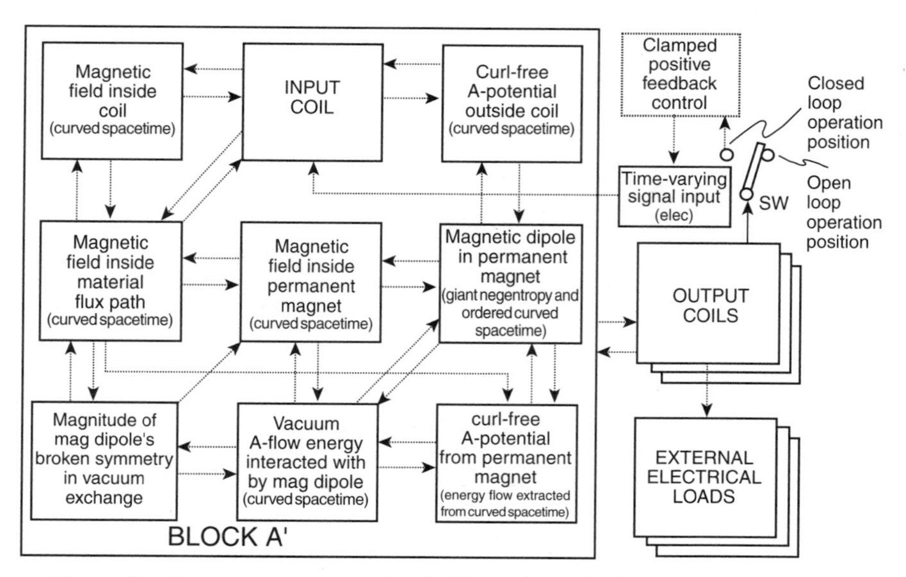

Figure 14. Energy gain process using feedforward and feedback subprocesses providing individual energy gain operations.

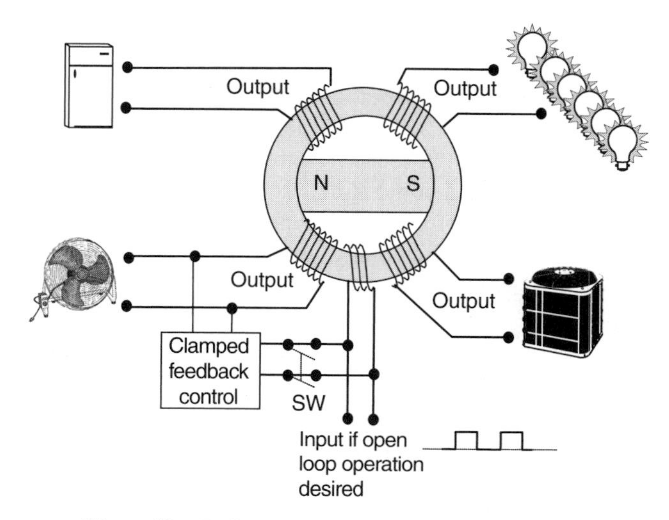

Figure 15. Typical embodiment system and application.

Figure 15 shows a type embodiment of the system and system process, perhaps at a home and powering a variety of home appliances and loads. The system as shown is "jump-started" initially in open-loop mode, and once in stable operation is disconnected from the jump starter (such as a battery) to run in closed-loop operational mode.

Figure 16 shows one of the former laboratory test buildups embodying the process of the invention. This test prototype was used for proof-of-principle and

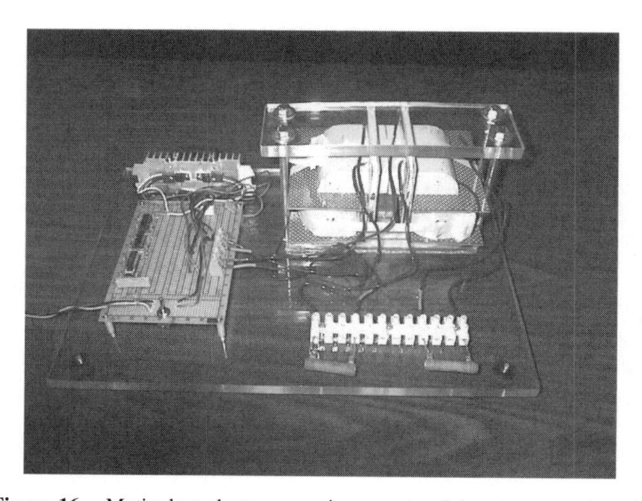

Figure 16. Motionless electromagnetic generator (laboratory experiment).

phenomenology testing. This experimental buildup has been substantially replicated by French researcher Naudin, who achieved COP = 1.76 with a less optimized core material (see footnote 10, above). Naudin is the first experimenter to replicate our MEG experimental apparatus. His key illustrations are used in our Figs. 26 and 28 (below) with his permission.

Figure 17 shows a simplified block diagram of a basic embodiment demonstrating the process. Many of these buildups were undertaken to test various core materials, observe phenomenology, and perform other tasks. The "square C's" of the flux path halves right and left, as shown in this Fig. 17, were actually made as "half-circle C-shaped flux path halves" right and left in Fig. 16 above.

Figure 18 shows the measurement of the input to the actuator coil of the test unit of Fig. 16 operated in open-loop mode.

Figure 19 shows the measurement of the output of one of the output coils of the test unit of Fig. 16 operated in open-loop mode.

Figure 20 shows the output power in watts as a function of the input potential in volts, thus indicating the output versus potentialization sensitivity. The circles indicate actual measurements, and the curve has been curve-fitted to them.

Figure 21 shows the COP of a single output coil's power divided by the input power, as a function of input potentialization. The circles indicate actual measurements, and the curve has been fitted to them. The second coil had the same power output and COP simultaneously, so the net unit COP of the unit is double what is shown in the figure.

Figure 22 shows the projected unit output power sensitivity versus voltage input, expected for the next prototype buildup now in progress.

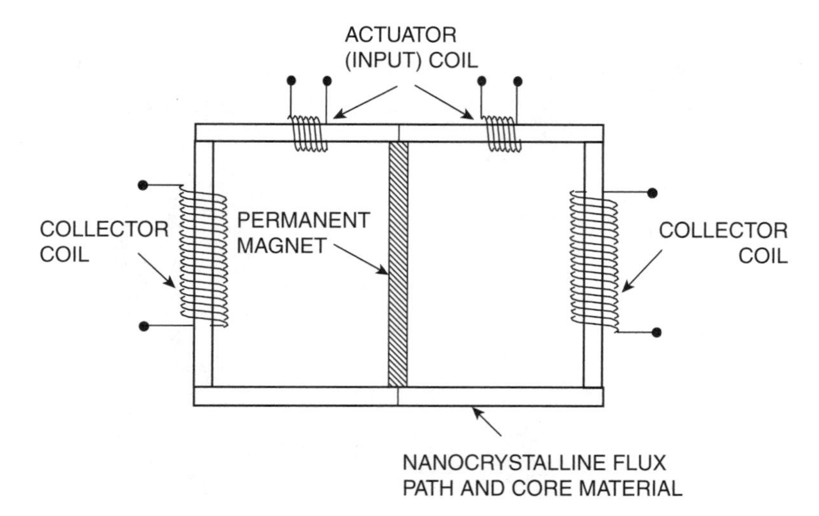

Figure 17. Diagram of laboratory test prototype.

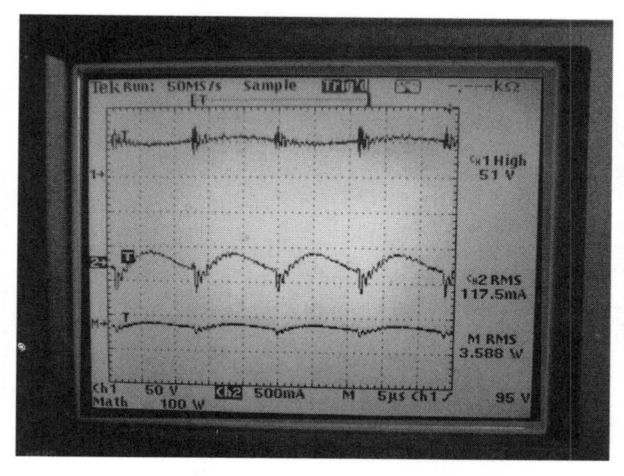

Figure 18. MEG input measurements.

Figure 23 shows the projected unit COP versus input potentialization expected for the next prototype buildup now in progress. We eventually expect this type of unit to easily operate at the COP = 30 or COP = 40 level, with multiple kilowatt output power.

Figure 24 shows the true operation of a typical coal-fired power plant and generator, and the associated power line. As can be seen, all that burning the coal accomplishes is to force the internal charges in the generator apart and form the source dipole. None of that adds a single watt to the power line. Instead, once

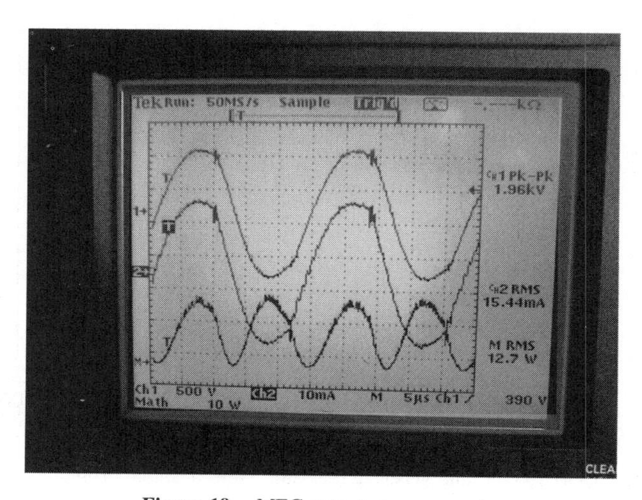

Figure 19. MEG output measurements.

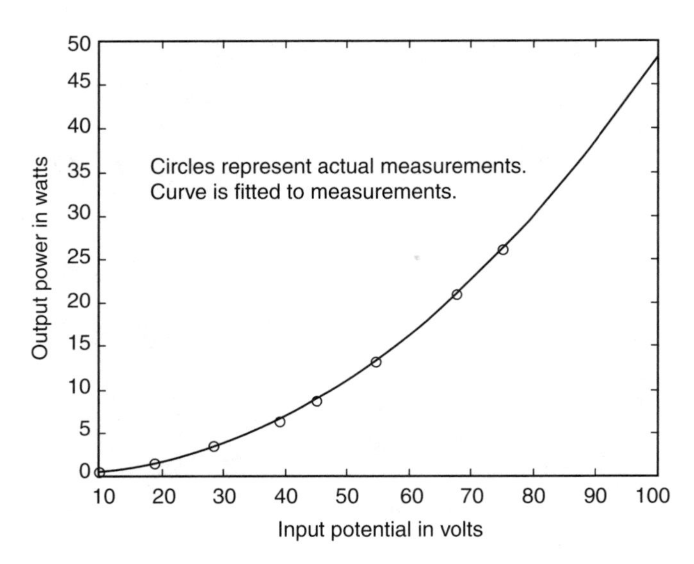

Figure 20. MEG prototype potentialization sensitivity.

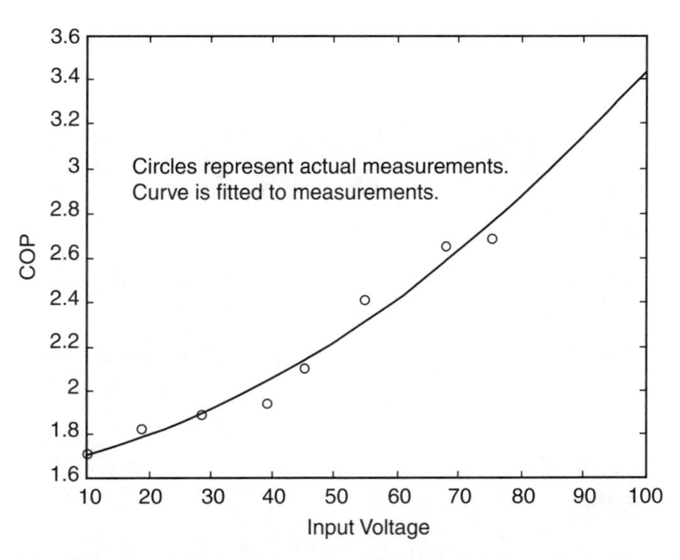

Figure 21. COP (power out versus power in), as a function of sensitivity (open-loop prototype).

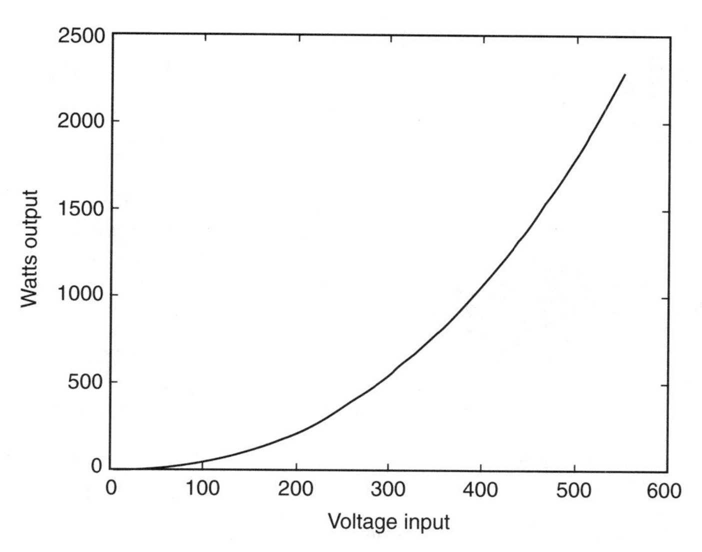

Figure 22. MEG projected sensitivity (test buildup in progress).

the dipole is formed, its broken 3-symmetry produces an inflow of energy from the complex plane (time domain) from the active vacuum. The dipole transduces the absorbed *reactive* power (in electrical engineering terms) and outputs it as real, observable EM energy flow. Note particularly the enormous amount of the

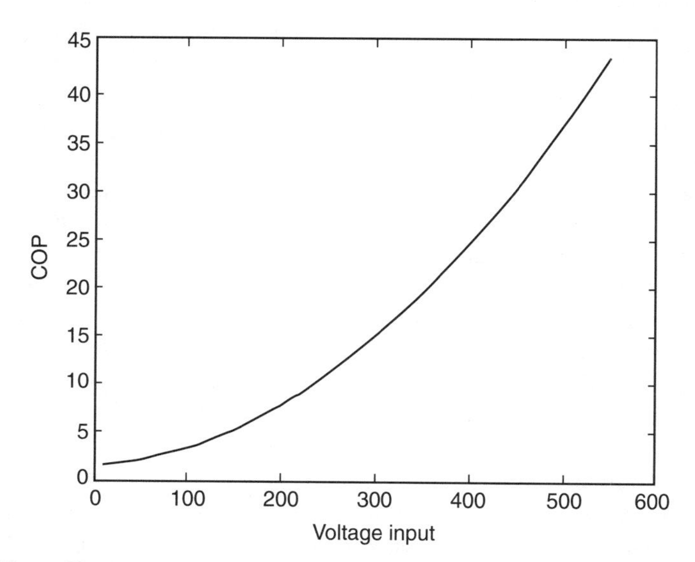

Figure 23. MEG projected COP versus input voltage (test buildup in process).

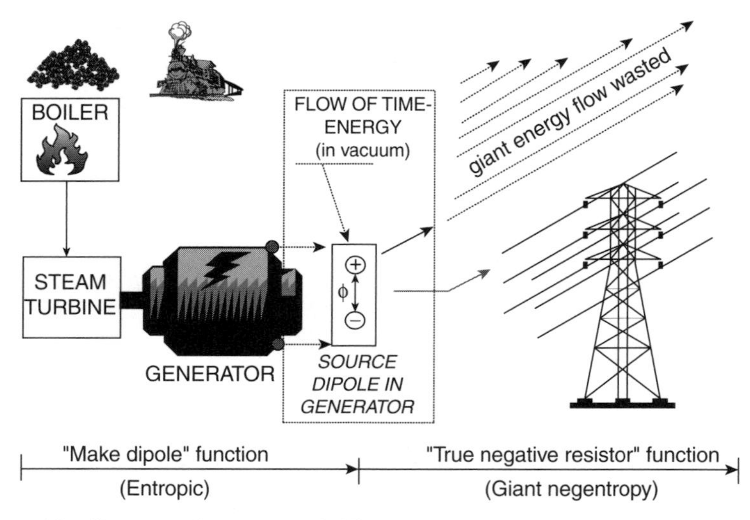

Figure 24. Generators do not power circuits and power lines; they make the source dipole.

energy flow pouring out of the terminals, most of which misses the power line entirely and is just wasted. Only a very tiny component of that giant Heaviside flow is intercepted by the power line and circuits, and diverged into the conductors to power the Drude electrons. In addition, half the Poynting energy collected in the circuit is dissipated in the dipole's back emf, scattering the charges and continually destroying the dipole—which is restored continually by additional shaft input energy that is continually input. This inane way of building power systems is what has required all the burning of hydrocarbons, use of nuclear fuel rods and dams, and so on. None of that powers the power grid, and we have been damaging and polluting our environment for no good reason at all. Every electrical power system is now and always has been powered by EM energy extracted directly from the active vacuum by the broken 3-symmetry of the source dipole in the generator or battery.

Figure 25 shows a waterwheel "material fluid flow analogy" of the "double-dipping" principle used by the MEG. In a fast-flowing stream with appreciable drop in elevation along the flow, we insert a waterwheel in the river past the bottom of a small waterfall. The "reaction cross section" of the bottom of the waterwheel immersed in the rushing stream is a certain amount, which determines the amount of energy diversion and hence the work done on the wheel if used in normal waterwheel fashion. However, we have inserted a pipe at the top of the waterfall, and "piped" a large flow of water down to the top of the waterwheel, turning its direction so that, when it flows out on the left side of the waterwheel, it strikes the same reaction cross section of the left side of the

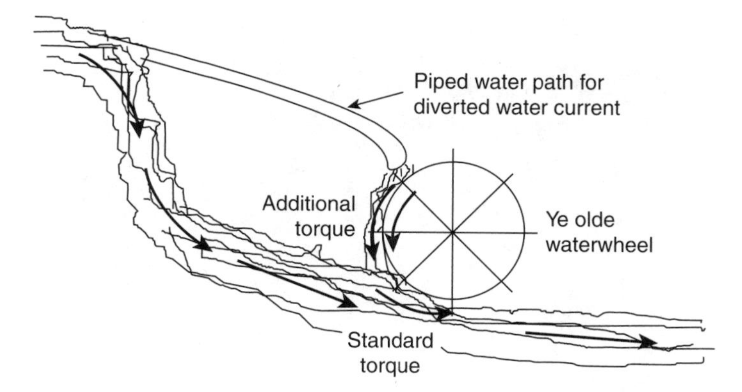

Figure 25. Double-dipping waterwheel analogy of MEG operation.

wheel. Hence a "couple" exists so that twice the energetic interaction occurs with the double-dipping waterwheel, as is normal for single-dipping waterwheels. In short, we get twice the energetic interaction on the waterwheel to power it, because the wheel interacts twice rather than once. If we do not press this simple analogy very far, it suffices to demonstrate that, in flowing stream of energy, we can divert part of the stream and then interact twice with the energy flow rather than just once. The difference is that, in the case of diversion of the energy flow in the magnetic vector potential as a magnetic field flux, the magnetic vector potential is replenished instantly and freely, so that the two streams are equal in flow, and hence equal in interaction. In the water analogy, of course, the two streams are not necessarily equal in flow magnitude.[31]

Figure 26 shows Naudin's second MEG variant buildup, in November 2000. His core material is less optimal than is the core material utilized by Magnetic Energy Ltd. Hence he achieved $COP = 1.76$ with this unit, instead of the full $COP = 2.5$ or 5.0. We are pleased that Naudin is the first researcher to have successfully replicated our MEG unit, including achieving $COP > 1.0$.

Figure 27 shows a simplified diagram of the double-dipping principle, applied by the MEG. From the permanent magnet's magnetic dipole, there pours forth an "energy stream" called the *magnetic vector potential.* We intercept this stream right at the poles of the magnetic dipole, with the special nanocrystalline core material. This material extracts and curls magnetic energy

[31] In the MEG, the *interception* of energy from the uncurled A-potential in space outside the core may be made much larger than energy intercepted from magnetic flux switched in the core. This is accomplished by adjusting the rise and decay times of the edges of rectangular pulses in the primary coil. The E-fields produced by dA/dt may be made very large, increasing the energy collected in a given set of external receiving circuits with loads.

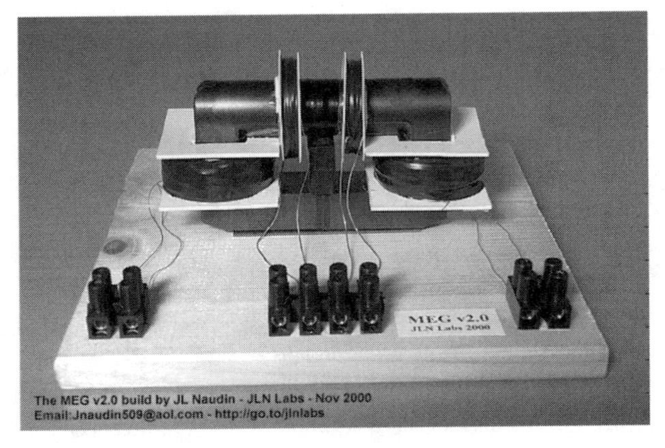

Figure 26. Naudin's replication of a MEG variant in Nov. 2000.

from the magnetic vector potential, and magnetic field energy is then separated from the curl-free magnetic vector potential river, which is instantly replenished. Thus the material acts as a "diverger" so that we conduct the extracted magnetic field energy through the core path and through the center of the input and output coils, in a closed path with the permanent magnet. Magnetic field measurements will show little or no magnetic field outside the core path, but of course the replenished **A** potential is there, but without curl. We use the input coil to perturb the magnetic flux and simultaneously perturb the **A** potential. So two equal energy perturbations—one a perturbation of the magnetic field flux in the core and the other a perturbation of the **A** potential in space outside the core material—are propagated. The **A**-potential perturbation propagates outside the core, and the **B**-field perturbation propagates inside the core. Both equal-energy perturbations strike the output coil simultaneously, where the rate of change of the **A** potential interacts with the coil's Drude electrons in **E**-field fashion,

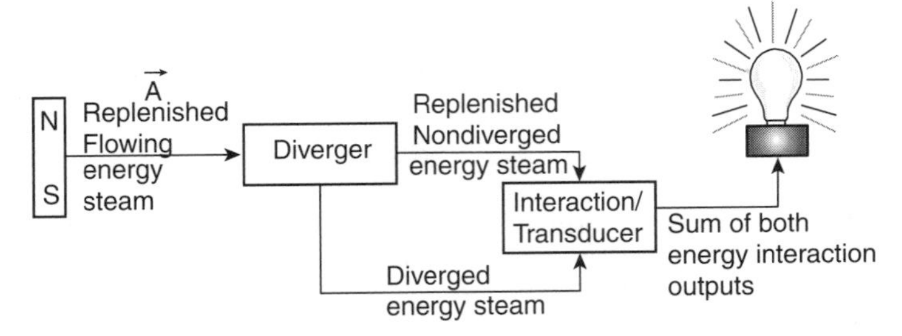

Figure 27. MEG "double dipping" from a replenished manetic vector potential energy stream.

while the magnetic field perturbation in the core in the center of the coil interacts with the Drude electrons in magnetic field fashion. A slight interaction within the coil occurs there between the two interactions, resulting in slight increase of the energy of the coil's interaction. Hence this double-acting coil is actually interacted with some 2.5 times the energy that normal magnetic field interaction would provide; that is, the output coil interacts and outputs 2.5 times as much EM energy as we input into the input coil (with two output coils, the COP = 5.0). No laws of physics or thermodynamics are violated, because the magnetic vector potential from the permanent magnet is freely and continuously replenished, regardless of how much energy we draw from it by "double-dipping," "triple-dipping," and so on. In theory, the COP is unlimited. In practice, a COP = 10 or even 20 can be achieved with some difficulty.

Figure 28 shows Naudin's actual measurements as he varied his input conditions to his MEG 2 variant. As can be seen, the COP varies as the efficiency of the separation process varies, and that varies as the conditions of the input vary

MEG v2.0 Tests by JL Naudin on 11-07-00 Email: JNaudin509@aol.com

Rload (ohms) = 100000

Time (sec)	VTG OUT (V)	PWR OUT (W)	VTG INP (V)	CUR INP (A)	PWR INP (W)	Efficiency (%)	COP
0	790	6.24	23.7	0.35	8.30	75%	0.76
30	909	8.26	23.7	0.35	8.30	100%	1.00
50	995	9.90	23.7	0.35	8.30	119%	1.19
60	1045	10.92	23.7	0.35	8.30	132%	1.32
90	1120	12.54	23.7	0.36	8.53	147%	1.47
120	1140	13.00	23.7	0.36	8.53	152%	1.52
150	1167	13.62	23.7	0.36	8.53	160%	1.60
180	1190	14.16	23.7	0.36	8.53	165%	1.66
210	1244	15.48	23.7	0.37	8.77	176%	1.76

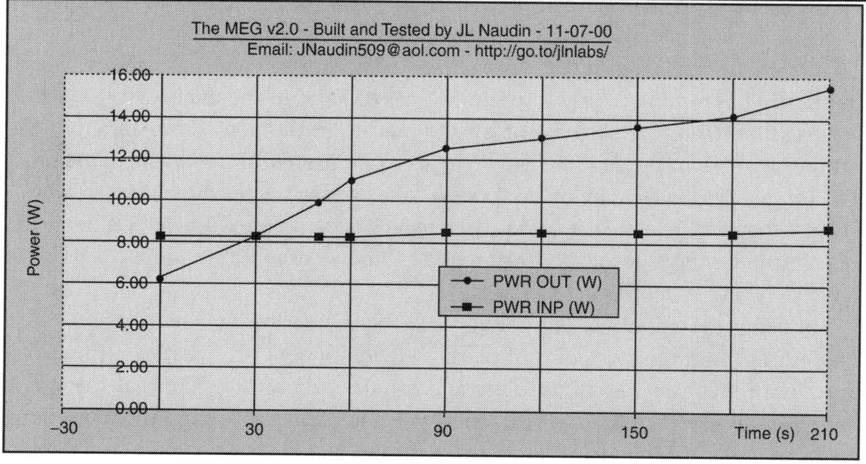

The MEG v2.0 - Built and Tested by JL Naudin - 11-07-00
Email: JNaudin509@aol.com - http://go.to/jlnlabs/

Figure 28. Naudin's Nov. 2000 results, reaching COP = 1.76.

for a given design. These results were obtained by Naudin shortly before the cutoff deadline of this paper, increasing his previous results of COP = 1.5. The point is that rigorous, independent replication has indeed been achieved.

VIII. RAMIFICATIONS

A. Importance of the Process and Its Subprocesses

A process has been provided whereby useful electromagnetic energy may be extracted from the dipole of a permanent magnet, via the giant negentropy process [16,20] associated with the magnetic dipole. In that process, an outflow of EM energy is continuously furnished by the magnet dipole in all directions in 3-space, and the energy to the dipole is freely furnished from the time domain of the active vacuum [1,16,20]. Whittaker [1] demonstrated this giant negentropy mechanism in 1903, but apparently failed to recognize its implications for electrical power systems. Recent recognition of the mechanism and its implications for electrical power systems was accomplished by one of the inventors, Bearden [16], and then more deeply examined by Evans and Bearden [20].

By using the principle that essentially unlimited energy can be withdrawn from (collected from) a potential, and the withdrawn energy will be replaced by the potential's negative-resistor action using the giant negentropy mechanism [1,16,20], a practical approach to free-energy sources for self-powering and COP > 1.0 electrical power systems anywhere in the universe is provided.

By using the principle that iterative transformation of its form energy in a replenishing potential environment can be repeatedly reused to do work, so long as the form of the energy resulting at the completion of each work phase is retained and reprocessed, one joule of energy can be utilized to do many joules of work, as precisely permitted by the energy conservation law with regauging. This is a major change to the work–energy theorem of electrodynamics, which implicitly has assumed only a single change of form of the energy, followed by loss (escape from the system) of all the energy in the new form. In short, *the present work–energy theorem is only a special case valid under those assumed special conditions*. The invented process takes advantage of the extended work–energy theorem where one joule of energy—accompanied by retention of the new form of energy resulting from work—can evoke multiple joules of work in a replenishing potential environment.

By using the principle that one joule in "iterative form changing mode with retention" can do many joules of work on a component of a system—to wit, on the Drude electron gas in an electrical circuit, where the potential energy is increased by the increased kinetic energy of the electrons having the work done on them—the extended work–energy theorem can be utilized to overpotentialize the receiving Drude electron gas, thereby regauging the system to add excess energy by gauge freedom and outputting more electrical energy to the load than is input to the system by the operator.

By then dissipating in loads this excess energy collected in the Drude electron gas in the output circuit, the invented process provides greater energy to be dissipated in the load than is input by the operator. The combination of processes thus allows an EM system freely functioning as an open system not in equilibrium with its active vacuum (due to the giant negentropy mechanism [1,16,20]), hence permitted to exhibit COP > 1.0. In this way, more work output can be accomplished by the system process than the work that the operator must perform on the system to operate it.

By using the principle of governed, clamped positive feedback[32] of a portion of the increased output back to the input, the system can be close-looped and can power itself and its load, with all the energy furnished by self-regauging from the active vacuum as an external energy source, furnishing excess energy to the magnetic dipole's magnetostatic potential and associated magnetic vector potential, thereby replenishing energy withdrawn from the magnetic vector potential by the subprocesses in the overall system process.

One system operating in closed-loop mode can also have one fraction of its output devoted to jump-starting another such system in tandem, then switching the second system into self-powering closed-loop mode, then jump-starting another such system, which is then switched to self-powering, and so on. In that way, multiple systems can be "piggybacked" so that an exceptionally large power system consisting of a group of such "piggy-backing" systems can be produced. In case of system failure, all can be started again in the same series, by furnishing only the initial small input required to jump-start the first system of the group. In this way, very large power systems such as necessary to power automobiles, trucks, ships, trains, and other vehicles can be produced, and yet the backup jump-starting source—such as a storage battery—can be very small, such as a simple flashlight battery.

B. Implications for the Crisis in Oil Supplies versus Energy Demands

The emerging overunity electrical power systems—including "self-powered" systems freely taking all their energy from the local vacuum—will produce a total revolution in transportation, electrical power systems, backup power systems, and so on [68,69]. In the process, the electrical power is obtained freely and cleanly from the vacuum, from permanent-magnet dipoles continuously replenished from the active vacuum via the giant negentropy process.

[32]We state, however, that at about COP $\approx$ 2.0 Special Dirac sea hole current phenomena are encountered in close-looping, as a new kind of decay mechanism from the disequilibrium state back to the Lorentz equilibrium. Bedini and Bearden have filed a patent application for energy transduction processes to overcome this effect and allow close-looping.

A more significant fraction of the electrical power system can thus be decentralized, and degradation in case of system failure will be graceful and local. Yet full use can still be made of the existing power grids and power systems. As an example, arrays of self-powering electrical heater systems can be developed and used to heat the boilers in many standard power systems, thereby stopping the burning of hydrocarbons in those plants, and drastically reducing the pollution of the biosphere and the lungs of living creature including humans. This would allow a graceful phase-in of new, clean, self-powering electrical power systems; reduction of hydrocarbon combustion for commercial electricity production; and ready increase in electrical power to meet increasing world demands, even in poor nations and developing countries, while capitalizing and using much of the very large "sunk costs" investment in present large power systems. The core material fabrication is labor-intensive, so it is made in developing nations where such jobs are sorely needed and greatly benefit both the individual people and the nation. The dramatically increased use of and demand for these materials would thus stimulate substantial economic growth in those nations by providing many more jobs.

The conversion of power systems and replacement of a fraction of them, can proceed vigorously, since production and scaleup of systems utilizing this process can be very rapid. Except for the cores, all fabrication, parts, techniques, tooling, and other procedures are simple and standard and very economical—and are already on hand and used by a great many manufacturing companies worldwide.

At this time of an escalating world oil crisis and particularly a shortage of refining facilities, a very rapid and permanent solution to the oil crisis and the rapidly increasing demand for electricity—and also much of the problem of the present pollution of the biosphere by combustion byproducts, and of the present global warming enhancement by the emitted CO_2 from the hydrocarbon combustion—can be provided cheaply, widely, and expeditiously.

The steady reduction and eventual near-elimination of hydrocarbon combustion in commercial power systems and transport, the dramatic reduction in nuclear fuel rod consumption, and other improvements will result in cleaner, cheaper, more easily maintained power systems and a reduction in the acreage required for these power systems.

The gradual decentralization and localization of a substantial fraction of the presently centralized power grid will eliminate a significant fraction of power transmission costs, thereby lowering the price of electrical energy to the consumers.

The scaleup weight-per-kilowatt of systems using this system process will be sufficiently low to enable rapid development of electrically powered transport media such as automobiles. These will have weight about the same as now, carry a small battery as a backup jump-starter, and have very agile performance suitable for modern driving in heavy traffic. With fuel costs zeroed, the cost to the citizens of owning and operating vehicles will be reduced. Costs to the trucking industry, for instance, will be dramatically reduced, since fuel is a major cost item. In turn, since most goods are moved via the trucking industry, the lower transport costs will mean more economical sales prices of the goods. These are very powerful and beneficial economic advantages of the new process.

C. Some Specific Advantages

The process ensures the following advantages for electrical power systems:

- The systems can have a high output power : weight ratio. Second- and third-generation equipment will have a very high output power : weight ratio.
- The systems can be highly portable for mobile applications.
- The size and output of the systems are easily scalable, and piggybacking is simple.
- The logistics burden for remote operation of MEG power systems such as on manned space capsules and satellites in orbit will be dramatically reduced.
- The systems will be rugged and reliable for use in hostile environments where conventional generators would fail or be extremely difficult to sustain. The systems can easily be environmentally shielded.
- The systems can function effectively in very wide operating temperature ranges and can be used where conventional batteries and fuel cells cannot function. As an example, a system can power a resistance heater to keep its own immediate environment continuously warm. It can also power electrostatic or magnetic cooling devices to keep the unit and its immediate environment cool in higher-temperature environments.
- The system will have an extremely long life cycle and high reliability, allowing it to be placed where frequent maintenance is not possible.
- The system uses no fuel or fuel transport, packaging, storage, and disposal systems and needs no intermediate refining facilities and operations. The resulting overhead and financial savings are vast and significant.
- Use of the systems in a combined centralized and decentralized electrical power system provides survival of electric power and graceful degradation, rather than catastrophic collapse, of electrical power in the presence of damage and destruction. This is particularly important since the greatest threat to the United States and many other nations is terrorist attacks against our cities, or against our fuel supplies, electrical power grids, and other structures and systems.

- The systems produce no harmful emission, harmful or radioactive byproducts, hazardous wastes, or biospheric pollutants. As usage is phased in worldwide, a significant reduction of environmental pollutants and hazardous wastes will result, as will a cleaner biosphere.
- The systems can produce AC or DC power directly by simple electrical additions, and provide shaft power simultaneously. Frequency can be changed by frequency conversion.
- Coupled with normal electric motors, the systems can provide attractive power system alternatives for automobiles, tractors, trucks, aircraft, boats, ships, submarines, trains, and other vehicles, again without exhaust emissions, pollutants, or harmful waste products and without fuel costs.
- The systems can be developed in small-system sizes, rugged and efficient, to replace the motors of hosts of small engine devices such as garden tractors, lawnmowers, power saws, and leaf blowers, which are presently recognized to be very significant biospheric polluters.

These descriptions provide illustrations of some of the presently envisioned preferred embodiments and applications of this invention.

D. Extension and Adaptation of the Process

For example, we have mentioned piggybacking arrays of such systems for easily assembled large power plants.

As another example, conversion to furnish either DC or AC, or combinations of either, at whatever frequencies are required, is easily accomplished by standard conversion techniques and add-on systems.

As another, less obvious, example, the process uses a multiplicity of positive energy feedforwards and feedbacks, and iterative change of the form of the energy between multiple states in a replenishing environment, to provide iterative gain by "pingpong." As the number of feedback and feedforward operations is increased, it is possible to advance the system process into a region where the regenerative feeds produce an exponentially *increasing* curve of regauging energy and potential energy increase, with concomitant exponentially increasing curve of output energy. Material characteristics, saturation levels of cores, and other variables provide "plateaus" where the exponentially rising output curve is damped, leveled off, and stabilized. By using spoiling and damping, such exponential increase in energy density of the system can be leveled off at specifically desired plateau regions, which can be easily adjusted at will, either manually or automatically in response to sensor inputs. By this means, these systems enable automatically self-regulating, self-adapting power grids and power systems, which automatically adjust their state and operation according to the exact needs and conditions, changes of these needs and conditions, and other requirements without impact on supporting fuel, transport, refining, storage, and other activities. These "exponential but plateau-curtailed" systems are capable of producing very large power-per-pound levels, and sustaining them without overheating, limited only by the saturation level of the core materials. Such new adaptations of the fundamental system process of this invention can be developed in a straightforward manner in the third generation.

The adaptations and alterations of the process are limited only by the ingenuity of the scientists and engineers and by the particular needs of a given application. The process uses the laws of nature in a novel and extended manner, such as using one joule of input energy together with automatic replenishing to cause many joules of output work to be done in the load. Many alternative subprocesses, embodiments, modifications, and variations will be apparent to those skilled in the art of conventional electrical power systems and magnetoelectric generators.

References

1. E. T. Whittaker, *Mathematishe Annalen* **57**, 333–335 (1903).
2. I. Prigogine, *Nature* **246**, 67–71 (1973).
3. G. Nicolis and I. Prigogine, *Exploring Complexity*, Piper, Munich, 1987 (a technical exposition of the thermodynamics of dissipative systems far from thermodynamic equilibrium).
4. G. Nicolis, in P. Davies (Ed.), *The New Physics*, Cambridge Univ. Press, Cambridge, UK, 1989, pp. 316–347 (a good overview of the thermodynamics of dissipative systems far from thermodynamic equilibrium).
5. A. Sommerfield, *Mechanics of Deformable Bodies* (includes discussion of the "perfect analogy between hydrodynamics and electrodynamics").
6. H. Lamb, *Hydrodynamics*, 1879.
7. L. V. Lorenz, *Phil. Mag.* **34**, 287–301 (1867) (discusses essentially what today is called "Lorentz symmetric regauging").
8. J. C. Maxwell, *Phil. Trans. Roy. Soc.* **155**, 71, 459 (1865) (presented orally in 1864; Maxwell's definitive presentation of his theory); also in W. D. Niven (Ed.), *The Scientific Papers of James Clerk Maxwell*, Dover, New York, 1952, Vol. 1, pp. 526–604.
9. M. W. Evans, P. K. Anastasovski, T. E. Bearden, et al., *Physica Scripta*, **61**(5), 513–517 (2000).
10. M. W. Evans, P. K. Anastasovski, T. E. Bearden, et al., *Found. Phys.* **39**(7), 1123 (2000).
11. M. W. Evans, P. K. Anastasovski, T. E. Bearden, et al., "Energy inherent in the pure gauge vacuum," *Physica Scripta* (in press).
12. M. W. Evans, P. K. Anastasovski, T. E. Bearden, et al., "Electromagnetic energy from curved space-time," *Optik* (in press).
13. M. W. Evans, P. K. Anastasovski, T. E. Bearden, et al., *Found. Phys. Lett.* **13**(3), 289–296 (2000).
14. M. W. Evans, P. K. Anastasovski, T. E. Bearden, et al., "The Aharonov-Bohm effect as the basis of electromagnetic energy inherent in the vacuum," *Optik* (in press).
15. J. D. Jackson, *Classical Electrodynamics*, 2nd ed., Wiley, New York, 1975, pp. 219–221; 811–812.
16. T. E. Bearden, "Giant negentropy from the common dipole," *Proc. IC-2000* (St. Petersburg, Russia, July 2000); *J. New Energy* **5**(1), 11–23 (summer 2000); also on http://www.ott.doe.gov/electromagnetic/papersbooks. html and www.cheniere.org.

17. T. E. Bearden, "Bedini's method for forming negative resistors in batteries," *Proc. IC-2000* (St. Petersburg, Russia, July 2000); *J. New Energy* **5**(1), 24–38 (summer 2000; this paper is also carried on DOE public Website http://www.ott.doe.gov/electromagnetic/papersbooks.html).
18. T. D. Lee., *Particle Physics and Introduction to Field Theory*, Harwood, New York, 1981.
19. T. E. Bearden, "On extracting electromagnetic energy from the vacuum," *IC-2000 Proc.* (St. Petersburg, Russia, 2000) (this paper is also published on DOE Website http://www.ott.doe.gov/electromagnetic/papersbooks.html and www.cheniere.org).
20. M. W. Evans and T. E. Bearden, "The most general form of the vector potential in electrodynamics," *Optik* (in press).
21. Y. Aharonov and D. Bohm, *Phys. Rev.* (2nd series) **115**(3), 485–491 (1959).
22. T. E. Bearden, J. C. Hayes, J. L. Kenny, K. D. Moore, and S. L. Patrick, "The motionless electromagnetic generator: Extracting energy from a permanent magnet with energy-replenishing from the active vacuum" (on DOE Website http://www.ott.doe.gov/electromagnetic/papersbooks.html).
23. S. Olariu and I. Iovitzu Popescu, *Reviews of Mod. Phy.* **57**(2), 339–436 (1985).
24. M. W. Evans, P. K. Anastasovski, T. E. Bearden, et al., "Spontaneous symmetry breaking as the source of the electromagnetic field," *Found. Phys. Lett.* (in press).
25. T. E. Bearden, "Extracting and using electromagnetic energy from the active vacuum," in M. W. Evans (Ed.), Modern Nonlinear Optics, Part 2, Second Edition, Wiley, New York, 3 vols. (in press), constituting a Special Topic issue as Vol. 114, I. Prigogine and S. A. Rice (Series Eds.), *Advances in Chemical Physics*, Wiley, ongoing.
26. M. W. Evans, P. K. Anastasovski, T. E. Bearden, et al., "On Whittaker's representation of the classical electromagnetic field in vacuo, Part II: Potentials without fields," *J. New Energy* **4**(3), 59–67 (winter 1999).
27. M. W. Evans, P. K. Anastasovski, T. E. Bearden, et al., "On Whittaker's F and G fluxes, Part III: The existence of physical longitudinal and timelike photons," *J. New Energy* **4**(3) (Special Issue), 68–71 (1999).
28. M. W. Evans, P. K. Anastasovski, T. E. Bearden et al., *Physica Scripta* **61**(5), 513–517 (2000).
29. M. W. Evans, P. K. Anastasovski, T. E. Bearden, et al., "Vacuum energy flow and Poynting theorem from topology and gauge theory," *Physica Scripta* (in press).
30. M. W. Evans, P. K. Anastasovski, T. E. Bearden, et al., *Found. Phys. Lett.* **13**(3), 289–296 (2000).
31. M. W. Evans, P. K. Anastasovski, T. E. Bearden et al., *Found. Phys.*, **39**(7), 1123 (2000).
32. M. W. Evans, P. K. Anastasovski, T. E. Bearden, et al., "Energy inherent in the pure gauge vacuum," *Physica Scripta* (in press).
33. M. W. Evans, P. K. Anastasovski, T. E. Bearden, et al., "Electromagnetic energy from curved space-time," *Optik* (in press).
34. M. W. Evans, P. K. Anastasovski, T. E. Bearden, et al., "Schrödinger equation with a Higgs mechanism: Inherentvacuum energy," *Found. Physics* (in press).
35. M. W. Evans, P. K. Anastasovski, T. E. Bearden et al., "The Aharonov-Bohm Effect as the Basis of Electromagnetic Energy Inherent in the Vacuum," submitted to *Optik*, 2000 (in press).
36. M. W. Evans, P. K. Anastasovski, T. E. Bearden et al., "Longitudinal Modes in Vacuo of the Electromagnetic Field in Riemannian Spacetime," submitted to *Optik*, 2000 (in press).
37. M. W. Evans, P. K. Anastasovski, T. E. Bearden et al., "O(3) Electrodynamics from the Irreducible Representations of the Einstein Group," submitted to *Optik*, 2000 (in press).
38. D. C. Cole and H. E. Puthoff, *Phys. Rev. E* **48**(2), 1562–1565 (1993).
39. J. O'M. Bockris, *J. Chem. Ed.* **48**(6), 352–358 (1971).

40. T. E. Bearden, *J. New Energy* **4**(4), 4–11 (2000) (also carried on the DOE Website listed in Ref. 22).
41. M. Sachs, *General Relativity and Matter*, Reidel, 1982.
42. M. W. Evans, "Precise statement on the importance and implications of O(3) electrodynamics as a special subset of Sachs' unified field theory,"(in press).
43. M. Sachs, *Physica B* **192**, 227, 237 (1992).
44. *Fragments of Science: Festschrift for Mendel Sachs*, Michael Ram (Ed.), World Scientific, Singapore, 1999.
45. M. Sachs, in T. W. Barrett and D. M. Grimes (Eds.), *Advanced Electromagnetism*, World Scientific, 1995, p. 551.
46. R. P. Feynman, R. B. Leighton, and M. Sands, *Lectures on Physics*, Addison-Wesley, Reading, MA, 1964, Vol. 1, p. 4-2.
47. Yilmaz, *Ann. Phys.* **81**, 179–200 (1973).
48. O. Heaviside, *Phil. Trans. Roy. Soc. Lond.* **183A**, 423–480 (1893).
49. O. Heaviside, *The Electrician* **27**, 270–272 (July 10, 1891).
50. O. Heaviside, *Electrical Papers*, Vol. 2, 1887, p. 94.
51. J. H. Poynting, *Phil. Trans. Roy. Soc. Lond.* **175**(Part II), 343–361 (1885).
52. H. A. Lorentz, *Vorlesungen über Theoretische Physik an der Universität Leiden*, Vol. V, *Die Maxwellsche Theorie (1900–1902)*, Akademische Verlagsgesellschaft M.B.H., Leipzig, 1931, pp. 179–186.
53. A. A. Logunov and Yu. M. Loskutov, *Sov. J. Part. Nucl.* **18**(3), 179–187 (1987).
54. D. Hilbert, *Gottingen Nachrichten*, Vol. 4, 1917, p. 21.
55. J. D. Jackson, *Classical Electrodynamics*, 2nd ed., Wiley, New York, 1975, pp. 219–221, 811–812.
56. E. T. Whittaker, *Proc. Lond. Math. Soc.* Vol. 1 (Series 2), 367–372 (1904).
57. I. Bloch and H. Crater, *Am. J. Phys.* **49**(1), (1981).
58. M. Phillips, in S. Flugge (Ed.), *Principles of Electrodynamics and Relativity*, Vol. IV of *Encyclopedia of Physics*, Springer-Verlag, 1962 (gives a useful overview of superpotential theory).
59. T. J. I'A. Bromwich, *An Introduction to the Theory of Infinite Series*, 2nd ed., rev., 1926, reprinted 1965; see also J. D. Zund and J. M. Wilkes, *Tensor* (NS), **55**(2), 192–196 (1994) (describes Bromwich's method for solving source-free Maxwell equations).
60. P. Debye, *Ann. Phys.* (Leipzig) **30**, 57 (1909).
61. A. Nisbet, *Physica* **21**, 799 (1955) (extends the Whittaker–Debye two-potential solutions of Maxwell's equations to points within the source distribution; a full generalization of the vector superpotentials, for media of arbitrary properties and their relations to such scalar potentials as those of Debye).
62. W. H. McCrea, *Proc. Roy. Soc. Lond. A* **240**, p. 447 (1957) (gives general properties in tensor form of superpotentials and their gauge transformations; treatment more concise than Nisbet's, but entirely equivalent when translated into ordinary spacetime coordinates).
63. M. W. Evans, "O(3) electrodynamics," a review in M. W. Evans (Ed.), Moderen Nonlinear Optics, Part 2, (this volume series).
64. M. W. Evans, P. K. Anastasovski, T. E. Bearden et al., "On Whittaker's representation of the classical electromagnetic field in vacuo, Part II: Potentials without fields," *J. New Energy* **4**(3), 59–67 (winter 1999).
65. M. W. Evans, P. K. Anastasovski, T. E. Bearden, et al., *J. New Energy* **4**(3) (Special Issue), 68–71 (1999).

66. R. B. Lindsay and H. Margenau, *Foundations of Physics*, Dover, New York, 1963, p. 217.
67. R. C. Gelinas, U.S. patents 4,429,280 (Jan 31, 1984); 4,429,288 (Jan. 31, 1984); 4,432,098 (Feb. 14, 1984); 4,447,779 (May 8, 1984); 4,605,897 (Aug. 12, 1986); 4,491,795 (Jan. 1, 1985).
68. T. E. Bearden, "The unnecessary energy crisis: How to solve it quickly" (Position Paper by the Association of Distinguished American Scientists; also carried on DOE Website http://www.ott.doe.gov/electromagnetic/papersbooks.html and on www.cheniere.org).
69. T. E. Bearden, "Toward solving the energy crisis: The emerging energy science for the new millennium," Oct. 17, 2000, presentation to senior staff assistants of the U.S. Senate Committee on Environment and Public Works (Senator Bob Smith, Chairman) and the U.S. Senate Armed Services Committee (Senator John Warner, Chairman) and technical representatives from other Congressional offices and government agencies.
70. J. D. Kraus, *Electromagnetics*, 4th ed., McGraw-Hill, New York, 1992.
71. T. E. Bearden, "Energy flow, collection, and dissipation in overunity EM devices," *Proc. 4th Int. Energy Conf.*, Academy for New Energy (Denver, CO, May 23–27, 1997), pp. 5–51.
72. W. K. H. Panofsky and M. Phillips, *Classical Electricity and Magnetism*, 2nd ed., Addison-Wesley, Reading, MA, 1962, p. 181; W. Gough and J. P. G. Richards, *Eur. J. Phys.* **7**, 195 (1986) (examples).
73. M. W. Evans, P. K. Anastasovski, T. E. Bearden et al., "Explanation of the Motionless Electromagnetic Generator with O(3) Electrodynamics," *Found. Phys. Lett.*, 14(1), Feb. 2000, p. 87–94.
74. M. W. Evans, P. K. Anastasovski, T. E. Bearden et al., "Explanation of the Motionless Electromagnetic Generator with Sachs Theory of Electrodynamics," *Found. Phys. Lett.* **14**(8), 387–393 (2001).
75. T. E. Bearden, "On Permissible COP $>$ 1.0 Maxwellian systems," *Found. Phys. Lett.* (2001) (in press) (a rigorous rebuttal to a charge that COP $>$ 1.0 circuits and systems are impossible.
76. T. E. Bearden, Letter to Dr. Bruce Alberts, President, U.S. National Academy of Sciences, "Solution for vast electrical energy, cleaner biosphere, and global warming reduction," May 29, 2001.

AUTHOR INDEX

Numbers in parentheses are reference numbers and indicate that the author's work is referred to although his name is not mentioned in the text. Numbers in *italic* show the pages on which the complete references are listed.

SUBJECT INDEX